Groups of Diffeomorphisms
in honor of Shigeyuki Morita
on the occasion of his 60th birthday

ADVANCED STUDIES
IN PURE MATHEMATICS 52

Chief Editor of the Series: Toshitake Kohno (The University of Tokyo)

Groups of Diffeomorphisms

in honor of Shigeyuki Morita on the occasion of his 60th birthday

Edited by

Robert Penner
(University of Southern California/Aarhus University)
Dieter Kotschick
(Ludwig-Maximilians-Universität München)
Takashi Tsuboi
(The University of Tokyo)
Nariya Kawazumi
(The University of Tokyo)
Teruaki Kitano
(Soka University)
Yoshihiko Mitsumatsu
(Chuo University, chief editor of the volume)

Mathematical Society of Japan

This book was typeset by AMS-TEX and AMS-LATEX, the TEX macro systems of the American Mathematical Society, together with the style files aspm.sty *and* aspmfm.sty *for AMS-TEX written by Dr. Chiaki Tsukamoto and* aspmproc.sty *for AMS-LATEX written by Dr. Akihiro Munemasa and* aspm.cls *for AMS-LATEX provided by Tokyo Shoseki Printing Co., Ltd.*

TEX is a trademark of the American Mathematical Society.

Edited by the Mathematical Society of Japan.

Published by the Mathematical Society of Japan.

Distributed by the Mathematical Society of Japan, the American Mathematical Society and the World Scientific Publishing Co., Ltd.
Distributed exclusively in North America by the American Mathematical Society.

Advanced Studies in Pure Mathematics 52
ISBN 978-4-931469-48-8

PRINTED IN JAPAN
by Tokyo Shoseki Printing Co., Ltd.

2000 Mathematics Subject Classification.
Primary 57–06.
Secondary 19D06, 19D10, 37C85, 37D40, 37E10, 37E25, 37E30, 53D05, 53D10, 53D12, 53D35, 53D40, 53D45, 55P35, 55P47, 55P62, 55R35, 55R40, 57R17, 57R19, 57R20, 57R30, 57R32, 57R50, 57S05, 57S25, 57S30, 58H10.

Shigeyuki Morita lecturing at BΓ school (2003)

This volume is dedicated to Professor Shigeyuki Morita on the occasion of his 60th birthday

Preface

This volume consists of selected paper on recent trends and results in the study of groups of diffeomorphisms and related topics. Most of the authors were invited speakers or participants of the *International Symposium on Groups of Diffeomorphisms 2006*, which was held at the University of Tokyo, Japan, during the period 11-15 September, 2006.

During the preceding week, Professor Shigeyuki Morita, one of the leading topologists of our time, saw his 60th birthday on September 5, 2006. A principal purpose of the symposium was to present and discuss mathematical results concerning various groups of diffeomorphisms, including mapping class groups, from the point of view of algebraic and differential topology, so that the lectures would form an appropriate birthday present for Professor Morita. This volume, which bears the same title as the symposium, has been prepared in the same spirit aiming in both cases for the highest possible mathematical quality. The program, the list of participants, and photos from the meeting can be found on the following web site: http://faculty.ms.u-tokyo.ac.jp/~gd2006/

The symposium was partially supported by JSPS Core-to-Core Program 18005 "New Developments of Arithmetic Geometry, Motive, Galois Theory, and Their Practical Applications", Grant-in-Aid for Scientific Research (A) 16204005, (A)17204007, (A)18204002, and (B)18340020. The publication of this volume was partially supported by Grant-in-Aid for Scientific Research (B)18340020 and by Chuo University as one of the 2008 Research Projects for Promotion of Advanced Research at Graduate School.

Several expository papers are included in this volume along with a large number of research papers. Centered around the study of mapping class groups of surfaces, the topics of the papers range from algebraic-topological studies to dynamical ones involving foliations and symplectic or contact diffeomorphisms. We believe that the scope of this volume well reflects Shigeyuki Morita's mathematical interests. It would be our great pleasure for this volume to inspire not only the specialists in these fields but also a wider audience of mathematicians.

We would like to express our deep gratitude to the contributors for their excellent articles as well as to the anonymous referees for their important contributions. In most cases, the manuscripts were greatly improved by the suggestions of the referees. We are also grateful to many people from Morita's school, including M. Fujii, Y. Kasahara, T. Morifuji, and K. Horie and also to younger mathematicians in the generation of his mathematical grandchildren who worked hard to make the symposium a success. Among other things, Masaaki Suzuki took care of most of the practical aspects of producing this volume. His devotion is highly appreciated.

However, our greatest thanks must be reserved for Shigeyuki Morita himself. His mathematical influence and his taste and style are surely evident from the contributions to this volume. His wit and frank and inspiring personality abundantly shone and reverberated at the symposium, during which the wide respect and love for the man were apparent. We know that Morita-sensei never stops developing his mathematics, and we hope this volume will help many other mathematicians do the same.

Editors
Robert Penner (University of Southern California/Aarhus University)
Dieter Kotschick (Ludwig-Maximilians-Universität München)
Takashi Tsuboi (The University of Tokyo)
Nariya Kawazumi (The University of Tokyo)
Teruaki Kitano (Soka University)
Yoshihiko Mitsumatsu (Chuo University)

All papers in this volume have been refereed and are in final form. No version of any of them will be submitted for publication elsewhere.

Photos from "Groups of Diffeomorphisms 2006"

from left to right:
Yukio Matsumoto, Akio Hattori (Morita's thesis advisor), Shigeyuki Morita

Shigeyuki Morita and Etienne Ghys, and other participants

CONTENTS

I. Expository Articles

II. Research Articles

Advanced Studies in Pure Mathematics 52, 2008
Groups of Diffeomorphisms
pp. 1–20

Asymptotic geometry of foliations and pseudo-Anosov flows — a survey

Sérgio R. Fenley

This expository article describes certain aspects of large scale geometry of foliations and flows in 3-manifolds with Gromov hyperbolic fundamental group. The first part is a survey of ideas and previous results about asymptotic behavior of leaves of foliations and flow lines. It reviews classical ideas such as quasi-isometries, quasigeodesics and ideal boundaries. This part includes a description of the seminal work of Cannon-Thurston on fibrations and suspension pseudo-Anosov flows as well as other important results concerning certain classes of foliations. The second part of the article describes more recent topics: What information can be obtained about the asymptotic structure of the universal cover of a manifold using only the dynamics of a pseudo-Anosov flow in the manifold? We explain how to create a dynamical systems ideal boundary for a certain class of such flows and a corresponding compactification of the universal cover. We then explain how these objects are strongly related to the large scale geometry of the manifolds, the large scale geometry of the flows themselves and also of some classes of foliations. Details and proofs of the results in the second part are found in [Fe8].

Let $\mathcal{F}$ be a Reebless foliation in M^3 with Gromov hyperbolic fundamental group, which is not virtually $\mathbf{Z}$. The lifted foliation to the universal cover $\widetilde{M}$ will be denoted by $\widetilde{\mathcal{F}}$. By Novikov's fundamental theorem, the leaves of $\mathcal{F}$ are incompressible and lift to simply connected leaves in $\widetilde{M}$ [No]. The leaves in $\widetilde{\mathcal{F}}$ cannot be spheres or else $\widetilde{M}$ would

Received July 13, 2007.
Revised October 9, 2007.
2000 *Mathematics Subject Classification.* Primary: 53C23, 57R30, 37C85, 37D20; Secondary: 57M50, 37D50, 58D19.
Research partially supported by NSF grant DMS-0305313.

be $\mathbf{S}^2 \times \mathbf{R}$ and $\pi_1(M)$ would be virtually $\mathbf{Z}$. Hence the leaves of $\widetilde{\mathcal{F}}$ are topological planes and they are properly embedded in $\widetilde{M}$ [No]. As $\mathcal{F}$ is Reebless then M is irreducible [Ros] and the universal cover of M is homeomorphic to $\mathbf{R}^3$ [Pa].

In terms of geometry, the universal cover $\widetilde{M}$ is Gromov hyperbolic and can be compactified with an ideal boundary, the Gromov ideal boundary of $\widetilde{M}$ [Gr]. Since M is irreducible and $\pi_1(M)$ is not virtually $\mathbf{Z}$, then Bestvina and Mess [Be-Me] proved that the ideal boundary is a topological 2-sphere S^2_∞. The *limit set* of a subset B of $\widetilde{M}$ is the set of accumulation points of B in S^2_∞. For such manifolds, the relationship between sets in $\widetilde{M}$ and their respective limit sets in S^2_∞ is of the utmost importance [Mo, Th1, Th2, Th3, Gr]. For background in Gromov hyperbolic groups and spaces see [Gr, Gh-Ha].

Notice that if M is closed, irreducible, atoroidal and with infinite fundamental group (which is not virtually $\mathbf{Z}$), then Perelman's results [Pe1, Pe2, Pe3] imply that M is hyperbolic. Here we do not make use of Perelman's results. In particular one goal of the second part of the article is to analyse which geometric information about M can be obtained solely from dynamical systems constructions.

The leaves of $\widetilde{\mathcal{F}}$ are properly embedded in $\widetilde{M}$ and hence only accumulate in the sphere at infinity. For such foliations there are two very important questions [Ga5]:

Property/Question 1) How do leaves approach their limit sets? Is it in a continuous manner? We explain continuity below;

Property/Question 2) Are leaves geometrically good in $\widetilde{M}$, that is, does distance along the leaves compare well with distance in the manifold?

Let us first explain question 2). Let $k \geq 1, C \geq 0$. A (k, C) *quasi-isometric embedding* between metric spaces (X, d), (X', d') is a map f from X to X' so that for any a, b in X then

$$\frac{1}{k} d(a,b) - C \ \leq \ d'(f(a), f(b)) \ \leq \ k d(a,b) + C.$$

The rough meaning of this is as follows: the constant C (which can be quite big) means that small distances ($\leq C$) do not matter. In particular the map f does not need to be continuous at all. The meaning of k is that for big distances the map f is essentially k-bilipschitz. A *quasi-isometric embedding* is a map which is a (k, C) quasi-isometric embedding for some (k, C). A *quasi-isometry* is a quasi-isometric embedding for which there is some positive C' so that $f(X)$ is C' dense in

X'. Then up to this type of error in distances, the metric spaces X, X' are the same. Quasi-isometries are extremely important in hyperbolic spaces: Mostow used quasi-isometries to prove a very strong rigidity result for hyperbolic manifolds [Mo]. Quasi-isometries were also extremely important in Thurston's proof of the geometrization conjecture in the Haken case [Th1, Th2, Th3, Mor].

For a Reebless foliation in such M^3 then one can also show that the leaves are uniformly Gromov hyperbolic [Pl, Su, Gr]. The meaning of question 2) above for foliations in M^3 with Gromov hyperbolic fundamental group is the following: Consider a leaf F of $\widetilde{\mathcal{F}}$ with its path metric and we ask whether the inclusion $i : F \to \widetilde{M}$ is a quasi-isometric embedding. If this is true then the limit set of F is a Jordan curve in S^2_∞ [Th1, Gr]. If every leaf F of $\widetilde{\mathcal{F}}$ is quasi-isometrically embedded in $\widetilde{M}$, then we say that $\mathcal{F}$ is a *quasi-isometric foliation.* Notice that uniformity of the quasi-isometry constants is not required. See remark below on the existence of these foliations.

We now explain question 1). In addition to the leaves being uniformly Gromov hyperbolic, in fact there is a metric in M so that leaves are hyperbolic [Ca] (meaning sectional curvature is constant $= -1$). Hence the inclusion $i : F \to \widetilde{M}$ of a leaf of $\widetilde{\mathcal{F}}$ can be thought of as a map from the hyperbolic plane $\mathbf{H}^2$ to $\widetilde{M}$. The hyperbolic plane has a canonical compactification to a closed disk with a circle at infinity S^1_∞. So the asymptotic behavior of leaves of $\widetilde{\mathcal{F}}$ or how they approach their limit sets can be phrased in the following way: given the inclusion $i : F \to \widetilde{M}$ does it extend to a continuous map between compactifications $F \cup S^1_\infty \to \widetilde{M} \cup S^2_\infty$? If this is the case for all leaves of $\widetilde{\mathcal{F}}$ then we say that $\mathcal{F}$ has the *continuous extension property.* The restriction to S^1_∞ expresses the limit set of the respective leaf as a continuous image of the circle. It is a classical result that for a given leaf F of $\widetilde{\mathcal{F}}$ property 2) is stronger than property 1) see [Mo, Th2, Gr]. In this way questions 1) and 2) are basic questions about the relationship between the foliation structure and the large scale geometric structure of the universal cover of these manifolds.

The level zero result in this area is that there are no quasi-isometric foliations in M^3 with $\pi_1(M)$ Gromov hyperbolic [Fe1]. In particular for every foliation there is at least one leaf of $\widetilde{\mathcal{F}}$ which is not quasi-isometrically embedded [Fe1]. This is easy to show if the leaf space of $\widetilde{\mathcal{F}}$ is not Hausdorff. In case this leaf space is Hausdorff, then it is homeomorphic to the real numbers $\mathbf{R}$. In that case the foliation is called an $\mathbf{R}$-*covered* foliation. Then the limit set of every leaf of $\widetilde{\mathcal{F}}$

is the whole sphere [Fe1]. It follows that in this case no leaf of $\widetilde{\mathcal{F}}$ is quasi-isometrically embedded. We remark that this particular result holds in higher dimensions: if $\pi_1(M^n)$ is Gromov hyperbolic and $\mathcal{F}$ is a codimension one foliation with Gromov hyperbolic leaves, then $\mathcal{F}$ is not a quasi-isometric foliation. For the remainder of the article we restrict to M being 3-dimensional.

By the the above remarks, question 2) always has a negative answer as stated. However if we consider singular foliations of a certain type, there are some examples of quasi-isometric foliations as we will see below.

As a result of the negative answer to question 2), then for general foliations one considers the continuous extension question. How can one think about this? The first situation that comes to mind is that of a compact leaf S of $\mathcal{F}$. What happens in this case? The compact leaf is incompressible or, in other words, π_1-injective (if S is two sided) [No]. It follows that M is Haken and therefore hyperbolic, by Thurston's proof of geometrization in the Haken case [Th1, Th2, Mor]. Here there is a fundamental dichotomy proved using results of Thurston [Th2] and Bonahon [Bon]: either the geometry of the compact leaf is very well behaved or is very badly distorted. The result is that either $\pi_1(S)$ corresponds to a quasi-Fuchsian group or S is a virtual fiber of a fibration over the circle. One of the equivalent ways to characterize the first situation is that any lift $\widetilde{S}$ to $\widetilde{M}$ is quasi-isometrically embedded in $\widetilde{M}$. In this case the limit set of any such $\widetilde{S}$ is a Jordan curve. Geometrically this is a very well behaved situation. The second possibility implies that the limit set of any such $\widetilde{S}$ is the whole sphere. This is the geometrically wild situation. Thurston wondered what could be said about virtual fibers, in particular whether their lifts to $\widetilde{M}$ extend continuously to the ideal boundary (In fact, before that, Thurston wondered if manifolds fibering over the circle could be hyperbolic. If that were true, then the limit set of the lift of a fiber is the whole sphere and that was completely mysterious at that point.. The double limit theorem was the main new tool to prove geometrization in that case [Th3]).

In a seminal work, Cannon and Thurston [Ca-Th] studied fibrations over the circle and proved the continuous extension property for them. This work was done in the early 1980's but only very recently was published.

How was this done in the fibration case? At this point we should introduce a fundamental tool in the study of foliations in 3-manifolds. In the case of a fibration in M^3 closed, atoroidal, it follows that the monodromy of the fiber is a *pseudo-Anosov* homeomorphism [Th4, Bl-Ca].

In particular it has stable and unstable singular 1-dimensional measured foliations in the fiber which are transverse to each other. These measured foliations are projectively invariant under a representative of the monodromy of the fibration. The singularities are of p-prong type. The suspension flow associated to this map preserves the suspended (2-dimensional) foliations and is an example of a pseudo-Anosov flow, see general definition below. Cannon and Thurston [Ca-Th] introduced a metric in M called the singular solv metric, which is quasi-isometric to the ambient metric when lifted to $\widetilde{M}$ and so that: given any stable/unstable leaf L in the universal cover then there is a distance decreasing retraction of $\widetilde{M}$ into L. This implies that L is quasi-isometrically embedded in the original metric of $\widetilde{M}$. So the leaves of the stable/unstable foliations are quasi-isometrically embedded. We stress that the stable/unstable foliations are <u>singular</u> 2-dimensional foliations. This shows that question 2) can have a positive answer if one considers certain types of singular 2-dim foliations in M^3 with $\pi_1(M)$ Gromov hyperbolic. The singular solv metric is obtained using the structure of pseudo-Anosov homeomorphisms of surfaces.

An important property here is that if L is quasi-isometrically embedded then L is a bounded distance (depending on the constants k, C) from the convex hull of the limit set of L. This also works for manifolds with $\pi_1(M)$ Gromov hyperbolic [Gr, Gh-Ha]. Hence if leaves L are far away from a base point it turns out that their limit sets are very small when seen from the base point in the visual measure. Cannon and Thurston use information on how stable/unstable leaves intersect leaves of $\widetilde{\mathcal{F}}$ to show that the leaves of $\widetilde{\mathcal{F}}$ (for $\mathcal{F}$ a fibration) extend continuously to the circle at infinity.

The main facts used here were the following:

I) A (suspension) pseudo-Anosov flow Φ transverse to $\mathcal{F}$ so that its stable and unstable foliations are quasi-isometric (leaves are quasi-isometrically embedded in $\widetilde{M}$),

II) Detailed information of how stable/unstable leaves in $\widetilde{M}$ intersect leaves of $\widetilde{\mathcal{F}}$. In particular considering the ideal compactification of a leaf F of $\widetilde{\mathcal{F}}$ with an ideal circle S^1_∞ then: intrinsic ideal points of F have neighborhood systems defined by intersections of F with stable/unstable leaves in the universal cover.

Information I) is strong metric information and information II) refers to topology and dynamical systems.

We now define a general pseudo-Anosov flow. Let Φ be a flow in M^3 closed. Then Φ is a *pseudo-Anosov flow* if the following conditions hold [Mo1, Fe8]: Φ has no point orbits and there are two dimensional foliations Λ^s, Λ^u, stable and unstable respectively , which are possibly singular and so that

a) There are at most finitely many "singular" orbits and off of the singular orbits Λ^s, Λ^u are non singular foliations, transverse to each other and intersect exactly along the flow lines of Φ;

b) A singular orbit α is a closed orbit and a leaf of Λ^s (or Λ^u) containing α is homeomorphic to $(P \times I)/f$ where P is a p-prong in the plane $(p \geq 3)$ and f is a homeomorphism from $P \times \{1\}$ to $P \times \{0\}$;

c) Finally all orbits in a leaf of Λ^s are forward asymptotic and likewise in a leaf Λ^u the orbits are asymptotic in the backward direction.

Pseudo-Anosov flows are extremely common amongst 3-manifolds: some classes of examples are: 1) Anosov flows. For example geodesic flows in the unit tangent bundle of closed surfaces of negative Gaussian curvature. Also suspensions of Anosov diffeomorphisms of the torus (here there are no singular orbits); 2) Suspension pseudo-Anosov flows; 3) Pseudo-Anosov flows survive most Dehn surgeries – perform Dehn surgery along a closed orbit of the flow (suppose orientation preserving curve), then outside of two lines in Dehn surgery space it follows that the resulting flow is still pseudo-Anosov [Fri]. In particular, using Thurston's Dehn surgery theorem [Th1], it follows that there is an enormous class of pseudo-Anosov flows where the underlying manifold is hyperbolic.

We should remark that foliations come in many different flavors. Some have compact leaves, others have all leaves dense in the manifold, some are **R**-covered, some have one sided branching (see definition below). Some are strongly connected with the topology of the manifold (finite depth foliations – see definition below); others have strong dynamical properties (Anosov foliations). A priori there is not a single method that allows one to study the continuous extension property for all of them at the same time.

We now describe a couple of important classes of foliations for which something is known concerning the continuous extension property. Both of these results were obtained right around the same time in the late 90's.

The first result concerns finite depth foliations. A foliation is *proper* if all its leaves are properly embedded in the manifold (here it is M, not $\widetilde{M}$). In particular they never limit in themselves (in the foliation sense). By minimality there are always compact leaves which are defined to have depth zero. The depth is then defined inductively: a leaf has depth i, if the maximum depth of the leaves where it limits on is $i-1$ (hence a depth one leaf only limits on compact leaves). The foliation $\mathcal{F}$ is said to be *finite depth* if it is proper and there is n positive integer which is an upper bound to the depths of all the leaves. The minimum such n is the depth of the foliation. In a fundamental paper Gabai [Ga1] proved that if M closed, oriented, irreducible, has non finite second homology then it has Reebless, finite depth foliations. These foliations are directly associated with a hierarchy of the manifold [Fe-Mo]: one cuts along leaves and finally arrives at a collection of balls. As such, these foliations have been successfully used by Gabai and others to obtain deep results in 3-dimensional topology [Ga1, Ga2, Ga3].

As in the fibration case one looks for a transverse pseudo-Anosov flow. This was studied by Mosher and Gabai [Mo2], but in general one can only obtain almost transversality: A pseudo-Anosov flow Φ is *almost transverse* to a foliation $\mathcal{F}$ if the following happens [Mo2]: after a possible blow up of some singular orbits of Φ into a collection of annuli (a type of flow blow up), then one obtains a flow which is transverse to the foliation $\mathcal{F}$. When there is a blow up, then the resulting flow is not pseudo-Anosov, but it still has stable/unstable singular foliations (which are not transverse to each other at the blown up annuli – they both contain the blown up annuli). This flow has several important properties. Given $\mathcal{F}$ a Reebless, transversely orientable, finite depth foliation in an atoroidal closed 3-manifold, then Mosher and Gabai [Mo2] constructed a pseudo-Anosov flow Φ which is almost transverse to $\mathcal{F}$. This also has consequences for manifolds obtained by Dehn surgeries on certain orbits of the flow [Mo2]. Concerning the asymptotic behavior of foliations the following happens: in some cases where $\mathcal{F}$ has depth one, we were able to prove that the stable/unstable foliations of Φ are quasi-isometric foliations [Fe4]. These were the first examples of quasi-isometric singular foliations which were not obtained from suspension pseudo-Anosov flows. In some situations we were also able to prove that in the universal cover, the intersections of leaves of $\widetilde{\mathcal{F}}$ with stable/unstable leaves have good properties. This allowed us to prove the continuous extension property for some depth one foliations [Fe4].

Another result concerning the continuous extension property was obtained by Thurston [Th5]. He considered uniform foliations and slitherings. A foliation is *uniform* if for any two leaves in $\widetilde{M}$ they are at a finite Hausdorff distance from each other. Assuming that the foliation is Reebless this implies that $\mathcal{F}$ is $\mathbf{R}$-covered. A related concept is that of slitherings: a *slithering* is a map f from the universal covering $\widetilde{M}$ of M to the circle, so that f is a submersion everywhere and f is group equivariant under the action of $\pi_1(M)$ by covering translations. See details in [Th5]. The slithering map f induces an $\mathbf{R}$-covered foliation in M which is uniform. In [Th5], Thurston proved, amongst other things, that if M is atoroidal, acylindrical with a transversely orientable, uniform foliation then it has a transverse pseudo-Anosov flow Φ. The proof is a very clever extension of the fibration situation. He produces the universal circle for such foliations as defined in [Th6, Th7] to relate the circles at infinity of different leaves of $\widetilde{\mathcal{F}}$. In the atoroidal case he produces 2 laminations transverse to $\mathcal{F}$ and transverse to each other, which blow down to a pseudo-Anosov flow transverse to $\mathcal{F}$ [Th5]. Once that is done, then using the uniform property it is not hard to show that there is a retraction from $\widetilde{M}$ to stable/unstable leaves which does not increase distances too much. The continuous extension follows as in the fibering case.

The following result is a much more recent result about the continuous extension property. In the Cannon-Thurston result they obtained a pseudo-Anosov flow with stable/unstable foliations which are quasi-isometric. There is a weaker property, namely that the flow Φ is quasigeodesic. A *quasigeodesic* in $\widetilde{M}$ is a quasi-isometric embedding $g : I \rightarrow \widetilde{M}$, where I is an interval in $\mathbf{R}$ or the integers. Given a flow Φ let $\widetilde{\Phi}$ be the lifted flow to $\widetilde{M}$. A flow Φ is quasigeodesic if all flow lines of $\widetilde{\Phi}$ in $\widetilde{M}$ are quasigeodesics. As for quasi-isometric objects, it is also true that quasigeodesics are extremely important in manifolds with Gromov hyperbolic fundamental group [Mo, Th1, Gr, Gh-Ha]. It is not too hard to prove that if Λ^s (or Λ^u) is a quasi-isometric foliation, then the flow Φ is a quasigeodesic flow [Fe4]. The result below is very recent and shows that the concept of quasigeodesic flows is extremely useful for studying asymptotic behavior of foliations:

Theorem 1. [Fe7] *Let $\mathcal{F}$ be a foliation in M^3 closed with $\pi_1(M)$ Gromov hyperbolic. Suppose that there is a pseudo-Anosov flow Φ which is almost transverse to $\mathcal{F}$ and so that Φ is a quasigeodesic flow. Then $\mathcal{F}$ has the continuous extension property.*

We first remark that if $\mathcal{F}$ is almost transverse to a pseudo-Anosov flow, then it follows that $\mathcal{F}$ is necessarily a Reebless foliation [Fe7]. For the purposes of this survey article we restrict ourselves to the situation when the flow is actually transverse to the foliation. In order to prove theorem 1, one first proves the following: given a leaf F of $\widetilde{\mathcal{F}}$ and a leaf L of $\widetilde{\Lambda}^s$ then the intersection $L \cap F$ is a 1-dimensional set in F, which may have prongs. We show that each ray of $L \cap F$ accumulates only in a single point in the ideal boundary $\partial_\infty F$ of F. This allows us to describe the neighborhood system of a point q in $\partial_\infty F$ using only intersections of F with leaves of $\widetilde{\Lambda}^s$ and $\widetilde{\Lambda}^u$. This gives the property II) which was mentioned in the Cannon-Thurston analysis. This description does not use any geometric hypothesis on M or Φ – it uses only dynamical properties of the flow. The next step is to use the geometry. The fact that flow lines of $\widetilde{\Phi}$ are quasigeodesics is weaker than the foliations Λ^s, Λ^u being quasi-isometric. But quasigeodesic behavior of Φ still has geometric consequences for the stable/unstable leaves of $\widetilde{\Lambda}^s, \widetilde{\Lambda}^u$ in $\widetilde{M}$. A quasigeodesic is a bounded distance (depending on the quasi-isometric constants (k, C)) from a minimal geodesic in $\widetilde{M}$ [Gr, Gh-Ha]. It follows that the convex hulls of the limit sets of the leaves of $\widetilde{\Lambda}^s, \widetilde{\Lambda}^u$ in $\widetilde{M}$ are at a bounded distance from the leaves themselves. This property together with fact II) and an extended analysis of the topological situation allows one to prove the continuous extension property for $\mathcal{F}$ under the hypothesis of theorem 1. See details in [Fe7].

One case where theorem 1 was successfully used is that of finite depth foliations. We previously proved jointly with Mosher that the pseudo-Anosov flows constructed by Mosher and Gabai which are almost transverse to finite depth foliations are quasigeodesic [Fe-Mo]. Theorem 1 then implies the continuous extension property for all Reebless, finite depth foliations.

Notice also that the condition on $\mathcal{F}$ in theorem 1 is an open condition: if it is satisfied by $\mathcal{F}$, then it is satisfied by any $\mathcal{F}'$ sufficiently near $\mathcal{F}$. This allows one to prove the continuous extension property for many other foliations.

The upshot is that in order to analyse the continuous extension property for foliations, one approach is to find an almost transverse pseudo-Anosov flow which is quasigeodesic. Notice that this has happened in all the situations analysed so far. The big question is: how does one find a quasigeodesic flow transverse to $\mathcal{F}$? In the fibration case it is extremely easy to prove that any transverse flow (pseudo-Anosov or not) is a quasigeodesic flow. This is because any such flow is a suspension flow and using geometric properties of the foliation it is easy to see that the flow is quasigeodesic. It is also very easy to prove quasigeodesic behavior for large classes of transverse flows in the uniform case.

How was the quasigeodesic behavior obtained for Φ in the finite depth foliation? Assuming that $\mathcal{F}$ is not a fake fibration over the circle, then it has compact leaves which are quasi-Fuchsian and therefore any lift F of this compact leaf A to $\widetilde{M}$ is quasi-isometrically embedded [Th1, Th2, Bon]. Notice that this is highly non trivial geometric information from the foliation. This implies that flow lines of Φ which keep intersecting compact leaves in M lift to flow lines of $\widetilde{\Phi}$ which keep intersecting these quasi-isometric leaves of $\widetilde{\mathcal{F}}$ in $\widetilde{M}$. Here is the key observation: because the leaf A is compact, these intersections with lifts of A form a discrete subset of $\widetilde{M}$. As the quasi-isometric leaves escape in $\widetilde{M}$ their limit sets shrink to a point u of S^2_∞. This is because lifts of compact leaves to $\widetilde{M}$ are quasi-isometrically embedded. The point u is the only possible limit point for the given flow line of $\widetilde{\Phi}$. This genesis of an argument can be carefully expanded, using the explicit structure of finite depth foliations to show that the flow (almost) transverse to the foliation is a quasigeodesic flow.

We summarize what we have obtained so far: in all of these cases we start with a foliation which has certain strong geometric properties and using these geometric properties, one can show that a transverse pseudo-Anosov flow is quasigeodesic. Then using the geometric properties of the flow, we can prove the continuous extension property for the foliation.

The difficulty with this approach is that we have to start with some geometric property of the foliation. So in the case that $\mathcal{F}$ does not have compact leaves and is not uniform there is no geometric information to start with. Notice that this is the generic situation. For example this occurs if $\mathcal{F}$ has dense leaves and is not $\mathbf{R}$-covered, or if $\mathcal{F}$ is $\mathbf{R}$-covered and not uniform, or if $\mathcal{F}$ is not finite depth. How does one deal with such cases?

We now describe a new, different approach: in this approach one starts with the pseudo-Anosov flow and its dynamic properties and obtains asymptotic information about $\widetilde{M}$ and geometric information about the flow, using only the dynamics of the flow. This can then be used to analyse certain classes of foliations and prove the continuous extension property. In order to apply this approach we will only use topological or dynamical information about the foliation and will not need geometric information. So for the time being we completely forget foliations and concentrate on pseudo-Anosov flows.

The first goal is to analyse what a pseudo-Anosov flow can say about the asymptotic structure of the universal cover of the manifold. We first study the orbit space of the flow $\widetilde{\Phi}$ in $\widetilde{M}$. A very important basic fact is that the lifted flow to the universal cover is topologically a product flow and the orbit space $\mathcal{O}$ of this flow is homeomorphic to an open disk (or the plane) [Fe2, Fe-Mo]. This is just a topological statement – the geometric or homotopic behavior of the flow is not a product at all. The fundamental group of M acts on $\widetilde{M}$ by covering translations which preserve the collection of flow lines in $\widetilde{M}$. Hence $\pi_1(M)$ acts naturally on $\mathcal{O}$. As the flow is a product flow in $\widetilde{M}$, then the lifted stable and unstable foliations to $\widetilde{M}$ project to one dimensional (perhaps singular) foliations in $\mathcal{O}$. These are denoted by $\mathcal{O}^s, \mathcal{O}^u$, the stable and unstable foliations in $\mathcal{O}$. They are transverse to each other and have only p-prong singularities. These are singular 1-dimensional foliations in the plane. The fundamental group also preserves the stable/unstable foliations in $\widetilde{M}$ and therefore acts on $\mathcal{O}^s, \mathcal{O}^u$. With this topological structure in the orbit space $\mathcal{O}$ one can construct a natural ideal boundary to $\mathcal{O}$:

Theorem 2. [Fe8] *Let Φ be a pseudo-Anosov flow in M^3 closed. Let $\widetilde{\Phi}$ be the lifted flow to the universal cover and $\mathcal{O}$ be the orbit space of $\widetilde{\Phi}$. Then there is an ideal circle boundary $\partial\mathcal{O}$ obtained using only the topological and dynamical structure of the stable and unstable foliations in $\mathcal{O}$. There is also a compactfication $\mathcal{D} = \mathcal{O} \cup \partial\mathcal{O}$ which is homeomorphic to a closed disk. The fundamental group acts by homeomorphism on $\mathcal{D}$.*

Notice there is absolutely no restriction on the pseudo-Anosov flow. We stress that so far we have a boundary for the orbit space $\mathcal{O}$, which is 2-dimensional and not a boundary for $\widetilde{M}$ which has dimension 3. The flow boundary of $\widetilde{M}$ will be constructed later.

The prototype here is a suspension pseudo-Anosov flow – then $\mathcal{O}$ is identified with a lift of a fiber of the fibration of M over the circle. The fiber is a hyperbolic surface. The lift of the fiber is identified with the hyperbolic plane $\mathbf{H}^2$. In this case the ideal circle boundary of $\mathcal{O}$ is identified with the circle at infinity S^1_∞ of the lift of the fiber (this is the ideal circle of $\mathbf{H}^2$). But this uses geometry in $\mathcal{O}$ and not just dynamics. Notice however that any point q in the circle boundary S^1_∞ has a neighborhood basis in $\mathbf{H}^2 \cup S^1_\infty$ which is bounded by leaves of $\mathcal{O}^s$ or $\mathcal{O}^u$. The topology in $\mathcal{O} \cup \partial\mathcal{O}$ can also be defined using only these neighborhood systems. In this way we can define the ideal boundary and ideal compactification of $\mathcal{O}$ using only $\mathcal{O}^s, \mathcal{O}^u$. We want to do something similar in the general case.

Before describing the general situation, notice that in general there is no geometry (even coarse geometry) in the space $\mathcal{O}$. It is only a topological space and the fundamental group acts by homeomorphisms on this space. We now explain ideal points of $\mathcal{O}$ in the general case:

For an arbitrary pseudo-Anosov flow, an ideal point of $\mathcal{O}$ is defined by a sequence of chains – a chain is a finite collection of segments alternatively in $\mathcal{O}^s, \mathcal{O}^u$ and 2 rays, forming a properly embedded path in $\mathcal{O}$. One can think of a chain as a properly embedded polygonal arc in $\mathcal{O}$. The sides of the polygonal arc are alternativately in $\mathcal{O}^s, \mathcal{O}^u$.

In the case of fibrations, the chains are very simple: they are just leaves of $\mathcal{O}^s$ or $\mathcal{O}^u$. In that case an ideal point of $\mathcal{O}$ is determined by a sequence of nested leaves of (say) $\mathcal{O}^s$ which escape compact sets in $\mathcal{O}$. This defines a neighborhood system of this ideal point of $\mathcal{O}$. In general it is necessary to use chains in $\mathcal{O}^s, \mathcal{O}^u$ rather than just leaves of $\mathcal{O}^s, \mathcal{O}^u$ because of the existence of perfect fits as defined below. Any ray in a leaf of $\mathcal{O}^s, \mathcal{O}^u$ is properly embedded and defines an ideal point in $\mathcal{O}$, but there are many other points.

Now we define perfect fits. An unstable leaf G of Λ^u makes a *perfect fit* with a stable leaf F of Λ^s if G and F do not intersect but any other unstable leaf sufficiently near G (and in the F side) will intersect F and vice versa. The same happens to projections to the orbit space $\mathcal{O}$. In the orbit space we can think of a perfect fit as a proper embedding of a rectangle minus a corner. Stable leaves are horizontal segments, unstable leaves are vertical. The 2 leaves which limit in the missing corner form a perfect fit. Roughly speaking, perfect fits are associated to freely homotopic closed orbits of the flow Φ: if there are freely homotopic closed orbits of Φ then there are such orbits α, β which are freely homotopic to the inverse of each other and so that [Fe3, Fe4]: If one lifts then coherently to orbits $\widetilde{\alpha}, \widetilde{\beta}$ of $\widetilde{\Phi}$ then $\widetilde{W^s}(\widetilde{\alpha}), \widetilde{W^u}(\widetilde{\beta})$ form a perfect fit. Notice one is stable and the other unstable. If there are perfect fits as above then the collection of unstable leaves intersecting $\widetilde{W^s}(\widetilde{\alpha})$ is nested, but does not escape in $\widetilde{M}$ – they limit on $\widetilde{W^u}(\widetilde{\beta})$ at least. It follows that when projected to the orbit space $\mathcal{O}$, the elements of the corresponding sequence of leaves of $\mathcal{O}^u$ are not "shrinking" to a single ideal point. In fact they limit at least to a leaf l of $\mathcal{O}^u$. This is why one needs to consider the more general object of chains in $\mathcal{O}^s, \mathcal{O}^u$.

The ideal points of $\mathcal{O}$ are defined as equivalence classes of escaping, nested sequences of chains which satisfy a convexity property. The convexity property is a technical property which guarantees for instance that the chains are properly embedded in $\mathcal{O}$. See [Fe8] for detailed proofs that $\mathcal{D}$ has a well defined topology making it homeormorphic to a closed disk. The proof is quite involved.

One main goal in introducing an ideal boundary for $\mathcal{O}$ is that it leads to an understanding of the asymptotic behavior of $\widetilde{M}$ in certain cases. The flow ideal boundary of $\widetilde{M}$ is constructed as a quotient of the ideal boundary of $\mathcal{O}$ as described in the next theorem:

Theorem 3. [Fe8] *Let Φ be a pseudo-Anosov flow without perfect fits and not topologically conjugate to a suspension Anosov flow. Let $\mathcal{O}$ be its orbit space and $\partial\mathcal{O}$ its ideal boundary. Let $\cong$ be the equivalence relation in $\partial\mathcal{O}$ generated by: two points are in the same class if they are the ideal points of the same stable or unstable leaf in $\mathcal{O}$. Let $\mathcal{R}$ be the set of equivalence classes with the quotient topology. Then $\mathcal{R}$ is a 2-sphere. The fundamental group of M acts on $\mathcal{R}$ by homeomorphisms. There is a natural topology in $\widetilde{M} \cup \mathcal{R}$ making it into a compactification of $\widetilde{M}$. The action of $\pi_1(M)$ on $\widetilde{M}$ extends to an action on $\widetilde{M} \cup \mathcal{R}$. The quotient map from $\partial\mathcal{O}$ to $\mathcal{R}$ is a group invariant Peano curve associated to the flow Φ. All of this uses only the dynamics of the flow Φ.*

How is the space $\mathcal{R}$ obtained from the ideal boundary of $\mathcal{O}$? Here is some information: If x, y points in $\partial\mathcal{O}$ are related under $\cong$ and x, y are ideal points of (say) the same stable leaf in $\mathcal{O}^s$, then the condition on perfect fits implies that there is no unstable leaf with ideal point x (or y). Hence if p is the maximum number of prongs in singular leaves of $\mathcal{O}^s$ (or $\mathcal{O}^u$), then any equivalence class of $\cong$ has at most p points. One can then use a theorem of Moore on classical topology and use the topological structure of $\mathcal{O}^s, \mathcal{O}^u$ [Fe4] to show that $\partial\mathcal{O}/\cong$ is a 2-dimensional sphere $\mathcal{R}$ where the group $\pi_1(M)$ acts naturally.

Unless otherwise stated, we assume from now on that Φ has no perfect fits and is not topologically conjugate to a suspension Anosov flow. One example of such a flow is a suspension pseudo-Anosov flow (with singularities!). Later we describe other classes of examples transverse to certain classes of foliations.

What good is the flow ideal boundary? Let us review what we have done so far: At this point one has an ideal compactification $\widetilde{M} \cup \mathcal{R}$ of $\widetilde{M}$ which is constructed using the dynamics of the pseudo-Anosov flow. Keeping in mind 3-dimensional topology and the Gromov ideal boundary if $\pi_1(M)$ is Gromov hyperbolic, one would certainly like to relate these two approaches to an ideal boundary of $\widetilde{M}$!

In order to do this, we want to analyse the properties of $\mathcal{R}$ and $\widetilde{M} \cup \mathcal{R}$. This is because, as we will see below, certain topological properties completely characterize the Gromov ideal boundary of a group. The fundamental group of M acts on $\mathcal{R}$ and $\widetilde{M} \cup \mathcal{R}$ and the action captures the essence of these objects, as M is compact and is $\widetilde{M}/\pi_1(M)$. One important question to ask is: which properties do the actions have? We have actions of $\pi_1(M)$ on a circle $\partial\mathcal{O}$ and a sphere $\mathcal{R}$ (besides the actions on $\widetilde{M}$ and $\mathcal{O}$ and the singular foliations $\widetilde{\Lambda}^s, \widetilde{\Lambda}^u$). Motivated by a lot of of previous work on 2 and 3-dimensional topology, one asks whether these actions are convergence group actions. For example a group that acts as a uniform convergence group in the circle is topologically conjugate to a Möbius group [Ga4, Ca-Ju]. with fundamental consequences for 3-manifold theory. In particular this is related to the Seifert fibered conjecture for 3-manifolds [Ga4, Ca-Ju]. Also a fundamental question of Cannon asks whether a uniform convergence group acting on a sphere is conjugate to a cocompact Kleinian group. This is an extremely important question which is related to the geometrization of 3-manifolds.

A compactum is a compact Hausdorff space. A group γ acts as a *convergence group* [Bo2] on a metrisable compactum Z if for any sequence $(\gamma_n), n \in \mathbf{N}$ of distinct elements in Γ, there is a subsequence $(\gamma_{n_i}), i \in \mathbf{N}$ and a source/sink pair y, x so that $(\gamma_{n_i}), i \in \mathbf{N}$ converges uniformly to x in compact sets of $Z - \{y\}$. This is equivalent to Γ acts properly discontinuously on the set of distinct triples $\Theta_3(Z)$ of Z. In addition the action is *uniform* if the quotient of $\Theta_3(Z)$ by the action of Γ is compact. If Z is perfect (no isolated points) then the additional condition is equivalent to every point of Z being a conical limit point for the action. A point x in Z is a *conical limit point* if there is a sequence $(\gamma_n), n \in \mathbf{N}$ in Γ and b, c distinct in Z, with $\gamma_n(x)$ converging to c, but for every other point y in Z then $\gamma_n(y)$ converges to b.

It is easy to see that the action of $\pi_1(M)$ on $\partial\mathcal{O}$ is not a convergence group (expected because $\pi_1(M)$ is not a surface group). But the action on $\mathcal{R}$ has excellent properties. This turns out to be a key result:

Theorem 4. [Fe8] *Let Φ be a pseudo-Anosov flow without perfect fits and not topologically conjugate to a suspension Anosov flow. Let $\mathcal{R}$ be the associated ideal boundary sphere with corresponding compactification $\widetilde{M} \cup \mathcal{R}$ of the universal cover. Then the action of $\pi_1(M)$ on $\mathcal{R}$ is a uniform convergence group action. The action of $\pi_1(M)$ on $\widetilde{M} \cup \mathcal{R}$ is also a convergence group action (not uniform).*

The main part of the proof is to prove uniform convergence action on $\mathcal{R}$. Here there is a strong interplay between 2-dimensional dynamics of the action of $\pi_1(M)$ on $\mathcal{R}$ and the action of $\pi_1(M)$ on $\partial\mathcal{O}$, which describes the other action. In other words 1-dimensional dynamics (action on a circle) completely encodes 2-dimensional dynamics.

We mention that when there are perfect fits it is not clear what will be the resulting structure of the quotient space $\mathcal{R}$. For example consider an **R**-covered Anosov flow. This means that the stable foliation of the flow is **R**-covered (and hence so is the unstable foliation [Ba, Fe2]). There are infinitely many examples where the manifold is hyperbolic [Fe2]. Then the quotient $\mathcal{R}$ (of the circle $\mathcal{O}$) is the union of a circle and two special points: each special point is non separated from every point in the circle. Hence $\mathcal{R}$ is not even Hausdorff and clearly in this case the quotient $\mathcal{R}$ does not provide the expected ideal boundary of $\widetilde{M}$ which should be a sphere!

In the situation of theorem 4, one can then use the excellent dynamical properties of the action on $\mathcal{R}$ and $\widetilde{M} \cup \mathcal{R}$ to relate them to the large scale geometry of the manifold. This will have geometric consequences for flows and foliations.

The key tool here is the following result: Bowditch [Bo1], following ideas of Gromov, proved the very interesting theorem that if Γ acts as a uniform convergence group on a perfect, metrisable compactum Z, then Γ is Gromov hyperbolic, Z is homeomorphic to the Gromov ideal boundary $\partial\Gamma$ of Γ and the action of Γ on Z is equivariantly topologically conjugate to the action of Γ on its Gromov ideal boundary $\partial\Gamma$. Theorem 4 then immediately implies the following:

Theorem 5. [Fe8] *Let Φ be a pseudo-Anosov flow without perfect fits and not topologically conjugate to a suspension Anosov flow. Let $\mathcal{R}$ be the associated flow ideal boundary of $\widetilde{M}$ and let $\widetilde{M} \cup \mathcal{R}$ be the flow ideal compactification. Then $\pi_1(M)$ acts as a uniform convergence group on $\mathcal{R}$ and Bowditch's theorem implies that $\pi_1(M)$ is Gromov hyperbolic and the action of $\pi_1(M)$ on $\mathcal{R}$ is topologically conjugate to the action on the Gromov ideal boundary S^2_∞. In addition the actions on $\widetilde{M} \cup \mathcal{R}$ and $\widetilde{M} \cup S^2_\infty$ are also topologically conjugate – by a homeomorphism which is the identity in $\widetilde{M}$.*

The first conclusion of theorem 5 is not new. The fact that $\pi_1(M)$ is Gromov hyperbolic can also be obtained by previous results of Gabai and Kazez [Ga-Ka] and the author [Fe6]. The reason is the following: if M is a toroidal manifold with a pseudo-Anosov flow, then either there is a free homotopy between closed orbits of the flow or the flow is topologically conjugate to a suspension Anosov flow [Fe6]. The second option is disallowed by hypothesis. If there is a free homotopy between closed orbits of the flow then there are perfect fits [Fe3, Fe4]. This is also disallowed by hypothesis, hence it follows that M is atoroidal. Start with the stable foliation of Φ and blow up each singular leaf to produce solid tori complementary components. This produces an essential lamination which is genuine [Ga-Ka]. Since M is atoroidal, the main result of [Ga-Ka] implies then that $\pi_1(M)$ is Gromov hyperbolic. The interest in theorem 5 is that the construction here gives a very explicit description of the Gromov ideal boundary of $\widetilde{M}$ – first as a purely dynamical systems object and a posteriori implying that $\pi_1(M)$ is Gromov hyperbolic and then totally relating the two boundaries. The result of [Ga-Ka] provides no direct relationship with the Gromov ideal boundary. Gabai and Kazez show that least area disks satisfy a linear isoperimetric inequality.

We stress that the flow ideal boundary constructed in theorem 3 and analysed in theorem 4 is obtained using only dynamical systems properties and has no geometric hypothesis. The geometric properties will come as a consequence of the flow behavior.

When Φ is a suspension (singular) pseudo-Anosov flow (fiber is hyperbolic), theorem 5 provides a new proof that if M fibers over the circle with pseudo-Anosov monodromy then $\pi_1(M)$ is Gromov hyperbolic.

These results have some important geometric consequences for flows and foliations. Flow objects (flowlines, stable/unstable leaves, foliations transverse to the flow) behave very well in the flow ideal compactification $\widetilde{M} \cup \mathcal{R}$. This is due to the explicit structure of the flow compactification $\widetilde{M} \cup \mathcal{R}$. We can show for example that flow lines of $\widetilde{\Phi}$ converge to unique points in $\mathcal{R}$ in both forward and backwards direction [Fe8]. One then uses the important fact that since the flow ideal compactification is equivalent to the Gromov compactification, then flow lines also have good asymptotic behavior in the Gromov compactification. Using results in [Fe-Mo] this implies the following:

Theorem 6. [Fe8] *Let Φ be a pseudo-Anosov flow not topologically conjugate to a suspension flow and which does not have perfect fits. By theorem 5, $\pi_1(M)$ is Gromov hyperbolic. Then Φ is a quasigeodesic flow in M. In addition Λ^s, Λ^u are quasi-isometric singular foliations in M.*

Finally we come back to full circle and apply these results to foliations and their asymptotic properties. This also will show that pseudo-Anosov flow without perfect fits are very common. We use the geometric tools developed above and prove:

Theorem 7. [Fe8] *Let $\mathcal{F}$ be an $\mathbf{R}$-covered foliation in M^3 acylindrical and atoroidal. Up to a double cover $\mathcal{F}$ is transverse to a pseudo-Anosov flow Φ without perfect fits and not conjugate to a suspension Anosov flow. Then Φ is quasigeodesic and consequently $\mathcal{F}$ satisfies the continuous extension property. In addition the stable/unstable foliations of Φ are quasi-isometric.*

In order to prove theorem 7, start with a pseudo-Anosov flow transverse to $\mathcal{F}$ as constructed previously by the author [Fe5] and independently by Calegari [Cal1]. We show that Φ is not conjugate to a suspension Anosov flow and has no perfect fits. By theorem 6, the flow Φ is quasigeodesic and its stable/unstable foliations are quasi-isometric. Theorem 1 then implies that $\mathcal{F}$ has the continuous extension property. This result includes the cases of fibrations and uniform foliations/slitherings.

Another class of foliations that can be analysed is the following. Given a Reebless foliation $\mathcal{F}$ in M^3 closed then the leaf space $\mathcal{H}$ of $\widetilde{\mathcal{F}}$ is a simply connected 1-manifold, which has a countable basis and no boundary. If $\mathcal{H}$ is Hausdorff then it is homeomorphic to the real numbers and $\mathcal{F}$ is $\mathbf{R}$-covered. On the other hand if $\mathcal{H}$ is not Hausdorff, then it is still orientable, as it is simply connected. Hence there is a notion of positive and negative sides of a leaf of $\widetilde{\mathcal{F}}$. The non separated points in $\mathcal{H}$ correspond to branching of $\widetilde{\mathcal{F}}$ in the negative direction if they are non separated from each other on their positive sides. Similarly for branching in the positive direction. A foliation $\mathcal{F}$ has *one sided branching* if the branching in $\widetilde{\mathcal{F}}$ is only in one direction (positive or negative). We prove the following:

Theorem 8. [Fe8] *Let $\mathcal{F}$ be a foliation with one sided branching in M^3 atoroidal, acylindrical. Then $\mathcal{F}$ is transverse to a pseudo-Anosov flow without perfect fits and not conjugate to a suspension Anosov flow. Then Φ is quasigeodesic, its stable/unstable foliations are quasi-isometric and consequently $\mathcal{F}$ has the continuous extension property. If F is a leaf of $\widetilde{\mathcal{F}}$ then the limit set of F is not the whole sphere.*

Under the conditions of this theorem, Calegari [Cal2] proved that $\mathcal{F}$ is transverse to a pseudo-Anosov flow Φ. We show that such Φ does not have perfect fits nor is conjugate to a suspension Anosov flow. By theorem 6, the flow Φ is quasigeodesic. As seen in theorem 7 this implies that $\mathcal{F}$ has the continuous extension property. The second statement follows from the metric properties of leaves of Λ^s, Λ^u.

In this way we one obtains the continuous extension property for these two classes of foliations: **R**-covered and with one sided branching assuming only topological conditions. These are conditions on the leaf spaces of $\widetilde{\mathcal{F}}$. Using the tools of flow ideal boundaries and uniform convergence groups we are able to analyse the continuous extension property for these foliations.

The open case for the continuous extension property now is foliations with two sided branching and not finite depth. We should remark that this is the generic case for foliations. It may be that the techniques mentioned here are useful in this situation.

References

[Ba] T. Barbot, Caractérization des flots d'Anosov pour les feuilletages, Ergodic Theory Dynam. Systems, **15** (1995), 247–270.

[Be-Me] M. Bestvina and G. Mess, The boundary of negatively curved groups, J. Amer. Math. Soc., **4** (1991), 469–481.

[Bl-Ca] S. Bleiler and A. Casson, Automorphism of surfaces after Nielsen and Thurston, Cambridge Univ. Press, 1988.

[Bon] F. Bonahon, Bouts des variétés hyperboliques de dimension 3, Ann. of Math. (2), **124** (1986), 71–158.

[Bo1] B. Bowditch, A topological characterization of hyperbolic groups, J. Amer. Math. Soc., **11** (1998), 643–667.

[Bo2] B. Bowditch, Convergence groups and configuration spaces, In: Group Theory Down Under, (eds. J. Cossey, C. F. Miller, W. D.Newmann and M. Shapiro), de Gruyter, Berlin, 1999, pp. 23–54.

[Cal1] D. Calegari, The geometry of **R**-covered foliations, Geom. Topol., **4** (2000), 457–515 (eletronic).

[Cal2] D. Calegari, Foliations with one-sided branching, Geom. Dedicata, **96** (2003), 1–53.

[Ca] A. Candel, Uniformization of surface laminations, Ann. Sci. École Norm. Sup. (4), **26** (1993), 489–516.

[Ca-Th] J. Cannon and W. Thurston, Group invariant Peano curves, Geom. Topol., **11** (2007), 1315–1355.

[Ca-Ju] A. Casson and D. Jungreis, Convergence groups and Seifert fibered 3-manifolds, Invent. Math., **118** (1994), 441–456.

[Fe1] S. Fenley, Quasi-isometric foliations, Topology, **31** (1992), 667–676.

[Fe2] S. Fenley, Anosov flows in 3-manifolds, Ann. of Math. (2), **139** (1994), 79–115.

[Fe3] S. Fenley, Quasigeodesic Anosov flows and homotopic properties of flow lines, J. Differential Geom., **41** (1995), 479–514.

[Fe4] S. Fenley, Foliations with good geometry, J. Amer. Math. Soc., **12** (1999), 619–676.

[Fe5] S. Fenley, Foliations, topology and geometry of 3-manifolds: **R**-covered foliations and transverse pseudo-Anosov flows, Comment. Math. Helv., **77** (2002), 415–490.

[Fe6] S. Fenley, Pseudo-Anosov flows and incompressible tori, Geom. Dedicata, **99** (2003), 61–102.

[Fe7] S. Fenley, Geometry of foliations and flows I: Almost transverse pseudo-Anosov flows and asymptotic behavior of foliations, eprint math.GT/0502330, to appear in J. Differential Geom.

[Fe8] S. Fenley, Ideal boundaries of pseudo-Anosov flows and uniform convergence groups with connections and applications to large scale geometry, eprint math.GT/0507153, submitted.

[Fe-Mo] S. Fenley and L. Mosher, Quasigeodesic flows in hyperbolic 3-manifolds, Topology, **40** (2001), 503–537.

[Fri] D. Fried, Transitive Anosov flows and pseudo-Anosov maps, Topology, **22** (1983), 299–303.

[Ga1] D. Gabai, Foliations and the topology of 3-manifolds, J. Differential Geom., **18** (1983), 445–503.

[Ga2] D. Gabai, Foliations and the topology of 3-manifolds II, J. Differential Geom., **26** (1987), 461–478.

[Ga3] D. Gabai, Foliations and the topology of 3-manifolds III, J. Differential Geom., **26** (1987), 479–536.

[Ga4] D. Gabai, Convergence groups are Fuchsian groups, Ann. of Math. (2), **136** (1992), 447–510.

[Ga5] D. Gabai, 8 problems in foliations and laminations, In: Geometric Topology, (ed. W. Kazez), Amer. Math. Soc., 1987, pp. 1–33.

[Ga-Ka] D. Gabai and W. Kazez, Group negative curvature for 3-manifolds with genuine laminations, Geom. Topol., **2** (1998), 65–77.

[Ga-Oe] D. Gabai and U. Oertel, Essential laminations and 3-manifolds, Ann. of Math. (2), **130** (1989), 41–73.

[Gh-Ha] E. Ghys and P. De la Harpe (eds.), Sur les groupes hyperboliques d'aprés Mikhael Gromov, Progr. Math., **83**, Birkhäuser, 1990.

[Gr] M. Gromov, Hyperbolic groups, In: Essays on Group theory, Math. Sci. Res. Inst. Publ., **8**, Springer-Verlag, 1987, pp. 75–263.

[Mor] J. Morgan, On Thurston's uniformization theorem for 3-dimensional manifolds, In: The Smith Conjecture, (eds. J. Morgan and H. Bass), Academic Press, New York, 1984, pp. 37–125.

[Mo1] L. Mosher, Dynamical systems and the homology norm of a 3-manifold II, Invent. Math., **107** (1992), 243–281.

[Mo2] L. Mosher, Laminations and flows transverse to finite depth foliations, Part I: Branched surfaces and dynamics – available from `http://newark.rutgers.edu:80/~mosher/`; Part II: in preparation.

[Mo] G. Mostow, Strong rigidity of locally symmetric spaces, Ann. of Math. Stud., **78**, Princeton Univ. Press, Princeton, NJ,1973.

[No] S. P. Novikov, Topology of foliations, Trans. Moscow Math. Soc., **14** (1963), 268–305.

[Pa] F. Palmeira, Open manifolds foliated by planes, Ann. of Math. (2), **107** (1978), 109–131.

[Pe1] G. Perelman, The entropy formulat for the Ricci flow and its geometric applications, eprint math.DG/0211159.

[Pe2] G. Perelman, Ricci flow with surgery on 3-manifolds, eprint math.DG/0303109.

[Pe3] G. Perelman, Finite extinction time for the solutions to the Ricci flow in certain 3-manifolds, eprint math.DG/0307245.

[Pl] J. Plante, Foliations with measure preserving holonomy, Ann. of Math. (2), **107** (1975), 327–261.

[Ros] H. Rosenberg, Foliations by planes, Topology, **7** (1968), 131–138.

[Su] D. Sullivan, Cycles for the dynamical study of foliated manifolds and complex manifolds, Invent. Math., **36** (1976), 225–255.

[Th1] W. Thurston, The geometry and topology of 3-manifolds, Princeton Univ. Lecture Notes, 1982.

[Th2] W. Thurston, Three dimensional manifolds, Kleinian groups and hyperbolic geometry, Bull. Amer. Math. Soc. (N.S.), **6** (1982), 357–381.

[Th3] W. Thurston, Hyperbolic structures on 3-manifolds II: Surface groups and 3-manifolds that fiber over the circle, preprint.

[Th4] W. Thurston, On the geometry and dynamics of diffeomorphisms of surfaces, Bull. Amer. Math. Soc. (N.S.), **19** (1988), 417–431.

[Th5] W. Thurston, 3-manifolds, foliations and circles I, preprint.

[Th6] W. Thurston, 3-manifolds, foliations and circles II: Transverse asymptotic geometry of foliations, preprint.

[Th7] W. Thurston, The universal circle for foliations and transverse laminations, lectures in M.S.R.I., 1997.

Florida State University
Tallahassee, FL 32306-4510, USA
and
Princeton University
Princeton, NJ 08540, USA
E-mail address: fenley@mail.math.fsu.edu

Advanced Studies in Pure Mathematics 52, 2008
Groups of Diffeomorphisms
pp. 21–29

Pontrjagin classes and higher torsion of sphere bundles

Kiyoshi Igusa

Abstract.

By a classical result of Schafer and Kahn ([12], [6]), oriented topological sphere bundles have well defined rational Pontrjagin classes. In the smooth case, we show that the corresponding Pontrjagin character is proportional to the higher Franz-Reidemeister torsion invariant in each degree when the fiber is even dimensional and we discuss the relationship in the odd dimensional case.

This short paper is intended to answer a question about oriented smooth sphere bundles that Shigeyuki Morita and Dieter Kotschick asked me at the AIM (American Institute of Mathematics in Palo Alto) conference in March, 2005 on the moduli space of curves, namely: Can higher Franz-Reidemeister torsion be used to define Pontrjagin classes for smooth oriented odd-dimensional sphere bundles?

If a smooth oriented sphere bundle $E \to B$ has a section then the vertical tangent bundle of E along the section can be used to define the Pontrjagin classes and therefore the Pontrjagin character of the bundle E. In the case when the fiber is an even dimensional sphere this Pontrjagin character is proportional to the higher Franz-Reidemeister torsion invariant. Therefore, Morita and Kotschick pointed out to me that (the appropriate scalar multiple of) this higher torsion invariant can be used as a generalization of the Pontrjagin character and therefore defines Pontrjagin classes for all oriented even dimensional smooth sphere bundles. In the case of an odd dimensional sphere bundle the analogous statement is false by a construction of Hatcher. If there is a section of the bundle, the higher Franz-Reidemeister torsion is equal to a multiple of the Pontrjagin character plus an exotic term which measures how far the bundle differs from the linear bundle given by the section. The question

Received April 30, 2007.
Revised December 28, 2007.
Research for this paper was supported by NSF Grant DMS 0309480.

is: Can we specify this decomposition in general, even when there is no section?

§1. Rational Pontrjagin classes

We recall that a *Euclidean bundle* is a fiber bundle with fiber $\mathbb{R}^n$ and structure group $\mathrm{Homeo}(\mathbb{R}^n, 0)$, the group of homeomorphisms of $\mathbb{R}^n$ preserving the basepoint 0 with the compact open topology. Euclidean bundles are equivalent to topological microbundles by [7]. We recall the following classical result.

Theorem 1.1 (Schafer[12], Kahn[6]). *Oriented Euclidean bundles over a finite cell complexes have natural and well defined rational Pontrjagin classes which agree, rationally, with the usual Pontrjagin classes if the bundle is a vector bundle.*

Corollary 1.2. *Oriented topological sphere bundles over finite complexes have well defined rational Pontrjagin classes.*

Proof. The fiberwise open cone of a topological sphere bundle is a Euclidean bundle. Q.E.D.

Lemma 1.3. *If an n-dimensional Euclidean bundle E over a finite complex contains an embedded n-disk bundle associated to a vector bundle V then E is fiberwise homeomorphic to V.*

Proof. Since $\mathbb{R}^n$ is contractible, we can move the base point to the center of the n-disk. This reduces the structure group of the bundle to the subgroup of $\mathrm{Homeo}(\mathbb{R}^n, 0)$ which is orthogonal in a small neighborhood of the origin. There is a deformation retraction of this group to the orthogonal group $O(n)$ by the one parameter family of continuous automorphisms ϕ_t given by $\phi_t(f)(x) = \frac{1}{t}f(tx)$ if $0 < t \leq 1$ and $\phi_0(f) = Df(0) \in O(n)$ is the derivative of f at 0. Q.E.D.

Corollary 1.4. *If $p : E \to B$ is a smooth oriented sphere bundle over a compact manifold B and $s : B \to E$ is a section, then the rational Pontrjagin classes of the sphere bundle E agree with the usual Pontrjagin classes of the pull-back s^*T^vE of the vertical tangent bundle T^vE of E.*

Proof. The Euclidean bundle given by coning off each fiber of E contains a linear disk bundle associated to the stabilization of s^*T^vE. By the lemma, these bundles are homeomorphic. So, they have the same rational Pontrjagin classes by the theorem. Q.E.D.

We will combine the Pontrjagin classes of a vector bundle E into the *Pontrjagin character* $ph(E) = \sum ph_k(E)$ where

$$ph_k(E) = (-1)^k ch_{2k}(E \otimes \mathbb{C})$$

is, up to sign, the degree $4k$ part of the Chern character of the complexification of E. Since $ph(E)$ is a polynomial with rational coefficients in the Pontrjagin classes, oriented Euclidean bundles have well-defined Pontrjagin characters. Also, it is well-known and easy to verify that the Pontrjagin character determines the rational Pontrjagin classes.

We will denote the Pontrjagin character of an oriented topological sphere bundle E by $ph^{top}(E)$.

§2. Higher torsion of sphere bundles

We recall the higher torsion invariants defined in [8], [3], [5], [4]. Given any smooth bundle $p : E \to B$ where $\pi_1 B$ acts trivially on the rational homology of the fiber, there are higher torsion invariants

$$\tau_{2k}^{FR}(E) \in H^{4k}(B; \mathbb{R})$$

called the *higher Franz-Reidemeister (FR) torsion invariants* of E which are invariants of the smooth bundle but, in general, are not topological invariants. Oriented smooth sphere bundles satisfy the trivial action assumption and therefore have well-defined higher FR-torsion invariants.

For oriented smooth bundles $E \to B$ with closed even dimensional fibers, the higher FR-torsion is proportional to *generalized Miller-Morita-Mumford classes* $M_{2k}(E)$ which can be defined rationally in terms of the Pontrjagin character as

$$M_{2k}(E) = tr_B^E \left(\frac{(-1)^k (2k)!}{2} ph_k(T^v E) \right) \in H^{4k}(B; \mathbb{Q})$$

where $tr_B^E : H^*(E) \to H^*(B)$ is the *transfer* [2]. The coefficients are chosen so that $M_{2k}(E)$ is equal to the usual Miller-Morita-Mumford classes, also called *tautological classes*, for oriented surface bundles ([11], [9], [10]).

Theorem 2.1. [3],[5] *If $E \to B$ is an oriented smooth bundle with closed even dimensional fibers so that $\pi_1 B$ acts trivially on the rational homology of the fiber then*

$$\tau_{2k}^{FR}(E) = \frac{(-1)^k \zeta(2k+1)}{2(2k)!} M_{2k}(E) = \frac{\zeta(2k+1)}{4} tr_B^E(ph_k(T^v E))$$

where $\zeta(s) = \sum \frac{1}{n^s}$ is the Riemann zeta function.

Corollary 2.2. *If $p : E \to B$ is an oriented smooth even dimensional sphere bundle then*

$$\tau_{2k}^{FR}(E) = \frac{1}{2}\zeta(2k+1)ph_k^{top}(E).$$

In particular, $\tau_{2k}^{FR}(E)$ is a topological invariant.

Proof. One of the basic properties of the transfer [2] is that the composition

$$H^*(B;\mathbb{Q}) \xrightarrow{p^*} H^*(E;\mathbb{Q}) \xrightarrow{tr_B^E} H^*(B;\mathbb{Q})$$

is equal to multiplication by the Euler characteristic of the fiber which in this case is 2. By naturality of ph^{top} and Corollary 1.4, $p^*(ph^{top}(E)) = ph^{top}(p^*E) = ph(T^vE)$. Transferring down to $H^{4k}(B)$ we get:

$$2ph_k^{top}(E) = tr_B^E(ph_k(T^vE)) = \frac{4}{\zeta(2k+1)}\tau_{2k}^{FR}(E)$$

proving the formula. Q.E.D.

For smooth oriented odd dimensional sphere bundles, the situation is not so clear. If the bundle is linear then we have the formula:

Theorem 2.3. [3],[5] *If $E = S^{2n-1}(\xi)$ is the S^{2n-1}-bundle associated to an $SO(2n)$-bundle ξ over B then*

$$\tau_{2k}^{FR}(S^{2n-1}(\xi)) = \frac{-\zeta(2k+1)}{2}ph_k(\xi)$$

However, when the bundle is not linear, there is an exotic component to the higher torsion. Hatcher gave a family of examples of such bundles. The first example has the following properties (Theorem 6.4.2 in [3]).

Theorem 2.4 (Hatcher's example). *There is a smooth bundle $E \to S^4$ with fiber S^{13} which has a section along which the vertical tangent bundle is trivial but so that $\tau_2^{FR}(E)$ is equal to $\pm 24\zeta(5)$ times the generator of $H^4(S^4)$.*

§3. Exotic torsion

The difference between the two expressions in Theorem 2.3 can be defined for all oriented smooth sphere bundles as follows.

Definition 3.1. For any oriented smooth S^n-bundle $E \to B$, we define the *exotic torsion* $\tau_{2k}^x(E) \in H^{4k}(B;\mathbb{R})$ by

$$\tau_{2k}^x(E) = \tau_{2k}^{FR}(E) - (-1)^n \frac{\zeta(2k+1)}{2} ph_k^{top}(E)$$

Exotic torsion is zero for all smooth oriented even dimensional sphere bundles by Corollary 2.2 and for all linear odd dimensional sphere bundles by Theorem 2.3. Therefore, it measures the extent to which E is not a linear bundle.

In [4], the general theory of higher torsion invariants is discussed. But the following proposition tells us that exotic torsion does not fit into this theory.

Proposition 3.2. *Exotic torsion, as defined above, is not the restriction to sphere bundles of a higher torsion theory as defined in* [4].

Proof. Higher torsion theories in degree $4k$ have an even and an odd component, each of which is unique up to a scalar multiple. However, exotic torsion is zero on all linear even and odd dimensional sphere bundles. So, both even and odd components would be zero making it identically zero if it were a higher torsion theory. Q.E.D.

This implies that exotic torsion is not an absolute higher torsion theory. However, it might be an example of a *relative theory*. As I explained in my lecture at the conference in honor of Professor Morita, there are different definitions of higher relative torsion, three of which agree according to my joint work with Sebastian Goette.

§4. Higher relative torsion

There are three definitions of relative smooth torsion: axiomatic relative torsion, higher relative Franz-Reidemeister (FR) torsion and relative Dwyer-Weiss-Williams (DWW) torsion. The axiomatic relative torsion is defined when we have a pair of smooth bundles $E \to B$, $E' \to B$ over the same base B with compact smooth manifold fibers M, M' and a fiber homotopy equivalence $f : E \to E'$. I.e., f commutes with the projection to B and induces a homotopy equivalence on fibers $M \simeq M'$. In this case we have "tangential" and "exotic" relative torsion invariants $\tau^T(f), \tau^X(f) \in H^{4k}(B;\mathbb{R})$ which measure the extent to which f is not a fiberwise diffeomorphism.

The *tangential relative torsion* measures the difference between the vertical tangent bundles of E and E'. It is the push-down of the Chern character of the difference bundle, i.e., it is the relative generalized

Miller-Morita-Mumford class. The *exotic relative torsion* is independent of the vertical tangent bundle. By an argument very similar to the one given in [4], it follows that tangential and exotic relative torsion are unique up to a scalar factor. This implies that exotic relative torsion is proportional to higher relative FR torsion.

What Goette and I proved is that the exotic relative torsion is also proportional to the relative Dwyer-Weiss-Williams torsion when the base and fiber are closed oriented manifolds. This extends to the case of arbitrary base spaces but the definitions become more complicated.

4.1. Dwyer-Weiss-Williams smoothing theory

Dwyer-Weiss-Williams smoothing theory works as follows. We take a topological manifold bundle $E \to B$ (with compact topological manifold fiber M) together with a linear vertical tangent bundle V^vE. This is a vector bundle whose total space is homeomorphic to a neighborhood of the diagonal ΔE in the fiberwise product $E \times_B E$ (the bundle over B with fiber $M \times M$). The question is: Given the pair (E, V^vE), can we find a smooth bundle $W \to B$ and a homeomorphism $f : W \to E$ which commutes with the projection to B so that f is covered by a nonsingular linear isomorphism of vector bundles $\tilde{f} : T^vW \to V^vE$ so that $\tilde{f}$ is compatible with the exponential maps to W and E? We call this a *fiberwise tangential smoothing* of (E, V^vE).

Let $\widetilde{\mathcal{S}}_B^{d/t}(E, V^vE)$ be the space of all fiberwise tangential smoothings of (E, V^vE). We want to know how many components this space has. In their paper [1] Dwyer, Weiss and Williams give a computation of the homotopy type of this space (and in particular of π_0) in the stable range. Stabilization is given by taking the direct limit with respect to all linear disk bundles $D(\xi)$ over E. This gives a space

$$s\widetilde{\mathcal{S}}_B^{d/t}(E, V^vE) = \varinjlim \widetilde{\mathcal{S}}^{d/t}(D(\xi), V^vE \oplus \xi).$$

Theorem 4.1 (Dwyer-Weiss-Williams [1]). *Assuming that this space is nonempty, we have a homotopy equivalence*

$$s\widetilde{\mathcal{S}}_B^{d/t}(E, V^vE) \simeq \Gamma_B\mathcal{H}^{\%}(E)$$

where $\Gamma_B\mathcal{H}^{\%}(E)$ is the space of sections of the bundle $\mathcal{H}^{\%}(E)$ over B whose fiber is the zero space $\Omega^\infty(M_+ \wedge \mathcal{H}(*))$ of the homology theory on the fiber of E with coefficients in the stable h-cobordism space $\mathcal{H}(*)$.

This theorem is not stated in this way in their paper [1]. So, Bruce Williams gave us (Goette and the author) handwritten notes proving this statement. Excerpts were shown in my lecture. This theorem leads to the concept of the *stable smooth structure class* of an exotic smooth structure. The idea is as follows.

First of all, this theorem implies that the set of stable tangential smooth structures on (E, V^vE) forms an abelian group since it is π_0 of an infinite loop space. However, it would be more accurate to say that it is an affine space which needs a choice of zero to become an additive group. This choice is given by a fixed tangential smoothing E_0 of (E, V^vE). Then, any other tangential smoothing E gives an element of this group:

$$\tilde{\theta}(E, E_0) \in \pi_0 \Gamma_B \mathcal{H}^{\%}(E)$$

We call $\tilde{\theta}(E, E_0)$ the *relative stable smooth structure class* of (E, E_0).

4.2. Results of Goette-I.

Sebastian Goette and I looked at the special case when both base and fiber are closed manifolds. In this case we have the following results. Details will appear elsewhere.

Theorem 4.2 (Goette-I). *If the fiber M and base B of the bundle $E \to B$ are closed oriented manifolds then*

$$\pi_0 \Gamma_B \mathcal{H}^{\%}(E) \otimes \mathbb{Q} \cong \bigoplus_{k>0} H_{\dim B - 4k}(E; \mathbb{Q}).$$

We call the image of $\tilde{\theta}(E, E_0)$ in $\bigoplus_{k>0} H_{\dim B-4k}(E;\mathbb{Q})$ the *relative rational stable smooth structure class* of (E, E_0) and denote it by $\theta(E, E_0)$. As a consequence of this calculation we can make the following definition. Again, this is not the same as the definition given in [1] but I claim that it is equivalent in the cases where both are defined.

Definition 4.3. Suppose that E, E_0 are smooth bundles over B which are tangentially fiberwise homeomorphic and suppose that the fiber M and base B are closed oriented manifolds. Then the degree $4k$ *relative Dwyer-Weiss-Williams torsion* $\tau_{2k}^{DWW}(E, E_0) \in H^{4k}(B; \mathbb{Q})$ is defined to be the Poincaré dual of the image of the relative rational stable smooth structure class $\theta(E, E_0)$ in $H_{\dim B-4k}(B; \mathbb{Q})$.

Theorem 4.4 (Goette-I). *In the situation above, the relative DWW torsion $\tau_{2k}^{DWW}(E, E_0)$ is a scalar multiple of the relative FR torsion $\tau_{2k}^{FR}(E, E_0)$.*

Corollary 4.5. *Suppose that $E \to B$ is a smooth oriented sphere bundle over a closed oriented manifold B. Suppose also that $E_0 \to B$ is a linear sphere bundle which is tangentially fiber homeomorphic to E. Then the relative DWW torsion $\tau_{2k}^{DWW}(E, E_0) \in H^{4k}(B; \mathbb{Q})$ is proportional to the exotic torsion $\tau_{2k}^{x}(E) \in H^{4k}(B; \mathbb{R})$.*

Proof. Let $c_{2k} \in \mathbb{R}$ be the proportionality constant between τ_{2k}^{DWW} and τ_{2k}^{FR}. Then

$$c_{2k}\tau_{2k}^{DWW}(E, E_0) = \tau_{2k}^{FR}(E, E_0) = \tau_{2k}^{FR}(E) - \tau_{2k}^{FR}(E_0).$$

Since E and E_0 are fiber homeomorphic, they have the same topological Pontrjagin classes. Therefore, the difference between their higher FR-torsions is equal to the difference between their exotic torsion invariants. Since E_0 is linear, its exotic torsion is zero. Therefore this difference is equal to the exotic torsion of E as claimed. Q.E.D.

§5. Questions

I will close with two questions, which arise from this correspondence between higher FR torsion and higher Dwyer-Weiss-Williams torsion.

Question 5.1. For a smooth oriented S^{2n-1}-bundle $p : E \to B$ where B is a smooth closed manifold, does there exist an "absolute rational smooth structure class"

$$\theta(E) = \sum \theta_k(E) \in \bigoplus H^{4k+2n-1}(E; \mathbb{R})$$

so that

$$p_*(\theta_k(E)) = \tau_{2k}^x(E)?$$

The answer to this question would be "Yes" if there were a unique or canonical linear sphere bundle E_0 which is tangentially fiber homeomorphic to E. Then we could define $\theta(E)$ to be $\theta(E, E_0)$. If E_0 does not exist, perhaps we could define $\theta(E)$ to be the average value of $\theta(E, E_0)$ using some canonically defined measure on the set of fiberwise smooth structures on $E \to B$.

By the Gysin sequence

$$\cdots \to H^{4k+2n-1}(E) \xrightarrow{p_*} H^{4k}(B) \xrightarrow{\cup e} H^{4k+2n}(B) \xrightarrow{p^*} H^{4k+2n}(E) \to \cdots$$

this question is almost the same as the question: Is $\tau_{2k}^x(E) \cup e = 0$? where $e \in H^{2n}(B)$ is the Euler class of E. This condition is certainly a necessary condition for the existence of $\theta(E)$.

For even dimensional sphere bundles, $\tau_{2k}^x(E) = 0$. So the analogous conjecture would be that $\theta(E)$ exists and is equal to zero. In the relative theory, $\theta(E, E_0) = \tilde{\theta}(E, E_0) \otimes \mathbb{Q}$ is the rational version of the integral obstruction $\tilde{\theta}(E, E_0)$ to fiberwise stable tangential diffeomorphism, i.e., $\tilde{\theta}(E, E_0) = 0$ if and only if $E \times D^N$ is fiberwise diffeomorphic to $E_0 \times D^N$ where E, E_0 are tangentially homeomorphic smooth bundles. So, the idea that $\theta(E)$ might be trivial would be expressed as follows.

Question 5.2. Are smooth oriented S^{2n}-bundles $E \to B$ "stably rationally rigid" in the sense that, for sufficiently large N, the smooth bundle $E \times D^N \to B$ is uniquely determined up to finite indeterminacy by the underlying topological bundle of E?

References

[1] W. Dwyer, M. Weiss and B. Williams, A parametrized index theorem for the algebraic K-theory Euler class, Acta Math., **190** (2003), 1–104.

[2] J. C. Becker and D. H. Gottlieb, The transfer and fiber bundles, Topology, **14** (1975), 1–12.

[3] K. Igusa, Higher Franz-Reidemeister Torsion, AMS/IP Stud. Adv. Math., **31**, International Press, 2002.

[4] ———, Axioms for higher torsion invariants of smooth bundles, J. Topol., **1** (2008), 159–186.

[5] ———, Higher complex torsion and the framing principle, Mem. Amer. Math. Soc., **177** (2005), no. 835, xiv+94.

[6] P. J. Kahn, A note on topological Pontrjagin classes and the Hirzebruch index formula, Illinois J. Math., **16** (1972), 243–256.

[7] J. M. Kister, Microbundles are fibre bundles, Ann. of Math. (2), **80** (1964), 190–199.

[8] J. R. Klein, The cell complex construction and higher R-torsion for bundles with framed Morse function, Ph.D. thesis, Brandeis Univ., 1989.

[9] S. Morita, Characteristic classes of surface bundles, Bull. Amer. Math. Soc. (N.S.), **11** (1984), 386–388.

[10] ———, Characteristic classes of surface bundles, Invent. Math., **90** (1987), 551–577.

[11] D. Mumford, Towards an enumerative geometry of the moduli space of curves, Arithmetic and geometry, Vol. II, Birkhäuser Boston, Boston, MA, 1983, pp. 271–328.

[12] J. A. Schafer, Topological Pontrjagin classes, Comment. Math. Helv., **45** (1970), 315–332.

Department of Mathematics
Brandeis University
Waltham, MA 02454
USA

Advanced Studies in Pure Mathematics 52, 2008
Groups of Diffeomorphisms
pp. 31–49

L^2-torsion invariants and the Magnus representation of the mapping class group

Teruaki Kitano[1] **and Takayuki Morifuji**[2]

Abstract.

In this paper, we study a series of L^2-torsion invariants from the viewpoint of the mapping class group of a surface. We establish some vanishing theorems for them. Moreover we explicitly calculate the first two invariants and compare them with hyperbolic volumes.

§1. Magnus representation

Let $\Sigma_{g,1}$ be a compact oriented smooth surface of genus g with a boundary $\partial\Sigma_{g,1} \cong S^1$. In this paper, we always assume that $g \geq 1$. We take and fix a base point $* \in \partial\Sigma_{g,1}$ of $\Sigma_{g,1}$. Let $\mathcal{M}_{g,1}$ be the mapping class group of $\Sigma_{g,1}$, namely, the group of all isotopy classes of orientation preserving diffeomorphisms of $\Sigma_{g,1}$ relative to the boundary. We denote $\pi_1(\Sigma_{g,1}, *)$ by Γ, which is a free group of rank $2g$, and fix a generating system $\Gamma = \langle x_1, \ldots, x_{2g} \rangle$. Let $\mathbb{Z}\Gamma$ be the group ring of Γ over $\mathbb{Z}$. We write $\varphi_* \in \mathrm{Aut}(\Gamma)$ to the automorphism induced from $\varphi \in \mathcal{M}_{g,1}$. The following result, usually called the Dehn-Nielsen-Baer theorem, is classical and fundamental to study the mapping class group $\mathcal{M}_{g,1}$ by using combinatorial group theories (see [9] Section 2.9).

Proposition 1.1 (Zieschang [27]). *The above induced homomorphism $\mathcal{M}_{g,1} \ni \varphi \mapsto \varphi_* \in \mathrm{Aut}(\Gamma)$ is injective.*

As a corollary, we see that φ can be determined by the words $\varphi_*(x_1), \ldots, \varphi_*(x_{2g}) \in \Gamma$. Since the fundamental formula $\gamma = 1 +$

Received May 5, 2007.
Revised April 4, 2008.
[1] This research was partially supported by the Grant-in-Aid for Scientific Research (No.17540064), the Ministry of Education, Culture, Sports, Science and Technology, Japan.
[2] This research was partially supported by the Grant-in-Aid for Scientific Research (No.17740032), the Ministry of Education, Culture, Sports, Science and Technology, Japan.

$\sum_{i=1}^{2g}(\partial\gamma/\partial x_i)(x_i - 1)$ holds in $\mathbb{Z}\Gamma$ for any $\gamma \in \Gamma$, the word $\varphi_*(x_j)$ is determined by $\{\partial\varphi_*(x_j)/\partial x_i\}$. Here $\partial/\partial x_i : \mathbb{Z}\Gamma \to \mathbb{Z}\Gamma$ denotes Fox's free differential. See [1] Section 3.1 for a systematic treatment of the subject. The Magnus representation of the mapping class group is defined as follows.

Definition 1.2. The Magnus representation of $\mathcal{M}_{g,1}$ is defined by the assignment

$$r : \mathcal{M}_{g,1} \ni \varphi \mapsto \left(\overline{\frac{\partial\varphi_*(x_j)}{\partial x_i}} \right)_{ij} \in GL(2g, \mathbb{Z}\Gamma),$$

where $\overline{\sum_g \lambda_g g} = \sum_g \lambda_g g^{-1}$ for any element $\sum_g \lambda_g g \in \mathbb{Z}\Gamma$.

Remark 1.3. By the expression $\gamma = 1 + \sum_i(\partial\gamma/\partial x_i)(x_i - 1)$, it follows that r is injective. However, it is not a group homomorphism, just a crossed homomorphism. According to the practice, we call it simply the Magnus representation of $\mathcal{M}_{g,1}$.

Now for a matrix $B \in M(n, \mathbb{C})$, let us recall that its characteristic polynomial

$$\det(tI - B)$$

is one of the fundamental tools in the linear algebra. Here I denotes the identity matrix of degree n. If we can define a characteristic polynomial of $r(\varphi)$, it may be useful tool to study the mapping class group. In order to define it for a Magnus matrix $r(\varphi)$, we need to clarify the following two points.

(1) What is the determinant over a non-commutative group ring?
(2) What is the meaning of a variable "t" in the group?

As an answer to these problems, we can formulate that

- the variable t lives in the fundamental group of the mapping torus of φ,
- a characteristic polynomial "det"$(tI - r(\varphi))$ with respect to the Fuglede-Kadison determinant.

In the later sections, we explain that the characteristic polynomial of $r(\varphi)$ is defined as a real number and it essentially gives the L^2-torsion and the hyperbolic volume of the mapping torus of φ. Moreover taking the lower central series of the surface group Γ, we obtain a family of Magnus representations, so that we can introduce a sequence of L^2-torsion invariants as an approximate sequence of the hyperbolic volume.

This paper is organized as follows. In the next section, we briefly recall the definition of the Fuglede-Kadison determinant. In Section 3, we summarize some properties of the L^2-torsion of 3-manifolds and explain a relation to the Magnus representation. We introduce a sequence of L^2-torsion invariants for a surface bundle over the circle in Section 4 and give some formulas for them in Section 5. In the last section, we discuss some vanishing theorems for L^2-torsion invariants.

§2. Fuglede-Kadison determinant

In this section, we review the combinatorial definition of the Fuglede-Kadison determinant over a non-commutative group ring and its basic properties (see [19] for details).

The idea to define a determinant over a group ring comes from the following observation. That is, for a matrix $B \in GL(n, \mathbb{C})$ with the (non-zero) eigenvalues $\lambda_1, \ldots, \lambda_n$, we can formally calculate

$$\begin{aligned}
\log|\det B|^2 &= \log \prod_{i=1}^{n} \lambda_i \overline{\lambda}_i = \sum_{i=1}^{n} \log \lambda_i \overline{\lambda}_i \\
&= \sum_{i=1}^{n} \left(\sum_{p=1}^{\infty} \frac{(-1)^{p+1}}{p} \left(\lambda_i \overline{\lambda}_i - 1\right)^p \right) \\
&= -\sum_{p=1}^{\infty} \left(\sum_{i=1}^{n} \frac{1}{p} \left(1 - \lambda_i \overline{\lambda}_i\right)^p \right) \\
&= -\sum_{p=1}^{\infty} \frac{1}{p} \mathrm{tr}\left((I - BB^*)^p\right)
\end{aligned}$$

by the power series expansion of log, where B^* is the adjoint matrix of B. More precisely, if we take a sufficiently large constant $K > 0$, we obtain

$$|\det B| = K^n \exp\left(-\frac{1}{2} \sum_{p=1}^{\infty} \frac{1}{p} \mathrm{tr}\left(\left(I - K^{-2} BB^*\right)^p \right) \right) \in \mathbb{R}_{>0}.$$

Thus if we can define a certain "trace" over a group ring, we get a determinant by using this formula.

Let π be a discrete group and $\mathbb{C}\pi$ denote its group ring over $\mathbb{C}$. For an element $\sum_{g \in \pi} \lambda_g g \in \mathbb{C}\pi$, we define the $\mathbb{C}\pi$-trace $\mathrm{tr}_{\mathbb{C}\pi} : \mathbb{C}\pi \to \mathbb{C}$ by

$$\mathrm{tr}_{\mathbb{C}\pi}\left(\sum_{g \in \pi} \lambda_g g \right) = \lambda_e \in \mathbb{C},$$

where e is the unit element in π. For an $n \times n$-matrix $B = (b_{ij}) \in M(n, \mathbb{C}\pi)$, we extend the definition of $\mathbb{C}\pi$-trace by means of

$$\mathrm{tr}_{\mathbb{C}\pi}(B) = \sum_{i=1}^{n} \mathrm{tr}_{\mathbb{C}\pi}(b_{ii}).$$

Next let us recall the definition of the L^2-Betti number of an $n \times m$-matrix $B \in M(n, m, \mathbb{C}\pi)$. We consider the bounded π-equivariant operator

$$R_B : \oplus_{i=1}^{n} l^2(\pi) \to \oplus_{i=1}^{m} l^2(\pi)$$

defined by the natural right action of B. Here $l^2(\pi)$ is the complex Hilbert space of the formal sums $\sum_{g\in\pi} \lambda_g g$ which are square summable. We fix a positive real number K so that $K \geq ||R_B||_\infty$ holds, where $||R_B||_\infty$ is the operator norm of R_B.

Definition 2.1. The L^2-Betti number of a matrix $B \in M(n, m, \mathbb{C}\pi)$ is defined by

$$b(B) = \lim_{p\to\infty} \mathrm{tr}_{\mathbb{C}\pi}\left(\left(I - K^{-2}BB^*\right)^p\right) \in \mathbb{R}_{\geq 0},$$

where $B^* = (\bar{b}_{ji})$ and $\overline{\sum \lambda_g g} = \sum \bar{\lambda}_g g^{-1}$ for each entry.

Roughly speaking, the L^2-Betti number $b(B)$ measures the size of the kernel of a matrix B. Hereafter we assume $b(B) = 0$. Then, for a matrix with coefficients in a non-commutative group ring, we can introduce the desired determinant as follows.

Definition 2.2. The Fuglede-Kadison determinant of a matrix $B \in M(n, m, \mathbb{C}\pi)$ is defined by

$$\mathrm{det}_{\mathbb{C}\pi}(B) = K^n \exp\left(-\frac{1}{2}\sum_{p=1}^{\infty}\frac{1}{p}\mathrm{tr}_{\mathbb{C}\pi}\left(\left(I - K^{-2}BB^*\right)^p\right)\right) \in \mathbb{R}_{>0},$$

if the infinite sum of non-negative real numbers in the above exponential converges to a real number.

Remark 2.3. It is shown that the L^2-Betti number $b(B)$ and the Fuglede-Kadison determinant $\mathrm{det}_{\mathbb{C}\pi}(B)$ are independent of the choice of the constant K (see [16] for example).

Here we consider the condition of the convergence. For any matrix $B \in M(n, \mathbb{C})$, the condition

$$\lim_{p\to\infty} \mathrm{tr}\left(\left(I - K^{-2}BB^*\right)^p\right) = 0$$

implies that B has no zero eigenvalues, and then $|\det B|$ converges. In the case of group rings, if $\det_{\mathbb{C}\pi}(B)$ converges, then $b(B) = 0$. But it is not a sufficient condition, so that we need additional one. It is a problem to decide when $\det_{\mathbb{C}\pi}(B)$ converges. Under the assumption that $b(B) = 0$, such a sufficient condition is given by the positivity of the Novikov-Shubin invariant $\alpha(B)$. Then the convergence of the infinite sum in the Fuglede-Kadison determinant is guaranteed. The Novikov-Shubin invariant of an operator R_B measures how concentrated the spectrum of $R_B^* R_B$ is. However, in general, it is hard to check the positivity of the Novikov-Shubin invariant.

To avoid the difficulty, we need to consider the determinant class condition for groups (see [19], [24] for details). A group π is of det ≥ 1-class if for any $B \in M(n, m, \mathbb{Z}\pi)$ the Fuglede-Kadison determinant of B satisfies $\det_{\mathbb{C}\pi}(B) \geq 1$. There are no known examples of groups which are not of det ≥ 1-class. Further recently it was proved that there is a certain large class $\mathcal{G}$ of groups for which they are of det ≥ 1-class. It includes amenable groups and countable residually finite groups. If we can see that π belongs to $\mathcal{G}$, namely it is of det ≥ 1-class, the convergence of the Fuglede-Kadison determinant is guaranteed when the L^2-Betti number is vanishing. See [18], [19], [24] for definitions and properties of these subjects.

§3. L^2-torsion of 3-manifolds

In this section, we quickly recall the definition of the L^2-torsion of 3-manifolds. It is an L^2-analogue of the Reidemeister and the Ray-Singer torsion and essentially gives Gromov's simplicial volume under certain general conditions [2], [3], [4], [8], [14], [15], [20], [21], [22]. See [19] and its references for historical background, related works and so on.

Let M be a compact connected orientable 3-manifold. We fix a CW-complex structure on M. We may assume that the action of $\pi_1 M$ on the universal covering $\widetilde{M}$ is cellular (if necessary, we have only to take a subdivision of the original structure). We consider the $\mathbb{C}\pi_1 M$-chain complex

$$0 \longrightarrow C_3(\widetilde{M}, \mathbb{C}) \xrightarrow{\partial_3} C_2(\widetilde{M}, \mathbb{C}) \xrightarrow{\partial_2} C_1(\widetilde{M}, \mathbb{C}) \xrightarrow{\partial_1} C_0(\widetilde{M}, \mathbb{C}) \longrightarrow 0$$

of $\widetilde{M}$. Since the boundary operator ∂_i is a matrix with coefficients in $\mathbb{C}\pi_1 M$, if we take the adjoint operator $\partial_i^* : C_{i-1}(\widetilde{M}, \mathbb{C}) \to C_i(\widetilde{M}, \mathbb{C})$ as in the previous section, we can define the ith (combinatorial) Laplace operator $\Delta_i : C_i(\widetilde{M}, \mathbb{C}) \to C_i(\widetilde{M}, \mathbb{C})$ by

$$\Delta_i = \partial_{i+1} \circ \partial_{i+1}^* + \partial_i^* \circ \partial_i.$$

Let us suppose that all the L^2-Betti numbers $b(\Delta_i)$ vanish and the fundamental group $\pi_1 M$ is of det ≥ 1-class. Thereby as a generalization of the classical Reidemeister torsion, the L^2-torsion $\tau(M)$ is defined by

Definition 3.1.

$$\tau(M) = \prod_{i=0}^{3} \det{}_{\mathbb{C}\pi_1 M}(\Delta_i)^{(-1)^{i+1}i} \in \mathbb{R}_{>0}.$$

As for the positivity of Novikov-Shubin invariants $\alpha(\Delta_i)$ for the Laplace operator Δ_i, it is known that $\alpha(\Delta_i) > 0$ holds under some general assumptions (see [15]). For example, if a compact connected orientable 3-manifold M satisfies

(1) $\pi_1 M$ is infinite,
(2) M is an irreducible 3-manifold or $S^1 \times S^2$ or $\mathbb{R}P^3 \sharp \mathbb{R}P^3$,
(3) if $\partial M \neq \phi$, it consists of tori,
(4) if $\partial M = \phi$, M is finitely covered by a 3-manifold which is a hyperbolic, Seifert or Haken 3-manifold,

then $b(\Delta_i) = 0$ and $\alpha(\Delta_i) > 0$ for each i. Therefore, we see that the L^2-torsion $\tau(M)$ is also well-defined in view of these conditions.

Remark 3.2. The above condition (4) is automatically satisfied by Perelman's proof of Thurston's Geometrization Conjecture.

As a notable property of the L^2-torsion, it is known that $\log \tau(M)$ can be interpreted as Gromov's simplicial volume $||M||$ and hyperbolic volume $\mathrm{vol}(M)$ (see [7]) of M. See [21] for the heart of the proof.

Theorem 3.3. *Let M be a compact connected orientable irreducible 3-manifold with an infinite fundamental group such that ∂M is empty or a disjoint union of incompressible tori. Then it holds that*

$$\log \tau(M) = C||M||,$$

where C is the universal constant not depending on M. In particular, if M is a hyperbolic 3-manifold, we obtain

$$\log \tau(M) = -\frac{1}{3\pi}\mathrm{vol}(M).$$

Next we review Lück's formula for the L^2-torsion of 3-manifolds ([16] Theorem 2.4). From this formula, we see that $\log \tau$ is a characteristic polynomial of the Magnus representation of the mapping class group.

Theorem 3.4. *Let M be as in the above theorem. We suppose that ∂M is non-empty and $\pi_1 M$ has a deficiency one presentation*

$$\langle s_1, \dots, s_{n+1} \mid r_1, \dots, r_n \rangle .$$

Put A to be the $n \times n$-matrix with entries in $\mathbb{Z}\pi_1 M$ obtained from the matrix $(\partial r_i / \partial s_j)$ by deleting one of the columns. Then the logarithm of the L^2-torsion of M is given by

$$\begin{aligned} \log \tau(M) &= -2 \log \det_{\mathbb{C}\pi_1 M}(A) \\ &= -2n \log K + \sum_{p=1}^{\infty} \frac{1}{p} \mathrm{tr}_{\mathbb{C}\pi_1 M} \left((I - K^{-2} AA^*)^p \right), \end{aligned}$$

where K is a constant satisfying $K \geq ||R_A||_\infty$.

To see a relation between the Magnus representation and the L^2-torsion, we describe the above Lück's formula for a surface bundle over the circle.

For an orientation preserving diffeomorphism φ of $\Sigma_{g,1}$, we form the mapping torus M_φ by taking the product $\Sigma_{g,1} \times [0,1]$ and gluing $\Sigma_{g,1} \times \{0\}$ and $\Sigma_{g,1} \times \{1\}$ via φ. This gives a surface bundle over S^1. Its diffeomorphism type is determined by the monodromy map φ, and conversely the monodromy map φ is determined by a given surface bundle up to conjugacy and isotopy. Here an isotopy fixes setwisely the points on the boundary $\partial \Sigma_{g,1}$. We take a deficiency one presentation of the fundamental group $\pi = \pi_1(M_\varphi, *)$,

$$\pi = \left\langle x_1, \dots, x_{2g}, t \mid r_i : t x_i t^{-1} = \varphi_*(x_i),\ 1 \leq i \leq 2g \right\rangle ,$$

where the base point $*$ of π and $\Gamma = \pi_1(\Sigma_{g,1}, *)$ is the same one on the fiber $\Sigma_{g,1} \times \{0\} \subset M_\varphi$ and $\varphi_* : \Gamma \to \Gamma$ is the automorphism induced by $\varphi : \Sigma_{g,1} \to \Sigma_{g,1}$. It should be noted that π is isomorphic to the semi-direct product of Γ and $\pi_1 S^1 \cong \mathbb{Z} = \langle t \rangle$.

Applying the free differential calculus to the relations r_i $(1 \leq i \leq 2g)$, we obtain the Alexander matrix

$$A = \left(\frac{\partial r_i}{\partial x_j} \right) \in M(2g, \mathbb{Z}\pi).$$

Then Lück's formula for a surface bundle over the circle is given by

$$\begin{aligned} \log \tau(M_\varphi) &= -2 \log \det_{\mathbb{C}\pi}(A) \\ &= -4g \log K + \sum_{p=1}^{\infty} \frac{1}{p} \mathrm{tr}_{\mathbb{C}\pi} \left((I - K^{-2} AA^*)^p \right), \end{aligned}$$

where K is a constant satisfying $K \geq ||R_A||_\infty$.

This formula enables us to interpret the L^2-torsion $\log \tau$ of a surface bundle over the circle as the characteristic polynomial of the Magnus representation $r(\varphi)$. In fact, an easy calculation shows that

$$A = \left(\frac{\partial r_i}{\partial x_j} \right) = tI - {}^t\overline{r(\varphi)}.$$

Then if we take the Fuglede-Kadison determinant in $M(2g, \mathbb{C}\pi)$, we have

$$\begin{aligned} \det_{\mathbb{C}\pi} \left(tI - {}^t\overline{r(\varphi)} \right) &= \det_{\mathbb{C}\pi} \left(tI - {}^t\overline{r(\varphi)} \right)^* \\ &= \det_{\mathbb{C}\pi} \left(t^{-1}I - r(\varphi) \right) \end{aligned}$$

because $\mathrm{tr}_{\mathbb{C}\pi}(BB^*) = \mathrm{tr}_{\mathbb{C}\pi}(B^*B)$ holds. Therefore the L^2-torsion is interpreted as the characteristic polynomial of $r(\varphi)$.

§4. Definition of L^2-torsion invariants

As was seen in Section 3, Lück's formula gives a way to calculate the simplicial volume from a presentation of the fundamental group. However, in general, it seems to be difficult to evaluate the exact values from the formula. In this section, we introduce a sequence of L^2-torsion invariants which approximates the original one for a surface bundle over the circle. See [12] for details.

In order to construct such a sequence of L^2-torsion invariants, we consider the lower central series of Γ. Namely, we take the descending infinite sequence

$$\Gamma_1 = \Gamma \supset \Gamma_2 \supset \cdots \supset \Gamma_k \supset \cdots,$$

where $\Gamma_k = [\Gamma_{k-1}, \Gamma_1]$ for $k \geq 2$. Let N_k be the kth nilpotent quotient $N_k = \Gamma/\Gamma_k$ and $p_k : \Gamma \to N_k$ be the natural projection.

In the previous section, we considered a chain complex $C_*(\widetilde{M_\varphi}, \mathbb{C})$ of $\mathbb{C}\pi$-modules. Instead of this complex, we can use the chain complex

$$C_*(M_\varphi, l^2(\pi)) = l^2(\pi) \otimes_{\mathbb{C}\pi} C_*(\widetilde{M_\varphi}, \mathbb{C})$$

to define the same L^2-torsion $\tau(M_\varphi)$. This point of view allows us to introduce a sequence of the L^2-torsion invariants.

The group Γ_k is a normal subgroup of π, so that we can take the quotient group $\pi(k) = \pi/\Gamma_k$. It should be noted that $\pi(k)$ is isomorphic to the semi-direct product $N_k \rtimes \mathbb{Z}$. We denote the induced projection

$\pi \longrightarrow \pi(k)$ by the same letter p_k. Thereby we can consider the chain complex

$$C_*\left(M_\varphi, l^2\left(\pi(k)\right)\right) = l^2\left(\pi(k)\right) \otimes_{\mathbb{C}\pi} C_*(\widetilde{M}_\varphi, \mathbb{C})$$

through the projection p_k. By using the Laplace operator

$$\Delta_i^{(k)} : C_i\left(M_\varphi, l^2\left(\pi(k)\right)\right) \to C_i\left(M_\varphi, l^2\left(\pi(k)\right)\right)$$

on this complex, we can formally define the kth L^2-torsion invariant $\tau_k(M_\varphi)$ as follows.

Definition 4.1.

$$\tau_k(M_\varphi) = \prod_{i=0}^{3} \det{}_{\mathbb{C}\pi(k)}(\Delta_i^{(k)})^{(-1)^{i+1}i}.$$

Of course, this definition is well-defined if every L^2-Betti number $b(\Delta_i^{(k)})$ vanishes and every $\pi(k)$ is of det $\geq$ 1-class. The next lemma is easily proved (see [12], [17]).

Lemma 4.2. *The L^2-Betti numbers of $\Delta_i^{(k)}$ are all zero.*

Recall the class $\mathcal{G}$ of groups. It is the smallest class of groups which contains the trivial group and is closed under the following processes: (i) amenable quotients, (ii) colimits, (iii) inverse limits, (iv) subgroups and (v) quotients with finite kernel (see [19], [24]). It is known that $\mathcal{G}$ contains all amenable groups. By definition, $N_k = \Gamma/\Gamma_k$ is a nilpotent group and in particular an amenable group. Hence every N_k belongs to $\mathcal{G}$. Further for any automorphism $\varphi_* : N_k \to N_k$, its mapping torus extension (*HNN*-extension) $N_k \rtimes \mathbb{Z}$ also belongs to $\mathcal{G}$. Therefore we have

Lemma 4.3. *The group $\pi(k)$ belongs to $\mathcal{G}$.*

As a result, we can conclude that our L^2-torsion invariants τ_k can be defined for any $k \geq 1$ and they are all homotopy invariants (see [19], [24]).

Now let us describe a formula of the kth L^2-torsion invariant $\tau_k(M_\varphi)$ and establish a relation to the Magnus representation of the mapping class group. Let $p_{k*} : \mathbb{C}\pi \to \mathbb{C}\pi(k)$ be an induced homomorphism over the group rings. For $k \geq 1$, we put

$$A_k = \left(p_{k*}\left(\frac{\partial r_i}{\partial x_j}\right)\right) \in M(2g, \mathbb{C}\pi(k)).$$

Moreover we fix a constant K_k satisfying $K_k \geq ||R_{A_k}||_\infty$. Then we have

$$\begin{aligned}\log \tau_k(M_\varphi) &= -2\log \det_{\mathbb{C}\pi(k)}(R_{A_k}) \\ &= -4g\log K_k + \sum_{p=1}^{\infty} \frac{1}{p} \mathrm{tr}_{\mathbb{C}\pi(k)}\left(\left(I - K_k^{-2}A_k{A_k}^*\right)^p\right),\end{aligned}$$

by virtue of the same argument as Theorem 3.4.

For the kth invariant τ_k, we have taken the lower central series $\{\Gamma_k\}$ of Γ and the nilpotent quotients $\{N_k\}$. These quotients induce a sequence of representations (more precisely, crossed homomorphisms)

$$r_k : \mathcal{M}_{g,1} \to GL(2g, \mathbb{Z}N_k)$$

for $k \geq 1$ (see [23]). They naively approximate the original Magnus representation $r : \mathcal{M}_{g,1} \to GL(2g, \mathbb{Z}\Gamma)$. By the similar observation as before, the kth invariant $\log \tau_k(M_\varphi)$ can be regarded as the characteristic polynomial of $r_k(\varphi)$ with respect to the Fuglede-Kadison determinant in $M(2g, \mathbb{C}\pi(k))$.

From the viewpoint of the Magnus representation of the mapping class group, it seems natural to raise the following problem.

Problem 4.4. *Show that the sequence* $\{\tau_k(M_\varphi)\}$ *converges to* $\tau(M_\varphi)$ *when we take the limit on* k.

In general, such an approximation problem for the L^2-torsion seems to be difficult. However, similar convergence results are known for the L^2-Betti numbers. In fact, Lück shows in [18] a theorem relating L^2-Betti numbers to ordinary Betti numbers of finite coverings. This result is generalized to more general settings by Schick in [24].

As for the Fuglede-Kadison determinant, Lück proves in [19] the following. Let $f : \mathbb{Q}[\mathbb{Z}] \to \mathbb{Q}[\mathbb{Z}]$ be the $\mathbb{Q}[\mathbb{Z}]$-map given by multiplication with $p(t) \in \mathbb{Q}[\mathbb{Z}]$ and $f_{(2)} : l^2(\mathbb{Z}) \to l^2(\mathbb{Z})$ be the linear operator obtained from f by tensoring with $l^2(\mathbb{Z})$ over $\mathbb{Q}[\mathbb{Z}]$. Further let $f_{[n]} : \mathbb{C}[\mathbb{Z}/n] \to \mathbb{C}[\mathbb{Z}/n]$ be the linear operator obtained from f by taking the tensor product with $\mathbb{C}[\mathbb{Z}/n]$ over $\mathbb{Q}[\mathbb{Z}]$. We then get an approximation result:

$$\log \det_{\mathbb{C}[\mathbb{Z}]}\left(f_{(2)}\right) = \lim_{n\to\infty} \frac{\log \det_{\mathbb{C}[\mathbb{Z}/n]}\left(f_{[n]}\right)}{n}$$

(see [11] for a similar statement). In [19] Lück also points out that there exists a purely algebraic example where Fuglede-Kadison determinants do not converge.

On the other hand, in general, we have at least an inequality for the Fuglede-Kadison determinant in the limit statement (see [24]). That is, for the operator R_{A_k} we see that

$$\log \det_{\mathbb{C}\pi}(R_A) \geq \limsup_k \log \det_{\mathbb{C}\pi(k)}(R_{A_k})$$

holds. In the last section, we shall discuss Problem 4.4 again and give an affirmative answer under certain conditions.

§5. Formulas of τ_1 and τ_2

In this section, we give explicit formulas of the first two invariants of a sequence of our L^2-torsion invariants. They are really computable formulas, so that we can make a systematic calculation for low genus cases. In particular, we compare them with hyperbolic volumes. The results discussed here are a summary of our previous paper [12] (see also [10], [11]).

First we consider the Magnus representation

$$r_1 : \mathcal{M}_{g,1} \to GL(2g, \mathbb{Z}N_1).$$

Here $N_1 = \Gamma/\Gamma_1$ is the trivial group and then the above representation is the same as the usual homological action of $\mathcal{M}_{g,1}$ on $H_1(\Sigma_{g,1}, \mathbb{Z})$. Namely we have the representation

$$r_1 : \mathcal{M}_{g,1} \to \mathrm{Aut}\,(H_1(\Sigma_{g,1}, \mathbb{Z}), \langle\ ,\ \rangle) \cong \mathrm{Sp}(2g, \mathbb{Z}),$$

where $\langle\ ,\ \rangle$ denotes the intersection form on the first homology group. Further $\pi(1) = \pi/\Gamma_1 \cong \mathbb{Z} = \langle t \rangle$ and its group ring $\mathbb{C}\langle t \rangle$ is a commutative Laurent polynomial ring $\mathbb{C}[t, t^{-1}]$. Then the matrix A_1 is nothing but the usual characteristic matrix of ${}^t r_1(\varphi)$. In this case, it is described by the usual determinant for a matrix with commutative entries.

In order to state the theorem, we recall a definition from number theory (see [6] and its references). For a Laurent polynomial $F(\mathbf{t}) \in \mathbb{C}[t_1^{\pm 1}, \dots, t_n^{\pm 1}]$, the Mahler measure of F is defined by

$$m(F) = \int_0^1 \cdots \int_0^1 \log \left| F(e^{2\pi\sqrt{-1}\theta_1}, \dots, e^{2\pi\sqrt{-1}\theta_n}) \right| d\theta_1 \cdots d\theta_n,$$

where we assume that undefined terms are omitted. Namely we define the integrand to be zero whenever we hit a zero of F.

Theorem 5.1 ([12]). *The logarithm of the first invariant τ_1 is given by*

$$\log \tau_1(M_\varphi) = -2m\left(\Delta_{r_1(\varphi)}\right),$$

where $\Delta_{r_1(\varphi)}(t) = \det A_1 = \det(tI - r_1(\varphi))$. *Moreover if* $\Delta_{r_1(\varphi)}(t)$ *has a factorization* $\Delta_{r_1(\varphi)}(t) = \prod_{i=1}^{2g}(t-\alpha_i)$ $(\alpha_i \in \mathbb{C})$, *then we have*

$$\log \tau_1(M_\varphi) = -2\sum_{i=1}^{2g} \log\max\{1, |\alpha_i|\}.$$

Remark 5.2. In other words, $\log \tau_1(M_\varphi)$ is given by the integral of the Alexander polynomial of M_φ over the circle S^1 (see [16], for the exterior of a knot K in the 3-sphere S^3). Further, $\log \tau_1(M_\varphi)$ can be described by the asymptotic behavior of the order of the first homology group of a cyclic covering (see [11]).

The point of the proof is to identify the Hilbert space $l^2(\mathbb{Z})$ with $L^2(\mathbb{R}/\mathbb{Z})$ in terms of the Fourier transforms. Then the $\mathbb{C}\langle t\rangle$-trace $\mathrm{tr}_{\mathbb{C}\langle t\rangle}$: $l^2(\mathbb{Z}) \to \mathbb{C}$ can be realized as the integration

$$L^2(\mathbb{R}/\mathbb{Z}) \ni f(\theta) \mapsto \int_0^1 f(\theta)d\theta \in \mathbb{C}$$

(see [12] for details). From this description and Kronecker's theorem ([6] Theorem 2), we obtain a certain vanishing theorem of the first invariant.

Corollary 5.3. *The logarithm of* $\tau_1(M_\varphi)$ *vanishes if and only if every eigenvalue of* $r_1(\varphi) \in \mathrm{Sp}(2g, \mathbb{Z})$ *is a root of unity.*

This corollary seems to be interesting from the viewpoint of Problem 4.4. Because in some case, we can say that the first invariant τ_1 already approximates the simplicial volume. In particular, Corollary 5.3 implies that a torus bundle M_φ $(g = 1)$ with a hyperbolic structure (namely, $|\mathrm{tr}\,(r_1(\varphi))| \geq 3$) has always non-trivial L^2-torsion invariant $\tau_1(M_\varphi)$. Summing up, we have

Corollary 5.4. *For any* $\varphi \in \mathcal{M}_{1,1}$, *its mapping torus* M_φ *admits a hyperbolic structure if and only if* M_φ *has a non-trivial* L^2*-torsion invariant* $\tau_1(M_\varphi)$.

Therefore, the first invariant τ_1 already approximates the simplicial volume in genus one case.

Remark 5.5. It is known that if the characteristic polynomial of $r_1(\varphi) \in \mathrm{Sp}(2g, \mathbb{Z})$ is irreducible over $\mathbb{Z}$, has no roots of unity as eigenvalues and is not a polynomial in t^n for any $n > 1$, then φ is pseudo-Anosov (see Casson-Bleiler [5]). In this case, $\mathrm{vol}(M_\varphi) \neq 0$ and further $\log \tau_1(M_\varphi) \neq 0$ by Corollary 5.3.

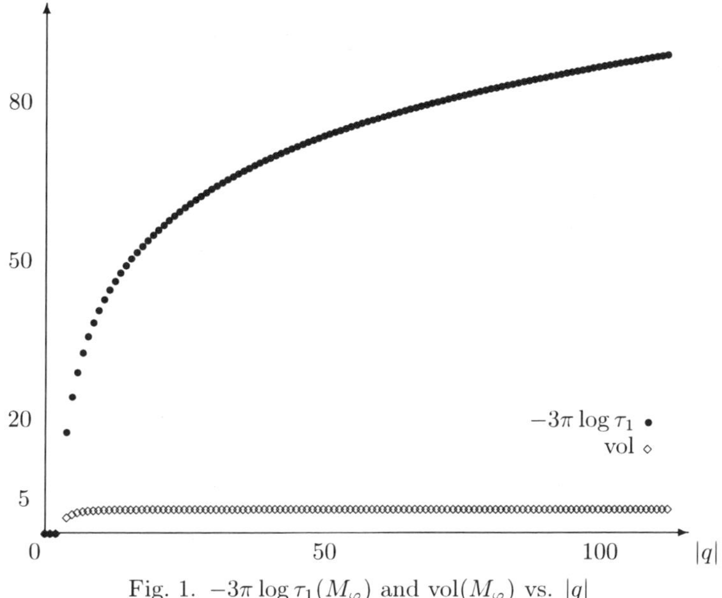

Fig. 1. $-3\pi \log \tau_1(M_\varphi)$ and $\mathrm{vol}(M_\varphi)$ vs. $|q|$

Example 5.6. It is well-known that the mapping class group of the two dimensional torus $T^2 = \mathbb{R}^2/\mathbb{Z}^2$ is isomorphic to $SL(2,\mathbb{Z})$. Taking a matrix $\begin{pmatrix} q & 1 \\ -1 & 0 \end{pmatrix} \in SL(2,\mathbb{Z})$, it gives a diffeomorphism φ on T^2. We may assume that it is the identity on some embedded disk by an isotopic deformation and it gives an element of $\mathcal{M}_{1,1}$. We use the same symbol φ for this mapping class. An easy calculation shows that

$$r_1(\varphi) = \begin{pmatrix} q & 1 \\ -1 & 0 \end{pmatrix}$$

and

$$\Delta_{r_1(\varphi)}(t) = \det(tI - r_1(\varphi)) = t^2 - qt + 1.$$

We put $\xi_\pm = (q \pm \sqrt{q^2 - 4})/2$ (the eigenvalues of the matrix $r_1(\varphi)$). If $|q| \leq 2$, then $|\xi_\pm| = 1$. Hence $\log \tau_1(M_\varphi) = 0$ in these cases. On the other hand, either $|\xi_+|$ or $|\xi_-|$ is greater than one when $|q| \geq 3$, so that M_φ has a non-trivial L^2-torsion invariant τ_1 in these cases. In fact, the logarithm of the first invariant is given by

$$\log \tau_1(M_\varphi) = -2 \log \max \{|\xi_+|, |\xi_-|\}.$$

The values of $\log \tau_1$ for the traces q and $-q$ are the same, so that it is a function of $|\mathrm{tr}(r_1(\varphi))|$. We put a graph of the L^2-torsion invariant

$-3\pi \log \tau_1(M_\varphi)$ and the hyperbolic volume $\mathrm{vol}(M_\varphi)$ as a function of $|q|$ in Fig 1.

Example 5.7. Next we consider the genus two case. Let $t_1, \ldots, t_5$ be the Lickorish-Humphries generators of $\mathcal{M}_{2,1}$. We take the element $\varphi = t_1 t_3 {t_5}^2 t_2^{-1} t_4^{-1} \in \mathcal{M}_{2,1}$. As was shown in [5], the characteristic polynomial of $r(\varphi)$ is

$$\begin{aligned}\Delta_{r_1(\varphi)}(t) &= \det(tI - r_1(\varphi)) \\ &= t^4 - 9t^3 + 21t^2 - 9t + 1\end{aligned}$$

and irreducible over $\mathbb{Z}$. Moreover it has no roots of unity as zeros. Hence, φ is pseudo-Anosov and M_φ has a non-trivial L^2-torsion invariant $\tau_1(M_\varphi)$. In fact, we have

$$-3\pi \log \tau_1(M_\varphi) = 52.954.... \quad \text{and} \quad \mathrm{vol}(M_\varphi) = 11.466....$$

Remark 5.8. In the above two examples, we used SnapPea [26] to compute the hyperbolic volumes.

Now in the following, we consider the second invariant τ_2. In the case of genus one, we can prove the vanishing of $\log \tau_2(M_\varphi)$.

Theorem 5.9 ([11]). $\log \tau_2(M_\varphi) = 0$ *for any* $\varphi \in \mathcal{M}_{1,1}$.

This follows from the fact that the group $\pi(2)$ is isomorphic to the fundamental group of a closed torus bundle over the circle. Such a 3-manifold admits no hyperbolic structures, so that the original L^2-torsion is trivial and we obtain the assertion.

On the other hand, in the case of $g \geq 2$, it is difficult to describe $\log \tau_2$ explicitly on the full mapping class group $\mathcal{M}_{g,1}$. However, we can do it on the Torelli group. Let φ be an element of the Torelli group $\mathcal{I}_{g,1}$, namely φ acts trivially on the first homology group $H_1(\Sigma_{g,1}, \mathbb{Z})$. Then we notice that $\log \tau_1(M_\varphi) = 0$ holds for any $\varphi \in \mathcal{I}_{g,1}$ (see Corollary 5.3). To give an explicit formula of $\log \tau_2$, we consider the Magnus representation

$$r_2 : \mathcal{M}_{g,1} \to GL(2g, \mathbb{Z}N_2),$$

where $N_2 = \Gamma/[\Gamma, \Gamma] \cong H_1(\Sigma_{g,1}, \mathbb{Z})$. If we restrict r_2 to the Torelli group $\mathcal{I}_{g,1}$, this is really a homomorphism (see [23] Corollary 5.4). Then our formula for the second L^2-torsion invariant is the following. The proof is similar to one for Theorem 5.1.

Theorem 5.10 ([12]). *For any mapping class* $\varphi \in \mathcal{I}_{g,1}$, *the logarithm of the second* L^2*-torsion invariant* $\tau_2(M_\varphi)$ *is given by*

$$\log \tau_2(M_\varphi) = -2m\left(\Delta_{r_2(\varphi)}\right),$$

where $\Delta_{r_2(\varphi)}(y_1,\ldots,y_{2g},t) = \det A_2 = \det(tI - \overline{r_2(\varphi)})$ *and* y_i *denotes the homology class corresponding to* x_i.

Now we suppose $F(\mathbf{t}) \in \mathbb{Z}[t_1^{\pm 1},\ldots,t_n^{\pm 1}]$ is primitive. We define F to be a generalized cyclotomic polynomial if it is a monomial times a product of one-variable cyclotomic polynomials evaluated at monomials.

The next corollary immediately follows from the theorem of Boyd, Lawton and Smyth (see [6] Theorem 4).

Corollary 5.11. *For any mapping class* $\varphi \in \mathcal{I}_{g,1}$, $\log \tau_2(M_\varphi) = 0$ *if and only if* $\Delta_{r_2(\varphi)}$ *is a generalized cyclotomic polynomial.*

As a typical element of the Torelli group $\mathcal{I}_{g,1}$, we first consider a BSCC-map φ_h $(1 \leq h \leq g)$ of genus h. That is, a Dehn twist along a bounding simple closed curve on $\Sigma_{g,1}$ which separates $\Sigma_{g,1}$ into $\Sigma_{h,1}$ and genus $g-h$ surface with two boundaries. We then see from [25] Corollary 4.3 that $\Delta_{r_2(\varphi_h)} = (t-1)^{2g}$. This is clearly a generalized cyclotomic polynomial, so that $\log \tau_2(M_{\varphi_h}) = 0$.

Second we consider a BP-map $\psi_h = D_c D_{c'}^{-1}$ of genus h $(1 \leq h \leq g-1)$, where c and c' are disjoint homologous simple closed curves on $\Sigma_{g,1}$ and D_c denotes the Dehn twist along c. Since

$$\Delta_{r_2(\psi_h)} = (t-1)^{2g-2h}(t - y_{g+h+1})^{2h}$$

holds (see [25]), where y_{g+h+1} denotes the homology class corresponding to the $(h+1)$th meridian of $\Sigma_{g,1}$, we also have $\log \tau_2(M_{\psi_h}) = 0$.

The next example shows the non-triviality of the second L^2-torsion invariant $\log \tau_2$.

Example 5.12. Let $\varphi = t_3 \varphi_1 t_3^{-1} \varphi_1 \in \mathcal{I}_{2,1}$. Then we see from a computation in [25] that

$$\Delta_{r_2(\varphi)} = (t-1)^4 + t(t-1)^2(y_1 - 2 + y_1{}^{-1})(y_2 - 2 + y_2{}^{-1}).$$

This is not a generalized cyclotomic polynomial, so that the mapping torus M_φ has a non-trivial L^2-torsion invariant $\tau_2(M_\varphi)$. In fact we can numerically compute it by means of Lawton's result (see [13]). More precisely we have

$$\begin{aligned} -3\pi \log \tau_2(M_\varphi) &= 6\pi m\left(\Delta_{r_2(\varphi)}\right) \\ &= 6\pi \lim_{r\to\infty} m\left(\Delta_{r_2(\varphi)}(u,u,u^r)\right) \\ &= 19.28.... \end{aligned}$$

§6. Vanishing of $\log \tau_k$ for reducible mapping classes

From the Nielsen-Thurston theory (see [5]), the mapping classes of a surface are classified into the following three types: (i) periodic, (ii) reducible and (iii) pseudo-Anosov. In our point of view, the most interesting object is a pseudo-Anosov map φ. Because the corresponding mapping torus M_φ has non-trivial hyperbolic volume.

In this final section, we show two vanishing theorems for $\log \tau_k$. We introduced an infinite sequence $\{\tau_k\}$ as an approximation of the hyperbolic volume. Thus if it behaves well with the index k, we ought to prove

$$\lim_{k\to\infty} \log \tau_k = 0$$

for non-hyperbolic 3-manifolds (see Problem 4.4). As a first step of this observation, we obtain the following.

Theorem 6.1. *If $\varphi \in \mathcal{M}_{g,1}$ is the product of Dehn twists along any disjoint non-separating simple closed curves on $\Sigma_{g,1}$ which are mutually non-homologous, then $\log \tau_k(M_\varphi) = 0$ for any $k \geq 1$.*

Remark 6.2. The mapping torus M_φ for $\varphi \in \mathcal{M}_{g,1}$ as above admits no hyperbolic structures, so that $\mathrm{vol}(M_\varphi) = 0$ holds.

Proof. At first, we prove the theorem for the genus one case. After that we give the outline of the proof in the higher genus case.

Let D_c be a Dehn twist along a non-separating simple closed curve c on $\Sigma_{1,1}$. Taking a conjugation, we can assume that the curve c is one of the standard generators of $\pi_1(\Sigma_{1,1})$. We then see that $\varphi = {D_c}^q$ is represented by a matrix $\begin{pmatrix} 1 & q \\ 0 & 1 \end{pmatrix} \in SL(2, \mathbb{Z})$. Thus we can choose a deficiency one presentation

$$\langle x, y, t \mid txt^{-1} = x,\ tyt^{-1} = x^q y \rangle$$

of $\pi_1(M_\varphi)$. Applying the free differential calculus to the relators $txt^{-1}x^{-1}$ and $tyt^{-1}(x^q y)^{-1}$, we obtain the Alexander matrix

$$A = \begin{pmatrix} t-1 & 0 \\ -\partial(x^q)/\partial x & t - x^q \end{pmatrix}.$$

Here we remark that the generators t and x can be commuted by the relation $txt^{-1} = x$. Hence in this case, the kth Alexander matrix A_k coincides with the original matrix A. In particular, t and x always commute. As we saw in Section 5, the L^2-torsion invariant $\tau_k(M_\varphi)$ ($k \geq$

1) can be computed by using the usual determinant and the Mahler measure in such a situation. Since

$$\det A = (t-1)(t-x^q)$$

is a generalized cyclotomic polynomial, we obtain $\log \tau_k(M_\varphi) = 0$ as desired (see Corollary 5.11).

In the higher genus case, we can assume that the mapping class φ is given by

$$\begin{aligned}
&\varphi_*(x_1) = x_1,\ \varphi_*(x_2) = x_1^{q_1}x_2, \ldots, \\
&\varphi_*(x_{2l-1}) = x_{2l-1},\ \varphi_*(x_{2l}) = x_{2l-1}^{q_l}x_{2l}, \\
&\varphi_*(x_{2l+1}) = x_{2l+1}, \ldots, \varphi_*(x_{2g}) = x_{2g}
\end{aligned}$$

by taking a conjugation, where $q_1, \ldots, q_l \in \mathbb{Z}$ and $1 \leq l \leq g-1$. We then obtain the following presentation of $\pi_1(M_\varphi)$:

$$\langle x_1, \ldots, x_{2g}, t \mid tx_it^{-1} = \varphi_*(x_i),\ 1 \leq i \leq 2g \rangle.$$

Since the Alexander matrix A is the direct sum of the 2×2-matrix in the genus one case, we obtain $\log \tau_k(M_\varphi) = 0$ by the similar arguments.
Q.E.D.

As another affirmative answer to Problem 4.4, we can show the vanishing of $\log \tau_k$ for the following mapping classes (see [12]). That is, we consider the case where there exists an integer n such that M_{φ^n} is topologically the product of $\Sigma_{g,1}$ and S^1. Here its bundle structure is non-trivial in general. Namely the nth power φ^n of a given monodromy φ is not trivial. A typical example is the Dehn twist along the simple closed curve on $\Sigma_{g,1}$ parallel to the boundary. The difference between an isotopy fixing the boundary pointwisely and such one setwisely, it gives birth to the difference between a bundle structure and a topological type. We then obtain

Theorem 6.3 ([12]). $\log \tau_k(M_\varphi) = 0$ *for any* $k \geq 1$.

It is easy to see that such a 3-manifold does not admit a hyperbolic structure. Hence it has trivial simplicial volume.

The above two examples are both non-hyperbolic cases, so that we conclude the present paper with the following problem.

Problem 6.4. *Show*

$$\lim_{k \to \infty} \log \tau_k(M_\varphi) = \log \tau(M_\varphi)$$

for a pseudo-Anosov diffeomorphism φ.

Acknowledgements. The authors are grateful to the referee for his/her numerous and helpful comments which greatly improved this paper.

References

[1] J. Birman, Braids, links, and mapping class groups, Ann. of Math. Stud., **82**, Princeton Univ. Press, Princeton, NJ, 1974.

[2] D. Burghelea, L. Friedlander, T. Kappeler and P. McDonald, Analytic and Reidemeister torsion for representations in finite type Hilbert modules, Geom. Funct. Anal., **6** (1996), 751–859.

[3] A. Carey, M. Farber and V. Mathai, Determinant lines, von Neumann algebras and L^2 torsion, J. Reine. Angew. Math., **484** (1997), 153–181.

[4] A. Carey and V. Mathai, L^2-torsion invariants, J. Funct. Anal., **110** (1992), 377–409.

[5] A. Casson and S. Bleiler, Automorphisms of surfaces after Nielsen and Thurston, Cambridge Univ. Press, Cambridge, 1988.

[6] G. Everest, Measuring the height of a polynomial, Math. Intelligencer, **20** (1998), 9–16.

[7] M. Gromov, Volume and bounded cohomology, Inst. Hautes Études Sci. Publ. Math., **56** (1982), 5–99.

[8] E. Hess and T. Schick, L^2-torsion of hyperbolic manifolds, Manuscripta Math., **97** (1998), 329–334.

[9] N. V. Ivanov, Mapping class groups, Handbook of geometric topology, North-Holland, Amsterdam, 2002, pp. 523–633.

[10] T. Kitano, T. Morifuji and M. Takasawa, Numerical calculation of L^2-torsion invariants, Interdiscip. Inform. Sci., **9** (2003), 35–42.

[11] T. Kitano, T. Morifuji and M. Takasawa, L^2-torsion invariants and homology growth of a torus bundle over S^1, Proc. Japan Acad. Ser. A Math. Sci., **79** (2003), 76–79.

[12] T. Kitano, T. Morifuji and M. Takasawa, L^2-torsion invariants of a surface bundle over S^1, J. Math. Soc. Japan, **56** (2004), 503–518.

[13] W. M. Lawton, A problem of Boyd concerning geometric means of polynomials, J. Number Theory, **16** (1983), 356–362.

[14] J. Lott, Heat kernels on covering spaces and topological invariants, J. Differential Geom., **35** (1992), 471–510.

[15] J. Lott and W. Lück, L^2-topological invariants of 3-manifolds, Invent. Math., **120** (1995), 15–60.

[16] W. Lück, L^2-torsion and 3-manifolds, Low-dimensional topology, Knoxville, TN, 1992, Conf. Proc. Lecture Notes Geom. Topology, III, Int. Press, Cambridge, MA, 1994, pp. 75–107.

[17] W. Lück, L^2-Betti numbers of mapping tori and groups, Topology, **33** (1994), 203–214.

[18] W. Lück, Approximating L^2-invariants by their finite-dimensional analogues, Geom. Funct. Anal., **4** (1994), 455–481.
[19] W. Lück, L^2-invariants: theory and applications to geometry and K-theory, Ergeb. Math. Grenzgeb. (3), **44**, Springer-Verlag, Berlin, 2002.
[20] W. Lück and M. Rothenberg, Reidemeister torsion and the K-theory of von Neumann algebras, K-Theory, **5** (1991), 213–264.
[21] W. Lück and T. Schick, L^2-torsion of hyperbolic manifolds of finite volume, Geom. Funct. Anal., **9** (1999), 518–567.
[22] V. Mathai, L^2-analytic torsion, J. Funct. Anal., **107** (1992), 369–386.
[23] S. Morita, Abelian quotients of subgroups of the mapping class group of surfaces, Duke Math. J., **70** (1993), 699–726.
[24] T. Schick, L^2-determinant class and approximation of L^2-Betti numbers, Trans. Amer. Math. Soc., **353** (2001), 3247–3265 (electronic).
[25] M. Suzuki, A class function on the Torelli group, Kodai Math. J., **26** (2003), 304–316.
[26] J. Weeks, SnapPea, `http://www.geometrygames.org/SnapPea` .
[27] H. Zieschang, E. Vogt and H-D. Coldewey, Surfaces and planar discontinuous groups, Translated from the German by John Stillwell, Lecture Notes in Math., **835**, Springer, Berlin, 1980.

Teruaki Kitano
Department of Information Systems Science
Soka University
Tokyo 192-8577, Japan

Takayuki Morifuji
Department of Mathematics
Tokyo University of Agriculture and Technology
Tokyo 184-8588, Japan

Advanced Studies in Pure Mathematics 52, 2008
Groups of Diffeomorphisms
pp. 51–75

Parameterized Gromov-Witten invariants and topology of symplectomorphism groups

Hông-Vân Lê[1] **and Kaoru Ono**[2]

Abstract.

In this note we introduce parameterized Gromov-Witten invariants for symplectic fiber bundles and study the topology of the symplectomorphism group. We also give sample applications showing the non-triviality of certain homotopy groups of some symplectomorphism groups.

§1. Introduction

Given a symplectic manifold (M, ω), one of the basic mathematical objects associated to (M, ω) is its automorphism group $\mathrm{Symp}(M, \omega)$. Since the group $\mathrm{Symp}(M, \omega)$ can be equipped with the C^∞ topology, we would like to know the homotopy type of this automorphism group. In this note we are interested in the following questions:

1) How large is the rank of the homotopy group $\pi_i(\mathrm{Symp}(M, \omega) \otimes \mathbf{Q})$?

2) What are the characteristic classes of $\mathrm{Symp}(M, \omega)$, that is, the cohomology ring of the classifying space $\mathrm{BSymp}(M, \omega)$.

Our approach to these problems uses symplectic fiber bundle setting (a similar setting is used in the study of homotopy type of diffeomorphism groups) and Gromov's technique of pseudoholomorphic curves. We would also like to remark that the Gromov technique of pseudoholomorphic curves has been developed and extended in different directions in the study of the topology of symplectomorphism groups. Recent developments in this direction can be found in, e.g., McDuff's survey [19].

Received April 30, 2007.
Revised October 24, 2007.
[1] Partially supported by grant of ASCR Nr IAA100190701.
[2] Partially supported by Grant-in-Aid for Scientific Research Nos. 08640093 and 18340014, Ministry of Education, Culture, Sports, Science and Technology, Japan.

This note consists of four sections. In section 2 we recall the definition of symplectic fiber bundles and we introduce the notion of fiber-wise (vertical) stable maps. In section 3 we show that the basic properties of the moduli space of stable maps also hold in the fiber-wise (family) version. As an immediate consequence we construct parameterized Gromov-Witten invariants for symplectic fiber bundles, which is a family version of the usual Gromov-Witten invariants for symplectic manifolds. In section 4 we apply this construction to problems 1, 2 mentioned above. We associate to each element $\pi_i(\mathrm{Symp}(M,\omega))$ a symplectic fiber bundle over S^{i+1} which is the union of two trivial symplectic bundles over a disk D^{i+1} glued along the boundary ∂D^{i+1} by this element $\pi_i(\mathrm{Symp}(M,\omega))$. We re-interpret a result by Gromov [8], Theorem 2.4.C_2 on the existence of an element of infinite order in the symplectomorphism group of non-monotone $S^2 \times S^2$ in terms of parameterized Gromov-Witten invariants, see Theorem 4.3. We also slightly generalize Gromov's result in the following cases. We denote by (X_1^4, ω_1) a non-monotone symplectic manifold which is diffeomorphic to $S^2 \times S^2$, and by (X_2^4, ω_2) a symplectic manifold which is diffeomorphic to $\mathbf{C}P^2 \# \overline{\mathbf{C}P^2}$.

Theorem 4.5. a) *Let* $(M_1, \Omega_1) = (X_1^4 \times N^{2k}, \omega_1 \oplus \omega_0)$ *be a symplectic manifold with* (X_1^4, ω_1) *as above and* (N, ω_0) *a compact symplectic manifold. Then we have* $rk(\pi_1(\mathrm{Symp}(M_1, \Omega_1)) \otimes \mathbf{Q}) \geq 1$.

b) *We also have* $rk(\pi_1(\mathrm{Symp}(X_2^4, \omega_2))) \geq 1$.

There are intensive studies on cohomology groups and homotopy types of symplectomorphism groups of rationally ruled symplectic 4-manifolds such as Abreu [1], Abreu-McDuff [2], Anjos [3], etc. In fact, we can improve Theorem 4.5 for the case of (M_1, Ω_1) without using "hard machinery'.

Theorem 4.8. a) *The rank of the homomorphism* $i_* : \pi_1(\mathrm{Symp}(M_1, \Omega_1)) \to \pi_1(\mathrm{Diff}(M_1))$ *is at least* 1.

b) *The rank of the homomorphism* $i_* : \pi_3(\mathrm{Symp}(M_1, \Omega_1)) \to \pi_3(\mathrm{Diff}(M_1))$ *is greater than or equal* 2.

In section 4 we also construct characteristic classes of the group $\mathrm{Symp}(M, \omega)$ by formulating the Gromov-Witten invariants in a dual way. We also include an Appendix, which contains an alternative proof of Theorem 4.3, a special version of Theorem 4.5.a.

After the preliminary version of this note was written [14], we learned several works on the topology of symplectomorphism groups, [10], [22], see also references in [19]. Since some results in [14] have been quoted in some literature e.g. [4], [5], [19], we feel a need to revise the version [14] to correct some errors as well as to add details to missing arguments.

Acknowledgement. We thank Forschungsinstitut Oberwolfach, where we exchanged some ideas during our stay in RIP program 1996. The first author is indebted to John Lott for his inspiring lecture [15] and his help in the reference in the homotopy type of diffeomorphism groups. We are also grateful to Dusa McDuff for her interest and critical helpful comments. A part of this note was written during the stay of the first author at the Max-Planck-Institute in Bonn and the Mathematical Institute of the University Leipzig during her Heisenberg fellowship. She thanks all these institutions for their hospitality. The second author thanks Professor Akira Kono, who suggested the way of simplifying the proof of Theorem 4.8.

§2. Symplectic fiber bundles and vertical stable maps

In this section we recall the notions of symplectic fiber bundles, stable maps and introduce the notion of vertical stable maps. We refer to [7] for more discussions on symplectic fiber bundles. The idea of counting fiber-wise holomorphic curves in symplectic fiber bundles is also suggested by Kontsevich (in his communication to us after a preliminary version of this note has been written in 1997) and by Lu-Tian[1].

2.1. Symplectic bundles and their fiber-wise compatible almost complex structures

A fibration $M \to E \xrightarrow{\pi} B$ is said to be a symplectic fiber bundle, if the fiber M is diffeomorphic to a symplectic manifold (M, ω) and the transition function takes its value in the group $\mathrm{Symp}(M, \omega)$. We denote by $\mathcal{J}_\pi(E)$ the associated bundle over B whose fiber is the space $\mathcal{J}(M)$ of smooth compatible almost complex structures on (M, ω). Since the fiber $\mathcal{J}(M)$ is contractible, the space of sections $J(E) : B \to \mathcal{J}_\pi(E)$ is also non-empty and contractible.

In what follows, we are interested in defining invariants which detect the non-triviality of symplectic fiber bundles. Associated to any symplectic fiber bundle $M \to E \xrightarrow{\pi} B$ we obtain the local system of fiberwise homology groups, resp. fiberwise cohomology groups, denoted by $\mathcal{H}_*(E)$, resp. $\mathcal{H}^*(E)$. In what follows, the coefficients of cohomology groups are in $\mathbf{R}$ or $\mathbf{Z}$, and the coefficients of homology groups are in $\mathbf{Z}$.

---

[1]We thank Dusa McDuff for informing us that they used this idea in order to construct equivariant Gromov-Witten invariants, which is a special case of the parameterized Gromov-Witten invariants for the symplectic fiber bundle associated with the Hamiltonian action of a compact Lie group on a symplectic manifold.

Clearly all the invariants of the associate local systems $\mathcal{H}_*(E), \mathcal{H}^*(E)$ are also invariants of symplectic fiber bundles. In particular, if a symplectic fiber bundle is trivial, then the associated local systems are simple, i.e., trivial. We also observe that for a symplectic fiber bundle E there is always a section $s^{[\omega]} : B \to \mathcal{H}^2(E)$ which takes a given value $[\omega] \in H^2(M; \mathbf{R})$ and there is also a section $s^{c_1} : B \to \mathcal{H}^2(E)$ which takes a given value $c_1(M, \omega) \in H^2(M; \mathbf{Z})$.

2.2. Stable maps and vertical stable maps.

Our notion of vertical stable maps is based on the notion of stable maps due to Kontsevich [12], [11], see also [6] whose exposition we follow closely.

Let g and m be nonnegative integers. A semistable curve with m marked points is a pair (Σ, z) of a connected space $\Sigma = \cup \pi_\nu(C_\nu)$, where C_ν is a Riemann surface and $\pi_\nu : C_\nu \to \Sigma$ is a continuous map, and $z = (z_1, \cdots, z_m)$ are m distinct points in Σ with the following properties.
(1) π_ν is the normalization of the irreducible component $\Sigma_\nu = \pi_\nu(C_\nu)$ of Σ for all ν.
(2) For each $p \in \Sigma$ we have $\sum_\nu \#\pi_\nu^{-1}(p) \leq 2$. Here # denotes the order of the set.
(3) $\sum_\nu \#\pi_\nu^{-1}(z_i) = 1$ for each z_i.
(4) The number of Riemann surfaces C_ν is finite.
(5) The set $\{p \in \Sigma | \sum_\nu \#\pi_\nu^{-1}(p) = 2\}$ is finite.

We denote by g_ν the genus of C_ν and by m_ν the number of points $\overline{p}$ on C_ν, which are the inverse image of nodes of Σ, i.e. $\sum_\gamma \#\pi_\gamma^{-1}(\pi_\nu(\overline{p})) = 2$, or marked points, i.e. $\pi_\nu(\overline{p}) = z_j$ for some j. The genus g of a semistable curve Σ is defined by by

$$g = \sum_\nu g_\nu + \dim H_1(T_\Sigma; \mathbf{Q}),$$

where T_Σ is a graph associated to Σ in the following way. The vertices of T_Σ correspond to the components of Σ. Denote by v_ν the vertex corresponding to Σ_ν. For a node $p \in \Sigma_\nu \cap \Sigma_{\nu'}$, we assign an edge e_p joining vertices v_ν and $v_{\nu'}$. (When p is a node of Σ_ν, the "edge" e_p becomes a loop based at v_ν.)

A homeomorphism $\theta : \Sigma \to \Sigma'$ between two semistable curves is called an isomorphism, if it restricts to a biholomorphic isomorphism $\theta_{\nu\nu'} : \Sigma_\nu \to \Sigma'_{\nu'}$ for each component Σ_ν of Σ and some component $\Sigma'_{\nu'}$. We also require that θ maps the marked points in Σ onto the corresponding marked points in Σ' bijectively.

Let $J(E)$ be a vertical compatible almost complex structure. A map $u : (\Sigma, z) \to E$ is called a vertical $J(E)$-stable map, if the composition map $\pi \circ u$ sends (Σ, z) to a point $b \in B$ and u is a stable map from (Σ, z) to $\pi^{-1}(b) = (M, J(E)|_{E_b})$. In other words, for each ν, the restriction of u to each component Σ_ν is either a non-constant map, or we have $m_\nu + 2g_\nu \geq 3$.

To define the moduli space of vertical stable maps, we assume first, for the sake of simplicity and a later application, that the local system $\mathcal{H}_2(E)$ is simple, i.e., the fundamental group $\pi_1(B)$ acts trivially on $\mathcal{H}_2(E)$ (e.g. it is the case if the base B is simply connected).

In this case, for a class $A \in H_2(M; \mathbf{Z})$, there is a global locally constant section $s_A : B \to \mathcal{H}_2(E)$ whose value is A at a reference fiber. We consider the moduli space of all vertical stable map $((\Sigma_g, z), u)$ such that (Σ_g, z) is of genus g with m marked points. We denote by $C\mathcal{M}_{g,m}(E, J(E), s_A)$ the moduli space of vertical stable maps representing the class $s_A(\pi(u))$:

$$C\mathcal{M}_{g,m}(E, J(E), s_A) := \cup_{b \in B}\{b\} \times C\mathcal{M}_{g,m}(E_b, J(E)|_{E_b}, s_A(b)),$$

which carries a Kuranishi structure in the sense of [6], see Lemma 3.1 below.

Here $C\mathcal{M}_{g,m}(E_b, J(E)|_{E_b}, s_A(b))$ is the moduli space of stable maps of genus g, with m marked points and representing the homology class $s_A(b)$. Here, two pairs $((\Sigma, z), h))$ and $((\Sigma', z'), h')$ are equivalent, if and only if there exists an isomorphism $\theta : (\Sigma, z) \to (\Sigma', z')$ satisfying $h' \circ \theta = h$.

If the action of $\pi_1(B)$ on the fiber $H_2(M; \mathbf{Z})$ is non-trivial, we can still define a notion of a moduli space of vertical stable maps by considering a multi-valued section $s_A : B \to \mathcal{H}_2(E)$ which is obtained by the locally constant continuation of A in a typical fiber to a multi-valued section. We note that the pairing $\langle s^{[\omega]}(b), s_A(b)\rangle$ as well as the pairing $\langle s^{c_1}(b), s_A(b)\rangle$ are constant functions on B, since they are locally constant functions and we assume that B is connected. The number $\langle s^{[\omega]}, s_A(b)\rangle$ is the "energy" of a holomorphic curve realizing any class in $s_A(b)$, and the second number $\langle s^{c_1}(b), s_A(b)\rangle$ enters in the expected dimension of the moduli space of stable maps representing any class in $s_A(b)$. Now using the Gromov compactness theorem it is easy to see that there is only a finite number of values of s_A in each fiber $H_2(M = \pi^{-1}(b); \mathbf{Z})$ such that there is a $J(E)|_{E_b}$-holomorphic curve representing a homology class in the set $s_A(b)$. The projection from $C\mathcal{M}_{g,m}(E, J(E), s_A)$ to B is proper. It follows from the Gromov compactness theorem: If u_i is a sequence of J_i-stable maps with energy bounded by a constant and J_i converges to J_∞ in the space of compatible almost complex structures, then there exists a subsequence u_{i_k}, which converges to a J_∞-stable map u_∞.

Finally we observe that any element in $s_A(b)$ induces the same class in $H_*(E;\mathbf{Z})$ by the inclusion.

Remark 2.1. It is sometimes more convenient to work with a tame almost complex structure (i.e. $\omega(X, JX) > 0$ for any non-zero tangent vector X). As in the non-parameterized case all the compactness and perturbation theorems for pseudo-holomorphic curves with respect to a compatible almost complex structure hold for a tame almost complex structure.

2.3. Examples of symplectic fiber bundles.

There are several ways to construct symplectic fiber bundles.

The first way is the associate bundle method. Suppose that a group G acts symplectically on a symplectic manifold (M,ω), i.e. there is a homomorphism $\rho : \mathrm{G} \to \mathrm{Symp}(M,\omega)$. Then we can associate to each principal G-bundle P over B a symplectic fiber bundle $P \times_{\mathrm{G}} (M,\omega)$. This symplectic fiber bundle is non-trivial, if and only if the image $\rho_*(\lambda)$ of the homotopy class $\lambda \in [B, \mathrm{BG}]$ defining the G-bundle is non-trivial in $[B, \mathrm{BSymp}(M,\omega)]$.

The second way is the pull-back method. Suppose that we are given a symplectic fiber bundle E over a base B. Then any map f from B' to B pulls the bundle E back to a symplectic fiber bundle E' over B'.

The third way is the reduction method. We begin with a differentiable fiber bundle $M \to E \to B$ with M being a symplectic manifold and ask if this fiber bundle also admits a structure of a symplectic fiber bundle. Of course, it is the case if the inclusion of $\mathrm{Symp}(M,\omega)$ to $\mathrm{Diff}^+(M)$ is a homotopy equivalence (e.g. if dim $M = 2$). In general, we can state the following criterion, see e.g., [7] for more information.

Lemma 2.2. *Let $\pi : E \to B$ be a fiber bundle and $\omega \in \Omega^2(E)$ be a closed form such that ω is non-degenerate along all fibers of E. Then $\pi : E \to B$ admits a structure of a symplectic fiber bundle, which is compatible with ω.*

The fourth way to construct symplectic fiber bundles is the gluing method. Suppose that we are given two symplectic bundles E_1 and E_2 over bases B_1 and B_2 respectively. Suppose that the restriction of E_1 over the boundary ∂B_1 is isomorphic to the restriction of E_2 over the boundary ∂B_2. Then we can glue the bundle E_1 with E_2 along the boundary $E_i|_{\partial B_i}$. In particular when the restriction of E_i over ∂E_i is trivial then the glued bundle is defined uniquely by a map $\partial B_1 \to \mathrm{Symp}(M,\omega)$. If B_i is closed, we can define the operation of fiber connected sum as follows. Choose a small disk D_i in B_i and take a trivialization of $E_i|_{D_i} \to D_i$. Then glue $E_i|_{B_i \setminus D_i}$, $i = 1, 2$, along ∂D_i.

Remark 2.3. Each element $g \in \pi_k(\mathrm{Symp}(M, \omega))$ defines a symplectic fiber bundle E with the fiber (M, ω) over a sphere S^{k+1} by gluing two trivial symplectic fiber bundles $D^{k+1} \times M$ along the boundary $M \times S^k$ by the element g. Conversely any symplectic fiber bundle over S^{k+1} is defined by such a method.

§3. Parameterized Gromov-Witten invariants

In this section we define parameterized Gromov-Witten invariants for symplectic fiber bundles over a closed oriented manifold B. The base B is assumed to be oriented in order to deal with the orientation of the moduli space of stable maps. The base B is also assumed to be a closed manifold in order to get the fundamental class of the moduli space of vertical stable maps.

3.1. Geometric picture

Recall that a semistable curve (Σ, z) with m marked points is called stable, if for all its component C_ν of the normalization of Σ we have $m_\nu + 2g_\nu \geq 3$. Let $C\mathcal{M}_{g,m}$ denote the Deligne-Mumford moduli space of stable curves, i.e. $C\mathcal{M}_{g,m}$ is the set of all isomorphism classes of stable curves with m marked points and of genus g. Let us denote by $E^{(m)}$ the "Whitney sum" (the multiple fiber product over B) of m copies of E. When $2g + m \geq 3$, as in the usual case (see e.g. [12], 2.4, [11], 1.5), there is the evaluation map

$$\Pi = pr \times ev_{g,m,s_A} : C\mathcal{M}_{g,m}(E, J(E), s_A) \to C\mathcal{M}_{g,m} \times E^{(m)},$$

$$((\Sigma, z), u) \mapsto ((\tilde{\Sigma}, \tilde{z}), u(z_1), \cdots, u(z_m)).$$

Here $(\tilde{\Sigma}, \tilde{z})$ is the stable curve with marked points obtained from (Σ, z) by consecutive contractions of non-stable components. When $2g + m = 0, 1$, we call ev_{g,m,s_A} the evaluation map.

We briefly recall the notion of Kuranishi structures, see [6], §5 for details. Roughly speaking a compact Hausdorff space X has a Kuranishi structure, if it is locally described as the zero set $s^{-1}(0)$ of a V-bundle over a V-manifold, namely for each $p \in X$, there exist a V-manifold U_p, a V-bundle E_p on it and a continuous section s of $E_p \to U_p$ such that the difference $\dim E_p - \dim U_p$ is independent of $p \in X$. Moreover, we assume that such local descriptions are compatible under coordinate changes $\{\phi_{pq}\}$ in a suitable sense. We also have the notion that X with Kuranishi structure has its tangent bundle. If a continuous map $f : X \to$

Y extends locally to $f_p : U_p \to Y$ for each p such that $f_p \circ \phi_{pq} = f_q$, we call f is a strongly continuous map. If X has an oriented Kuranishi structure and f is a strongly smooth map from X to a topological space Y, then we can define the image $f_*([X])$ of the fundamental class of X as the image $f_*[(s')^{-1}(0)]$ of the fundamental class of the zero set $(s')^{-1}(0)$ of a perturbed smooth multi-section s' of E which is transversal to zero. Taking into account of multiplicity in an appropriate way, this fundamental class gives a well-defined element in $H_*(Y; \mathbf{Q})$.

Lemma 3.1. *The space* $C\mathcal{M}_{g,m}(E, J(E), s_A)$ *has a Kuranishi structure with oriented tangent bundle. This space is compact and of dimension* $\dim B + 2m + 2\langle c_1(M), s_A\rangle + (6 - \dim M)(g-1)$. *Moreover,* Π *is strongly continuous map in the sense of Kuranishi structure.*

By this Lemma, we can define the virtual fundamental cycle of the moduli space of vertical stable maps. Denote by $\Pi_*([C\mathcal{M}_{g,m}(E, J(E), s_A)])$ the induced class in $H_*(C\mathcal{M}_{g,m} \times E^{(m)}; \mathbf{Q})$. The map Π induces a map in cohomologies

$$I^E_{g,m,s_A} : H^*(E^{(m)}; \mathbf{Q}) \to H^{*+\mu}(C\mathcal{M}_{g,m}; \mathbf{Q}) \tag{1}$$

by

$$I^E_{g,m,s_A}(\gamma) = PD(\gamma \backslash \Pi_*(C\mathcal{M}_{g,m}(E, J(E), s_A))) \tag{2}$$

If $2g + m = 0, 1$, we define

$$I^E_{g,m,s_A}(\gamma) = \langle \gamma, (ev_{g,m,s_A})_*(C\mathcal{M}_{g,m}(E, J(E), s_A))\rangle \in \mathbf{Q}. \tag{3}$$

The parameterized Gromov-Witten invariants, as in the usual case, are the collection of maps I^E_{g,m,s_A} defined in (2), (3). The shift of grading is given by $\mu = -\langle 2c_1(M), A\rangle + (g-1)\dim M - \dim B$.

Remark 3.2. The parametrized Gromov-Witten invariant $I^E_{g,0,s_A}$ is said to be of relative degree 0, if the expected dimension of $C\mathcal{M}_{g,0}(E, J(E), s_A)$ is zero, i.e., $\mu = 6(g-1)$. For symplectic fiber bundles E with the local system $\mathcal{H}_2(E)$ being simple, we can interpret the invariants $I^E_{g,0,s_A}$ of relative degree 0 in term of "counting vertical holomorphic curves" of genus g representing any class in s_A. Here "counting" means the "order" of the space with Kuranishi structure of expected dimension 0. The invariant

3.2. Proof of Lemma 3.1

To define rigorously the parametrized Gromov-Witten invariants for all compact symplectic fiber bundles, we need to prove Lemma 3.1 and moreover, to show that the map I^E_{g,m,s_A} does not depend on the choice of $J(E)$. Since the base space B is assumed to be compact, the moduli space $C\mathcal{M}_{g,m}(E, J(E), s_A)$ is compact, cf. section 2.2.

Proof of Lemma 3.1. Our proof of Lemma 3.1 is an adaptation of the proof of the corresponding results concerning Gromov-Witten invariants [6], Theorems 7.10 and 7.11.

Let u be a vertical stable map over $b_0 \in B$. For simplicity, we assume that the domain of u is irreducible. (The general case is handled in a similar way.) Pick a finite dimensional space $\mathcal{E}_0 \subset L^p\Omega^1(u^*TE_{b_0})$ such that

$$\mathrm{Im}\ D_u\overline{\partial}_{J_{b_0}} + \mathcal{E}_0 = L^p\Omega^1(u^*TE_{b_0}).$$

If $b \in B$ is in a neighborhood D_0 of b_0, we can identify fibers E_b and E_{b_0} by a local trivialization of E. Thus J_b is considered as an almost complex structure on E_{b_0}. Then there is a smaller neighborhood $D_0' \subset D_0$ of b_0 such that

$$\mathrm{Im}\ D_u\overline{\partial}_{J_b} + \mathcal{E}_0 = L^p\Omega^1(u^*TE_{b_0}).$$

Using this observation, the argument in [6] implies that there exists a Kuranishi neighborhood $(U, \mathcal{E}, s)$ of u on $C\mathcal{M}_{g,m}(E, J(E), s_A)$ so that a V-manifold U is fibered over an open subset of B.

All arguments in [6] can be directly adapted to the parametrized case or they even imply the corresponding statements in the parametrized case. For details of the construction of Kuranishi structure, see [6], §12. Q.E.D.

In order to show that the usual Gromov-Witten invariants do not depend on the choice of compatible almost complex structures, perturbations, we needed to construct a bordism between the moduli spaces corresponding to two almost complex structures J_1 and J_2 (with perturbations). This bordism is a version of the moduli space of vertical stable pseudo-holomorphic curves parametrized by the interval $[0, 1]$.

3.3. Properties of parameterized Gromov-Witten invariants

The following theorem immediately follows from Lemma 3.1 applied to the case that the base space is $B \times [0, 1]$.

Theorem 3.3. *The parameterized Gromov-Witten invariants are invariants of symplectic fiber bundles.*

Remark 3.4. The bordism type invariants of the moduli space of pseudoholomorphic curves may have more informations than the (cohomological) Gromov-Witten invariants. Such examples of finer Gromov-invariants of symplectic manifolds can be found in [17].

In order to distinguish a symplectic fiber bundle from the trivial one by parameterized Gromov-Witten invariants, we need to compute those for trivial symplectic fiber bundles $E = B \times M$. By the Künneth formula, the algebra $H^*(B \times (M)^{(m)}; \mathbf{Q})$ is isomorphic to $H^*(B; \mathbf{Q}) \otimes (H^*(M; \mathbf{Q}))^{\otimes m}$. Let us denote by α_i elements in $H^*(M; \mathbf{Q})$ and by β an element in $H^*(B; \mathbf{Q})$.

Proposition 3.5. *The Gromov-Witten invariants of a trivial symplectic fiber bundle satisfy*

$$I^E_{g,m,s_A}(\beta \otimes \alpha_1 \otimes \cdots \otimes \alpha_m) = I^M_{g,m,A}(\alpha_1 \otimes \cdots \otimes \alpha_m) \int_B \beta. \tag{4}$$

Proof. We choose a vertical compatible almost complex structure $J(E = B \times M)$ such that it is constant in the B-direction. Clearly the moduli space of vertical stable maps $C\mathcal{M}_{g,m}(E = B \times M, \{J_b \equiv J\}, s_A)$ is the direct product $B \times C\mathcal{M}_{g,m}(M, J, A)$. We take a multi-valued perturbation of Kuranishi map for $C\mathcal{M}_{g,m}(M, J, A)$ to define the virtual fundamental cycle of $B \times C\mathcal{M}_{g,m}(M, J, A)$. Denote by Π^{pt} the evaluation map in section 3.1 for $C\mathcal{M}_{g,m}(M, J, A)$, i.e. the case with base $B = pt$. By (2), the left hand side of (4) equals

$$\begin{aligned} &PD(\beta \otimes \alpha_1 \otimes \cdots \otimes \alpha_m \backslash \Pi_*[B \times C\mathcal{M}_{g,m}(M, J, A)]) \\ =&PD(\beta \otimes \alpha_1 \otimes \cdots \otimes \alpha_m \backslash ([B] \times \Pi^{pt}_*[C\mathcal{M}_{g,m}(M, J, A)]) \end{aligned} \tag{5}$$

Clearly the right hand side of (4) equals the right hand side of (5).

Q.E.D.

Now let us compute parameterized Gromov-Witten invariants of a pull-back symplectic fiber bundle. Let $p : B_1 \to B_2$ be a k-fold covering space and $E_2 \to B_2$ a symplectic fiber bundle. Then the pull-back $E_1 = p^* E_2 \to B_1$ is also a symplectic fiber bundle. For a single section s_A of $\mathcal{H}_2(E_2)$, denote by $p^* s_A$ its pull back. We get immediately the following

Proposition 3.6. *We have*

$$I^{E_1}_{g,m,p^* s_A}(p^*_{(m)} \alpha) = k I^{E_2}_{g,m,s_A}(\alpha),$$

where $p_{(m)} : E_1^{(m)} \to E_2^{(m)}$.

Parameterized Gromov-Witten invariants of relative degree 0 and without marked points satisfy the following additivity.

Proposition 3.7. *Let $E = E_1 \# E_2$ be a fiber connected sum of symplectic fiber bundles E_1 and E_2. Then we have the following formula for parameterized Gromov-Witten invariants of relative degree 0.*

$$I^E_{g,0,s_A} = I^{E_1}_{g,0,s_A} + I^{E_2}_{g,0,s_A}.$$

Proof. By the dimension assumption of the Gromov-Witten invariants, we take perturbation, if necessary, so that there is no vertical stable curves representing the class A over a small disk $D_i(\varepsilon)$ in the base B_i for $i = 1, 2$. We can assume further that our fiberwise almost complex structures on $D_i(\varepsilon)$, $i = 1, 2$, are constant and isomorphic each other. Now we perform the connected sum of symplectic fiber bundles using these disks. The almost complex structures on E_i can be glued together identifying their restrictions on $D_i(\varepsilon) \times M$. Hence we obtain the proposition. Q.E.D.

Remark 3.8. For symplectic fiber bundles $\pi_i : E_i \to B_i$ with simple local systems $\mathcal{H}^*(E_i)$, $\alpha_j \in H^{q_j}(M; \mathbf{Q})$ defines the locally constant sections of $\mathcal{H}^{q_j}(E_i)$, $i = 1, 2$, and $\mathcal{H}^{q_j}(E_1 \# E_2)$. Suppose that there exist cohomology classes $\widetilde{\alpha}^{(i)}_j \in H^{q_j}(E_i; \mathbf{Q})$ such that their restrictions to typical fibers coincide with α_j. Let $\widetilde{\alpha}_j \in H^{q_j}(E_1 \# E_2; \mathbf{Q})$ be a cohomology class, which is equal to $\widetilde{\alpha}^{(i)}_j$ after restricting to $\pi_i^{-1}(B_i \setminus D_i(\varepsilon))$. When $\mu + \sum_j q_j = 6(g-1)$, we have

$$I^E_{g,m,s_A}(\prod_j \widetilde{\alpha}_j) = I^{E_1}_{g,m,s_A}(\prod_j \widetilde{\alpha}^{(1)}_j) + I^{E_2}_{g,m,s_A}(\prod_j \widetilde{\alpha}^{(2)}_j).$$

At the end of this section we would like to suggest that many properties of the Gromov-Witten invariants (e.g. the Kontsevich-Manin axioms) should be valid in the family version. Specially interesting seems to us an analog of Taubes' theorem on the relation of Gromov-Witten invariants and Seiberg-Witten invariants in dimension 4. It would imply that the parametrized Gromov-Witten invariants also bring information on the homotopy type of the diffeomorphism group of 4-dimensional symplectic manifolds.

§4. Homotopy groups of symplectomorphism groups.

In this section we combine Remark 2.3, Propositions 3.5, 3.6, 3.7 and other observations to estimate the rank of homotopy groups of symplectomorphism groups.

Proposition 4.1. *Parameterized Gromov-Witten invariants of relative degree* 0 *and without marked points* $I^E_{g,0,s_A}$ *over sphere* S^{i+1} *define elements in* $Hom(\pi_i(\mathrm{Symp}(M,\omega)), \mathbf{Q})$, $i \ge 1$.

Proof. Denote by E_g the symplectic fiber bundle over S^{i+1} by gluing $D_1^{i+1} \times M \cup_g D_2^{i+1} \times M$ with the help of a map $g : \partial D_1^{i+1} \to \mathrm{Symp}(M)$, i.e. we identify a pair $(x, y) \in \partial D_1^{i+1} \times M$ with $(x, g(x)\cdot y) \in \partial D_2^{i+1} \times M$. Then we have $E_{g\cdot f} \cong E_g \# E_f$. Now we get Proposition 4.1 immediately from Remark 2.3 and Proposition 3.7. Q.E.D.

Remark 4.2. Taking into account Remark 3.8 we can get a similar statement for certain parameterized Gromov-Witten invariants. We use such invariants in the proof of Theorem 4.5.a.

As an application of Proposition 4.1, Remark 4.2, we shall prove Theorem 4.3 and Theorem 4.5.

Let $(M, \omega) = (S^2 \times S^2, \omega^{(1)} \oplus \omega^{(2)})$ be a product of symplectic manifolds. We denote by A_i the generators of $H_2(M; \mathbf{Z})$ realizing by the i-th sphere S^2.

Theorem 4.3. *If* $\omega^{(1)}(A_1) > \omega^{(2)}(A_2)$ *then there is an* $S^2 \times S^2$*-symplectic fiber bundle over* S^2 *with non-vanishing parameterized Gromov-Witten invariants of relative degree* 0 *and* $m = 0$. *In particular the rank of* $\pi_1(\mathrm{Symp}(S^2 \times S^2, \omega))$ *is at least* 1.

We recall that the last statement in Theorem 4.3 is established in Theorem 2.4. C_2 in [8].

Proof of Theorem 4.3. There are several ways of describing the proof of this theorem (see also Appendix, which follows the original idea of Gromov [8]). We present here a proof following the idea of McDuff in [17], Lemma 3.1, which uses the deformation space of complex structures of Hirzebruch's surfaces of even degree, which is diffeomorphic to $S^2 \times S^2$, see [16]. Thus we can apply technique in complex analytic geometry for our computation.

Denote by $\mathcal{O}(\ell)$ the holomorphic line bundle of degree ℓ on $\mathbf{C}P^1$. Let us recall that the Hirzebruch surface F_k is the projectivization of a rank 2 holomorphic vector bundle $W_k = \mathcal{O}(0) \oplus \mathcal{O}(k)$ with $k \ge 0$ over $\mathbf{C}P^1$. The line subbundles $\mathcal{O}(0) \oplus 0$ and $0 \oplus \mathcal{O}(k)$ define sections of $F_k = \mathbf{P}(W_k) \to \mathbf{C}P^1$ with self-intersection number k and $-k$, respectively. We denote by J^{F_k} the complex structure on the Hirzebruch surface F_k. It is known that all Hirzebruch surfaces with even degree k are diffeomorphic to $S^2 \times S^2$, and F_0 is biholomorphic to $\mathbf{C}P^1 \times \mathbf{C}P^1$ (see e.g., [16], chapter 1).

We define a family of holomorphic vector bundles $\{V_a\}$ of rank 2 on $\mathbf{C}P^1$ as follows. Write $U_0 = \mathbf{C}P^1 \setminus \{\infty\}$ and $U_\infty = \mathbf{C}P^1 \setminus \{0\}$. Consider the transition function $f_a : (U_0 \setminus \{0\}) \times \mathbf{C}^2 \to (U_\infty \setminus \{\infty\}) \times \mathbf{C}^2$ by

$$f_a(z, v_1, v_2) = (z, zv_1 + av_2, z^{-1}v_2).$$

Denote by X_a the projectivization of V_a and J_a the complex structure on it. Note that $V_{a=0}$ is isomorphic to $\mathcal{O}(-1) \oplus \mathcal{O}(1)$ and that $\mathbf{P}(V_{a=0})$ is isomorphic to $\mathbf{P}(W_2)$. Thus $\{X_a\}$ is a complex one-dimensional deformation of $X_0 = F_2$. The complex structure J_0 is the complex structure J^{F_2}. All J_a, $a \neq 0$, are isomorphic to the complex structure J^{F_0}, i.e., the product complex structure.

Note that

$$\mathcal{X} = \cup_{a \in \mathbf{C}^1} \{a\} \times X_a \to \mathbf{C}^1 \times \mathbf{C}P^1$$

is the projectivization of the vector bundle

$$\mathcal{V} = \cup_{a \in \mathbf{C}^1} \{a\} \times V_a \to \mathbf{C}^1 \times \mathbf{C}P^1.$$

Since the restriction of the vector bundle $\mathcal{V}$ to $(\mathbf{C} \setminus \{0\}) \times \mathbf{C}P^1$ is holomorphically trivialized by the following 2 sections σ_1 and σ_2, which are everywhere linearly independent:

$$\sigma_1(a, z) = \begin{cases} (z, 0, 1) \in U_0 \times \mathbf{C}^2, & \text{if } z \in U_0 \\ (z, a, \frac{1}{z}) \in U_\infty \times \mathbf{C}^2, & \text{if } z \in U_\infty \end{cases}$$

$$\sigma_2(a, z) = \begin{cases} (z, 1, -\frac{z}{a}) \in U_0 \times \mathbf{C}^2, & \text{if } z \in U_0 \\ (z, 0, -\frac{1}{a}) \in U_\infty \times \mathbf{C}^2, & \text{if } z \in U_\infty \end{cases}$$

Note that

$$f_a(z, 0, 1) = (z, a, \frac{1}{z}), \quad f_a(z, 1, -\frac{z}{a}) = (z, 0, -\frac{1}{a}).$$

Thus σ_1, σ_2 are well-defined holomorphic sections. Using this trivialization, we extend $\mathcal{V}$ and $\mathcal{X}$ across $\{\infty\} \times \mathbf{C}P^1$ to $\mathbf{C}P^1 \times \mathbf{C}P^1$. We denote these extensions by the same symbols. $\mathcal{X}$ is also considered as an $S^2 \times S^2$ bundle with the projection to the first factor $\mathbf{C}P^1$ parameterized by a. Denote this fiber bundle by $E \to \mathbf{C}P^1$. The complex structure on $\mathcal{X}$ induces the fiberwise complex structure $J(E)$.

Clearly the parameterized Gromov-Witten invariant $I_{0,0,A_1-A_2}$ is of relative degree 0. We shall show that its value $I^E_{0,0,A_1-A_2}$ is 1. Note that there is no J^{F_0}-holomorphic sphere realizing class $(A_1 - A_2)$, since $J^{F_0} = \mathbf{C}P^1 \times \mathbf{C}P^1$. Further, there is exactly one J^{F_2}-holomorphic sphere realizing class $(A_1 - A_2)$, which is the section of self-intersection number -2 in the $\mathbf{C}P^1$ bundle over $\mathbf{C}P^1$. Thus the moduli space $C\mathcal{M}_{0,0,A_1-A_2}(E, J(E))$ consists of one point at $a = 0$.

To prove the transversality of this moduli space, we argue as follows. Let N be the normal bundle of the unique J^{F_2}-holomorphic sphere in the class $(A_1 - A_2)$ in the total space of fibration. In order to show the transversality in the case of an integrable complex structure, it suffices to prove the following

Lemma 4.4. *We have* $H^1(\mathbf{C}P^1; \mathcal{O}(N)) = 0$. *Hence,* $N = \mathcal{O}(-1) \oplus \mathcal{O}(-1)$.

Proof. Consider the cohomology exact sequence associated to

$$0 \to \mathcal{O}(-2) \to N \to \mathcal{O}_D \to 0,$$

where $\mathcal{O}(-2)$ is the normal bundle of the (-2)-curve in the central fiber $X_{a=0} \subset \mathcal{X}$ with the complex structure J^{F_2}, and the third term $\mathcal{O}_D$ is the quotient, which is nothing but the pull back of the normal bundle of the origin in $\mathbf{C}P^1$ parameterized by a. We shall show that the connecting homomorphism $H^0(\mathcal{O}_D) \to H^1(\mathcal{O}(-2))$ is surjective, (hence isomorphism). As a consequence we get $H^1(N) = 0$.

Here we regard X_a as the projectivization of $\mathcal{O}(-1) \otimes V_a$. When $a = 0$, it is isomorphic to $\mathcal{O}(-2) \oplus \mathcal{O}(0)$. The vector bundle $\mathcal{O}(-1) \otimes V_a$ is written as the gluing $U_0 \times \mathbf{C}^2$ and $U_\infty \times \mathbf{C}^2$ by

$$(z, v_1, v_2) \mapsto (z, z^2 v_1 + azv_2, v_2).$$

Note that the (-2)-curve representing $A_1 - A_2$ is the image of $\{0\} \otimes \mathcal{O}(0)$, hence the image of the section $(0, 1)$ in $\mathbf{P}(\mathcal{O}(-2) \oplus \mathcal{O}(0))$ over $a = 0$.

Consider the restrictions of the section $(0, 1)$ of $\mathcal{O}(-1) \otimes V_{a=0}$ over U_0 and U_∞, respectively. For $a \in \mathbf{C}$, the pair $z \in U_0 \mapsto (z, 0, 1) \in U_0 \times \mathbf{C}^2$ and $z \in U_\infty \mapsto (z, 0, 1) \in U_\infty \times \mathbf{C}^2$ gives a deformation as a Cech 0-cocycle. Taking the Cech differential, we get a Cech 1-cocycle $z \in U_0 \cap U_\infty \subset U_\infty \mapsto (z, az, 0) \in (U_\infty \setminus \{\infty\}) \times \mathbf{C}^2$. Differentiate in a, then we find that the last component of the Cech 1-cocycle vanishes and obtain a Cech 1-cocycle $z \in U_0 \cap U_\infty \subset U_\infty \mapsto (z, z) \in (U_\infty \setminus \{\infty\}) \times \mathbf{C}$ of $\mathcal{O}(-2)$, which represents a non-zero element in $H^1(\mathbf{C}P^1; \mathcal{O}(-2))$. Hence we conclude that the connecting homomorphism $H^0(\mathcal{O}_D) \to H^1(\mathcal{O}(-2))$ is surjective. Q.E.D.

Let us generalize Theorem 4.3. We denote by (X_1^4, ω_1) a non-monotone symplectic manifold which is diffeomorphic to $S^2 \times S^2$ and by (X_2^4, ω_2) a symplectic manifold which is diffeomorphic to $\mathbf{C}P^2 \# \overline{\mathbf{C}P^2}$. We are going to prove the following

Theorem 4.5. a) *Let* $(M_1, \Omega_1) = (X_1^4 \times N^{2k}, \omega_1 \oplus \omega_0)$ *be a symplectic manifold with* (X_1, ω_1) *as above and* (N^{2k}, ω) *a compact symplectic manifold. Then we have* $rk(\pi_1(\mathrm{Symp}(M_1, \Omega_1)) \otimes \mathbf{Q}) \geq 1$.
b) *We also have* $rk(\mathrm{Symp}(M_2^4, \omega_2)) \geq 1$.

Proof. a) To prove the statement for (M_1, Ω_1) we shall construct a symplectic fiber bundle E with fiber (M_1, Ω_1) over $B = S^2$ and compute a certain parameterized Gromov-Witten invariant with one marked point of E. For a symplectic fiber bundle E with fiber (M_1, Ω_1), we consider the moduli space of vertical holomorphic mappings $f : S^2 \to E$ whose image represents the class $A := A_1 - A_2$. Note that the local system $\mathcal{H}_2(E)$ is simple for any symplectic fiber bundle over the base space $B = S^2$.

The dimension computation shows us that the moduli space of vertical stable maps of genus 0 in the class A on an (M, ω)-bundle over S^2 has dimension $2k = \dim N^{2k}$. We will compare symplectic fiber bundles over S^2, each of which is the product of a symplectic fiber bundle E' with the fiber X_1^4 over S^2 and $N = N^{2k}$, i.e., $E = E' \times N \to S^2$. We shall count the number of vertical pseudo-holomorphic spheres u with one marked point z in the class A so that $ev(u; z) \in \Gamma$. Here Γ is a cycle represented by a submanifold $E' \times \{y_0\} \subset E$, for some arbitrary chosen point $y_0 \in N^{2k}$. This number is the parameterized Gromov-Witten invariant $I^E_{0,1,A_1-A_2}(PD[\Gamma])$.

Note that the restriction of $PD[\Gamma] \in H^*(E)$ to $E|_b$, $b \in B$, is $PD[X_1^4 \times \{y_0\}] \in H^*(X_1^4 \times N)$. But this condition does not characterize $PD[\Gamma] \in H^*(E)$. Let $\Gamma' \in E$ be another cycle such that the restriction of $PD[\Gamma']$ to $E|_b$ is equal to $PD[X_1^4 \times \{y_0\}] \in H^*(X_1^4 \times N)$. Since the base space B is S^2, we find that $PD[\Gamma'] - PD[\Gamma] \in H^2(B) \otimes H^*(X_1^4 \times N)$. In other words, $[\Gamma'] - [\Gamma]$ is represented by some cycle contained in a single fiber $E|_{b_0}$. By dimensional counting argument, we can take a fiberwise almost complex structure such that there are no pseudo-holomorphic $(A_1 - A_2)$-spheres contained in $E|_{b_0}$. Thus we have $I^E_{0,1,A_1-A_2}(PD[\Gamma]) = I^E_{0,1,A_1-A_2}(PD[\Gamma'])$. Hence $I^E_{0,1,A_1-A_2}(PD[\Gamma])$ gives an invariant for symplectic $X_1^4 \times N$-bundles over S^2.

We claim that this number of the trivial bundle $S^2 \times (M_1, \Omega_1)$ equals zero. To show it we consider a vertical almost complex structure J^{prod} on the trivial bundle $(M, \omega) \times S^2$ such that on each fiber (M, ω) we have $J^{\mathrm{prod}} = J^0 \times J^N$, where J^0 is the standard product complex structure on $S^2 \times S^2$ and J^N is an almost complex structure on (N^{2k}, ω_0). Clearly the projection on the first factor of any J^{prod}-sphere is also a J^0-sphere in $S^2 \times S^2$. Hence the moduli space of J^{prod}-sphere in class $A_1 - A_2$ is empty.

Now we construct a symplectic (M_1, Ω_1)-bundle E over S^2 by gluing two trivial symplectic (M_1, Ω_1)-bundles, one over a disk B^2 and the other over a disk D^2, using the loop $\tilde{g}_t : S^1 \to \mathrm{Symp}(M_1, \Omega_1)$ of the form $\tilde{g}_t = (g_t \times \{Id\}) \in (\mathrm{Symp}(X_1^4, \omega_1) \times \{Id\}) \subset \mathrm{Symp}(M_1, \Omega_1)$. Here g_t is the transition function for $\mathcal{X} \to \mathbf{C}P^1$. (See also Appendix.)

To compute the Gromov-Witten invariant $I^E_{0,1,A_1-A_2}(PD[\Gamma])$, we construct a fiberwise compatible almost complex structure $J(E)$ by gluing two fiberwise compatible almost complex structures on the restriction of E to disks D_0^2 and D_1^2. The first fiberwise compatible almost complex structure is defined as follows: $J(z) = (J^0 \times J^N)$ for $z \in D_0^2$. The second one is defined as follows: $J(a) = (J_a \times J^N)$, $a \in D_1^2 \subset \mathbf{C}$. They are glued by using the symplectomorphism loop $\tilde{g}_t$.

We claim that the constructed vertical almost complex structure is $A = A_1 - A_2$-generic. Clearly outside the singular point $a = 0$ in the disk D_1^2 of the base S^2, where the vertical almost complex structure take value $(J^{F_2} \times J^N)$, the moduli space of $J(E)$-holomorphic spheres realizing A is empty. At the singular point $a = 0$ the moduli space is diffeomorphic to N^{2k}, namely it consists of maps $u_y(t) = \{u_1(t) \times y\}, y \in N^{2k}$, where $t \in S^2$ and u_1 is the J^{F_2}-holomorphic (-2)-sphere in X_1^4. Clearly the transversality of the constructed $J(E)$ is equivalent to the surjectivity of the linearization map $D_u\bar{\partial}_{J(E)} : T_{\pi(u)}S^2 \times L_1^p(u^*T_{ver}E) \to L^p(\Lambda^{0,1}S^2 \otimes_{J(E)} u^*T_{ver}E)$. Since $u = (u_1, y)$, we have $u^*T_{ver}E = (u_1^*TX_1^4) \times T_yN^{2k}$, so the surjectivity of $D\bar{\partial}_{J(E)}$ follows from the surjectivity of the linearization map considered in the proof of Theorem 4.3, see also the proof of Lemma 5.1 in Appendix. This proves the first statement in Theorem 4.5 for the case (M_1, Ω_1).

Now let us prove the statement b). Denote by $\mathrm{Symp}(\mathbf{C}P^2, \omega, pt)$ the subgroup of the symplectomorphisms of $(\mathbf{C}P^2, \omega)$ which preserve a point pt. Clearly Theorem 4.5.b follows from the following Lemmas 4.6, 4.7. Q.E.D.

Lemma 4.6. *The fundamental group of* $\mathrm{Symp}(\mathbf{C}P^2, \omega, pt)$ *contains a subgroup* $\mathbf{Z}$.

Lemma 4.7. *There is an injective homomorphism α from the infinite cyclic subgroup of* $\pi_1(\mathrm{Symp}(\mathbf{C}P^2, \omega, pt))$ *in Lemma* 4.6 *to* $\pi_1(\mathrm{Symp}(X_2^4, \omega_2))$.

Lemma 4.6 is a direct consequence of Gromov's theorem, which states that $\mathrm{Symp}(\mathbf{C}P^2, \omega)$ is homotopy equivalent to $PU(3)$, since the quotient space $\mathrm{Symp}(\mathbf{C}P^2, \omega)/\mathrm{Symp}(\mathbf{C}P^2, \omega, pt)$ is isomorphic to $\mathbf{C}P^2$. We can also see this fact as follows. It suffices to show that the inclusion $U(2) \to \mathrm{Symp}(\mathbf{C}P^2, \omega, pt)$ induces an injective homomorphism on the corresponding fundamental groups. To see it we consider the evaluation map $ev : \mathrm{Symp}(\mathbf{C}P^2, \omega, pt) \to Sp(4) : g \mapsto Dg(pt, v)$, where v is an element in the frame $Sp(4)$ over the fixed point in $\mathbf{C}P^2$. The restriction of this evaluation map to $U(2)$ is injective, and we know that the image $ev(U(2))$ is a deformation retract of $Sp(4)$. Hence follows the Lemma.

Proof of Lemma 4.7. Denote by E_a the symplectic fiber bundle over S^2 with fiber $(\mathbf{C}P^2, \omega, pt)$ corresponding to element $a \in \pi_1(U(2))(\cong \mathbf{Z}) \subset \pi_1(\mathrm{Symp}(\mathbf{C}P^2, \omega, pt))$. Note that each fiber has a base point, hence there is a canonical section s of E_a.

Pick a $U(2)$-invariant Darboux ball B such that (X_2^4, ω_2) is symplectomorphic to the symplectic manifold $\mathbf{C}P^2 \setminus B$ with symplectic reduction applied to the boundary (symplectic cutting construction). Then we have the homomorphism $\rho : \pi_1(U(2)) \subset \pi_1(\mathrm{Symp}(\mathbf{C}P^2, \omega, pt)) \to \pi_1(\mathrm{Symp}(X_2^4, \omega_2))$. Denote by $\widetilde{E}_a$ the symplectic fiber bundle corresponding to $\rho(a)$. Then $\widetilde{E}_a$ is the fiberwise blow-up of E_a along the section s.

In fact, E_a carries a symplectic structure and $\widetilde{E}_a$ can be obtained by symplectic blowing-up of E_a along the symplectic submanifold $s(S^2)$ as follows. Using the spectral sequence for E_a we see that there is a closed 2-form Ω such that the restriction of Ω to the fiber $\mathbf{C}P^2$ equals ω. Thus we can apply the Thurston construction (see e.g. [21], p. 193) to conclude that E_a is a symplectic manifold with a symplectic form Ω_K in a class $K\pi^*(\omega_0) + \Omega$, where ω_0 is a symplectic form on the base S^2 and K is a sufficiently large real number. Moreover, all the fibers $\mathbf{C}P^2$ are symplectic submanifolds of (E_a, Ω_K) and we may assume that $s(S^2)$ is a symplectic submanifold. Recall that $\widetilde{E}_a$ is the symplectic fiber bundle over S^2 by fiber-wise blowing-up a symplectic fiber bundle with the fiber $(\mathbf{C}P^2, \omega)$ at point $s(x), x \in S^2$. This is exactly the blow-up (E_a, Ω_K) along the submanifold $s(S^2)$. Clearly the fiber of this new fiber bundle is $\mathbf{C}P^2 \# \overline{\mathbf{C}P^2}$. Now we apply Lemma 2.2 to conclude that $\widetilde{E}_a$ is a symplectic fiber bundle with the fiber (X_2^4, ω_2).

Now assume that α is not injective. Then, for some integer $a \neq 0$, we can find a trivialization of symplectic fiber bundle $\widetilde{E}_a$ with a constant compatible complex structure $J^0_{\widetilde{E}_a}$. We denote by $\{J^t_{\widetilde{E}_a}\}_{0 \le t \le 1}$ a family of compatible almost complex structures on $\widetilde{E}_a$ with $J^1_{\widetilde{E}_a}$ being the almost complex structure resulting from the blow-up process along $s(S^2)$. Note that the space of compatible almost complex structures, which admit $(-k)$-curve, is of real codimension $2(k-1)$. (See Appendix for the proof in the case that $k = 2$. The argument can be generalized for general

$k > 2$.) Note also that $\mathbf{C}P^2\#\overline{\mathbf{C}P^2}$ does not contain any cycle of self-intersection number -2. In our case, $\{J^t_{\widetilde{E}_a}\}$ gives a three parameter family of compatible almost complex structures on $\mathbf{C}P^2\#\overline{\mathbf{C}P^2}$. Thus we can take $\{J^t_{\widetilde{E}_a}\}_{0\le t\le 1}$ such that there are no $(-k)$-curves, $k \ge 2$, in the fibers of $(\widetilde{E}_a, J^t_{\widetilde{E}_a})$ over $\mathbf{C}P^1$. Then, for each $t \in [0,1]$ and $b \in S^2$, there is a unique (-1) embedded curve in each fiber $(\widetilde{E}_a|_b \cong X^4_2, \omega, J^t)$. Thus the families of $J^t_{\widetilde{E}_a}$-holomorphic $(A_1 - A_2)$-spheres in the fibers of $\widetilde{E}_a$ parametrized by $b \in S^2$ are isotopic, when t varies. Therefore the process of blowing-down of all $J^t_{\widetilde{E}_a}$-holomorphic $(A_1 - A_2)$-spheres in the fibers of $\widetilde{E}_a$ is unique up to isomorphisms of symplectic fiber bundles. Hence follows that our symplectic fiber bundle (E_a, Ω_K) is also trivial, which is a contradiction. Q.E.D.

We shall improve Theorem 4.5 in the following statement[2].

Theorem 4.8. a) *The rank of the homomorphism* $\pi_1(\mathrm{Symp}(M_1, \Omega_1)) \to \pi_1(\mathrm{Diff}(M_1))$ *is at least* 1.

b) *The rank of the homomorphism* $\pi_3(\mathrm{Symp}(M_1, \Omega_1)) \to \pi_3(\mathrm{Diff}(M_1))$ *is at least* 2.

Proof. a) It is enough to show that the symplectic fiber bundle constructed by the loop $(\tilde{g}_t)^k$, $k \neq 0$, in the proof of Theorem 4.5.a is a non-trivial differentiable fibration for all k. To do so, it suffices to compute the cohomology ring of this differentiable bundle. Since the loop $\widetilde{g}_t$ by our choice is the product of g_t and the identity element in $\mathrm{Symp}(N, \omega_0)$, our cohomology is also the tensor product of the corresponding rings.

Here, we consider the projectivization E of the bundle $\mathcal{V}$ over $\mathbf{C}P^1 \times \mathbf{C}P^1$, which is given in the proof of Theorem 4.3. We compute the cohomology ring of E. We regard E as a family of $\mathbf{C}P^1$-bundles on $\mathbf{C}P^1$ parametrized by $\mathbf{C}P^1$, namely a $\mathbf{C}P^1$-bundle over $S^2 \times S^2$.

Let us compute $H^*(E; \mathbf{Z})$ by using the Leray-Hirsch Theorem. To see that the first Chern class $c_1(\mathcal{V})$ vanishes, it suffices to compute its evaluation on $S^2 \times \{pt\}$ and $\{pt\} \times S^2$. The second Chern class $c_2(\mathcal{V})$ equals the Poincare dual of the class of the zero section of σ_1, which extends to a section on $\mathbf{C}P^1 \times \mathbf{C}P^1$. This zero locus consists of the only point $(z = \infty, a = 0)$. So we find that $c_2(\mathcal{V})$ is the generator $\{S^2 \times S^2\}$

[2]This proof was suggested by Professor A. Kono, when K.O. gave a proof of this result at his seminar in 1996.

of $H^4(S^2 \times S^2; \mathbf{Z})$. Applying the Leray-Hirsch-Theorem we get

$$H^*(E; \mathbf{Z}) = \frac{H^*(S^2 \times S^2; \mathbf{Z})[t]}{t^2 - \{S^2 \times S^2\}}.$$

In this ring t^2 cannot be divided by 2. But in the ring $H^*(S^2 \times S^2 \times S^2; \mathbf{Z})$ any element t^2 can be divided by 2. Hence E is not homotopic to $S^2 \times S^2 \times S^2$. That proves the non-triviality of the loop $\{g_t\}$ in the diffeomorphism group.

Similarly, we compare the modules consisting elements of degree 2 of $H^*(E \times N; \mathbf{Z})$ and $H^*(S^2 \times S^2 \times S^2 \times N; \mathbf{Z})$, whose square is divisible by 2 to conclude that the loop $\{\tilde{g}_t\}$ is not null homotopic in the diffeomorphism group.

To prove that the loop $\{g_t^n\}$, $n \neq 0$, also realizes a non-trivial element in the diffeomorphism group we proceed similarly. Namely the loop $\{g_t^n\}$ corresponds to the n-time fiber connected sum $E^{\langle n \rangle}$. We have the following formula

$$H^*(E^{\langle n \rangle}; \mathbf{Z}) = \frac{H^*(S^2 \times S^2; \mathbf{Z})[t]}{t^2 - n\{S^2 \times S^2\}}.$$

Now we note that the set of element $x \in H^*(E^{\langle n \rangle}; \mathbf{Z})$ such that $x^2 = 0$ is the union $\mathbf{Z}(\{S^2\} \times 1) \cup \mathbf{Z}(1 \times \{S^2\})$. In particular from any 3 elements in this set we can get 2 linearly dependent elements. This implies that $E^{\langle n \rangle}$ and $S^2 \times S^2 \times S^2$ not homotopic, because there are 3 linearly independent elements of the cohomology ring of $S^2 \times S^2 \times S^2$, namely the generators $\{S^2\} \times 1 \times 1$, $1 \times \{S^2\} \times 1$, $1 \times 1 \times \{S^2\}$, whose square vanish.

Similarly, the ranks of the modules generated by elements of square zero in $H^2(E \times N; \mathbf{Z})$ and $H^2(S^2 \times S^2 \times S^2 \times N; \mathbf{Z})$ are different. Hence $\{\tilde{g}_t\}$ is not null homotopic in the diffeomorphism group.

b) It suffices to show that the two subgroups $SO(3) \times Id \subset \mathrm{Symp}(M_1, \omega)$ and $Id \times SO(3) \subset \mathrm{Symp}(M_1, \omega)$ realize two linearly independent elements in $\pi_3(\mathrm{Diff}(M_1))$. We denote by E_1 and E_2 two differentiable bundles with fiber M_1 over S^4, which correspond to the elements in $\pi_3(\mathrm{Diff}(M_1))$ realized by these subgroups $SO(3)$. Let us consider the homotopy exact sequences of these fibration E_i, which give us two connecting homomorphisms $h_i : \pi_4(S^4) \to \pi_3(M_1)$. We observe that $\pi_4(S^4) = \mathbf{Z}$, $\pi_3(M_1) = \mathbf{Z} \oplus \mathbf{Z} \oplus \pi_3(N)$. Now it is easy to check that the homomorphism h_i are linearly independent, and hence the image of the two subgroups are also linearly independent elements in $\pi_3(E)$. Q.E.D.

We can also describe parametrized Gromov-Witten invariants in a different way using the Poincaré duality. Define

$$\begin{aligned} CGW^E_{g,m,s_A}: \quad & H^*(C\mathcal{M}_{g,m};\mathbf{R}) \to H^{*+\nu}(E^{(m)};\mathbf{R}) \\ & \delta \mapsto PD(\delta \backslash \Pi_*(C\mathcal{M}_{g,m,s_A}(E,J(E))). \end{aligned}$$

Here $\nu = m \dim M - 2\langle c_1(M), A\rangle + (\dim M - 6)(g-1) - 2m$.

Assume that the local system $\mathcal{H}_2(E)$ for the symplectic fiber bundle $p: E \to B$ is trivial. Using the moduli space with $m = 0$, we define the following homomorphism

$$\begin{aligned} CGW_{g,A}(E): \quad & H^*(C\mathcal{M}_{g,0};\mathbf{R}) \to H^{*+\nu}(B;\mathbf{R}) \\ & \delta \mapsto PD(p_*(\delta \backslash \Pi_*(C\mathcal{M}_{g,0,s_A}(E,J(E)))), \end{aligned}$$

where $p: C\mathcal{M}_{g,0,s_A}(E,J(E))) \to B$ is the projection.

We call the image of the map $CGW_{g,A}(E)$ the Gromov-Witten characteristic classes.

Theorem 4.9. *The Gromov-Witten characteristic classes are invariants of symplectic fiber bundles E with simple local system $\mathcal{H}_2(E)$. All the Gromov-Witten characteristic classes with positive degree vanish for trivial symplectic fiber bundles.*

Proof. The first statement is obvious (cf. Theorem 3.3). To prove the second statement we compute the Gromov-Witten characteristic classes on a trivial bundle $B \times M$. Let $[T]$ be a cycle in B. As before we denote also by p the projection $E^{(m)} \to B$.

$$\begin{aligned} \langle CGW_{g,A}(E)(\delta), [T]\rangle &= \langle PD(p_*(\delta \backslash \Pi_*(B \times C\mathcal{M}_{g,0}(J,A)))), ([T])\rangle \\ &= \langle PD([B]), [T]\rangle \langle \delta, \Pi_*(C\mathcal{M}_{g,0}(J,A))\rangle. \end{aligned}$$

Clearly $\langle PD([B]), [T]\rangle = 0$, if $\dim[T] \geq 1$. Q.E.D.

The word "characteristic class" is explained by the following functoriality of these classes. In particular we see that the Gromov-Witten characteristic classes are cohomology classes of the classifying space $\mathrm{BSymp}_0(M,\omega)$. (Note that $\mathrm{BSymp}_0(M,\omega)$ is simply connected, hence any local systems on it are simple. If we consider the moduli space of vertical stable maps with a fixed energy and a fixed Chern number, we can obtain characteristic classes for $\mathrm{Symp}(M,\omega)$-bundles.)

Theorem 4.10. *Let E be a symplectic fiber bundle over B with the simple local system $\mathcal{H}_2(E)$ and $f^*(E)$ be the induced symplectic fiber bundle by a map $f: B_1 \to B$. Then we find that*

$$\widetilde{f}^*(CGW^E_{g,m,A}) = CGW^{f^*E}_{g,m,A},$$

where $\widetilde{f} : f^*E \to E$ is the tautological bundle map of symplectic fiber bundles and

$$CGW_{g,A}(f^*E) = f^* \circ (CGW_{g,A})(E).$$

Proof. By the construction, we have

$$C\mathcal{M}_{g,m,s_A}(f^*(E)) = (C\mathcal{M}_{g,m,s_A}(E))_p \times_f B_1,$$

where we take the fiber product in the sense of spaces with Kuranishi structures and p is the projection from $C\mathcal{M}_{g,m,s_A}(E))$ to B. Hence, we find that

$$\widetilde{f}^*(PD(\Pi_*(C\mathcal{M}_{g,m,s_A}(E)))) = PD(\Pi_*(C\mathcal{M}_{g,m,s_A}(f^*E))).$$

Hence we can see immediately that

$$(id. \otimes (\widetilde{f}^*)^{\otimes m})(CGW^E_{g,m,A}) = CGW^{f^*E}_{g,m,A}.$$

When $m = 0$, it implies that

$$CGW_{g,A}(f^*E) = f^* \circ CGW_{g,A}(E)$$

Q.E.D.

It is an easy exercise to interpret Theorem 4.3 and Theorem 4.5 in term of Gromov-Witten characteristic classes.

§5. Appendix. An alternative proof of Theorem 4.3

Let $\omega^{(1)}$, $\omega^{(2)}$ be symplectic forms on S^2 such that $\int_{S^2} \omega^{(1)} > \int_{S^2} \omega^{(2)}$. According to Proposition 4.1 it suffices to find a symplectic bundle E over S^2 with fiber $(S^2 \times S^2, \omega = \omega^{(1)} \oplus \omega^{(2)})$ and a parameterized Gromov-Witten invariant whose value on E is non-trivial. We shall construct the bundle E by finding its transition function g, i.e., a loop in $\mathrm{Symp}(S^2 \times S^2, \omega)$. The existence of such element g was shown by Gromov [8], and in what follows we shall give a detailed proof. First we need the following lemma (compare with [8], 2.4.C). Denote by A_1, resp. A_2 the homology classes $[S^2 \times \{pt\}]$, resp. $[\{pt\} \times S^2]$.

Lemma 5.1. *The subspace $\mathcal{J}_0$ of compatible almost complex structures J on $S^2 \times S^2$, for which there exists a J-holomorphic sphere in a class $A_1 - \ell A_2$ for some $\ell \geq 1$, is a non-empty closed subset of codimension 2 in $\mathcal{J}(S^2 \times S^2, \omega)$. For each $J \in (\mathcal{J}(S^2 \times S^2, \omega) \setminus \mathcal{J}_0)$ and for each point in $S^2 \times S^2$ there is a unique J-holomorphic sphere representing class A_i and passing through x. Moreover, these spheres are embedded.*

Proof. First of all, we note that for any $J \in \mathcal{J}(S^2 \times S^2, \omega)$ there exists a unique embedded J-holomorphic sphere representing the class A_2 and passing through each point x, cf. [18]. We include here the proof of this fact for the reader's convenience.

For a generic compatible almost complex structure J, there exists such a J-holomorphic sphere. For $J_\infty \in \mathcal{J}(S^2 \times S^2, \omega)$, pick a sequence $\{J_i\}$ of generic compatible almost complex structures, which converges to J_∞. Let u_i be the J_i-holomorphic sphere representing the class A_2 and passing through x. Suppose that there exists a subsequence u_{i_k} converging to a J_∞-holomorphic sphere u_∞. Since A_2 is a primitive class, the adjunction formula implies that u_∞ is embedded. Clearly it passes through x. Hence we obtain the desired existence. If it is not the case, a subsequence of $\{u_i\}$ converges to a J_∞-stable map and there appears a J_∞-holomorphic map v representing the class $kA_2 - \ell A_1$ for some integers k and ℓ such that k is positive and $(k, \ell) \neq (1, 0)$. If v is multiply covered, factorize it as $v = p \circ v'$, where v' is a simple map and p is a ramified covering of $\mathbf{C}P^1$. Replace v by v', if necessary, we may assume that v is simple. Since $\int_{A_1} \omega > \int_{A_2} \omega$ and v is a J_∞-holomorphic map with the symplectic area smaller than $\int_{A_2} \omega$, we have $k \geq \ell + 1$ and $\ell \geq 1$. By the adjunction formula, the virtual genus of $C = v(\mathbf{C}P^1)$ is

$$\begin{aligned} g_v(C) &= 1 + \frac{1}{2}(C \cdot C - c_1(C)) \\ &= 1 - k\ell - k + \ell \\ &\leq 1 - (\ell + 1)\ell - (\ell + 1) + \ell \\ &< 0. \end{aligned}$$

However, the virtual genus $g_v(C)$ is a non-negative integer, which is a contradiction. Therefore we obtain the existence of J_∞-holomorphic sphere representing the class A_2 and passing through the given point x.

Next we prove that $\mathcal{J}_0$ is a non-empty closed subset of codimension 2. The subspace $\mathcal{J}_0$ is non-empty because the sphere $(x, -x)$ is symplectic. To prove the closedness of $\mathcal{J}_0$ we first notice that the energy of a holomorphic sphere in class $A_1 - \ell A_2$ is less than $\omega^{(1)}(A_1)$. Thus we can apply the Gromov compactness argument to the following situation. Let a sequence of J_i-holomorphic spheres u_i representing $A_1 - \ell A_2$. Suppose that J_i converges to a compatible almost complex structure J_∞. Then there is a subsequence $\{u_{i_k}\}$, which converges to a J_∞-stable map u_∞. If u_∞ is a J_∞-holomorphic sphere, we find that $J_\infty \in \mathcal{J}_0$. (In this case, by the adjunction formula, u_∞ is an embedding.) Otherwise, u_∞ consists of at least two irreducible components, which represent the classes $k_i A_1 - \ell_i A_2$ such that $\sum k_i = 1$ and $\sum \ell_i = \ell$. As we mentioned above, there is always a J_∞-holomorphic sphere in class A_2. Taking into account of posivity of intersection in dimension 4, we conclude that these J_∞-holomorphic spheres must be of type $A_1 - \ell_i A_2$ and $m_j A_2$ such that $\sum(-\ell_i)+\sum m_j = -\ell$. Since m_j, if exists, must be positive, we conclude that there must be a bubble of type $A_1 - \ell' A_2$, $\ell' \geq 1$, that proves the closedness of $\mathcal{J}_0$.

The codimension of $\mathcal{J}_0$ is at least 2 by a similar argument as in [13], [14]. (Namely for a fixed ℓ we consider the universal moduli space of the pairs (J, J-holomorphic sphere in class $A_1 - lA_2$). The Fredholm index of the projection of this moduli space on the first factor is equal $4 + 2(1 - \ell) - 6 \leq -2$.) To prove that the codimension is precisely 2, we use the uniqueness of J-holomorphic sphere in class $A_1 - A_2$ if it exists. (cf. with the argument in [9]. It follows that the kernel of the linearization of the projection from the universal moduli space to the first factor, i.e., the space of compatible almost complex structures equals zero.).

Finally we prove the existence of J-holomorphic sphere in class A_1 for $J \in (\mathcal{J} \setminus \mathcal{J}_0)$ and use again the bubbling-off argument. For a generic compatible almost complex structure J, there exist J-holomorphic spheres in the class A_1, see [18]. Note that such J-holomorphic spheres are automatically embedded by the adjunction formula and that the class A_1 is primitive. If there is no J_0-holomorphic curve in the class A_1, we pick a sequence of generic compatible almost complex structures converging to J_0. Then the bubbling-off argument implies that there must be a J_0-holomorphic sphere in a class $A_1 - \ell A_2$ for $\ell \geq 1$. Q.E.D.

Now let us find an element $g \in \mathrm{Symp}(S^2 \times S^2, \omega^{(1)} \oplus \omega^{(2)})$ by studying the action of the group of symplectomorphisms on $\mathcal{J} \setminus \mathcal{J}_0$. Since $\mathcal{J}_0$ is a closed subset of codimension 2 we can choose a small disk D in $\mathcal{J}(S^2 \times S^2)$ such that this disk intersects $\mathcal{J}_0$ transversally at exactly one interior point. By results of Gromov [8] and McDuff [18], for any compatible almost complex structure J_θ, $\theta \in \partial D$, $S^2 \times S^2$ is foliated by A_1-curves and A_2-curves, respectively. In particular, J_θ is pointwisely positive on these A_1-curves and A_2-curves. It implies that there is a loop g_t in the group $\mathrm{Symp}(S^2 \times S^2, x)$ such that the image $g_t(J_0)$ is homotopic to the loop ∂D in $\mathcal{J} \setminus \mathcal{J}_0$, where J_0 is the complex structure on $\mathbf{C}P^1 \times \mathbf{C}P^1$. Thus, we can deform D along the boundary so that D intersects $\mathcal{J}_0$ transversally at one point and $\partial D = \{g_t(J_0)\}$.

Now we construct our bundle E by gluing two trivial $S^2 \times S^2$ bundles over another disk D' using this loop g_t. Since the base space of E is S^2, which is simply connected, $\mathcal{H}_2(E)$ is a simple local system. We claim that the parametrized Gromov-Witten invariant $I^E_{0,0,A_1-A_2}$ is 1.

Since $c_1(A_1 - A_2) = 0$, the moduli space $C\mathcal{M}_{0,0}(E, J(E), A_1 - A_2)$ is 0-dimensional. To compute the invariant for our bundle E, we choose a generic fiberwise compatible almost complex structure $J(E)$ on E as follows. Note that it is the case of weakly monotone symplectic manifolds and we are working with Gromov-Witten invariants of genus 0, $c_1(A_1 - A_2) = 0$ and the dimension of the base of E is 2. Therefore it suffices to perturb J to get the fundamental class of the corresponding moduli space. We observe that the standard product complex structure J_0 on $S^2 \times S^2$ is a $(A_1 - A_2)$-regular. Then we take $J(E)$ being the gluing of the constant complex structure J_0 over D' and the compatible almost complex structure parametrized by D along the boundary ∂D by g_t. By the transversality of the intersection of $J(E)$ with $\mathcal{J}_0$, we find that the vertical almost complex structure $J(E)$ is $(A_1 - A_2)$-regular for the symplectic fiber bundle. By the construction the parametrized moduli space $C\mathcal{M}_{0,0,A_1-A_2}(E, J(E))$ consists of one point over the point $D \cap \mathcal{J}_0$. Therefore the value $I^E_{0,0,A_1-A_2} = 1$. By Proposition 4.1, this nontrivial parametrized Gromov-Witten invariant defines a non-trivial element in $Hom(\pi_2(\mathrm{BSymp}(M, \omega), \mathbf{Q})$.

References

[1] M. Abreu, Topology of symplectomorphism groups of $S^2 \times S^2$, Invent. Math., **131** (1998), 1–23.

[2] M. Abreu and D. McDuff, Topology of symplectomorphism groups of rational ruled surfaces, J. Amer. Math. Soc., **13** (2000), 971–1009.

[3] S. Anjos, Homotopy type of symplectomorphism groups of $S^2 \times S^2$, Geom. Topol., **6** (2002), 195–218.

[4] O. Buse, Relative family Gromov-Witten invariants and symplectomorphisms, Pacific. J. Math., **218** (2005), 315–341.

[5] O. Buse, Whitehead products in symplectomorphism groups and Gromov-Witten invariants, arXiv: math.SG/0411108.

[6] K. Fukaya and K. Ono, Arnold conjecture and Gromov-Witten invariant, Topology, **38** (1999), 933–1048.

[7] V. Guillemin, E. Lerman and S. Sternberg, Symplectic fibrations and multiplicity diagrams, Cambridge Univ. Press, Cambrdige, 1996.

[8] M. Gromov, Pseudo holomorphic curves in symplectic manifolds, Invent. Math., **82** (1985), 307–334.

[9] H. Hofer, V. Lizan and J.-C. Sikorav, On genericity for holomorphic curves in four-dimensional almost-complex manifolds, J. Geom. Anal., **7** (1997), 149–159.

[10] J. Kędra, Restrictions on symplectic fibrations, Differential Geom. Appl., **21** (2004), 93–112.

[11] M. Kontsevich, Enumeration of rational curves via torus actions, In: Moduli space of surfaces, 1995, Birkhäuse, pp. 335–368.
[12] M. Kontsevich and Yu. Manin, Gromov-Witten classes, quantum cohomology and enumerative geometry, Comm. Math. Physics, **164** (1994), 525–562.
[13] H. V. Lê and K. Ono, Perturbation of pseudo-holomorphic curves, Internat. J. Math., **7** (1996), 771–774.
[14] H. V. Lê and K. Ono, Parameterized Gromov-Witten invariants and topology of symplectomorphism groups, preprint, MPIMIS 28/2001.
[15] J. Lott, To see the group of diffeomorphisms by using analysis, talk at the Univ. of Bonn in 1996.
[16] J. Morrow and K. Kodaira, Complex manifolds, Holt, Rinehalt and Winston, 1971.
[17] D. McDuff, Examples of symplectic structures, Invent. Math., **89** (1987), 13–36.
[18] D. McDuff, Blow ups and symplectic embedding in dimension 4, Topology, **30** (1991), 409–421.
[19] D. McDuff, A survey of the topological properties of symplectomorphism groups, In: Topology, geometry and quantum field theory, London Math. Soc. Lecture Note Ser., **308**, Cambridge Univ. Press, 2004, pp. 173–193.
[20] D. McDuff and D. Salamon, J-holomorphic Curves and Quantum Cohomology, Univ. Lecture Ser., **6**, Amer. Math. Soc., 1994.
[21] D. McDuff and D. Salamon, Introduction in Symplectic Topology, Oxford Sci. Publ., 1995.
[22] T. Nishinou, Some nontrivial homology classes on the space of symplectic forms, J. Math. Kyoto Univ., **42** (2002), 599–606.
[23] P. Seidel, π_1 of symplectic automorphism groups and invertible element in quantum cohomology ring, Geom. Funct. Anal., **7** (1997), 1046–1095.

Hông-Vân Lê
Mathematical Institute, Zitna 25,
CZ-11567 Praha, Czech Republic

Kaoru Ono
Department of Mathematics
Hokkaido University
Sapporo, 060-0810, Japan

Advanced Studies in Pure Mathematics 52, 2008
Groups of Diffeomorphisms
pp. 77–92

Mapping class actions on surface group completions

Robert C. Penner

Abstract.

Recent projects on pronilpotent and profinite completions of surface groups are discussed, and common threads are then compared in the context of general group completions. In particular, a profinite version of the Torelli groups and a pronilpotent version of the punctured solenoid are introduced.

§1. Introduction

I would like to describe here both the union and then the intersection of two recent projects:

- on pronilpotent completions of surface groups, which is joint work with Shigeyuki Morita [14];
- on profinite completions of surface groups, which is joint work with Dragomir Šarić [19] and with Dragomir Šarić and Sylvain Bonnot [3].

Background for both of these projects is the decorated Teichmüller theory of punctured surfaces [17, 18], which I shall first briefly recall.

§2. Decorated Teichmüller Theory

Let F_g^s denote a fixed smooth oriented surface of genus g with $s \geq 1$ punctures, where $2g - 2 + s > 0$. The *mapping class group* $MC(F_g^s)$ of isotopy classes of orientation-preserving diffeomorphisms of F_g^s acts on the *Teichmüller space* $\mathcal{T}(F_g^s)$ of isotopy classes of hyperbolic metrics on F_g^s by push-forward of metric with quotient *Riemann's moduli space*

Received March 11, 2007.
Revised September 27, 2007.
It is a pleasure for the author to thank the Center for the Topology and Quantization of Moduli Spaces, Aarhus University, Denmark, for kind hospitality when this paper was written.

$\mathcal{M}(F_g^s)$. There is a trivial $\mathbb{R}_{>0}^s$-bundle over $\mathcal{T}(F_g^s)$ called the *decorated Teichmüller space* $\tilde{\mathcal{T}}(F_g^s)$, where the fiber over a point is identified with the collection of all s-tuples of horocycles in F_g^s, one horocycle about each puncture. Indeed, since horocycles in F_g^s are closed curves, we may take the coordinate on the fiber to be simply the tuple of hyperbolic lengths of the (not necessarily embedded) horocycles. $MC(F_g^s)$ acts on $\tilde{\mathcal{T}}(F_g^s)$ by permuting these numbers, and we thus have the diagram

$$\begin{array}{rl} MC(F_g^s) \circlearrowright & \tilde{\mathcal{T}}(F_g^s) - \mathbb{R}_{>0}^s \\ & \downarrow \\ MC(F_g^s) \circlearrowright & \mathcal{T}(F_g^s) \\ & \downarrow \\ & \mathcal{M}(F_g^s) \end{array}$$

relating these spaces and mapping class group actions.

Uniformize in Minkowski three-space with its pairing

$$< (x, y, z), (x', y', z') > = \; xx' + yy' - zz',$$

where (x, y, z) are the usual Cartesian coordinates, and with its group $SO(2, 1)$ of isometries. We may regard a point of $\mathcal{T}(F_g^s)$ as the conjugacy class of a discrete and faithful representation of the fundamental group of F_g^s in the component $SO^+(2, 1) \approx PSL_2(\mathbb{R})$ of the identity, where the representation is required to map peripheral elements of the fundamental group to parabolic elements of $SO^+(2, 1)$. Letting

$$\mathbb{H} = \{w = (x, y, z) : < w, w > = \; -1 \text{ and } z > 0\}$$

denote the upper sheet of the hyperboloid and

$$L^+ = \{w = (x, y, z) : < w, w > = \; 0 \text{ and } z > 0\}$$

denote the subspace of isotropic vectors with positive height, we may identify the collection of horocycles in $\mathbb{H}$ with L^+ via affine duality $L^+ \ni v \leftrightarrow h(v) = \{w \in \mathbb{H} : < v, w > = \; -2^{-\frac{1}{2}}\}$.

There are two basic ingredients to the decorated Teichmüller theory, the first of which gives coordinates on $\tilde{\mathcal{T}}(F_g^s)$, as follows. Given a pair of horocycles $h(u), h(v)$, a natural invariant is given by

$$\sqrt{- < u, v >} = \exp \delta/2,$$

where δ is the signed hyperbolic distance between $h(u), h(v)$ taken with positive sign if and only if the horocycles are disjoint. We promote this invariant to the setting of decorated hyperbolic surfaces in the natural way: If $\tilde{G} \in \tilde{\mathcal{T}}(F_g^s)$ is a decorated hyperbolic structure on F_g^s and α is an isotopy class of arcs connecting punctures of F_g^s, then a lift to $\mathbb{H}$ of α is asymptotic to a pair of rays in L^+, each of which contains a well-defined point determined by the decoration, say these points are $u, v \in L^+$. Define the *lambda length* of α with respect to $\tilde{G}$ to be $\lambda(\alpha; \tilde{G}) = \sqrt{- < u, v >}$, which is independent of the choice of lift since the inner product is invariant by $SO^+(2,1)$.

The second main ingredient of decorated Teichmüller theory is the "convex hull construction" [6], which is described as follows. A point $\tilde{G} \in \tilde{\mathcal{T}}(F_g^s)$ determines not only (the conjugacy class of) a discrete group $G < SO^+(2,1)$ of isometries, but it also determines $s \geq 1$ many G-orbits of points in L^+, namely, the set $\mathcal{B} \subset L^+$ of points corresponding via affine duality to the collection of horocycles in the decoration of F_g^s. The closed convex hull of $\mathcal{B}$ in the underlying vector space structure of Minkowski three-space is a G-invariant convex body since $G < SO^+(1,2)$ acts linearly, and the *convex hull construction* associates to $\tilde{G}$ the projection $\Delta(\tilde{G})$ to F_g^s of the edges in the frontier of this closed convex hull of $\mathcal{B}$ which meet L^+ only in their endpoints, and we may straighten these arcs connecting punctures to geodesics for the hyperbolic structure underlying $\tilde{G}$.

Define an *ideal triangulation* Δ of F_g^s to be (the isotopy class of) a collection of disjointly embedded arcs connecting punctures which decompose F_g^s into triangles with vertices at the punctures. Given an arc $\alpha \in \Delta$ so that α triangulates a quadrilateral complementary to $\Delta - \{\alpha\}$ in F_g^s, let β denote the other diagonal of this quadrilateral, and define the *flip* along α in Δ to produce the ideal triangulation $\Delta \cup \{\beta\} - \{\alpha\}$. Finally, define an *ideal cell decomposition* (i.c.d) of F_g^s to be (the isotopy class of) a collection of disjointly embedded arcs connecting punctures which decompose F_g^s into ideal polygons with vertices at the punctures. We have the following "omnibus" theorem:

Theorem 2.1. [17, 18] **Part 1** *If Δ is an ideal triangulation of F_g^s, then*

$$\begin{aligned} \tilde{\mathcal{T}}(F_g^s) &\to \mathbb{R}_{>0}^{\Delta} \\ \tilde{G} &\mapsto (\alpha \mapsto \lambda(\alpha; \tilde{G})) \end{aligned}$$

is a real-analytic surjective homeomorphism. The action of $MC(F_g^s)$ *on these global coordinates is given by permutation followed by finite compositions of "Ptolemy transformations"* $ef = ac + bd$, *which describe the lambda lengths* e, f *of the pair of diagonals of a quadrilateral involved in a flip whose opposite sides have lambda lengths* a, c *and* b, d. *Furthermore, the Weil-Petersson Kähler two-form on* $\mathcal{T}(F_g^s)$ *pulls back to*

$$2\sum \mathrm{dlog}\ a \wedge \mathrm{dlog}\ b + \mathrm{dlog}\ b \wedge \mathrm{dlog}\ c + \mathrm{dlog}\ c \wedge \mathrm{dlog}\ a,$$

where the sum is over all triangles in F_g^s *complementary to* Δ *with lambda lengths* a, b, c *in this clockwise cyclic order as determined by the orientation of* F_g^s.

Part 2 *The convex hull construction* $\tilde{G} \mapsto \Delta(\tilde{G})$ *provides a canonical i.c.d.* $\Delta(\tilde{G})$ *for each* $\tilde{G} \in \tilde{\mathcal{T}}(F_g^s)$. *Furthermore,*

$$\{\{\tilde{G} \in \mathcal{T}(F_g^s) : \Delta(\tilde{G}) \text{ is isotopic to } \Delta\} : \Delta \text{ is an i.c.d of } F_g^s\}$$

is an $MC(F_g^s)$*-invariant ideal cell decomposition of* $\tilde{\mathcal{T}}(F_g^s)$. *If* Δ' *is any ideal triangulation containing the i.c.d.* Δ, *then the corresponding cell in decorated Teichmüller space is described in lambda length coordinates on* Δ' *by the coupled inequalities*

$$0 \leq \frac{a^2 + b^2 - e^2}{abe} + \frac{c^2 + d^2 - e^2}{cde} = E,$$

where the inequality is strict for each arc in Δ, *equality holds for each arc in* $\Delta' - \Delta$, *and the notation is as above with the edge of lambda length* e *separating those with lengths* a, b *from those with lengths* c, d.

The quantity E is called the *simplicial coordinate* of the edge with lambda length e, and it turns out that $2^{\frac{3}{2}}abcdE$ is the signed volume of the corresponding tetrahedron in Minkowski three-space, which is taken with a positive sign if the edge of lambda length e lies below that with lambda length f. (Note that elements of $SO^+(2,1)$ have unit determinant and hence preserve signed Euclidean volume.)

As a first corollary, notice that taking all the lambda lengths on an ideal triangulation to have value one gives positive simplicial coordinates, so every isotopy class of ideal triangulation of F_g^s actually arises from the convex hull construction. Since $\tilde{\mathcal{T}}(F_g^s)$ is connected, it follows from general position that finite compositions of flips act transitively on ideal triangulations of F_g^s, thus giving a new proof of this classical fact due to Whitehead.

This leads to two faithful representations of $MC(F_g^s)$. The first algebraic one given by the action on coordinates with respect to a fixed ideal triangulation follows from the evident naturality of lambda lengths $\lambda(\alpha, \tilde{G}) = \lambda(\phi(\alpha), \phi(\tilde{G}))$, for any $\phi \in MC(F_g^s)$, and represents an element of $MC(F_g^s)$ as a permutation followed by a finite composition of Ptolemy transformations. By the way, one easily directly checks that the two-form described in Part 1 of the Omnibus Theorem is invariant by Ptolemy transformations, which in particular shows directly that it is invariant under $MC(F_g^s)$.

For the second combinatorial representation of $MC(F_g^s)$, let us consider *labeled ideal triangulations*, by which we mean an ideal triangulation Δ of F_g^s together with an enumeration of the arcs in Δ by $1, 2, \ldots, 6g - 6 + 3s$. Define the *mapping class groupoid* $MD(F_g^s)$ to be the category whose objects are $MC(F_g^s)$-orbits of labeled ideal triangulations and whose morphisms are pairs of labeled ideal triangulations modulo the diagonal action of $MC(F_g^s)$ together with the obvious composition. The mapping class group $MC(F_g^s)$ is then just the group of self-morphisms in $MD(F_g^s)$ of any object.

Consider the flip on an arc α in an ideal triangulation supported in some ideal quadrilateral. Given a labeling on the ideal triangulation where α has label j, the corresponding *labeled flip* f_j along α simply assigns the labeling of α to the other diagonal of this quadrilateral and leaves invariant the labeling of the other arcs in the ideal triangulation.

Corollary 2.2. *The mapping class groupoid $MD(F_g^s)$ of $F_g^s \neq F_0^3$ admits the following presentation. Generators are given by labeled flips and permutations (i, j) of labels i and j. Relations are given by those of the symmetric group together with the following:*

Involutivity *for any labeled flip f_i, we have $f_i \circ f_i =$ identity;*

Commutativity *for any labeled flips f_i, f_j where the arcs with labels i, j do not lie in the frontier of a common complementary triangle, we have $f_i \circ f_j = f_j \circ f_i$;*

Pentagon *for any labeled flips f_i, f_j where the arcs with labels i, j triangulate a pentagon with frontier in the ideal triangulation, we have $f_i \circ f_j \circ f_i \circ f_j \circ f_i = (i, j)$;*

Naturality *for any labeled flip f_i and any permutation σ, we have $\sigma \circ f_i = f_{\sigma(i)} \circ \sigma$.*

We have included this result here because of its independent interest and because it is a paradigm for several of the subsequent results. The main point of its proof is that a homotopy of paths in $\tilde{\mathcal{T}}(F_g^s)$ can be put into general position with respect to the codimension-two skeleton of our ideal cell decomposition, and there are precisely two kinds of codimension-two cells: either five points in $\mathcal{B} \subset L^+$ become coplanar, which leads to the pentagon relation, or two sets of four points in $\mathcal{B} \subset L^+$ become coplanar, which leads to the commutativity relation. In the context of unlabeled ideal triangulations, there are thus corresponding relations among unlabeled flips (where the right-hand side of the pentagon relation is replaced by the identity), and we shall continue to refer to these identities as involutivity, commutativity, and the pentagon relation.

In fact, it is convenient in the sequel to reformulate the combinatorics of i.c.d.'s as follows. The Poincaré dual of an i.c.d. of F_g^s is a *fatgraph* embedded as spine of F_g^s, namely, a graph in the usual sense of the term together with a cyclic ordering on the half-edges about each vertex, where the cyclic ordering is determined by the clockwise cyclic ordering in F_g^s. It is often more convenient to employ the fatgraph formalism rather than the entirely equivalent formalism of i.c.d.'s. The dual of a flip on an ideal triangulation is a *Whitehead move* on the dual trivalent fatgraph, namely, contract an edge of the fatgraph with distinct endpoints (corresponding to removing its dual arc), and then expand the resulting four-valent vertex differently (corresponding to adding the other diagonal of the corresponding quadrilateral). We shall also in this context of Whitehead moves refer to the identities discussed above as involutivity, commutativity, and the pentagon relation.

§3. Pronilpotent Case

We shall think of the surface F_g^1 as the unpunctured surface F_g^0 of genus g with a basepoint $*$ that plays the role of the puncture. Define the *lower central series* of the fundamental group recursively by setting $\Gamma_0 = \pi_1(F_g^0, *)$ and $\Gamma_{k+1} = [\Gamma_k, \Gamma_0]$, where the square brackets denote the commutator group. There are the corresponding *kth nilpotent quotients* defined by $N_k = \Gamma_0/\Gamma_k$ related by the basic exact sequence

$$0 \to \Gamma_k/\Gamma_{k+1} \to N_{k+1} \to N_k \to 1,$$

and in particular, N_1 is the integral first homology group of F_g^0. The mapping class group $MC(F_g^1)$ is identified with the mapping class group of F_g^0 which fixes $*$, there are induced actions of $MC(F_g^1)$ on each N_k, and we define the *kth Torelli group*

$$MC_{g,*}[k] \;=\; ker(MC(F_g^1) \to Aut(N_k))$$

to be those mapping classes that act identically on N_k. Dennis Johnson defined homomorphisms

$$\tau_k : MC_{g,*}[k] \to \mathrm{Hom}(N_1, \Gamma_k/\Gamma_{k+1}),$$

which arise naturally from the basic exact sequence.

Johnson himself proved that $MC_g[1] = ker(MC(F_g^0) \to Sp(2g, \mathbb{Z}))$ and that $MC_{g,*}[1]$ is finitely generated [7] by an explicit set of homeomorphisms for $g \geq 3$, he calculated $ker(\tau_1)$ in [8], and he computed the abelianization of $MC_{g,*}[1]$ in [9]. Geoff Mess proved [12] that $MC_2[1]$ is not finitely generated. Daniel Biss and Benson Farb have shown [2] that $MC_{g,*}[2]$ are not finitely generated for $g \geq 2$ together with the analogous result for closed surfaces.

Here we shall give infinite presentations of all of the groups $MC_{g,*}[k]$ and indeed finite presentations for certain related groupoids and groups. These results should be compared with recent work by Andrew Putman, who has given different infinite presentations for $MC_g[1]$ and $MC_{g,*}[1]$ when $g \geq 2$ which have a more efficient and geometrical generating set but more complicated relations. We shall also describe an explicit combinatorial cocycle expression for the first Johnson homomorphism.

The elementary but potent remarks are that the cell decomposition of $\tilde{\mathcal{T}}(F_g^1)$ descends to an ideal cell decomposition of $\mathcal{T}(F_g^1)$ itself (since changing the unique horocycle in the decoration simply scales the set $\mathcal{B}$ of points in Minkowski space and the convex hull construction is linear), and furthermore, this decomposition of $\mathcal{T}(F_g^1)$ is invariant under any subgroup of the mapping class group, and in particular is invariant under each $MC_{g,*}[k]$. Thus, there is a "Torelli tower"

$$\mathcal{T}(F_g^1) \to \cdots \to T_{k+1} \to T_k \to T_{k-1} \to \cdots \to T_1 \to \mathcal{M}(F_g^1),$$

where the *kth* "Torelli space" defined by

$$T_k = \mathcal{T}(F_g^1)/MC_{g,*}[k], \text{ for each } k \geq 1,$$

is a manifold and an Eilenberg-MacLane space $K(MC_{g,*}[k], 1)$, each mapping $T_{k+1} \to T_k$ is an unbranched covering of manifolds, and each $T_k \to \mathcal{M}(F_g^1)$ is an orbifold covering. The point is that the ideal cell decomposition of $\mathcal{T}(F_g^1)$ descends to each of these spaces and gives a combinatorial realization of this tower of spaces and maps.

Just as it is useful to consider the Poincaré dual of the cell decomposition of a surface to get a fatgraph, so too is it useful to pass to Poincaré duals here. Define $\hat{\mathcal{G}}$ to be the dual cell complex to the ideal cell decomposition of $\mathcal{T}(F_g^1)$, so $\hat{\mathcal{G}}$ is an *honest* cell complex (not an ideal cell complex) of dimension $4g-3$, and is contractible since its dual is. We shall also consider the quotients $\hat{\mathcal{G}}_k = \hat{\mathcal{G}}/MC_{g,*}[k]$, for $k \geq 1$.

Define an N_k*-marking* on a fatgraph Γ to be a function

$$\mu : \{\text{oriented edges of } \Gamma\} \to N_k$$

so that:

- $\mu(\bar{e}) = [\mu(e)]^{-1}$ if $e, \bar{e}$ denote the two orientations on a common underlying edge of Γ;

- if $a_1, \ldots, a_k$ are oriented edges pointing towards a common vertex of Γ in this counter-clockwise cyclic ordering, then

$$\mu(a_1)\mu(a_2)\cdots\mu(a_k) = 1 \in N_k$$

  where we compose from left to right by convention;

- the image of μ generates N_k.

Suppose that the trivalent fatgraph Γ' arises from the trivalent fatgraph Γ by a Whitehead move along the edge e of Γ producing the edge f of Γ'. Suppose that the oriented edge e points from the vertex with incident oriented edges $c, d, \bar{e}$ towards the vertex with incident oriented edges a, b, e in these correct counter-clockwise cyclic orders, where the oriented edges a, b, c, d point towards the corresponding vertex. Identify the oriented edges of Γ' with those of Γ in the natural way, where the oriented edge f points from the vertex with incident oriented edges $d, a, \bar{f}$ towards the vertex with incident oriented edges b, c, f in these correct counter-clockwise cyclic orders with these oriented edges pointing towards the corresponding vertices as before. The N_k-marking μ' on Γ' induced by an N_k-marking μ on Γ agrees with μ on oriented edges other than e, f, and

$$\mu'(f) = \mu(d)\mu(a) = [\mu(b)\mu(c)]^{-1}.$$

Theorem 3.1. [14] **Part 1** *There is a $MC(F_g^1)$-invariant cocycle*

$$j \in Z^1(\hat{\mathcal{G}}; \Lambda^3 N_1)$$

defined for a Whitehead move in the notation of the previous paragraph by $j(\Gamma \to \Gamma') = a \wedge b \wedge c = c \wedge d \wedge a$, *where* $\Lambda^3 N_1$ *denotes the third exterior power of* N_1 *and* $\wedge$ *denotes the exterior product, and the associated group homomorphism*

$$[j] \in H^1(\hat{\mathcal{G}}_1; \Lambda^3 N_1) \approx \mathrm{Hom}(MC_{g,*}[1], \Lambda^3 N_1)$$

coincides with $6\tau_1$.

Part 2 *For each* $k \geq 1$, *cells in* $\hat{\mathcal{G}}_k$ *are indexed by isotopy classes of suitable* N_k*-marked fatgraph spines of* F_g^1, *and coordinates on a given cell are provided by simplicial coordinates. Furthermore,* $MC_{g,*}[k]$ *is generated (in fact enumerated) by sequences of Whitehead moves beginning and ending on the same* $MC(F_g^1)$*-orbit which leave invariant the* N_k*-marking. Relations in* $MC_{g,*}[k]$ *are given by involutivity, commutativity, and the pentagon relation as before.*

In fact, one can use Part 2 together with Johnson's generating set for $MC_{g,*}[1]$ to give *finite presentations* for the fundamental path *groupoid* of the Torelli spaces T_1 and T_2 and also finite presentations for the "level N classical Torelli *groups*", i.e., the subgroup of $MC(F_g^1)$ that acts trivially on homology with coefficients mod N, for any $N \geq 1$.

Furthermore, one can use Part 1 with the Alexander-Whitney approximation to the diagonal together with results of Morita [13] and Kawazumi-Morita [10, 11] to give new cocycles generating the tautological algebra of $\mathcal{M}(F_g^1)$. Please see [14] for details on Theorem 3.1 as well as these further applications.

In fact, one uses the description of relations in Part 2 to show that the expression in Part 1 for the Johnson homomorphism is well-defined. The more general point is that there is a kind of "machine" here for producing $MC(F_g^1)$-invariant cocycles with coefficients in general modules by solving for coefficients so that the required relations of involutivity, commutativity, and the pentagon relation hold. Together with Shigeyuki Morita and Alex Bene, we have produced roughly ten or so such cocycles with values in various modules, so far without success in writing down an explicit combinatorial formula for the second Johnson homomorphism. We take this opportunity just to mention one further very simple such cocycle with values in the second symmetric power of the second exterior power of N_1, namely in the notation of Part 1, assigning the expression

$$(a \wedge c) \otimes (b \wedge d) \ + \ (b \wedge d) \otimes (a \wedge c)$$

to a Whitehead move gives another invariant cocycle as one can check.

Let us finally mention that recent joint work with Alex Bene and Nariya Kawazumi has shown that the higher Johnson homomorphisms lift to the mapping class groupoid in the sense of Part 1 of Theorem 3.1 for the case of surfaces with a single boundary component. Furthermore, this work also provides an algorithm for the explicit calculation of representative combinatorial cocycles in analogy to the cocyle j, and explicit formulae for the first several Johnson homomorphisms have been obtained.

§4. Profinite Case

Consider the collection of all finite pointed unbranched covers of some fixed surface $F = F_g^s$ of negative Euler characteristic with $s \geq 1$. Put another way, we may consider all finite pointed branched covers, where the branching occurs only at the removed punctures of F. This collection of covers is naturally inverse directed, where $\pi_1 : \tilde{F}_1 \to F$ is greater than or equal to $\pi_2 : \tilde{F}_2 \to F$ if there is a finite pointed unbranched cover $\pi_{12} : \tilde{F}_1 \to \tilde{F}_2$ so that $\pi_1 = \pi_2 \circ \pi_{12}$ (here reading composition functionally from right to left).

Define the *punctured solenoid* S to be the inverse limit of this collection of covers, and note that the homeomorphism type of S is independent of the initial choice of surface F subject to the stated conditions since any two such surfaces admit a common finite unbranched cover. For an explicit model of a space homeomorphic to S, take the classical modular group $G = PSL_2(\mathbb{Z})$, or any finite-index subgroup of it, let $\hat{G}$ denote its profinite completion, i.e., $\hat{G}$ is the inverse limit of G/H over all normal subgroups $H < G$ of finite index, and set

$$S_G = (\mathbb{D} \times \hat{G})/G, \text{ where } \gamma(z,t) = (\gamma z, t\gamma^{-1}) \text{ for } \gamma \in G,$$

$\mathbb{D}$ denoting the Poincaré disk with its usual action of G by hyperbolic isometries. Thus, $S \approx S_G$ has local neighborhoods modeled on the product of an open set in $\mathbb{D}$ with a Cantor set, and furthermore, these local foliations by open sets in $\mathbb{D}$ combine to give a canonical foliation of S_G by leaves which are identified with $\mathbb{D}$, here using that surface groups are residually finite.

Following Sullivan's work [20] in the analogous case of closed surfaces, we may define a Teichmüller theory of S in the spirit of Ahlfors-Bers theory [1], namely: We say that a homeomorphism from S_G is *quasiconformal* if it is quasiconformal in the usual sense in the $\mathbb{D}$ direction and varies continuously (technical point: for both the C^∞ and quasiconformal topologies) in the transverse Cantor set direction.

The *Teichmüller space* $\mathcal{T}(S)$ is then defined in perfect analogy to the classical case, namely, the space of equivalence classes of quasiconformal homeomorphisms from S_G, where two such $S_G \to S_i$, for $i = 1, 2$ are equivalent if there is a conformal map (or in other words, hyperbolic isometry) $S_1 \to S_2$, which is defined analogously, so that the obvious diagram commutes up to a homotopy (technical point: the homotopy must be bounded for the hyperbolic metric). By Sullivan's work in the closed case, $\mathcal{T}(S)$ has the natural structure of a Banach space in analogy to the classical case. It is easy to see that there are embeddings of $\mathcal{T}(F_g^s) \subset \mathcal{T}(S)$ by demanding that the covering maps over some F_g^s are conformal. Sullivan called these the "TLC" for transversely locally constant structures on S, and he proved [20] that these TLC structures on S are dense in $\mathcal{T}(S)$. Thus, the solenoid is a natural universal object in the sense that its Teichmüller space is a natural completion of all of the classical ones.

One may then define the *mapping class group* $MC(S)$ of the punctured solenoid as the group of all such equivalence classes of quasiconformal self-homeomorphisms of S_G. This is a very mysterious uncountable group which contains the more tractable discrete "baseleaf preserving" subgroup $MC_{BLP}(S) < MC(S)$ comprised of those mapping classes which fix some leaf (called the "baseleaf") of the natural foliation of S. Essentially by definition *as a topological space*, we have

$$MC(S) \approx MC_{BLP}(S) \times_G \hat{G},$$

but really nothing is known about $MC(S)$ as a group. There is the feeling that perhaps it is related to the absolute Galois group.

Chris Odden showed in his thesis [15, 16] that $MC_{BLP}(S)$ is isomorphic to the "virtual automorphism group" of G, namely, equivalence classes of isomorphisms between finite-index subgroups of G, which may have different indexes, where two such are equivalent if they agree on a common finite-index subgroup of their domains. Furthermore, Sullivan showed that the "Ehrenpreis Conjecture" (namely, for any points of $\mathcal{T}(F_{g_1}^{s_1})$ and $\mathcal{T}(F_{g_2}^{s_2})$, there are finite unbrached covers of $F_{g_1}^{s_1}$ and $F_{g_2}^{s_2}$ by a common surface F_g^s so that the induced structures on $\mathcal{T}(F_g^s)$ are arbitrarily close) is equivalent to the statement that $MC_{BLP}(S)$ has dense orbits on $\mathcal{T}(S)$.

The *Farey tesselation* Δ_* is the $PSL_2(\mathbb{Z})$-orbit of any ideal triangle in $\mathbb{D}$, and in a sense we shall make precise in Theorem 4.1, Δ_* plays the role for the punctured solenoid of an ideal triangulation for a punctured surface. To apply the decorated Teichmüller theory in the current context, we note that there are natural *punctures* of S corresponding to equivalence class of wandering rays, where two rays are equivalent if they are asymptotic in their common leaf. We may furthermore decorate these punctures by choosing a horocycle in each leaf centered at each puncture and requiring these choices of horocycles to vary continuously in the Cantor set direction. There is thus a natural *decorated Teichmüller space* $\tilde{\mathcal{T}}(S)$ of the solenoid, a natural continuous forgetful mapping $\tilde{\mathcal{T}}(S) \to \mathcal{T}(S)$, and given a homotopy class of arc in S connecting punctures and a point of $\tilde{\mathcal{T}}(S)$, there is again a lambda length defined as before.

Say that an ideal triangulation Δ of $\mathbb{D}$ is TLC if it is the lift to $\mathbb{D}$ of some ideal triangulation on $\mathbb{D}/K$ for some finite-index torsion free group $K < G$. Given a TLC tesselation Δ corresponding to $K < G$ and any arc $e \in \Delta$ whose projection $\bar{e}$ to $\mathbb{D}/K$ triangulates a quadrilateral complementary to $\Delta/K - \{\bar{e}\}$, we may define the result of the *K-equivariant flip on e* to be the lift to $\mathbb{D}$ of the ideal triangulation of $\mathbb{D}/K$ that arises by flipping $\bar{e}$ in Δ/K.

Theorem 4.1. [19] **Part 1** *The assignment of lambda lengths*

$$\tilde{\mathcal{T}}(S) \to Cont^G(\hat{G}, \mathbb{R}_{>0}^{\Delta_*})$$

gives a homeomorphism onto the space of all G-equivariant continuous functions with the compact-open topology from the profinite completion $\hat{G}$ to $\mathbb{R}_{>0}$-valued functions on the Farey tesselation Δ_* with the strong topology. The same formula for Ptolemy transformations describes the action of $MC_{BLP}(S)$ on these global coordinates. Furthermore, the same formula for the Weil-Petersson two-form gives a $MC_{BLP}(S)$-invariant two-form on $\tilde{\mathcal{T}}(S)$, which descends to a non-degenerate two-form on the Teichmüller space $\mathcal{T}(S)$.

Part 2 *Again, there is a convex hull construction associating to a point $\tilde{G}' \in \tilde{\mathcal{T}}(S)$ a collection $\Delta(\tilde{G}')$ of disjointly embedded arcs connecting punctures in S. The subspace*

$$\cup_{\mathrm{TLC}\Delta}\ \{\tilde{G}' \in \tilde{\mathcal{T}}(S) : \Delta(\tilde{G}') \simeq \Delta \text{ on the baseleaf}\}$$

of $\tilde{\mathcal{T}}(S)$ is open and dense, where $\simeq$ means homotopic rel endpoints at infinity, and the quotient of this subspace by $MC_{BLP}(S)$ is Hausdorff. Furthermore, $PSL_2(\mathbb{Z})$ together with all K-equivariant flips generate $MC_{BLP}(S)$, and a complete set of relations for these generators of $MC_{BLP}(S)$ can be written [3].

Thus, whereas lambda length numbers on the arcs in an ideal triangulation of F_g^s give global coordinates for $\tilde{\mathcal{T}}(F_g^s)$, continuous G- equivariant lambda length *functions*

$$\lambda_e : \hat{G} \to \mathbb{R}_{>0}, \text{ for } e \in \Delta_*,$$

on the edges of the Farey tesselation give global coordinates for $\tilde{\mathcal{T}}(S)$.

The same formula for the Weil-Petersson two-form gives an invariant two-form on $\tilde{\mathcal{T}}(S)$ in two senses: either fix a fundamental domain for $K < G$ torsion free of finite index, sum over the finitely many triangles in this fundamental domain the formula from before, and then integrate against Haar measure on $\hat{K}$, or equivalently choose a sequence of groups G_n, for $n \geq 1$, that are cofinal in the set of all finite-index normal subgroups of G (see the next section for such a cofinal set), choose nested fundamental domains D_n for each G_n, and take the limit as $n \to \infty$ of the reciprocal of the index of G_n in G times the two-form defined by summing the formula from before only over triangles in D_n.

It is entirely unclear which polyhedral decompositions of the baseleaf arise from the convex hull construction on $\tilde{\mathcal{T}}(S)$, let alone what is the topology of the corresponding subset of $\tilde{\mathcal{T}}(S)$. The presentation of $MC_{BLP}(S)$ therefore cannot be derived in analogy to the classical case, and a different complex on which $MC_{BLP}(S)$ suitably acts must be introduced to derive this presentation; again, the relations are tantamount to involutivity, commutativity, and the pentagon relation plus the obvious further relation corresponding to the case of K_1- and K_2-equivariant flips when $K_1 < K_2$ is finite-index. See [3] for details.

Since the union over all TLC tesselations Δ of

$$\{\tilde{G}' \in \tilde{\mathcal{T}}(S) : \Delta(\tilde{G}') \simeq \Delta \text{ on the baseleaf}\}$$

is open dense with Hausdorff quotient by $MC_{BLP}(S)$, the Ehrehpreis conjecture is dramatically false for decorated structures on the punctured solenoid.

§5. Group Completions

The previous two sections have surveyed applications of the decorated Teichmüller theory to pronilpotent and profinite completions of surface groups. We now turn our attention briefly to general group completions citing [4] for generalities and [5] for details on the profinite case in particular.

In the general set-up given a group G, we may consider any family $\mathcal{N}$ of normal subgroups of G where the intersection of any two groups in $\mathcal{N}$ again contains a group in $\mathcal{N}$, so that $\mathcal{N}$ is naturally a directed set. The canonical projections $G/N \to G/M$ for $N \subset M$ with $M, N \in \mathcal{N}$ thus provide an inverse system of groups, and the inverse limit of this system is the *pro-$\mathcal{N}$-completion* of G, to be denoted $\hat{\mathcal{N}}(G)$, which is a totally disconnected and complete topological group. The canonical homomorphism $G \to \hat{\mathcal{N}}(G)$ has dense image with kernel $\cap\mathcal{N}$, the canonical projection $G \to G/N$ extends uniquely to a homomorphism $\hat{\mathcal{N}}(G) \to G/N$, for each $N \in \mathcal{N}$, and the right action of G on itself extends uniquely to a continuous right action of G on $\hat{\mathcal{N}}(G)$. In case $\mathcal{N}$ is countable, we may assume that $\mathcal{N}$ is given by a nested sequence $N_1 \supset N_2 \supset \cdots$ and define the functions $\nu_i : G \to \{0,1\}$ by $\nu_i(g) = 0$ if and only if $g \in N_i$. The topology on G with local neighborhoods of the identity given by the elements of $\mathcal{N}$ is induced by the metric $\mu : G \times G \to [0,1]$ given by $\mu(g_1, g_2) = \sum_{i \geq 1} 2^{-i} \nu_i(g_1^{-1} g_2)$, and $\hat{\mathcal{N}}(G)$ is the metric completion of G for this metric comprised of equivalence classes of convergent Cauchy sequences. In the special case that $\cap\mathcal{N}$ is trivial, the canonical homomorphism $G \to \hat{\mathcal{N}}(G)$ is injective, and $\hat{\mathcal{N}}(G)$ is Hausdorff.

For instance, the two examples we have considered correspond to fixing an appropriate surface group G and taking $\mathcal{N}$ to be all normal subgroups $\mathcal{N}_f$ with finite index, giving the profinite completion, or all normal subgroups $\mathcal{N}_n$ with nilpotent quotient, giving the pronilpotent completion. Since fundamental groups of surfaces are residually finite and residually nilpotent, we have $\cap\mathcal{N}_f = \{id\} = \cap\mathcal{N}_p$, so the group injects into its Hausdorff completion in these cases. Furthermore, in the profinite case, we have the cofinal set

$$G_k = \cap\{K < G : [G : K] \leq k\} \in \mathcal{N}_f,$$

of characteristic subgroups, for $k \geq 1$, and in the pronilpotent case, we have the cofinal set of characteristic subgroups given by the terms of the lower central series for G.

There are many other interesting possibilities for surface groups where little is known including the prosolvable completion, the proLie completion, and the extreme case of the pronormal completion (corresponding to all normal subgroups).

In addition to simply pointing out here that these seem to be interesting directions for future research, we wish also to remark that in each case of the described projects on pronilpotent and profinite completions, there are aspects that are more or less functorial and are applicable in other cases of completions of a surface group G. Specifically for any completion, we may construct a corresponding "pro-$\mathcal{N}$ solenoid"

$$S(\mathcal{N}) = (\mathbb{D} \times \hat{\mathcal{N}}(G))/G, \text{ where } \gamma(z,t) = (\gamma z, t\gamma^{-1}) \text{ for } \gamma \in G,$$

we may define quasiconformal mappings as before, and we may ask for a description of the corresponding mapping class group, or more explicitly the relationship between this mapping class group and the groups $MC(S)$ and $MC_{BLP}(S)$ for $S = S(\mathcal{N}_f)$ discussed in the previous section.

Likewise, for any completion with a cofinal set $\{U_k\}_{k\geq 1}$ of characteristic groups in $\mathcal{N}$, we may consider the "$\mathcal{N}$-Torelli groups" defined by

$$\mathcal{I}_{\mathcal{N}}[k] = \{\phi \in Aut(G) : \phi \text{ acts identically on } G/U_k\},$$

and ask the obvious questions, for instance, are these groups finitely generated?, what are their abelianizations?, and so on, in analogy to the case $MC_{g,*}[k] = \mathcal{I}_{\mathcal{N}_p}[k]$ discussed already for G the fundamental group of a closed surface with basepoint. In particular, it certainly seems viable to study the profinite Torelli groups and the pronilpotent solenoid again using the decorated Teichmüller theory.

References

[1] L. V. Ahlfors and L. Bers, Riemann's mapping theorem for variable metrics, Ann. of Math. (2), **72** (1960), 385–404.

[2] D. Biss and B. Farb, $\mathcal{K}_g$ is not finitely generated, Invent. Math., **163** (2006), 213–226.

[3] S. Bonnot, R. C. Penner and D. Šarić, A presentation for the baseleaf preserving mapping class group of the punctured solenoid, preprint, math.DS/0608066, to appear Algebr. Geom. Topol.

[4] N. Bourbaki, Topologie générale, Hermann, 1960.

[5] J. D. Dixon, M. du Sautoy, A. Mann and D. Sega, Analytic pro-p groups, Cambridge Univ. Press, 1991.

[6] D. B. A. Epstein and R. C. Penner, Euclidean decompositions of noncompact hyperbolic manifolds, J. Differential Geom., **27** (1988), 67–80.

[7] D. Johnson, The structure of the Torelli group I: A finite set of generators for $\mathcal{I}_g$, Ann. of Math. (2), **118** (1983), 423–442.

[8] D. Johnson, The structure of the Torelli group II: A characterization of the group generated by twists on bounding simple closed curves, Topology, **24** (1985), 113–126.

[9] D. Johnson, The structure of the Torelli group III: The abelianization of $\mathcal{I}_g$, Topology, **24** (1985), 127–144.

[10] N. Kawazumi and S. Morita, The primary approximation to the cohomology of the moduli space of curves and cocycles for the stable characteristic classes, Math. Res. Lett., **3** (1996), 629–641.

[11] N. Kawazumi and S. Morita, The primary approximation to the cohomology of the moduli space of curves and cocycles for the Mumford-Morita-Miller classes, preprint.

[12] G. Mess, The Torelli groups for genus 2 and 3 surfaces, Topology, **31** (1992), 775–790.

[13] S. Morita, A linear representation of the mapping class group of orientable surfaces and characteristic classes of surface bundles, In: Proceedings of the Taniguchi Symposium on Topology and Teichmüller Spaces, Finland, July 1995, World Scientific, 1996, pp. 159–186.

[14] S. Morita and R. C. Penner, Torelli groups, extended Johnson homomorphisms, and new cycles on the moduli space of curves, preprint, math.GT/0602461, to appear Math. Proc. Camb. Phil. Soc.

[15] C. Odden, Virtual automorphism group of the fundamental group of a closedsurface, Ph.D. thesis, Duke Univ., Durham, 1997.

[16] ———, The baseleaf preserving mapping class group of the universal hyperbolic solenoid, Trans. Amer. Math Soc., **357** (2005), 1829–1858.

[17] R. C. Penner, The decorated Teichmüller space of punctured surfaces, Comm. Math. Phys., **113** (1987), 299–339.

[18] ———, Weil-Petersson volumes, J. Differential Geom., **35** (1992), 559–608.

[19] R. C. Penner and D. Šarić, Teichmüller theory of the punctured solenoid, preprint, math.DS/0508476, to appear in Geom. Dedicata.

[20] D. Sullivan, Linking the universalities of Milnor-Thurston, Feigenbaum and Ahlfors-Bers, In: Milnor Festschrift, Topological methods in modern mathematics, (eds. L. Goldberg and A. Phillips), Publish or Perish, 1993, pp. 543–563.

Departments of Mathematics and Physics
University of Southern California
Los Angeles, CA 90089
USA

Advanced Studies in Pure Mathematics 52, 2008
Groups of Diffeomorphisms
pp. 93–109

Johnson's homomorphisms and the rational cohomology of subgroups of the mapping class group

Takuya Sakasai

Abstract.

The Torelli group and the Johnson kernel are important subgroups of the mapping class group of a surface. Using Johnson's homomorphisms as their free abelian quotients together with the representation theory of the symplectic groups, we will give some descriptions of cup products of rational cohomology classes of degree one obtained from the dual of Johnson's homomorphisms.

§1. Introduction

Let Σ_g be a closed oriented surface of genus $g \geq 2$ and let $\mathrm{Diff}_+\Sigma_g$ be the topological group of orientation preserving diffeomorphisms of Σ_g with the C^∞-topology. The mapping class group $\mathcal{M}_g$ is defined as the group of all connected components of $\mathrm{Diff}_+\Sigma_g$. By a result of Earle-Eells [7], we have $\mathrm{BDiff}_+\Sigma_g = K(\mathcal{M}_g, 1)$, so that cohomology classes of $\mathcal{M}_g$ give characteristic classes of oriented Σ_g-bundles and we can study the theory of oriented Σ_g-bundles from the algebraic point of view. We refer [20] for generalities of characteristic classes of oriented Σ_g-bundles.

To understand the structure of $\mathcal{M}_g$, we consider the following two subgroups. The first one is the *Torelli group* $\mathcal{I}_g$ consisting of all elements which act trivially on the first homology group $H := H_1(\Sigma_g; \mathbb{Z})$ of Σ_g, namely it is the kernel of the classical representation $\rho_0 : \mathcal{M}_g \to \mathrm{Sp}(2g, \mathbb{Z})$, where $\mathrm{Sp}(2g, \mathbb{Z})$ denotes the integral symplectic group. The second one is the group $\mathcal{K}_g$ generated by Dehn twists along bounding simple closed curves. Johnson [14] showed that $\mathcal{K}_g$ coincides with the kernel of what is now called the first Johnson homomorphism $\tau_g(1) : \mathcal{I}_g \to (\wedge^3 H)/H$ of $\mathcal{I}_g$ (see [11]). For this reason, $\mathcal{K}_g$ is often called the

Received April 27, 2007.
Revised April 10, 2008.

Johnson kernel. Summarizing, we have the exact sequences

$$1 \longrightarrow \mathcal{I}_g \longrightarrow \mathcal{M}_g \xrightarrow{\rho_0} \mathrm{Sp}(2g, \mathbb{Z}) \longrightarrow 1,$$

$$1 \longrightarrow \mathcal{K}_g \longrightarrow \mathcal{I}_g \xrightarrow{\tau_g(1)} (\wedge^3 H)/H \longrightarrow 1.$$

Generally, the structures of $\mathcal{I}_g$ and $\mathcal{K}_g$ are more difficult to understand than the structure of $\mathcal{M}_g$. In fact, it is known that $\mathcal{M}_g$ is finitely presentable. On the other hand, it is not known whether $\mathcal{I}_g$ is finitely presentable or not, while Johnson [13] showed that it is finitely generated for $g \geq 3$. As for $\mathcal{K}_g$, Biss-Farb [5] showed that it is not even finitely generated for $g \geq 2$. However, we can use Johnson's homomorphisms $\tau_g(1)$ and $\tau_g(2)$, defined by Johnson [11, 12] and Morita [22, 27], to obtain non-trivial finitely generated free abelian quotients of $\mathcal{I}_g$ and $\mathcal{K}_g$. They will play important roles as primary approximations of $\mathcal{I}_g$ and $\mathcal{K}_g$, from which we can extract several pieces of information. This situation should be compared with the fact that $\mathcal{M}_g$ is perfect for $g \geq 3$, namely it has no abelian quotients except the trivial one.

Now we are particularly interested in the rational cohomology of $\mathcal{I}_g$ and $\mathcal{K}_g$, which give characteristic classes of oriented Σ_g-bundles with restricted holonomies. By passing to the dual over $\mathbb{Q}$ of $\tau_g(k)$ $(k = 1, 2)$, we can regard the rational images $\mathrm{Im}\,\tau_g^{\mathbb{Q}}(k)$ of $\tau_g(k)$ as subspaces of $H^1(\mathcal{I}_g; \mathbb{Q})$ and $H^1(\mathcal{K}_g; \mathbb{Q})$. Next we consider cup products of them. In [10], Hain determined the kernel of the cup product map

$$\cup : \wedge^2(\mathrm{Im}\,\tau_g^{\mathbb{Q}}(1)) \longrightarrow H^2(\mathcal{I}_g; \mathbb{Q}),$$

which is stable under the natural action of the rational symplectic group $\mathrm{Sp}(2g, \mathbb{Q})$ on $\wedge^2(\mathrm{Im}\,\tau_g^{\mathbb{Q}}(1))$, in terms of the representation theory of $\mathrm{Sp}(2g, \mathbb{Q})$. Continuing his result, in Sections 3 and 4, we will summarize two previous results of the author:

- ([30]) the kernel of the cup product map

$$\cup : \wedge^2(\mathrm{Im}\,\tau_g^{\mathbb{Q}}(2)) \longrightarrow H^2(\mathcal{K}_g; \mathbb{Q}),$$

- ([29]) the kernel (modulo one irreducible summand) of the triple cup product map

$$\cup^3 : \wedge^3(\mathrm{Im}\,\tau_g^{\mathbb{Q}}(1)) \longrightarrow H^3(\mathcal{I}_g; \mathbb{Q}).$$

Note that the collections of the cup product maps $\wedge^*(\mathrm{Im}\,\tau_g^{\mathbb{Q}}(1)) \to H^*(\mathcal{I}_g; \mathbb{Q})$ and $\wedge^*(\mathrm{Im}\,\tau_g^{\mathbb{Q}}(2)) \to H^*(\mathcal{K}_g; \mathbb{Q})$ are far from surjective, since Akita [1] showed that $H^*(\mathcal{I}_g; \mathbb{Q})$ and $H^*(\mathcal{K}_g; \mathbb{Q})$ are infinitely generated vector space (while $\mathcal{I}_g$ and $\mathcal{K}_g$ have finite cohomological dimensions). It is an interesting (and difficult) problem to find cohomology classes of $\mathcal{I}_g$ and $\mathcal{K}_g$ not obtained from Johnson's homomorphisms.

We also mention some similar research. Brendle-Farb [6] studied the second cohomology of $\mathcal{I}_g$ and $\mathcal{K}_g$ by using the Birman-Craggs-Johnson homomorphism, and Pettet [28] studied the second cohomology of the (outer-)automorphism group of a free group by using its first Johnson homomorphism.

§2. Preliminaries

2.1. Surfaces and their mapping class groups

Let $H_1(\Sigma_g)$ be the first integral homology group of a closed oriented surface Σ_g of genus g. $H_1(\Sigma_g)$ has a natural intersection form $\mu : H_1(\Sigma_g) \otimes H_1(\Sigma_g) \to \mathbb{Z}$ which is non-degenerate and skew symmetric. We fix a symplectic basis $\langle a_1, \ldots, a_g, b_1, \ldots, b_g \rangle$ of $H_1(\Sigma_g)$ with respect to μ as in Figure 1.

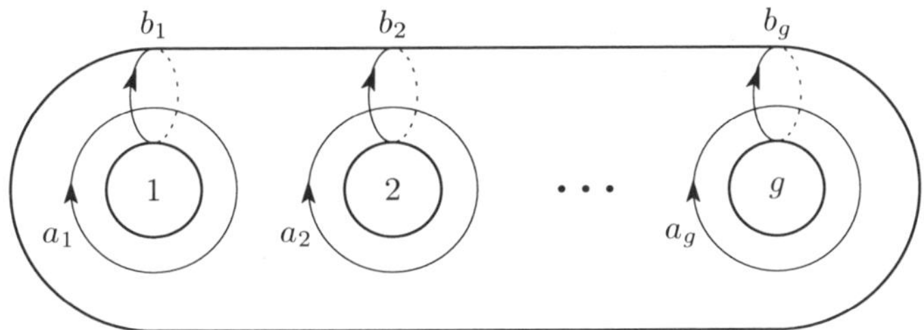

Fig. 1. A symplectic basis of $H_1(\Sigma_g)$

Then we have

$$\mu(a_i, a_j) = 0, \quad \mu(b_i, b_j) = 0, \quad \mu(a_i, b_j) = \delta_{ij}.$$

Poincaré duality gives a canonical isomorphism between $H_1(\Sigma_g)$ and its dual $\mathrm{Hom}(H_1(\Sigma_g), \mathbb{Z}) = H^1(\Sigma_g)$, the first integral cohomology group of Σ_g. In this isomorphism, a_i (resp. b_i) $\in H_1(\Sigma_g)$ corresponds to $-b_i^*$ (resp. a_i^*) $\in H^1(\Sigma_g)$ where $\langle a_1^*, \ldots, a_g^*, b_1^*, \ldots, b_g^* \rangle$ is the dual basis of $H^1(\Sigma_g)$. We use the same symbol H for these canonically isomorphic abelian groups.

We also use a compact oriented surface $\Sigma_{g,1}$ of genus g with a connected boundary. $H_1(\Sigma_{g,1})$ and $H^1(\Sigma_{g,1})$ can be naturally identified with H. The fundamental group $\pi_1\Sigma_{g,1}$ of $\Sigma_{g,1}$, where we take a base point of $\Sigma_{g,1}$ on $\partial\Sigma_{g,1}$, is known to be a free group of rank $2g$. We write

$\zeta \in \pi_1\Sigma_{g,1}$ for the boundary loop of $\Sigma_{g,1}$. Then the fundamental group $\pi_1\Sigma_g$ of Σ_g is given by $\pi_1\Sigma_{g,1}/\langle\zeta\rangle$ where $\langle\zeta\rangle$ is the normal closure of the subgroup generated by ζ.

Let $\mathcal{M}_g, \mathcal{M}_{g,*}, \mathcal{M}_{g,1}$ be the mapping class group of Σ_g, of Σ_g relative to the base point, of $\Sigma_{g,1}$, respectively. They are related by the exact sequences

$$0 \longrightarrow \mathbb{Z} \longrightarrow \mathcal{M}_{g,1} \longrightarrow \mathcal{M}_{g,*} \longrightarrow 1, \tag{1}$$

$$1 \longrightarrow \pi_1\Sigma_g \longrightarrow \mathcal{M}_{g,*} \longrightarrow \mathcal{M}_g \longrightarrow 1, \tag{2}$$

where $\mathbb{Z}$ corresponds to the Dehn twist along a loop which is parallel to $\partial\Sigma_{g,1}$, and $\pi_1\Sigma_g$ is embedded in $\mathcal{M}_{g,*}$ as spin-maps (see Theorem 4.3 in [4]). The former sequence is a central extension.

The natural action of $\mathcal{M}_g$ on H gives the classical representation

$$\rho_0 : \mathcal{M}_g \longrightarrow \mathrm{Sp}(2g, \mathbb{Z}),$$

and we also have similar ones for $\mathcal{M}_{g,*}$ and $\mathcal{M}_{g,1}$. The kernels of these representations are denoted by $\mathcal{I}_g$, $\mathcal{I}_{g,*}$ and $\mathcal{I}_{g,1}$, respectively and called the *Torelli group* for each case. Note that among $\mathcal{I}_g$, $\mathcal{I}_{g,*}$ and $\mathcal{I}_{g,1}$, we have exact sequences similar to (1) and (2). Indeed the Dehn twist along $\partial\Sigma_{g,1}$ and spin-maps act on H trivially.

Let $\mathcal{K}_g$ (resp. $\mathcal{K}_{g,1}$) be the subgroup of $\mathcal{M}_g$ (resp. $\mathcal{M}_{g,1}$) generated by Dehn twists along bounding simple closed curves on Σ_g (resp. $\Sigma_{g,1}$). We define $\mathcal{K}_{g,*} \subset \mathcal{M}_{g,*}$ to be the image of $\mathcal{K}_{g,1}$ by the map $\mathcal{M}_{g,1} \to \mathcal{M}_{g,*}$. Then we have the exact sequences

$$0 \longrightarrow \mathbb{Z} \longrightarrow \mathcal{K}_{g,1} \longrightarrow \mathcal{K}_{g,*} \longrightarrow 1,$$

$$1 \longrightarrow [\pi_1\Sigma_g, \pi_1\Sigma_g] \longrightarrow \mathcal{K}_{g,*} \longrightarrow \mathcal{K}_g \longrightarrow 1,$$

where the former sequence is the pull-back of the central extension of $\mathcal{M}_{g,*}$, and the latter one follows from a result of Asada-Kaneko [2].

2.2. Johnson's homomorphisms

Here, we recall what we call Johnson's homomorphisms defined by Johnson [11, 12] and Morita [22, 24, 27].

By results of Dehn, Nielsen and many people, we have

$$\mathcal{M}_g \cong \mathrm{Out}_+\,\pi_1\Sigma_g, \tag{3}$$

$$\mathcal{M}_{g,*} \cong \mathrm{Aut}_+\,\pi_1\Sigma_g, \tag{4}$$

$$\mathcal{M}_{g,1} \cong \{\varphi \in \mathrm{Aut}\,\pi_1\Sigma_{g,1} \mid \varphi(\zeta) = \zeta\}, \tag{5}$$

where $\mathrm{Out}_+\,\pi_1\Sigma_g := \mathrm{Ker}(\mathrm{Out}\,\pi_1\Sigma_g \to \mathrm{Aut}\,H_2(\pi_1\Sigma_g))$, and $\mathrm{Aut}_+\,\pi_1\Sigma_g$ is similar.

For a group G, let $\{\Gamma^k G\}_{k\geq 1}$ be the lower central series of G inductively defined by $\Gamma^1 G = G$ and $\Gamma^i G = [\Gamma^{i-1}G, G]$ for $i \geq 2$. The collection $\{(\Gamma^k G)/(\Gamma^{k+1}G)\}_{k\geq 1}$ forms a graded Lie algebra whose bracket map is induced from taking commutators. It is well known that the Lie algebra $\{(\Gamma^k \pi_1\Sigma_{g,1})/(\Gamma^{k+1}\pi_1\Sigma_{g,1})\}_{k\geq 1}$ is isomorphic to the free Lie algebra $\mathcal{L}_{g,1} = \{\mathcal{L}_{g,1}(k)\}_{k\geq 1}$ generated by H. Furthermore, by a result of Labute [17], the Lie algebra $\{(\Gamma^k \pi_1\Sigma_g)/(\Gamma^{k+1}\pi_1\Sigma_g)\}_{k\geq 1}$ is given by $\mathcal{L}_g := \mathcal{L}_{g,1}/I$ where I is the ideal of $\mathcal{L}_{g,1}$ generated by $\omega_0 := \sum_{i=1}^g [a_i, b_i]$.

The isomorphisms (3), (4) and (5) induce the homomorphisms

$$\begin{aligned} \sigma_k \ : \mathcal{M}_g \ &\longrightarrow \mathrm{Out}\big(\pi_1\Sigma_g/(\Gamma^k\pi_1\Sigma_g)\big),\\ \sigma_{k,*} : \mathcal{M}_{g,*} &\longrightarrow \mathrm{Aut}\big(\pi_1\Sigma_g/(\Gamma^k\pi_1\Sigma_g)\big),\\ \sigma_{k,1} : \mathcal{M}_{g,1} &\longrightarrow \mathrm{Aut}\big(\pi_1\Sigma_{g,1}/(\Gamma^k\pi_1\Sigma_{g,1})\big) \end{aligned}$$

for each $k \geq 2$, and we define filtrations of $\mathcal{M}_g, \mathcal{M}_{g,*}, \mathcal{M}_{g,1}$ by

$$\begin{aligned} \mathcal{M}_g[1] &:= \mathcal{M}_g, & \mathcal{M}_g[k] &:= \mathrm{Ker}\,\sigma_k \quad (k \geq 2),\\ \mathcal{M}_{g,*}[1] &:= \mathcal{M}_{g,*}, & \mathcal{M}_{g,*}[k] &:= \mathrm{Ker}\,\sigma_{k,*} \quad (k \geq 2),\\ \mathcal{M}_{g,1}[1] &:= \mathcal{M}_{g,1}, & \mathcal{M}_{g,1}[k] &:= \mathrm{Ker}\,\sigma_{k,1} \quad (k \geq 2). \end{aligned}$$

For each $\varphi \in \mathcal{M}_{g,1}[k+1]$ and $\gamma \in \pi_1\Sigma_{g,1}$, we have $\varphi(\gamma)\gamma^{-1} \in \Gamma^{k+1}\pi_1\Sigma_{g,1}$. This induces a map

$$\begin{aligned} \tau_{g,1}(k) : \mathcal{M}_{g,1}[k+1] \to \mathrm{Hom}(\pi_1\Sigma_{g,1}, (\Gamma^{k+1}\pi_1\Sigma_{g,1})/(\Gamma^{k+2}\pi_1\Sigma_{g,1}))\\ = \mathrm{Hom}(H, \mathcal{L}_{g,1}(k+1)), \end{aligned}$$

and it is in fact a homomorphism.

In [23, 24], Morita showed the following. By taking commutators, we can endow $\{\mathcal{M}_{g,1}[k+1]/\mathcal{M}_{g,1}[k+2]\}_{k\geq 1} = \{\mathrm{Im}\,\tau_{g,1}(k)\}_{k\geq 1} =: \mathrm{Im}\,\tau_{g,1}$ with a Lie algebra structure. We can also endow $\mathrm{Hom}(H, \mathcal{L}_{g,1}) := \{\mathrm{Hom}(H, \mathcal{L}_{g,1}(k+1))\}_{k\geq 1}$ with a Lie algebra structure, so that $\tau_{g,1} := \{\tau_{g,1}(k)\}_{k\geq 1}$ becomes a Lie algebra inclusion of $\mathrm{Im}\,\tau_{g,1}$ into $\mathrm{Hom}(H, \mathcal{L}_{g,1})$. Moreover, he showed that $\mathrm{Im}\,\tau_{g,1}$ is contained in the Lie subalgebra $\mathfrak{h}_{g,1} = \{\mathfrak{h}_{g,1}(k)\}_{k\geq 1}$ defined by

$$\mathfrak{h}_{g,1}(k) := \mathrm{Ker}\left(\mathrm{Hom}(H, \mathcal{L}_{g,1}(k+1)) \xrightarrow{\cong} H \otimes \mathcal{L}_{g,1}(k+1) \xrightarrow{[\cdot,\cdot]} \mathcal{L}_{g,1}(k+2)\right),$$

where the maps in the right hand side are given by Poincaré duality and the bracket operation.

A similar argument gives a homomorphism $\tau_{g,*}(k) : \mathcal{M}_{g,*}[k+1] \to \mathfrak{h}_{g,*}(k) \subset \mathrm{Hom}(H,\ \mathcal{L}_g(k+1))$, where

$$\mathfrak{h}_{g,*}(k) := \mathrm{Ker}\left(\mathrm{Hom}(H, \mathcal{L}_g(k+1)) \xrightarrow{\cong} H \otimes \mathcal{L}_g(k+1) \xrightarrow{[\cdot,\cdot]} \mathcal{L}_g(k+2)\right),$$

and the corresponding Lie algebra inclusion. Moreover, Asada-Kaneko [2] showed that $\pi_1\Sigma_g \cap \mathcal{M}_{g,*}[k+1] = \Gamma^k\pi_1\Sigma_g$. Hence we have an inclusion

$$\Psi_k : (\Gamma^k\pi_1\Sigma_g)/(\Gamma^{k+1}\pi_1\Sigma_g) \cong \mathcal{L}_g(k) \hookrightarrow \mathfrak{h}_{g,*}(k).$$

If we set $\mathfrak{h}_g(k) := \mathfrak{h}_{g,*}(k)/\mathcal{L}_g(k)$, we obtain a homomorphism $\tau_g(k) : \mathcal{M}_g[k+1] \to \mathfrak{h}_g(k)$ and the corresponding Lie algebra inclusion. We call the homomorphisms $\tau_g(k)$, $\tau_{g,*}(k)$, $\tau_{g,1}(k)$ the k-th Johnson homomorphism for each case. Note that $\tau_g(k)$ is $\mathcal{M}_g$-equivariant, where $\mathcal{M}_g$ acts on $\mathcal{M}_g[k+1]$ by conjugation and acts on the target through the classical representation $\rho_0 : \mathcal{M}_g \to \mathrm{Sp}(2g,\mathbb{Z})$. Similar results hold for $\tau_{g,*}(k)$ and $\tau_{g,1}(k)$.

We have $\mathcal{M}_{g,1}[2] = \mathcal{I}_{g,1}$, $\mathcal{M}_{g,*}[2] = \mathcal{I}_{g,*}$ and $\mathcal{M}_g[2] = \mathcal{I}_g$ by definition. Johnson [14] showed that $\mathcal{M}_{g,1}[3] = \mathcal{K}_{g,1}$ and $\mathcal{M}_g[3] = \mathcal{K}_g$. Combining the fact that the first Johnson homomorphisms for $\mathcal{I}_{g,1}$ and $\mathcal{I}_{g,*}$ have the same target $\wedge^3 H$, we can see that $\mathcal{M}_{g,*}[3] = \mathcal{K}_{g,*}$.

2.3. The representation theory of $\mathrm{Sp}(2g,\mathbb{Q})$

Here we summarize the notation and general facts concerning the representation theory of $\mathrm{Sp}(2g,\mathbb{Q})$ from [8], [10] and [27]. First we consider the Lie group $\mathrm{Sp}(2g,\mathbb{C})$ and its Lie algebra $\mathfrak{sp}(2g,\mathbb{C})$. It is known that finite dimensional representations of $\mathrm{Sp}(2g,\mathbb{C})$ coincide with those of $\mathfrak{sp}(2g,\mathbb{C})$, and their common irreducible representations (up to isomorphisms) are parameterized by Young diagrams whose numbers of rows are less than or equal to g. These representations are all defined over $\mathbb{Q}$ so that we can consider them as irreducible representations of $\mathrm{Sp}(2g,\mathbb{Q})$ and $\mathfrak{sp}(2g,\mathbb{Q})$. We follow the notation in [27] to describe Young diagrams as in Figure 2.

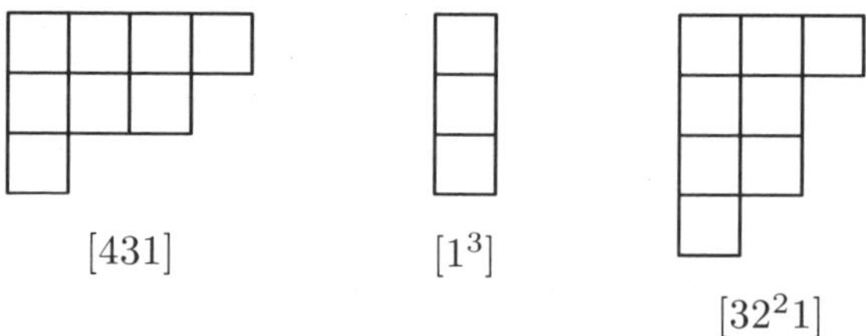

Fig. 2. Notation for Young diagrams

For example, the trivial representation $\mathbb{Q}$ is denoted by $[0]$ and the fundamental representation $H_{\mathbb{Q}} := H \otimes \mathbb{Q}$ is denoted by $[1]$. We fix a symplectic basis $\langle a_1, \ldots, a_g, b_1, \ldots, b_g \rangle$ of $H_{\mathbb{Q}}$ with respect to the non-degenerate skew symmetric bilinear form, denoted by μ again, on $H_{\mathbb{Q}}$ induced from the intersection form μ on H. In general, the Young diagram $[n_1 n_2 \cdots n_l]$, where n_i are integers satisfying $n_1 \geq n_2 \geq \cdots \geq n_l \geq 1$ and $l \leq g$, corresponds to the $\mathrm{Sp}(2g, \mathbb{Q})$-vector space V given as follows. Let $[m_1 m_2 \cdots m_k]$ be the Young diagram obtained by transposing $[n_1 n_2 \cdots n_l]$. Then V is explicitly defined to be the irreducible $\mathrm{Sp}(2g, \mathbb{Q})$-subspace of

$$(\wedge^{m_1} H_{\mathbb{Q}}) \otimes (\wedge^{m_2} H_{\mathbb{Q}}) \otimes \cdots \otimes (\wedge^{m_k} H_{\mathbb{Q}})$$

containing the vector

$$(a_1 \wedge a_2 \wedge \cdots \wedge a_{m_1}) \otimes (a_1 \wedge a_2 \wedge \cdots \wedge a_{m_2}) \otimes \cdots \otimes (a_1 \wedge a_2 \wedge \cdots \wedge a_{m_k}),$$

which is called the *highest weight vector of* $[n_1 n_2 \cdots n_l]$.

2.4. Johnson's homomorphisms from the view point of the representation theory of $\mathrm{Sp}(2g, \mathbb{Q})$

Let $\tau_g^{\mathbb{Q}}(k)$ denote the Johnson homomorphism $\tau_g(k)$ tensored by $\mathbb{Q}$, namely

$$\tau_g^{\mathbb{Q}}(k) : (\mathcal{M}_g[k+1]/(\Gamma^2 \mathcal{M}_g[k+1])) \otimes \mathbb{Q} \longrightarrow \mathfrak{h}_g^{\mathbb{Q}}(k) := \mathfrak{h}_g(k) \otimes \mathbb{Q}.$$

As mentioned in Section 2.2, $\tau_g^{\mathbb{Q}}(k)$ is $\mathcal{M}_g$-equivariant, so that $\mathrm{Im}\, \tau_g^{\mathbb{Q}}(k)$ is an $\mathrm{Sp}(2g, \mathbb{Z})$-vector space. Moreover, $\mathrm{Im}\, \tau_g^{\mathbb{Q}}(k)$ turns out to be an $\mathrm{Sp}(2g, \mathbb{Q})$-vector space by Lemma 2.2.8 of Asada-Nakamura [3]. In particular, $\mathrm{Im}\, \tau_g^{\mathbb{Q}}(k)$ and $\mathfrak{h}_g^{\mathbb{Q}}(k)$ can be written in terms of the representation theory of $\mathrm{Sp}(2g, \mathbb{Q})$. Similar results hold for $\tau_{g,*}^{\mathbb{Q}}(k) := \tau_{g,*}(k) \otimes \mathbb{Q}$ and $\tau_{g,1}^{\mathbb{Q}}(k) := \tau_{g,1}(k) \otimes \mathbb{Q}$. By results of Johnson [11] for $k = 1$, Hain [10] and Morita [22] for $k = 2$, Hain [10] and Asada-Nakamura [3] for $k = 3$, we have the following.

$$\begin{aligned}
\mathrm{Im}\, \tau_{g,1}^{\mathbb{Q}}(1) &= \mathrm{Im}\, \tau_{g,*}^{\mathbb{Q}}(1) = \mathfrak{h}_{g,1}^{\mathbb{Q}}(1) = \mathfrak{h}_{g,*}^{\mathbb{Q}}(1) = [1^3] + [1] = \wedge^3 H_{\mathbb{Q}}, \\
\mathrm{Im}\, \tau_g^{\mathbb{Q}}(1) &= \mathfrak{h}_g^{\mathbb{Q}}(1) = [1^3] = (\wedge^3 H_{\mathbb{Q}})/H_{\mathbb{Q}}, \\
\mathrm{Im}\, \tau_{g,1}^{\mathbb{Q}}(2) &= \mathfrak{h}_{g,1}^{\mathbb{Q}}(2) = [2^2] + [1^2] + [0], \\
\mathrm{Im}\, \tau_{g,*}^{\mathbb{Q}}(2) &= \mathfrak{h}_{g,*}^{\mathbb{Q}}(2) = [2^2] + [1^2], \\
\mathrm{Im}\, \tau_g^{\mathbb{Q}}(2) &= \mathfrak{h}_g^{\mathbb{Q}}(2) = [2^2], \\
\mathrm{Im}\, \tau_{g,1}^{\mathbb{Q}}(3) &= \mathrm{Im}\, \tau_{g,*}^{\mathbb{Q}}(3) = [31^2] + [21] \\
&\qquad \subset \mathfrak{h}_{g,1}^{\mathbb{Q}}(3) = \mathfrak{h}_{g,*}^{\mathbb{Q}}(3) = [31^2] + [21] + [3], \\
\mathrm{Im}\, \tau_g^{\mathbb{Q}}(3) &= [31^2] \subset \mathfrak{h}_g^{\mathbb{Q}}(3) = [31^2] + [3]
\end{aligned}$$

for $g \geq 3$, where we write $+$ for the direct sum. Moreover, in [27], Morita announced that

$$\begin{aligned}\operatorname{Im}\tau_{g,1}^{\mathbb{Q}}(4) &= \operatorname{Im}\tau_{g,*}^{\mathbb{Q}}(4) = [42]+[31^3]+2[31]+[2^3]+[21^2]+2[2],\\ \operatorname{Im}\tau_{g}^{\mathbb{Q}}(4) &= [42]+[31^3]+[31]+[2^3]+[2]\end{aligned}$$

in

$$\begin{aligned}\mathfrak{h}_{g,1}^{\mathbb{Q}}(4) &= [42]+[31^3]+2[31]+[2^3]+2[21^2]+3[2],\\ \mathfrak{h}_{g,*}^{\mathbb{Q}}(4) &= [42]+[31^3]+2[31]+[2^3]+2[21^2]+2[2],\\ \mathfrak{h}_{g}^{\mathbb{Q}}(4) &= [42]+[31^3]+[31]+[2^3]+[21^2]+[2]\end{aligned}$$

for $g \geq 4$.

Remark 2.1. Hain showed in [10] that as Lie algebras, $\operatorname{Im}\tau_{g,1}^{\mathbb{Q}}$, $\operatorname{Im}\tau_{g,*}^{\mathbb{Q}}$, $\operatorname{Im}\tau_{g}^{\mathbb{Q}}$ are generated by their degree one part for $g \geq 3$, and that $\operatorname{Im}\tau_{g,1}^{\mathbb{Q}}(k) \cong \operatorname{Im}\tau_{g,*}^{\mathbb{Q}}(k)$ for $k \geq 3$. The proof, on which Hain kindly informed the author, uses Lemmas and Propositions 4.5–4.8 in [10] with some general facts about mixed Hodge structures.

The bracket operation of $\mathfrak{h}_{g,1}$ is explicitly given in [23, 24]. However, here we use an alternative description given by Garoufalidis-Levine [9], Levine [18, 19], which will be easier to handle. We recall the following Lie algebra of labeled unitrivalent trees. Let $\mathcal{A}_k^t(H)$ be the abelian group generated by unitrivalent trees with $k+2$ univalent vertices labeled by elements of H and a cyclic order of each trivalent vertex modulo relations of AS and IHX together with linearity of labels. We can endow $\mathcal{A}^t(H) := \{\mathcal{A}_k^t(H)\}_{k\geq 1}$ with a bracket operation

$$[\cdot,\cdot] : \mathcal{A}_k^t(H) \otimes \mathcal{A}_l^t(H) \longrightarrow \mathcal{A}_{k+l}^t(H)$$

given below, and $\mathcal{A}^t(H)$ becomes a quasi Lie algebra. For labeled trees $T_1, T_2 \in \mathcal{A}^t(H)$, we define

$$[T_1, T_2] := \sum_{i,j} \mu(c_i, d_j) T_1 *_{i,j} T_2$$

where μ is the intersection form on H and the sum is taken over all pairs of a univalent vertex of T_1, labeled by c_i, and one of T_2, labeled by d_j, and $T_1 *_{i,j} T_2$ is the tree given by welding T_1 and T_2 at the pair. We define a map $\eta_k : \mathcal{A}_k^t(H) \to H \otimes \mathcal{L}_{g,1}(k+1)$ by

$$\eta_k(T) := \sum_v c_v \otimes T_v,$$

where the sum is over all univalent vertices of T, and for each univalent vertex v, c_v denotes the label of v and T_v denotes the rooted labeled planar binary tree obtained from T by removing the label c_v and considering v to be an unlabeled root, which can be regarded as an element of $\mathcal{L}_{g,1}(k+1)$ by a standard method. It is shown that $\eta := \{\eta_k\}_{k\geq 1} : \mathcal{A}^t(H) \to H \otimes \mathcal{L}_{g,1}$ is a quasi Lie algebra homomorphism and $\operatorname{Im}\eta \subset \mathfrak{h}_{g,1}$. Moreover $\eta \otimes \mathbb{Q} : \mathcal{A}^t(H) \otimes \mathbb{Q} \to \mathfrak{h}_{g,1}^{\mathbb{Q}}$ becomes an isomorphism of Lie algebras. In what follows, we identify $\mathcal{A}^t(H) \otimes \mathbb{Q}$ with $\mathfrak{h}_{g,1}^{\mathbb{Q}}$ by $\eta \otimes \mathbb{Q}$.

Using $\mathcal{A}^t(H) \otimes \mathbb{Q}$, we now give a graphical description of the map

$$\Psi_k : \mathcal{L}_g^{\mathbb{Q}}(k) := \mathcal{L}_g(k) \otimes \mathbb{Q} \hookrightarrow \operatorname{Im}\tau_{g,*}^{\mathbb{Q}}(k) \subset \mathfrak{h}_{g,*}^{\mathbb{Q}}(k),$$

which was mentioned in Section 2.2 and is explicitly given by

$$\mathcal{L}_g^{\mathbb{Q}}(k) \ni X \mapsto \sum_{i=1}^{g} \big(a_i \otimes [b_i, X] - b_i \otimes [a_i, X]\big) \in H \otimes \mathcal{L}_g^{\mathbb{Q}}(k+1).$$

For each rooted labeled planar binary tree T as an element of $\mathcal{L}_{g,1}^{\mathbb{Q}}(k)$, we can construct an element of $\mathcal{A}_k^t(H) \otimes \mathbb{Q} \cong \mathfrak{h}_{g,1}^{\mathbb{Q}}(k)$ by gluing T to the rooted labeled planar binary tree $T_{\omega_0} \in \mathcal{L}_{g,1}^{\mathbb{Q}}(2) \cong \wedge^2 H_{\mathbb{Q}}$ corresponding to $\omega_0 = \sum_{i=1}^{g} a_i \wedge b_i$ at their roots as depicted in Figure 3.

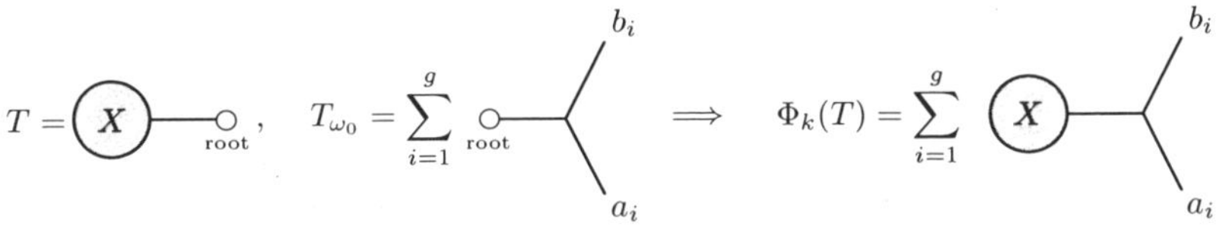

Fig. 3. The map $\Phi_k : \mathcal{L}_{g,1}^{\mathbb{Q}}(k) \to \mathfrak{h}_{g,1}^{\mathbb{Q}}(k)$

We can easily check that this construction gives an $\mathrm{Sp}(2g, \mathbb{Q})$-equivariant homomorphism $\Phi_k : \mathcal{L}_{g,1}^{\mathbb{Q}}(k) \to \mathfrak{h}_{g,1}^{\mathbb{Q}}(k)$, and it induces the desired map $\Psi_k : \mathcal{L}_g^{\mathbb{Q}}(k) \hookrightarrow \mathfrak{h}_{g,*}^{\mathbb{Q}}(k)$ by using Labute's result [17].

§3. The second cohomology of the Torelli group and the Johnson kernel

We start our investigation of cup products of cohomology classes of degree one obtained from Johnson's homomorphisms.

By passing to the dual over $\mathbb{Q}$ of $\tau_g(1) : \mathcal{I}_g \to [1^3]$, we obtain an injection

$$\tau_g^{\mathbb{Q}}(1)^* : [1^3] \hookrightarrow H^1(\mathcal{I}_g; \mathbb{Q})$$

and more generally, we have the cup product map

$$\cup^n : \wedge^n [1^3] \longrightarrow H^n(\mathcal{I}_g; \mathbb{Q})$$

for each $n \geq 2$. In [15], Johnson showed that $\tau_g^{\mathbb{Q}}(1)$ is an isomorphism, so that studying the map $\cup^n$ is equivalent to determining the subring of $H^*(\mathcal{I}_g; \mathbb{Q})$ generated by its degree one part.

The map $\cup^n$ is $\mathcal{M}_g$-equivariant, since $\tau_g(1)$ is so. Hence $\operatorname{Ker} \cup^n$ is an $\mathcal{M}_g$-submodule (in fact, an $\mathrm{Sp}(2g, \mathbb{Z})$-submodule) of $\wedge^n[1^3]$. Moreover, by Asada-Nakamura's argument [3] mentioned before, it is stable under the natural action of $\mathrm{Sp}(2g, \mathbb{Q})$ extending that of $\mathrm{Sp}(2g, \mathbb{Z})$.

The kernel of the cup product map $\cup : \wedge^2[1^3] \to H^2(\mathcal{I}_g; \mathbb{Q})$ was determined by Hain as follows. (We only mention the case in the stable range $g \geq 6$.)

Lemma 3.1 (Hain [10]). *For $g \geq 6$, the irreducible decomposition of $\wedge^2[1^3]$ is given by*

$$\wedge^2[1^3] = [2^2 1^2] + [2^2] + [1^6] + [1^4] + [1^2] + [0].$$

Theorem 3.2 (Hain [10]). *For $g \geq 6$, the kernel of the cup product map $\cup : \wedge^2[1^3] \to H^2(\mathcal{I}_g; \mathbb{Q})$ is $[2^2] + [0]$.*

This theorem is obtained as a corollary of his argument about the finite presentation of Torelli Lie algebra and it is not mentioned directly in [10]. Here we sketch the proof of the theorem, which is divided into the following two steps. Note that the idea of this proof will be applicable to general cases.

Step 1. (Lower bound of $\operatorname{Ker} \cup$) Since $\wedge^2[1^3] = H^2([1^3]; \mathbb{Q})$, the map $\cup : \wedge^2[1^3] \to H^2(\mathcal{I}_g; \mathbb{Q})$ coincides with the homomorphism

$$\tau_g(1)^* : H^2([1^3]; \mathbb{Q}) \longrightarrow H^2(\mathcal{I}_g; \mathbb{Q}).$$

Hence, by passing to the dual, our task is equivalent to observing the cokernel of the map

$$\tau_g(1)_* : H_2(\mathcal{I}_g; \mathbb{Q}) \longrightarrow H_2([1^3]; \mathbb{Q}).$$

By applying Stallings' exact sequence [31] to the group extension

$$1 \longrightarrow \Gamma^2 \mathcal{I}_g \longrightarrow \mathcal{I}_g \longrightarrow H_1(\mathcal{I}_g) \longrightarrow 1$$

and observing the homomorphisms, we obtain the exact sequence

$$H_2(\mathcal{I}_g;\mathbb{Q}) \longrightarrow \wedge^2(H_1(\mathcal{I}_g;\mathbb{Q})) = \wedge^2[1^3] \xrightarrow{[\cdot,\cdot]} (\Gamma^2\mathcal{I}_g/\Gamma^3\mathcal{I}_g)\otimes\mathbb{Q} \longrightarrow 0$$

where the first map is the coproduct on the rational homology and the second one is the Lie bracket

$$[\cdot,\cdot]: \wedge^2(H_1(\mathcal{I}_g)) = \wedge^2(\Gamma^1\mathcal{I}_g/\Gamma^2\mathcal{I}_g) \longrightarrow \Gamma^2\mathcal{I}_g/\Gamma^3\mathcal{I}_g.$$

By observing the image of this map, we obtain $[2^2] \subset \operatorname{Ker}\cup$.

We can see $[0] \subset \operatorname{Ker}\cup$ from the fact that the first Morita-Miller-Mumford class vanishes on $H^2(\mathcal{I}_g;\mathbb{Q})$ (see [26] for details).

Step 2. (Upper bound of $\operatorname{Ker}\cup$) We can obtain summands in $\wedge^2[1^3]$ which survive in $H^2(\mathcal{I}_g;\mathbb{Q})$ when we take Kronecker products with elements in $H_2(\mathcal{I}_g;\mathbb{Q})$ and the results are non-trivial. We can construct an element of $H_2(\mathcal{I}_g;\mathbb{Q})$ by constructing a homomorphism $\mathbb{Z}^2 \to \mathcal{I}_g$ and considering the image of $1 \in \mathbb{Z} \cong H_2(\mathbb{Z}^2)$ in $H_2(\mathcal{I}_g)$. Such classes are called *abelian cycles*. Note that what we need to construct a homomorphism $\mathbb{Z}^2 \to \mathcal{I}_g$ is only a choice of a pair of commuting elements in $\mathcal{I}_g$. By an explicit computation, we can see that $[2^21^2]+[1^6]+[1^4]+[1^2]$ is not in $\operatorname{Ker}\cup$.

Since the upper bound and the lower one coincides, Theorem 3.2 is proved. □

Remark 3.3. While some additional arguments are needed, the cases of $\mathcal{I}_{g,*}$ and $\mathcal{I}_{g,1}$ will be settled similarly. See Section 6 of [26] for details.)

Next we consider the case of $\mathcal{K}_g$. By passing to the dual of $\tau_g(2):\mathcal{K}_g \to [2^2]$, we obtain an injection

$$\tau_g^{\mathbb{Q}}(2)^*: [2^2] \hookrightarrow H^1(\mathcal{K}_g;\mathbb{Q})$$

and the cup product map

$$\cup^n: \wedge^n[2^2] \longrightarrow H^n(\mathcal{K}_g;\mathbb{Q})$$

for each $n \geq 2$. We now observe the map $\cup: \wedge^2[2^2] \to H^2(\mathcal{K}_g;\mathbb{Q})$. In this case, the stability range is given by $g \geq 4$.

Lemma 3.4 ([30]). *For $g \geq 4$, the irreducible decomposition of $\wedge^2[2^2]$ is given by*

$$\wedge^2[2^2] = [431]+[42]+[32^21]+[321]+[31^3]+[31]+[2^3]+[21^2]+[2].$$

Theorem 3.5 ([30]). *For $g \geq 4$, the kernel of the cup product map $\cup : \wedge^2[2^2] \to H^2(\mathcal{K}_g;\mathbb{Q})$ is*

$$[42] + [31^3] + [31] + [2^3] + [2],$$

which is, as an $\mathrm{Sp}(2g,\mathbb{Q})$-vector space, isomorphic to the rational image of the fourth Johnson homomorphism $\tau_g(4)$.

(*Sketch of Proof*) The proof goes parallel to that of Theorem 3.2. The point is the relationship to the fourth Johnson homomorphism.

Step 1. (Lower bound of $\mathrm{Ker} \cup$) A lower bound of $\mathrm{Ker} \cup$ is obtained by observing the map

$$[\cdot,\cdot] : \wedge^2(H_1(\mathcal{K}_g)) = \wedge^2(\Gamma^1\mathcal{K}_g/\Gamma^2\mathcal{K}_g) \longrightarrow \Gamma^2\mathcal{K}_g/\Gamma^3\mathcal{K}_g.$$

To see the image of this map, we can use the diagram

$$\begin{array}{ccccc}
\wedge^2(H_1(\mathcal{K}_g)) & \xrightarrow{[\cdot,\cdot]} & \Gamma^2\mathcal{K}_g/\Gamma^3\mathcal{K}_g & \longrightarrow & \mathcal{M}_g[5]/\mathcal{M}_g[6] \\
{\scriptstyle \wedge^2\tau_g(2)}\downarrow & & & & \cong\downarrow{\scriptstyle \tau_g(4)} \\
\wedge^2(\mathrm{Im}\,\tau_g(2)) & \xrightarrow{[\cdot,\cdot]} & \mathrm{Im}\,\tau_g(4) & = & \mathrm{Im}\,\tau_g(4),
\end{array}$$

whose commutativity follows from the fact that the collection $\{\tau_g(k)\}_{k\geq 1}$ forms a Lie algebra homomorphism. By direct computations, we can show that $[42] + [31^3] + [31] + [2^3] + [2] \subset \mathrm{Ker} \cup$.

Step 2. (Upper bound of $\mathrm{Ker} \cup$) By computations using abelian cycles in $H_2(\mathcal{K}_g)$, we can see that $[431] + [32^21] + [321] + [21^2]$ is not in $\mathrm{Ker} \cup$.

Since the upper bound and the lower one coincides, Theorem 3.5 is proved. □

Remark 3.6. From Theorems 3.2 and 3.5, we can give lower bounds of the ranks of $H^2(\mathcal{I}_g)$ and $H^2(\mathcal{K}_g)$, which may be infinite, by using Weyl's character formula (see Section 24.2 of [8]). For example, the summand $[32^21] \subset \wedge^2[2^2]$ survives in $H^2(\mathcal{K}_g;\mathbb{Q})$, so that the rational dimension

$$\frac{1}{36}(g-3)(g-2)(g-1)(g+2)(2g-1)(2g+1)^2(2g+3)$$

of this summand gives a lower bound of the rank of $H^2(\mathcal{K}_g)$.

§4. The third cohomology of the Torelli group

Finally, we consider triple cup products of $[1^3] = H^1(\mathcal{I}_g;\mathbb{Q})$. In this case the stability range is given by $g \geq 9$.

Lemma 4.1 ([29]). *For $g \geq 9$, the irreducible decomposition of $\wedge^3[1^3]$ is given by*

$$\begin{aligned}\wedge^3[1^3] = \;& [3^2 1^3] + [3^2 1] + [32^3] + [321^2] + [32] \\ &+ [2^3 1^3] + [2^3 1] + [2^2 1^5] + 2[2^2 1^3] + 2[2^2 1] + [21^5] + 2[21^3] + [21] \\ &+ [1^9] + [1^7] + 2[1^5] + 3[1^3] + [1].\end{aligned}$$

Theorem 4.2 ([29]). *For $g \geq 9$, the kernel of the cup product map $\cup^3 : \wedge^3[1^3] \to H^3(\mathcal{I}_g;\mathbb{Q})$ contains the direct sum*

$$[3^2 1] + [321^2] + [32] + [2^2 1^3] + [2^2 1] + [21^3] + [21] + 2[1^3]$$

which is equal to $\mathrm{Im}(\cup : [1^3] \otimes ([2^2] + [0]) \to \wedge^3[1^3])$. *Moreover, one of the following two possibilities holds:*

a) $\mathrm{Ker}\cup^3$ *coincides with the above.*

b) $\mathrm{Ker}\cup^3$ *coincides with the direct sum of the above summands and one more summand* $[1]$.

(*Sketch of Proof*) Recall that $\mathrm{Ker}(\cup : \wedge^2[1^3] \to H^2(\mathcal{I}_g;\mathbb{Q})) = [2^2] + [0]$. Hence a lower bound of $\mathrm{Ker}\cup^3$ is given by observing the image of the cup product map $\cup : [1^3] \otimes ([2^2] + [0]) \to \wedge^3[1^3]$. On the other hand, an upper bound of $\mathrm{Ker}\cup^3$ is given by computations using abelian cycles in $H_3(\mathcal{I}_g)$. By these computations, we can give the bounds which coincide modulo the summand $[1]$. □

As for the summand $[1]$, we can relate it with the Euler class $e \in H^2(\mathcal{M}_{g,*};\mathbb{Q})$ and the pull-back of the second Morita-Miller-Mumford class $e_2 \in H^4(\mathcal{M}_g;\mathbb{Q})$ to $\mathcal{M}_{g,*}$ (see [20] for the definitions). We have the following.

Theorem 4.3 ([29]). *For $g \geq 5$,*

$$[1] \subset \mathrm{Ker}\cup^3 \iff e_2 - (2-2g)e^2 = 0 \in H^4(\mathcal{I}_{g,*};\mathbb{Q}).$$

(*Sketch of Proof*) First we can see that $[1] \subset \mathrm{Ker}\cup^3$ if and only if the pull-back $p \circ \tau_g(1) : H_3(\mathcal{I}_g) \to [1]$ of the unique (up to scalar) $\mathrm{Sp}(2g,\mathbb{Q})$-equivariant projection $p : \wedge^3[1^3] \to [1]$ to $H_3(\mathcal{I}_g)$ is trivial. The map $p \circ \tau_g(1)$ belongs to $\mathrm{Hom}(H_3(\mathcal{I}_g),[1]) \cong H^3(\mathcal{I}_g;[1])$, which can be embedded in $H^4(\mathcal{I}_{g,*};\mathbb{Q})$ by using a canonical inclusion given by Kawazumi-Morita [16]. On the other hand, using the natural embedding

$$\mathrm{Hom}(\wedge^3[1^3],[1]) \cong \wedge^3[1^3] \otimes [1] \hookrightarrow \wedge^4(\wedge^3[1])$$

and the first Johnson homomorphism $\tau_{g,*} : \mathcal{I}_{g,*} \to \wedge^3[1] = [1^3] + [1]$ for $\mathcal{I}_{g,*}$, we obtain a commutative diagram

$$\begin{array}{ccc} \mathrm{Hom}(\wedge^3[1^3], [1]) & \xrightarrow{\tau_g(1)^*} & H^3(\mathcal{I}_g; [1]) \\ \downarrow & & \downarrow \\ \wedge^4(\wedge^3[1]) & \xrightarrow{\tau_{g,*}(1)^*} & H^4(\mathcal{I}_{g,*}; \mathbb{Q}). \end{array}$$

Therefore we see that $[1] \subset \operatorname{Ker} \cup^3$ if and only if $\tau_{g,*}(1)^*(p) \in H^4(\mathcal{I}_{g,*}; \mathbb{Q})$ is trivial.

Since $p : \wedge^3[1^3] \to [1]$ is $\mathrm{Sp}(2g, \mathbb{Q})$-equivariant, as an element of $\wedge^4(\wedge^3[1])$, it belongs to the $\mathrm{Sp}(2g, \mathbb{Q})$-invariant part $(\wedge^4(\wedge^3[1]))^{\mathrm{Sp}}$. In [26], Morita constructed a commutative diagram

$$\begin{array}{ccc} (\wedge^4(\wedge^3[1]))^{\mathrm{Sp}} & \xrightarrow{\rho_1^*} & H^4(\mathcal{M}_{g,*}; \mathbb{Q}) \\ \downarrow & & \downarrow \\ \wedge^4(\wedge^3[1]) & \xrightarrow{\tau_{g,*}(1)^*} & H^4(\mathcal{I}_{g,*}; \mathbb{Q}), \end{array}$$

where the upper horizontal map is induced from the extended Johnson homomorphism $\rho_1 : \mathcal{M}_{g,*} \to \wedge^3[1^3] \rtimes \mathrm{Sp}(2g, \mathbb{Z})$ (see [25]). Finally, using an explicit description of ρ_1^* given by Kawazumi-Morita [16], we can compute that $\tau_{g,*}(1)^*(p) = e_2 - (2 - 2g)e^2$ up to scalar. Theorem 4.3 follows from this. □

At present, it is not known whether powers of the Euler class e^i $(i \geq 2)$ and even Morita-Miller-Mumford classes e_{2i} $(i \geq 1)$ are trivial or not when they are restricted to the Torelli group, and we have little information about it. (As for odd Morita-Miller-Mumford classes e_{2i-1} $(i \geq 1)$, it is known that they are trivial in $H^*(\mathcal{I}_g, \mathbb{Q})$. See [20] for details.) Theorem 4.3 can be read that there is a method to attack the non-triviality problem for $e^2, e_2 \in H^4(\mathcal{I}_{g,*}; \mathbb{Q})$ from the theory of *three*-dimensional manifolds. That is, by the well-known fact about the realization of a homology class of degree three, it follows that if $\cup^3([1])$ is non-trivial, it must be evaluated non-trivially by the fundamental class of an oriented closed three-dimensional manifold. However this approach has the following difficulty. The condition in Theorem 4.3 is compatible with the pull-back of the universal Σ_g-bundle. Therefore comparing the result of Morita in [21] that the pull-back of $e^2 \in H^4(\mathcal{M}_{g,*}; \mathbb{Q})$ to an amenable group always vanishes, we obtain the following.

Corollary 4.4 ([29]). *For every amenable group G and every group homomorphism $f : G \to \mathcal{I}_g$,*

$$f^* \circ \cup^3([1]) = \{0\} \subset H^3(G;\mathbb{Q}).$$

Since abelian groups are amenable, this corollary implies that we cannot evaluate $\cup^3([1])$ by using abelian cycles even if $\cup^3([1])$ is non-trivial in $H^3(\mathcal{I}_g;\mathbb{Q})$. Therefore the first step to determine whether $\cup^3([1])$ is trivial or not is to study the following problem.

Problem 4.5. *Construct a non-trivial homomorphism $G \to \mathcal{I}_g$ where G is not amenable and is given as the fundamental group of an oriented closed three-dimensional manifold.*

§5. Acknowledgement

The author would like to express his gratitude to Professor Shigeyuki Morita for his encouragement and helpful suggestions. The author also would like to thank Professor Richard Hain, who kindly informed the author on the proof of his result mentioned in Remark 2.1, the referee of [30], who corrected an insufficient point in the original argument and the referee of this manuscript for his/her useful comments.

The author was supported by JSPS Research Fellowships for Young Scientists, KAKENHI (No. 19840009) and 21st century COE program at Graduate School of Mathematical Sciences, The University of Tokyo.

References

[1] T. Akita, Homological infiniteness of Torelli groups, Topology, **40** (2001), 213–221.

[2] M. Asada and M. Kaneko, On the automorphism group of some pro-l fundamental groups, Adv. Stud. Pure Math., **12** (1987), 137–159.

[3] M. Asada and H. Nakamura, On graded quotient modules of mapping class groups of surfaces, Israel J. Math., **90** (1995), 93–113.

[4] J. Birman, Braids, Links and Mapping Class Groups, Ann. of Math. Stud., **82**, Princeton Univ. Press, 1974.

[5] D. Biss and B. Farb, $\mathcal{K}_g$ is not finitely generated, Invent. Math., **163** (2006), 213–226.

[6] T. Brendle and B. Farb, The Birman-Craggs-Johnson homomorphism and abelian cycles in the Torelli group, Math. Ann., **338** (2007), 33–53.

[7] C. J. Earle and J. Eells, The diffeomorphism group of a compact Riemann surface, Bull. Amer. Math. Soc., **73** (1967), 557–559.

[8] W. Fulton and J. Harris, Representation Theory, Grad. Texts in Math., **129**, Springer-Verlag, 1991.

[9] S. Garoufalidis and J. Levine, Tree-level invariants of three-manifolds, Massey products and the Johnson homomorphism, In: Graphs and patterns in mathematics and theoretical physics, Proc. Sympos. Pure Math., **73**, Amer. Math. Soc., 2005, pp. 173–205.
[10] R. Hain, Infinitesimal presentations of the Torelli groups, J. Amer. Math. Soc., **10** (1997), 597–651.
[11] D. Johnson, An abelian quotient of the mapping class group $\mathcal{I}_g$, Math. Ann., **249** (1980), 225–242.
[12] D. Johnson, A survey of the Torelli group, Contemp. Math., **20** (1983), 165–179.
[13] D. Johnson, The structure of the Torelli group I: A finite set of generators for $\mathcal{I}$, Ann. of Math. (2), **118** (1983), 423–442.
[14] D. Johnson, The structure of the Torelli group II: A characterization of the group generated by twists on bounding curves, Topology, **24** (1985), 113–126.
[15] D. Johnson, The structure of the Torelli group III: The abelianization of $\mathcal{I}_g$, Topology, **24** (1985), 127–144.
[16] N. Kawazumi and S. Morita, The primary approximation to the cohomology of the moduli space of curves and cocycles for the Mumford-Morita-Miller classes, preprint (the Preprint Series of the Graduate School of Mathematical Sciences, The University of Tokyo, 2001–13) (2001).
[17] J. Labute, On the descending central series of groups with a single defining relation, J. Algebra, **14** (1970), 16–23.
[18] J. Levine, Homology cylinders: an enlargement of the mapping class group, Algebr. Geom. Topol., **1** (2001), 243–270.
[19] J. Levine, Addendum and correction to: Homology cylinders: an enlargement of the mapping class group, Algebr. Geom. Topol., **2** (2002), 1197–1204.
[20] S. Morita, Characteristic classes of surface bundles, Invent. Math., **90** (1987), 551–577.
[21] S. Morita, Characteristic classes of surface bundles and bounded cohomology, A Fête of Topology, Academic Press, 1988, pp. 233–257.
[22] S. Morita, Casson's invariant for homology 3-spheres and characteristic classes of surface bundles I, Topology, **28** (1989), 305–323.
[23] S. Morita, On the structure of the Torelli group and the Casson invariant, Topology, **30** (1991), 603–621.
[24] S. Morita, Abelian quotients of subgroups of the mapping class group of surfaces, Duke Math. J., **70** (1993), 699–726.
[25] S. Morita, The extension of Johnson's homomorphism from the Torelli group to the mapping class group, Invent. Math. **111** (1993), 197–224.
[26] S. Morita, A linear representation of the mapping class group of orientable surfaces and characteristic classes of surface bundles, In: Proceeding of the Taniguchi Symposium on Topology and Teichmüller Spaces, Finland, July 1995, World Scientific, 1996, pp. 159–186.

[27] S. Morita, Structure of the mapping class groups of surfaces: a survey and a prospect, In: Proceedings of the Kirbyfest, Geom. Topol. Monogr., **2**, Geom. Topol. Publ., 1999, pp. 349–406.
[28] A. Pettet, The Johnson homomorphism and the second cohomology of IA_n, Algebr. Geom. Topol., **5** (2005), 725–740.
[29] T. Sakasai, The Johnson homomorphism and the third rational cohomology group of the Torelli group, Topology Appl., **148** (2005), 83–111.
[30] T. Sakasai, The second Johnson homomorphism and the second rational cohomology of the Johnson kernel, Math. Proc. Cambridge Philos. Soc., **143** (2007), 627–648.
[31] J. Stallings, Homology and central series of groups, J. Algebra, **2** (1965), 170–181.

Department of Mathematics
Tokyo Institute of Technology
2-12-1 Oh-okayama, Meguro-ku
Tokyo 152-8551
Japan

Advanced Studies in Pure Mathematics 52, 2008
Groups of Diffeomorphisms
pp. 111–118

On mod p Riemann-Roch formulae for mapping class groups

Toshiyuki Akita

Abstract.

We provide affirmative evidences for conjectural mod p Riemann-Roch formulae for mapping class groups by considering Steenrod operations on mod p Morita-Mumford classes.

§1. Introduction

Let Σ_g be a closed oriented surface of genus $g \geq 2$ and let Γ_g be its mapping class group. Namely, it is the group consisting of path components of $\mathrm{Diff}_+\Sigma_g$, which is the group of orientation preserving diffeomorphisms of Σ_g. Any cohomology class of Γ_g can be considered as a characteristic class of oriented surface bundles. Indeed, by a theorem of Earle and Eells [4], the classifying space $B\mathrm{Diff}_+\Sigma_g$ of oriented Σ_g-bundles is an Eilenberg-MacLane space $K(\Gamma_g, 1)$ so that we have a natural isomorphism

$$H^*(B\mathrm{Diff}_+\Sigma_g;\mathbb{Z}) \cong H^*(\Gamma_g;\mathbb{Z}).$$

Morita [11] and Mumford [12] independently introduced a series of cohomology classes $e_k \in H^{2k}(\Gamma_g;\mathbb{Z})$ of Γ_g which are called Morita-Mumford classes (or Mumford-Morita-Miller classes in the literature). Over the rationals, the natural homomorphism

$$\mathbb{Q}[e_1, e_2, e_3, \dots] \to H^*(\Gamma_g;\mathbb{Q})$$

is an isomorphism in dimensions less than $2g/3$ by the proof of Mumford conjecture [9]. On the contrary, less is known about integral or mod p Morita-Mumford classes of Γ_g. In [1], the author proposed a conjecture

Received April 30, 2007.
Revised January 10, 2008.
The research is supported in part by the Grant-in-Aid for Scientific Research (C) (No. 17560054) from the Japan Society for Promotion of Sciences.

concerning of integral Morita-Mumford classes. To be precise, let B_{2k} be the $2k$-th Bernoulli number, and define N_{2k}, D_{2k} to be coprime integers satisfying $B_{2k}/2k = N_{2k}/D_{2k}$. Let $s_k \in H^{2k}(\Gamma_g;\mathbb{Z})$ be the k-th Newton class of Γ_g which will be defined in §3.

Conjecture 1 (integral Riemann-Roch formulae for Γ_g).

$$N_{2k}e_{2k-1} = D_{2k}s_{2k-1} \in H^*(\Gamma_g;\mathbb{Z})$$

holds for all $k \geq 1$ *and* $g \geq 2$.

With rational coefficients, it is deduced from the Grothendieck-Riemann-Roch theorem. See [11, 12]. The conjecture is affirmative for $k = 1$ (i.e. $e_1 = 12s_1 \in H^2(\Gamma_g;\mathbb{Z})$ for all $g \geq 2$), since $H^2(\Gamma_g,\mathbb{Z}) \cong \mathbb{Z}$ for $g \geq 3$ as was proved by Harer [6] (see [1] for the case $g = 2$). The author and Kawazumi [2] showed that the conjecture is affirmative for any cyclic subgroup of Γ_g. Kawazumi [8] showed that a slightly weaker version of the conjecture holds for hyperelliptic mapping class groups. In addition, a result of Galatius, Madsen and Tillmann [5, Theorem 1.2] can be regarded as an affirmative evidence of the conjecture for the stable mapping class group (see [1, Section 7] for the detail). Now let p be a prime and $\mathbb{F}_p$ the field consisting of p elements. Passing to the mod p cohomology, Conjecture 1 leads to the following conjecture:

Conjecture 2 (mod p Riemann-Roch formulae for Γ_g).

$$N_{2k}e_{2k-1} = D_{2k}s_{2k-1} \in H^*(\Gamma_g;\mathbb{F}_p)$$

holds for all $k \geq 1$ *and* $g \geq 2$.

The purpose of this paper is to provide affirmative evidences of Conjecture 2. Our main result is the following:

Theorem 1.1. *Let* p *be an odd prime and* $g \geq 2$. *If*

$$N_{2k}e_{2k-1} = D_{2k}s_{2k-1} \in H^*(\Gamma_g;\mathbb{F}_p)$$

holds for some $k \geq 1$, *and if* $\binom{2k-1}{i}$ *is prime to* p, *then*

$$N_{2k+i(p-1)}e_{2k-1+i(p-1)} = D_{2k+i(p-1)}s_{2k-1+i(p-1)} \in H^*(\Gamma_g;\mathbb{F}_p).$$

In other words, the affirmative solution of Conjecture 2 for some k implies that for $k + i(p-1)/2$, provided $\binom{2k-1}{i}$ is prime to p. In particular, since Conjecture 1 and hence Conjecture 2 are affirmative for $k = 1$ as was mentioned earlier, one has the following result:

Corollary 1.2. *Let p be an odd prime. Then*

$$N_{p^n+1}e_{p^n} = D_{p^n+1}s_{p^n} \in H^*(\Gamma_g; \mathbb{F}_p)$$

for all $n \geq 0$ and $g \geq 2$.

Theorem 1.1 is proved by considering reduced power operations on mod p Morita-Mumford and Newton classes of Γ_g, together with Kummer's congruences on Bernoulli numbers. Similar considerations are possible for $p = 2$ by using squaring operations in place of reduced power operations.

The rest of the paper is organized as follows. In §2, we will recall the definition of Morita-Mumford classes, and compute the action of Steenrod operations on them. In §3, we will recall the definition of Newton classes. The proof of Theorem 1.1 will be given in §4.

Notation. There are some conflicting notations of Bernoulli numbers. We define B_{2k} $(k \geq 1)$ by a power series expansion

$$\frac{z}{e^z - 1} + \frac{z}{2} = 1 + \sum_{k=1}^{\infty} \frac{B_{2k}}{(2k)!} z^{2k}.$$

Thus $B_2 = 1/6, B_4 = -1/30, B_6 = 1/42, B_8 = -1/30, B_{10} = 5/66$, and so on. Our notation is consistent with [2, 7] but differs from [1, 8].

§2. Morita-Mumford classes

2.1. Definition

Let $\pi : E \to B$ be an oriented Σ_g-bundle, $T_{E/B}$ the tangent bundle along the fiber of π, and $e \in H^2(E; \mathbb{Z})$ the Euler class of $T_{E/B}$. Then the k-th Morita-Mumford class $e_k \in H^{2k}(B; \mathbb{Z})$ of π is defined by

$$e_k := \pi_!(e^{k+1})$$

where $\pi_! : H^*(E; \mathbb{Z}) \to H^{*-2}(B; \mathbb{Z})$ is the Gysin homomorphism (or the integration along the fiber). The structure group of oriented Σ_g-bundles is $\mathrm{Diff}_+\Sigma_g$. Hence passing to the universal Σ_g-bundle

$$E\mathrm{Diff}_+\Sigma_g \times_{\mathrm{Diff}_+\Sigma_g} \Sigma_g \to B\mathrm{Diff}_+\Sigma_g$$

we obtain the cohomology classes $e_k \in H^*(B\mathrm{Diff}_+\Sigma_g; \mathbb{Z})$. As was mentioned in Introduction, the classifying space $B\mathrm{Diff}_+\Sigma_g$ is an Eilenberg-MacLane space $K(\Gamma_g, 1)$ so that we obtain the *k-th Morita-Mumford class $e_k \in H^{2k}(\Gamma_g; \mathbb{Z})$ of Γ_g.*

2.2. Steenrod operations

For an odd prime p, let

$$\mathrm{P}^i : H^k(-;\mathbb{F}_p) \to H^{k+2i(p-1)}(-;\mathbb{F}_p)$$

be the i-th reduced power operation. We will compute the action of P^i's on mod p Morita-Mumford classes. To this end let us return to an oriented smooth Σ_g-bundle $\pi : E \to B$ and choose a smooth embedding $E \to \mathbb{R}^n$ of E in some Euclidean space $\mathbb{R}^n$. The normal bundle $N^f E$ of the resulting embedding $f : E \to B \times \mathbb{R}^n$ is called the *normal bundle along the fiber*:

$$T_{E/B} \oplus N^f E \cong E \times \mathbb{R}^n \quad \text{(product bundle)}.$$

Let $q_\bullet(N^f E) \in H^*(E, \mathbb{F}_p)$ be the total Wu class of $N^f E$ defined by $q_\bullet(N^f E) = \phi^{-1} \circ \mathrm{P} \circ \phi(1)$ where ϕ is the Thom isomorphism for $N^f E$ (see [10, p.228] for instance). Applying the generalized Riemann-Roch theorem [3, p.65 Theorem 9] to the total reduced power operation $\mathrm{P} = \sum_i \mathrm{P}^i$, one has

$$\mathrm{P}(\pi_!(u)) = \pi_!(\mathrm{P}(u) \cdot q_\bullet(N^f E)) \tag{1}$$

for every $u \in H^*(E;\mathbb{F}_p)$. For the mod p Euler class $e \in H^2(E;\mathbb{F}_p)$ of $T_{E/B}$, one has $\mathrm{P}(e) = \mathrm{P}^0(e) + \mathrm{P}^1(e) = e + e^p$ and hence

$$\mathrm{P}(e^{k+1}) = \mathrm{P}(e)^{k+1} = (e + e^p)^{k+1} = e^{k+1}(1 + e^{p-1})^{k+1}$$

by Cartan formula. On the other hand, one has

$$\begin{aligned} &q_\bullet(T_{E/B}) \cdot q_\bullet(N^f E) = q_\bullet(T_{E/B} \oplus N^f E) = q_\bullet(E \times \mathbb{R}^n) = 1 \\ &q_\bullet(T_{E/B}) = 1 + e^{p-1} \quad \text{(see [10, p.228])}. \end{aligned}$$

Applying $u = e^{k+1}$ to (1), one has

$$\begin{aligned} \mathrm{P}(e_k) = \mathrm{P}(\pi_!(e^{k+1})) &= \pi_!(e^{k+1}(1 + e^{p-1})^{k+1} \cdot q_\bullet(N^f E)) \\ &= \pi_!(e^{k+1}(1 + e^{p-1})^k \cdot q_\bullet(T_{E/B}) \cdot q_\bullet(N^f E)) \\ &= \pi_!(e^{k+1}(1 + e^{p-1})^k) \end{aligned}$$

and hence

$$\mathrm{P}(e_k) = \pi_!\left(\sum_{i=0}^{k} \binom{k}{i} e^{k+i(p-1)+1}\right) = \sum_{i=0}^{k} \binom{k}{i} e_{k+i(p-1)}. \tag{2}$$

Since reduced power operations are natural with respect to bundle maps, the last equality (2) is valid in $H^*(\Gamma_g;\mathbb{F}_p)$. Thus we have proved the following proposition:

Proposition 2.1. *For $e_k \in H^*(\Gamma_g; \mathbb{F}_p)$, one has*

$$\mathrm{P}^i(e_k) = \binom{k}{i} e_{k+i(p-1)} \in H^*(\Gamma_g; \mathbb{F}_p).$$

Now let $\mathrm{Sq}^i : H^k(-, \mathbb{F}_2) \to H^{k+i}(-, \mathbb{F}_2)$ be the i-th squaring operation. Applying the generalized Riemann-Roch theorem to the total squaring operation $\mathrm{Sq} = \sum_i \mathrm{Sq}^i$, one has

$$\mathrm{Sq}(\pi_!(u)) = \pi_!(\mathrm{Sq}(u) \cdot \omega_\bullet(N^f E)) \tag{3}$$

for every $u \in H^*(E; \mathbb{F}_2)$, where $\omega_\bullet(N^f E) \in H^*(E; \mathbb{F}_2)$ is the total Stiefel-Whitney class of $N^f E$. With the equation (3) in mind, the proof of the following proposition is similar to that of Proposition 2.1, and is left to the reader:

Proposition 2.2. *For $e_k \in H^*(\Gamma_g; \mathbb{F}_2)$, one has*

$$\mathrm{Sq}^{2i}(e_k) = \binom{k}{i} e_{k+i}, \ \mathrm{Sq}^{2i+1}(e_k) = 0 \in H^*(\Gamma_g; \mathbb{F}_2).$$

§3. Newton classes

Let $U(n)$ be the n-dimensional unitary group. The k-th Newton class $s_k \in H^*(BU(n); \mathbb{Z})$ is the characteristic class associated to the formal sum $\sum_{l=1}^n x_l^k$. Steenrod operations on mod p Newton classes $s_k \in H^*(BU(n); \mathbb{F}_p)$ can be computed quite easily, as in the proposition below.

Proposition 3.1. *For an odd prime p, one has*

$$\mathrm{P}^i(s_k) = \binom{k}{i} s_{k+i(p-1)} \in H^*(BU(n); \mathbb{F}_p).$$

For $p = 2$, one has

$$\mathrm{Sq}^{2i}(s_k) = \binom{k}{i} s_{k+i}, \ \mathrm{Sq}^{2i+1}(s_k) = 0 \in H^*(BU(n); \mathbb{F}_2).$$

Proof. By the definition of the Newton class s_k, it suffices to prove the case $n = 1$. Since $s_k = c_1^k \in H^*(BU(1); \mathbb{Z})$ where c_1 is the first Chern class, for an odd prime p, one has

$$\mathrm{P}^i(s_k) = \mathrm{P}^i(c_1^k) = \binom{k}{i} c_1^{k+i(p-1)} = \binom{k}{i} s_{k+i(p-1)}$$

as desired. The case $p = 2$ is similar. Q.E.D.

Now we recall the definition of Newton classes of Γ_g. The natural action of Γ_g on the first real homology $H_1(\Sigma_g;\mathbb{R})$ induces a homomorphism $\Gamma_g \to Sp(2g,\mathbb{R})$. The homomorphism yields a continuous map

$$\eta : K(\Gamma_g, 1) \to BU(g),$$

for the maximal compact subgroup of $Sp(2g,\mathbb{R})$ is isomorphic to $U(g)$. The *k-th Newton class* $s_k \in H^{2k}(\Gamma_g;\mathbb{Z})$ *of* Γ_g is defined to be the pullback of $s_k \in H^*(BU(g);\mathbb{Z})$ by η (we use the same symbol).

§4. Proof of the main results

The proof of Theorem 1.1 is based on the following facts concerning of number theoretic properties of Bernoulli numbers:

Theorem 4.1. *Let p be a prime number.*

(1) $p|D_{2k}$ *if and only if* $p-1|2k$.

(2) *If* $p-1 \not| \, 2k$ *and* $k \equiv h \pmod{p-1}$ *then*

$$\frac{B_{2k}}{2k} \equiv \frac{B_{2h}}{2h} \pmod{p}.$$

The last congruence is considered in $\mathbb{Z}_{(p)} = \{m/n \in \mathbb{Q} \mid (n,p) = 1\}$. Namely, for $r, s \in \mathbb{Z}_{(p)}$, we write $r \equiv s \pmod{p}$ if $r-s = m/n, (n,p) = 1$, and $p|m$. The first statement (1) is called von Staudt's theorem, while the second statement (2) is called the Kummer's congruence. See [7] for the proof of Theorem 4.1.

Proof of Theorem 1.1. Applying the i-th reduced power operation to the both sides of equality $N_{2k}e_{2k-1} = D_{2k}s_{2k-1}$, one has

$$N_{2k}e_{2k-1+i(p-1)} = D_{2k}s_{2k-1+i(p-1)} \in H^*(\Gamma_g;\mathbb{F}_p) \tag{4}$$

by Proposition 2.1 and 3.1.

Suppose first $p-1$ divides $2k$. It follows from von Staudt's theorem that $D_{2k} \equiv D_{2k+i(p-1)} \equiv 0 \pmod{p}$. Consequently, the equality $N_{2k}e_{2k-1} = D_{2k}s_{2k-1}$ implies the condition $e_{2k-1} = 0$, and the equality (4) implies $e_{2k-1+i(p-1)} = 0$. Since

$$N_{2k+i(p-1)}e_{2k-1+i(p-1)} = D_{2k+i(p-1)}s_{2k-1+i(p-1)}$$

is equivalent to $e_{2k-1+i(p-1)} = 0$, theorem follows.

Now suppose $p-1$ does not divide $2k$. By von Staudt's theorem, D_{2k} and $D_{2k+i(p-1)}$ are prime to p. Choose integers I_{2k} and $I_{2k+i(p-1)}$

satisfying $I_{2k}D_{2k} \equiv 1$ and $I_{2k+i(p-1)}D_{2k+i(p-1)} \equiv 1 \pmod p$. Then the equality (4) is equivalent to

$$N_{2k}I_{2k} \cdot e_{2k-1+i(p-1)} = s_{2k-1+i(p-1)}. \tag{5}$$

Now the Kummer's congruence implies

$$N_{2k}I_{2k} \equiv N_{2k+i(p-1)}I_{2k+i(p-1)} \pmod p.$$

Consequently, one has

$$N_{2k+i(p-1)}I_{2k+i(p-1)} \cdot e_{2k-1+i(p-1)} = s_{2k-1+i(p-1)}$$

and hence

$$N_{2k+i(p-1)}e_{2k-1+i(p-1)} = D_{2k+i(p-1)}s_{2k-1+i(p-1)}$$

as desired. Q.E.D.

References

[1] T. Akita, Nilpotency and triviality of mod p Morita-Mumford classes of mapping class groups of surfaces, Nagoya Math. J., **165** (2002), 1–22.

[2] T. Akita and N. Kawazumi, Integral Riemann-Roch formulae for cyclic subgroups of mapping class groups, Math. Proc. Cambridge Phil. Soc., **144** (2008), 411–421.

[3] E. Dyer, Cohomology Theories, Benjamin, 1969.

[4] C. J. Earle and J. Eells, A fibre bundle description of Teichmüller theory, J. Differential Geom., **3** (1969), 19–43.

[5] S. Galatius, Ib Madsen and U. Tillmann, Divisibility of the stable Miller-Morita-Mumford classes, J. Amer. Math. Soc., **19** (2006), 759–779.

[6] J. L. Harer, The second homology group of the mapping class group of an oriented surface, Invent. Math., **72** (1983), 221–239.

[7] K. Ireland and M. I. Rosen, A classical Introduction to Modern Number Theory, Grad. Texts in Math., **84**, Springer-Verlag, 1990.

[8] N. Kawazumi, Weierstrass points and Morita-Mumford classes on hyperelliptic mapping class groups, Topology Appl., **125** (2002), 363–383.

[9] Ib Madsen and M. S. Weiss, The stable moduli space of Riemann surfaces: Mumford's conjecture, Ann. of Math. (2), **165** (2007), 843–941.

[10] J. W. Milnor and J. D. Stasheff, Characteristic Classes, Princeton Univ. Press, 1974.

[11] S. Morita, Characteristic classes of surface bundles, Invent. Math., **90** (1987), 551–577.

[12] D. Mumford, Towards an enumerative geometry of the moduli space of curves, In: Arithmetic and Geometry Vol. II, Birkhäuser, 1983, pp. 271–328.

Department of Mathematics
Hokkaido University
Sapporo 060-0810
Japan

Advanced Studies in Pure Mathematics 52, 2008
Groups of Diffeomorphisms
pp. 119–134

Calculating the image of the second Johnson-Morita representation

Joan S. Birman[1], Tara E. Brendle[2] and Nathan Broaddus[3]

Abstract.

Johnson has defined a surjective homomorphism from the *Torelli subgroup* of the mapping class group of the surface of genus g with one boundary component to $\wedge^3 H$, the third exterior product of the homology of the surface. Morita then extended Johnson's homomorphism to a homomorphism from the entire mapping class group to $\frac{1}{2} \wedge^3 H \rtimes \mathrm{Sp}(H)$. This *Johnson-Morita homomorphism* is not surjective, but its image is finite index in $\frac{1}{2} \wedge^3 H \rtimes \mathrm{Sp}(H)$ [11]. Here we give a description of the exact image of Morita's homomorphism. Further, we compute the image of the *handlebody subgroup* of the mapping class group under the same map.

§1. Introduction

Let S_g be a closed surface of genus g. We fix a closed disk D in S_g, and by deleting its interior, obtain $S_{g,1}$, a genus g surface with one boundary component, as illustrated in Figure 1. Let $\mathcal{M}_g$ (resp. $\mathcal{M}_{g,1}$) denote the mapping class group of the surface S_g (resp. $S_{g,1}$). In the case of $\mathcal{M}_{g,1}$ we assume the boundary component is fixed pointwise.

We choose a base point on $\partial S_{g,1}$, and let $\alpha_1, \ldots, \alpha_g, \beta_1, \ldots, \beta_g$ denote the based loops illustrated in Figure 1(b). Let $a_1, \ldots, a_g, b_1, \ldots, b_g$ denote the corresponding homology classes, as in Figure 1(a). It will sometimes be convenient to denote these same homology classes by $x_1, \ldots, x_{2g}$ with the understanding that $x_i = a_i$ and $x_{i+g} = b_i$ for

Received April 30, 2007.
Revised August 21, 2007.

[1]The first author was supported in part by NSF grant DMS-0405586.
[2]The second author was supported in part by NSF grant DMS-0606882.
[3]The third author was supported in part by an NSF Postdoctoral Fellowship.

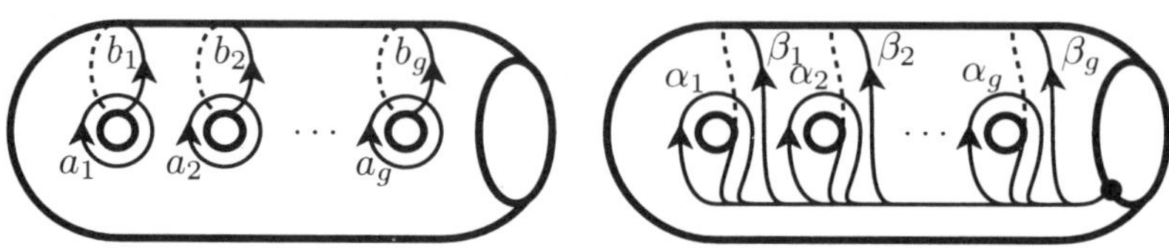

Fig. 1. (a) A basis for $H_1(S_{g,1})$ (b) Generators for $\pi_1(S_{g,1})$

$1 \leq i \leq g$. Likewise, we will sometimes refer to the based loops $\alpha_1, \ldots, \alpha_g, \beta_1, \ldots, \beta_g$ by $\xi_1, \ldots, \xi_{2g}$ with the understanding that $\xi_i = \alpha_i$ and $\xi_{i+g} = \beta_i$ for $1 \leq i \leq g$.

Now, let $H = H_1(S_{g,1})$ be the free abelian group with generating set $\{a_1, \ldots, a_g, b_1, \ldots, b_g\}$ and $\pi = \pi_1(S_{g,1})$ which is a free group on the generating set $\{\alpha_1, \ldots, \alpha_g, \beta_1, \ldots, \beta_g\}$. The action of $\mathcal{M}_{g,1}$ on π gives an injection $\mathcal{M}_{g,1} \hookrightarrow \mathrm{Aut}(\pi)$. More generally, we can compose with the homomorphism $\mathrm{Aut}(\pi) \to \mathrm{Aut}(\pi/\chi)$ for any characteristic subgroup $\chi \subset \pi$. The *lower central series* of the free group π is a sequence of characteristic subgroups defined inductively by setting $\pi^{(0)} = \pi$ and $\pi^{(k+1)} = [\pi, \pi^{(k)}]$. We define the kth *Johnson-Morita representation* to be the map

$$\rho_k : \mathcal{M}_{g,1} \to \mathrm{Aut}(\pi/\pi^{(k)})$$

We note that these maps were first studied by Johnson in [7, 6] and subsequently developed by Morita in a series of papers [11, 12, 13, 14].

Observe that the first Johnson-Morita map is just the classical symplectic representation $\rho_1 : \mathcal{M}_{g,1} \to \mathrm{Sp}(H)$ which is surjective ([4], in particular pp. 209-212). In [11, Theorem 4.8] Morita shows that the image of ρ_2 is isomorphic to a subgroup of finite index in $\frac{1}{2} \wedge^3 H \rtimes \mathrm{Sp}(H)$. Our first main result in this paper, given in Theorem 2.4, is to identify the precise image $\rho_2(\mathcal{M}_{g,1})$ using a formulation due to Perron [16].

Let us now consider S_g as ∂X_g, where X_g is a genus g handlebody. Let $\mathcal{H}_g$ denote the *handlebody subgroup* of $\mathcal{M}_g$, that is, the subgroup consisting of maps of S_g which extend to the handlebody X_g. There is a natural surjection $\mathcal{M}_{g,1} \to \mathcal{M}_g$ obtained by extending via the identity map along D. The kernel of this surjection is generated by two kinds of elements: the Dehn twist along the boundary curve, and "push" maps along elements of $\pi_1(S_{g,1})$ [1]. Note that any map in this kernel extends to X_g. Hence, we are justified in defining the handlebody subgroup $\mathcal{H}_{g,1}$ of $\mathcal{M}_{g,1}$ as the pullback of $\mathcal{H}_g$.

The handlebody group arises naturally in a number of applications in 3-manifold topology, particularly through Heegaard splittings of 3-manifolds. Our second result in this paper is to compute $\rho_2(\mathcal{H}_{g,1})$, given in Theorem 3.5.

The authors would like to thank the referee for helpful comments and suggestions.

§2. The second Johnson-Morita map

In this section we will describe Perron's formulation [16] of the second Johnson-Morita representation. We will give a precise characterization of the image of the mapping class group under this map. First, it will be useful to review the image of the first Johnson-Morita representation, i.e., the symplectic group.

2.1. The symplectic group

The group $H = H_1(S_{g,1})$ is free abelian with free basis $a_1, \ldots, a_g$, $b_1, \ldots, b_g$, as in Figure 1(a), and has a symplectic intersection form given by signed intersection of curves which is preserved by every mapping class $f \in \mathcal{M}_{g,1}$. In the basis above, the intersection form is given by the the matrix J with $g \times g$ block form

$$J = \begin{pmatrix} 0 & -I \\ I & 0 \end{pmatrix} \tag{1}$$

The intersection form got by acting by the linear transformation M on an intersection form with matrix L is given by $ML\overline{M}$ where $\overline{M}$ denotes the transpose of M. Hence for every M in the image of the mapping class group

$$MJ\overline{M} = J, \qquad \text{or equivalently} \qquad \overline{M}JM = J \tag{2}$$

In fact (2) is a sufficient condition for M to be in the image of the mapping class group under ρ_1. It is sometimes useful to write a symplectic matrix M in $g \times g$ block form as

$$M = \begin{pmatrix} S & T \\ P & Q \end{pmatrix}$$

A convenient consequence of (2) is that $M^{-1} = J\overline{M}J^{-1}$. In block form this becomes

$$\begin{pmatrix} S & T \\ P & Q \end{pmatrix}^{-1} = \begin{pmatrix} \overline{Q} & -\overline{T} \\ -\overline{P} & \overline{S} \end{pmatrix}$$

The group of such matrices form the *symplectic group*. Writing M and $\overline{M}$ in $g \times g$ block form

$$M = \begin{pmatrix} S & T \\ P & Q \end{pmatrix}, \qquad \overline{M} = \begin{pmatrix} \overline{S} & \overline{P} \\ \overline{T} & \overline{Q} \end{pmatrix}$$

we derive the *symplectic constraints*, which follow directly from the condition in (2):

(3) (i) $Q\overline{S} - P\overline{T} = I$, (ii) $S\overline{T}$ symmetric, (iii) $P\overline{Q}$ symmetric.

2.2. Perron's formulation of ρ_2

The *Torelli group* $\mathcal{I}_{g,1}$ is the kernel of the symplectic representation $\rho_1 : \mathcal{M}_{g,1} \to \mathrm{Sp}(H)$. Johnson proved, in [5], that the image of the Torelli group under ρ_2 is $\wedge^3 H$. In the next section we will go a step further, and describe, in Theorem 2.4, the image of the full mapping class group $\mathcal{M}_{g,1}$ under ρ_2 noting that Morita [11, Theorem 4.8] has already identified this image as being finite index in $\frac{1}{2} \wedge^3 H \rtimes \mathrm{Sp}(H)$. We begin by summarizing Morita's explicit description of ρ_2 as given in [11, Section 4]. Consider the 2-step nilpotent group

$$\Phi_2 = \left\{ (\eta, y) \middle| \eta \in \frac{1}{2} \wedge^2 H, \ y \in H \right\}$$

with multiplication in Φ_2 given by $(\eta, y)(\nu, z) = (\eta + \nu + \frac{1}{2} y \wedge z, y + z)$. It contains a subgroup of finite index which can be identified (see [8, Sec. 5.5]) with the second nilpotent quotient $\pi/\pi^{(2)} = \pi/[\pi, [\pi, \pi]]$ of our surface group via the homomorphism $\phi_2 : \pi \to \Phi_2$

$$\phi_2(\xi_i) = (0, x_i)$$

where $\{\xi_1, \cdots, \xi_{2g}\}$ generate $\pi = \pi_1(S_{g,1})$ and $\{x_1, \cdots, x_{2g}\}$ is our basis for $H = H_1(S_{g,1})$ (see Figure 1(a-b)). The group Φ_2 can be viewed as a subgroup of the Mal'cev completion of the nilpotent group $\pi/\pi^{(2)}$. Any automorphism of $\pi/\pi^{(2)}$ extends to the Mal'cev completion and preserves Φ_2 so we may think of $\mathcal{M}_{g,1}$ as acting on Φ_2 [11, Proposition 2.5].

In [11, Section 3] Morita describes a function $\mathcal{M}_{g,1} \to \mathrm{Hom}(H, \frac{1}{2} \wedge^2 H)$. An automorphism f of Φ_2 coming from an automorphism of the Mal'cev completion of $\pi/\pi^{(2)}$ can be specified by the images

$$f(0, x_i) = (w_i, h_i) \qquad w_i \in \frac{1}{2} \wedge^2 H, \ h_i \in H$$

for each x_i. The homomorphism $\rho_1(f) : H \to H$ given by $\rho_1(f)(x_i) = h_i$ is just the image of f under the symplectic representation. Johnson looks at the homomorphism $\tilde{\tau}_2(f) : H \to \frac{1}{2} \wedge^2 H$ given by

$$\tilde{\tau}_2(f)(x_i) = w_i$$

The function $\tilde{\tau}_2 : \mathcal{M}_{g,1} \to \mathrm{Hom}(H, \frac{1}{2} \wedge^2 H)$ is a homomorphism when restricted to the kernel $\mathcal{I}_{g,1}$ of the symplectic representation. Johnson [5, Theorem 1] identifies its image as $\wedge^3 H \subset \mathrm{Hom}(H, \frac{1}{2} \wedge^2 H)$, where $x_i \wedge x_j \wedge x_k \in \wedge^3 H$ is understood to be the homomorphism

$$(x_i \wedge x_j \wedge x_k)(y) = \langle y, x_k \rangle x_i \wedge x_j + \langle y, x_i \rangle x_j \wedge x_k + \langle y, x_j \rangle x_k \wedge x_i \tag{4}$$

where $\langle , \rangle$ gives the symplectic pairing for vectors in H. The map $\mathcal{I}_{g,1} \to \wedge^3 H \subset \mathrm{Hom}(H, \frac{1}{2} \wedge^2 H)$ is usually referred to as the *Johnson homomorphism.*

Morita [11, Section 3] begins by considering this map $\tilde{\tau}_2 : \mathcal{M}_{g,1} \to \mathrm{Hom}(H, \frac{1}{2} \wedge^2 H)$ (in Morita's notation this is the map $\tilde{k}$). While not a homomorphism it is a crossed homomorphism with respect to the symplectic action of the mapping class group on $\mathrm{Hom}(H, \frac{1}{2} \wedge^2 H)$. In other words, the map $\tilde{\tau}_2$ satisfies:

$$\tilde{\tau}_2(fg) = \tilde{\tau}_2(f) + \rho_1(f)\tilde{\tau}_2(g) \qquad f, g \in \mathcal{M}_{g,1}$$

Choose $R \in \mathrm{Sp}(H)$, $y \in H$, and $m \in \mathrm{Hom}(H, \frac{1}{2} \wedge^2 H)$. We note that the action of $\mathrm{Sp}(H)$ on $\mathrm{Hom}(H, \frac{1}{2} \wedge^2 H)$ in the equation above (and in the remainder of this paper) is the natural "change-of-basis" action:

$$(Rm)(y) = Rm(R^{-1}y) \tag{5}$$

The crossed homomorphism property is exactly what is needed for the map $\tilde{\rho}_2 : \mathcal{M}_{g,1} \to \mathrm{Hom}(H, \frac{1}{2} \wedge^2 H) \rtimes \mathrm{Sp}(H)$ given by

$$\tilde{\rho}_2(f) = (\tilde{\tau}_2(f), \rho_1(f))$$

to be a homomorphism. The homomorphism $\tilde{\rho}_2$ gives the action of $\mathcal{M}_{g,1}$ on $\phi_2(\pi) \subset \Phi_2$, via the action of $(r, R) \in \mathrm{Hom}(H, \frac{1}{2} \wedge^2 H) \rtimes \mathrm{Sp}(H)$ on Φ_2:

$$(r, R) * (\eta, y) = (r(Ry) + R\eta, Ry) \tag{6}$$

Morita shows that by modifying the crossed homomorphism $\tilde{\tau}_2 : \mathcal{M}_{g,1} \to \mathrm{Hom}(H, \frac{1}{2} \wedge^2 H)$, one obtains a crossed homomorphism $\tilde{\tau}_2'$ (Morita denotes this map by $\tilde{k}'$ in [11, Section 4] and $\tilde{k}$ in [11, Section 5]) from $\mathcal{M}_{g,1}$ to the submodule $\frac{1}{2} \wedge^3 H$ of $\mathrm{Hom}(H, \frac{1}{2} \wedge^2 H)$ which extends the Johnson homomorphism. We will modify $\tilde{\tau}_2$ to get a different crossed homomorphism $\tau_2 : \mathcal{M}_{g,1} \to \frac{1}{2} \wedge^3 H$ extending the Johnson homomorphism. Our map τ_2 is a trivial modification of Morita's map $\tilde{\tau}_2'$ which will lend itself to later calculations.

For any $m \in \mathrm{Hom}(H, \frac{1}{2} \wedge^2 H)$, the map $\sigma_m : \mathcal{M}_{g,1} \to \mathrm{Hom}(H, \frac{1}{2} \wedge^2 H)$ given by

$$\sigma_m(f) = m - \rho_1(f)m$$

is a crossed homomorphism. Such a crossed homomorphism is called *principal*; two crossed homomorphisms are cohomologous if they differ by a principal crossed homomorphism [3, Chapter IV.2].

Let $\kappa \in \mathrm{Hom}(H, \frac{1}{2} \wedge^2 H)$ be the homomorphism

$$\kappa(a_i) = \frac{1}{2} a_i \wedge b_i \qquad \kappa(b_i) = -\frac{1}{2} a_i \wedge b_i$$

or equivalently

$$\kappa(x_i) = \frac{1}{2} x_i \wedge C x_i \tag{7}$$

where C is the $2g \times 2g$ matrix with $g \times g$ block form $\begin{pmatrix} 0 & I \\ I & 0 \end{pmatrix}$. Define

$$\tau_2(f) = \tilde{\tau}_2(f) + \kappa - \rho_1(f)\kappa \tag{8}$$

This is the crossed homomorphism that Perron [16, Remark 5.5] denotes $-\frac{1}{6}\widetilde{A_1}$. We note that by comparing the above with [11, Proposition 4.7], it is straightforward to see that Morita's crossed homomorphism $\tilde{\tau}_2'$ can be expressed as

$$\tilde{\tau}_2'(f) = \tau_2(f) + m - \rho_1(f)m$$

where $m = -\frac{1}{2}(\sum_{i=1}^{g} a_i + b_i) \wedge (\sum_{i=1}^{g} a_i \wedge b_i)$. In other words, the map τ_2 and Morita's original map $\tilde{\tau}_2'$ are cohomologous, that is, they represent the same element of $H^1(\mathcal{M}_{g,1}, \frac{1}{2} \wedge^3 H)$.

We can now define a homomorphism $\rho_2 : \mathcal{M}_{g,1} \to \frac{1}{2} \wedge^3 H \rtimes \mathrm{Sp}(H)$ as follows:

$$\rho_2(f) = (\tau_2(f), \rho_1(f))$$

Using (8), (6), (5), and (4), we obtain the correct action of $\rho_2(\mathcal{M}_{g,1})$ on Φ_2:

$$\begin{aligned} &\Big(\sum r_{ijk} x_i \wedge x_j \wedge x_k, R\Big) * (\eta, y) \\ (9)\qquad &= (R\eta - \kappa(Ry) + R(\kappa(y)) + r(y), Ry) \\ (10)\qquad &= \left(R\eta - \kappa(Ry) + R(\kappa(y)) + \sum r_{ijk} \begin{pmatrix} \langle Ry, x_k\rangle x_i \wedge x_j \\ +\langle Ry, x_i\rangle x_j \wedge x_k \\ +\langle Ry, x_j\rangle x_k \wedge x_i \end{pmatrix}, Ry \right) \end{aligned}$$

where $\langle , \rangle$ is the symplectic pairing on H and the sums are taken over $1 \leq i < j < k \leq 2g$.

2.3. Calculating the image of the mapping class group

In this section we compute $\rho_2(\mathcal{M}_{g,1})$. See Theorem 2.4 below.

Recall the map $\phi_2 : \pi \to \Phi_2$ given in the previous section. It will be helpful for us to identify $\phi_2(\pi) \subset \Phi_2$ precisely. The gist of the following lemma is that for pairs in the image of ϕ_2, the second coordinate determines the first coordinate modulo 1.

Lemma 2.1. *The image of π under the map ϕ_2 is given as follows.*

$$\phi_2(\pi) = \left\{ \left(\sum_{1<i<j<2g} \left(n_{ij} + \frac{l_i l_j}{2} \right) x_i \wedge x_j \,,\, \sum_{i=1}^{2g} l_i x_i \right) \,\middle|\, n_{ij},\ l_i \in \mathbb{Z} \right\}$$

Proof. Let $G \subset \Phi_2$ denote the set on the right-hand side of the equation in the lemma. We claim that the set G is a subgroup of Φ_2. First, G is closed under inversion since $(\eta, y)^{-1} = (-\eta, -y)$. For closure under products consider

$$\begin{aligned} &\left(\sum_{1<i<j<2g} \left(n_{ij} + \frac{l_i l_j}{2} \right) x_i \wedge x_j \,,\, \sum_{i=1}^{2g} l_i x_i \right) \\ &\qquad \cdot \left(\sum_{1<i<j<2g} \left(n'_{ij} + \frac{l'_i l'_j}{2} \right) x_i \wedge x_j \,,\, \sum_{i=1}^{2g} l'_i x_i \right) \\ &= \left(\sum_{1<i<j<2g} \begin{pmatrix} n_{ij} + n'_{ij} + \frac{l_i l_j}{2} \\ + \frac{l'_i l'_j}{2} + \frac{l_i l'_j}{2} - \frac{l_j l'_i}{2} \end{pmatrix} x_i \wedge x_j \,,\, \sum_{i=1}^{2g} (l_i + l'_i) x_i \right) \end{aligned}$$

This product is in G because $l_i l_j + l'_i l'_j + l_i l'_j - l_j l'_i \equiv (l_i + l'_i)(l_j + l'_j) \bmod 2$.

Clearly, G contains each generator $\phi_2(\xi_i) = (0, x_i)$ of $\phi_2(\pi)$. For the reverse inclusion, note that any element of the form

$$(0, x_i)(0, x_j)(0, -x_i)(0, -x_j) = (x_i \wedge x_j, 0)$$

lies in $\phi_2(\pi)$. In fact such an element is in the center of G. Now, any element of G can be written as a product of $(0, x_i)$'s to get the correct second coordinate, followed by a product of $(x_i \wedge x_j, 0)$'s to get the correct first coordinate. Hence $G \subset \phi_2(\pi)$. Q.E.D.

We are almost ready to characterize the subgroup $\rho_2(\mathcal{M}_{g,1}) \subset \frac{1}{2} \wedge^3 H \rtimes \mathrm{Sp}(H)$. We begin with a simple yet fundamental observation.

Remark 2.2. Suppose R is a symplectic matrix and (r_1, R), $(r_2, R) \in \rho_2(\mathcal{M}_{g,1})$. Then $(r_1, R)^{-1} = (-R^{-1}r_1, R^{-1}) \in \rho_2(\mathcal{M}_{g,1})$ so

$$(r_2, R)(-R^{-1}r_1, R^{-1}) = (r_2 - r_1, I) \in \rho_2(\mathcal{M}_{g,1}).$$

In other words, we have that $(r_2 - r_1, I) \in \rho_2(\mathcal{I}_{g,1})$. Using Johnson's characterization of $\tau_2(\mathcal{I}_{g,1})$ [5, Theorem 1] we conclude that if two elements of $\rho_2(\mathcal{M}_{g,1})$ have identical symplectic matrices, then their $\frac{1}{2} \wedge^3 H$ coordinate must differ by an *integral* element of $\wedge^3 H$.

As a consequence of this observation, we expect that the symplectic matrix R will determine the coefficients of r_1 and r_2 modulo 1. Theorem 2.4 makes this precise and gives the characterization of $\rho_2(\mathcal{M}_{g,1})$. First we give a short definition.

Definition 2.3. Given three n-dimensional vectors $\vec{w} = (w_1, \ldots, w_n)$, $\vec{y} = (y_1, \ldots, y_n)$, $\vec{z} = (z_1, \ldots, z_n)$ in basis $\mathcal{B}$, their *$\mathcal{B}$-triple dot product* is the scalar

$$\bullet_{\mathcal{B}}(\vec{w}, \vec{y}, \vec{z}) = \sum_{i=1}^{n} w_i y_i z_i.$$

When the basis $\mathcal{B}$ is clear, we will write $\bullet(\vec{w}, \vec{y}, \vec{z})$.

Recall that J is the matrix given in (1).

Theorem 2.4. *Let $R \in \mathrm{Sp}(2g, \mathbb{Z})$ be an arbitrary symplectic matrix. Let r be any element of $\frac{1}{2} \wedge^3 H$ with $r = \sum_{1 \le i < j < k \le 2g} r_{ijk} x_i \wedge x_j \wedge x_k$. Then $(r, R) \in \rho_2(\mathcal{M}_{g,1})$ if and only if*

$$r_{ijk} \equiv \frac{E_{ijk}}{2} \bmod 1$$

where

$$\begin{aligned} E_{ijk} &= \bullet(\mathrm{row}_i(RJ), \mathrm{row}_j(R), \mathrm{row}_k(R)) \\ &\quad - \bullet(\mathrm{row}_i(R), \mathrm{row}_j(RJ), \mathrm{row}_k(R)) \\ &\quad + \bullet(\mathrm{row}_i(R), \mathrm{row}_j(R), \mathrm{row}_k(RJ)) \end{aligned}$$

for all $1 \le i < j < k \le 2g$.

Proof. Let $(r,R) \in \rho_2(\mathcal{M}_{g,1})$, and let

$$r = \sum_{1\le i<j<k\le 2g} r_{ijk} x_i \wedge x_j \wedge x_k.$$

For $1 \le i,j,k \le 2g$ we set $r_{ijk} = 0$ unless $i<j<k$. The group $\rho_2(\mathcal{M}_{g,1})$ preserves $\phi_2(\pi)$, described in Lemma 2.1. Let x_n be an arbitrary basis element of H, and consider the action of (r,R) on $(0,x_n)$. We will use the standard notation M_{ij} to denote the entry in the ith row and j^{th} column of a matrix M throughout. By (10), we get that the second coordinate of $(r,R)*(0,x_n)$ is simply Rx_n, which we can write as $\sum_{i=1}^{2g} R_{in}x_i$, with an eye on eventually applying Lemma 2.1. Using (10) and (7), we obtain the following for the first coordinate of $(r,R)*(0,x_n)$:

$$-\kappa(Rx_n) + R(\kappa(x_n)) + \sum_{1\le i<j<k\le 2g} r_{ijk}\begin{pmatrix} \langle Rx_n, x_k\rangle x_i\wedge x_j \\ +\langle Rx_n, x_i\rangle x_j \wedge x_k \\ -\langle Rx_n, x_j\rangle x_i\wedge x_k\end{pmatrix}$$

Notice that under the symplectic pairing $\langle Rx_n, x_k\rangle = (JR)_{kn}$ so the above can be rewritten:

$$\begin{aligned}
&-\kappa\left(\sum_{i=1}^{2g} R_{in}x_i\right) + R\left(\frac{1}{2}x_n\wedge Cx_n\right) \\
&+ \sum_{1\le i<j<k\le 2g} r_{ijk}\begin{pmatrix} ((JR)_{kn})x_i\wedge x_j \\ +((JR)_{in})x_j\wedge x_k \\ -((JR)_{jn})x_i\wedge x_k\end{pmatrix} \\
= &-\left(\sum_{i=1}^{2g}\frac{R_{in}}{2}x_i\wedge Cx_i\right) + \left(\sum_{1\le i,j\le 2g}\frac{R_{in}(RC)_{jn}}{2}x_i\wedge x_j\right) \\
&+ \sum_{1\le i<j<k\le 2g} r_{ijk}\begin{pmatrix} ((JR)_{kn})x_i\wedge x_j \\ +((JR)_{in})x_j\wedge x_k \\ -((JR)_{jn})x_i\wedge x_k\end{pmatrix} \\
= &\left(\sum_{i=1}^{g}\frac{(CR)_{in}-R_{in}}{2}x_i\wedge x_{i+g}\right) \\
&+ \left(\sum_{1\le i<j\le 2g}\frac{R_{in}(RC)_{jn}-R_{jn}(RC)_{in}}{2}x_i\wedge x_j\right) \\
&+ \sum_{1\le i<j<k\le 2g} r_{ijk}\begin{pmatrix} ((JR)_{kn})x_i\wedge x_j \\ +((JR)_{in})x_j\wedge x_k \\ -((JR)_{jn})x_i\wedge x_k\end{pmatrix}
\end{aligned}$$

Now, applying Lemma 2.1 to the coefficient of $x_p \wedge x_q$, where $p < q$, gives

$$\frac{\delta_{q,p+g}((CR)_{pn} - R_{pn}) + R_{pn}(RC)_{qn} - R_{qn}(RC)_{pn}}{2} + \sum_{i=1}^{2g} (r_{ipq}(JR)_{in} - r_{piq}(JR)_{in} + r_{pqi}(JR)_{in}) \equiv \frac{R_{pn}R_{qn}}{2} \bmod 1$$

Note that for fixed i, p, q, at most one of the r-coefficients in the above summation is nonzero. For bookkeeping purposes, when $1 \leq j < r \leq 2g$ we define $\vec{r}_{jk}$ be the $2g$-dimensional column vector whose i^{th} entry is r_{ijk} if $i < j$, $-r_{jik}$ if $j < i < k$, r_{jki} if $k < i$, and 0 otherwise. If $\mathrm{col}_n(M)$ denotes the nth column vector of M, we may rewrite this to obtain that $\mathrm{col}_n(JR) \cdot \vec{r}_{pq}$ is congruent $(\bmod 1)$ to

$$\frac{\delta_{q,p+g}(R_{pn} - (CR)_{pn}) + R_{pn}R_{qn} - R_{pn}(RC)_{qn} + R_{qn}(RC)_{pn}}{2}$$

In order to write this a bit more compactly, for $1 \leq j < k \leq 2g$, we define $\vec{t}_{jk}$ to be the $2g$-dimensional column vector whose i^{th} entry is $\delta_{k,j+g}(R_{ji} - (CR)_{ji}) + R_{ji}R_{ki} - R_{ji}(RC)_{ki} + R_{ki}(RC)_{ji}$. Combining the equations above for all $1 \leq n \leq 2g$ we get:

$$\overline{JR}\vec{r}_{pq} \equiv \frac{\vec{t}_{pq}}{2} \bmod 1 \qquad \forall 1 \leq p < q \leq 2g$$

Solving for $\vec{r}_{pq}$, we obtain:

$$\vec{r}_{pq} \equiv \frac{(\overline{JR})^{-1}\vec{t}_{pq}}{2} \bmod 1$$

Since R is assumed to be symplectic, we can rewrite this as:

$$\vec{r}_{pq} \equiv \frac{RJ\vec{t}_{pq}}{2} \bmod 1$$

Observe that the ith entry of the vector on the right-hand side is

$$\begin{aligned} &\frac{1}{2}\delta_{q,p+g}\mathrm{row}_i(RJ) \cdot (\mathrm{row}_p(R) - \mathrm{row}_p(CR)) \\ &+\frac{1}{2} \bullet (\mathrm{row}_i(RJ), \mathrm{row}_p(R), \mathrm{row}_q(R)) \\ &-\frac{1}{2} \bullet (\mathrm{row}_i(RJ), \mathrm{row}_p(R), \mathrm{row}_q(RC)) \\ &+\frac{1}{2} \bullet (\mathrm{row}_i(RJ), \mathrm{row}_p(RC), \mathrm{row}_q(R)) \end{aligned} \tag{11}$$

We are interested in calculating the coefficients r_{ipq} for $1 \leq i < p < q \leq 2g$. Thus we are interested in the ith entry of $\vec{r}_{pq}$ when $1 \leq i < p < q \leq 2g$. If $q \neq p+g$ then $\delta_{q,p+g} = 0$. Assume that $q = p+g$. Then $1 \leq i < p \leq g$, and if we write $R = \begin{pmatrix} S & T \\ P & Q \end{pmatrix}$, we have

$$\begin{aligned} &\mathrm{row}_i(RJ) \cdot (\mathrm{row}_p(R) - \mathrm{row}_p(CR)) \\ &\quad = \mathrm{row}_i(T) \cdot \mathrm{row}_p(S) - \mathrm{row}_i(S) \cdot \mathrm{row}_p(T) \\ &\quad\quad -\mathrm{row}_i(T) \cdot \mathrm{row}_p(P) + \mathrm{row}_i(S) \cdot \mathrm{row}_p(Q) \\ &\quad = (T\overline{S})_{ip} - (S\overline{T})_{ip} - (T\overline{P})_{ip} + (S\overline{Q})_{ip} \\ &\quad = 0 - 0 \end{aligned}$$

The last equality results from using the symplectic conditions (3 i,ii) and by our assumption that $i \neq p$. Thus we may drop the first term of (11). In other words, for $1 \leq i < p < q \leq 2g$ the ith entry of $\vec{r}_{pq}$ (mod 1) is given by

$$\begin{aligned} &\frac{1}{2} \bullet (\mathrm{row}_i(RJ), \mathrm{row}_p(R), \mathrm{row}_q(R)) \\ &-\frac{1}{2} \bullet (\mathrm{row}_i(RJ), \mathrm{row}_p(R), \mathrm{row}_q(RC)) \\ &+\frac{1}{2} \bullet (\mathrm{row}_i(RJ), \mathrm{row}_p(RC), \mathrm{row}_q(R)) \mod 1 \end{aligned}$$

For aesthetic reasons we rewrite the expression above more symmetrically to show that i^{th} entry of $\vec{r}_{pq}$ (mod 1) is:

$$\begin{aligned} &\frac{1}{2} \bullet (\mathrm{row}_i(RJ), \mathrm{row}_p(R), \mathrm{row}_q(R)) \\ &-\frac{1}{2} \bullet (\mathrm{row}_i(R), \mathrm{row}_p(RJ), \mathrm{row}_q(R)) \\ &+\frac{1}{2} \bullet (\mathrm{row}_i(R), \mathrm{row}_p(R), \mathrm{row}_q(RJ)) \mod 1 \end{aligned}$$

We have just shown that the $\binom{2g}{3}$ equations in the statement of the lemma are necessary for (r, R) to be an element of $\rho_2(\mathcal{M}_{g,1})$. Since the symplectic representation ρ_1 is surjective, $\rho_2(\mathcal{M}_{g,1})$ contains an element of the form (r, R) for any given R. Johnson [5, Theorem 1] showed that any element of the form (w, I) with $w \in \wedge^3 H$ is in $\rho_2(\mathcal{M}_{g,1})$. Then if $(r, R) \in \rho_2(\mathcal{M}_{g,1})$, so is $(w, I)(r, R) = (w+r, R)$ for any $w \in \wedge^3 H$. Hence we can hit any other possible choice of the coefficients r_{ijk} satisfying the "mod 1" conditions imposed by R by composing our map with different choices of Torelli elements. This shows sufficiency. Q.E.D.

§3. The handlebody group

Our primary goal in this section is to compute $\rho_2(\mathcal{H}_{g,1})$ explicitly. We will begin with some known algebraic characterizations of $\mathcal{H}_{g,1}$ and of $\rho_1(\mathcal{H}_{g,1})$ which will be helpful to us, and use them to derive an analogous characterization at the second level. Thus equipped, we derive an explicit formulation of $\rho_2(\mathcal{H}_{g,1})$ in Section 3.2.

3.1. Algebraic characterizations of the handlebody subgroup

Let $\mathfrak{b}$ denote the normal closure in π of $\{\beta_1, \ldots, \beta_g\}$. Note that $\mathfrak{b}$ is also the kernel of the homomorphism $\pi \to \pi_1(X_g)$ induced by inclusion.

The following proposition was first proved by McMillan [9]. The proof given here was suggested to the authors by Saul Schleimer.

Proposition 3.1. *The handlebody subgroup $\mathcal{H}_{g,1}$ of the mapping class group $\mathcal{M}_{g,1} \subset \mathrm{Aut}(\pi_1(S_{g,1}))$ is precisely the subgroup which preserves $\mathfrak{b}$.*

Proof. One direction is immediate; in order for a mapping class in $\mathcal{M}_{g,1}$ to extend to the X_g it must preserve $\mathfrak{b}$. Now suppose f is a mapping class which preserves $\mathfrak{b}$. Then f sends each β_i to a loop that can be represented by a simple closed curve which is trivial in $\pi_1(X_g)$. Dehn's Lemma [15] shows that these curves bound disks in X_g that can be made disjoint. By matching these disks to the ones bounded by each β_i we may construct a homeomorphism from X_g to itself restricting to f on its boundary. Q.E.D.

Moving on to level one of the Johnson-Morita representations, Birman has shown that the image of the handlebody group in $\mathrm{Sp}(2g, \mathbb{Z})$ is particularly nice [2, Lemma 2.2]. All subblocks are $g \times g$ matrices.

Proposition 3.2 (Birman). *The image of the handlebody group under the symplectic representation is characterized by a $g \times g$ block of zeroes in the upper-right corner. That is,*

$$\rho_1(\mathcal{H}_{g,1}) = \left\{ M \in \mathrm{Sp}(2g; \mathbb{Z}) \middle| M \text{ has block form } \begin{pmatrix} * & 0 \\ * & * \end{pmatrix} \right\}$$

Sufficiency is shown in [2] by exhibiting generators for $\rho_1(\mathcal{H}_{g,1})$ which are in the image of the handlebody group. The necessity of this condition for membership in $\rho_1(\mathcal{H}_{g,1})$ follows from the observation that in the handlebody X_g, the homology classes of the generators of type b_i are all 0. Any homeomorphism of S_g which extends to X_g must take trivial elements in the homology of the handlebody to trivial elements in the homology of the handlebody. In other words, $\rho_1(\mathcal{H}_{g,1})$ is characterized by the property that its elements must preserve the subgroup of H generated by the b_i's.

We will now give a second-level analogue of these characterizations by describing a subgroup of $\pi/\pi^{(2)}$ which must be preserved by $\rho_2(\mathcal{H}_{g,1})$, thus giving a restriction on the image of the handlebody group.

The second Johnson-Morita homomorphism is given by the action of $\mathcal{M}_{g,1}$ on the nilpotent quotient $\pi/\pi^{(2)}$. Let $\mathfrak{b} \subset \pi$ be as above, and recall from Section 2.2 the map $\phi_2 : \pi \to \Phi_2$ be as above. The following lemma computes $\phi_2(\mathfrak{b})$.

Lemma 3.3.

$$\phi_2(\mathfrak{b}) = \left\{ \left(\begin{array}{c} \sum_{1\le i,j\le g} m_{ij} a_i \wedge b_j \\ +\sum_{1\le i<j\le g} \left(n_{ij} + \frac{l_i l_j}{2}\right) b_i \wedge b_j \end{array}, \sum_{i=1}^{g} l_i b_i \right) \middle| m_{ij}, n_{ij}, l_i \in \mathbb{Z} \right\}$$

Proof. In light of Lemma 2.1, the right-hand side above is clearly the kernel of the quotient homomorphism $\pi/\pi^{(2)} \to \pi_1(X_g)/\pi_1(X_g)^{(2)}$. Q.E.D.

Now that we have identified $\phi_2(\mathfrak{b})$ we will describe $\rho_2(\mathcal{H}_{g,1})$.

3.2. Image of the handlebody subgroup under ρ_2

Theorem 2.4 above gives $\rho_2(\mathcal{M}_{g,1})$. The missing ingredient for a characterization of $\rho_2(\mathcal{H}_{g,1})$ is $\rho_2(\mathcal{I}_{g,1} \cap \mathcal{H}_{g,1})$ which was computed by Morita.

Proposition 3.4 ([10, Lemma 2.5]). *$\rho_2(\mathcal{I}_{g,1} \cap \mathcal{H}_{g,1})$ is the free abelian group with free basis:*

$$(b_i \wedge b_j \wedge b_k, I), \quad (a_i \wedge b_j \wedge b_k, I), \quad \text{and} \quad (a_i \wedge a_j \wedge b_k, I) \quad 1 \le i,j,k \le g.$$

Now we have the tools to assemble a description of $\rho_2(\mathcal{H}_{g,1})$. The following theorem gives a complete characterization of $\rho_2(\mathcal{H}_{g,1})$; it says that an element is in this image if and only if its first factor has no "triple-a" terms and its second factor has the form of Proposition 3.2.

Theorem 3.5. *Let $R \in \mathrm{Sp}(2g, \mathbb{Z})$ be an arbitrary symplectic matrix. Let r be any element of $\frac{1}{2} \wedge^3 H$ with $r = \sum_{1\le i<j<k\le 2g} r_{ijk} x_i \wedge x_j \wedge x_k$. Then $(r, R) \in \rho_2(\mathcal{H}_{g,1})$ if and only if all of the following three conditions hold:*

(1) *R has $g \times g$ block form* $\begin{pmatrix} * & 0 \\ * & * \end{pmatrix}$

(2) $r_{ijk} \equiv \frac{1}{2} E_{ijk} \bmod 1$ *for all* $1 \le i < j < k \le 2g$.

(3) $r_{ijk} = 0$ *for all* i, j, k *with* $0 \leq i < j < k \leq g$. *(i.e.* r *contains no terms of the form* $a_i \wedge a_j \wedge a_k$.*)*

We refer the reader to Theorem 2.4 for the definition of E_{ijk}, which depends on the matrix R.

Proof. The necessity of condition 1 has already been established in [2, Lemma 2.2]. We claim that only elements of $\frac{1}{2}\wedge^3 H \rtimes \mathrm{Sp}(H)$ satisfying condition 3 above preserve $\phi_2(\mathfrak{b})$ under the action of (10). Suppose R is symplectic with the required block form and r contains a term of the form $ca_i \wedge a_j \wedge a_k$. Since R^{-1} must satisfy condition 1 above and using Lemma 3.3, there is an element $(\nu, R^{-1}b_i) \in \phi_2(\mathfrak{b})$ where ν has only terms of the form $\frac{1}{2}b_n \wedge b_m$. Applying (9) we get

$$\begin{aligned}
&(r, R) * (\nu, R^{-1}b_i) = \\
&= \big(R(\nu) + \kappa(RR^{-1}b_i) + R\kappa(R^{-1}b_i) + r(RR^{-1}b_i), RR^{-1}b_i\big) \\
&= \big(R(\nu) + \kappa(b_i) + R\kappa(R^{-1}b_i) + r(b_i), b_i\big)
\end{aligned}$$

Consider each of the terms in the first coordinate of the ordered pair above. Since ν only has terms of the form $\frac{1}{2}b_n \wedge b_m$ and the matrix R has the block form given in condition 1, we must have that $R(\nu)$ contains no terms of the form $a_j \wedge a_k$. The image of the homomorphism κ has no $a_j \wedge a_k$ terms so neither $\kappa(b_i)$ nor $\kappa(R^{-1}b_i)$ contains any $a_j \wedge a_k$ terms. Application of the matrix R preserves this quality; hence $R\kappa(R^{-1}b_i)$ contains no $a_j \wedge a_k$ terms. We can see using (4) that $r(b_i)$ will contain a term of the form $-ca_j \wedge a_k$ by construction. Then Lemma 3.3 implies that $c = 0$. It follows that the two conditions of the corollary are necessary.

For each R satisfying 1 there is some mapping class $f \in \mathcal{H}_{g,1}$ with $\rho_1(f) = R$ as shown in [2, Lemma 2.2]. We have shown that $\rho_2(f)$ satisfies conditions 1 and 2. Applying Proposition 3.4 we can get every other element of the form (w, R) satisfying 1 and 2 as a product $(z, I)\rho_2(f)$ where $(z, I) \in \rho_2(\mathcal{I}_{g,1} \cap \mathcal{H}_{g,1})$. This establishes sufficiency. Q.E.D.

References

[1] J. Birman, Braids, Links and Mapping Class Groups, Ann. of Math. Stud., **82**, 1974.

[2] J. Birman, On the equivalence of Heegaard splittings of closed, orientable 3-manifolds, In: Knots, Groups and 3-manifolds, Ann. of Math. Stud., **84**, 1975, pp. 137–164.

[3] K. Brown, Cohomology of Groups, Grad. Texts in Math., **87**, Springer-Verlag, New York, 1994.

[4] H. Burkhardt, Grundzüge einer allgemeinen Systematik der hyperelliptischen Funktionen erster Ordnung, Math. Ann., **35** (1890), 198–296.

[5] D. Johnson, An abelian quotient of the mapping class group $\mathcal{I}_g$, Math. Ann., **249** (1980), 225–242.

[6] D. Johnson, The structure of the Torelli group II: A characterization of the group generated by twists on bounding curves, Topology, **24** (1985), 113–126.

[7] D. Johnson, A survey of the Torelli group, Contemp. Math., **20** (1983), 165–179.

[8] W. Magnus, A. Karrass and D. Solitar, Combinatorial Group Theory, Interscience, John Wiley, 1966.

[9] D.R. McMillan, Jr., Homeomorphisms on a solid torus, Proc. Amer. Math. Soc., **14** (1963), 386–390.

[10] S. Morita, Casson's invariant for homology 3-spheres and characteristic classes of surface bundles. I, Topology, **28** (1989), 305–323.

[11] S. Morita, The extension of Johnson's homomorphism from the Torelli group to the mapping class group, Invent. Math., **111** (1993), 197–224.

[12] S. Morita, Abelian quotients of subgroups of the mapping class group of surfaces, Duke Math. J., **70** (1993), 699–726.

[13] S. Morita, A linear representation of the mapping class group of orientable surfaces and characteristic classes of surface bundles, In: Topology and Teichmüller spaces, Katinkulta, 1995, World Sci. Publ., River Edge, NJ, 1996, pp. 159–186.

[14] S. Morita, Structure of the mapping class group and symplectic representation theory, In: Essays on geometry and related topics, Vol. 1, 2, Monogr. Enseign. Math., **38**, Enseignement Math., Geneva, 2001, pp. 577–596.

[15] C. Papakyriakopoulos, On Dehn's lemma and the asphericity of knots, Ann. of Math. (2), **66** (1957), 1–26.

[16] B. Perron, Homomorphic extensions of Johnson homomorphisms via Fox calculus, Ann. Inst. Fourier (Grenoble), **54** (2004), 1073–1106.

Joan S. Birman
Department of Mathematics
Barnard College of Columbia University
New York, New York 10027
USA

Tara E. Brendle
Department of Mathematics
Louisiana State University
Lockett Hall
Baton Rouge, LA 70803
USA

Nathan Broaddus
Department of Mathematics
University of Chicago
5734 S. University Ave.
Chicago, IL 60637
USA

Advanced Studies in Pure Mathematics 52, 2008
Groups of Diffeomorphisms
pp. 135–220

Symplectic Heegaard splittings and linked abelian groups

Joan S. Birman, Dennis Johnson and Andrew Putman

Contents

§1. Introduction

1.1. The Johnson-Morita filtration of the mapping class group

Let M_g be a closed oriented 2-manifold of genus g and $\tilde{\Gamma}_g$ be its mapping class group, that is, the group of isotopy classes of orientation-preserving diffeomorphisms of M_g. Also, let $\pi = \pi_1(M_g)$, and denote by $\pi^{(k)}$ the kth term in the lower central series of π, i.e. $\pi^{(1)} = \pi$ and $\pi^{(k+1)} = [\pi, \pi^{(k)}]$. Then $\tilde{\Gamma}_g$ acts on the quotient groups $\pi/\pi^{(k)}$, and that action yields a representation $\rho^k : \tilde{\Gamma}_g \to \Gamma_g^k$, where $\Gamma_g^k < \text{Aut}\ (\pi/\pi^{(k)})$.

Received November 14, 2007.
Revised March 11, 2008.

Supported in part by NSF Grant DMS-040558

With these conventions, ρ^1 is the trivial representation and ρ^2 is the symplectic representation. The kernels of these representations make what has been called the 'Johnson filtration' of $\tilde{\Gamma}_g$, because they were studied by Johnson in [21, 20]. Subsequently they were developed by Morita in a series of papers [30, 31, 32, 33]. In particular, Morita studied the extensions of Johnson's homomorphisms to $\tilde{\Gamma}$, and so we refer to our representations as the *Johnson-Morita* filtration of $\tilde{\Gamma}$.

Our work in this article is motivated by the case when M_g is a Heegaard surface in a 3-manifold W and elements of $\tilde{\Gamma}_g$ are 'gluing maps' for the Heegaard splitting. One may then study W by investigating the image under the maps ρ^k of the set of all possible gluing maps that yield W. Among the many papers which relate to this approach to 3-manifold topology are those of Birman [2], Birman and Craggs [3], Brendle and Farb [5], Broaddus, Farb and Putman [6], Cochran, Gerges and Orr [10], Garoufalidis and Levine [14], Johnson [17, 18, 19, 20], Montesinos and Safont [27], Morita [29], Pitsch [37], [38], and Reidemeister [40]. The papers just referenced relate to the cases k =1-4 in the infinite sequence of actions of $\tilde{\Gamma}_g$ on the quotient groups of the lower central series, but the possibility is there to study deeper invariants of W, obtainable in principle from deeper quotients of the lower central series. The foundations for such deeper studies have been laid in the work of Morita[30, 31], who introduced the idea of studying higher representations via crossed homomorphisms. It was proved by Day [11] that the crossed product structure discovered by Morita in the cases $k = 3$ and 4 can be generalized to all k, enabling one in principle to separate out, at each level, the new contributions.

The invariants of 3-manifolds that can be obtained in this way are known to be closely related to finite type invariants of 3-manifolds [9, 14], although as yet this approach to finite type invariants opens up many more questions than answers. For example, it is known that the Rochlin and Casson invariants of 3-manifolds appear in this setting at levels 3 and 4, respectively. It is also known that in general there are finitely many linearly independent finite-type invariants of 3-manifolds at each fixed order (or, in our setting, fixed level) k, yet at this moment no more than one topological invariant has been encountered at any level.

The simplest non-trivial example of the program mentioned above is the case $k = 2$. Here $\tilde{\Gamma}_g$ acts on $H_1(M_g) = \pi/[\pi,\pi]$. The information about W that is encoded in $\rho^2(\phi)$, where $\phi \in \mathrm{Diff}^+(M_g)$ is the Heegaard gluing map for a Heegaard splitting of W of minimum genus, together with the images under ρ^2 of the Heegaard gluing maps of all 'stabilizations' of the given splitting, is what we have in mind when we refer to a 'symplectic Heegaard splitting'.

The purpose of this article is to review the literature on symplectic Heegaard splittings of 3-manifolds and the closely related literature on linked abelian groups, with the goal of describing what we know, as completely and explicitly and efficiently as possible, in a form in which we hope will be useful for future work. At the same time, we will add a few new things that we learned in the process. That is the broad outline of what the reader can expect to find in the pages that follow.

This article dates back to 1989. At that time, the first two authors had discussed the first author's invariant of Heegaard splittings, in [2], and had succeeded in proving three new facts : first, that the invariant in [2] could be improved in a small way; second, that the improved invariant was essentially the only invariant of Heegaard splittings that could be obtained from a symplectic Heegaard splitting; and third, that the index of stabilization of a symplectic Heegaard splitting is one. That work was set aside, in partially completed form, to gather dust in a filing cabinet. An early version of this paper had, however, been shared with the authors of [27] (and was referenced and acknowledged in [27]). Alas, it took us 18 years to prepare our work for publication! Our work was resurrected, tentatively, at roughly the time of the conference on *Groups of Diffeomorphisms* that was held in Tokyo September 11-15, 2006. As it turned out, the subject still seemed to be relevant, and since a conference proceedings was planned, we decided to update it and complete it, in the hope that it might still be useful to current workers in the area. When that decision was under discussion, the manuscript was shared with the third author, who contributed many excellent suggestions, and also answered a question posed by the first author (see §8). Soon after that, he became a coauthor.

1.2. Heegaard splittings of 3-manifolds

Let W be a closed, orientable 3-dimensional manifold. A *Heegaard surface* in W is a closed, orientable surface M of genus $g \geqslant 0$ embedded in W which divides W into homeomorphic handlebodies $N \cup \bar{N}$, where $N \cap \bar{N} = \partial N = \partial \bar{N} = M$. For example, if W is the 3-sphere

$$\mathbb{S}^3 = \left\{(x_1, x_2, x_3, x_4) \in \mathbb{R}^4 \mid x_1^2 + x_2^2 + x_3^2 + x_4^2 = 1\right\},$$

then the torus

$$M = \left\{(x_1, x_2, x_3, x_4) \in \mathbb{S}^3 \mid x_1^2 + x_2^2 = x_3^2 + x_4^2 = \frac{1}{2}\right\}$$

is a Heegaard surface.

Proposition 1.1. *Every closed orientable 3-manifold W admits Heegaard splittings.*

See [41], for example, for a proof. One will also find there related notions of Heegaard splittings of non-orientable 3-manifolds, of open 3-manifolds such as knot complements, and of 3-manifolds with boundary, and also an excellent introduction to the topic and its many open problems from the viewpoint of geometric topology.

Since any Heegaard splitting clearly gives rise to others under homeomorphisms of W, an equivalence relation is in order.

Definition 1.2. Assume that W is an oriented 3-manifold, and write $W = N \cup \bar{N} = N' \cup \bar{N}'$. These two Heegaard splittings will be said to be *equivalent* if there is a homeomorphism $F : W \to W$ which restricts to homeomorphisms $f : N \to N'$ and $\bar{f} : \bar{N} \to \bar{N}'$. Observe that our particular way of defining equivalent Heegaard splittings involve a choice of the initial handlebody N and a choice of an orientation on W. The *genus* of the splitting $W = N \cup \bar{N}$ is the genus of N. ||

There are 3-manifolds and even prime 3-manifolds which admit more than one equivalence class of splittings (for example, see [12, 4]), there are also 3-manifolds which admit unique equivalence classes of splittings of minimal genus (*e.g.* lens spaces and the 3-torus $\mathbb{S}^1 \times \mathbb{S}^1 \times \mathbb{S}^1$), and there are also 3-manifolds which admit unique equivalence classes of Heegaard splittings of every genus. A very fundamental example was studied by Waldhausen in [46], who proved:

Theorem 1.3 ([46]). *Any two Heegaard splittings of the same, but arbitrary, genus of the 3-sphere $\mathbb{S}^3$ are equivalent.*

After that important result became known, other manifolds were investigated. At the present writing, it seems correct to say that 'most' 3-manifolds admit exactly one equivalence class of minimal genus Heegaard splittings. On the other hand, many examples are known of manifolds that admit more than one equivalence class of splittings. See, for example, [28], where all the minimal genus Heegaard splittings of certain Seifert fiber spaces are determined.

If a three-manifold admits a Heegaard splitting of genus g, then it also admits of one genus g' for every $g' > g$. To see why this is the case, let $N_g \cup \bar{N}_g$ be a Heegaard splitting of W of genus g, and let $T_1 \cup \bar{T}_1$ be a Heegaard splitting of the 3-sphere $\mathbb{S}^3$ of genus 1. Remove a 3-ball from W and a 3-ball from $\mathbb{S}^3$, choosing these 3-balls so that they meet the respective Heegaard surfaces in discs. Using these 3-balls to form the connected sum $W\#\mathbb{S}^3$, we obtain a new Heegaard splitting $(N_g\#T_1) \cup (\bar{N}_g\#\bar{T}_1)$ of $W \cong W\#\mathbb{S}^3$ of genus $g+1$. This process is called *stabilizing* a Heegaard splitting. Note that Theorem 1.3 implies that the equivalence class of the new genus $g+1$ splitting is independent of the choice of T_1 and of $\bar{T}_1$, as subsets of $\mathbb{S}^3$, since all splittings of $\mathbb{S}^3$ of genus 1, indeed of any genus, are equivalent. Iterating the procedure, we obtain splittings $(N_g\#T_1\#\cdots\#T_1) \cup (\bar{N}_g\#\bar{T}_1\#\cdots\#\bar{T}_1)$ of W of each genus $g+k, k>0$.

Heegaard splittings of genus g and g' of a 3-manifold W are said to be *stably equivalent* if they have equivalent stabilizations of some genus $g+k = g'+k'$. In this regard, we have a classical result, proved in 1933 by Reidemeister [39] and (simultaneously and independently) by James Singer [44]:

Theorem 1.4 ([39], [44]). *Any two Heegaard splittings of any closed, orientable 3-manifold W are stably equivalent.*

Remark 1.5. We distinguish two types of candidates for inequivalent minimum genus Heegaard splittings of a 3-manifold. The first (we call it *ordinary*) is always present: two splittings which differ in the choice of ordering of the two handlebodies, i.e. $N \cup \bar{N}$ in one case and $\bar{N} \cup N$ in the other. Two ordinary Heegaard splittings may or may not be equivalent. The second are *all examples which are not ordinary*, e.g. the 'horizontal' and 'vertical' Heegaard splittings of certain Seifert fibered spaces [28]. In view of the fact that Theorem 1.4 was proved in 1933, it seems remarkable that the following situation exists, as we write in 2008:

- The only examples of inequivalent minimal genus Heegaard splittings of genus g of the same 3-manifold which can be proved to require more than one stabilization before they become equivalent are ordinary examples;
- The discovery of the first ordinary examples which can be proved to require more than one stabilization was made in 2008 [15].
- While non-ordinary examples have been known for some time, at this writing there is no known pair which do not not become equivalent after a single stabilization. For example, the inequivalent minimal genus Heegaard splittings of Seifert fiber spaces which were studied in [28] were proved in [42] to become equivalent after a single stabilization.

Note that ordinary examples can be ruled out by a small change in the definition of equivalence, although we have chosen not to do so, because the situation as regards ordinary examples is still far from being understood. ||

Keeping Remark 1.5 in mind, we have several classical problems about Heegaard splittings:

- How many stabilizations are needed before two inequivalent Heegaard splittings of a 3-manifold become equivalent, as they must because of Theorem 1.4? Is there a uniform bound, which is independent of the choice of W and of the Heegaard surface $\partial N = \partial \bar{N}$ in W?
- Can we use *stabilized* Heegaard splittings to find topological invariants of 3-manifolds?

An example of a 3-manifold invariant which was discovered with the help of Heegaard splittings is Casson's invariant [1].

A Heegaard splitting of a 3-manifold W is said to have *minimal genus* (or simply to be *minimal*) if there do not exist splittings of W which have smaller genus. Our second problem involves Heegaard splittings which are not stabilized. Since it can happen that a Heegaard splitting of a 3-manifold W is non-minimal in genus, but is not the stabilization of a Heegaard splitting of smaller genus (a complication which we wish to avoid), we assume from now on that wherever we consider *unstabilized* Heegaard splittings, we assume the genus to be minimal over all Heegaard splittings of the particular manifold. This brings us to another problem:

- Can we find invariants of *unstable* Heegaard splittings, and so reach a better understanding of the classification of Heegaard splittings?

Surprisingly, such invariants are very hard to come by, and little is known.

1.3. Symplectic Heegaard splittings

We begin by setting up notation that will be used throughout this paper. We will use a standard model for a symplectic space and for the symplectic group $\mathrm{Sp}(2g, \mathbb{Z})$. Let N_g be a handlebody. Then $H_1(\partial N_g)$ is a free abelian group of rank $2g$. Thinking of it as a vector space, the free abelian group $H_1(N_g; \mathbb{Z})$ is a subspace. We choose as basis elements for the former the ordered array of homology classes of the loops

$a_1, \dots, a_g, b_1, \dots, b_g$ which are depicted in Figure 1. With our choices, the images of the a_i under the inclusion map $\partial N_g \to N_g$ are a basis for $H_1(N_g; \mathbb{Z})$. The algebraic intersection pairing $(\cdot, \cdot)$ defines a symplectic

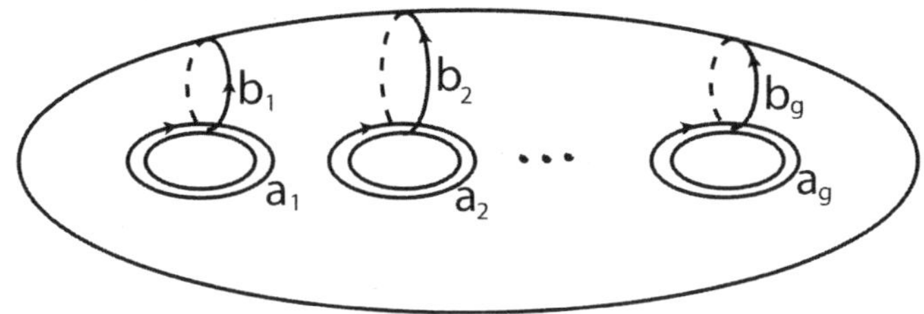

Fig. 1. Curves representing a canonical basis for $H_1(\partial N_g)$

form on $H_1(\partial N_g; \mathbb{Z})$, making it into a symplectic space. The matrix of intersection numbers for our canonical basis is $\mathcal{J} = \begin{pmatrix} 0_g & \mathcal{I}_g \\ -\mathcal{I}_g & 0_g \end{pmatrix}$, where 0_g and $\mathcal{I}_g$ are the $g \times g$ zero and identity matrices.

Definition 1.6. $\mathrm{Sp}(2g, \mathbb{Z})$ is the group of all $2g \times 2g$ matrices $\mathcal{H} = \begin{pmatrix} \mathcal{R} & \mathcal{P} \\ \mathcal{S} & \mathcal{Q} \end{pmatrix}$ over $\mathbb{Z}$ which satisfy

$$\hat{\mathcal{H}}\, \mathcal{J}\, \mathcal{H} = \mathcal{J} \tag{1}$$

where $\hat{\mathcal{H}}$ denotes the transpose of $\mathcal{H}$. Hence $\mathcal{H} \in \mathrm{Sp}(2g, \mathbb{Z})$ if and only if its $g \times g$ blocks $\mathcal{R}, \mathcal{P}, \mathcal{S}, \mathcal{Q}$ satisfy

(2)
$\hat{\mathcal{R}}\mathcal{S}, \hat{\mathcal{P}}\mathcal{Q}, \mathcal{R}\hat{\mathcal{P}}$ and $\mathcal{S}\hat{\mathcal{Q}}$ are symmetric, and $\hat{\mathcal{R}}\mathcal{Q} - \hat{\mathcal{S}}\mathcal{P} = \mathcal{R}\hat{\mathcal{Q}} - \mathcal{P}\hat{\mathcal{S}} = \mathcal{I}$.

Note that $\mathcal{H} \in \mathrm{Sp}(2g, \mathbb{Z})$ if and only if $\hat{\mathcal{H}} \in \mathrm{Sp}(2g, \mathbb{Z})$. ||

Lemma 1.7. *The group Γ_g (i.e. the image of the mapping class group under ρ^2) coincides with* $\mathrm{Sp}(2g, \mathbb{Z})$.

Proof. The fact that elements of Γ_g satisfy the constraints in (2) comes from the fact that topological mappings preserve algebraic intersection numbers. The fact that *every* symplectic matrix is in the image of ρ^2 can be proven by combining the classical fact that $\mathrm{Sp}(2g, \mathbb{Z})$ is generated by symplectic transvections with the fact that every such symplectic transvection is the image of a Dehn twist. This fact was used by Humphries, in his famous paper [16], to find a lower bound on the number of Dehn twists needed to generate the mapping class group. He used the known fact that Γ_g cannot be generated by fewer than $2g+1$ transvections. Q.E.D.

Lemma 1.8. *Let Λ_g be the subgroup of matrices in Γ_g which are induced by topological mappings of ∂N_g which extend to homeomorphisms of N_g (the so-called handlebody subgroup). Then Λ_g coincides with the subgroup of all elements in Γ_g with a $g \times g$ block of zeros in the upper right corner.*

Proof. By our choice of a basis for $H_1(\partial N_g;\mathbb{Z})$, a topologically induced automorphism of $H_1(\partial N_g;\mathbb{Z})$ extends to an automorphism of $H_1(N_g;\mathbb{Z})$ only if it preserves the kernel of the inclusion-induced homomorphism $H_1(\partial N_g) \to H_1(N_g)$, i.e. the subgroup generated by $b_1,\ldots,b_g$. Sufficiency is proved by finding generators for the group Λ_g, given in [34], and showing that each comes from a topological mapping on ∂N_g which extends to a homeomorphism of N_g. Explicit lifts are given in [2]. Q.E.D.

In § 1.2, we saw that every closed orientable 3-manifold admits Heegaard splittings. Let us now choose coordinates to make this more explicit. Let $N = N_g$ be a standard model for an oriented handlebody of genus g, and let $\bar{N} = \phi(N)$ be a copy of N, where ϕ is a fixed orientation-reversing homeomorphism. (Note that representative diffeomorphisms are always required to be orientation-preserving.) Choosing any element $\tilde{h} \in \mathrm{Diff}^+(\partial N_g)$, we may then construct a 3-manifold W as the disjoint union of N_g and $\bar{N}_g$, glued together by the rule $\phi \circ \tilde{h}(x) = x,\ x \in \partial N_g$. To stress the role of $\tilde{h}$ we will write $W = N_g \cup_{\phi\circ\tilde{h}} \bar{N}_g$. With these conventions, if we choose $\tilde{h}$ to be the identity map, the manifold W will be the connect sum of g copies of $\mathbb{S}^2 \times \mathbb{S}^1$. The mapping class group $\tilde{\Gamma}_g$ now means $\pi_0\mathrm{Diff}^+(\partial N_g)$.

Now let $\tilde{\Lambda} = \tilde{\Lambda}_g$ denote the subgroup of $\tilde{\Gamma}_g$ consisting of mapping classes which have a representative which extends to a homeomorphism of N_g. Note that every map of ∂N_g which is isotopic to the identity extends, hence if one representative extends then so does every other representative, so $\tilde{\Lambda}_g$ is well-defined.

Proposition 1.9. *Equivalence classes of genus g Heegaard splittings of 3-manifolds are in 1-1 correspondence with double cosets in the sequence of groups $\tilde{\Gamma}_g$ mod $\tilde{\Lambda}_g$.*

Proof. Each Heegaard splitting of a 3-manifold determines a (non- unique) $\tilde{h} \in \tilde{\Gamma}_g$ for some g, and each $\tilde{h} \in \tilde{\Gamma}_g$ determines a 3-manifold $W = N_g \cup_{\phi \circ \tilde{h}} \bar{N}_g$. Suppose $N_g \cup_{\phi\circ\tilde{h}} \bar{N}_g$ and $N_g' \cup_{\phi\circ\tilde{h}'} \bar{N}_g{}'$ are equivalent splittings of a 3-manifold W. Then there is an equivalence F which restricts to equivalences $f, \bar{f}$ on $N_g, \bar{N}_g$ and then to $f_0 = f|_{\partial N_g}, \bar{f}_0 = \bar{f}|_{\partial \bar{N}_g}$. There is thus a commutative diagram

$$\begin{array}{ccccc} \partial N_g & \xrightarrow{\tilde{h}} & \partial N_g & \xrightarrow{\phi} & \partial \bar{N}_g \\ \big\downarrow{f_0} & & & & \big\downarrow{\bar{f}_0} \\ \partial N_g & \xrightarrow{\tilde{h}'} & \partial N_g & \xrightarrow{\phi} & \partial \bar{N}_g \end{array}$$

Then $\tilde{h}' f_0 = \phi^{-1} \bar{f}_0 \phi \tilde{h}$, hence $\tilde{h}' \in \tilde{\Lambda}\tilde{h}\tilde{\Lambda}$. Conversely, if $\tilde{h}' \in \tilde{\Lambda}\tilde{h}\tilde{\Lambda}$ then $\tilde{h}' f_0 = \phi^{-1} \bar{f}_0 \phi \tilde{h}$ for some $f_0, \phi^{-1}\bar{f}_0\phi \in \tilde{\Lambda}$. Let $f, \phi^{-1}\bar{f}\phi$ be an extension of $f_0, \phi^{-1} f_0 \phi$ to N_g. Define $F|_{N_g} = f$, $F|_{\bar{N}_g} = \bar{f}$. Q.E.D.

For convenience, we will not distinguish between the diffeomorphism $\tilde{h}$ and the mapping class it determines in $\tilde{\Gamma}_g$.

Corollary 1.10. *Let $W = N_g \cup_{\phi\circ\tilde{h}} \bar{N}_g$ and let $W' = N_{g'} \cup_{\phi\circ\tilde{h}'} \bar{N}_{g'}$. Let $\tilde{s}$ be any choice of gluing map for a genus 1 splitting of $\mathbb{S}^3$. Then W is homeomorphic to W' if and only if there are integers k, k' with $g+k = g'+k'$ so that $\tilde{h}\#_k\tilde{s}$ is in the same double coset of $\tilde{\Gamma}_{g+k}$* mod $\tilde{\Lambda}_{g+k}$ *as $\tilde{h}'\#_{k'}\tilde{s}$.*

Proof. This follows directly from Theorem 1.4. Q.E.D.

Corollary 1.11. *Let W be a closed, orientable 3-dimensional manifold which is defined by any Heegaard splitting of genus g with Heegaard gluing map $\tilde{h}$. Then invariants of the stable double coset of $\tilde{h}$ in $\tilde{\Gamma}_g$ are topological invariants of the 3-manifold W.*

Proof. This is a direct consequence of Proposition 1.1, Proposition 1.9, and Corollary 1.10. Q.E.D.

We pass to the action of $\tilde{\Gamma}_g$ on $\pi_1(\partial N_g)/[\pi_1(\partial N_g), \pi_1(\partial N_g)]$, i.e. to the representation $\rho^2 : \tilde{\Gamma}_g \to \Gamma_g$. What information might we expect to detect about Heegaard splittings from the image $\rho^2(\tilde{h})$ of our gluing map $\tilde{h}$ in Γ_g?

Definition 1.12. A *stabilization of index k of $\mathcal{H}$* is the image of $\mathcal{H} \in \Gamma_g$ under the embedding $\Gamma_g \to \Gamma_{g+k}$ defined by bordering $\mathcal{R}, \mathcal{P}, \mathcal{S}, \mathcal{Q}$ according to the rule

$$\mathcal{R} \mapsto 0_k \oplus \mathcal{R}, \quad \mathcal{P} \mapsto \mathcal{I}_k \oplus \mathcal{P}, \quad \mathcal{S} \mapsto -\mathcal{I}_k \oplus \mathcal{S}, \quad \mathcal{Q} \mapsto 0_k \oplus \mathcal{Q}.$$

This is a particular way of taking the direct sum of $\mathcal{H} \in \Gamma_g$ with the matrix $\mathcal{J} \in \Gamma_1$, which is the image under ρ^2 of a Heegaard gluing map that defines $\mathbb{S}^2$.

Define $\mathcal{H}, \mathcal{H}' \in \Gamma_g$ to be *equivalent* ($\mathcal{H} \simeq \mathcal{H}'$) if $\mathcal{H}' \in \Lambda_g \mathcal{H} \Lambda$ and *stably equivalent* ($\mathcal{H} \simeq_s \mathcal{H}'$) if $\mathcal{H}$ and $\mathcal{H}'$ have equivalent stabilizations for some index $k \geqslant 0$. Equivalence classes are then double cosets in Γ_g mod Λ_g and stable equivalence classes are double cosets in Γ_{g+k} modulo Λ_{g+k}.

A *stabilized* symplectic Heegaard splitting is the union of all stabilizations of the double coset $\Lambda_g \mathcal{H} \Lambda_g$. ||

This brings us to the main topic of this article. Choose any $\tilde{h} \in \tilde{\Gamma}_g$ and use it to construct a 3-manifold W as above. Let $\mathcal{H}$ be the symplectic matrix that is induced by the action of $h = \rho^2(\tilde{h})$.

Definition 1.13. A *symplectic Heegaard splitting* of the 3-manifold $W = N_g \cup_{\phi \circ \tilde{h}} \bar{N}_g$ is the double coset $\Lambda_g \mathcal{H} \Lambda_g \subset \mathrm{Sp}(2g, \mathbb{Z})$, together with the double cosets of all stabilizations of $\mathcal{H}$. A symplectic Heegaard splitting is *minimal* if it is not the stabilization of a symplectic Heegaard splitting of lower genus which is in the same double coset. ||

1.4. Survey of the literature

The earliest investigation of Heegaard splittings were the proofs, by Singer [44] and Reidemeister [39] that all Heegaard splittings of an arbitrary 3-manifold are stably equivalent. Shortly after the publication of [39] Reidemeister asked about invariants of 3-manifolds that can be determined from a Heegaard splittings. His invariants are given in the paper [40]. He proves by an example (the Lens spaces) that the invariants he discovered distinguish manifolds which have the same fundamental group $\pi_1(W)$, and so are independent of the rank and torsion coefficients of W. Reidemeister's invariants are determined from the action of a Heegaard gluing map on $H_1(W; \mathbb{Z})$. We will explain exactly what he proved at the end of §6.4.

Essentially simultaneously and independently of Reidemeister's work, Seifert [43] introduced the concept of a linking form on a 3-manifold whose homology group has a torsion subgroup T, and studied the special case when T has no 2-torsion, obtaining a complete set of invariants for linked abelian groups in this special case. His very new idea was that linking numbers could be defined not just in homology spheres, but also in 3-manifolds whose $\mathbb{Z}$-homology group has torsion. Let W be a closed, oriented 3-manifold and suppose that the torsion subgroup T of $H_1(W;\mathbb{Z})$ is non-trivial. Let a, b be simple closed curves in W which represent elements of T of order α, β respectively. Since $\alpha a, \beta b$ are homologous to zero they bound surfaces $A, B \subset W$. Let $A \cdot b$ denote the algebraic intersection number of A with b, similarly define $B \cdot a$. The *linking number* $\lambda(a, b)$ of a with b is the natural number

$$\lambda(a,b) = \frac{1}{\alpha} A \cdot b = \frac{1}{\beta} a \cdot B.$$

Seifert's invariants are defined in terms of an array of integer determinants associated to the p-primary cyclic summands of T. The invariant depends upon whether each determinant in the array is or is not a quadratic residue mod p^k. His work is, however, restricted to the case when there is no 2-torsion. In the appendix to [43], and also at the end of [40], both Seifert and Reidemeister noted that their invariants are in fact closely related, although neither makes that precise. Both [40] and [43] are, at this writing very well known but it takes some work to pin down the precise relationship so that one can move comfortably between them. See §6.4.

In [7] Burger reduced the problem of classifying linked p-groups ($p \geqslant 2$) to the classification of symmetric bilinear forms over $\mathbb{Z}_{p^n}$. His procedure, together with Minkowski's work on quadratic forms [26] gives a complete set of invariants for the case $p = 2$, but they are inconveniently cumbersome. Our contribution here is to reduce Burger's invariants to a simple and useful set. Most of what we do is probably obtainable from Burger's work together with the work of O'Meara [36]; however, our presentation is unified and part of a systematic study, hence it may be more useful than the two references [7] and [36]. We note that Kawauchi and Kojima [22] *also* studied linked abelian groups with 2-torsion. They obtained a solution of the problem which is similar to ours, however, their goal was different and the intersection between their paper and ours is small.

Invariants of Heegaard splittings, rather than of the manifold itself, were first studied in the context of symplectic Heegaard splittings, in [2]. Later, the work in [2] was further investigated in [27], from a slightly different perspective, with two motivations behind their work. The first is that they thought that linking forms in 3-manifold might give more information than intersection forms on a Heegaard surface, but that is not the case. Second, they thought that, because a finite abelian group can be decomposed as a direct sum of cyclic groups of prime power order, whereas in [2] the decomposition was as a direct sum of a (in general smaller) set of cyclic groups which are not of prime power order, that perhaps there were invariants of unstabilized Heegaard splittings that were missed in [2]. The main result in [27] is that, with one small exception in the case when there is 2-torsion, the Heegaard splitting invariants in [2] cannot be improved.

See [23] for an invariant of Heegaard splittings which is related in an interesting way to our work in this paper. The relationship will be discussed in §8 of this paper.

1.5. Six problems about symplectic Heegaard splittings

In this article we will consider six problems about symplectic Heegaard splittings, giving complete solutions for the first five and a partial solution for the sixth:

Problem 1: Find a complete set of invariants for stabilized symplectic Heegaard splittings.
The full solution is in Corollary 5.16, which asserts the well-known result that a complete set of invariants are the rank of $H_1(W;\mathbb{Z})$, its torsion coefficients, and the complete set of linking invariants.

Problem 2: Knowing the invariants which are given in the solution to Problem 1 above, the next step is to learn how to compute them. Problem 2 asks for a constructive procedure for computing the invariants in Problem 1 for particular $\mathcal{H} \in \Gamma$. The easy part of this, i.e. the computation of invariants which determine $H_1(W;\mathbb{Z})$, is given in Theorem 2.4. The hard part is in the analysis of the linking invariants associated to the torsion subgroup of $H_1(W;\mathbb{Z})$. See §6.2 for the case when p is odd and §6.3 for the case where there is 2-torsion.

Problem 3: Determine whether there is a bound on the stabilization index of a symplectic Heegaard splitting. We will prove that there is a uniform bound, and it is 1. See Corollary 5.22.

Problem 4: Find a complete set of invariants which characterize minimal (unstabilized) symplectic Heegaard splittings and learn how to compute them. In Theorem 7.5 we will prove that the only invariant is a strengthened form of the invariant which was discovered in [2], using very different methods. Example 7.14 shows that we have, indeed, found an invariant which is stronger than the one in [2].

Problem 5: Count the number of equivalence classes of minimal (unstabilized) symplectic Heegaard splittings. The answer is given in Theorem 7.7.

Problem 6: This problem asks for a normal form which allows one to choose a unique representative for the collection of matrices in an unstabilized double coset in Γ_g (mod Λ_g). We were only able to give a partial solution to this problem. In §7.5 we explain the difficulty.

In §8 we go a little bit beyond the main goal of this paper, and consider whether the work in §3 of this paper can be generalized to the higher order terms in the Johnson-Morita filtration. As we shall see, the approach generalizes, but it does not yield anything new.

§2. Symplectic matrices : a partial normalization

Our task in this section is the proof of Theorem 2.4, which gives a partial solution to Problem 2 and tells us how to recognize when a symplectic Heegaard splitting is stabilized.

2.1. Preliminaries

We follow the notation that we set up §1. Let $\tilde{h}$ be the gluing map for a Heegaard splitting of a 3-manifold W. We wish to study the double coset $\Lambda_g h \Lambda_g \subset \Gamma$. For that it will be helpful to learn a little bit more about the subgroup Λ_g. Recall that, by Lemma 1.8, the group Λ_g is the subgroup of elements in Γ_g with a $g \times g$ block of zeros in the upper right corner.

Lemma 2.1. (i) *The group Λ_g is the semi-direct product of its normal subgroup*

$$\Omega = \left\{ \begin{pmatrix} \mathcal{I} & 0 \\ \mathcal{Z} & \mathcal{I} \end{pmatrix} \;\middle|\; \mathcal{Z} \textit{ symmetric} \right\}$$

and its subgroup

$$\Sigma = \left\{ \begin{pmatrix} \mathcal{A} & 0 \\ 0 & \hat{\mathcal{A}}^{-1} \end{pmatrix} \;\middle|\; \mathcal{A} \textit{ unimodular} \right\}.$$

(ii) *Every element in Ω and every element in Σ is induced by a homeomorphism of ∂X_g which extends to a homeomorphism of X_g.*

Proof. (i) Since a general matrix $\left(\begin{smallmatrix} \mathcal{A} & 0 \\ \mathcal{C} & \mathcal{D} \end{smallmatrix}\right) \in \Lambda_g$ is symplectic, it follows from (2) that $\hat{\mathcal{A}}\mathcal{D} = \mathcal{I}$, hence $\hat{\mathcal{A}} = \mathcal{D}^{-1}$, so $\mathcal{A} \in \mathrm{GL}(g, \mathbb{Z})$. Since

$\left(\begin{smallmatrix}\mathcal{A} & 0\\ 0 & \mathcal{A}^{-1}\end{smallmatrix}\right) \in \Lambda_g$, it follows that the most general matrix in Λ_g has the form:

$$\begin{pmatrix}\mathcal{A} & 0\\ \mathcal{C} & \mathcal{A}^{-1}\end{pmatrix} = \begin{pmatrix}\mathcal{A} & 0\\ 0 & \mathcal{A}^{-1}\end{pmatrix}\begin{pmatrix}\mathcal{I} & 0\\ \mathcal{Z} & 0\end{pmatrix} = \begin{pmatrix}\mathcal{I} & 0\\ \hat{\mathcal{A}}^{-1}\mathcal{Z}\mathcal{A}^{-1} & \mathcal{I}\end{pmatrix}\begin{pmatrix}\mathcal{A} & 0\\ 0 & \hat{\mathcal{A}}^{-1}\end{pmatrix},$$

with $\mathcal{Z} = \mathcal{AC}$. But then (by (2) again) $\mathcal{Z}$ must be symmetric. A simple calculation reveals that the conjugate of any element in Ω by an element in Λ is in Ω. The semi-direct product structure follows from the fact that both Σ and Ω embed naturally in Λ, and that they generate Λ.

(ii) The reader is referred to [2] for explicit lifts of generators of Ω and Σ to the mapping class group. Q.E.D.

In several places in this article it will be necessary to pass between the two canonical ways of decomposing a finite abelian group T into cyclic summands. We record here the following well-known theorem:

Theorem 2.2 (The fundamental theorem for finitely generated abelian groups). *Let G be a finitely generated abelian group. Then the following hold:*

(i) *G is a direct sum of r infinite cyclic groups and a finite abelian group T. The group T is a direct sum of t finite cyclic subgroups $T^{(1)} \oplus \cdots \oplus T^{(t)}$, where $T^{(i)}$ has order τ_i. Each τ_i divides $\tau_{i+1}, 1 \leq i \leq t-1$. The integers $r, t, \tau_1, \ldots, \tau_t$ are a complete set of invariants of the isomorphism class of G.*

Let $p_1, \ldots, p_k$ be the prime divisors of τ_t. Then each integer τ_i, $1 \leq i \leq t$ has a decomposition as a product of primes:

(3)
$$\tau_i = p_1^{e_{i,1}} p_2^{e_{i,2}} \cdots p_k^{e_{i,k}}, \quad 0 \leqslant e_{1,d} \leqslant e_{2,d} \leqslant \cdots \leqslant e_{t,d}, \text{ for each } 1 \leq d \leq k.$$

(ii) *T is also a direct sum of p-primary groups $T(p_1) \oplus \cdots \oplus T(p_k)$. Here each $T(p_d)$ decomposes in a unique way as a direct sum of cyclic groups, each of which has order a power of p_d. Focusing on one such prime p_d, $1 \leq d \leq k$, the group $T(p_d)$ is a sum of cyclic groups of orders $p_d^{e_{1,d}}, p_d^{e_{2,d}}, \ldots, p_d^{e_{t,d}}$, where the powers $e_{i,d}$ that occur are not necessarily distinct. That is, we have:*

(4)
$$\begin{aligned} e_{1,d} = e_{2,d} = \cdots = e_{t_1,d} < e_{t_1+1,d} = \cdots \\ = e_{t_2,d} < \cdots < e_{t_r+1,d} = e_{t_r+2,d} = \cdots = e_{t_{r+1},d}. \end{aligned}$$

(iii) *Let y_i be a generator of the cyclic group of order τ_i in* (i) *above. Let $g_{i,d}$ be a generator of the cyclic group of order $p_d^{e_{i,d}}$ in* (ii) *above. Note that there may be more than one group with this* *order. Then the generators $g_{i,d}$ and y_i, where $1 \leq i \leq t$ and $1 \leq d \leq k$ are related by:*

$$g_{i,d} = \left(\frac{\tau_i}{(p_d^{e_{i,d}})}\right) y_i = (p_1^{e_{i,1}} p_2^{e_{i,2}} \cdots p_{d-1}^{e_{i,d-1}} p_{d+1}^{e_{i,d+1}} \cdots p_k^{e_{i,k}}) y_i. \tag{5}$$

The following corollary to statement (i) of Theorem 2.2 allows us to transform a presentation matrix for a finitely generated abelian group into a particularly simple form.

Corollary 2.3 (Smith normal form, see, e.g., [34, Theorem II.9]). *Let $\mathcal{P}$ be any $g \times g$ integer matrix. Then there exist $\mathcal{U}, \mathcal{V} \in \mathrm{GL}(g, \mathbb{Z})$ so that $\mathcal{U}\mathcal{P}\mathcal{V} = \mathrm{Diag}(1, \ldots, 1, \tau_1, \ldots, \tau_t, 0, \ldots, 0)$, where the τ_i are non-negative integers which are different from 1 and satisfying $\tau_i | \tau_{i+1}$ for all $1 \leq i < g$. The diagonal matrix is called the Smith normal form of $\mathcal{P}$. Additionally, the Smith normal form of a matrix is unique, so that in particular the torsion free rank r (the number of zeros in the diagonal) and the torsion rank t are unique. The number of $1'$s is the index of stabilization of the symplectic Heegaard splitting, which can vary.*

2.2. A partial normal form

Theorem 2.4. *Let*

$$\mathcal{H} = \rho^2(\tilde{h}) = \begin{pmatrix} \mathcal{R} & \mathcal{P} \\ \mathcal{S} & \mathcal{Q} \end{pmatrix} \tag{6}$$

be the symplectic matrix associated to a given Heegaard splitting of a 3-manifold W, where $\tilde{h}$ is the Heegaard gluing map. Then:

(i) *The g-dimensional matrix P is a relation matrix for $H = H_1(W; \mathbb{Z})$. This is true, independent of the choice of $\mathcal{H}$ in its double coset modulo Λ_g. Different choices correspond to different choices of basis for H.*

(ii) *The double coset $\Lambda_g \mathcal{H} \Lambda_g$ has a representative:*

$$\mathcal{H}' = \left(\begin{array}{ccc|ccc} 0 & 0 & 0 & \mathcal{I} & 0 & 0 \\ 0 & \mathcal{R}^{(2)} & 0 & 0 & \mathcal{P}^{(2)} & 0 \\ 0 & 0 & \mathcal{I} & 0 & 0 & 0 \\ \hline -\mathcal{I} & 0 & 0 & 0 & 0 & 0 \\ 0 & \mathcal{S}^{(2)} & 0 & 0 & \mathcal{Q}^{(2)} & 0 \\ 0 & 0 & 0 & 0 & 0 & \mathcal{I} \end{array}\right), \tag{7}$$

where $\mathcal{P}^{(2)} = Diag`(\tau_1, \ldots, \tau_t)$ with the τ_i positive integers satisfying $\tau_i|\tau_{i+1}$ for $1 \leq i < t$. In this representation the submatrix

$$\mathcal{H}^{(2)} = \left(\begin{array}{c|c} \mathcal{R}^{(2)} & \mathcal{P}^{(2)} \\ \hline \mathcal{S}^{(2)} & \mathcal{Q}^{(2)} \end{array}\right) \tag{8}$$

is symplectic.

(iii) *The $t \times t$ matrix $\mathcal{P}^{(2)}$ is a relation matrix for the torsion subgroup T of H, which is a direct sum of cyclic groups of orders $\tau_1, \ldots, \tau_t$. The number r of zeros in the lower part of the diagonal of $P^{(1)} = \mathrm{Diag}(1, \ldots, 1, \tau_1, \ldots, \tau_t, 0, \ldots 0)$ is the free rank of H and the number of $1'$s is the index of stabilization of the splitting. In particular, a symplectic Heegaard splitting with defining matrix $\mathcal{H} \in \mathrm{Sp}(2g, \mathbb{Z})$ is unstabilized precisely when the diagonal matrix $\mathcal{P}^{(1)}$ contains no unit entries.*

(iv) *We may further assume that every entry $q_{ij} \in \mathcal{Q}^{(2)}$ and every entry $r_{ij} \in \mathcal{R}^{(2)}$ is constrained as follows. Assume that $i \leq j$. Then:*

$$0 \leq q_{ji} < \tau_j, \quad q_{ij} = (\tau_j/\tau_i)q_{ji}, \quad \text{and} \quad 0 \leq r_{ij} < \tau_i, \quad r_{ji} = (\tau_j/\tau_i)r_{ij}.$$

Proof. Proof of (i). Apply the Mayer-Vietoris sequence to the decomposition of the 3-manifold W that arises through the Heegaard splitting $N_g \cup_{\phi \circ \tilde{h}} \bar{N}_g$.

Proof of (ii). The proof is a fun exercise in manipulating symplectic matrices, but without lots of care the proof will not be very efficient.

In view of Lemma 2.1, the most general element in the double coset of $\mathcal{H} = \left(\begin{smallmatrix} \mathcal{R} & \mathcal{P} \\ \mathcal{S} & \mathcal{Q} \end{smallmatrix}\right)$ has the form

$$\begin{aligned} \mathcal{M} &= \begin{pmatrix} \mathcal{I} & 0 \\ \mathcal{Z}_1 & \mathcal{I} \end{pmatrix} \begin{pmatrix} \mathcal{U} & 0 \\ 0 & \hat{\mathcal{U}}^{-1} \end{pmatrix} \begin{pmatrix} \mathcal{R} & \mathcal{P} \\ \mathcal{S} & \mathcal{Q} \end{pmatrix} \begin{pmatrix} \hat{\mathcal{V}}^{-1} & 0 \\ 0 & \mathcal{V} \end{pmatrix} \begin{pmatrix} \mathcal{I} & 0 \\ \mathcal{Z}_2 & \mathcal{I} \end{pmatrix} \\ &= \begin{pmatrix} * & \mathcal{UPV} \\ * & * \end{pmatrix} \end{aligned} \tag{9}$$

where $\mathcal{U}, \mathcal{V}$ are arbitrary matrices in GL(2,$\mathbb{Z}$).

Choose $\mathcal{U}, \mathcal{V} \in \mathrm{GL}(t, \mathbb{Z})$ so that

$$\begin{aligned} \mathcal{P}^{(1)} &= (I \oplus \mathcal{U} \oplus I)(\mathcal{P})(I \oplus \mathcal{V} \oplus I) \\ &= \mathrm{Diag}(1, 1, \ldots, 1, \tau_1, \tau_2, \ldots, \tau_t, 0, \ldots, 0) \in \mathrm{GL}(g, \mathbb{Z}). \end{aligned}$$

By Corollary 2.3, this is always possible. Let $\mathcal{P}^{(2)} = \mathrm{Diag}(\tau_1, \tau_2, \ldots, \tau_t) \in \mathrm{GL}(t, \mathbb{Z})$. Using (9). we have shown that $\mathcal{H}$ is in the same double coset as

(10)

$$\mathcal{H}^{(1)} = \begin{pmatrix} \mathcal{R}^{(1)} & \mathcal{P}^{(1)} \\ \mathcal{S}^{(1)} & \mathcal{Q}^{(1)} \end{pmatrix} = \left(\begin{array}{ccc|ccc} \mathcal{R}_{11} & \mathcal{R}_{12} & \mathcal{R}_{13} & \mathcal{I} & 0 & 0 \\ \mathcal{R}_{21} & \mathcal{R}_{22} & \mathcal{R}_{23} & 0 & \mathcal{P}^{(2)} & 0 \\ \mathcal{R}_{31} & \mathcal{R}_{32} & \mathcal{R}_{33} & 0 & 0 & 0 \\ \hline \mathcal{S}_{11} & \mathcal{S}_{12} & \mathcal{S}_{13} & \mathcal{Q}_{11} & \mathcal{Q}_{12} & \mathcal{Q}_{13} \\ \mathcal{S}_{21} & \mathcal{S}_{22} & \mathcal{S}_{23} & \mathcal{Q}_{21} & \mathcal{Q}_{22} & \mathcal{Q}_{23} \\ \mathcal{S}_{31} & \mathcal{S}_{32} & \mathcal{S}_{33} & \mathcal{Q}_{31} & \mathcal{Q}_{32} & \mathcal{Q}_{33} \end{array}\right)$$

This is the first step in our partial normal form.

It will be convenient to write $\mathcal{H}^{(1)}$ in several different ways in block form. The first one is the block decomposition in (10). In each of the other cases, given below, the main decomposition is into square $g \times g$ blocks, and these blocks will not be further decomposed (although much later they will be modified):

$$(11) \qquad \mathcal{H}^{(1)} = \left(\begin{array}{cc|cc} \mathcal{A}_{11} & \mathcal{A}_{12} & \mathcal{B}_{11} & 0 \\ \mathcal{A}_{21} & \mathcal{A}_{22} & 0 & 0 \\ \hline \mathcal{C}_{11} & \mathcal{C}_{12} & \mathcal{D}_{11} & \mathcal{D}_{12} \\ \mathcal{C}_{21} & \mathcal{C}_{22} & \mathcal{D}_{21} & \mathcal{D}_{22} \end{array}\right) = \left(\begin{array}{c|c} \mathcal{A} & \mathcal{B} \\ \hline \mathcal{C} & \mathcal{D} \end{array}\right)$$

where

$$\mathcal{A}_{11} = \begin{pmatrix} \mathcal{R}_{11} & \mathcal{R}_{12} \\ \mathcal{R}_{21} & \mathcal{R}^{(2)} \end{pmatrix}, \quad \mathcal{B}_{11} = \begin{pmatrix} \mathcal{I} & 0 \\ 0 & \mathcal{P}^{(2)} \end{pmatrix},$$

$$\mathcal{C}_{11} = \begin{pmatrix} \mathcal{S}_{11} & \mathcal{S}_{12} \\ \mathcal{S}_{21} & \mathcal{S}^{(2)} \end{pmatrix}, \ldots$$

$$\mathcal{A}_{12} = \begin{pmatrix} \mathcal{R}_{13} \\ \mathcal{R}_{23} \end{pmatrix}, \quad \mathcal{A}_{21} = \begin{pmatrix} \mathcal{R}_{31} & \mathcal{R}_{32} \end{pmatrix}, \quad \mathcal{A}_{22} = (\mathcal{R}_{33}), \ldots$$

In general, the blocks $\mathcal{R}_{ij}, \mathcal{A}_{ij}, \ldots$ are *not* square, however $\mathcal{R}^{(2)}, \mathcal{S}^{(2)}, \mathcal{P}^{(2)}, \mathcal{Q}^{(2)}$ are square $t \times t$ matrices.

Now $\mathcal{H}^{(1)} \in \Gamma_g$, hence its $g \times g$ block satisfy the conditions (1) and (2). Working with the decomposition of $\mathcal{H}^{(1)}$ into the block form given in (10), one sees that because of the special form of $\mathcal{B} = \begin{pmatrix} \mathcal{B}_{11} & 0 \\ 0 & 0 \end{pmatrix}$, the $2(g-r) \times 2(g-r)$ matrix $\begin{pmatrix} \mathcal{A}_{11} & \mathcal{B}_{11} \\ \mathcal{C}_{11} & \mathcal{D}_{11} \end{pmatrix}$ *also* satisfies (1) and (2), now with respect to its $(g-r) \times (g-r)$ block. From there it follows (using Definition 1.6) that the matrix given in (10) is in the group $\Gamma_{2(g-r)}$. One may then verify without difficulty that the augmented matrix

$$(12)\qquad \mathcal{M}^{(2)} = \left(\begin{array}{cc|cc} \mathcal{A}_{11} & 0 & \mathcal{B}_{11} & 0 \\ 0 & \mathcal{I} & 0 & 0 \\ \hline \mathcal{C}_{11} & 0 & \mathcal{D}_{11} & 0 \\ 0 & 0 & 0 & \mathcal{I} \end{array}\right)$$

$$= \left(\begin{array}{ccc|ccc} \mathcal{R}_{11} & \mathcal{R}_{12} & 0 & \mathcal{I} & 0 & 0 \\ \mathcal{R}_{21} & \mathcal{R}^{(2)} & 0 & 0 & \mathcal{P}^{(2)} & 0 \\ 0 & 0 & \mathcal{I} & 0 & 0 & 0 \\ \hline \mathcal{S}_{11} & \mathcal{S}_{12} & 0 & \mathcal{Q}_{11} & \mathcal{Q}_{12} & 0 \\ \mathcal{S}_{21} & \mathcal{S}^{(2)} & 0 & \mathcal{Q}_{21} & \mathcal{Q}^{(2)} & 0 \\ 0 & 0 & 0 & 0 & 0 & \mathcal{I} \end{array}\right),$$

which has dimension $2g$ again, *also* satisfies the conditions (1) and (2), now with respect to its $g \times g$ block, and so $\mathcal{H}^{(2)}$ is in Γ_g.

We will need further information about $\mathcal{H}^{(1)}$ and $\mathcal{M}^{(2)}$. Returning to (10), and using the right decomposition of $\mathcal{H}^{(1)}$, we now verify that conditions (1) and (2) imply the following relations between the subblocks:

$$(13)\qquad \mathcal{Q}_{13} = \mathcal{Q}_{23} = \mathcal{R}_{31} = \mathcal{R}_{32} = 0$$

$$(14)\qquad \mathcal{P}^{(2)}\mathcal{Q}_{21} = \hat{\mathcal{Q}}_{12}$$

$$(15)\qquad \mathcal{R}_{12}\mathcal{P}^{(2)} = \hat{\mathcal{R}}_{21}$$

$$(16)\qquad \mathcal{P}^{(2)}\mathcal{Q}^{(2)} \text{ symmetric}$$

$$(17)\qquad \mathcal{R}^{(2)}\mathcal{P}^{(2)} \text{ symmetric}$$

$$(18)\qquad \mathcal{Q}_{11}, \mathcal{R}_{11} \text{ symmetric}$$

Now observe that if $\left(\begin{smallmatrix} \mathcal{A} & \mathcal{B} \\ \mathcal{C} & \mathcal{D} \end{smallmatrix}\right) \in \Gamma_g$ then (1) and (2) imply that

$$(19)\qquad \begin{pmatrix} \mathcal{A} & \mathcal{B} \\ \mathcal{C} & \mathcal{D} \end{pmatrix}^{-1} = \begin{pmatrix} \hat{\mathcal{D}} & -\hat{\mathcal{B}} \\ -\hat{\mathcal{C}} & \hat{\mathcal{A}} \end{pmatrix}.$$

Using equation (19) to compute $(\mathcal{M}^{(2)})^{-1}$, and making use of the conditions in (13)-(18), one may then verify that the product matrix $(\mathcal{M}^{(2)})^{-1}\mathcal{H}^{(1)}$ has a $g \times g$ block of zeros in the upper right corner. But then $(\mathcal{M}^{(2)})^{-1}\mathcal{H}^{(1)} \in \Lambda_g$, hence $\mathcal{M}^{(2)}$ and $\mathcal{H}^{(1)}$ are in the same double coset.

Further normalizations are now possible. Since $\mathcal{R}_{11}$ and $\mathcal{Q}_{11}$ are symmetric (by the symplectic constraints (2)) the following matrices are in the subgroup $\Omega \subset \Lambda_g$ defined in Lemma 2.1:

$$\mathcal{N}_1 = \left(\begin{array}{ccc|ccc} \mathcal{I} & 0 & 0 & 0 & 0 & 0 \\ 0 & \mathcal{I} & 0 & 0 & 0 & 0 \\ 0 & 0 & \mathcal{I} & 0 & 0 & 0 \\ \hline -\mathcal{Q}_{11} & -\hat{\mathcal{Q}}_{21} & 0 & \mathcal{I} & 0 & 0 \\ -\mathcal{Q}_{21} & 0 & 0 & 0 & \mathcal{I} & 0 \\ 0 & 0 & 0 & 0 & 0 & \mathcal{I} \end{array}\right)$$

$$\mathcal{N}_2 = \left(\begin{array}{ccc|ccc} \mathcal{I} & 0 & 0 & 0 & 0 & 0 \\ 0 & \mathcal{I} & 0 & 0 & 0 & 0 \\ 0 & 0 & \mathcal{I} & 0 & 0 & 0 \\ \hline -\mathcal{R}_{11} & -\mathcal{R}_{21} & 0 & \mathcal{I} & 0 & 0 \\ -\hat{\mathcal{R}}_{12} & 0 & 0 & 0 & \mathcal{I} & 0 \\ 0 & 0 & 0 & 0 & 0 & \mathcal{I} \end{array}\right)$$

Computing, we find that

$$\mathcal{N}_1\mathcal{M}^{(2)}\mathcal{N}_2 = \left(\begin{array}{ccc|ccc} 0 & 0 & 0 & \mathcal{I} & 0 & 0 \\ 0 & \mathcal{R}^{(2)} & 0 & 0 & \mathcal{P}^{(2)} & 0 \\ 0 & 0 & \mathcal{I} & 0 & 0 & 0 \\ \hline * & * & 0 & 0 & 0 & 0 \\ * & \mathcal{S}^{(2)} & 0 & 0 & \mathcal{Q}^{(2)} & 0 \\ 0 & 0 & 0 & 0 & 0 & 0 \end{array}\right).$$

Since this matrix is in Γ_g, its entries satisfy the conditions (1) and (2). An easy check shows that the lower left $g \times g$ box necessarily agrees with the entries in the matrix defined in the statement of Theorem 2.4. Thus $\mathcal{H}^{(2)} = \mathcal{N}_1\mathcal{M}^{(2)}\mathcal{N}_2$ is in the same double coset as $\mathcal{M}^{(2)}$, $\mathcal{H}^{(1)}$ and $\mathcal{H}$. This completes the proof of (ii).

Proof of (iii). In (i) we saw that in the partial normal form the matrix $\mathcal{I}_{g-r-t} \oplus \mathcal{P}^{(2)} \oplus 0_r$ is a relation matrix for H. Since H is a finitely generated abelian group, it is a direct sum of t cyclic groups of order $\tau_1, \dots, \tau_t$ and r infinite cyclic groups and $g-r-t$ trivial groups. The $g-r-t$ trivial groups indicate that the symplectic Heegaard splitting has been stabilized $g-r-t$ times. That is, (iii) is true.

Proof of (iv). We consider additional changes in the submatrix $\mathcal{H}^{(2)}$ which leave the $\mathcal{P}^{(2)}$-block unchanged. Note that by Lemma 2.1, any changes in the double coset of $\mathcal{H}^{(2)}$ in Γ_t can be lifted canonically to corresponding changes in the double coset of $\mathcal{H}'$ in Γ_g, and therefore it suffices to consider modifications to the double coset of $\mathcal{H}^{(2)}$ in Γ_t. To simplify notation for the remainder of this proof, we set $\mathcal{H}^{(2)} = \left(\begin{smallmatrix} \mathcal{R} & \mathcal{P} \\ \mathcal{S} & \mathcal{Q} \end{smallmatrix}\right)$

Choose any element $\left(\begin{smallmatrix} \mathcal{I}_t & 0_t \\ \mathcal{Z} & \mathcal{I}_t \end{smallmatrix}\right)$ in the subgroup Ω_t of $\mathrm{Sp}(2t, \mathbb{Z})$. Then:

$$\begin{pmatrix} \mathcal{I}_t & 0_t \\ \mathcal{Z} & \mathcal{I}_t \end{pmatrix} \begin{pmatrix} \mathcal{R} & \mathcal{P} \\ \mathcal{S} & \mathcal{Q} \end{pmatrix} = \begin{pmatrix} \mathcal{R} & \mathcal{P} \\ \star & \mathcal{Q} + \mathcal{Z}\mathcal{P} \end{pmatrix},$$

$$\begin{pmatrix} \mathcal{R} & \mathcal{P} \\ \mathcal{S} & \mathcal{Q} \end{pmatrix} \begin{pmatrix} \mathcal{I}_t & 0_t \\ \mathcal{Z} & \mathcal{I}_t \end{pmatrix} = \begin{pmatrix} \mathcal{R} + \mathcal{P}\mathcal{Z} & \mathcal{P} \\ \star & \mathcal{Q} \end{pmatrix}$$

Let $\mathcal{Q} = (q_{ij})$, $\mathcal{R} = (r_{ij}), \mathcal{Z} = (z_{ij})$. Then $\mathcal{Z}\mathcal{P} = (\tau_j z_{ij})$ and $\mathcal{P}\mathcal{Z} = (\tau_i z_{ij})$. Therefore we may perform multiplications as above so that if $i \leq j$ then $0 \leq q_{ji} < \tau_j$ and $0 \leq r_{ij} < \tau_i$. The fact that $\mathcal{H}^{(2)}$ is symplectic shows that $\mathcal{P}\mathcal{Q}$ and $\mathcal{R}\mathcal{P}$ are symmetric. Therefore $q_{ij} = (\tau_j/\tau_i)q_{ji}$, $r_{ji} = (\tau_j/\tau_i)r_{ij}$. Thus the matrices $\mathcal{Q}$ and $\mathcal{R}$ are completely determined once we fix the entries q_{ji} and r_{ij} which satisfy $i \leq j$. This completes the proof of (iv), and so of Theorem 2.4. Q.E.D.

Remark 2.5. As noted in §1.2, we made two choices when we defined equivalence of Heegaard splittings: the choice of one of the two handlebodies as the preferred one, and the choice of a preferred orientation on the 3-manifold W. When we allow for all possible choices, we see that the symplectic matrix $\mathcal{H}$ of Theorem 2.4 is replaced by 4 possible symplectic matrices, related by the operations of taking the transpose and the inverse and the inverse of the transpose:

$$\begin{pmatrix} \mathcal{R}^{(2)} & \mathcal{P}^{(2)} \\ \mathcal{S}^{(2)} & \mathcal{Q}^{(2)} \end{pmatrix}, \begin{pmatrix} \hat{\mathcal{R}}^{(2)} & \hat{\mathcal{S}}^{(2)} \\ \hat{\mathcal{P}}^{(2)} & \hat{\mathcal{Q}}^{(2)} \end{pmatrix}, \begin{pmatrix} \hat{\mathcal{Q}}^{(2)} & -\hat{\mathcal{P}}^{(2)} \\ -\hat{\mathcal{S}}^{(2)} & \hat{\mathcal{R}}^{(2)} \end{pmatrix}, \begin{pmatrix} \mathcal{Q}^{(2)} & -\mathcal{S}^{(2)} \\ -\mathcal{P}^{(2)} & \mathcal{R}^{(2)} \end{pmatrix}.$$

Any one of the four could equally well have been chosen as a representative of the Heegaard splitting. These four matrices may or may not be in the same double coset. ||

2.3. Uniqueness questions

There is a source of non-uniqueness in the partial normal form of Theorem 2.4. It lies in the fact that further normalizations are possible after those in (iv) of Theorem 2.4, but they are difficult to understand. By Lemma 2.1, we know that Λ_t is the semi-direct product of the normal subgroup Ω_t and the subgroup Σ_t that were defined there. We already determined how left and right multiplication by elements in Ω_t change $\mathcal{H}'$ in the proof of part (iv) of Theorem 2.4. We now investigate further changes, using left (resp right) multiplication by matrices in Σ_t.

Lemma 2.6. *Assume that* $\mathcal{P}^{(2)} = \mathrm{Diag}(\tau_1, \ldots, \tau_t)$ *is fixed, and that* $\mathcal{P}^{(2)}\mathcal{Q}^{(2)} = \hat{\mathcal{Q}}^{(2)}\mathcal{P}^{(2)}$. *Then there is a well-defined subgroup* G *of* $\Sigma_t \times \Sigma_t$, *determined by the condition that there exist matrices $\mathcal{U}, \mathcal{V} \in \mathrm{GL}(t, \mathbb{Z})$ such that $\mathcal{U}\mathcal{P}^{(2)} = \mathcal{P}^{(2)}\mathcal{V}$. Equivalently, there exist symplectic matrices $U, V \in \Sigma_t$ such that $(U,V) \in G \Longleftrightarrow$*

$$(20)\quad U\begin{pmatrix}\mathcal{R}^{(2)} & \mathcal{P}^{(2)}\\ \mathcal{S}^{(2)} & \mathcal{Q}^{(2)}\end{pmatrix}V = \begin{pmatrix}\mathcal{U} & 0\\ 0 & \hat{\mathcal{U}}^{-1}\end{pmatrix}\begin{pmatrix}\mathcal{R}^{(2)} & \mathcal{P}^{(2)}\\ \mathcal{S}^{(2)} & \mathcal{Q}^{(2)}\end{pmatrix}\begin{pmatrix}\mathcal{V} & 0\\ 0 & \hat{\mathcal{V}}^{-1}\end{pmatrix}$$
$$= \begin{pmatrix}\star & \mathcal{P}^{(2)}\\ \star & \hat{\mathcal{U}}^{-1}\mathcal{Q}^{(2)}\hat{\mathcal{V}}^{-1}\end{pmatrix}.$$

For later use, we also have that if $(\mathcal{P}^{(2)})^{-1}$ is the diagonal matrix whose i^{th} entry is the rational number $1/\tau_i$, then $\mathcal{Q}^{(2)}(\mathcal{P}^{(2)})^{-1}$ will be replaced by $\hat{\mathcal{U}}^{-1}(\mathcal{Q}^{(2)}(\mathcal{P}^{(2)})^{-1})\mathcal{U}^{-1}$.

Proof. The statement in (20) is a simple calculation. We need to prove that it determines a group. Suppose that (U_1, V_1), $(U_2, V_2) \in G$. Then $\mathcal{U}_i\mathcal{P}^{(2)} = \mathcal{P}^{(2)}\mathcal{V}_i$ for $i = 1, 2$, so $\mathcal{U}_1\mathcal{U}_2\mathcal{P}^{(2)} = \mathcal{U}_1\mathcal{P}^{(2)}\hat{\mathcal{V}}_2 = \mathcal{P}^{(2)}\hat{\mathcal{V}}_1\hat{\mathcal{V}}_2$. Therefore $(U_1U_2, V_1V_2) \in G$. Also, $(\mathcal{P}^{(2)})^{-1}\mathcal{U}_1^{-1} = \hat{\mathcal{V}_1}^{-1}(\mathcal{P}^{(2)})^{-1}$, which implies that $(U_1^{-1}, V_1^{-1}) \in G$, so G is a group. It is immediate that $\mathcal{P}^{(2)}$ remains unchanged and that $\mathcal{Q}^{(2)}(\mathcal{P}^{(2)})^{-1}$ changes in the stated way. Q.E.D.

Remark 2.7. The condition $\mathcal{U}\mathcal{P}^{(2)} = \mathcal{P}^{(2)}\hat{\mathcal{V}}$ means that $\mathcal{U}$ is restricted to $t \times t$ unimodular matrices which satisfy the condition: $\mathcal{U} = (u_{ij})$, where u_{ji} is divisible by τ_j/τ_i whenever $j < i$. There are no restrictions on u_{ij} when $j \geq i$ other than that the determinant $|u_{ij}| = \pm 1$. ||

Remark 2.8. We were unable to find a nice way to choose U and V so as to obtain a unique representative of the double coset of a symplectic Heegaard splitting, in the case when H is not torsion-free. The reason will become clear in §6: invariants of the matrix $\mathcal{Q}^{(2)}(\mathcal{P}^{(2)})^{-1}$, and so also a normal form, depend crucially on whether or not there is 2-torsion in the torsion subgroup T of H, and so a general rule cannot be easily stated. See also the discussion in §7.5. ||

§3. Presentation theory for finitely generated abelian groups

We are ready to begin the main work in this article. In Section 1 we described the topological motivation that underlies the work in this paper, namely we were interested in understanding all topological invariants of a 3-manifold W and of its Heegaard splittings that might arise through symplectic Heegaard splittings. In Theorem 2.4 we saw that the matrix associated to a symplectic Heegaard splitting gives a natural presentation of $H_1(W;\mathbb{Z})$. Therefore it is natural to begin our work by investigating the theory of presentations of abelian groups. Our goal is to understand, fully, that part of the obstruction to stable equivalence coming from a symplectic Heegaard splitting.

We begin by introducing the two concepts of isomorphism and equivalence of presentations. The rank of a presentation is defined and a concept of stabilizing a presentation (thereby increasing the rank) is introduced. Most of this is aimed at Theorem 3.15, which gives necessary and sufficient conditions for an automorphism of H to lift to an automorphism of the free group of a presentation. Theorem 3.15 implies Corollaries 3.16, 3.17 and 3.18, which assert (in various ways) that any two presentations of the same finitely generated abelian group which are of non-minimal equal rank are equivalent. From this it follows that at most a single stabilization is required to remove any obstruction to equivalence between two presentations. This does not solve Problem 3, but it is a first step in the direction of this problem's solution. We remark that, by contrast, the usual proof of Tietze's Theorem on equivalence of two particular presentations of an arbitrary finitely generated but in general non-abelian group shows that presentations of rank r, r' become equivalent after stabilizations of index r', r, respectively [25].

The latter half of the section focuses on presentations of minimal rank of a finitely generated abelian group. An "orientation" and a "volume" on H are defined. The determinant of an endomorphism h of abelian groups, and hence of a presentation π of an abelian group, is introduced. The key result is Theorem 3.26, which gives necessary and sufficient conditions for an isomorphism between two *minimal* presentations to lift to the presentation level. Corollary 3.27 follows: two minimal presentations of H are equivalent if and only if they have the same volume on H. The section closes with two examples which illustrate the application of Theorem 3.26 and Corollary 3.27 to explicit group presentations.

In §7, we will apply the notion of "volume" to obtain invariants of Heegaard splittings. It turns out that associated to a symplectic Heegaard splitting is a natural presentation of the first homology group, and thus an induced volume. We will use the interplay between this volume and a linking form on the first homology group to find invariants of Heegaard splittings.

3.1. Equivalence classes of not-necessarily minimal presentations

We begin our work with several definitions which may seem unnecessary and even pedantic; however, extra care now will help to make what follows later seem natural and appropriate.

Definition 3.1. A *free pair* is a pair of groups (F, R) with $R \subset F$ and F free abelian and finitely generated; its *quotient* is F/R. If H is a finitely generated abelian group, a *presentation* of H is a surjection $\pi : F \to H$, with F again a finitely generated free abelian group. The *rank* of a free pair and of a presentation is the rank of F. Direct sums of these objects are defined in the obvious way. The *index k stabilization of a free pair* (F, R) is the free pair $(F \oplus \mathbb{Z}^k, R \oplus \mathbb{Z}^k)$, and the *index k stabilization of a presentation* $\pi : F \to H$ is the presentation $\pi \oplus 0 : F \oplus \mathbb{Z}^k \to H$. ||

Definition 3.2. An *isomorphism of free pairs* $(F, R), (F', R')$ is an isomorphism $f : F \to F'$ such that $f(R) = R'$. An *isomorphism of presentations* is a commutative diagram

$$\begin{array}{ccc} F & \xrightarrow{\pi} & H \\ {\scriptstyle f}\downarrow & & \downarrow{\scriptstyle h} \\ F' & \xrightarrow{\pi'} & H' \end{array}$$

with f, h isomorphisms. If $H = H'$, then we have the stronger notion of an *equivalence of presentations*, which is a commutative diagram

$$\begin{array}{ccc} F & & \\ & \searrow^{\pi} & \\ {\scriptstyle f}\downarrow & & H. \\ & \nearrow_{\pi'} & \\ F' & & \end{array}$$

with f an isomorphism. Two free pairs (resp. two presentations) are *stably isomorphic* (resp. *stably equivalent*) if they have isomorphic (resp. equivalent) stabilizations. If $\pi : F \to H$, $\pi' : F' \to H$ are both of minimal rank and stably equivalent, then we define the *stabilization index* of π, π' to be the smallest index k such that π, π' have equivalent stabilizations of index k. ||

Example 3.3. To see that equivalence and isomorphism of presentations are distinct concepts, let $F = \mathbb{Z}$ and let $H = \mathbb{Z}_5$ with $\pi : \mathbb{Z} \to \mathbb{Z}_5$ defined by $\pi(1) = 1$ and $\pi' : \mathbb{Z} \to \mathbb{Z}_5$ defined by $\pi'(1) = 2$. Then π and

π' are isomorphic because the automorphism $h : \mathbb{Z}_5 \to \mathbb{Z}_5$ defined by $h(1) = 2$ lifts to the identity automorphism of $\mathbb{Z}$. However, it is easy to see that π and π' are not equivalent. ||

The standard 'elementary divisor theorem' concerning presentations of abelian groups may be phrased as follows:

Proposition 3.4. *Two free pairs are isomorphic if and only if they have the same rank and isomorphic quotients. For any pair (F, R) there is a basis f_i of F and integers m_i so that $\{m_i f_i \mid m_i \neq 0\}$ is a basis for R. The f_i and m_i may be chosen so that $m_i | m_{i+1}$ for all i.*

Since we can always stabilize two pairs to the same rank, two pairs are stably isomorphic if and only if their quotients are isomorphic.

We now investigate equivalence classes of presentations of a finitely generated abelian group H.

Definition 3.5. If H is a finitely generated abelian group, its *rank* is, equivalently,

- the minimal number of infinite and finite cyclic direct summands required to construct H
- the minimal rank of a presentation of H
- the number of torsion coefficients of the torsion subgroup T of H plus the rank of H/T.

A presentation of minimal rank is simply called a minimal presentation. ||

Lemma 3.6. *Every non-minimal presentation $F \xrightarrow{\pi} H$ is equivalent to a presentation of the form $F' \oplus \mathbb{Z} \xrightarrow{\pi' \oplus 0} H$, where π' is a presentation of H. Every presentation of H is a stabilization of a minimal one.*

Proof. Clearly, by stabilizing, H has a presentation of every rank $\geqslant$ rank H. Let $F' \xrightarrow{j} H$ be a presentation of rank (rank $F - 1$). By Proposition 3.4, the stabilization of j is isomorphic to $F \xrightarrow{\pi} H$, say by a diagram of the form

$$\begin{array}{ccc} F & \xrightarrow{\pi} & H \\ {\scriptstyle f}\downarrow & & \downarrow{\scriptstyle h} \\ F' \oplus \mathbb{Z} & \xrightarrow{j \oplus 0} & H \end{array}$$

Hence

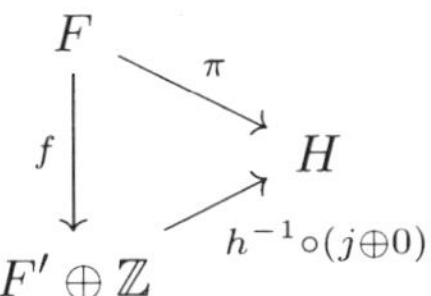

is an equivalence, as desired. By induction, we conclude that every presentation of H is a stabilization of a minimal one. Q.E.D.

In the next few lemmas, we will show that all presentations of H are stably equivalent and that the index of stabilization required is at most one.

Lemma 3.7. *Let n be the rank of* rank (H/T) *and let* $T = \operatorname{Tor}(H)$. *Then any presentation of H is equivalent to one of the form $F \oplus \mathbb{Z}^n \xrightarrow{\pi} H$, where $\pi|_F$ is a presentation of T and $\pi|_{\mathbb{Z}^n}$ is injective.*

Proof. A presentation of the required type certainly exists and may be of any rank $\geqslant$ rank H. The proof that any presentation is equivalent to one of this form is similar to the proof of Lemma 3.6. Q.E.D.

Lemma 3.8. *Let h be an automorphism of H which acts trivially on T. Then for any presentation $\pi : F \to H$ there is an automorphism f of F so that*

$$\begin{array}{ccc} F & \xrightarrow{\pi} & H \\ {\scriptstyle f}\downarrow & & \downarrow{\scriptstyle h} \\ F & \xrightarrow{\pi} & H \end{array}$$

commutes.

Proof. By Lemma 3.7, our presentation is equivalent to the direct sum of presentations $F_0 \to T$ and $\mathbb{Z}^n \xrightarrow{id} \mathbb{Z}^n$, where H has been decomposed as $T \oplus \mathbb{Z}^n$. Representing elements of H by column vectors $\left(\begin{smallmatrix} t \\ z \end{smallmatrix}\right)$ with $t \in T$, $z \in \mathbb{Z}^n$, any automorphism h of H must be of the form

$$\left(\begin{smallmatrix} t \\ z \end{smallmatrix}\right) \mapsto \left(\begin{smallmatrix} A & B \\ 0 & C \end{smallmatrix}\right) \left(\begin{smallmatrix} t \\ z \end{smallmatrix}\right),$$

where $A : T \to T$ and $C : \mathbb{Z}^n \to \mathbb{Z}^n$ are automorphisms and $B : \mathbb{Z}^n \to T$ is a homomorphism; by hypothesis, $A = 1$. If we lift B to a homomorphism $\bar{B} : \mathbb{Z}^n \to F_0$, then the endomorphism $\left(\begin{smallmatrix} 1 & \bar{B} \\ 0 & C \end{smallmatrix}\right)$ of $F_0 \oplus \mathbb{Z}^n$ is an automorphism which clearly induces $\left(\begin{smallmatrix} 1 & B \\ 0 & C \end{smallmatrix}\right)$ on $T \oplus \mathbb{Z}^n = H$, as desired. Q.E.D.

Definition 3.9. In the situation of Lemma 3.8, we say that f *lifts* h. ||

Corollary 3.10. *Using the notation of the proof of Lemma 3.8, if h is any automorphism of H so that $h|_T$ lifts to F_0, then h lifts to F.*

Proof. Let $g = (h|_T) \oplus 1_{\mathbb{Z}^n}$. Clearly g lifts to $F_0 \oplus \mathbb{Z}^n$. Hence $h \circ g^{-1} = 1$ on T, so hg^{-1} also lifts. We conclude $(hg^{-1}) \circ g$ lifts, as desired. Q.E.D.

Definition 3.11. Let $f : F \to F$ be an endomorphism. Since F is free abelian, we may represent f by a matrix with respect to any basis for F. We define the *determinant* of F to be the determinant of any such matrix. Clearly $\det f$ is well-defined, up to sign, independent of the choice of basis. ||

Lemma 3.12. *Let T be an abelian p-group for some prime p. Let $\pi : F \to T$ be a presentation and let $h : T \to T$ be an automorphism. Then there is an endomorphism (which we are not claiming is an automorphism) $f : F \to F$ lifting h so that p does not divide* $\det f$.

Proof. Since F is a *free* abelian group, it is easy to construct an endomorphism f of F that lifts h, so the key point is to construct one so that p does not divide $\det(f)$.

By Lemma 3.6, the presentation π is equivalent to $\pi_0 \oplus 0^k : F_0 \oplus \mathbb{Z}^k \to T$ with π_0 minimal (here possibly $k = 0$). Choose $f_0 \in \operatorname{End} F_0$ so that $\pi_0 f_0 = h\pi_0$, and let $f = f_0 \oplus 1_{\mathbb{Z}^k}$. By construction f lifts h, and we claim that p does not divide $\det f$.

Consider the canonical map $T \to T/pT$ and the composite $\pi_p : F_0 \to T \to T/pT$. Since T is an abelian p-group and since π_0 is minimal, we have rank T/pT = rank T = rank F_0. Hence $\ker \pi_p = pF_0$. Now h induces an automorphism h_p of T/pT and we have

$$\begin{array}{ccc} F_0/pF_0 & \xrightarrow{\cong} & T/pT \\ {\scriptstyle f_p}\downarrow & & {\scriptstyle h_p}\downarrow \\ F_0/pF_0 & \xrightarrow{\cong} & T/pT \end{array}$$

Hence f_p must be an isomorphism. This implies that p does not divide $\det f = \det f_0$, for p divides $\det f_0$ if and only if the induced map on F_0/pF_0 is *not* an isomorphism. Q.E.D.

Lemma 3.13. *Let T be any finite abelian group, $\pi : F \to T$ be a presentation of T, and $h : T \to T$ be any automorphism. Then there is an endomorphism (which we are again not claiming is an automorphism) $f : F \to F$ lifting h so that* $(\det f, |T|) = 1$.

Proof. Let p be a prime divisor of $|T|$. Suppose that p^k is the highest power of p which divides $|T|$. Then $T_p := T/p^kT$ is an abelian p-group isomorphic to the p-component of T, and h induces an automorphism h_p of T_p. By the previous lemma, there is an endomorphism f_p of F with $p \nmid \det f_p$ so that

$$\begin{array}{ccc} F & \xrightarrow{\pi_p} & T_p \\ {\scriptstyle f_p}\downarrow & & \downarrow{\scriptstyle h_p} \\ F & \xrightarrow{\pi_p} & T_p \end{array}$$

commutes. Fixing a basis of F and representing f_p as a matrix, we note that any matrix congruent to f_p mod p^k also induces h_p on T_p. By the Chinese remainder theorem, there is a single matrix f so that $f \equiv f_p \bmod p^k$ for all primes p which divide $|T|$, that is to say, a single endomorphism f of F inducing h_p on T_p for each such prime. Since $\det f \equiv \det f_p \bmod p^k$, we have $p \nmid \det f$ for all such p, *i.e.* $(\det f, |T|) = 1$.

It remains to prove that f induces h on T. Since $T_p = T/p^kT$ is a p-group isomorphic to the p-component of T, it follows that the kernel of $\pi : F \to T$ is precisely

$$\ker \pi = \bigcap_{p \text{ divides } |T|} \ker(\pi_p : F \to T_p).$$

Hence

$$f(\ker \pi) = f\left(\bigcap_p \ker \pi_p\right) \subset \bigcap_p f(\ker \pi_p) = \bigcap_p \ker \pi_p = \ker \pi.$$

This shows that f induces *some* automorphism of T. But this automorphism induces h_p on T_p for every p, so it must be h. Q.E.D.

Lemma 3.14. *Let T be a torsion group and let $h \in \mathrm{Aut}\,(T)$. Then for any non-minimal presentation $\pi : F \to T$ there is some $f \in \mathrm{Aut}\,(F)$ which lifts h.*

Proof. Since π is non-minimal, by Lemma 3.6 we may assume that $\pi = \pi' \oplus 0 : F' \oplus \mathbb{Z} \to T$. By Lemma 3.13, we may lift h to $f \in \text{End } F'$ with $(\det f, m) = 1$, where $m = |T|$. Choose an integer δ such that $\delta \cdot \det f \equiv 1 \bmod m$. Let d denote the endomorphism of $\mathbb{Z}$ defined by $d(1) = \delta$. Then $f \oplus \delta \in \text{End } (F' \oplus \mathbb{Z})$ and $\det(f \oplus \delta) \equiv 1 \bmod m$. Since the canonical homomorphism $\text{SL}(r, \mathbb{Z}) \to \text{SL}(r, \mathbb{Z}_m)$ is surjective, we may lift $f \oplus \delta$ to $f' \in \text{Aut } (F' \oplus \mathbb{Z})$. More precisely, we can find some $f' \in \text{Aut}(F' \oplus \mathbb{Z})$ which (when considered as a matrix over $\mathbb{Z}$) is equal to $f \oplus \delta \bmod m$. Since $f \oplus \delta$ lifts h, the diagram

$$\begin{array}{ccc} F' \oplus \mathbb{Z} & \xrightarrow{\pi \oplus 0} & T \\ {\scriptstyle f\oplus\delta}\downarrow & & \downarrow{\scriptstyle h} \\ F' \oplus \mathbb{Z} & \xrightarrow{\pi \oplus 0} & T \end{array}$$

commutes. Now $m(F' \oplus \mathbb{Z}) \subset \ker(\pi \oplus 0)$, since $mx = 0$ for all $x \in T$. Note that by construction, $f' \equiv f \oplus \delta \bmod m$, *i.e.* for each $x \in F' \oplus \mathbb{Z}$ there is a $y \in F' \oplus \mathbb{Z}$ such that $f'(x) = (f \oplus \delta)(x) + my$. Hence f' also lifts h. Q.E.D.

Theorem 3.15. *If h is any automorphism of H and if $\pi : F \to H$ is any non-minimal presentation of H, then h lifts to F.*

Proof. By Lemma 3.7, we may decompose F as $F = F_0 \oplus \mathbb{Z}^n$, where $\pi|_{F_0} : F_0 \to T$ is a presentation and $\pi|_{\mathbb{Z}^n} = 1 : \mathbb{Z}^n \to \mathbb{Z}^n$. The non-minimality of F then implies the non-minimality of F_0. Hence by Lemma 3.14, we may lift $h|_T$ to F_0. Corollary 3.10 then implies that h lifts to F, as desired. Q.E.D.

The following three results will be important later. They are immediate consequences of Theorem 3.15.

Corollary 3.16. *If $h : H \to H'$ is an isomorphism and $\pi : F \to H$, $\pi' : F' \to H'$ are non-minimal of equal rank, then h lifts to a presentation isomorphism.*

Corollary 3.17. *All presentations of H are stably equivalent, and any of two presentations of non-minimal, equal rank are equivalent.*

Corollary 3.18. *If $\pi : F \to H$, $\pi' : F \to H$ are two minimal presentations, then π, π' have stabilization index* 0 *or* 1.

Example 3.19. To illustrate Corollary 3.18, recall Example 3.3. Two rank 1 presentations π, π' of $H = \mathbb{Z}_5$ were defined by $\pi(1) = 1$ and $\pi'(1) = 2$. These are obviously inequivalent. We claim that they have equivalent index 1 stabilizations $\pi \oplus 0, \pi' \oplus 0$, *i.e.* there exists some f so that the following diagram commutes:

$$\begin{array}{ccc} \mathbb{Z} \oplus \mathbb{Z} & & \\ & \searrow^{\pi \oplus 0} & \\ f \downarrow & & H. \\ & \nearrow_{\pi' \oplus 0} & \\ \mathbb{Z} \oplus \mathbb{Z} & & \end{array}$$

For example, we may define f by

$$\begin{pmatrix} z_1 \\ z_2 \end{pmatrix} \mapsto \begin{pmatrix} 3 & 5 \\ 1 & 2 \end{pmatrix} \begin{pmatrix} z_1 \\ z_2 \end{pmatrix} = \begin{pmatrix} 3z_1 + 5z_2 \\ z_1 + 2z_2 \end{pmatrix}$$

We then have

$$(\pi' \oplus 0) \circ f(z_1, z_2) = 6z_1 + 10z_2 \equiv z_1 \pmod 5 = (\pi \oplus 0)(z_1, z_2). \quad \|$$

3.2. Equivalence classes of minimal presentations

We continue our study of presentations of finitely generated abelian groups by investigating equivalence classes of *minimal* presentations of finitely generated abelian groups. The main results are Theorem 3.26 and Corollary 3.27, which give a complete invariant of equivalence of minimal presentations of H.

First we recall the definition of the exterior powers of an abelian group H. From the kth tensor power $H^k = H \otimes \cdots \otimes H$ we form a quotient by dividing out by the subgroup generated by all $x_1 \otimes \cdots \otimes x_k$ in which two x_i's are equal. This quotient is the kth *exterior power of* H, denoted by $\Lambda^k H$. The image of an arbitrary tensor product $x_1 \otimes \cdots \otimes x_k$ in $\Lambda^k H$ is denoted by $x_1 \wedge \cdots \wedge x_k$, and we have the usual law

$$x_1 \wedge \cdots x_i \wedge x_{i+1} \cdots \wedge x_k = -x_1 \wedge \cdots x_{i+1} \wedge x_i \cdots \wedge x_k.$$

Also as usual, $x_1 \wedge \cdots \wedge x_k = 0$ if any x_i is a linear combination of the other terms, which implies that $\Lambda^k H = 0$ if $k >$ rank H.

Lemma 3.20. *Let $r = \text{rank } H$ and τ be the smallest elementary divisor of the torsion subgroup T of H. If $T = 0$, we put $\tau = 0$. Then $\Lambda^r H$ is cyclic of order τ if $T \neq 0$, whereas if $T = 0$, then $\Lambda^r H$ is infinite cyclic. If $x_1, \ldots, x_r$ generate H, then $x_1 \wedge \cdots \wedge x_r$ generates $\Lambda^r H$.*

Proof. The case when $T = 0$ is well known, so we assume that $\tau > 0$. We prove the last statement first. Now, $\Lambda^r H$ is generated by all $y_1 \wedge \cdots \wedge y_r$ as the y_i range over H. But let $y_i = \sum_{j=1}^r \alpha_{ij} x_j$

for integers α_{ij}. A straightforward check shows that $y_1 \wedge \cdots \wedge y_r = \det(\alpha_{ij})(x_1 \wedge \cdots \wedge x_r)$.

Thus $\Lambda^r H$ is cyclic. Elementary divisor theory tells us that H is the direct sum of r cyclic groups $\mathbb{Z}_{\tau_i}$, where the τ_i are the elementary divisors ($\mathbb{Z}_0$ means $\mathbb{Z}$ here). If these cyclic summands have generators x_i, with x_1, say, of order $\tau = \tau_1$, then $\theta = x_1 \wedge \cdots \wedge x_r$ generates $\Lambda^r H$, and $\tau\theta = (\tau x_1) \wedge x_2 \wedge \cdots \wedge x_r = 0$ since $\tau x_1 = 0$. Thus $|\Lambda^r H| \leqslant \tau$.

We must therefore prove that $|\Lambda^r H| \geqslant \tau$. Consider the map $d : H^r \to \mathbb{Z}_\tau$ defined by

$$d(y_1 \otimes \cdots \otimes y_r) = \det(\alpha_{ij}) \bmod \tau,$$

where $y_i = \sum_j \alpha_{ij} x_j$. This is well defined, since if for some i we had $\sum_j \alpha_{ij} x_j = 0$ in H, then we must have $\alpha_{ij} \equiv 0 \bmod \tau$ for all j, and hence $\det(\alpha_{ij}) \equiv 0 \bmod \tau$. The map d is also clearly onto (let $y_i = x_i$). Finally, d kills all terms having two y_i's equal, so it induces a map of $\Lambda^r H$ onto $\mathbb{Z}_\tau$. Q.E.D.

Definition 3.21. If H has rank r, an *orientation* of H is a selection of a generator θ of $\Lambda^r H$. A *volume* of H is a pair $\pm\theta$ of orientations of H. *i.e.* an orientation of H, determined up to sign. Observe that a free abelian group of rank r has $\Lambda^r \simeq \mathbb{Z}$ and hence two orientations and only one volume, but if H has torsion, then it will in general have many volumes. ||

If $f : H \to H'$ is a homomorphism between groups of the same rank r, then f induces a homomorphism $\Lambda^r f : \Lambda^r H \to \Lambda^r H'$ in the standard way; we will write simply f for $\Lambda^r f$.

Lemma 3.22. *Assume that H and H' both have rank r and that $f : H \to H'$ is surjective. Let τ and τ' be the smallest elementary divisors of H and H', respectively. Then $\tau' \mid \tau$ and if θ is any orientation of H, then $f(\theta)$ is an orientation of H'.*

Proof. Let $x_1, \ldots, x_r$ generate H, so $\varphi = x_1 \wedge \cdots \wedge x_r$ generates $\Lambda^r H$. There is thus some generator m of $\mathbb{Z}_\tau$ so that $\theta = m\varphi$. But since f is onto, H' is generated by $f(x_1), \ldots, f(x_r)$ and $\varphi' = f(x_1) \wedge \cdots \wedge f(x_r) = f(\varphi)$ generates $\Lambda^r H'$. This shows that $f : \Lambda^r H \to \Lambda^r H'$ is also surjective and hence that $\tau' \mid \tau$. We conclude that m is also a generator of $\mathbb{Z}_{\tau'}$, and hence that $\theta' = f(\theta) = mf(\varphi) = m\varphi'$ generates $\Lambda^r H'$. Q.E.D.

If H, H' have specific orientations θ, θ' and $f : H \to H'$ is a homomorphism, then since θ' generates $\Lambda^r H'$ we have $f(\theta) = m\theta'$ for a unique $m \in \mathbb{Z}_{\tau'}$. We call m the *determinant* of f (with respect to the orientations θ, θ') and write $f(\theta) = \det f \cdot \theta'$. If H, H' have only *volumes* specified, then $\det f$ is determined up to sign. If, however, $H = H'$ and $\theta = \theta'$, then $\det f$ is independent of θ; in fact, $f : \Lambda^r H \to \Lambda^r H'$ is just multiplication by $\det f \in \mathbb{Z}_\tau$. Thus endomorphisms of H have a well-defined determinant, and it is easy to see that this definition is the classical one when H is free. More generally, we have:

Lemma 3.23. *Suppose*

$$\begin{array}{ccc} F & \xrightarrow{\pi} & H \\ {\scriptstyle f}\downarrow & & \downarrow{\scriptstyle h} \\ F' & \xrightarrow{\pi'} & H' \end{array}$$

commutes, where π, π' are presentations and all groups have the same rank. Then if φ, φ' are orientations of F, F' inducing orientations θ, θ' of H, H', we have $\det h \equiv \det f \bmod \tau'$. *If no orientations are specified, then we measure* $\det h$ *with respect to the canonical induced volumes, and the above congruence holds up to sign.*

Proof. Observe that $\det f \cdot \varphi' = f(\varphi)$ and that

$$\det h \cdot \theta' = h(\theta) = h\pi(\varphi) = \pi'(\det f \cdot \varphi') \equiv_{\mathrm{mod}\ \tau'} \det f \cdot \pi'(\varphi') = \det f \cdot \theta'.$$

The final statement is obvious. Q.E.D.

Note that if $H = H'$, $F = F'$ and $\pi = \pi'$, then $\det f$, $\det h$ and the congruence are independent of the orientations.

The following two lemmas show that det behaves like the classical determinant.

Lemma 3.24. *If $f : (H_1, \theta_1) \to (H_2, \theta_2)$ and $g : (H_2, \theta_2) \to (H_3, \theta_3)$ are homomorphisms of oriented groups so that all the H_i have the same rank, then if τ_3 is the smallest elementary divisor of H_3 we have* $\det(gf) \equiv (\det g) \cdot (\det f) \bmod \tau_3$.

Proof. We calculate:

$$\det(gf) \cdot \theta_3 = gf(\theta_1) = g(\det f \cdot \theta_2) = \det f \cdot g(\theta_2) = \det f \cdot \det g \cdot \theta_3.$$

Q.E.D.

Lemma 3.25. *Let F and G be abelian groups with F free, and let h be an endomorphism of $F \oplus G$ so that $h(G) < G$. Let $g = h|_G$ and let f be the map on $F = \frac{F \oplus G}{G}$ induced by h. Then* $\det h = \det f \cdot \det g \bmod \tau$, *where τ is the smallest elementary divisor of G (and hence of $F \oplus G$). In particular, if h is an automorphism, then* $\det h = \pm \det g$.

Proof. Let $m = \operatorname{rank} F$ and $n = \operatorname{rank} G$; then $m + n = \operatorname{rank} F \oplus G$ holds because F is free. Let $x_1, \ldots, x_m$ and $y_1, \ldots, y_n$ be minimal sets of generators of F and G; their union is then a minimal set of generators of $F \oplus G$. By hypothesis, $h(y_i) = g(y_i)$; also, $h(x_i) = f(x_i) + e(x_i)$ is the direct sum decomposition of $h(x_i)$, where e is some homomorphism $F \to G$. Hence

$$\begin{aligned}
&\det h \cdot (x_1 \wedge \cdots \wedge x_m) \wedge (y_1 \wedge \cdots \wedge y_n) \\
&\quad = (f(x_1) + e(x_1)) \wedge \cdots \wedge (f(x_m) + e(x_m)) \wedge (g(y_1) \wedge \cdots \wedge g(y_n)) \\
&\quad = \det g \cdot (f(x_1) + e(x_1)) \wedge \cdots \wedge (f(x_m) + e(x_m)) \wedge (y_1 \wedge \cdots \wedge y_n)\,.
\end{aligned}$$

But since $e(x_i)$ is a linear combination of the y_i's, the above reduces to just

$$\det g \cdot (f(x_1) \wedge \cdots \wedge f(x_m)) \wedge (y_1 \wedge \cdots \wedge y_n) = \det f \cdot \det g \cdot (x_1 \wedge \cdots \wedge y_n)$$

as desired. If h is an automorphism, then $\det f$ must be ± 1, proving the last statement. Q.E.D.

Suppose now that $F \xrightarrow{\pi} H$ is a minimal presentation, so that $\operatorname{rank} F = \operatorname{rank} H = r$. Let $\pm\varphi$ be the unique volume on F, and let $\pm\theta \in \wedge^r H$ be $\pm\pi(\varphi')$; we call the volume $\pm\theta$ the *volume of* (or *induced by*) *the presentation* π. Suppose now that $F \xrightarrow{\pi'} H$ is an equivalent presentation, *i.e.* there exists a diagram

$$\begin{array}{ccc}
F & & \\
& \searrow^{\pi} & \\
{\scriptstyle f}\downarrow & & H \\
& \nearrow_{\pi'} & \\
F' & &
\end{array}$$

with f an isomorphism. If $\pm\theta'$ is the volume induced by π', we then have

$$\pm\theta = \pi(\pm\varphi) = \pi' f(\pm\varphi) = \pi'(\pm \det f \cdot \varphi') = \pm \det f \cdot \theta'.$$

But the fact that f is an isomorphism implies that $\det f = \pm 1$, and hence, *equivalent presentations have the same volume.* This argument generalizes in the obvious way to prove the necessity in the following:

Theorem 3.26. *Let $\pi : F \to H$, $\pi' : F' \to H'$ be two minimal presentations with volumes θ, θ' and let $h : H \to H'$ be an isomorphism. Then h lifts to an isomorphism $f : F \to F'$ if and only if $h(\pm\theta) = \pm\theta'$; that is, if and only if* $\det h = \pm 1 \bmod \tau$ *where τ is the smallest elementary divisor of $H \cong H'$.*

Proof. We first claim that it suffices to consider the special case when the two presentations are identical. Indeed, by Proposition 3.4 the two presentations are isomorphic; *i.e.* there exists a commutative diagram

$$\begin{array}{ccc} F & \xrightarrow{\pi} & H \\ {\scriptstyle f'}\downarrow & & \downarrow{\scriptstyle h'} \\ F' & \xrightarrow{\pi'} & H' \end{array}$$

with both f' and h' isomorphisms. Since f' is an isomorphism, the map h' has determinant 1 and hence so does $h \circ h'^{-1}$. If we could lift the automorphism $h \circ h'^{-1}$ to an automorphism f'' of F', then $f := f'' \circ f' : F \to F'$ would be the desired lift of h.

Hence let h be an automorphism of H with $\det h = \pm 1$ and let $F \xrightarrow{\pi} H$ be any minimal presentation. The proof proceeds just as the proof of Theorem 3.15: if T is the torsion subgroup of H and $F_0 = \pi^{-1}(T)$ and $h_0 = h|_T$, then it still suffices to lift h_0 to F_0. Since H is the direct sum of T and a free abelian group, the presentation $F_0 \to T$ is also minimal. Furthermore, the conditions of Lemma 3.25 hold here, so $\det h_0 = \pm 1$ also. Thus it suffices to prove the theorem when $H = T$ is a torsion group.

Let $|H| = m$. By Lemma 3.12 we may lift h to an endomorphism f_0 of F such that $(\det f_0, m) = 1$; by Lemma 3.14, it follows that $\det f_0 \equiv \det h \equiv \pm 1 \bmod \tau$. Choose k such that $k \cdot \det f_0 \equiv \pm 1 \bmod m$, where the sign here is to be the same as the one above, so that $k \equiv 1 \bmod \tau$. We choose a basis $e_1, \ldots, e_r$ of F (as in Proposition 3.4) so that H is the direct sum of the cyclic subgroups generated by $x_i = \pi(e_i)$ and x_1 has order $\tau = m_1$. The endomorphism f_1 of F defined by $e_1 \mapsto ke_1$, $e_i \mapsto e_i$ for all $i > 1$ clearly induces the identity map on H, since $k \equiv 1 \bmod \tau$. Hence $f_0 f_1$ still induces h on H, and its determinant is now $\det f_0 \cdot \det f_1 \equiv k \det f_0 \equiv \pm 1 \bmod m$. Just as in the proof of Lemma 3.14, we conclude that there exists an *isomorphism* f of F, with determinant ± 1 (same sign!) such that $f \equiv f_0 f_1 \bmod m$ (we are here using the fact that $\mathrm{GL}(r, \mathbb{Z})$ maps onto all elements of $\mathrm{GL}(r, \mathbb{Z}_m)$ with determinant ± 1). As in Lemma 3.14, f still induces h on H, and we are done. Q.E.D.

Lifting the identity automorphism gives:

Corollary 3.27. *Two minimal presentations of H are equivalent if and only if they induce the same volume on H.*

Here are some examples to show that calculations can actually be done with this machinery.

Example 3.28. In Example 3.3 we gave an example of inequivalent minimal presentations, namely $\mathbb{Z} \xrightarrow{1\mapsto 1} \mathbb{Z}_5$ and $\mathbb{Z} \xrightarrow{1\mapsto 2} \mathbb{Z}_5$. Now observe that $r = 1$, $\tau = 5$, $\Lambda^1\mathbb{Z}_5 = \mathbb{Z}_5$; $\theta = \pm 1$, $\theta' = 2(\pm 1) = \pm 2 \not\equiv \pm 1 \bmod 5$. ||

Example 3.29. Let $H = \mathbb{Z}_{2^{n-1}} \oplus \mathbb{Z}^2_{2^n}$ with standard generators e_1, e_2, e_3 and standard presentation $\mathbb{Z}^3 \to H$ taking $(1,0,0) \mapsto e_1$, etc. Let the second presentation be given by

$$(1,0,0) \mapsto e_1 + 2e_2 - 2e_3 \quad ; \quad (0,1,0) \mapsto e_1 + e_2 \quad ; \quad (0,0,1) \mapsto e_1 - e_3$$

(it is easily seen that this map is onto H). The former volume is $\pm e_1 \wedge e_2 \wedge e_3$, the latter

$$\pm(e_1+2e_2-2e_3)\wedge(e_1+e_2)\wedge(e_1-e_3) = \pm\theta\cdot\det\begin{pmatrix} 1 & 2 & -2 \\ 1 & 1 & 0 \\ 1 & 0 & -1 \end{pmatrix} = \pm 3\theta.$$

Since $\tau = 2^{n-1}$ here, the presentations are equivalent if and only if $\pm 3 \equiv \pm 1 \bmod 2^{n-1}$, *i.e.* if and only if $n \leqslant 3$ (the signs on 3 and 1 are independent).||

§4. Symplectic spaces, Heegaard pairs and symplectic Heegaard splittings

As we noted at the start of the previous section, when a 3-manifold W is defined by a Heegaard splitting, then we have, in a natural way, a presentation of $H_1(W;\mathbb{Z})$. In fact we have more, because there is also a natural symplectic form associated to the presentation. In this section, our goal is to begin to broaden the concept of a presentation by placing additional structure on the free group of the presentation, and then to extend the results of Section 3 to include the symplectic structure. With that goal in mind we introduce symplectic spaces and their lagrangian subspaces, leading to the concept of a *Heegaard pair.* There are equivalence relations on Heegaard pairs analogous to those on free pairs (F, R). Just as we stabilized free pairs by taking their direct sums with $(\mathbb{Z}^k, \mathbb{Z}^k)$, we will see that there is an analogous concept of stabilization of Heegaard pairs, only now we need direct sums with a *standard* Heegaard pair. At the end of the section (see Theorem 4.8) we will relate our Heegaard pairs to the symplectic Heegaard splittings that were introduced in §1.

4.1. Symplectic spaces and Heegaard pairs

To begin, we reinterpret the free group F of Definition 3.1, introducing new notation, ideas and structure in the process.

Definition 4.1. A *symplectic space* is a finitely generated free abelian group V which is endowed with a non-singular antisymmetric bilinear pairing, written here as a dot product. *Non-singular* means that for each homomorphism $\alpha : V \to \mathbb{Z}$ there is an $x_\alpha \in V$ (necessarily unique) such that $\alpha(y) = x_\alpha \cdot y$ $(\forall y \in V)$. A *symplectic* or Sp-*basis* for V is a basis $\{a_i, b_i; 1 \leqslant i \leqslant g\}$ such that $a_i \cdot a_j = b_i \cdot b_j = 0, a_i \cdot b_j = \delta_{ij}, 1 \leqslant i, j \leqslant g$. Every symplectic space has such a basis, and so is of even rank, say $2g$. As our *standard model* of a rank $2g$ symplectic space we have $X_g = \mathbb{Z}^{2g}$ with basis $\{a_1, \ldots, a_g, b_1, \ldots, b_g\}$ the $2g$ unit vectors, given in order. An isomorphism $V \to V$ which is form-preserving is a *symplectic isomorphism*. The group of all symplectic isomorphisms of V is denoted $\mathrm{Sp}(V)$. $\|$

Definition 4.2. Let $B \subset V$ be a subset of a symplectic space V and define $B^\perp = \{v \in V \mid v \cdot b = 0 \ (\forall b \in B)\}$.

- A subspace $B \subset V$ is *symplectic* if, equivalently,
  a) the symplectic form restricted to B is non-singular, or
  b) $V = B \oplus B^\perp$.
- A subspace $B \subset V$ is *isotropic* if, equivalently,
  a) $x \cdot y = 0$ for all $x, y \in B$, or
  b) $B \subset B^\perp$.
- A subspace $B \subset V$ is *lagrangian* if, equivalently,
  a) B is maximal isotropic, or
  b) $B = B^\perp$, or
  c) B is isotropic, a direct summand of V, and rank $B = \frac{1}{2}$rank V.

We shall omit the proof that these various conditions are indeed equivalent. $\|$

Our next definition is motivated by the material in §1.2, where we defined symplectic Heegaard splittings. We will see very soon that our current definitions lead to the identical concept.

Definition 4.3. A *Heegaard pair* is a triplet $(V; B, \bar{B})$ consisting of a symplectic space V and an ordered pair $B, \bar{B}$ of lagrangian subspaces. The *genus* of the pair is rank $B =$ rank $\bar{B} = \frac{1}{2}$rank V. An *isomorphism* of Heegaard pairs $(V_i; B_i, \bar{B}_i), i = 1, 2$ is a symplectic isomorphism $f : V_1 \to V_2$ such that $f(B_1) = B_2, f(\bar{B}_1) = \bar{B}_2$. ‖

We now want to define a concept of "stabilization" for Heegaard pairs. If V has Sp-basis $\{a_i, b_i \mid i = 1, \ldots, g\}$, then the a_i's (and also the b_i's) generate a lagrangian subspace. These two subspaces A, B have the following properties:

(a) $A \oplus B = V$
(b) the symplectic form induces a dual pairing of A and B, *i.e.* $a_i \cdot a_j = b_i \cdot b_j = 0$, $a_i \cdot b_j = \delta_{ij}$, $1 \leqslant i, j \leqslant g$.

Any pair of lagrangian subspaces of V satisfying these two properties with respect to some basis will be called a *dual pair*, and either space will be called the *dual complement* of the other.

If X_g is the standard model for a symplectic space, then the lagrangian subspaces E_g spanned by $a_1, \ldots, a_g$ and F_g spanned by $f_1, \ldots, f_g$ are a dual pair. We will refer to $(X_g; E_g, F_g)$ as the *standard Heegaard pair*. Note that in an arbitrary Heegaard pair $(V; B, \bar{B})$ the lagrangian subspaces $B, \bar{B}$ need not be dual complements.

If V_1 and V_2 are symplectic spaces, then $V_1 \oplus V_2$ has an obvious symplectic structure, and V_1 and V_2 are Sp-subspaces of $V_1 \oplus V_2$ with $V_1 = V_2^{\perp}$ and $V_2 = V_1^{\perp}$. This induces a natural direct sum construction for Heegaard pairs, with $(V_1; B_1, \bar{B}_1) \oplus (V_2; B_2, \bar{B}_2) = (V_1 \oplus V_2; B_1 \oplus B_2, \bar{B}_1 \oplus \bar{B}_2)$. The *stabilization of index* k of a Heegaard pair is its direct sum with the standard Heegaard pair $(X_k; E_k, F_k)$ of genus k. Two Heegaard pairs $(V_i; B_i, \bar{B}_i)$, $i = 1, 2$, are then *stably isomorphic* if they have isomorphic stabilizations.

These concepts will soon be related to topological ideas. First, however, we will show that stable isomorphism classes and isomorphism classes of Heegaard pairs are in 1-1 correspondence with stable double cosets and double cosets in the symplectic modular group Γ, with respect to its subgroup Λ.

Note that if A, B is a dual pair of V and $\mathcal{U} : B \to B$ is a linear automorphism, then the adjoint map $(\mathcal{U}^*)^{-1}$ is an isomorphism of $A = B^*$. Moreover $(\mathcal{U}^*)^{-1} \oplus \mathcal{U}$ is a symplectic automorphism of $V = A \oplus B$.

Lemma 4.4. *If $B \subset V$ is lagrangian, $A \subset V$ is isotropic and $A \oplus B = V$, then A is lagrangian and A, B is a dual pair of V. Every lagrangian subspace has a dual complement.*

Proof. Since A is an isotropic direct summand of V and rank $A =$ rank V – rank $B = \frac{1}{2}$rank V, it follows that A is lagrangian. Let now $f : B \to \mathbb{Z}$ be linear, and extend it to $\alpha : V \to \mathbb{Z}$ by setting $\alpha(A) = 0$. Then $\alpha(v) = x \cdot v$ for some $x \in V$. Since $x \cdot A = 0$ and A is maximal isotropic, we must have $x \in A$, showing that A, B are dually paired and hence a dual pair of V. To prove the second statement, let b_i be a basis of B. Since B is lagrangian it is a direct summand of V, so we may choose a homomorphism $f_1 : V \to \mathbb{Z}$ such that $f_1(b_1) = 1$ and $f_1(b_i) = 0$ for $i > 1$. Let $a_1 \in V$ be such that $a_1 \cdot v = f_1(v)$ for all v. Clearly the subgroup generated by a_1 and B is still a direct summand of V, so choose $f_2 : V \to \mathbb{Z}$ such that $f_2(b_2) = 1$, $f_2(a_1) = f_2(b_i) = 0$ $(i \neq 2)$ and $a_2 \in V$ realizing this map. Continuing in this way, we get finally $a_1, \ldots, a_g$ such that a_i, b_i satisfy the laws of a symplectic basis and generate a direct summand of V. This direct summand has the same rank as V, so it equals V, and the group A generated by the a_i's is then a dual complement of B. Q.E.D.

Proposition 4.5. *Let A, B be a dual splitting and b_i a basis of B. If a_i is the dual basis of A defined by $a_i \cdot b_j = \delta_{ij}$, then a_i, b_i is a symplectic basis of V.*

Corollary 4.6. *If A, B and A', B' are two dual pairs of V, then there is an $f \in \mathrm{Sp}(V)$ such that $f(A) = A'$, $f(B) = B'$.*

Proof. Choose symplectic bases a_i, b_i adapted to A, B and a_i', b_i' adapted to A', B'. Then the map defined by $a_i \mapsto a_i'$ and $b_i \mapsto b_i'$ is symplectic. Q.E.D.

Corollary 4.7. *If B, B' are lagrangian, there is an $f \in \mathrm{Sp}(V)$ such that $f(B) = B'$.*

Proof. By Lemma 4.4, B and B' have dual complements A and A'. By Corollary 4.6 we may find $f \in \mathrm{Sp}(V)$ such that $f(B) = B'$. Q.E.D.

4.2. Heegaard pairs and symplectic Heegaard splittings

We are now ready to relate our work on Heegaard pairs to the double cosets introduced in §1. We follow notation used there.

Theorem 4.8. *The following hold:*

(1) *Isomorphism classes of Heegaard pairs are in 1-1 correspondence with double cosets in* Γ mod Λ. *Stable isomorphism classes of Heegaard pairs are in 1-1 correspondence with stable double cosets in* Γ mod Λ.

(2) *Let* $j : \partial N_g \to N_g$, $\bar{j} : \partial N_g \to \bar{N}_g$, *and let* $j_*, \bar{j}_*$ *be the induced actions on homology.*
Then the triplet $(H_1(M; \mathbb{Z}); \ker j_*, \ker \bar{j}_*)$ *is a Heegaard pair.*

(3) *Every Heegaard pair is topologically induced as the Heegaard pair associated to a topological Heegaard splitting of some* 3-*manifold. Moreover, equivalence classes and stable equivalence classes of Heegaard pairs are topologically induced by equivalence classes and stable equivalence classes of Heegaard splittings.*

Proof. We begin with assertion (1). Let $(V; B, \bar{B})$ be a Heegaard pair of genus g. Then by Corollary 4.6 we may find a symplectic isomorphism $f : V \to X_g$ such that $f(B) = F_g$. Putting $\bar{F}_g = f(\bar{B})$, we then have $(V; B, \bar{B})$ isomorphic to $(X_g; F_g, \bar{F}_g)$. If $f' : (V, B) \to (X_g, F_g)$ is another choice, with $f'(\bar{B}) = \bar{F}'_g$, then $f'f^{-1}(F_g) = F_g$, hence $f'f^{-1} \in \Lambda$. Then we see that the isomorphism classes of genus g Heegaard pairs correspond to equivalence classes of lagrangian subspaces $\bar{F}_g \subset X_g$, with $\bar{F}_g, \bar{F}'_g$ equivalent if there is a map $m \in \Lambda$ such that $m(\bar{F}_g) = \bar{F}'_g$. Now, we have seen that there is a map $h \in \Gamma$ such that $\bar{F}_g = h(F_g)$, and $h_1(F_g) = h_2(F_g)$ if and only if $h_2 = h_1 f$ for some $f \in \Lambda$. Then each $\bar{F}_g$ can be represented by an element $h \in \Gamma$ and h, h' give equivalent subspaces $\bar{F}_g = h(F_g)$, $\bar{F}'_g = h'(F_g)$ if and only if there are $f_1, f_2 \in \Lambda$ such that $h' = f_1 h f_2$. The set of all $f_1 h f_2$, $f_i \in \Lambda$, is a double coset of Γ mod Λ. Then the isomorphism classes of Heegaard pairs of genus g are in 1-1 correspondence with the double cosets of Γ mod Λ.

Direct sums and stabilizations of Heegaard pairs corresponds to a topological construction. If $(W_i; N_i, \bar{N}_i)$ $(i = 1, 2)$ are Heegaard splittings, their connected sum $(W_1 \# W_2; N_1 \# N_2, \bar{N}_1 \# \bar{N}_2)$ is a Heegaard splitting whose abelianization to a Heegaard pair is readily identifiable as the direct sum of the Heegaard pairs associated to the summands. Moreover, if $(\mathbb{S}^3; Y_k, \bar{Y}_k)$ is a standard Heegaard splitting of genus k for $\mathbb{S}^3$, its Heegaard pair may be identified with the standard Heegaard pair $(X_k; E_k, F_k)$ of index k. This stabilization of Heegaard pairs is induced by the topological construction $(W; N, \bar{N}) \to (W \# \mathbb{S}^3; N \# Y_k, \bar{N} \# \bar{Y}_k)$.

In an entirely analogous manner to the proof just given for (1), stable isomorphism classes of Heegaard pairs correspond to stable double cosets in Γ mod Λ.

Proof of (2): There is a natural symplectic structure on the free abelian group $H_1(M; \mathbb{Z})$, with the bilinear pairing defined by intersection numbers of closed curves which represent elements of $H_1(M; \mathbb{Z})$ on M. In fact, as claimed in (2) above, the triplet $(H_1(M; \mathbb{Z}); \ker j_*, \ker \bar{j}_*)$ is a

Heegaard pair. To see this, let $B = \ker j_*$. Since $x \cdot y = 0 \; \forall x, y \in B$, the subspace B is isotropic. Also, rank $B = \frac{1}{2}$rank $H_1(M;\mathbb{Z}) =$ genus $N =$ genus M. Hence B is lagrangian. Similarly, $\bar{B}$ is lagrangian. Therefore the assertion is true.

Proof of (3): It remains to show that every Heegaard pair is topologically induced as the Heegaard pair associated to a topological Heegaard splitting of some 3-manifold, and also that equivalence classes and stable equivalence classes of Heegaard pairs are topologically induced by equivalence classes and stable equivalence classes of Heegaard splittings. To see this, let $(V; B, \bar{B})$ be a Heegaard pair. By Theorem 4.8 we may without loss of generality assume that $(V; B, \bar{B})$ is $(X_g; F_g, \bar{F}_g)$. Choose a standard basis for $H_1(M;\mathbb{Z})$, with representative curves as illustrated in Figure 1. We may without loss of generality take one of these (say $w_1, \ldots, w_g$) to be standard and cut M open along $w_1, \ldots, w_g$ to a sphere with $2g$ boundary components $w_i, \bar{w}_i$ $(i = 1, \ldots, g)$. Choose $2g$ additional curves $V_1, \ldots, V_g, W_1, \ldots, W_g$ on M such that each pair w_i, W_i is a canceling pair of handles, *i.e.* $w_i \cdot W_i = 1$ point, $w_i \cdot W_j = W_i \cdot W_j = \emptyset$ if $i \neq j$, and similarly for the V_i's. Then the matrices of algebraic intersection numbers

$$\left\|\left|v_i \cdot w_j\right|\right\| , \quad \left\|\left|v_i \cdot W_j\right|\right\|$$

uniquely determine a symplectic Heegaard splitting. This gives a natural symplectic isomorphism from $H_1(M;\mathbb{Z})$ to X_g. Also, since M is pictured in Figure 1 as the boundary of a handlebody N, our map sends $H_1(N;\mathbb{Z})$ to F_g. By Corollary 4.7 we may find $h \in \Gamma$ such that $h(F_g) = \bar{F}_g$. By [8] each $h \in \Gamma$ is topologically induced by a homeomorphism $\tilde{h} : M \to M$. Let $\bar{N}$ be a copy of N, and let W be the disjoint union of N and $\bar{N}$, identified along $\partial N = M$ and $\partial \bar{N} = M$ by the map $\tilde{h}$. Then $(W; N, \bar{N})$ is a Heegaard splitting of W which induces the Heegaard pair $(V; B, \bar{B})$.

In an entirely analogous manner, the correspondence between (stable) isomorphism classes of Heegaard pairs and symplectic Heegaard splittings may be established, using the method of proof of Theorem 4.8 and the essential fact that each $h \in \Lambda$ is topologically induced by a homeomorphism $\tilde{h} : M \to M$. Q.E.D.

§5. Heegaard pairs and their linked abelian groups

In this section we meet linked groups for the first time in our investigations of Heegaard pairs. We show that the problem of classifying stable isomorphism classes of Heegaard pairs reduces to the problem of classifying linked abelian groups. This is accomplished in Theorem 5.15

and Corollary 5.16. In Theorem 5.18 and Corollary 5.21 we consider the question: how many stabilizations are needed to obtain equivalence of minimal, stably equivalent Heegaard pairs? Corollary 5.22 asserts that a single stabilization suffices, generalizing the results of Theorem 3.15 and Corollary 3.18. This solves Problem 3.

The final part of the section contains partial results about classifying Heegaard pairs of minimal rank. Theorem 5.20 is a first step. The complete solution to that problem will be given, later, in Theorem 7.5.

We will not be able to address the issue of computing the linking invariants in this section. Later, after we have learned more, we will develop a set of computable invariants for both stable and unstable double cosets.

5.1. The quotient group of a Heegaard pair and its natural linking form. Solution to Problem 1

We now introduce the concept of the quotient group of a Heegaard pair. We will prove (see Theorem 5.5) that the quotient group of a Heegaard pair has a natural non-singular linking form. This leads us to the concept of a 'linked abelian group'. In Corollary 5.9 we show that, as a consequence of Theorem 5.5, the linked abelian group that is associated to a Heegaard pair is an invariant of its stable isomorphism class. Corollary 5.16 solves Problem 1.

Lemma 5.1. *Let* $(V; B, \bar{B})$ *be a Heegaard pair and let* $C = \{x \in V \mid x \cdot (B + \bar{B}) = 0\}$. *Then* $C = B \cap \bar{B}$.

Proof. We have $x \in C$ if and only if $x \cdot (B + \bar{B}) = 0$, which is true if and only if $x \cdot B = 0$ and $x \cdot \bar{B} = 0$. Since $B, \bar{B}$ are maximally isotropic, this is true if and only if $x \in B$ and $x \in \bar{B}$. Q.E.D.

This lemma implies that, for lagrangian subspaces $B, \bar{B}, B', \bar{B}'$, if $B + \bar{B} = B' + \bar{B}'$, then $B \cap \bar{B} = B' \cap \bar{B}'$.

Lemma 5.2. *Every Heegaard pair* $(V; B, \bar{B})$ *is a direct sum of Heegaard pairs of the form* $(V_1; C, C) \oplus (V_2; D, \bar{D})$ *where* $C = B \cap \bar{B}$ *and* $D \cap \bar{D} = 0$.

Proof. Since $B/C = B/(B \cap \bar{B}) \cong (B + \bar{B})/\bar{B} \subset V/\bar{B}$, the group B/C is free and thus C is a direct summand of B. Thus $B = C \oplus D$ for some subgroup D of B. Let now B^* be a dual complement of B in V; the splitting $C \oplus D$ of B then induces, dually, a splitting $C^* \oplus D^*$ of B^*, where $C \perp D^*$ and $D \perp C^*$ and where C and C^* are dually paired (by the symplectic form) and likewise D, D^*. Thus $V_1 = C \oplus C^*$ and $V_2 = D \oplus D^*$ are *symplectic* subspaces of V, with $V_1 = V_2^{\perp}$ and $V = V_1 \oplus V_2 = C \oplus C^* \oplus D \oplus D^*$. We claim that $\bar{B} \subset C \oplus D \oplus D^*$. Indeed, express $\bar{b} \in \bar{B}$ as $\bar{b} = c + c^* + d + d^*$ with $c \in C$, etc. Since $\bar{B} \supset C$ and $\bar{B}$ is isotropic, we have $\bar{b} \cdot c' = 0$ for all $c' \in C$, *i.e.* $(c + c^* + d + d^*) \cdot c' = c^* \cdot c' = 0$ for all $c' \in C$. But C, C^* are dually paired, so c^* must be zero.

Hence we have $C \oplus D \oplus D^* \supset \bar{B} \supset C$. But this implies that $\bar{B} = C \oplus \bar{D}$, where $\bar{D} = \bar{B} \cap (D \oplus D^*) = \bar{B} \cap V_2 = V_2$. We have now shown that:

a) $V = V_1 \oplus V_2$
b) $B = C \oplus D$ with $C = B \cap \bar{B} \subset V_1$, $D \subset V_2$
c) $\bar{B} = C \oplus \bar{D}$ with $\bar{D} \subset V_2$

Note that $C = B \cap \bar{B} = (C \oplus D) \cap (C \cap \bar{D}) = C \oplus (D \cap \bar{D})$, so $D \cap \bar{D} = 0$. To finish the proof, it suffices then to show that $(V_1; C, C)$ and $(V_2; D, \bar{D})$ are Heegaard pairs. The former is trivially so since $V_1 = C \oplus C^*$, and for the same reason, D is lagrangian in V_2. We must then show that $\bar{D}$ is lagrangian in V_2. It is certainly isotropic, since $\bar{B} = C \oplus \bar{D}$ is so. But let $x \in V_2$ be such that $x \cdot \bar{D} = 0$. We also have $x \cdot V_1 = 0$ since $V_1 \perp V_2$ and hence $x \cdot \bar{B} = x \cdot (C \oplus \bar{D}) = 0$. Since $\bar{B}$ is maximally isotropic, $x \in \bar{B}$ and hence $\bar{B} \cap V_2 = \bar{D}$, showing $\bar{D}$ to be maximally isotropic in V_2. Q.E.D.

Lemma 5.3. *Let $(V; B, \bar{B})$ and $(V; B, \bar{B}')$ be two Heegaard pairs such that $B + \bar{B} = B + \bar{B}'$. If the first is split as in the previous lemma, then the second has a splitting of the form $(V_1; C, C) \oplus (V_2; D, \bar{D}')$, where $D \cap \bar{D}' = 0$ and $D + \bar{D}' = D + \bar{D}$.*

Proof. By Lemma 5.1, the fact that $B + \bar{B} = B + \bar{B}'$ implies that $C = B \cap \bar{B} = B \cap \bar{B}'$. Examining the construction of Lemma 5.2, we see that since B and C are the same for both pairs, we may choose B^* and D the same, and hence C^*, D^* and $V_1 = C \oplus C^*$, $V_2 = D \oplus D^*$ will also be the same. Thus the second pair has a splitting satisfying all the requirements except possibly the last. But we have $D + \bar{D}' = (B + \bar{B}') \cap V_2 = (B + \bar{B}) \cap V_2 = D + \bar{D}$. Q.E.D.

We define the *quotient* of a Heegaard pair $(V; B, \bar{B})$ to be the group $H = V/(B + \bar{B})$. Clearly, isomorphic Heegaard pairs have isomorphic quotients. Furthermore, stabilization does not change this quotient either, since

$$(V \oplus X_k)/[(B \oplus E_k) + (\bar{B} \oplus F_k)] = (V \oplus X_k)/[(B + \bar{B}) \oplus X_k] \cong V/(B + \bar{B}).$$

Thus the isomorphism class of the quotient is an invariant of stable isomorphism classes of pairs. We cannot conclude, however, that two Heegaard pairs are stably isomorphic if they have isomorphic quotients; there are further invariants. To pursue these, we need the following concepts.

Definition 5.4. If T is a finite abelian group, a *linking form* on T is a symmetric bilinear $\mathbb{Q}/\mathbb{Z}$-valued form on T, where $\mathbb{Q}/\mathbb{Z}$ is the group of rationals mod 1. More generally, a linking form on any finitely generated abelian group H means a linking form on its torsion subgroup T. A linking form λ is *non-singular* if for every homomorphism $\varphi : T \to \mathbb{Q}/\mathbb{Z}$, there is a (necessarily unique) $x \in T$ such that $\varphi(y) = \lambda(x, y)$ for all $y \in T$. A group H will be called a *linked group* if its torsion subgroup is endowed with a non-singular linking form. $\|$

Theorem 5.5. *The quotient group of a Heegaard pair has a natural non-singular linking form.*

Proof. Let the pair be $(V; B, \bar{B})$, the quotient be H, and its torsion subgroup be T. Consider $x, y \in T$ and suppose that $mx = 0$. Lift x, y to $u, v \in V$; then $mx = 0$ implies that $mu \in B + \bar{B}$, say $mu = b + \bar{b}$. Define $\lambda(x, y)$ to be $\frac{1}{m}(b \cdot v)$ mod 1. Note that if $ny = 0$ and hence $nv \in B + \bar{B}$, say $nv = c + \bar{c}$ $(c \in B, \bar{c} \in \bar{B})$ then we have

$$\lambda(x,y) \equiv \frac{1}{m}(b \cdot v) \equiv \frac{1}{mn}(b \cdot nv) \equiv \frac{1}{mn} b \cdot (c + \bar{c}) \equiv \frac{1}{mn} b \cdot \bar{c} \bmod 1,$$

which gives a more symmetric definition of $\lambda(x, y)$. We now verify the necessary facts about λ.

a) Independent of the choice of $b, \bar{b}$: if $b + \bar{b} = b' + \bar{b}'$, then we have $b' = b + \delta$, $\bar{b}' = \bar{b} - \delta$, with $\delta \in B \cap \bar{B}$. But then $\frac{1}{mn} b' \cdot \bar{c} = \frac{1}{mn}(b \cdot \bar{c} + \delta \cdot \bar{c}) = \frac{1}{mn} b \cdot \bar{c}$, since $\delta \cdot \bar{c} = 0$. Similarly, $\lambda(x, y)$ is independent of the choice of $c, \bar{c}$.
b) Independent of the lifting u, v: a different lifting u' satisfies $u' = u + b_1 + \bar{b}_1$, so

$$mu' = mu + m(b_1 + b_1') = (b + mb_1) + (\bar{b} + m\bar{b}_1),$$

and hence

$$\lambda(x,y) \equiv \frac{1}{m} b \cdot v \equiv \frac{1}{m}(b + mb_1) \cdot v \bmod 1.$$

Similarly, $\lambda(x, y)$ does not depend on the lifting v of y.

c) Independent of the choice of m: if $m'x = 0$ also, with $m'u = b' + \bar{b}'$, then $mm'u = m'b + m'\bar{b} = mb' + m\bar{b}'$ and

$$\frac{1}{m'}(b' \cdot v) \equiv \frac{1}{mm'}(mb' \cdot v) \equiv_{\text{by a)}} \frac{1}{mm'}(m'b \cdot v) \equiv \frac{1}{m}(b \cdot v) \bmod 1.$$

d) Bilinearity and symmetry: the first follows immediately from a). Then for symmetry, we have $\lambda(x,y) = \frac{1}{mn}(b \cdot \bar{c})$ and $\lambda(y,x) = \frac{1}{mn}(c \cdot \bar{b})$. But

$$mn(u \cdot v) = (mu) \cdot (nv) = (b + \bar{b}) \cdot (c + \bar{c}) = b \cdot \bar{c} + \bar{b} \cdot c = b \cdot \bar{c} - c \cdot \bar{b} \equiv 0 \bmod mn,$$

so $\frac{1}{mn}(b \cdot \bar{c}) \equiv \frac{1}{mn}(c \cdot \bar{b}) \bmod 1$.

e) Non-singularity: this is equivalent to the statement that $\lambda(x,y) \equiv 0 \bmod 1$ for all $y \in T$ implies that $x = 0$ in T.

Suppose then $x \in T$ and $\lambda(x,y) \equiv 0$ for all $y \in T$, and let u be a lifting of x to V. Now by Lemma 5.2, our Heegaard pair is a direct sum $(V_1; C, C) \oplus (V_2; D, \bar{D})$ with $C = B \cap \bar{B}$ and $D \cap \bar{D} = 0$. Since the quotient $V_1/(C + C)$ is *free*, V_2 projects onto T and so the lifting of any torsion element may always be chosen in V_2. If E is a dual complement of D in V_2, then in fact the projection $V \to H$ will take E onto T. Thus we may assume that $u \in E$. If $mx = 0$ in T then $mu = d + \bar{d}$ for some $d \in D$, $\bar{d} \in \bar{D}$. The hypothesis that $\lambda(x,y) \equiv 0 \bmod 1$, all $y \in T$ is equivalent to $d \cdot v \equiv 0 \bmod m$ for all $v \in V_2$. Since V_2 is symplectic, this implies that d is *divisible by* m in V_2, that is, $d = md'$ for some $d' \in V_2$. Now clearly $d' \cdot D = 0$, and hence $d' \in D$ since D is maximally isotropic. Thus we have $mu = md' + \bar{d}$, $\bar{d} = m(u - d')$ and we conclude similarly that $u - d' \in \bar{D}$, say $u - d' = \bar{d}'$. Thus $u = d' + \bar{d}' \in D + \bar{D}$, which implies that $x = 0$.

Q.E.D.

Remark 5.6. Note that the *maximal* isotropic nature of $B, \bar{B}$ was used only in proving e); the weaker assumption that they are only isotropic still suffices to prove a)–d) and thus construct a natural linking on H. ||

Lemma 5.7. *Let $B \subset V$ be lagrangian, let $\bar{B}$ be isotropic of rank $\frac{1}{2}\mathrm{rank}\ V$, and suppose that $B \cap \bar{B} = 0$. Then $\bar{B}$ is lagrangian if and only if the induced linking form λ on H is non-singular.*

Proof. We have already proved the necessity, so suppose that λ is non-singular. By Definition 4.2, we need only show that $\bar{B}$ is a direct summand of V. This is equivalent to showing that $V/\bar{B}$ is torsion free, *i.e.* that for $u \in V$, if $mu \in \bar{B}$ for some nonzero m, then $u \in \bar{B}$. If then

$mu \in \bar{B}$ and x is the image of u in H, then $mx = 0$ in H, so x is in the torsion group T. But then u lifts x and mu decomposes in $B + \bar{B}$ as $0 + mu$. Hence if $y \in T$ is lifted to $v \in V$, we get $\lambda(x, y) = \frac{1}{m}(0 \cdot v) \equiv 0 \bmod 1$, *i.e.* $\lambda(x, y) \equiv 0$ for all $y \in T$. By hypothesis, we have $x = 0$ in T, and hence $u \in B + \bar{B}$, say $u = b + \bar{b}$. Since $mu = mb + m\bar{b}$ is in $\bar{B}$, mb is also in $\bar{B}$. But it is also in B, so must be zero, *i.e.* $b = 0$. Thus $u = \bar{b} \in \bar{B}$. Q.E.D.

We have shown that the quotient of a Heegaard pair has the structure of a linked group in a natural way; clearly, isomorphic Heegaard pairs have isomorphic *linked* quotients: the Heegaard isomorphism induces an isomorphism on the quotients which preserves the linking. Let us see how the linked quotient behaves under stabilization.

Lemma 5.8. *The linked quotient of a stabilization of* $(V; B, \bar{B})$ *is canonically isomorphic to the unstabilized quotient.*

Proof. The canonical isomorphism of the quotients is induced by the inclusion $V \hookrightarrow V \oplus X_k$, and we identify the two quotients in this way. To see that the linking defined by the two pairs are equal, let $x, y \in T$. Their liftings u, v in $V \oplus X_k$ may be chosen to lie in $V \oplus 0$, since X_k projects to 0 in the quotient, and the splitting of mu may then be chosen to be $(b, 0) + (\bar{b}, 0)$. The stabilized linking number, defined thus, is then obviously the same as the unstabilized one. Q.E.D.

Corollary 5.9. *The linked abelian group is an invariant of the stable isomorphism class of a Heegaard pair.*

The remainder of this section is devoted to strengthening Corollary 5.9 by showing that, in fact, two Heegaard pairs are stably isomorphic if and only if their linked quotients are isomorphic (see Corollary 5.16). It is easily verified that two linked groups are link-isomorphic if and only if they have link-isomorphic torsion groups and, mod their torsion groups, the same (free) rank.

Lemma 5.10. *Let* $(V; B, \bar{B})$ *be a Heegaard pair with* $B \cap \bar{B} = 0$, *and let* A *be a dual complement of* B. *If* $\bar{A}$ *is the direct projection of* $\bar{B}$ *into* A, *then there is a symplectic basis* a_i, b_i *of* V $(a_i \in A,\ b_i \in B)$ *such that:*

a) $m_i a_i$ *is a basis for* $\bar{A}$, *for some integers* $m_i \neq 0$;
b) $m_i a_i + \sum_j n_{ij} b_j$ *is a basis for* $\bar{B}$, *for some integers* n_{ij} *such that* $n_{ij}/m_i = n_{ji}/m_j$.

Remark 5.11. Note that the hypothesis $B \cap \bar{B} = 0$ is equivalent to the fact that the rank of $B + \bar{B}$ is equal to the rank of V, *i.e.* that the quotient is *finite.* ||

Proof. Since $B \cap \bar{B} = 0$, the projection of $\bar{B}$ into A is 1-1, *i.e.*

$$\operatorname{rank} \bar{A} = \operatorname{rank} \bar{B} = \frac{1}{2}\operatorname{rank} V = \operatorname{rank} A.$$

By Proposition 3.4, there is a basis a_i of A such that $m_i a_i$ is a basis of $\bar{A}$, and $m_i \neq 0$ because rank $\bar{A}$ = rank A. Let b_i be the dual basis of $B = A^*$; then a_i, b_i is a symplectic basis of V. The inverse of the projection $\bar{B} \to \bar{A}$ takes $m_i a_i$ into a basis of $\bar{B}$, which must then be of the form $\bar{b}_i = m_i a_i + \sum_k n_{ik} b_k$ for $n_{ij} \in \mathbb{Z}$. But $\bar{b}_i \cdot \bar{b}_j = 0$ for all j, *i.e.* $m_i n_{ji} - m_j n_{ij} = 0$ for all i, j. Q.E.D.

Lemma 5.12. *Let $(V; B, \bar{B})$ and $(V; B, \bar{B}')$ be two Heegaard pairs such that $B + \bar{B} = B + \bar{B}'$ and $B \cap \bar{B} = B \cap \bar{B}' = 0$. Then the linkings λ, λ' induced on the common quotient H are identical if and only if there is an $f \in \mathrm{Sp}(V)$ such that $f(B) = B$, $f(\bar{B}) = \bar{B}'$, and such that the automorphism h of H induced by f is the identity map.*

Proof. Certainly the condition is sufficient. To prove the necessity, let A be a dual complement of B and let $\bar{A}$, $\bar{A}'$ be the projections of $\bar{B}$, $\bar{B}'$ into A. Note that $\bar{A} = A \cap (B + \bar{B})$ and $\bar{A}' = A \cap (B + \bar{B}')$. Hence $\bar{A} = \bar{A}'$. As in the previous lemma we choose an Sp-basis a_i, b_i with $m_i a_i$ a basis of $\bar{A} = \bar{A}'$, and corresponding bases

$$\bar{b}_i = m_i a_i + \sum_j n_{ij} b_j \text{ of } \bar{B} \qquad \text{and} \qquad \bar{b}'_i = m_i a_i + \sum_j n'_{ij} b_j \text{ of } \bar{B}'.$$

Let $\beta_{ij} = \frac{n'_{ij} - n_{ij}}{m_i}$; the "symmetry" conditions of the previous lemma on the n_{ij}, n'_{ij} imply that $\beta_{ij} = \beta_{ji}$. Let x_i be the image of a_i in H. Since $m_i a_i \in B + \bar{B}$, when calculating $\lambda(x_i, x_k)$ we may choose the B-part of the lift of x_i to be $-\sum_j n_{ij} b_j$, and we get

$$\lambda(x_i, x_k) = \frac{1}{m_i}(-\sum n_{ij} b_j) \cdot a_k = \frac{n_{ik}}{m_i}.$$

Likewise we get $\lambda'(x_i, x_k) = \frac{n'_{ik}}{m_i}$. By hypothesis, for all i, k we have $\frac{n'_{ik}}{m_i} \equiv \frac{n_{ik}}{m_i} \bmod 1$, *i.e.* $\beta_{ik} = \frac{n'_{ik} - n_{ik}}{m_i} \equiv 0 \bmod 1$; that is, β_{ik} is *integral* as well as symmetric. Thus the transformation $f : V \to V$ which fixes the b_i and takes a_i to $a_i + \sum_j \beta_{ij} b_j$ is easily seen to be *symplectic.* We have $f(B) = B$ obviously, and

$$\begin{aligned} f(\bar{b}_i) &= m_i(a_i + \sum_j \beta_{ij} b_j) + \sum n_{ij} b_j = m_i a_i + \sum_j (m_i \beta_{ij} + n_{ij}) b_j \\ &= m_i a_i + \sum_j n'_{ij} b_j = \bar{b}'_i, \end{aligned}$$

so $f(\bar{B}) = \bar{B}'$. Finally, $h(x_i)$ is the image of $f(a_i) = a_i + \sum_j \beta_{ij} b_j$, which is just x_i; this shows that $h = 1$. Q.E.D.

Lemma 5.13. *With hypotheses as in the preceding lemma, but omitting the assumption that $B \cap \bar{B} = B \cap \bar{B}' = 0$, the conclusion remains valid.*

Proof. Again we need only prove the necessity. By Lemma 5.3, we split $(V; B, \bar{B})$ as $(V_1; C, C) \oplus (V_2; D, \bar{D})$, where $C = B \cap \bar{B} = B \cap \bar{B}'$, and $(V; B, \bar{B}')$ as $(V_1; C, C) \oplus (V_2; D, \bar{D}')$. Since $D + \bar{D} = (B + \bar{B}) \cap V_2 = (B + \bar{B}') \cap V_2 = D + \bar{D}'$, both the V_2 pairs have the same quotient, namely the torsion subgroup T, and both these pairs define the same linking form. By the preceding lemma we have a map $f_2 \in \mathrm{Sp}(V_2)$ such that $f_2(D) = D$, $f_2(\bar{D}) = \bar{D}'$ and so that

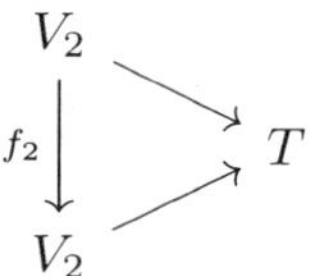

commutes. Let $f = 1_{V_1} \oplus f_2$; then f satisfies the requirements. Q.E.D.

Lemma 5.14. *Let $(V; B, \bar{B})$ and $(V'; B', \bar{B}')$ be Heegaard pairs of the same genus g, and let $h : H \to H'$ be an isomorphism of the quotients (not necessarily linking-preserving). If $g >$ rank H, then there is a symplectic isomorphism $f : V \to V'$ lifting h such that $f(B) = B'$ and $f(B + \bar{B}) = B' + \bar{B}'$.*

Proof. Let A, A' be dual complements of B, B'. Then the projections π, π' map A, A' onto H, H'. Since A, A' are free of rank $g >$ rank H, we have non-minimal presentations $A \to H$, $A' \to H'$, and hence by Theorem 3.15 there is an isomorphism $p : A \to A'$ lifting h. Let q be the adjoint map of p on $B = A^*$ (*i.e.* $q = (p^*)^{-1}$); then $f = p \oplus q$ is a symplectic isomorphism of $A \oplus B = V$ to $A' \oplus B' = V'$. It clearly still lifts h, which implies that $f(B+\bar{B}) = f(\ker \pi) = \ker \pi' = B' + \bar{B}'$. Finally, $f(B) = B'$ by the construction of f. Q.E.D.

Theorem 5.15. *Let $(V; B, \bar{B})$ and $(V'; B', \bar{B}')$ be two Heegaard pairs of genus g and $h : H \to H'$ a link-isomorphism of their linked quotient groups. If $g >$ rank H, then h lifts to a Heegaard isomorphism $j : (V; B, \bar{B}) \to (V'; B', \bar{B}')$.*

Proof. Lift h to f as in the previous lemma, and put $\bar{B}_1 = f^{-1}(\bar{B}')$. Then $(V; B, \bar{B}_1)$ is a Heegaard pair and f maps it isomorphically to $(V'; B', \bar{B}')$. Note that

$$B + \bar{B}_1 = f^{-1}(B' + \bar{B}') = f^{-1}f(B + \bar{B}) = B + \bar{B},$$

so the quotient of $(V; B, \bar{B}_1)$ is also H. Moreover, by construction the linking form on H induced by $(V, B, \bar{B}_1)$ is identical to the linking form induced by $(V, B, \bar{B})$. By Lemma 5.13, there is a map $g \in \mathrm{Sp}(V)$ such that $g(B) = B$, $g(\bar{B}) = \bar{B}_1$, and g induces the identity map on H. Hence the map fg also induces $h : H \to H'$, and $fg(B) = B'$, $fg(\bar{B}) = f(\bar{B}_1) = \bar{B}'$. Q.E.D.

We are now ready to give our solution to Problem 1 of the introduction to this paper.

Corollary 5.16. *Two Heegaard pairs are stably isomorphic if and only if they have the same torsion free ranks, and their torsion groups are link-isomorphic; that is, if and only if they have isomorphic linked quotients.*

5.2. The stabilization index. Solution to Problem 3

In this subsection we introduce the notion of a Heegaard presentation and define the genus of a Heegaard presentation. We return to the concept of the volume of a presentation of an abelian group, relating it now to Heegaard presentations. See Theorem 5.20. At the end of this section we give the solution to Problem 3 of the Introduction to this article. See Corollary 5.22.

Definition 5.17. Let H be a linked group. A *Heegaard presentation* of H consists of a Heegaard pair $(V; B, \bar{B})$ and a surjection $\pi : V \to H$ such that:

a) $\ker \pi = B + \bar{B}$
b) the linking induced on H by means of π is the given linking on H.

The *genus* of the presentation is the given genus of the pair. We will use the symbol $(V; B, \bar{B}; \pi)$ to denote a Heegaard presentation. ||

Theorem 5.18. *Every linked group H has a Heegaard presentation of genus equal to the rank of H.*

Proof. Let $H = F_k \oplus T$, where F_k is free of rank k and T is torsion. If $(V; B, \bar{B}; \pi)$ is a Heegaard presentation of T with genus equal to rank T, then taking the direct sum with $(X_k; E_k, E_k; \rho)$ where $\rho : X_k \to F_k$ is a surjection with kernel E_k gives the required presentation for H. Thus we need only prove the theorem for torsion groups T. Let V be symplectic rank of 2rank T and let A, B be a dual pair in V. Let $\pi_A : A \to T$ be a presentation of T, which is possible since rank A = rank T. We may, by Proposition 3.4, choose a basis a_i of A such that $m_i a_i$ is a basis of $\ker \pi_A$, and $m_i \neq 0$ since T is a torsion group. If b_i is the dual basis of B, then a_i, b_i is a symplectic basis of V. Let now $x_i = \pi_A(a_i)$; the x_i's generate T, and the order of x_i in T is m_i.

If now λ is the linking form on T, choose rational numbers q_{ij} representing $\lambda(x_i, x_j)$ mod 1, which, since $\lambda(x_i, x_j) \equiv \lambda(x_j, x_i)$, may be assumed to satisfy $q_{ij} = q_{ji}$. Note that

$$m_i q_{ij} \equiv m_i \lambda(x_i, x_j) \equiv \lambda(m_i x_i, x_j) \equiv \lambda(0, x_j) \equiv 0 \bmod 1,$$

that is, $m_i q_{ij} = n_{ij}$ is *integral.*

We now define $\bar{B} \subset V$ to be generated by $\bar{b}_i = m_i a_i + \sum_j n_{ij} b_j$. Clearly the map $\pi : a_i \mapsto x_i, b_i \mapsto 0$ is a surjection of V onto T, and its kernel is generated by $m_i a_i$ and b_i, or just as well by $\bar{b}_i$ and b_i. In other words, $\ker \pi = B + \bar{B}$. Observe that $\bar{B}$ is isotropic since

$$\bar{b}_i \cdot \bar{b}_k = m_i n_{ki} - m_k n_{ik} = m_i m_k q_{ki} - m_k m_i q_{ik} = 0.$$

Hence we have a linking λ' induced on T, as in Lemma 5.7, by the isotropic pair $B, \bar{B}$. An easy calculation shows that $\lambda' = \lambda$, and hence is non-singular by hypothesis. Now rank $\bar{B}$ is obviously = rank A = $\frac{1}{2}$rank V, and $B \cap \bar{B} = 0$: for $\sum_i r_i \bar{b}_i \in B$ if and only if $\sum_i r_i m_i a_i = 0$, *i.e.* if and only if $r_i = 0$ all i (since $m_i \neq 0$). We now apply Lemma 5.7 to conclude that $\bar{B}$ is lagrangian and so $(V; B, \bar{B}; \pi)$ is a Heegaard presentation of T with genus equal to the rank of T. Q.E.D.

Our next goal is to show that a minimal Heegaard pair has a natural volume in the sense of §3.2.

Lemma 5.19. *Let $(V; B, \bar{B})$ be a minimal Heegaard pair of genus g with quotient H. Then for any two dual complements A_i of B $(i = 1, 2)$ the two presentations $A_i \to H$ induce the same volumes. We shall call this volume the volume induced by the Heegaard pair.*

Proof. The direct sum projection of $V = A_2 \oplus B$ onto A_2 gives a map $j : A_1 \to A_2$, and j is an *isomorphism.* Clearly,

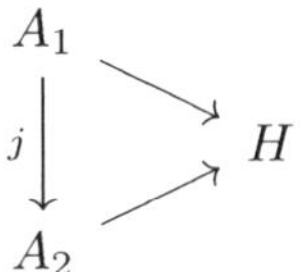

commutes, so the presentations are equivalent and have the same volume. Q.E.D.

We can strengthen Theorem 5.18 in the minimal volume case.

Theorem 5.20. *Let $(V_i; B_i, \bar{B}_i)$, $(i = 1, 2)$ be minimal Heegaard pairs of genus g with quotients H_i and induced volumes $\pm\theta_i$, and let $h : H_1 \to H_2$ be a linking isomorphism. Then h lifts to a Heegaard isomorphism if and only if h is also volume preserving.*

Proof. Assume that $g = \operatorname{rank} H$. The proof of Theorem 5.15 goes through exactly as is whenever we can lift h to f as in Lemma 5.14, and examining the proof of this lemma, we see that it also goes through as is if we can only lift h to an isomorphism $p : A \to A'$ such that

$$\begin{array}{ccc} A & \longrightarrow & H \\ {\scriptstyle p}\downarrow & & \downarrow{\scriptstyle h} \\ A' & \longrightarrow & H' \end{array}$$

commutes. Since $\operatorname{rank} A = \operatorname{rank} A' = g = \operatorname{rank} H$, the abelian groups H and H' have volumes θ, θ' induced by these presentations, and Theorem 3.26 tells us that h lifts if and only if $h(\pm\theta) = \pm\theta'$, as desired. Q.E.D.

Corollary 5.21. *Every Heegaard pair is isomorphic to a stabilization of a Heegaard pair whose genus is equal to the rank of the quotient.*

Proof. Let $(V; B, \bar{B})$ be of genus g and let its quotient be H of rank r. If $g = r$ we are done. If $g > r$, then by Theorem 5.18 there is a Heegaard presentation of H of genus r, and then by Corollary 5.16, its stabilization of index $k = g - r > 0$ is isomorphic to $(V; B, \bar{B})$. Q.E.D.

Problem 3 asked whether there is a bound, or even more a uniform bound on the stabilization index for arbitrary minimal inequivalent but stably equivalent pairs $\mathcal{H}_1, \mathcal{H}_2 \in \Gamma_g$.

Corollary 5.22. *If two Heegaard splittings of the same 3-manifold W have the same genus, then their associated symplectic Heegaard splittings are either isomorphic or become isomorphic after at most single stabilization. In particular, if the genus of the Heegaard splitting is greater than the rank of $H_1(W;\mathbb{Z})$, then the symplectic Heegaard splittings are always isomorphic.*

Proof. Let $\tilde{h}, \tilde{h}'$ be Heegaard gluing maps of genus g for the same 3-manifold. Let h, h' be their images in $\mathrm{Sp}(2g,\mathbb{Z})$. By Theorem 4.8 we know that the stable double cosets which characterize their stabilized symplectic Heegaard splittings are in 1-1 correspondence with isomorphism classes of associated stabilized Heegaard pairs. Let $(V;B,\bar{B})$, $(V';B',\bar{B}')$ be the Heegaard pairs determined by $h = \rho^2(\tilde{h}), h' = \rho^2(\tilde{h}')$. The fact that $\tilde{h}, \tilde{h}'$ determine the same 3-manifold W shows that there is a linking isomorphism $H \to H'$ of their linked quotient groups. Theorem 5.15 then asserts that, if $g > \mathrm{rank} H_1(W;\mathbb{Z})$, then there is a Heegaard isomorphism $(V;B,\bar{B}) \to (V';B',\bar{B}')$. In particular, the Heegaard splittings are equivalent. By Theorem 5.18, every linked group H has a Heegaard presentation of genus equal to the rank of H. Thus, at most a single stabilization is required, and that only if the genus is minimal and the Heegaard pairs are not isomorphic. Q.E.D.

§6. The classification problem for linked abelian groups

We have reduced the problem of classifying symplectic Heegaard splittings to the problem of the classification of linked abelian groups. It remains to find a system of invariants that will do the job, and that is our goal in this section.

6.1. Direct sum decompositions

We begin by showing that the problem of finding a complete system of invariants for a linked group (H, λ) reduces to the problem of studying the invariants on the p-primary summands of the torsion subgroup T of H.

Theorem 6.1 ([43]). *Every linking form on T splits as a direct sum of linkings associated to the p-primary summands of T, and two linking* *forms are equivalent if and only if the linkings on the summands are equivalent.*

Proof. Let $x, y \in T$ where x has order m and y has order n. Then $\lambda(x,y) = \lambda(x, y+0) = \lambda(x,y) + \lambda(x,0)$, hence $\lambda(x,0) = 0$. From this it follows that $n\lambda(x,y) = \lambda(x, ny) = \lambda(x,0) = 0 (\text{mod } 1)$ because $ny = 0$. By the symmetry of linking numbers, we also have $m\lambda(x,y) = 0(\text{mod } 1)$. Therefore $\lambda(x,y) = \frac{r}{m} = \frac{s}{n} (\text{mod } 1)$ for some integers r, s. This implies that $\lambda(x,y) = \frac{t}{(m,n)} (\text{mod } 1)$ for some integer t, where (m,n) is the greatest common divisor of m and n. Thus if x, y have order p^a, q^b where p, q are distinct primes, then $\lambda(x,y) = 0$. Thus the linking on T splits into a direct sum of linkings on the p-primary summands, as claimed. Q.E.D.

In view of Theorem 6.1, we may restrict our attention to a summand $T(p_j)$ of T of prime power order p_j^t, where $p_j \in \{p_1, p_2, \ldots, p_k\}$, the set of prime divisors of the largest torsion coefficient τ_t. This brings us, immediately, to a very simple question: how do we find the linking form on $T(p_j)$ from a symplectic Heegaard splitting? Our next result addresses this issue.

Corollary 6.2. *Let T be the torsion subgroup of $H_1(W;\mathbb{Z})$. Let $\mathcal{Q}^{(2)} = ||q_{ij}||$ and $\mathcal{P}^{(2)} = \text{Diag}(\tau_1, \tau_2, \ldots, \tau_t)$ be the matrices that are given in Theorem 2.4.*

(1) *The $t \times t$ matrix $\mathcal{Q}^{(2)}(\mathcal{P}^{(2)})^{-1} = \|\lambda(y_r, y_j)\| = ||\frac{q_{ij}}{\tau_j}||$ determines a linking on H.*

(2) *The linking matrices that were studied by Seifert in [43] are the direct sum of k distinct $t \times t$ matrices, one for each prime divisor p_d of τ_t. Each summand represents the restriction of the linking in (1) to the cyclic summands of T whose order is a fixed power of p_d. The one that is associated to the prime p_d is a matrix of dimension at most $t \times t$:*

$$\lambda(g_{id}, g_{jd}) = ||\frac{\tau_i \tau_j q_{ij}}{(p_d^{e_{i,d}})(p_d^{e_{jd}})\tau_j}|| = ||\frac{\tau_i q_{ij}}{(p_d^{e_{id}})(p_d^{e_{jd}})}|| \tag{21}$$

Proof. (1) The easiest way to see that $\mathcal{Q}^{(2)}(\mathcal{P}^{(2)})^{-1}$ is a linking on H is from the geometry. The matrix $\mathcal{Q}^{(2)}$ records the number of appearances (algebraically) of each homology basis element b_i in $h(b_j)$. The curves which represent the b_i's bound discs D_i in N_g, and the curves which represent the $h(b_j)$'s bounds discs $\bar{D}_j$ in $\bar{N}_g$. The g_{ij}'s are intersection numbers of $h(b_j)$ with D_i or of b_i with $\bar{D}_j$. Dividing by τ_j, intersection numbers go over to linking numbers. Note that the submatrix

$\mathcal{P}^{(2)}\mathcal{Q}^{(2)}$ is symmetric by (2), hence $\mathcal{Q}^{(2)}(\mathcal{P}^{(2)})^{-1}$ is likewise symmetric, as it must be because $\mathcal{Q}^{(2)}(\mathcal{P}^{(2)})^{-1}$ is a linking.

(2) By Theorem 6.1, T is a direct sum of p-primary groups $T(p_1) \oplus \cdots \oplus T(p_k)$. From this it follows that the linking on T also splits as a direct sum of the linkings associated to $T(p_1), \ldots, T(p_k)$.

We focus on one such prime $p = p_d$. By Theorem 2.2 the p-primary group $T(p)$ splits in a unique way as a direct sum of cyclic groups whose orders are powers of p, moreover the powers of p which are involved occur in a non-decreasing sequence, as in (4) of Theorem 2.2. The generators of these groups are ordered in a corresponding way, as $g_{1,d}, g_{2,d}, \ldots, g_{t,d}$, where distinct generators $g_{i,d}, g_{i+1,d}$ may generate cyclic groups of the same order, and where it is possible that the first q of these groups are trivial. As in the statement of Theorem 2.2, the generators $g_{i,d}$ and y_i are related by (5). The expression on the right in (21) follows immediately. There are k such matrices in all, where k is the number of distinct prime divisors of τ_t. Q.E.D.

6.2. Classifying linked p-groups when p is odd. Solution to Problem 2, odd p.

In this section we describe Seifert's classification theorem for the case when all of the torsion coefficients are odd. Seifert studied the $t \times t$ matrix $\lambda(g_i, g_j)$ in (21) belonging to a fixed prime $p = p_d$. To explain what he did, we start with the inequalities in (4), but restrict to a subsequence of cyclic groups all of which have the same prime power order. We simplify the notation, using the symbols

$$\varepsilon_1 = \cdots = \varepsilon_1 < \varepsilon_2 = \cdots = \varepsilon_2 < \cdots < \varepsilon_r = \varepsilon_r = \cdots \varepsilon_r$$

in place of the powers $e_{i,d}$ which appear in (21). The linking matrix then divides into blocks whose size is determined by the number of times, denoted t_i, that a given power, say ε_i, is repeated. Among these, the blocks that interest us are the square blocks whose diagonals are along the main diagonal of the linking matrix. There will be r such blocks of dimension $t_1, \ldots, t_r$ if r distinct powers p^{ε_i} occur in the subgroups of T that are cyclic with order a power of p:

$$\|\lambda(g_i, g_j)\| = \begin{pmatrix} \frac{\mathcal{A}_1}{p^{\varepsilon_1}} & * & \cdots & * \\ * & \frac{\mathcal{A}_2}{p^{\varepsilon_2}} & \cdots & * \\ \vdots & \vdots & \ddots & \vdots \\ * & \cdots & \cdots & \frac{\mathcal{A}_r}{p^{\varepsilon_r}} \end{pmatrix} \tag{22}$$

The stars relate to linking numbers that we shall not consider further.

Theorem 6.3 ([43]). *Two linkings of $T(p)$ are equivalent if and only if the corresponding box determinants $|\mathcal{A}_1|, |\mathcal{A}_2|, \dots, |\mathcal{A}_r|$ have the same quadratic residue characters* mod p.

Summarizing, we can now give the promised solution to Problem 2 in the case when T has no 2-torsion.

Theorem 6.4. (1) *We are given the gluing map $\tilde{h} \in \tilde{\Gamma}_g$ for a Heegaard splitting of genus g of a 3-manifold W. Let*

$$\mathcal{H} = \rho^2(\tilde{h}) = \begin{pmatrix} \mathcal{R} & \mathcal{P} \\ \mathcal{S} & \mathcal{Q} \end{pmatrix} \in \mathrm{Sp}(2g, \mathbb{Z}).$$

(2) *Use the methods described in the proof of Theorem* II.9 *of* [34] *to find matrices $\mathcal{U}, \mathcal{V}$ such that $\mathcal{UPV} = \mathcal{P}^{(1)} = \mathrm{Diag}(1, 1, \dots, 1, \tau_1, \tau_2, \dots, \tau_t, 0, \dots, 0)$.*

(3) *Use the methods described in the proof of Theorem* 2.4 *to find the equivalent matrix $\mathcal{H}'$.*

(4) *Let y_i be a generator of the subgroup of order τ_i of T. Let $p_1 < p_2 < \cdots < p_k$ be the primes divisors of $\tau_1, \tau_2, \dots, \tau_t$. Compute the elements g_{ij}, using* (5)*. Then compute the k symmetric matrices $\|\lambda(g_i, g_j)\|$, using* (21) *above.*

(5) *Using the matrices $\|\lambda(g_i, g_j)\|$, determine the submatrices $\mathcal{A}_1, \dots, \mathcal{A}_r$ that are shown in* (22) *above. Here each block matrix $\mathcal{A}_q$ belongs to a sequence (possibly of length* 1*) of cyclic subgroups of $T(p)$ of like prime power order. The matrix $\mathcal{A}_q$ might be the identity matrix. Compute the quadratic residue characters* mod p *of the determinants $|\mathcal{A}_1|, |\mathcal{A}_2|, \dots, |\mathcal{A}_r|$. Repeat this for each p.*

(6) *By Corollary* 5.16 *and Theorem* 6.3*, the rank r, the torsion coefficients $\tau_1, \dots, \tau_t$ and the complete array of quadratic reside characters* mod p *are the complete set of topological invariants of the associated symplectic Heegaard splitting of W that is determined by $\mathcal{H}$.*

6.3. Classifying linked p-groups, $p = 2$. Solution to Problem 2, $p = 2$.

If (H, λ) is a linked p-group, we have seen that H may be decomposed into orthogonal summands B_j, each of which is a free $\mathbb{Z}_{p^j}$-module, that is, a direct sum of some number r_j of copies of $\mathbb{Z}_{p^j}$; r_j is the *rank* of B_j, and B_j is the jth *block*. The r_j's are as usual invariants of H alone. For odd p, the linking type of $\lambda|_{B_j}$ is an invariant of λ, and these types (which are determined by a Legendre symbol) give a complete set of invariants. For $p = 2$, however, no such decomposition is possible; this is due to the fact that the linking type of B_j *is no longer an invariant.* Here is a typical example of the kind of thing that can happen.

Definition 6.5. An element $x \in H$ is said to be *primitive* if it generates a direct summand of H, or, equivalently, if $x \notin 2H$. ||

Example 6.6. Let $H = \mathbb{Z}_{2^{n-1}} \oplus \mathbb{Z}_{2^n} \oplus \mathbb{Z}_{2^n}$, so H has rank 3. Let

$$\mathcal{C} = \begin{pmatrix} 0 & 1 \\ 1 & 0 \end{pmatrix} \quad \text{and} \quad \mathcal{D} = \begin{pmatrix} 2 & 1 \\ 1 & 2 \end{pmatrix},$$

and consider the two linking forms λ and λ' on H whose matrices with respect to the standard basis are

$$(\frac{1}{2^{n-1}}) \oplus (\frac{1}{2^n}\mathcal{C}) \quad \text{and} \quad (\frac{-3}{2^{n-1}}) \oplus (\frac{1}{2^n}\mathcal{D}), \tag{23}$$

respectively, where $(\frac{1}{2^{n-1}})$ and $(\frac{-3}{2^{n-1}})$ denote 1×1 matrices. We claim that $\lambda \simeq \lambda'$. Indeed, let $\vec{e} = (e_1, e_2, e_3)$ be the standard basis of H and put

$$e_1' = e_1 + 2e_2 - 2e_3, \qquad e_2' = e_1 + e_2, \qquad e_3' = e_1 - e_3.$$

Now, we have:

$$3e_1 = -e_1' + 2e_2' + 2e_3', \qquad 3e_2 = e_1' + e_2' - 2e_3', \qquad 3e_3 = -e_1' + 2e_2' - e_3'.$$

Since 3 is a unit in $\mathbb{Z}_{2^n}$, it follows that $\vec{e'} = (e_1', e_2', e_3')$ is also a basis. Now, observe that

$$\begin{aligned}
\lambda(e_1', e_1') &= \lambda(e_1, e_1) - 8\lambda(e_2, e_3) = \frac{-3}{2^{n-1}} \\
\lambda(e_2', e_2') &= \lambda(e_1, e_1) = \frac{2}{2^n} \\
\lambda(e_3', e_3') &= \lambda(e_1, e_1) = \frac{2}{2^n} \\
\lambda(e_1', e_2') &= \lambda(e_1, e_1) - 2\lambda(e_3, e_2) = 0 \\
\lambda(e_1', e_3') &= \lambda(e_1, e_1) - 2\lambda(e_3, e_2) = 0 \\
\lambda(e_2', e_3') &= \lambda(e_1, e_1) - \lambda(e_2, e_3) = \frac{1}{2^n}
\end{aligned}$$

Thus λ with respect to the basis $\vec{e'}$ is equal to λ' with respect to the basis $\vec{e}$. But $\frac{1}{2^n}\mathcal{C}$ is *not* isomorphic to $\frac{1}{2^n}\mathcal{D}$ as a linking on $\mathbb{Z}_{2^n}^2$ when $n \geqslant 2$. For example, $\lambda(x, x) = 0$ has no primitive solutions for the second linking, whereas it does for the first. ||

Our example shows that we must approach the case $p = 2$ in a different manner. Burger (see [7]) reduced the classification of linked p-groups to the classification of quadratic forms (more precisely, symmetric bilinear forms) over $\mathbb{Z}_{p^n}$, and his procedure gives in theory at least a complete set of invariants, but for $p = 2$ they are inconveniently cumbersome. Our goal in this section is to reduce his invariants to a simple, manageable set and to demonstrate how to calculate them. We will proceed as follows:

- In §6.3.1 we show how to decompose a linking form (in a non-canonical way) into a direct sum of 3 types of basic forms.
- In §6.3.2 we discuss a variant of the Burger's numerical invariants.
- Our variant of Burger's invariants cannot achieve arbitrary values. To determine the values they can achieve, we calculate the values of our numerical invariants on the basic forms. This calculation will be the basis of the next step, given in §6.3.3.
- Finally, in §6.3.4 we show that all the information in Burger's numerical invariants is contained in a simpler set of mod 8 *phase invariants* together with the ranks of the various blocks.

The section ends with several examples.

6.3.1. *Decomposing a linked abelian 2-group* Strengthening a result of Burger [7], Wall [47] showed how to decompose a linked abelian 2-group into an orthogonal direct sum of certain basic linking forms. In this section, we review this result and give a proof of Wall's theorem which builds on Burger's original ideas and which is better suited for computations.

Definition 6.7. The *basic linking forms* on a finite abelian 2-group are the following

- The *unary forms*. These are the forms on $\mathbb{Z}_{2^j}$ for $j \geq 1$ whose matrices are $\left(\frac{a}{2^j}\right)$ for odd integers a.
- The two *binary forms*. These are the forms on $(\mathbb{Z}_{2^j})^2$ for $j \geq 1$ whose matrices are either $\frac{1}{2^j}\mathcal{C}$ or $\frac{1}{2^j}\mathcal{D}$, where $\mathcal{C}$ and $\mathcal{D}$ were defined in Example 6.6. ||

We can now state Wall's theorem.

Theorem 6.8 (Wall, [47]). *Let (H, λ) be a linked abelian 2-group. Then (H, λ) is isomorphic to an orthogonal direct sum $(H_1, \lambda_1) \oplus \cdots \oplus (H_n, \lambda_n)$, where the (H_i, λ_i) are basic linking forms.*

Proof. We will use the following well-known result about solving congruences mod p^k.

Lemma 6.9 (Hensel's Lemma [35, Theorem 2.23]). *Let $f(x)$ be a polynomial with integer coefficients, let p be a prime and let $k \geq 2$ be an integer. Assume that the integer r is a solution to the equation $f(x) = 0 \pmod{p^{k-1}}$, and moreover assume that $f'(r) \neq 0 \bmod p$. Then $f(r+tp^{k-1}) = 0 \pmod{p^k}$ for some (unique) integer t with $0 \leq t \leq p-1$.*

The proof of Theorem 6.8 begins with a lemma:

Lemma 6.10. *Let $H = \mathbb{Z}_{p^j} \oplus \mathbb{Z}_{p^j}$ and let λ be the linking form on H whose matrix with respect to the standard basis (e_1, e_2) for H is $\frac{1}{2^j}\left(\begin{smallmatrix} 2m & 1 \\ 1 & 2n \end{smallmatrix}\right)$. Then (H, λ) is isomorphic to the linking form whose matrix is $\frac{1}{2^j}C$ if both m and n are odd and to the linking form whose matrix is $\frac{1}{2^j}D$ otherwise.*

Proof. Assume first that one of m and n (say m) is even. Observe that if a is an integer, then since λ is defined by the matrix $\frac{1}{2^j}\left(\begin{smallmatrix} 2m & 1 \\ 1 & 2n \end{smallmatrix}\right)$, we have:

$$\lambda((1,a),(1,a)) = 2m + 2a^2n + 2a = 2(na^2 + a + m).$$

Now, since m is even it follows that $a = 1$ is a solution to the equation $na^2 + a + m = 0 \bmod 2$, so Hensel's lemma implies that there is some odd integer $\tilde{a}$ so that $n\tilde{a}^2 + \tilde{a} + m = 0 \bmod 2^j$. Set $\vec{v} = (1, a)$. Since λ is nondegenerate, there is some $\vec{w} \in H$ so that $\lambda(\vec{v}, \vec{w}) = \frac{1}{2^j}$. Let k be so that $\lambda(\vec{w}, \vec{w}) = 2k$. The pair $\{\vec{v}, \vec{w} - k\vec{v}\}$ is then a new basis, and with respect to this basis λ has the matrix $\frac{1}{2^j}\left(\begin{smallmatrix} 0 & 1 \\ 1 & 0 \end{smallmatrix}\right)$, as desired.

Now assume that both m and n are odd. By the same argument as in the previous paragraph (but solving $na^2 + a + m = 1 \bmod 2^j$), we can assume that $m = 1$, so the matrix for λ is $\frac{1}{2^j}\left(\begin{smallmatrix} 2 & 1 \\ 1 & 2n \end{smallmatrix}\right)$. Set $\vec{v} = (1, 0)$, and for an arbitrary integer c define $\vec{w}_c = (c, 1 - 2c)$. Observe that

$$\lambda(\vec{v}, \vec{w}_c) = 2c + (1 - 2c) = 1.$$

Moreover,

$$\lambda(\vec{w}_c, \vec{w}_c) = 2c^2 + 2c(1-2c) + 2n(1-2c)^2 = 2((4n-1)c^2 + (1-4n)c + n).$$

Now, since n is odd it follows that $c = 1$ is a solution to

$$(4n-1)c^2 + (1-4n)c + n = 1 \pmod 2,$$

so Hensel's lemma implies that there is some odd integer $\tilde{c}$ so that

$$(4n-1)\tilde{c}^2 + (1-4n)\tilde{c} + n = 1 \pmod{2^j}.$$

It follows that $\lambda(\vec{w}_{\tilde{c}}, \vec{w}_{\tilde{c}}) = 2$. Observe that since $\tilde{c}$ is odd the vectors $\vec{v}$ and $\vec{w}_{\tilde{c}}$ form a basis, and with respect to this basis λ has the matrix $\frac{1}{2^j}\left(\begin{smallmatrix} 2 & 1 \\ 1 & 2 \end{smallmatrix}\right)$, as desired. Q.E.D.

We are ready to prove Theorem 6.8. For $x \in H$, define the *order* of x to be the smallest nonnegative integer n so that $2^n x = 0$. Assume first that there is some primitive $x \in H$ of order j so that $\lambda(x,x) = \frac{a}{2^j}$ with a odd. Construct a basis $x = x_1, x_2, \ldots, x_N$ for H. Observe now that for any $2 \leq i \leq N$, we have $\lambda(x, x_i) = \frac{b_i}{2^j}$ for some integer b_i. Since a is odd, we can write $b_i = c_i a \bmod 2^j$ for some integer c_i. Set $x_i' = x_i - c_i x$. Observe that $x = x_1, x_2', \ldots, x_N'$ is a new basis for H, and that moreover (H, λ) has an orthogonal direct sum decomposition

$$\langle x \rangle \oplus \langle x_2', \ldots, x_N' \rangle.$$

Since λ restricted to $\langle x \rangle$ is the basic form with matrix $\left(\frac{a}{2^j}\right)$, we are done by induction.

We can therefore assume that for all $x \in H$, if j is the order of x then $\lambda(x,x) = \frac{2n}{2^j}$ for some integer n. Fixing some primitive $x \in H$ of order j and letting n be the integer with $\lambda(x,x) = \frac{2n}{2^j}$, the fact that λ is non-singular implies that there is some $y \in H$ so that $\lambda(x,y) = \frac{1}{2^j}$. Necessarily y has order j, so we can write $\lambda(y,y) = \frac{2m}{2^j}$ for some integer m. Construct a basis $x = z_1, y = z_2, z_3, \ldots, z_N$ for H, and for $3 \leq i \leq N$ write $\lambda(x, z_i) = \frac{h_i}{2^j}$ and $\lambda(y, z_i) = \frac{k_i}{2^j}$. Since the matrix $\left(\begin{smallmatrix} 2n & 1 \\ 1 & 2m \end{smallmatrix}\right)$ is invertible mod 2^j, we can find integers a_i and b_i so that

$$\begin{pmatrix} 2n & 1 \\ 1 & 2m \end{pmatrix} \cdot \begin{pmatrix} a_i \\ b_i \end{pmatrix} = \begin{pmatrix} -h_i \\ -k_i \end{pmatrix}$$

mod 2^j. For $3 \leq i \leq N$, set $z_i' = z_i + a_i x + b_i y$, and observe that $\lambda(x, z_i') = \lambda(y, z_i') = 0$. The linked group (H, λ) now has an orthogonal direct sum decomposition

$$\langle x, y \rangle \oplus \langle z_3', \ldots, z_N' \rangle.$$

Since λ restricted to $\langle x, y \rangle$ is the linking form with matrix $\left(\begin{smallmatrix} 2n & 1 \\ 1 & 2m \end{smallmatrix}\right)$, Lemma 6.10 says that it is one of the basic forms, and we are done by induction. This completes the proof of Wall's theorem. Q.E.D.

6.3.2. *Burger's numerical invariants* In [7], Burger constructs a complete set of numerical invariants of a linked abelian p-group (H, λ). The following formulation of Burger's result was suggested by Fox in [13], and is manifestly equivalent to Burger's original formulation. Define the *degree* of an abelian p-group to be the smallest nonnegative integer n so that $p^n H = 0$.

Definition 6.11. Let (H, λ) be a linked abelian p-group of degree n, and let $a = \frac{k}{p^n}$ for some $0 \leq k < p^n$. We then define $\mathbf{N}_a(\lambda)$ to be the number of solutions $x \in H$ to the equation $\lambda(x, x) = a$. $\|$

Theorem 6.12 (Burger, [7]). *Fix an abelian p-group H of degree n. Two linking forms λ and λ' on H are then isomorphic if and only if $\mathbf{N}_a(\lambda) = \mathbf{N}_a(\lambda')$ for all $a = \frac{k}{p^n}$ with $0 \leq k < p^n$.* $\|$

We now confine ourselves to the case $p = 2$. Fix a linked abelian 2-group (H, λ) of degree n. We shall use the notation $E(z)$ for $e^{2\pi i z}$. Also, to simplify our formulas we shall denote $\lambda(x, x)$ by x^2. Observe now that if $x \in H$, then $x^2 = \frac{k}{2^n}$ where k is an integer which is well-defined mod 2^n. Hence if $b \in \mathbb{Z}_{2^n}$, then the expression $bx^2 = \frac{bk}{2^n}$ is a well-defined real number mod 1, so the number $E(bx^2)$ is well-defined. The following definition therefore makes sense.

Definition 6.13. For $b \in \mathbb{Z}_{2^n}$, define $\overline{\Gamma}_b(\lambda) := \sum_{x \in H} E(bx^2)$. $\|$

Observe that we clearly have the identity

$$\overline{\Gamma}_b(\lambda) = \sum_{k=0}^{2^n-1} \mathbf{N}_{\frac{k}{p^n}}(\lambda) E(\frac{bk}{2^n}).$$

In other words, the numbers $\overline{\Gamma}_b(\lambda)$ are the result of applying the *discrete Fourier transform* (see [45]) to the numbers $\mathbf{N}_a(\lambda)$. In particular, since the discrete Fourier transform is invertible, it follows that the numbers $\overline{\Gamma}_b(\lambda)$ for $b \in \mathbb{Z}_{2^n}$ also form a complete set of invariants for linking forms on H. In fact, since $\overline{\Gamma}_0(\lambda) = \sum_{x \in H} 1 = |H|$, we only need $\overline{\Gamma}_b(\lambda)$ for $b \neq 0$.

However, there is a certain amount of redundancy among the $\overline{\Gamma}_b(\lambda)$. Indeed, set $\theta = E(1/2^n)$. Observe that θ is a $(2^n)^{\text{th}}$ root of unity and that the numbers $\overline{\Gamma}_b(\lambda)$ lie in $\mathbb{Q}(\theta)$. If a is any odd integer, then θ^a is another $(2^n)^{\text{th}}$ root of unity, and $\theta \mapsto \theta^a$ defines a Galois automorphism of $\mathbb{Q}(\theta)$, which we will denote by g_a. Consider a nonzero $b \in \mathbb{Z}_{2^n}$. Write

$b = 2^k b_0$, where b_0 is odd and $k < n$. Since $2^n x^2$ is an integer for all $x \in H$, we have

$$\begin{aligned}\overline{\Gamma}_b(\lambda) &= \sum_{x\in H} e^{\frac{2\pi i(2^k b_0)(2^n x^2)}{2^n}} = \sum_{x\in H}\left(e^{\frac{2\pi i b_0}{2^n}}\right)^{2^{k+n}x^2} \\ &= g_{b_0}\left(\sum_{x\in H}\left(e^{\frac{2\pi i}{2^n}}\right)^{2^{k+n}x^2}\right) = g_{b_0}\left(\sum_{x\in H} E(2^k x^2)\right) = g_{b_0}(\overline{\Gamma}_{2^k}(\lambda)).\end{aligned}$$

We conclude that the set of all $\overline{\Gamma}_b(\lambda)$ for $b \in \mathbb{Z}_{2^n}$ can be calculated from the set of all $\overline{\Gamma}_{2^k}(\lambda)$ for $0 \le k < n$. This discussion is summarized in the following definition and lemma.

Definition 6.14. Let λ be a linking form on H. For $0 \le k < n$, define $\Gamma_k(\lambda) := \overline{\Gamma}_{2^k}(\lambda)$. $\|$

Lemma 6.15. *The set of numbers $\Gamma_k(\lambda)$ for $0 \le k < n$ form a complete set of invariants for linking forms on H.*

6.3.3. *Calculating the numerical invariants for the basic forms* By Theorem 6.8, we can decompose any linked abelian group into a direct sum of basic linking forms. The following lemma shows how this reduces the calculation of the numbers $\Gamma_k(\lambda)$ to the knowledge of the $\Gamma_k(\cdot)$ for the basic forms.

Lemma 6.16. *If (H, λ) is the orthogonal direct sum $A_1 \oplus A_2$ with linkings λ_i on A_i, then for $0 \le k < n$ we have $\Gamma_k(\lambda) = \Gamma_k(\lambda_1) \cdot \Gamma_k(\lambda_2)$.*

Proof. For $x = (x_1, x_2)$:

$$\begin{aligned}\Gamma_k(\lambda) &= \sum_{x\in H} E(2^k x^2) = \sum_{x_1\in A_1}\sum_{x_2\in A_2} E(2^k(x_1^2 + x_2^2)) \\ &= \left(\sum_{x_1\in A_1} E(2^k x_1^2)\right)\cdot\left(\sum_{x_2\in A_2} E(2^k x_2^2)\right) = \Gamma_k(\lambda_1)\cdot\Gamma_k(\lambda_2).\end{aligned}$$

Q.E.D.

In the remainder of this section, we calculate the numbers $\Gamma_k(\cdot)$ for the basic forms. Recall that $\mathcal{C} = \left(\begin{smallmatrix}0&1\\1&0\end{smallmatrix}\right)$ and $\mathcal{D} = \left(\begin{smallmatrix}2&1\\1&2\end{smallmatrix}\right)$. The calculation is summarized in the following proposition.

Proposition 6.17. *Define $\rho = E(1/8)$ and*

$$\varepsilon(a) = \begin{cases} 1 \bmod 8 & \text{if } a \equiv 1 \bmod 4 \\ -1 \bmod 8 & \text{if } a \equiv -1 \bmod 4 \end{cases}$$

Then for $j \geq 1$ and $k \geq 0$, we have

- $\Gamma_k\left(\frac{a}{2^j}\right) = \begin{cases} 2^j & \textit{if } k \geqslant j \\ 0 & \textit{if } k = j-1 \\ 2^{\frac{j+k+1}{2}}\rho^{\varepsilon(a)} & \textit{if } j-k \geqslant 2 \textit{ and is even} \\ 2^{\frac{j+k+1}{2}}\rho^{a} & \textit{if } j-k \geqslant 2 \textit{ and is odd.} \end{cases}$
- $\Gamma_k\left(\frac{\mathcal{C}}{2^j}\right) = \begin{cases} 2^{2j} & \textit{if } k \geqslant j-1 \\ 2^{j+k+1} & \textit{if } k < j-1. \end{cases}$
- $\Gamma_k\left(\frac{\mathcal{D}}{2^j}\right) = \begin{cases} 2^{2j} & \textit{if } k \geqslant j-1 \\ (-1)^{j+k+1}2^{j+k+1} & \textit{if } k < j-1. \end{cases}$

Proof of Proposition 6.17 *for* $\left(\frac{a}{2^j}\right)$. We begin with the case $k \geq j$. For these values of k, the number $2^k x^2$ is an integer for all $x \in H$, so $E(2^k x^2) = 1$ for all $x \in H$. This implies that

$$\Gamma_k\left(\frac{a}{2^j}\right) = |H| = 2^j,$$

as desired.

The next step is to prove the following formula, which reduces the remaining cases to the case $a = 1$ and $k = 0$:

$$\Gamma_k\left(\frac{a}{2^j}\right) = 2^k g_a\left(\Gamma_0\left(\frac{1}{2^{j-k}}\right)\right) \qquad \text{for } k < j \tag{24}$$

We calculate:

$$\Gamma_k\left(\frac{a}{2^j}\right) = \sum_{x \bmod 2^j} E\left(\frac{2^k a x^2}{2^j}\right) = g_a\left(\sum_{x \bmod 2^j} E\left(\frac{x^2}{2^{j-k}}\right)\right).$$

Since $E(\frac{x^2}{2^{j-k}})$ depends only on x mod 2^{j-k}, this equals

$$g_a\left(2^k \sum_{x \bmod 2^{j-k}} E\left(\frac{x^2}{2^{j-k}}\right)\right) = 2^k g_a\left(\Gamma_0\left(\frac{1}{2^{j-k}}\right)\right),$$

as desired.

To simplify our notation, define $\mathsf{T}_k = \Gamma_0(\frac{1}{2^k})$. We will prove that

$$\mathsf{T}_k = \begin{cases} 0 & \text{if } k = 1 \\ 2^{\frac{k+1}{2}}\rho & \text{if } k \geq 2 \end{cases} \tag{25}$$

First, however, we observe that the proposition follows from Equations (25) and (24). Indeed, for $k = j - 1$ this is immediate. For $k < j - 1$ with $j - k$ odd, we have

$$\Gamma_k\left(\frac{a}{2^j}\right) = 2^k g_a(2^{\frac{j-k+1}{2}}\rho) = 2^{\frac{j+k+1}{2}}\rho^a.$$

Finally, for $k < j-1$ with $j-k$ even, using the fact that $\sqrt{2}\rho^{\varepsilon(\pm 1)} = 1 \pm i$ we have

$$\begin{aligned}\Gamma_k\left(\frac{a}{2^j}\right) &= 2^k g_a(2^{\frac{j-k+1}{2}}\rho) = 2^{\frac{j+k}{2}} g_a(\sqrt{2}\rho) = 2^{\frac{j+k}{2}} g_a(1+i)\\ &= 2^{\frac{j+k}{2}}(1+i^a) = 2^{\frac{j+k+1}{2}}\rho^{\varepsilon(a)},\end{aligned}$$

as desired.

The proof of Equation (25) will be by induction on k. The base cases $1 \le k \le 3$ are calculated as follows:

$$\mathsf{T}_1 = E(0) + E(\frac{1}{2}) = 1 - 1 = 0$$
$$\mathsf{T}_2 = E(0) + E(\frac{1}{4}) + E(\frac{4}{4}) + E(\frac{9}{4}) = 1 + i + 1 + i = 2(1+i) = 2\sqrt{2}\rho$$
$$\mathsf{T}_3 = 2(E(0) + E(\frac{1}{8}) + E(\frac{4}{8}) + E(\frac{1}{8})) = 2(1 + \rho - 1 + \rho) = 4\rho$$

Assume now that $k \geq 4$. We must prove that $\mathsf{T}_k = 2\mathsf{T}_{k-2}$. To see this, note first that $2^{k-1}+1$ is a *square* mod 2^k. Indeed,

$$(2^{k-2}+1)^2 = (2^{k-2})^2 + 2^{k-1} + 1 \equiv 2^{k-1} + 1 \bmod 2^k$$

when $k \geqslant 4$. Hence $x^2 \mapsto (2^{k-1}+1)x^2$ defines a bijection of the set of squares modulo 2^k. This implies that

$$\begin{aligned}\mathsf{T}_k &= \sum_{x \bmod 2^k} E\left(\frac{x^2}{2^k}\right) = \sum_{x \bmod 2^k} E\left(\frac{(2^{k-1}+1)x^2}{2^k}\right)\\ &= \sum_{x \bmod 2^k} E\left(\frac{x^2}{2^k}\right) E\left(\frac{x^2}{2}\right) = \sum_{x \bmod 2^k} (-1)^x E\left(\frac{x^2}{2^k}\right),\end{aligned}$$

and hence

$$\begin{aligned}2\mathsf{T}_k &= \sum_{x \bmod 2^k} E\left(\frac{x^2}{2^k}\right) + \sum_{x \bmod 2^k} (-1)^x E\left(\frac{x^2}{2^k}\right)\\ &= 2 \sum_{\text{even } x \bmod 2^k} E\left(\frac{x^2}{2^k}\right).\end{aligned}$$

We conclude that

$$\mathsf{T}_k = \sum_{y \bmod 2^{k-1}} E\left(\frac{(2y)^2}{2^k}\right) = \sum_{y \bmod 2^{k-1}} E\left(\frac{y^2}{2^{k-2}}\right)$$
$$= 2 \sum_{y \bmod 2^{k-2}} E\left(\frac{y^2}{2^{k-2}}\right) = 2\mathsf{T}_{k-2}.$$

Q.E.D.

Proof of Proposition 6.17 *for* $\left(\frac{\mathcal{C}}{2^j}\right)$. Here H consists of all pairs (x, y) with $x, y \in \mathbb{Z}_{2^j}$. We begin with the case $k \geq j-1$. For these values of k, the number $2^k(x,y)^2 = 2^k \frac{2xy}{2^j}$ is an integer for all $(x, y) \in H$, so $E(2^k(x,y)^2) = 1$ for all $(x, y) \in H$. This implies that

$$\Gamma_k\left(\frac{a}{2^j}\right) = |H| = 2^{2j},$$

as desired.

The next step is to prove the following formula:

$$\Gamma_k\left(\frac{\mathcal{C}}{2^j}\right) = 2^{2k+2} \sum_{x,y \bmod 2^{j-k-1}} E\left(\frac{xy}{2^{j-k-1}}\right) \qquad \text{for } k < j-1 \tag{26}$$

We calculate:

$$\Gamma_k\left(\frac{\mathcal{C}}{2^j}\right) = \sum_{x,y \bmod 2^j} E\left(\frac{2^k(2xy)}{2^j}\right) = \sum_{x,y \bmod 2^j} E\left(\frac{xy}{2^{j-k-1}}\right).$$

Since $E(\frac{xy}{2^{j-k-1}})$ depends only on x and y mod 2^{j-k-1}, this equals

$$2^{2k+2} \sum_{x,y \bmod 2^{j-k-1}} E\left(\frac{xy}{2^{j-k-1}}\right),$$

as desired.

Define

$$\mathsf{U}_k = \sum_{x,y \bmod 2^k} E\left(\frac{xy}{2^k}\right).$$

By Formula (26), to prove the proposition it is enough to prove that $\mathsf{U}_k = 2^k$ for $k \geq 0$. The proof of this will be by induction on k. The base cases $k = 0, 1$ are as follows:

$$\mathsf{U}_0 = E(0) = 1,$$
$$\mathsf{U}_1 = E(0) + E(0) + E(0) + E(\frac{1}{2}) = 1 + 1 + 1 - 1 = 2.$$

Assume now that $k \geq 2$. We will show that $\mathsf{U}_k = 4\mathsf{U}_{k-2}$. First, we first fix some y with $0 \leq y \leq 2^k$. Write y as $2^r y_0$ with y_0 odd, and suppose that y is not equal to 0 or 2^{k-1}. This implies that $r < k-1$ and that $2^{k-r+1}+1$ is odd. Using the fact that in the following sum the numbers xy_0 are odd, we have

$$\begin{aligned}\sum_{\text{odd } x \bmod 2^k} E\left(\frac{xy}{2^k}\right) &= \sum_{\text{odd } x} E\left(\frac{(2^{k-r-1}+1)xy}{2^k}\right)\\ &= \sum_{\text{odd } x} E\left(\frac{xy}{2^k}\right) E\left(\frac{2^{k-r-1}x \cdot 2^r y_0}{2^k}\right)\\ &= \sum_{\text{odd } x} E\left(\frac{xy}{2^k}\right) E\left(\frac{xy_0}{2}\right) = -\sum_{\text{odd } x} E\left(\frac{xy}{2^k}\right)\end{aligned}$$

Thus

$$\sum_{\text{odd } x} E\left(\frac{xy}{2^k}\right) = 0 \text{ for all } y \neq 0, 2^{k-1}.$$

In particular, since $k \geqslant 2$, the number 2^{k-1} is even and hence

$$\sum_{\text{odd } x, \text{odd } y} E\left(\frac{xy}{2^k}\right) = 0.$$

Also,

$$\begin{aligned}\sum_{\text{odd } x, \text{even } y} E\left(\frac{xy}{2^k}\right) &= \sum_{\text{odd } x} \left(E\left(\frac{x \cdot 0}{2^k}\right) + E\left(\frac{x \cdot 2^{k-1}}{2^k}\right)\right) = \sum_{\text{odd } x} (1-1)\\ &= 0;\end{aligned}$$

by symmetry,

$$\sum_{\text{even } x, \text{ odd } y} E\left(\frac{xy}{2^k}\right) = 0.$$

We have thus shown that U_k is equal to

$$\begin{aligned}\sum_{\text{even } x, \text{ even } y} E\left(\frac{xy}{2^k}\right) &= \sum_{x,y \bmod 2^{k-1}} E\left(\frac{(2x)(2y)}{2^k}\right) = \sum_{x,y \bmod 2^{k-1}} E\left(\frac{xy}{2^{k-2}}\right)\\ &= 4 \sum_{x,y \bmod 2^{k-2}} E\left(\frac{xy}{2^{k-2}}\right) = 4\mathsf{U}_{k-2},\end{aligned}$$

as desired. Q.E.D.

Proof of Proposition 6.17 *for* $\left(\frac{\mathcal{D}}{2^j}\right)$. The cases $k \geq j-1$ are identical to the analogous cases for $\left(\frac{\mathcal{C}}{2^j}\right)$. For $k < j-1$, we use Lemma 6.16 together with the isomorphism of Example 6.6 to conclude that

$$\Gamma_k\left(\frac{1}{2^{j-1}}\right)\Gamma_k\left(\frac{\mathcal{C}}{2^j}\right) = \Gamma_k\left(\frac{-3}{2^{j-1}}\right)\Gamma_k\left(\frac{\mathcal{D}}{2^j}\right).$$

But by the previously proven cases of Proposition 6.17, we have

$$\frac{\Gamma_k\left(\frac{1}{2^{j-1}}\right)}{\Gamma_k\left(\frac{-3}{2^{j-1}}\right)} = \begin{cases} 1 & \text{if } j-k \text{ is odd (since } \varepsilon(-3) \equiv 1 \bmod 8) \\ \frac{\rho}{\rho^{-3}} = \rho^4 = -1 & \text{if } j-k \text{ is even.} \end{cases}$$

The proposition follows. Q.E.D.

6.3.4. *Reduction to the phase invariants* By Theorem 6.8, Lemma 6.16, and Proposition 6.17, for any linked abelian group (H, λ) the invariants $\Gamma_k(\lambda)$ are either equal to 0 or to $(\sqrt{2})^m \rho^\varphi$ for some $m \geq 0$ and some $\varphi \in \mathbb{Z}_8$. The point of the following definition and theorem is that the $\sqrt{2}$-term is purely an invariant of the abelian group H, and thus is unnecessary for the classification of linking forms (the re-indexing is done to simplify the formulas in Theorem 6.20.

Definition 6.18. Let (H, λ) be a linked abelian group of degree n. Then for $1 \leq k \leq n$ the kth *phase invariant* $\varphi_k(\lambda) \in \mathbb{Z}_8 \cup \{\infty\}$ of (H, λ) is defined to equal ∞ if $\Gamma_{k-1}(\lambda) = 0$ and to equal $\varphi \in \mathbb{Z}_8$ with

$$\frac{\Gamma_{k-1}(\lambda)}{|\Gamma_{k-1}(\lambda)|} = \rho^\varphi$$

if $\Gamma_{k-1}(\lambda) \neq 0$. The *phase vector* $\varphi(\lambda)$ of λ is the vector $(\varphi_n(\lambda), \ldots, \varphi_1(\lambda))$. ‖

Theorem 6.19. *Fix a finite abelian group H of degree n. Then two linking forms λ_1 and λ_2 on H are isomorphic if and only if $\varphi_k(\lambda_1) = \varphi_k(\lambda_2)$ for all $1 \leq k \leq n$; i.e. if and only if λ_1 and λ_2 have identical phase vectors.*

Proof. Let r_j be the ranks of the blocks B_j of H, and consider a linking form λ on H. We must show that for $0 \leq k < n$ the number $\Gamma_k(\lambda)$ is determined by the r_j and the phase invariants of λ. First, we have $\Gamma_k(\lambda) = 0$ if and only if $\varphi_k(\lambda) = 0$. We can thus assume that $\Gamma_k(\lambda) \neq 0$, so

$$\Gamma_k(\lambda) = (\sqrt{2})^{p_k} \rho^{\varphi_{k+1}(\lambda)}.$$

We must determine p_k. By Theorem 6.8, we can write (H, λ) as an orthogonal direct sum of copies of the basic linking forms. Fixing such a decomposition, Lemma 6.16 and Proposition 6.17 say that the orthogonal components of (H, λ) make the following contributions to p_k :

- Every unary summand $\frac{a}{2^j}$ with $j \leqslant k$ contributes $2j$; binary summands contribute $4j$. Thus the block B_j contributes $2jr_j$ for all $j \leqslant k$.
- There are no unary summands when $j = k+1$, and each binary summand contributes $4(k+1)$; thus B_{k+1} contributes $2(k+1)r_{k+1} = 2jr_j$ also.
- For $j > k+1$, each unary summand contributes $j+k+1$ and each binary summand $2(j+k+1)$, so B_j contributes $r_j(j+k+1)$.

Adding these three sets of contributions gives us

$$p_k = \sum_{j \leqslant k+1} 2jr_j + \sum_{j > k+1} (j+k+1)r_j,$$

so p_k is a function of the ranks r_j alone, as desired. Q.E.D.

The computation of the phase invariants is facilitated by the following result, whose proof is immediate from Lemma 6.16 and Proposition 6.17.

Theorem 6.20. *The phase invariants are additive under direct sums, and have the following values on the basic linking forms:*

$$\varphi_k\left(\frac{a}{2^j}\right) = \begin{cases} 0 & \text{if } k > j \\ \infty & \text{if } k = j \\ \varepsilon(a) & \text{if } k < j \text{ and } k - j \text{ is odd} \\ a & \text{if } k < j \text{ and } k - j \text{ is even} \end{cases}$$

$$\varphi_k\left(\frac{\mathcal{C}}{2^j}\right) = 0, \text{ all } k,$$

$$\varphi_k\left(\frac{\mathcal{D}}{2^j}\right) = \begin{cases} 0 & \text{if } k \geq j \\ 4(j+k) & \text{if } k < j \end{cases}$$

Before doing some examples, we record a definition we will need later.

Definition 6.21. A linking form λ on a free $\mathbb{Z}_{2^n}$ module is *even* if $\lambda(x,x) = \frac{2k}{2^n}$ for some integer k. Otherwise, is *odd.* We remark that λ is even if and only if any orthogonal decomposition of it into basic forms has no unary summands. ||

Example 6.22. Let us test the method on the linkings of Example 6.6.

- $\left(\frac{\mathcal{C}}{2^n}\right) \oplus \left(\frac{1}{2^{n-1}}\right)$ gives $\varphi = (0, \infty, 1, 1, \ldots)$ since $a = \varepsilon(a) = 1$ and $\varphi\left(\frac{\mathcal{C}}{2^n}\right) = 0$.
- $\left(\frac{\mathcal{D}}{2^n}\right) \oplus \left(\frac{-3}{2^{n-1}}\right)$ gives

$$\varphi = (0, 4, 0, 4, \ldots) + (0, \infty, 1, -3, \ldots) = (0, \infty, 1, 1, \ldots),$$

as desired. ||

Example 6.23. $\frac{1}{2^n} \oplus \frac{3}{2^{n-1}} \not\simeq \frac{3}{2^n} \oplus \frac{1}{2^{n-1}}$. The reason is: φ of the left hand side is $\varphi = (\infty, 1, 1, 1, \ldots) + (0, \infty, -1, 3, \ldots) = (\infty, \infty, 0, 4, \ldots)$, and φ of the right hand side is $\varphi = (\infty, -1, 3, -1, \ldots) + (0, \infty, 1, 1, \ldots) = (\infty, \infty, 4, 0, \ldots)$. ||

Example 6.24.

$$\begin{aligned}
\varphi\left(\frac{1}{2^n} \oplus \frac{3}{2^{n-1}} \oplus \frac{5}{2^{n-2}}\right) &= (\infty, \infty, 0, 4, 0, \ldots) + (0, 0, \infty, 1, 5, \ldots) \\
&= (\infty, \infty, \infty, 5, 5, \ldots), \\
\varphi\left(\frac{3}{2^n} \oplus \frac{5}{2^{n-1}} \oplus \frac{1}{2^{n-2}}\right) &= (\infty, -1, 3, -1, 3, \ldots) + (0, \infty, 1, 5, 1, \ldots) \\
&\quad + (0, 0, \infty, 1, 1, \ldots) \\
&= (\infty, \infty, \infty, 5, 5, \ldots) \\
\varphi\left(\frac{5}{2^n} \oplus \frac{1}{2^{n-1}} \oplus \frac{3}{2^{n-2}}\right) &= (\infty, \infty, \infty, 1, 1, \ldots)
\end{aligned}$$

Therefore the first two forms are isomorphic, and the third form is different from the first two. ||

Example 6.25. If B_j is odd for all j from 1 to n, then $\varphi(\lambda) = (\infty, \infty, \ldots, \infty)$; in particular, the class of λ does not depend at all on the particular form of the individual blocks B_j. ||

6.4. Reidemeister's invariants

As we have seen, the Seifert-Burger viewpoint gives us a family of topological invariants of a 3-manifold that are associated to $H = H_1(W, \mathbb{Z})$, yet are not determined by H. Moreover, their determination depends crucially on whether there is, or is not, 2-torsion in H. We have learned how to compute them from a symplectic Heegaard splitting.

In 1933, the year that [43] was published, a paper by Reidemeister [40] also appeared. Moreover, there are remarks at the end of both [43] and [40] pointing to the work of the other. In particular, Reidemeister notes in his paper that Seifert's invariants are more general than his, as they must be because a quick scan of Reidemeister's paper [40] does not reveal any dependence of his results on whether there is 2-torsion. We describe Reidemeister's invariants briefly.

Referring to Theorem 2.2 and using the notation adopted there, Reidemeister defines the integers $\tau_{i,j} = \tau_j/\tau_i$, where $1 \leq i < j \leq t$. Let $\mathcal{Q}^{(2)} = (q_{ij})$ be the $t \times t$ matrix in (8), where we defined the partial normal form of Theorem 2.4. Let p_{im} be a non-trivial prime factor of the greatest common divisor of $(\tau_{1,2}, \tau_{2,3}, \ldots, \tau_{i,i+1})$. His invariants are a set of symbols which he calls ε_{im}, defined as follows:

$$\begin{aligned} \varepsilon_{im} &= 0 && \text{if } p_{im} \text{ divides } q_{ii}, \\ &= q_{ii}/p_{im} && \text{if } p_{im} \text{ does not divide } q_{ii}. \end{aligned} \tag{27}$$

Thus ε_{im} is defined for every non-trivial prime divisor of the greatest common divisor of $(\tau_{1,2}, \tau_{2,3}, \ldots, \tau_{i,i+1})$, and for every $i = 1, \ldots, t-1$. He proves that his symbols are well-defined, independent of the choice of the representative $\mathcal{H}^{(2)}$ within the double coset of $\mathcal{H}^{(2)}$ in Γ_t, by proving that remain unaltered under the changes which we described in §2.3. The fact that they are undefined when the greatest common divisor of $(\tau_{1,2}, \tau_{2,3}, \ldots, \tau_{i,i+1})$ is equal to 1 show that they do not change under stabilization.

Remark 6.26. We remark that Reidemeister's invariants are invariants of both the Heegaard splitting and of the stabilized Heegaard splitting. Therefore, if one happened to be working with a manifold which admitted two inequivalent Heegaard splittings, it would turn out that their associated Reidemeister symbols would coincide. This illustrates the very subtle nature of the Heegaard splitting invariants that are, ipso facto, encoded in the higher order representations of the mapping class group in the Johnson-Morita filtration. Any such Heegaard splitting invariant is either a topological invariant (as is the case for Reidemeister's invariant), or an invariant which vanishes after sufficiently many stabilizations. We will uncover an example of the latter type in the next section.

§7. The classification problem for minimal (unstabilized) symplectic Heegaard splittings.

Referring the reader back to Remark 1.5, it is clear that knowledge of a complete set of invariants of minimal symplectic Heegaard splittings, and the ability to compute them, are of interest in their own right. This was our motivation when we posed Problems 4, 5 and 6 in §1.5. In this section we will solve these problems.

Given two minimal Heegaard pairs with isomorphic linked quotient groups H_i and canonical volumes $\pm\theta_i$, Theorem 5.20 tells us that the pairs are isomorphic if and only if there is a volume preserving linking isomorphism $H_1 \to H_2$. By hypothesis there is a linking isomorphism h, but it may not be volume preserving — $\det h$ may not equal ± 1. Suppose $f : H_1 \to H_1$ is a linking automorphism, that is, an *isometry*, of H_1. Then hf is still a linking isomorphism, and $\det(hf) = \det h \cdot \det f$. Thus if $\det h \neq \pm 1$, we may hope to change it to ± 1 by composing it with some isometry of H_1. This will be our approach, and it will give us a complete set of invariants for *minimal* Heegaard pairs.

7.1. Statement of Results. Solutions to Problems 4, 5, 6.

Let $(V; B, \bar{B})$ be a minimal Heegaard pair with quotient H_0. Choose any dual complement A of B and let $\pi : A \to H_0$ be the projection. Let $H < H_0$ be the torsion subgroup with associated linking form λ, and set $F = \pi^{-1}(H)$. We thus have a minimal presentation $\pi : F \to H$. If e_i $(i = 1, \ldots, r)$ is a basis for F, put $x_i = \pi(e_i)$. We define a *linking matrix* for λ by choosing rational numbers λ_{ij} which are congruent mod 1 to $\lambda(x_i, x_j)$ for each i, j, subject to the symmetry condition $\lambda_{ij} = \lambda_{ji}$; we call such a choice a *lifting* of $\lambda(x_i, x_j)$.

Though the linking matrix for λ depends on choices, we can extract an invariant from it. The first step is the following theorem, which will be proven in §7.2. Let τ be the smallest elementary divisor of H.

Theorem 7.1. *The number $|H|\det(\lambda_{ij})$ is an integer, and its reduction modulo τ is a unit in $\mathbb{Z}_\tau$ which depends only on the isomorphism class of (H, λ) and the isomorphism class of the presentation $\pi : F \to H$.*

This theorem was first proven (by different methods) in [2]. In some cases, we can do better. We will need the following definition.

Definition 7.2. The linking on H is *even* if for all $x \in H$ with $\tau x = 0$, we have $x^2 \in \left(\frac{2}{\tau}\right)$ (here we have abbreviated $x \cdot x$ to x^2 and $\left(\frac{2}{\tau}\right)$ is the subgroup of $\mathbb{Q}/\mathbb{Z}$ generated by $\frac{2}{\tau}$). Otherwise the linking is *odd*.

Remark 7.3. See Lemma 7.18 below to relate this to the definition of an even linking form on a 2-group defined in Definition 6.21.

We will prove the following refinement of Theorem 7.1 in §7.3.

Theorem 7.4. *Let $\overline{\tau} = \tau$ if λ is odd and 2τ if λ is even. Then the reduction modulo $\overline{\tau}$ of $|H|\det(\lambda_{ij})$ depends only on the isomorphism class of (H, λ) and the isomorphism class of the presentation $\pi : F \to H$.*

At the end of §7.3 we give an example which shows that we have indeed found a stronger invariant than the one that was given in [2].

Corollary 3.27 and Lemma 5.19 say that the reduction modulo $\overline{\tau}$ of $|H|\det(\lambda_{ij})$ is actually a well-defined invariant of $(V; B, \bar{B})$, which we will denote by $\det(V; B, \bar{B})$. Our next theorem say that it is a complete invariant of minimal Heegaard pairs, solving Problem 4 of §1.5.

Theorem 7.5. *Let $(V_i; B_i, \bar{B}_i)$ $(i = 1, 2)$ be minimal Heegaard pairs with linked quotients (H_i, λ_i). Then the pairs are isomorphic if and only if the linked quotients are isomorphic and $\det(V_1; B_1, \bar{B}_1) = \det(V_2; B_2, \bar{B}_2)$.*

Theorem 7.5 will be proven in §7.4. With Theorem 7.5 in hand, we will be able to count the number of isomorphism classes of minimal Heegaard pairs with linked quotients (H, λ), solving Problem 5 of §1.5. To make sense of that result, we will need the following lemma.

Lemma 7.6. *Consider an integer $n \in \mathbb{Z}_\tau$. Then n^2 is well defined mod $\overline{\tau}$.*

Proof. We may assume that our linking is even, so $\overline{\tau} = 2\tau$. Then for $a, b \in \mathbb{Z}$ we have

$$(a + b\tau)^2 = a^2 + 2ab\tau + b^2\tau^2,$$

which equals a^2 modulo 2τ since 2τ divides τ^2. Hence knowledge of a modulo τ sufficies to determine a^2 modulo $\overline{\tau}$, as desired. Q.E.D.

Denote the group of units in $\mathbb{Z}_\tau$ by $\mathbb{U}$. In light of Lemma 7.6, it makes sense to define

$$\sqrt{1} = \{x \in \mathbb{U} \mid x^2 = 1 \text{ modulo } \overline{\tau}\}.$$

Our result is the following; it will be proven in §7.4 as a byproduct of the proof of Theorem 7.5.

Theorem 7.7. *The number of isomorphism classes of distinct minimal Heegaard pairs with linked quotients (H, λ) is $\frac{|\mathbb{U}|}{|\sqrt{1}|}$.*

We close this chapter in §7.5 with a discussion of how our techniques give normal forms for symplectic gluing matrices, in certain situations. This was the problem that was posed in Problem 6 of §1.5.

7.2. Proof of Invariance 1

This section contains the proof of Theorem 7.1. We will need several definitions.

Let F be a free abelian group and $R \subset F$ be a subgroup of equal rank. If φ, ψ are orientations of F, R, then the inclusion map of R into F has an integral determinant, and this determinant's absolute value is well known to be $(F : R) = |F/R|$ (for example, choose a basis for F as in Proposition 3.4). Thus, given either φ or ψ there is a unique choice of the other so that this determinant is *positive.* Orientations of F, R so chosen will be called *compatible.* Note that a change in the sign of one orientation necessitates a change in the other also to maintain compatibility. An orientation φ of F also induces a canonical orientation φ^* on the dual group $F^* = \mathrm{Hom}(F, \mathbb{Z})$: choose any basis $\{e_i\}$ of F with $e_1 \wedge \cdots \wedge e_r = \varphi$, then use the dual basis of F^* to define φ^*.

If H is any finite group, its *character group* is the additive group $H^* = \mathrm{Hom}(H, \mathbb{Q}/\mathbb{Z})$. It is well known that H^* is isomorphic to H, but not canonically so. This mirrors the relationship between a free (finitely generated) abelian group F and its dual (in the following, the * on a finite group indicates its character group, but on a free group indicates its dual). Also as in the free case, H^{**} is *canonically* isomorphic to H.

Suppose that $F \xrightarrow{\pi} H$ is a presentation of H with kernel R; we construct from it a canonical presentation of H^* which we call the *dual* presentation. The group F^* is a subgroup of $F^* \otimes \mathbb{Q} = \mathrm{Hom}(F, \mathbb{Q})$, namely, all maps $f : F \to \mathbb{Q}$ such that $f(F) \subset \mathbb{Z}$. The fact that H is finite and hence rank $R =$ rank F implies that R^* is precisely the subgroup of maps $f \in F^* \otimes \mathbb{Q}$ such that $f(R) \subset \mathbb{Z}$. Note that $R^* \supset F^*$ in $F^* \otimes \mathbb{Q}$. If $f \in R^*$, then f is a map of F into $\mathbb{Q}$ taking R into $\mathbb{Z}$ and so induces a map of $F/R = H$ to $\mathbb{Q}/\mathbb{Z}$. This element of H^* we denote by $\pi^*(f)$; we have then that π^* is a map $R^* \to H^*$. If $v \in H^*$, *i.e.* $v : F/R \to \mathbb{Q}/\mathbb{Z}$, we can lift v to a map $f : F \to \mathbb{Q}$ since F is free, and clearly $f(R) \subset \mathbb{Z}$, so $f \in R^*$ and $\pi^*(f) = v$. This shows that $R^* \xrightarrow{\pi^*} H^*$ is a presentation of H^*. The kernel of π^* consists of all f such that $f(F) \subset \mathbb{Z}$, that is, precisely F^*. Note thus that the index $(R^* : F^*) = |H^*| = |H| = (F : R)$.

Just as a choice of symmetric isomorphism $F \to F^*$ is the same as an "inner product" on F, so the choice of a symmetric isomorphism $H \xrightarrow{\lambda} H^*$ is the same as a *linking* on H: if we write $x \cdot y$ for the linking of x, y, then $x \cdot y$ is defined to be $\lambda(x)(y) \in \mathbb{Q}/\mathbb{Z}$ and conversely. The fact that λ is an isomorphism corresponds to the non-singularity of the linking.

If H is a linked group and $F \xrightarrow{\pi} H$ is a presentation of H, consider the diagram

$$\begin{array}{ccccccccc} 0 & \longrightarrow & R & \longrightarrow & F & \longrightarrow & H & \longrightarrow & 0 \\ & & \downarrow{\scriptstyle K} & & \downarrow{\scriptstyle L} & & \downarrow{\scriptstyle \lambda} & & \\ 0 & \longrightarrow & F^* & \longrightarrow & R^* & \longrightarrow & H^* & \longrightarrow & 0. \end{array}$$

The fact that F is free implies the existence of a map L making the right square commute, and L induces K on R. We call L a *lifting* of λ, and it is well defined up to the addition of a map $X : F \to F^*$. If now we choose an orientation φ of F, it induces φ^* on F^* and compatible orientations ψ, ψ^* on R, R^*; it is easy to see that ψ, ψ^* are also dual orientations. Furthermore, if $F \to H$ is *minimal*, then so is $R^* \to H^*$ and we get induced orientations θ, θ^* on H, H^*.

A change in the sign of φ uniformly changes the sign of all the other orientations. Thus the determinants $\det K, \det L \in \mathbb{Z}$ and $\det \lambda \in \mathbb{Z}_\tau$ are all well defined and independent of the orientations. Furthermore, $\det \lambda$ depends only on λ and the presentation π. The connection between these determinants and that of Theorem 7.1 is given by the following lemma.

Lemma 7.8. a) $\det L \equiv \det \lambda \bmod \tau$

b) $\det K = \det L$

c) *If* (λ_{ij}) *is a linking matrix as in Theorem* 7.1, *then it determines a lifting* $L : F \to R^*$ *and* $|H| \det(\lambda_{ij}) = \det K = \det L$.

Proof. a) is proved in Lemma 3.23; b) follows from the commutativity of the left square and the fact that the determinant of both (compatibly oriented) $R \to F$, $F^* \to R^*$ is $|H|$. It remains to prove c). Now the linking matrix (λ_{ij}) is clearly just the matrix of a map $L : F \to F^* \otimes \mathbb{Q}$ in terms of the basis $\{e_i\}$ of F used to define (λ_{ij}) and its dual basis in $F^* \otimes \mathbb{Q}$, namely, $L(e_i) = \sum_j \lambda_{ij} e_j^*$. Put $x_i = \pi(e_i)$ and denote the linking in H by the inner product dot; if then $s = \sum \alpha_i e_i$ is in R, we find

$$\begin{aligned} L(e_i)(s) &= \sum_j \lambda_{ij} e_j^*(s) = \sum_j \lambda_{ij} \alpha_j \equiv_{\text{mod } 1} \sum (x_i \cdot x_j) \alpha_j \\ &= x_i \cdot \left(\sum_j \alpha_j x_j \right) = x_i \cdot \pi(s) \equiv 0 \bmod 1. \end{aligned}$$

In other words, $L(e_i)$ takes R into $\mathbb{Z}$ for all i, that is, $L(e_i) \in R^*$ for all i, which means that L is actually a map from F to R^*. By its very definition it is a lifting of λ. Let now $s_1, \ldots, s_r$ be a basis of R compatible with $e_1, \ldots, e_r$ of F, and let $s_i = \sum_j A_{ij} e_j$; thus $\det(A_{ij}) = |H|$. We then find that

$$K(s_i) = L(s_i) = \sum_{j,k} A_{ij} \lambda_{jk} e_k^*,$$

and since $K(s_i)$ is in F^*, the matrix $A \cdot (\lambda_{ij})$ is integral and its determinant is $\det K = \det A \det(\lambda_{ij}) = |H| \det(\lambda_{ij})$. Q.E.D.

This lemma proves that $|H| \det(\lambda_{ij})$ is an integer whose mod τ reduction only depends on the isomorphism class of $\pi : F \to H$ and the linking. The fact that it is a unit in $\mathbb{Z}_\tau$ follows from the fact that $\lambda : H \to H^*$ is an isomorphism.

7.3. Proof of Invariance 2

We can assume that the linking form is even. Let the notation be as in the previous section. The first step is to prove that the lifts L which come from the symmetric linking matrices are *symmetric* in the sense that the induced map $L^* : R \to F^*$ is the map K. Let the e_i, the s_i, and the matrix A be as in the proof of Lemma 7.8. We will determine the matrix of L in terms of the bases $\{e_i\}$ of F and $\{s_i^*\}$ of R^*. Since $s_i = \sum_j A_{ij} e_j$, we have $e_j^* = \sum_i s_i^* A_{ij}$ so

$$L(e_i) = \sum_j \lambda_{ij} e_j^* = \sum_{j,k} \lambda_{ij} A_{kj} s_k^*.$$

Thus the matrix of L is λA^t, and in the dual bases s_i and e_i^* the operator L^* has matrix $A\lambda^t$. Since (λ_{ij}) was chosen to be symmetric, we have finally $L^* = A\lambda$, which is the matrix of K. Note that two symmetric liftings of λ differ by a *symmetric* map $F \to F^*$.

Our goal is to prove that modulo 2τ the number $\det L$ is independent of the choice of a symmetric lifting of λ. By Theorem 7.1 and the fact that τ is even, $(\det L, 2\tau) = 1$ and hence the matrix of L has a mod 2τ inverse, that is, there is an integral matrix L^{-1} such that $LL^{-1} \equiv \mathcal{I} \bmod 2\tau$. Any other symmetric lifting of L is of the form $L+X$ where X is a symmetric map $F \to F^*$. But note that $R \subset \tau F$ and so $F^* \subset \tau R^*$; hence $X = \tau Y$ for some map $Y : F \to R^*$. Modulo 2τ we then have

$$\det(L + X) \equiv \det L \cdot \det(\mathcal{I} + L^{-1}X) = \det L \cdot \det(\mathcal{I} + \tau L^{-1}Y).$$

Lemma 7.9. *If A is any square matrix and $\tau > 1$, then* $\det(\mathcal{I} + \tau A) \equiv 1 + \tau \mathrm{Tr}(A) \bmod \tau^2$, *where* $\mathrm{Tr}(A)$ *is the trace of* A.

Proof. In the expansion of $\det(1 + \tau A)$ only those monomials involving at most one off-diagonal factor are non-zero mod τ^2. A single off-diagonal factor cannot occur, however, in any monomial, and so $\det(1 + \tau A) \equiv \prod_i (1 + \tau A_{ii}) \bmod \tau^2$. The product is clearly equal to $1 + \tau(\sum A_{ii}) \bmod \tau^2$. Q.E.D.

This lemma shows that

$$\det(L + X) \equiv \det L + \tau \det L \cdot \mathrm{Tr}(L^{-1}Y) \bmod \tau^2.$$

But since $\det L \equiv 1 \bmod 2$, it follows that $\tau \det L \equiv \tau \bmod 2\tau$. Hence since $2\tau | \tau^2$, we have

$$\det(L + X) \equiv \det L + \tau \mathrm{Tr}(L^{-1}Y) \bmod 2\tau.$$

Thus to prove the desired result it suffices to show that $\mathrm{Tr}(L^{-1}Y) \equiv 0 \bmod 2$. We must now take a closer look at the matrices L and X.

We choose the basis of F as in Proposition 3.4, so that $s_i = m_i e_i$ $(i = 1, \ldots, r)$ with $m_1 = \tau$ and $m_i | m_{i+1}$. Let 2^n be the highest power of 2 dividing τ (we notate this in the future by $2^n \parallel \tau$) and suppose that the same is true for m_1 through m_b; that is, $\frac{m_i}{2^n}$ is odd for $1 \leqslant i \leqslant b$ and even for $i > b$. Let (λ_{ij}) be a (symmetric) linking matrix lifting $x_i \cdot x_j$ as before. The matrix A describing the basis $\{s_i\}$ in terms of $\{e_i\}$ is now the diagonal matrix $\mathrm{Diag}(m_i)$, and so the matrix of L is of the form $(L_{ij}) = (\lambda_{ij})A^t = (\lambda_{ij} m_j)$. We claim that L_{ij} is even for $i \leqslant b$ and $j > b$. Indeed, since $x_i = \pi(e_i)$ has order $m_i < m_j$ we must have $\lambda_{ij} = \frac{N}{m_i}$ for some integer N, and thus $L_{ij} = \frac{N}{m_i} m_j = N \frac{m_j}{m_i}$. But $\frac{m_j}{m_i}$ is even whenever $i \leqslant b$ and $j > b$. If we divide the coordinate indices into two blocks with $1 \leqslant i \leqslant b$ in the first block and $i > b$ in the second, then mod 2, the matrix L takes the form $\left(\begin{smallmatrix} B & 0 \\ C & D \end{smallmatrix}\right)$. The fact that $\det L \equiv 1 \bmod 2$ implies that $\det B \equiv \det D \equiv 1 \bmod 2$.

Lemma 7.10. *B is symmetric* mod 2 *and has zero diagonal* mod 2.

Proof. If $i < j$, then $B_{ij} = N \frac{m_j}{m_i}$. But $2^n \parallel m_i$ and $2^n \parallel m_j$, so $\frac{m_j}{m_i}$ is odd and $B_{ij} \equiv N \bmod 2$. On the other hand, $B_{ji} = \lambda_{ji} m_i = \frac{N}{m_i} m_i = N$. Thus B is symmetric mod 2. Its diagonal term B_{ii} is $\lambda_{ii} m_i$ where $\lambda_{ii}^2 \equiv x_i^2 \bmod 1$. Now x_i is of order m_i so $\frac{m_i}{\tau} x_i$ is of order τ, and $\frac{m_i}{\tau}$ is *odd.* Thus $\left(\frac{m_i}{\tau}\right)^2 \lambda_{ii} \equiv \left(\frac{m_i}{\tau} x_i\right)^2 \bmod 1$, and the latter is in $\left(\frac{2}{\tau}\right)$ by the assumption that λ is even, so we have $\left(\frac{m_i}{\tau}\right)^2 \lambda_{ii} = \frac{2N}{\tau}$ for some integer

N. Multiplying by τ we get $\frac{m_i}{\tau}(m_i\lambda_{ii}) = \frac{m_i}{\tau}B_{ii} \equiv 0 \bmod 2$, which by the oddness of $\frac{m_i}{\tau}$ implies that $B_{ii} \equiv 0 \bmod 2$. Q.E.D.

Lemma 7.11. *Let $X : F \to F^* \subset R^*$ be symmetric. Then its matrix, written in the bases e_i, s_i^*, is congruent* mod 2τ *to a matrix of the block form $\tau\left(\begin{smallmatrix} U & 0 \\ V & 0 \end{smallmatrix}\right)$, where U is symmetric.*

Proof. In the bases e_i, e_i^* the matrix of X is $\left(\begin{smallmatrix} U & V^t \\ V & W \end{smallmatrix}\right)$ where U and W are symmetric, but in the bases e_i, s_i^* it is $\left(\begin{smallmatrix} U & V \\ V^t & W \end{smallmatrix}\right)\mathrm{Diag}(m_i)$. Since $\frac{m_i}{\tau}$ is odd for $i \leqslant b$ and even for $i > b$, the block form of $\mathrm{Diag}(m_i)$ is congruent mod 2τ to $\left(\begin{smallmatrix} \tau\mathcal{I}_b & 0 \\ 0 & 0 \end{smallmatrix}\right)$ where $\mathcal{I}_b$ is the identity matrix. Hence $X \equiv_{\mathrm{mod}\ 2\tau} \left(\begin{smallmatrix} \tau U & 0 \\ \tau V & 0 \end{smallmatrix}\right) = \tau\left(\begin{smallmatrix} U & 0 \\ V & 0 \end{smallmatrix}\right)$. Q.E.D.

Recall the map $Y = \frac{1}{\tau}X$; by the above it is congruent mod 2 to $\left(\begin{smallmatrix} U & 0 \\ V & 0 \end{smallmatrix}\right)$. We now calculate

$$
\begin{aligned}
L^{-1}Y \equiv_{\mathrm{mod}\ 2} \begin{pmatrix} B & 0 \\ C & D \end{pmatrix}^{-1} \begin{pmatrix} U & 0 \\ V & 0 \end{pmatrix} &\equiv \begin{pmatrix} B^{-1} & 0 \\ C' & D^{-1} \end{pmatrix} \begin{pmatrix} U & 0 \\ V & 0 \end{pmatrix} \\
&\equiv \begin{pmatrix} B^{-1}U & 0 \\ C'' & 0 \end{pmatrix},
\end{aligned}
$$

where the precise calculation of the matrices C', C'' is unimportant to us. We are interested only in $\mathrm{Tr}(L^{-1}Y) \equiv \mathrm{Tr}(B^{-1}U) \bmod 2$, where B, U are symmetric and B has zero diagonal.

Lemma 7.12. *Let B be a nonsingular symmetric matrix over $\mathbb{Z}_2$ with zero diagonal; then its inverse has the same properties.*

Proof. We may lift B to an *integral* matrix $\tilde{B}$ which is *antisymmetric*, that is, $\tilde{B}^t = -\tilde{B}$. Since $\det \tilde{B} \equiv 1 \bmod 2$, the matrix $\tilde{B}^{-1}$ is rational and antisymmetric and $\tilde{C} = \tilde{B}^{-1} \cdot \det \tilde{B}$ is *integral* and antisymmetric. Reduced mod 2, it is also an inverse of B, since $\tilde{B} \cdot \tilde{C} = \det \tilde{B}\mathcal{I} \equiv \mathcal{I} \bmod 2$. This proves the lemma. Q.E.D.

We can now see that the desired result follows from the above results and the following:

Lemma 7.13. *Let C, U be symmetric matrices over $\mathbb{Z}_2$ such that C has zero diagonal; then* $\mathrm{Tr}(CU) = 0$.

Proof. $\operatorname{Tr}(CU) = \sum_{i,j} C_{ij}U_{ji}$. We split this sum into three parts: $\sum_{i<j} + \sum_{i>j} + \sum_{i=j}$. Now

$$\begin{aligned}\sum_{i<j} C_{ij}U_{ji} &= \sum_{i<j} C_{ji}U_{ji} && \text{by symmetry of } C \text{ and } U \\ &= \sum_{j<i} C_{ij}U_{ji} && \text{interchanging } i,j;\end{aligned}$$

thus $\sum_{i<j} + \sum_{i>j}$ cancel over $\mathbb{Z}_2$. But the last summand is $\sum_i C_{ii}U_{ii} = 0$ since $C_{ii} = 0$ for all i. Q.E.D.

We close this section with an example showing that we have indeed found a stronger invariant.

Example 7.14. Consider the matrices

$$\mathcal{U} = \begin{pmatrix} \begin{pmatrix} 0 & -15 \\ -15 & 0 \end{pmatrix} & \begin{pmatrix} 8 & 0 \\ 0 & 8 \end{pmatrix} \\ \begin{pmatrix} -2 & 0 \\ 0 & -2 \end{pmatrix} & \begin{pmatrix} 0 & 1 \\ 1 & 0 \end{pmatrix} \end{pmatrix} \quad \text{and} \quad \mathcal{V} = \begin{pmatrix} \begin{pmatrix} 0 & -5 \\ -5 & 0 \end{pmatrix} & \begin{pmatrix} 8 & 0 \\ 0 & 8 \end{pmatrix} \\ \begin{pmatrix} -2 & 0 \\ 0 & -2 \end{pmatrix} & \begin{pmatrix} 0 & 3 \\ 3 & 0 \end{pmatrix} \end{pmatrix}.$$

They are easily seen to be symplectic and thus define Heegaard pairs as discussed in §4, namely $(X_2; F_2, \mathcal{U}(F_2)$ or $\mathcal{V}(F_2))$. The quotient groups are $\mathbb{Z}_8 \oplus \mathbb{Z}_8$ in both cases, with respective linking matrices $\frac{1}{8}\left(\begin{smallmatrix} 0 & 1 \\ 1 & 0 \end{smallmatrix}\right)$ and $\frac{1}{8}\left(\begin{smallmatrix} 0 & 3 \\ 3 & 0 \end{smallmatrix}\right)$, which are even, so $\bar{\tau} = 16$. Multiplication by 3 in $\mathbb{Z}_8^2$ gives an isomorphism between the two linkings, so the pairs are stably isomorphic; furthermore, their determinants are both $\equiv -1 \bmod \tau (= 8)$. But these pairs are *not* isomorphic, since their respective determinants mod $\bar{\tau}$ are $-1 \bmod 16$ and $-9 \bmod 16$. ||

7.4. Proof of Completeness and a Count.

We now prove Theorem 7.5, which says that our mod $\bar{\tau}$ determinantal invariant is a complete invariant of minimal Heegaard pairs. A byproduct of our proof will be a proof of Theorem 7.7. As we observed in §5, we can assume that the the Heegaard pairs in question have finite quotients. Now, we have shown that our invariant is really an invariant of the linked quotient together with its induced volume, and it is easy to see that all such volumes occur. For a linked finite abelian group (H, λ) with a volume θ, denote this determinantal invariant by $\det(\lambda, \theta)$. Our first order of business is determine which volumes on (H, λ) have the same determinant, so fix a linked finite abelian group (H, λ) together with a minimal presentation $F \to H$, and let τ and $\bar{\tau}$ be defined as before. We begin with a lemma.

Lemma 7.15. *In the commutative diagram*

$$\begin{array}{ccc} F & \xrightarrow{\pi} & H \\ {\scriptstyle f}\downarrow & & \downarrow{\scriptstyle h} \\ F' & \xrightarrow{\pi'} & H' \end{array}$$

let π, π' *be minimal presentations of the linked groups* (H, λ) *and* (H', λ'), *and let* h *be a linking isomorphism (we do not assume that* f *is an isomorphism). Then if* $\det h$ *is measured with respect to the induced volumes on* H, H', *we have* $\det(\lambda, \pi) = (\det h)^2 \det(\lambda', \pi') \bmod \bar{\tau}$.

Proof. Lemma 7.6 shows that the statement is meaningful. Let now $\{e_i\}$, $\{e'_i\}$ be bases of F, F' respectively, and (λ'_{ij}) be a linking matrix for H' in the basis $\{e'_i\}$. If f has matrix A, then the fact that h is a linking isomorphism implies easily that $A\lambda' A^t$ is a linking matrix for H in the basis $\{e_i\}$. Hence

$$\begin{aligned} \det(\lambda, \pi) &\equiv_{\text{mod } \bar{\tau}} |H| \det(\lambda_{ij}) = |H| \det A^2 \det(\lambda'_{ij}) \\ &\equiv_{\text{mod } \bar{\tau}} (\det f)^2 \det(\lambda', \pi'). \end{aligned}$$

But by lemma 3.23 we have $\det f \equiv \pm \det h \bmod \tau$, so

$$(\det h)^2 \equiv (\det f)^2 \bmod \bar{\tau}.$$

Q.E.D.

Corollary 7.16. *Let* h *be an isometry of* H; *then* $(\det h)^2 \equiv 1 \bmod \bar{\tau}$.

This corollary restricts the determinant of an isometry to lie in $\sqrt{1}$. The following is an immediate corollary of Lemma 7.15.

Lemma 7.17. *For any volume* θ *on* (H, λ), *we have*

$$\det(\lambda, m\theta) \equiv m^2 \det(\lambda, \theta) \bmod \bar{\tau}$$

for any m *in* $\mathbb{U}$.

In particular, $\det(\lambda, m\theta) = \det(\lambda, \theta)$ if and only if $m \in \sqrt{1}$. Observe that this immediately implies Theorem 7.7 : the set of volumes is in bijection with $\mathbb{U}/\{\pm 1\}$, and two volumes define isomorphic Heegaard pairs if and only if they are in the same coset of $\mathbb{U}/\sqrt{1}$.

We conclude that to prove Theorem 7.5, it is enough by Theorem 5.20 to prove that everything in $\sqrt{1}$ is the determinant of an isometry.

We shall see that this problem can be solved in each p-component independently and pieced together to get the general solution. As a technical tool, we will need the following result.

Lemma 7.18. *Let H be a linked group whose smallest elementary divisor τ is even. Then the following statements are equivalent:*

a) *Every $x \in H$ such that $\tau x = 0$ satisfies $x^2 \in \left(\frac{2}{\tau}\right)$ (here we have abbreviated $x \cdot x$ to x^2 and $\left(\frac{2}{\tau}\right)$ is the subgroup of $\mathbb{Q}/\mathbb{Z}$ generated by $\frac{2}{\tau}$).*

b) *The lowest block of the 2-component of H is even (in the sense of Definition 6.21).*

Proof. Let $\{p_i\}$ be the primes dividing $|H|$, with $p_1 = 2$, and let $\tau = \prod_i p_i^{n_i}$ (some of the n_i's may be zero here, but $n_1 > 0$). The group H splits as an orthogonal direct sum of its p_i components H_i, and every $x \in H$ can be written uniquely as $x = \sum_i x_i$ with $x_i \in H_i$. If x_i has order $p_i^{r_i}$ then x has order $\prod_i p_i^{r_i}$. Thus if $\tau x = 0$ we must have $r_i \leqslant n_i$ for all i. Furthermore, $x^2 = \sum_i x_i^2$ and

$$x_i^2 \in \left(\frac{1}{p_i^{r_i}}\right) \subset \left(\frac{1}{p_i^{n_i}}\right) = \left(\frac{M_i}{\tau}\right)$$

where $M_i = \prod_{i \neq j} p_j^{n_j}$. For odd primes (*i.e.* $i > 1$), the number M_i is even and so $x_i^2 \in \left(\frac{2}{\tau}\right)$, but for $i = 1$ the number M_i is odd. Hence $x^2 \in \left(\frac{2}{\tau}\right)$ for all x satisfying $\tau x = 0$ if and only if $x_1^2 \in \left(\frac{2}{\tau}\right)$ for all $x_1 \in H_1$ satisfying $2^{n_1} x_1 = 0$. The lowest block B_1 of H_1 consists of elements all of which satisfy $2^{n_1} x_1 = 0$; their self linkings x_1^2 are in $\left(\frac{1}{2^{n_1}}\right)$, and are in $\left(\frac{2}{\tau}\right)$ if and only if they are in $\left(\frac{2}{2^{n_1}}\right)$. Thus in this case B_1 must be even. Conversely, suppose B_1 is even. The group H_1 splits orthogonally into blocks $B_1 \oplus B_2 \oplus \cdots \oplus B_k$ where each B_i is a free $\mathbb{Z}_{2^{s_i}}$-module, with $n_1 = s_1 < s_2 < \cdots < s_k$. If $x_1 \in H_1$, write $x_1 = \sum_{i=1}^k y_i$, with $y_i \in B_i$. We then have $2^{n_1} x_1 = 0$ if and only if $2^{n_1} y_i = 0$ for all i, which is true if and only if $y_i \in 2^{s_i - n_1} B_i$ for all i. Then $x_1^2 = \sum y_i^2$ (by orthogonality) and

$$y_i^2 \in \left(\frac{2^{2(s_i - n_1)}}{2^{n_1}}\right) = \left(\frac{2^{s_i - n_1}}{2^{n_1}}\right) \subset \left(\frac{2}{2^{n_1}}\right) \text{ for all } i > 1;$$

but $y_1^2 \in \left(\frac{2}{2^{n_1}}\right)$ also since B_1 is even. This concludes the proof. Q.E.D.

Lemma 7.19. *$\Lambda^r H$ is naturally isomorphic to the direct sum of $\Lambda^r H_p$ over all p which divide τ. If θ is an orientation (volume) on H then its projection θ_p in $\Lambda^r H_p$ is an orientation (volume) on H_p.*

Proof. The tensor power H^r splits into the direct sum H_p^r over all p since $H_p \otimes H_q = 0$ if $p \neq q$. Likewise, the kernel of $H^r \to \Lambda^r H$ splits into its p-component parts, and so we get a natural direct sum $\Lambda^r H = \bigoplus_{\text{all } p} \Lambda^r H_p$. But if $p \nmid \tau$ then rank $H_p < r$ and so $\Lambda^r H_p = 0$, proving the first statement. Note that $\Lambda^r H_p$ is precisely the p-component of $\Lambda^r H$, which is $\simeq \mathbb{Z}_{p^n}$ if p^n is the largest power of p dividing τ. If now $\theta \in \Lambda^r H$, we may write $\theta = \sum_{p|\tau} \theta_p$. Thus if θ generates $\Lambda^r H$, then θ_p must generate $\Lambda^r H_p$. Finally, if θ is determined up to sign, so is θ_p. Q.E.D.

Suppose now that $h : H \to H$ is an isometry. Hence h takes each p-component into itself, so h splits into a direct sum of maps h_p on H_p; conversely the maps h_p define h on H. Furthermore, the action of h on $\Lambda^r H$ is just multiplication by $\det h \bmod \tau$ and hence the action of h_p on $\Lambda^r H_p$ is also multiplication by $\det h$. But it is also multiplication by $\det h_p \bmod p^n$ (where $p^n \parallel \tau$), since $\Lambda^r H_p \simeq \mathbb{Z}_{p^n}$; in other words:

Lemma 7.20. *If h is an endomorphism of H, then* $\det h_p \equiv \det h \bmod p^n$ *for every p dividing τ, where $p^n \parallel \tau$. Thus* $\det h_p$ *is determined by* $\det h$. *Conversely* $\det h$ *is determined by the values of* $\det h_p$ *(the proof of this is by the Chinese Remainder Theorem).*

If H has linking λ then, by virtue of the orthogonality of distinct primary components, λ splits into the direct sum of linkings λ_p on H_p. Thus h is an isometry of H if and only if h_p is so for each p. Let now $\tau = \prod_i p_i^{n_i}$.

Lemma 7.21. $\sqrt{1} \bmod \tau$ *splits into the direct product of the groups* $\sqrt{1} \bmod p_i^{n_i}$.

Proof. Observe that $e \in \sqrt{1} \bmod \tau$ means that $e \in \mathbb{Z}_\tau$ and $e^2 \equiv 1 \bmod \bar{\tau}$. This means that $e^2 \equiv 1 \bmod p_i^{\bar{n}_i}$, where $\bar{n}_i = n_i$ if p_i is odd or if $p_i = 2$ and λ is odd, but $\bar{n}_i = n_i + 1$ if $p_i = 2$ and λ is even, which by Lemma 7.18 is true if and only if λ_2 is even. Hence e reduced mod $p_i^{n_i}$ is in $\sqrt{1} \bmod p_i^{n_i}$. Conversely let $e_i \in \sqrt{1} \bmod p_i^{\bar{n}_i}$, that is $e_i \in \mathbb{Z}_{p_i^{n_i}}$ be such that $e_i^2 \equiv 1 \bmod p_i^{\bar{n}_i}$; by the Chinese Remainder Theorem again, there is a unique $e \in \mathbb{Z}_\tau$ such that $e \equiv e_i \bmod p_i^{n_i}$, and we find $e^2 \equiv e_i^2 \equiv 1 \bmod p_i^{\bar{n}_i}$ which implies $e^2 \equiv 1 \bmod \bar{\tau}$. Q.E.D.

Corollary 7.22. *An element $a \in \sqrt{1} \bmod \tau$ is the determinant of an isometry of (H, λ) if and only if its reduction* $\bmod p_i^{n_i}$ *is the determinant of an isometry of (H_{p_i}, λ_{p_i}).*

We have now reduced the proof of Theorem 7.5 to the proof of:

Lemma 7.23. *Let H be a linked p-group with $\tau = p^n$ and $e \in \sqrt{1} \bmod p^n$. Then there is an isometry of H with determinant e.*

Proof. If λ is odd, by [7] the linked group H has an orthogonal splitting of the form $(x) \oplus H_0$, where (x) is the cyclic subgroup generated by an element x of order p^n satisfying $x^2 = \frac{u}{p^n}$ with $p \nmid u$. The map h which takes x to ex and which is the identity on H_0 is then an *isometry* and its determinant is clearly e. So now let $p = 2$ and λ be even. In this case, $\sqrt{1}$ consists of all $e \bmod 2^n$ such that $e^2 \equiv 1 \bmod 2^{n+1}$. There are four square roots of 1 mod 2^{n+1} (when $n \geqslant 2$), namely ± 1 and $2^n \pm 1$, but mod 2^n these give only two distinct elements ± 1 in $\sqrt{1}$. Thus we must simply exhibit an isometry with determinant $-1 \bmod 2^n$. By Lemma 7.18, the even nature of λ implies that the 2^n-block of H is even. By the classification of linked 2-groups in §6.3, the linked group H has an orthogonal splitting of the form $Q \oplus H_0$, where $Q \cong \mathbb{Z}_{2^n} \oplus \mathbb{Z}_{2^n}$, generated by let us say x, y of order 2^n, and where Q has a linking matrix equal to one of $\frac{1}{2^n}\left(\begin{smallmatrix} 0 & 1 \\ 1 & 0 \end{smallmatrix}\right)$ or $\frac{1}{2^n}\left(\begin{smallmatrix} 2 & 1 \\ 1 & 2 \end{smallmatrix}\right)$ mod 1. Clearly interchanging x and y is an isometry on either form, and extending by the identity on H_0 gives an isometry of H with determinant $\equiv -1 \bmod 2^n$. This proves the lemma and concludes the proof of Theorem 7.5. Q.E.D.

7.5. Problem 6

In Theorem 2.4 we found a partial normal form for the double coset associated to a symplectic matrix $\mathcal{H}$, and in §2.3 we investigated its non-uniqueness. We raised the question of whether the submatrix $\mathcal{Q}^{(2)}$ could be diagonalized. In fact, the following is true:

Proposition 7.24. *Let W be a 3-manifold which is defined by a Heegaard splitting. Let $H = H_1(W;\mathbb{Z})$ and let T be the torsion subgroup of H. Let t be the rank of T, so that T is a direct sum of cyclic groups of order $\tau_1, \dots, \tau_t$, where each τ_i divides τ_{i+1}. Assume that every τ_i is odd. Then T is an abelian group with a linking, and there is a choice of basis for T such that the linking form for T is represented by a $t \times t$ diagonal matrix.*

Proof. Consider, initially, a fixed p-primary component $T(p)$ of T and its splitting $T(p) = T_1 \oplus \cdots \oplus T_\nu$ into cyclic groups T_j of prime power order p^{e_j}. The T_j's may be collected into subsets consisting of groups of like order. Keeping notation adopted earlier, consider a typical such subset $T_{\rho+1}, \dots, T_{\rho+k}$ containing all cyclic summands of $T(p)$ of order $p^{\mu+1}$. Let $g_{\rho+1}, \dots, g_{\rho+k}$ generate these summands. Define a new linking

λ' on $T_{\rho+j}$ $(j = 1, \ldots, k)$ by the rule:

$$(28) \qquad \lambda'(g_{\rho+i}, g_{\rho+j}) = \begin{cases} \frac{|\mathcal{A}_{\rho+1}|}{p^{e_{\rho+1}}} \bmod 1 & \text{if } i = j = 1 \\ \frac{1}{p^{e_{\rho+1}}} \bmod 1 & \text{if } i = j = 2, \ldots, k \\ 0 & \text{if } i \neq j. \end{cases}$$

There is an induced linking λ' on $T(p)$ obtained by direct summing the linkings on all of the cyclic summands, and thus an induced linking on T obtained by taking the orthogonal direct sum of all the p-primary summands. We will also denote this by λ'. By Theorem 6.3, the linking on T is determined entirely by the quadratic residue characters of linkings on the p-primary summands of T. It follows that (T, λ') is equivalent as a linked group to (T, λ).

To complete the proof we need only note that by Theorem 2.2 the generators g_{ij} of the cyclic summands of prime power order determine the generators y_i of the cyclic summands of order $\tau_1, \ldots, \tau_t$. This follows from (5) of Theorem 2.2. Therefore there is a $t \times t$ matrix which also defines λ', and λ' is equivalent to λ. The proof is complete. Q.E.D.

Remark 7.25. One might be tempted to think that Proposition 7.24 implies that there is a matrix in the same double coset as the matrix $\mathcal{H}'$ in (7) of Theorem 2.4 in which the blocks $\mathcal{P}^{(2)}, \mathcal{Q}^{(2)}$ are both diagonal. Suppose we could prove that. Then for each $j = 1, \ldots, t$ choose r_j, s_j so that $r_j q_j - \tau_j s_j = 1$. Define $\mathcal{R}^{(3)} = \mathrm{Diag}(r_1, \ldots, r_t)$ and $\mathcal{S}^{(3)} = \mathrm{Diag}(s_1, \ldots, s_t)$. With these choices it is easy to verify that $\begin{pmatrix} \mathcal{R}^{(3)} & \mathcal{P}^{(2)} \\ \mathcal{S}^{(3)} & \mathcal{Q}^{(3)} \end{pmatrix}$ is symplectic. If so, that would imply that $\mathcal{R}^{(2)}$ and $\mathcal{S}^{(2)}$ also are diagonal. However, while we have learned that there is a change in basis for T in which $\mathcal{Q}^{(2)}$ is diagonal, we do not know whether this change in basis preserves the diagonal form of the matrix $\mathcal{P}^{(2)}$. Therefore we do not know whether it is possible to find a representative of the double coset in which all four blocks are diagonal. Proposition 7.24 tells us that there is no reason to rule this out. The discussion in §2.3 also tells us that it might be possible. On the other hand, the fact that such a diagonalization cannot always be achieved when there is 2-torsion tells us that the proof would have to be deeper than the work we have already done. ||

Example 7.26. In spite of the difficulties noted in Remark 7.25, we are able to construct a very large class of examples for which all four blocks are diagonal, even when there is 2-torsion. We construct our examples in stages:

- First, consider the case where our 3-manifold $W(h_{p,q})$ is a lens space of type (p,q). Then $W(h_{p,q})$ admits a genus 1 Heegaard splitting with gluing map that we call $h_{p,q}$, where p is the order of $\pi_1(W(h_{p,q}))$. The symplectic image of $h_{p,q}$ will be $\left(\begin{smallmatrix} r & p \\ s & q \end{smallmatrix}\right)$, where $rq - ps = \pm 1$.
- Next, consider the case when our 3-manifold $W(\tilde{h})$ is the connect sum of g lens spaces of types $(p_1, q_1), \ldots, (p_g, q_g)$, so that it admits a Heegaard splitting of genus g. Think of the Heegaard surface as the connect sum of g tori. The restriction of the gluing map $\tilde{h}$ to the ith handle will be h_{p_i,q_i}, so that the symplectic image of the gluing map will be $M = \left(\begin{smallmatrix} R & P \\ S & Q \end{smallmatrix}\right)$, where $P = \text{diag}(p_1, \ldots, p_g)$, $R = \text{diag}(r_1, \ldots, r_g)$, $S = \text{diag}(s_1, \ldots, s_g)$, $Q = \text{diag}(q_1, \ldots, q_g)$.
- Finally, consider the class of 3-manifolds $W(\tilde{f}\tilde{h})$ of Heegaard genus g which are defined by the gluing map $\tilde{f}\tilde{h}$, where $\tilde{f}$ is any element in kernel of the homomorphism $\rho^{(2)} : \tilde{\Gamma}_g \to Sp(2g, \mathbb{Z})$. The fact that $\tilde{f}$ has trivial image in $Sp(2g, \mathbb{Z})$ shows that the symplectic image of the gluing map for $W(\tilde{f}\tilde{h})$ will still be M. Thus we obtain an example for every element in the Torelli group, i.e. the kernel of $\rho^{(2)}$, for every choice of integers $(p_1, q_1), \ldots, (p_g, q_g)$. ||

§8. Postscript : Remarks on higher invariants

In this section, we make a few comments about the search for invariants of Heegaard splittings coming from the action of the mapping class group on the higher nilpotent quotients of the surface group (*i.e.* the higher terms in the Johnson-Morita filtration). In this paper, our invariants have come from 3 sources:

(1) The abelian group $H_1(W)$ of the 3-manifold W.
(2) The linking form on the torsion subgroup of $H_1(W)$.
(3) The presentation of $H_1(W)$ arising from the Heegaard splitting.

With regard to (1), It is easy to see that there is a natural generalization. The classical Van Kampen Theorem shows that the Heegaard gluing map $\tilde{h}$ determines a canonical presentation for $G = \pi_1(W)$ which arises via the action of h on π. This action determines in a natural way a presentation for $G/G^{(k)}$, the kth quotient group in the lower central series for G. We do not know of systematic studies of these invariants of the fundamental groups of closed, orientable 3-manifolds.

With regard to (2) and (3), if π_1 is the fundamental group of the Heegaard surface (which in this section we will consider to be a surface with 1 boundary component corresponding to a disc fixed by the gluing map – this will make π_1 a free group), then $H_1(W)$ is the quotient of the abelian group $\pi_1/\pi_1^{(2)}$ by the two lagrangians arising from the handlebodies. The obvious generalization of this is a quotient of the free nilpotent group $\pi_1/\pi_1^{(k)}$. Since it is unclear what the appropriate generalization of the linking form to this situation would be, one's first impulse might be to search for presentation invariants.

Now, it is easy to see that the quotient of $\pi_1/\pi_1^{(k)}$ by one of the "nilpotent lagrangians" is another free nilpotent group. Our presentation is thus a surjection $\pi : N_1 \to N_2$, where N_1 is a free nilpotent group. The invariants of presentations of abelian groups arise from the fact that automorphisms of the presented group may not lift to automorphisms of the free abelian group. Unfortunately, the following theorem says that no further obstructions exist:

Theorem 8.1. *Let $\pi : N_1 \to N_2$ be a surjection between finitely generated nilpotent groups, where N_1 is a free nilpotent group. Also, let ϕ be an automorphism of N_2. Then ϕ may be lifted to an automorphism of N_1 if and only if the induced automorphism ϕ_* of N_2^{ab} can be lifted to an automorphism of N_1^{ab}.*

The key to proving Theorem 8.1 is the following criterion for an endomorphism of a nilpotent group to be an automorphism. It is surely known to the experts, but we were unable to find an appropriate reference.

Theorem 8.2. *Let N be a finitely generated nilpotent group and let $\psi : N \to N$ be an endomorphism. Then ψ is an isomorphism if and only if the induced map $\psi_* : N^{ab} \to N^{ab}$ is an isomorphism.*

Proof. The forward implication being trivial, we prove the backward implication. The proof will be by induction on the degree n of nilpotency. If $n = 1$, then N is abelian and there is nothing to prove. Assume, therefore, that $n > 1$ and that the theorem is true for all smaller n. We begin by observing that since finitely generated nilpotent groups are Hopfian, it is enough to prove that ψ is surjective. Letting

$$N = N^{(1)} \rhd N^{(2)} \rhd \cdots \rhd N^{(n)} \rhd N^{(n+1)} = 1$$

be the lower central series of N, we have an induced commutative diagram

$$\begin{array}{ccccccccc} 1 & \longrightarrow & N^{(n)} & \longrightarrow & N & \longrightarrow & N/N^{(n)} & \longrightarrow & 1 \\ & & \downarrow \psi & & \downarrow \psi & & \downarrow & & \\ 1 & \longrightarrow & N^{(n)} & \longrightarrow & N & \longrightarrow & N/N^{(n)} & \longrightarrow & 1 \end{array}$$

Since $N/N^{(n)}$ is an $(n-1)$-step nilpotent group, the inductive hypothesis implies that the induced endomorphism of $N/N^{(n)}$ is an isomorphism. The five lemma therefore says that to prove that the map

$$\psi : N \longrightarrow N$$

is surjective, it is enough to prove that the map

$$\psi : N^{(n)} \longrightarrow N^{(n)}$$

is surjective. Now, $N^{(n)}$ is generated by commutators of weight n in the elements of N. Let β be a bracket arrangement of weight n and let $\beta(g_1, \dots, g_n) \in N^{(n)}$ with $g_i \in N$ be some commutator of weight n. Since ψ induces an isomorphism of $N/N^{(n)}$, we can find some $\tilde{g}_1, \dots, \tilde{g}_n \in N$ and $h_1, \dots, h_n \in N^{(n)}$ so that

$$\psi(\tilde{g}_i) = g_i h_i$$

for all i. Hence ψ maps $\beta(\tilde{g}_1, \dots, \tilde{g}_n)$ to $\beta(g_1 h_1, \dots, g_n h_n)$. However, since $N^{(n)}$ is central we have that

$$\beta(g_1 h_1, \dots, g_n h_n) = \beta(g_1, \dots, g_n),$$

so we conclude that $\beta(g_1, \dots, g_n)$ is in $\psi(N^{(n)})$, as desired. Q.E.D.

We now prove Theorem 8.1.

Proof of Theorem 8.1. Let $\{g_1, \dots, g_k\}$ be a free nilpotent generating set for N_1, and let ρ be an automorphism of N_1^{ab} lifting ϕ_*. Also, let $\overline{g}_i \in N_1^{\mathrm{ab}}$ be the image of g_i. Now, pick any lift $h_i \in N_1$ of $\rho(\overline{g}_i)$. Observe that by assumption $\pi(h_i)$ and $\phi(\pi(g_i))$ are equal modulo $[N_2, N_2]$. Since the restricted map $\pi : [N_1, N_1] \to [N_2, N_2]$ is easily seen to be surjective, we can find some $k_i \in [N_1, N_1]$ so that $\pi(h_i k_i) = \phi(\pi(g_i))$. Since N_1 is a free nilpotent group, the mapping

$$g_i \mapsto h_i k_i$$

induces an endomorphism $\tilde{\phi}$ of N_1 which by construction lifts ϕ. Moreover, Theorem 8.2 implies that $\tilde{\phi}$ is actually an automorphism, as desired.
Q.E.D.

Remark 8.3. Theorem 8.1 does not destroy all hope for finding invariants of presentations, as there may be obstructions to lifting automorphisms to automorphisms which arise "geometrically". However, it makes the search for obstructions much more subtle. Moreover, we note that in [23] Y. Moriah and M. Lustig used the presentation of $\pi_1(W)$ arising from a Heegaard splitting to prove that certain Heegaard splittings of Seifert fibered spaces are in fact inequivalent. Their subsequent efforts to generalize what they did [24] show that the problem is difficult, and the final word has not been said on invariants of Heegaard splittings that arise from the associated presentation of $\pi_1(W)$. ||

References

[1] S. Akbulut and J. D. McCarthy, Casson's invariant for oriented homology 3-spheres, Princeton Univ. Press, Princeton, NJ, 1990.

[2] J. S. Birman, On the equivalence of Heegaard splittings of closed, orientable 3-manifolds, In: Knots, groups, and 3-manifolds (Papers dedicated to the memory of R. H. Fox), Ann. of Math. Stud., **84**, Princeton Univ. Press, Princeton, NJ, pp. 137–164.

[3] J. S. Birman and R. Craggs, The μ-invariant of 3-manifolds and certain structural properties of the group of homeomorphisms of a closed, oriented 2-manifold, Trans. Amer. Math. Soc., **237** (1978), 283–309.

[4] J. S. Birman, F. González-Acuña and J. M. Montesinos, Heegaard splittings of prime 3-manifolds are not unique, Michigan Math. J., **23** (1976), 97–103.

[5] T. E. Brendle and B. Farb, The Birman-Craggs-Johnson homomorphism and abelian cycles in the Torelli group, Math. Ann., **338** (2007), 33–53.

[6] N. Broaddus, B. Farb and A. Putman, The Casson invariant and the word metric in the Torelli group, C. R. Acad. Sci. Paris, Ser. I Math., **345** (2007), 449–452.

[7] E. Burger, Über Gruppen mit Verschlingungen, J. Reine Angew. Math., **188** (1950), 193–200.

[8] H. Burkhardt, Grundzüge einer allgemeinen Systematik der hyperelliptischen Funktionen erster Ordnung, Math. Ann., **35** (1890), 198–296 (esp. pp. 209–212).

[9] T. D. Cochran and P. Melvin, Finite type invariants of 3-manifolds, Invent. Math., **140** (2000), 45–100.

[10] T. D. Cochran, A. Gerges and K. Orr, Dehn surgery equivalence relations on 3-manifolds, Math. Proc. Cambridge Philos. Soc., **131** (2001), 97–127.

[11] M. Day, Extending Johnson's and Morita's homomorphisms to the mapping class group, Algebr. Geom. Topol., **7** (2007), 1297–1326.

[12] R. Engmann, Nicht-homöomorphe Heegaard-Zerlegungen vom Geschlecht 2 der zusammenhängenden Summe zweier Linsenräume, Abh. Math. Sem. Univ. Hamburg, **35** (1970), 33–38.

[13] R. H. Fox, review of [7], Math Reviews, MR0043089 (13,204a).

[14] S. Garoufalidis and J. Levine, Finite type 3-manifold invariants, the mapping class group and blinks, J. Differential Geom., **47** (1997), 257–320.

[15] J. Hass, A. Thompson and W, Thurston, Stabilization of Heegaard splittings, preprint, arXiv:0802.2145v2 [math.GT] 6 Mar 2008.

[16] S. P. Humphries, Generators for the mapping class group, In: Topology of low-dimensional manifolds, Proc. Second Sussex Conf., Chelwood Gate, 1977, Lecture Notes in Math., **722**, Springer, Berlin, pp. 44–47.

[17] D. Johnson, An abelian quotient of the mapping class group $\mathcal{I}_g$, Math. Ann., **249** (1980), 225–242.

[18] D. Johnson, Conjugacy relations in subgroups of the mapping class group and a group-theoretic description of the Rochlin invariant, Math. Ann., **249** (1980), 243–263.

[19] D. Johnson, The structure of the Torelli group. I. A finite set of generators for $\mathcal{I}$, Ann. of Math. (2), **118** (1983), 423–442.

[20] D. Johnson, The structure of the Torelli group. II. A characterization of the group generated by twists on bounding curves, Topology, **24** (1985), 113–126.

[21] D. Johnson, A survey of the Torelli group, In: Low-dimensional topology, San Francisco, Calif., 1981, Contemp. Math., **20**, Amer. Math. Soc., Providence, RI, pp. 165–179.

[22] A. Kawauchi and S. Kojima, Algebraic classification of linking pairings on 3-manifolds, Math. Ann., **253** (1980), 29–42.

[23] M. Lustig and Y. Moriah, Nielsen equivalence in Fuchsian groups and Seifert fibered spaces, Topology, **30** (1991), 191–204.

[24] M. Lustig and Y. Moriah, Generating systems of groups and Reidemeister-Whitehead torsion, J. Algebra, **157** (1993), 170–198.

[25] W. Magnus, A. Karrass and D. Solitar, Combinatorial group theory: Presentations of groups in terms of generators and relations, Interscience Publishers, John Wiley & Sons, Inc., New York, 1966.

[26] H. Minkowski, Grundlagen für eine Theorie der quadratischen Formen mit ganzzahligen koeffiziente, Werke, Bd. I, 3–143.

[27] J. M. Montesinos and C. Safont, On the Birman invariants of Heegaard splittings, Pacific J. Math., **132** (1988), 113–142.

[28] Y. Moriah and J. Schultens, Irreducible Heegaard splittings of Seifert fibered spaces are either vertical or horizontal, Topology, **37** (1998), 1089–1112.

[29] S. Morita, On the structure of the Torelli group and the Casson invariant, Topology, **30** (1991), 603–621.

[30] S. Morita, The extension of Johnson's homomorphism from the Torelli group to the mapping class group, Invent. Math., **111** (1993), 197–224.

[31] S. Morita, Abelian quotients of subgroups of the mapping class group of surfaces, Duke Math. J., **70** (1993), 699–726.

[32] S. Morita, A linear representation of the mapping class group of orientable surfaces and characteristic classes of surface bundles, In: Topology and Teichmüller spaces, Katinkulta, 1995, World Sci. Publ., River Edge, NJ, pp. 159–186.

[33] S. Morita, Structure of the mapping class group and symplectic representation theory, In: Essays on geometry and related topics, Vol. 1, 2, Monogr. Enseign. Math., **38**, Enseignement Math., Geneva, 2001, pp. 577–596.

[34] M. Newman, Integral matrices, Academic Press, New York, 1972.

[35] I. Niven, H. S. Zuckerman and H. L. Montgomery, An introduction to the theory of numbers, Fifth edition, Wiley, New York, 1991.

[36] O. T. O'Meara, Introduction to quadratic forms, Academic Press, Publishers, New York, 1963.

[37] W. Pitsch, Trivial cocycles and invariants of homology 3-spheres, preprint, math.GT/0605725.

[38] W. Pitsch, Integral homology 3-spheres and the Johnson filtration, to appear in Trans. Amer. Math. Soc.

[39] K. Reidemeister, Zur Dreidimensionalen Topologie, Abh. Math. Sem. Univ. Hamburg, **9** (1933), 189–194.

[40] K. Reidemeister, Heegaarddiagramme und Invarianten von Mannigfaltighesten, Abh. Math. Sem. Univ. Hamburg, **10** (1934), 109–118.

[41] M. Scharlemann, Heegaard splittings of compact 3-manifolds, In: Handbook of geometric topology, North-Holland, Amsterdam, pp. 921–953.

[42] J. Schultens, The stabilization problem for Heegaard splittings of Seifert fibered spaces, Topology Appl., **73** (1996), 133–139.

[43] H. Seifert, Verschlingungsinvarianten, Sitzungsber. Preu. Akad. Wiss., **26-29** (1933), 811–828.

[44] J. Singer, Three-dimensional manifolds and their Heegaard diagrams, Trans. Amer. Math. Soc., **35** (1933), 88–111.

[45] D. Sundararajan, The discrete Fourier transform, World Sci. Publishing, River Edge, NJ, 2001.

[46] F. Waldhausen, Heegaard-Zerlegungen der 3-Sphäre, Topology, **7** (1968), 195–203.

[47] C. T. C. Wall, Quadratic forms on finite groups, and related topics, Topology, **2** (1963), 281–298.

Advanced Studies in Pure Mathematics 52, 2008
Groups of Diffeomorphisms
pp. 221–250

Conjugation-invariant norms on groups of geometric origin

Dmitri Burago[a], Sergei Ivanov[b] and Leonid Polterovich[c]

Abstract.

A group is said to be bounded if it has a finite diameter with respect to any bi-invariant metric. In the present paper we discuss boundedness of various groups of diffeomorphisms.

§1. Introduction and main results

1.1. The main phenomenon

A group G is said to be *bounded* if it is bounded with respect to any bi-invariant metric (that is, as a metric space, it has a finite diameter).

A *conjugation-invariant norm* $\nu : G \to [0; +\infty)$ is a function which satisfies the following axioms:

(i) $\nu(1) = 0$;
(ii) $\nu(f) = \nu(f^{-1})\ \ \forall f \in G$;
(iii) $\nu(fg) \le \nu(f) + \nu(g)\ \ \forall f, g \in G$;
(iv) $\nu(f) = \nu(gfg^{-1})\ \ \forall f, g \in G$;
(v) $\nu(f) > 0$ for all $f \ne 1$.

Thus a group is bounded iff every conjugation-invariant norm is bounded.

Convention: In this paper we work only with conjugation-invariant norms, so by default a *norm* is a conjugation-invariant norm.

If one drops condition (v), ν is said to be a *pseudo-norm*. It can immediately be converted into a norm by adding 1 to all elements except the

---

Received October 4, 2007.
Revised February 26, 2008.
[a]Partially supported by the NSF Grant DMS-0412166.
[b]Partially supported by RFBR grant 05-01-00939.
[c]Partially supported by the Israel Science Foundation grant # 509/07.

unity. Hence a group is unbounded if it admits an unbounded pseudo-norm. Observe that on a simple group every non-trivial pseudo-norm is automatically a norm: Indeed, the set of all elements with vanishing pseudo-norm forms a normal subgroup. Hence in the sequel condition (v) can be dropped everywhere when we deal with simple groups such as various groups of smooth diffeomorphisms.[d]

Two norms on a group are called *equivalent* if their ratio is bounded away from 0 and ∞. *The trivial norm*, which exists on any group, equals 1 on every element except the identity.

Given a connected manifold M, denote by $\mathrm{Diff}_0(M)$ the identity component of the group of C^∞ smooth compactly supported diffeomorphisms. This group is simple due to a theorem by Thurston [34]. The central phenomenon discussed in this paper is as follows: *in all known to us examples any norm on* $\mathit{Diff}_0(M)$ *is equivalent to the trivial one.* Below we confirm this phenomenon for spheres, all closed connected three-manifolds and the annulus. However we have neither a proof nor a counter-example for closed surfaces of genus ≥ 1 and the Möbius strip.

1.2. Setting the stage

1.2.1. *Conjugation-generated norms.* Many interesting norms come from the following construction: Let G be a group, and let $K \subset G$ be a symmetric subset, that is $x \in K$ whenever $x^{-1} \in K$. We say that the set K *conjugation-generates* (or, for brevity, *c-generates*) G if every element $h \in G$ can be represented as a product

$$h = \tilde{h}_1 \tilde{h}_2 \dots \tilde{h}_N \tag{1}$$

where each $\tilde{h}_i$ is conjugate to some element $h_i \in K$: $\tilde{h}_i = \alpha_i h_i \alpha_i^{-1}$, $\alpha_i \in G$. In this case define a norm $q_K(h)$ as the minimal N for which such a representation exists. We shall say that *the norm q_K is c-generated by the subset K*. If K is finite, G is said to be *finitely c-generated.* For instance, every simple group G is finitely c-generated by $K = \{x, x^{-1}\}$ with an arbitrary $x \neq 1$.

Note that the norm q_K has the following extremal property: for any norm q bounded on K there is a constant λ such that $q \leq \lambda q_K$. Hence, if K is finite, the group G is bounded if and only if q_K is bounded.

---

[d] A group is called simple if it has no non-trivial normal subgroups. In the 1970-ies, simplicity of various interesting groups of diffeomorphisms was established by highly non-trivial methods in works of Herman [16], Thurston [34], Mather [21, 22], Banyaga [2]. We refer to Banyaga's book [3] for a detailed discussion.

Example 1.1. Groups $SL(n,\mathbb{R})$ for $n \geq 2$ and $SL(n,\mathbb{Z})$ for $n \geq 3$ are finitely c-generated by the set K of all elementary matrices whose off-diagonal term equals ± 1. Moreover we claim that the number of terms in the decomposition (1) is bounded by a constant which does not depend on h.

In the case of $SL(n,\mathbb{R})$ the claim follows from an appropriate version of the Gauss elimination process.

As for $SL(n,\mathbb{Z})$, denote by $\mathcal{E}$ the set of all elementary matrices whose only non-zero off-diagonal element equals to 1. There exists $N = N(n) \in \mathbb{N}$ so that every element from $SL(n,\mathbb{Z})$ can be written as a product of $\leq N$ matrices of the form E^p, where $E \in \mathcal{E}$ and $p \in \mathbb{Z}$ (in other words, $SL(n,\mathbb{Z})$ possesses a bounded generation by elements from $\mathcal{E}$), see [9, 37]. The claim readily follows from the fact that each $E^p = [A, B^p]$ for some $A, B \in \mathcal{E}$. Let us prove this identity: let E_{ij} (where $i \neq j$) denotes the elementary matrix from $\mathcal{E}$ whose only non-zero off-diagonal element stands in the i-th raw and j-th column. Without loss of generality, put $i = 1, j = 3$. Then $E_{13}^p = [E_{12}, E_{23}^p]$ as required.

It follows from the claim that the "extremal" norm q_K is bounded, and hence the groups in question are bounded in view of extremality of q_K.

Example 1.2. The commutator length. Given a group G, denote by G' its commutator subgroup. The norm on G' c-generated by the set of all simple commutators $[a,b] = aba^{-1}b^{-1}$ is called the *commutator length* and is denoted by cl_G. This norm has a long history and has been intensively studied in various contexts, see e.g. [5].

1.2.2. *The role of the commutator subgroup.* The next observations suggest that the commutator subgroup plays a significant role in the study of boundedness.

Proposition 1.3. *If $H_1(G) := G/G'$ is infinite then G is unbounded.*

In particular, *an abelian group is bounded if and only if it is finite.*

Note that unbounded norms maybe non-extendable from a normal subgroups to the ambient group. Consider, for instance, the group $Aff(\mathbb{Z})$ of transformations of the real line of the form $u \mapsto \epsilon u + z$ with $\epsilon = \pm 1$ and $z \in \mathbb{Z}$. It can be considered as an extension of $\mathbb{Z}$ (the group of integer translations) by an element t of order 2 (the reflection over the origin) and with one additional relation $tz = z^{-1}t$. Thus $\mathbb{Z}$ is a normal subgroup of index 2 in $Aff(\mathbb{Z})$. Of course, $\mathbb{Z}$ has an unbounded norm, while $Aff(\mathbb{Z})$ admits no unbounded norms since t is conjugate to tz^{2n} (by z^n) for all integers n. However, the situation changes when one deals with the commutator length on the commutator subgroup:

Proposition 1.4. *Let G be any group. If the commutator length on G' is unbounded then G itself is unbounded.*

Propositions 1.3 and 1.4 are proved in Section 1.2.5 below.

1.2.3. *Stably unbounded norms.* Given a conjugation-invariant norm ν on a group G, we define its *stabilization* by

$$\nu_\infty(f) = \lim_{n\to\infty} \frac{\nu(f^n)}{n} .$$

Let us emphasize that stabilization of a norm is not in general a norm. An unbounded norm ν is called *stably unbounded* if $\nu_\infty(f) \neq 0$ for some $f \in G$.

For instance, an infinite abelian torsion group is unbounded by Proposition 1.3 but never stably unbounded.

Example 1.5. Consider a group $\mathbb{Z}_2^\infty$ of all finite words over $\{0, 1\}$ with componentwise addition mod 2 (that is, a direct sum of countably many copies of $\mathbb{Z}_2$). This group admits no stably unbounded norms since the order of every element is 2. On the other hand, the length of a word is an unbounded norm. There is a natural action of $\mathbb{Z}_2^\infty$ on $\mathbb{Z} \times \mathbb{Z}_2$: the i-th generator swaps $(i, 0)$ and $(i, 1)$. Thus the norm in our example can be interpreted as "the size of support".

Open Problem. Does there exist a group that does not admit a stably unbounded norm and yet admits a norm unbounded on some cyclic subgroup?

1.2.4. *Stable commutator length and quasi-morphisms.* In what follows we shall focus on the stable commutator length. Let G be any group. The commutator length cl_G on G' is stably unbounded if and only if G admits *non-trivial homogeneous quasi-morphisms* [5]. Recall that a function $r : G \to R$ is called a quasi-morphism if there exists $C > 0$ so that

$$|r(ab) - r(a) - r(b)| \leq C \;\; \forall a, b \in G .$$

A quasi-morphism is called *homogeneous* if $r(a^n) = nr(a)$ for all $a \in G$ and $n \in \mathbb{Z}$. A quasi-morphism is called non-trivial if it is not a morphism.

Convention: In this paper we deal with homogeneous quasi-morphisms only, so by default quasi-morphism means a homogeneous quasi-morphism.

Example 1.6. $G = SL(2, \mathbb{Z})$ carries an abundance of quasi-morphisms (cf. e.g. [4]) and hence the commutator norm on $SL(2, \mathbb{Z})$ is stably unbounded. Thus G is unbounded in view of Proposition 1.4, in contrast with $SL(n, \mathbb{Z})$ for $n \geq 3$ (see Example 1.1 above).

Introduce the class $\mathcal{G}$ of groups G with finite $H_1(G) = G/G'$ (we wish to rule out conjugation-invariant stably unbounded norms coming from the first homology, see Proposition 1.3 above). Note that various interesting groups of diffeomorphisms are simple (see footnote in Section 1.1 above) and hence belong to this class.

Open Problem. Does there exist a finitely presented group $G \in \mathcal{G}$ whose commutator length is unbounded but stably bounded?

Open Problem. Does there exist an unbounded finitely presented group which admits no unbounded quasi-morphisms?

A. Muranov informed us that he has an example of a finitely generated, but not finitely presented, group from $\mathcal{G}$ whose commutator length is unbounded but stably bounded. The existence of an infinitely generated group with this property readily follows from Muranov's work [26], who constructed a sequence of simple groups $G_i, i \in \mathbb{N}$ of finite commutator length diameter n_i, where $n_i \to \infty$. The infinite direct product $G = \prod_i G_i$ is as required.

A mystery related to the notion of stable unboundedness is as follows.

Open Problem. Does there exist a group $G \in \mathcal{G}$ whose commutator length is stably bounded, but which admits a stably unbounded norm? In other words, does the existence of a stably unbounded norm on G yields existence of non-trivial quasi-morphisms? In fact, we do not know even a single example of a group from $\mathcal{G}$ that admits no non-trivial quasi-morphisms but carries a norm that is unbounded on some cyclic subgroup.

Here is a (somewhat artificial) example of groups for which existence of a stably unbounded norm yields existence of non-trivial quasi-morphisms. Start with an arbitrary group $G \in \mathcal{G}$ and set $\bar{G}$ to be the extension of $G \times G$ by an element t so that

$$t^2 = 1, \quad \text{and} \quad t(g_1, g_2)t^{-1} = (g_2, g_1) \ \forall g_1, g_2 \in G.$$

Proposition 1.7. *The group $\bar{G}$ lies in $\mathcal{G}$ for every $G \in \mathcal{G}$.*

Proposition 1.8. *Suppose that for some $G \in \mathcal{G}$, the group $\bar{G}$ admits a stably unbounded norm. Then $\bar{G}$ admits a non-trivial quasi-morphism.*

1.2.5. *Quasi-norms.*

Definition 1.9. *Let G be a group. We say that a function $q : G \to [0; +\infty)$ is a a quasi-norm (for brevity, a q-norm) if:*

(i) *q is quasi-subadditive: there is a constant c such that*

$$q(ab) \leq q(a) + q(b) + c\,;$$

(ii) *q is quasi-conjugation-invariant: there is a constant c such that*

$$|q(b^{-1}ab) - q(a)| \leq c\,;$$

(iii) *q is unbounded.*

One can see that in fact the existence of a q-norm implies the existence of an unbounded norm: This norm can be constructed by (i) symmetrization: taking the maximum of the norm of a and a^{-1} for each a, (ii) redefining the norm of a to be the maximum of norms of its conjugates $b^{-1}ab$, and (iii) by adding a sufficiently large constant to the norm of all elements excluding the identity.

Hence a group is unbounded if it admits a q-norm; in other words, the existence of unbounded norms and q-norms are equivalent. However q-norms are often defined in a more natural way: A motivating example is provided by the absolute value of a non-trivial homogeneous quasi-morphism. Another advantage of q-norms is that they behave nicely under epimorphisms:

Lemma 1.10. *The pull-back of a q-norm under an epimorphism is a q-norm. In particular, if a group G admits a homomorphism onto an unbounded group, G itself is unbounded.*

This follows immediately from the definitions and discussion above. Let us apply the lemma for proving results stated in 1.2.2:

Proof of Proposition 1.3:

STEP 1: Let us show that any infinite abelian group G admits an unbounded norm.

If G is finitely generated, than by the classification theorem it has a $\mathbb{Z}$ as a direct factor, and hence it admits an epimorphism onto $\mathbb{Z}$. Thus G admits an unbounded norm by Lemma 1.10.

For a countably generated G, let us enumerate its generators g_1, $g_2, \ldots$. Define the norm of g to be the smallest k such that g lies in the subgroup generated by $g_1, g_2, \ldots, g_k$. This norm is unbounded.

In general, any infinite abelian group contains an infinite finitely or countably generated subgroup, and the above construction provides us with a norm on this subgroup H. Now choose any element g from $G \setminus H$ and consider a subgroup H' generated by the union of H and g. Combining the easily verifiable fact that the norm extends from H to H' with Zorn's lemma completes the proof.

STEP 2: Assume now that G/G' is infinite. By Step 1, it admits an unbounded norm. Look at the epimorphism $G \to G/G'$. Applying Lemma 1.10 we conclude that G is unbounded. ■

Proof of Proposition 1.4: If $[G, G]$ has infinite index, look at the epimorphism $G \to H := G/G'$. The group H is an infinite abelian group, thus by Proposition 1.3 H is unbounded, and hence G is unbounded in view of Lemma 1.10.

Otherwise, if H is finite, one can check that the commutator norm can be extended from the commutator to the whole group (even though in general q-norms cannot be extended from finite index subgroups, see an example above). Indeed, pick a (finite!) set S of representatives from cosets of G'. Then every element of G can be uniquely written as hs where $h \in G'$, $s \in S$. Define a q-norm of such an element $g = hs$ by $q(g) = cl_G(h)$. The approximate conjugation invariance of this norm follows from the fact that conjugation can be written as a multiplication by a commutator (and hence it changes the norm by at most 1). To prove the approximate triangle inequality, note that for $g_1 = h_1 s_1$ and $g_2 = h_2 s_2$

$$g_1 g_2 = h_1 h_2 [h_2^{-1}, s_1] s_1 s_2 \ .$$

Write

$$s_1 s_2 = h(s_1, s_2) t(s_1, s_2) \ ,$$

where $h(s_1, s_2) \in G'$ and $t(s_1, s_2) \in S$. Thus

$$q(g_1 g_2) = cl_G(h_1 h_2 [h_2^{-1}, s_1] h(s_1, s_2)) \ .$$

Put $C = \max_{s_1, s_2 \in S} cl_G(h(s_1, s_2))$. Applying the triangle inequality for the commutator length, we get

$$q(g_1 g_2) \le cl_G(h_1) + cl_G(h_2) + 1 + C = q(g_1) + q(g_2) + 1 + C \ .$$

Thus q is indeed a q-norm. ■

1.2.6. *Fine norms.* A norm ν on G is called *fine* if 0 is a limit point of $\nu(G)$. Otherwise the norm is called, following a suggestion by Yehuda Shalom, *discrete.* For instance, conjugation-generated norms assume integer values only and hence are discrete. On the other hand a bi-invariant Riemannian metric on a compact Lie group gives rise to a bounded fine norm on the group.

1.2.7. *Meager groups.* A norm ν on a group is **not** equivalent to the trivial norm if it is either unbounded or fine. A group G is called *meager* if every conjugation-invariant norm on G is equivalent to the trivial one (i.e. is bounded and discrete).

1.3. Norms on diffeomorphism groups

1.3.1. *Smooth diffeomorphisms.* In this section we present the main results of the paper which deal with norms on groups $\mathrm{Diff}_0(M)$, where M is a smooth connected manifold. We start with the case of closed manifolds.

Theorem 1.11 (Main Theorem)**.**

(i) *The group $\mathit{Diff}_0(M)$ does not admit a fine conjugation-invariant norm for all connected manifolds M.*

(ii) *The group $\mathit{Diff}_0(S^n)$ is meager (where S^n is a sphere);*

(iii) *The group $\mathit{Diff}_0(M)$ is meager for any closed connected 3-dimensional manifold M.*

After the first draft of this paper appeared, T.Tsuboi [35] generalized this result and, remarkably, established meagerness of $Diff_0(M)$ for all odd-dimensional closed manifolds.

Let us give two important examples of conjugation-invariant norms on $\mathrm{Diff}_0(M)$.

Example 1.12. The commutator length: Since $\mathrm{Diff}_0(M)$ is a simple group [34] it coincides with its commutator subgroup and hence the commutator length (see Example 1.2) is a well-defined invariant norm on $\mathrm{Diff}_0(M)$. Introduce the *commutator length diameter* $cld(M) \in \mathbb{N} \cup \infty$ as $\max cl(f)$ over all $f \in \mathrm{Diff}_0(M)$.

Theorem 1.13.

(i) *For the sphere, $cld(S^n) \leq 4$;*

(ii) *For any closed connected 3-dimensional manifold M, $cld(M) \leq 10$.*

Example 1.14. The fragmentation norm: Every element $f \in \mathrm{Diff}_0(M)$ can be represented as a finite product of diffeomorphisms supported in an embedded open ball (this is the famous fragmentation lemma, see e.g. [3]). The *fragmentation norm* $frag(f)$ is the minimal number of factors required to represent an element $f \in \mathrm{Diff}_0(M)$. Clearly, $frag$ is an conjugation-invariant norm on $\mathrm{Diff}_0(M)$. The next result shows that the fragmentation norm is responsible for meagerness of $\mathrm{Diff}_0(M)$.

Proposition 1.15. *The group $\mathit{Diff}_0(M)$ is meager if and only if the fragmentation norm is bounded.*

Open Problem. Is the fragmentation norm bounded for the case of closed surfaces?

Let us now turn to open manifolds.

Definition 1.16. We say that a smooth connected open manifold M is *portable*[e] if it admits a complete vector field X and a compact subset M_0 with the following properties:

- M_0 is an attractor of the flow X^t generated by X: for every compact subset $K \subset M$ there exists $\tau > 0$ so that $X^\tau(K) \subset M_0$.
- There exists a diffeomorphism $\theta \in \mathrm{Diff}_0(M)$ so that $\theta(M_0) \cap M_0 = \emptyset$.

The set M_0 is called *the core* of a portable manifold M.

For instance, any manifold M which splits as $P \times \mathbb{R}^n$, where P is a closed manifold, is portable. Indeed, the vector field $X(p,z) = -z\frac{\partial}{\partial z}$ and the compact $M_0 = P \times \{|z| \le 1\}$ satisfy the conditions above. Furthermore, M is portable if it admits an exhausting Morse function with finite number of critical points so that all the indices are strictly less than $\frac{1}{2}\dim M$. This implies, for example, that every 3-dimensional handlebody is a portable manifold.

The next result is the main "local" block in the proof of Theorem 1.11(ii) and (iii).

Theorem 1.17. *The group $\mathit{Diff}_0(M)$ is meager provided M is portable.*

For instance, any norm on Diff_0 of an open ball is bounded. Together with Theorem 1.11(i) this immediately yields Proposition 1.15. Furthermore, Diff_0 of a 2-dimensional annulus is meager (as well as for any product $\mathbb{R} \times M$). However, it is still unknown whether the same holds for the open Möbius band!

Our next result deals with the commutator length diameter of a portable manifold.

Theorem 1.18. *For a portable manifold M, $\mathit{cld}(M) \le 2$.*

---

[e]This notion is a mock version of subcritical Liouville manifolds in symplectic topology.

1.3.2. *Volume-preserving and symplectic diffeomorphisms: examples and problems.* In contrast to groups Diff_0, the identity components of groups of compactly supported volume preserving and symplectic diffeomorphisms, as well as their commutator subgroups, are **never** meager: they admit a fine norm.

Example 1.19. The size-of-support norm: The counterpart of Example 1.5 above for diffeomorphism groups is as follows. Consider the identity component $\mathrm{Diff}_0(M, \mathrm{vol})$ of the group of compactly supported volume-preserving diffeomorphisms of a smooth manifold M of dimension > 0. Define the norm of a diffeomorphism as the volume of its support. This norm is necessarily fine, and it is unbounded whenever the volume of M is infinite. However this norm is never stably unbounded: in fact, it is bounded on all cyclic subgroups.

In some situations, stably unbounded norms on the commutator subgroup of $\mathrm{Diff}_0(M, \mathrm{vol})$ can be "induced" from the fundamental group of M even when the volume of M is finite:

Example 1.20. Suppose that M is a closed manifold equipped with a volume form. Suppose that $H := \pi_1(M)$ has trivial center. Then the commutator length on the commutator subgroup of $\mathrm{Diff}_0(M, \mathrm{vol})$ is stably unbounded provided the commutator length on H' is stably unbounded, see [15, 29].

However, in dimension ≥ 3 no unbounded norms on volume-preserving diffeomorphisms are known so far in the cases when the manifold has simple topology and finite volume.

Open Problem. Assume that $n \geq 3$. Does the identity component of the group of volume preserving diffeomorphisms of the sphere S^n admit an unbounded conjugation-invariant norm? Does the identity component of the group of compactly supported volume preserving diffeomorphisms of the ball of finite volume admit an unbounded conjugation-invariant norm?

In the symplectic category, interesting norms inhabit the group $\mathrm{Ham}(M, \omega)$ of compactly supported Hamiltonian diffeomorphisms of a symplectic manifold (M, ω).

Example 1.21. The Hofer norm on $\mathrm{Ham}(M, \omega)$ ([17], see also [28]) is fine. Its unboundedness is a long-standing conjecture in symplectic topology. Nowadays it is confirmed for various symplectic manifolds including for instance surfaces, complex projective spaces with the Fubini-Studi symplectic form and closed manifolds with $\pi_2 = 0$. Further, the Hofer norm on groups of Hamiltonian diffeomorphisms is known to be stably unbounded for various closed symplectic manifolds. However it is unbounded, but not stably unbounded, for the standard symplectic vector space $\mathbb{R}^{2n}$ (Sikorav, [33]).

Example 1.22. The commutator length on $\mathrm{Ham}(M,\omega)$ is known to be stably unbounded for various symplectic manifolds (see [4, 12, 13, 7, 15, 30, 31]) including all surfaces and complex projective spaces of arbitrary dimension.

Example 1.23. The group $\mathrm{Ham}(\mathbb{R}^{2n})$ admits the Calabi homomorphism (the average Hamiltonian) to $\mathbb{R}$. The kernel of the Calabi homomorphism coincides with the commutator subgroup of $\mathrm{Ham}(\mathbb{R}^{2n})$, which is known to be simple [2]. This group is stably bounded with respect to the commutator length. This is proved by D. Kotschick in [18]. Alternatively, this readily follows from the algebraic packing inequality given by Theorem 2.7 below. In contrast to this, the commutator length on $[\mathrm{Ham}(B^{2n}), \mathrm{Ham}(B^{2n})]$, where B^{2n} is the standard symplectic ball, is stably unbounded, see [4].

Example 1.24. A somewhat less understood example is *the fragmentation norm* (cf. Example 1.14 above). Let (M,ω) be a closed symplectic manifold and let $U \subset M$ be an open subset. The Hamiltonian fragmentation lemma (see [2]) states that every Hamiltonian diffeomorphism f can be written as a product $h_1 \circ \ldots \circ h_N$, where each h_i is conjugate to an element from $\mathrm{Ham}(U)$. Define the fragmentation norm $frag_U(f)$ as the minimal number of factors in such a decomposition. Using methods of [14], one can show that for certain symplectic manifolds $frag_U$ is unbounded on $\mathrm{Ham}(M)$ provided the subset U is *displaceable by a Hamiltonian diffeomorphism* (e.g. U is a ball of a small diameter). Let us elaborate this statement.

First of all recall [32, 27, 25] that for elements of *the universal cover* $\widetilde{\mathrm{Ham}}(M)$ of the group of Hamiltonian diffeomorphisms one can define *spectral invariants* which come from Floer homology of the action functional. Denote by $\widetilde{\mu} : \widetilde{\mathrm{Ham}}(M) \to \mathbb{R}$ the asymptotic spectral invariant as defined in [14]. For various interesting symplectic manifolds $\widetilde{\mu}$ descends to a function μ on $\mathrm{Ham}(M)$. This is for instance the case for symplectic manifolds with $\pi_2(M) = 0$ (due to M. Schwarz [32]) and for standard complex projective spaces (see [13]). [f] We continue discussion on the fragmentation norm assuming that $\widetilde{\mu}$ does descend to μ.

Second, if U is displaceable, Theorem 7.1 in [14] guarantees that

$$|\mu(\phi\psi) - \mu(\phi) - \mu(\psi)| \leq \min(frag_U(\phi), frag_U(\psi)) \quad (2)$$

for all $\phi, \psi \in \mathrm{Ham}(M)$. At this point there is a dichotomy which roughly speaking depends on the algebraic structure of the quantum homology ring of (M, ω):

Possibility 1: The left hand side of (2) is a bounded function on $\mathrm{Ham}(M) \times \mathrm{Ham}(M)$, and thus μ is a quasi-morphism on $\mathrm{Ham}(M)$. For instance, this is the case for the complex projective spaces [13].

Possibility 2: The left hand side of (2) is unbounded, and thus *a fortiori* the fragmentation norm on $\mathrm{Ham}(M)$ is unbounded. For instance this is the case for the standard symplectic tori $(\mathbb{R}^{2n}/\mathbb{Z}^{2n}, dp \wedge dq)$.

Let us explain why Possibility 2 holds for the two-torus: Take a pair of disjoint meridians L and K on the torus. Let Φ, Ψ be two smooth cut off functions on the torus with disjoint supports which equal 1 near L and K respectively. Let $\{\phi_t\}$ and $\{\psi_t\}$ be the Hamiltonian flows generated by Φ and Ψ. A standard calculation in Floer homology shows that the left hand side of (2) with $\phi = \phi_t, \psi = \psi_t$ goes to infinity as $t \to \infty$. This proves unboundedness of the Hamiltonian fragmentation norm $frag_U$ for the 2-torus.

We conclude with an open problem. In spite of the fact that the complex projective spaces enjoy Possibility 1 above, the question on unboundedness of the Hamiltonian fragmentation norm in this case is widely open even for $\mathbb{C}P^1 = S^2$.

Organization of the paper: In the next section we introduce algebraic packing and displacement technique which is used for the proof of the main results stated in the introduction. As an illustration, we deduce there Theorem 1.11(i) and Proposition 1.8. Theorems 1.17 and 1.18 are proved in Section 3.1. These theorems, combined with topological decomposition technique (which is standard in the case of spheres, and less trivial in the case of three-manifolds) are applied to the proof of Theorems 1.11(ii),1.13(i) in Section 3.2 and of Theorems 1.11(iii),1.13(ii) in Section 3.3.

[f] In general, $\widetilde{\mu}$ may descend to $\mathrm{Ham}(M)$ and may not. We refer to a recent paper [23] by D. McDuff for new results and a detailed discussion of the current state of art in this problem. We thank D. McDuff for an illuminating discussion on this topic.

§2. Algebraic tools: packing and displacement

Here we present the algebraic tools used for proving Theorems 1.11 (i), 1.17 and 1.18. We use a number of tricks which imitate displacement of supports of diffeomorphisms and decomposition of diffeomorphisms into products of commutators in a more general algebraic setting. The tricks of this nature appear in the context of transformation groups at least since the beginning of 1960-ies (see e.g. [1]). The system of notions introduced below in parts imitates and extends the one arising in the study of Hofer's geometry on the group of Hamiltonian diffeomorphisms. Note also that various interesting results on infinitely displaceable subgroups were obtained in a recent work of D. Kotschick [18].

2.1. Algebraic packing and displacement energy

Let G be any group. We say that two subgroups $H_1, H_2 \subset G$ commute if $h_1 h_2 = h_2 h_1$ for all $h_1 \in H_1, h_2 \in H_2$. We denote by $Conj_\phi$ the automorphism of G given by $g \mapsto \phi g \phi^{-1}$. A subgroup $H \subset G$ is called *m-displaceable* (where $m \geq 1$ is an integer) if there exist elements $\phi_0 := 1, \phi_1, ..., \phi_m \in G$ so that the subgroups $\mathrm{Conj}_{\phi_i}(H), \mathrm{Conj}_{\phi_j}(H)$ pair-wise commute for all distinct $i, j \in \{0; ...; m\}$. A subgroup H is called *strongly* m-displaceable if in the previous definition one can choose ϕ_k's to be consecutive powers of the same element $\phi \in G$: $\phi_k = \phi^k$. In this case we shall say that ϕ m-displaces H.

Note that for $m = 1$ both notions coincide, and, for brevity, we refer to a 1-displaceable subgroup as to displaceable.

Introduce two numerical invariants related to the above notions. The *algebraic packing number* $p(G, H) = m + 1$, where m is the maximal integer such that H is m-displaceable. This is a purely algebraic invariant. The second quantity involves a conjugation-invariant norm, say ν on G. Define the *order m displacement energy* [g] of H with respect to ν as $e_m(H) = \inf \nu(\phi)$ where the infimum is taken over all $\phi \in G$ which m-displace H. We put $e_m(H) = +\infty$ if H is not strongly m-displaceable.

While speaking on displaceability, we tacitly assume that the subgroup H is non-abelian. Indeed, every abelian subgroup H is m-displaceable by 1 for every $m \in \mathbb{N}$ and hence $e_m(H) = 0$.

Example 2.1. Let M be a smooth connected manifold. Put $G = \mathrm{Diff}_0(M)$. Take any open ball $B \subset M$. Let H be the subgroup of G

---

[g]This notion is an algebraic counterpart of the symplectic displacement energy introduced by Hofer in [17].

consisting of all diffeomorphisms supported in B. Choose any diffeomorphism $\phi \in \mathrm{Diff}_0(M)$ which displaces B: $B \cap \phi(B) = \emptyset$. Then H commutes with $Conj_\phi(H)$, so H is displaceable.

Theorem 2.2. *Let $H \subset G$ be a strongly m-displaceable subgroup of G. Assume that G is endowed with a conjugation-invariant norm ν.*

(i) *For every element $x \in H'$ with $cl_H(x) = m$ the following inequalities hold:*

$$\nu(x) \leq 14e_m(H) \tag{3}$$

and

$$cl_G(x) \leq 2; \tag{4}$$

(ii) *In the case $cl_H(x) = 1$, that is $x = [f, g]$ for some $f, g \in H$, we have that*

$$\nu(x) \leq 4e_1(H)\ ; \tag{5}$$

Corollary 2.3. *Assume that an element $F \in G$ m-displaces H for every $m \geq 1$. Then $cl_G(h) \leq 2$ for all $h \in H'$.*

This follows immediately from inequality (4).

Theorem 2.2(ii) is proved in [11]. The argument is very short: indeed, assume that $\mathrm{Conj}_\phi(H)$ commutes with H. Then $[f, g] = [f \cdot \phi f^{-1} \phi^{-1}, g]$. Using bi-invariance of ν we get that

$$\nu([f, g]) \leq 2\nu([f, \phi]) \leq 4\nu(\phi)\ .$$

Taking the infimum over all ϕ displacing H we get inequality (5). The proof of Theorem 2.2(i) is more involved, see Section 2.2 below.

Let us give some sample applications of Theorem 2.2. First, we deduce from inequality (5) the fact that the group $\mathrm{Diff}_0(M)$ does not admit a fine norm.

Proof of Theorem 1.11(i): Assume on the contrary that $\mathrm{Diff}_0(M)$ admits a fine norm, say ν. Take any ball $B \subset M$ and pick two non-commuting diffeomorphisms f and g supported in B. For any $\epsilon > 0$ take $h \in \mathrm{Diff}_0(M)$ with $0 < \nu(h) < \epsilon$. Note that since $h \neq \mathbb{1}$ there exists a ball $C \subset M$ so that h displaces C. Since all balls in M are isotopic, there is a diffeomorphism $\psi \in \mathrm{Diff}_0(M)$ with $\psi(C) = B$. Therefore $\phi := \psi h \psi^{-1}$

displaces B, and hence ϕ displaces the subgroup $\mathrm{Diff}_0(B) \subset \mathrm{Diff}_0(M)$. Applying inequality (5) we get that

$$\nu([f,g]) \le 4\nu(\phi) = 4\nu(h) < 4\epsilon \,.$$

Sending ϵ to zero, we conclude that $\nu([f,g]) = 0$, a contradiction with the non-degeneracy of a norm. ■

Next, we apply Theorem 2.2 to proving that for a class of groups introduced in Section 1.2.4 existence of stably unbounded norms yields existence of quasi-morphisms.

Proof of Propositions 1.7 and 1.8: First of all note that every element $h \in \bar{G}$ can be uniquely written in the following *normal form*: either $h = (g_1, g_2)$ or $h = (g_1, g_2)t$. This readily yields Proposition 1.7. Second, we claim that it suffices to show that G has a non-trivial homogeneous quasi-morphism, say r. Indeed, put $\bar{r}(h) = r(g_1) + r(g_2)$, where h is in the normal form as above. A straightforward analysis shows that $\bar{r}$ is a (not necessarily homogeneous!) quasi-morphism on $\bar{G}$. For instance, if $h = (h_1, h_2)t$ and $f = (f_1, f_2)$ then $hf = (h_1 f_2, h_2 f_1)t$ and hence

$$|\bar{r}(hf) - \bar{r}(h) - \bar{r}(f)| \le |r(h_1 f_2) - r(h_1) - r(f_2)| + |r(h_2 f_1) - r(h_2) - r(f_1)|$$

and hence is uniformly bounded. The other cases are considered similarly. Finally note that the stabilization $\bar{r}_\infty(h) := \lim_{n\to\infty} \bar{r}(h^n)/n$ does not vanish on $h = (g, 1)$ provided $r(g) \ne 0$. Since $\bar{r}_\infty$ is a homogeneous quasi-morphism, the claim follows.

Let ν be a stably unbounded norm on $\bar{G}$. Assume that $\nu_\infty(w) > 0$ for some $w \in \bar{G}$.

CASE 1: $w = (g_1, g_2)$. Put $w_1 = (g_1, 1)$ and $w_2 = (1, g_2)$. We claim that either $\nu_\infty(w_1) > 0$ or $\nu_\infty(w_2) > 0$. Indeed, $w^k = w_1^k w_2^k$ and hence

$$0 < \nu_\infty(w) \le \nu_\infty(w_1) + \nu_\infty(w_2) \,,$$

which yields the claim.

CASE 2: $w = (g_1, g_2)t$. Put $w_1 = (g_1 g_2, 1)$ and $w_2 = (1, g_2 g_1)$. We claim that either $\nu_\infty(w_1) > 0$ or $\nu_\infty(w_2) > 0$. Indeed, $w^{2k} = w_1^k w_2^k$ and hence

$$0 < \nu_\infty(w) \le \frac{1}{2}(\nu_\infty(w_1) + \nu_\infty(w_2)) \,,$$

which yields the claim.

Looking at elements w_1 and tw_2t above we conclude that there exists an element $u = (g, 1)$ with $\nu_\infty(u) > 0$. Replacing, if necessary, u by its power we can assume that $g \in G'$ (here we use that $H_1(G)$ is finite). Denote by $H \subset \bar{G}$ the subgroup consisting of all elements of the form $(f, 1)$ where $f \in G$. Clearly, H is isomorphic to G and $u \in H'$. Furthermore, t displaces H. Thus inequality (5) yields that

$$\nu(z) \leq 4\nu(t) \cdot cl_H(z) \ \ \forall z \in H' .$$

Substituting $z = u^k$, dividing by k and passing to the limit as $k \to \infty$ we get that

$$0 < \nu_\infty(u) \leq 4\nu(t) \cdot scl_H(u) .$$

Thus $scl_G(g) = scl_H(u) > 0$. Therefore Bavard's theorem [5] yields existence of a non-trivial homogeneous quasi-morphism on G. ■

2.2. Inequalities with commutators

Here we prove Theorem 2.2(i). For an element $F \in G$, we say that $g \in G$ is an *F-commutator* if $g = \mathrm{Conj}_f[F, h]$ for some $f, h \in G$. Note that the inverse of an F-commutator is again an F-commutator.

Fix $F \in G$ such that the subgroups

$$H_0 := H, \ \ H_1 := \mathrm{Conj}_F H, \ \ ..., \ \ H_m := \mathrm{Conj}_{F^m} H$$

pair-wise commute. We shall show that every element x from the commutator subgroup H' with $cl_H(x) = m$ can be represented as a product of seven F-commutators. Note that given a conjugation-invariant norm ν on G, for every F-commutator g we have $\nu(g) \leq 2\nu(F)$. Thus we shall get that $\nu(x) \leq 14\nu(F)$, which yields inequality (3).

We shall consider products $\prod_0^m \mathrm{Conj}_{F^i}(g_i)$, where $g_i \in H$, $i = 0, ..., m$. Since H_i's pair-wise commute, the product of such elements $\prod_0^m \mathrm{Conj}_{F^i}(f_i)$ and $\prod_0^m \mathrm{Conj}_{F^i}(g_i)$ can be computed component-wise: it equals $\prod_0^m \mathrm{Conj}_{F^i}(f_i g_i)$.

Lemma 2.4. *Let a collection of $g_i \in H$, $i = 0, 1, \ldots, m$ be such that $\prod_0^m g_i = 1$. Then the product $g = \prod_0^m \mathit{Conj}_{F^i}(g_i)$ is an F-commutator.*

Proof. We will show that $g = [F, \phi^{-1}]$ where $\phi = \prod_0^{m-1} \mathrm{Conj}_{F^i}(\phi_i)$, $\{\phi_i\}_{i=0}^{m-1}$ is a collection of elements of H which will be defined later. We set $\phi_m = 1$ for convenience of notation.

Note that $[F, \phi^{-1}] = \mathrm{Conj}_F(\phi^{-1})\phi$ and $\mathrm{Conj}_F(\phi^{-1})$ equals the product $\prod_0^{m-1} \mathrm{Conj}_{F^{i+1}}(\phi_i^{-1}) = \prod_1^m \mathrm{Conj}_{F^i}(\phi_{i-1}^{-1})$ whose terms lie in $H_1, ..., H_m$. Hence

$$[F, \phi^{-1}] = \mathrm{Conj}_F(\phi^{-1})\phi = \phi_0 \cdot \prod_1^m \mathrm{Conj}_{F^i}(\phi_{i-1}^{-1}\phi_i)$$

and the equation $[F, \phi^{-1}] = g$ is equivalent to the system

$$\begin{cases} \phi_0 = g_0 \\ \phi_0^{-1}\phi_1 = g_1 \\ \phi_1^{-1}\phi_2 = g_2 \\ \quad \dots \\ \phi_{m-1}^{-1}\phi_m = g_m \end{cases}$$

The solution of this system is $\phi_k = \prod_0^k g_i$, $k = 0, 1, \dots, m$. The equation $\phi_m = 1$ is satisfied by the assumption $\prod g_i = 1$. ■

Lemma 2.5. *Let $g_1, g_2, \dots, g_m$ be a collection of elements of H. Then $g = \prod_m^1 g_i$ equals an F-commutator times the product $\prod_1^m \mathrm{Conj}_{F^i}(g_i)$.*

Proof. Introduce $g_0' = g$ and $g_i' = g_i^{-1}$. Note that $\prod_0^m g_i' = 1$. Then apply the previous lemma. ■

Lemma 2.6. *Any commutator from H is a product of two F-commutators.*

Proof. Consider a commutator $[f, g]$ with $f, g \in H$. Then by Lemma 2.4, the elements

$$(fg)\mathrm{Conj}_F(g^{-1})\mathrm{Conj}_{F^2}(f^{-1})$$

and

$$(f^{-1}g^{-1})\mathrm{Conj}_F(g)\mathrm{Conj}_{F^2}(f)$$

are F-commutators. Their product is $[f, g]$. ■

End of the proof of Theorem 2.2(i): Consider $h = \prod_m^1 [f_i, g_i]$ with $f_i, g_i \in H$. By Lemma 2.5, h equals an F-commutator times a product $\theta := \prod_1^m \mathrm{Conj}_{F^i}([f_i, g_i])$. The latter in its turn is equal to the commutator of two products $\phi := \prod_1^m \mathrm{Conj}_{F^i}(f_i)$ and $\psi := \prod_1^m \mathrm{Conj}_{F^i}(g_i)$ since the subgroups H_i and H_j commute for $i \neq j$. This proves inequality (4).

Applying again Lemma 2.5 we have that $\phi = fx$ and $\psi = gy$ where $f = f_m \dots f_1$ and $g = g_m \dots g_1$ and x, y are F-commutators. We write

$$\theta = [fx, gy] = [f, g] \cdot \mathrm{Conj}_g\{\mathrm{Conj}_f(g^{-1}xg \cdot y \cdot x^{-1}) \cdot y^{-1}\}\,.$$

Since $f, g \in H$, we have by Lemma 2.6 that $[f, g]$ equals a product of two F-commutators. Hence θ is a product of six F-commutators and therefore h is a product of seven F-commutators. As we explained in the beginning of this section, this completes the proof of the theorem.■

2.3. Packing and distortion of subgroups

Let G be a group and $H \subset G$ a subgroup. Consider the embedding of metric spaces $(H', cl_H) \mapsto (G', cl_G)$. Obviously $cl_G(w) \le cl_H(w)$ for all $w \in H'$. It turns out that, after stabilization, this inequality can be refined provided H is m-displaceable in G: the larger m is, the stronger H' is distorted in G' with respect to the stable commutator lengths.

Theorem 2.7.

$$scl_G(w) \le \frac{1}{p(G,H)} scl_H(w) \;\; \forall w \in H' .$$

Example 2.8. Let $G = \widetilde{Sp(2n,R)}$ be the universal cover of the linear symplectic group and let $H = \widetilde{Sp(2,R)} \subset G$. Here we fix the splitting $\mathbb{R}^{2n} = \mathbb{R}^2 \oplus \mathbb{R}^{2n-2}$. The monomorphism $Sp(2,\mathbb{R}) \to Sp(2n,\mathbb{R})$ which sends a matrix A to $A \oplus \mathbf{1}_{2n-2}$ induces the isomorphism of the fundamental groups $\pi_1(Sp(2,\mathbb{R})) = \pi_1(Sp(2n,\mathbb{R})) = \mathbb{Z}$, and hence H naturally embeds into G. Let $(p_1, q_1, ..., p_n, q_n)$ be the standard symplectic coordinates on $\mathbb{R}^{2n}$. Denote by I_j the symplectic transformation which permutes (p_1, q_1) and (p_j, q_j)-coordinates. Write $\widetilde{I_j}$ for a lift of I_j to G. Then the subgroups $\mathrm{Conj}_{I_j}(H)$ pairwise commute, and hence $p(G,H) \ge n$. Denote by $e \in H$ the generator of the center of H. One can show (see Remark 2.11 below) that

$$scl_H(e) = n \cdot scl_G(e) . \tag{6}$$

Thus the inequality in Theorem 2.7 yields $p(G,H) \le n$. We conclude that $p(G,H) = n$ and the inequality is sharp.

Example 2.9. Let (M,ω) be a symplectic manifold, and let $U \subset M$ be an open subset. Let $G = \mathrm{Ham}(M,\omega)$ and let $H = \mathrm{Ham}(U,\omega)$. In this case the algebraic packing number $p(G,H)$ has a simple geometric meaning: It equals to the *geometric packing number* $p_{geom}(M,U)$ which is defined as the maximal number of diffeomorphisms from G which take U to pairwise disjoint subsets of M. In the case when U is a standard symplectic ball the geometric packing number was intensively studied in the framework of the symplectic packing problem (see [6] for a survey). For instance, assume that M and U are $2n$-dimensional symplectic balls. In the case $n = 1$ the geometric packing number is simply the integer part of the ratio of the areas. In the case $n = 2$ the situation is more complicated: For instance, if the ratio of volumes of M and U lies in the interval $(8; (1 + 1/288) \cdot 8)$, the geometric packing number equals 7 (see [24]). It would be interesting to explore the sharpness of the inequality in Theorem 2.7 in these examples.

The proof of Theorem 2.7 is based on the following observation (thanks to Sasha Furman for help). For a subgroup $H \subset G$ write $Q(H)$ for the set of homogeneous quasi-morphisms on H modulo morphisms, and for $\phi \in Q(H)$ put

$$||\phi||_H = \sup_{x,y \in H} \phi([x, y]) .$$

Proposition 2.10. *Let $H_1, ..., H_N$ be subgroups of G so that H_i and H_j commute for $i \neq j$. Put $K = H_1 \cdot ... \cdot H_N$. Then for every $\phi \in Q(K)$*

$$||\phi||_K = \sum_{i=1}^{N} ||\phi||_{H_i} .$$

Proof of Proposition 2.10: Take any $x, y \in K$ and write

$$x = x_1 \cdot ... \cdot x_N,\ y = y_1 \cdot ... \cdot y_N ,$$

where $x_i, y_i \in H_i$. Then

$$[x, y] = [x_1, y_1] \cdot ... \cdot [x_N, y_N] .$$

Since the commutators in the right hand side pair-wise commute we get that for every quasi-morphism $\phi \in Q(K)$

$$\phi([x, y]) = \sum_{i=1}^{N} \phi([x_i, y_i]) .$$

Since pairs x_i, y_i can be chosen in an arbitrary way we get the desired equality. ■

Proof of Theorem 2.7: Suppose that $p(G, H) \geq N$. Then there exist elements $g_1 = 1, g_2, ..., g_N$ so that subgroups $H_i := g_i H g_i^{-1}$ pair-wise commute. For every $\phi \in Q(G)$ we have $||\phi||_{H_i} = ||\phi||_H$. Put $K = H_1 \cdot ... \cdot H_N$. Applying Proposition 2.10 we have

$$||\phi||_G \geq ||\phi||_K = N||\phi||_H . \tag{7}$$

Denote by $Q_*(H)$ the set of non-trivial quasi-morphisms from $Q(H)$, and by $Q_*(G, H)$ the set of quasi-morphisms from $Q_*(G)$ which restrict to a non-trivial quasi-morphism on H. Apply now Bavard's theorem [5]: given $w \in H'$ we have

$$scl_H(w) = \frac{1}{2} \sup_{\phi \in Q_*(H)} \frac{\phi(w)}{||\phi||_H} \geq \frac{1}{2} \sup_{\phi \in Q_*(G,H)} \frac{\phi(w)}{||\phi||_H} .$$

Using inequality (7) above and applying the same Bavard's theorem we have

$$\begin{aligned} scl_H(w) &\geq N \cdot \frac{1}{2} \sup_{\phi \in Q_*(G,H)} \frac{\phi(w)}{||\phi||_G} \\ &= N \cdot \frac{1}{2} \sup_{\phi \in Q_*(G)} \frac{\phi(w)}{||\phi||_G} = N scl_G(w). \end{aligned} \tag{8}$$

The equality in the middle follows from the fact that for $\phi \in Q_*(G) \setminus Q_*(G,H)$ and $w \in H'$ one has $\phi(w) = 0$. Using inequality (8), we readily complete the proof. ■

Remark 2.11. Denote by G_n the universal cover of the group $Sp(2n, \mathbb{R})$ and by $e_n \in G_n$ the generator of $\pi_1(Sp(2n, \mathbb{R}))$ with Maslov index 2. The group G_n carries unique homogeneous quasi-morphism μ_n with $\mu_n(e_n) = 1$ (see [4]). Put

$$I_n := \frac{||\mu_n||_{G_n}}{||\mu_1||_{G_1}} .$$

One can show that $I_n = n$. The only known to us proof of this innocently looking fact is surprisingly involved: it can be extracted from [8] (thanks to A. Iozzi and A. Wienhard for illuminating consultations). By the above-cited theorem due to Bavard

$$\frac{scl_{G_1}(e_1)}{scl_{G_n}(e_n)} = I_n ,$$

which proves equality (6) above.

§3. Topological arguments

3.1. Portable manifolds

Let M be a portable manifold. We shall use notations of Definition 1.16.

Lemma 3.1. *There exists a neighborhood U of the core M_0 of M and a diffeomorphism $\phi \in Diff_0(M)$ so that the sets $\phi^i(U)$, $i \geq 1$ are pair-wise disjoint.*

Proof. Choose a sufficiently small neighbourhood U of the core so that $\theta(U) \cap \text{Closure}(U) = \emptyset$. Put $V = \theta(U)$ and consider the vector field $Y = \theta_* X$ on M. Note that V is an attractor of Y. In particular there exists $\tau > 0$ large enough so that the closure of $Y^\tau(U \cup V)$ is contained in V. Cutting off Y^τ outside a sufficiently large compact set, we get that there exists a diffeomorphism $\phi \in \text{Diff}_0(M)$ so that

$$\text{Closure } \phi(U \cup V) \subset V .$$

Observe that $\phi^i(U) \subset \phi^{i-1}(V) \setminus \phi^i(V)$. Thus the sets $\phi^i(U)$, $i \geq 1$ are pair-wise disjoint. ■

Proof of Theorem 1.17: Let ν be any conjugation-invariant norm on $\text{Diff}_0(M)$. It suffices to show that ν is bounded.

We shall use notations of Definition 1.16 of a portable manifold. Look at the neighborhood U of the core and at the diffeomorphism ϕ from Lemma 3.1. Note that ϕ m-displaces the subgroup $\text{Diff}_0(U)$ for any m. Take any diffeomorphism $h \in \text{Diff}_0(U)$. Since the group $\text{Diff}_0(U)$ is perfect, it follows from inequality (3) that $\nu(h) \leq 14\nu(\phi)$.

Further, take any diffeomorphism $f \in \text{Diff}_0(M)$. The first item of the Definition 1.16 guarantees that for $\tau > 0$ large enough $X^\tau(\text{support} f) \subset U$. Applying the ambient isotopy theorem, we can find a diffeomorphism $\psi \in \text{Diff}_0(M)$ with $\psi(\text{support} f) \subset U$. Thus $\psi f \psi^{-1}$ lies in $\text{Diff}_0(U)$. We conclude that

$$\nu(f) = \nu(\psi f \psi^{-1}) \leq 14\nu(\phi)$$

which implies that ν is bounded. This completes the proof. ■

Proof of Theorem 1.18: The proof above shows that the diffeomorphism ϕ m-displaces the subgroup $H := \text{Diff}_0(U)$ for any m. Corollary 2.3 above implies that $cl_G(h) \leq 2$ for all $h \in \text{Diff}_0(U)$, where $G = \text{Diff}_0(M)$. But every element $f \in G$ is conjugate to an element from H. Thus $cld(M) \leq 2$. ■

Remark 3.2. Theorem 1.17 admits the following straightforward generalization. Let G be any group acting by homeomorphisms on a topological space X. Assume that there exist two disjoint open subsets $U, V \subset X$ and an element $\phi \in G$ which satisfy the following two easily verifiable properties:

(i) Closure $\phi(U \cup V) \subset V$;

(ii) For every finite collection of elements $\psi_1, ..., \psi_k \subset G$ there exists $h \in G$ so that

$$h\Big(\bigcup_{i=1}^{k} \text{support}(\psi_i)\Big) \subset U .$$

Then any invariant norm on G is bounded on the commutator subgroup G'.

3.2. Spheres

Lemma 3.3. *Every diffeomorphism $f \in \mathrm{Diff}_0(S^n)$ can be written as $f = gh$ where $g \in \mathrm{Diff}_0(S^n \setminus \{z\})$ and $h \in \mathrm{Diff}_0(S^n \setminus \{w\})$ for some points $z, w \in S^n$.*

Since $S^n \setminus \{\text{point}\} = \mathbb{R}^n$ is a portable manifold, Theorem 1.11(ii) follows from Theorem 1.17 and Theorem 1.13(i) follows from Theorem 1.18.

Proof of Lemma 3.3: This fact is standard: Let $\{f_t\}$, $t \in [0;1]$ be a path in $\mathrm{Diff}_0(S^n)$ with $f_0 = \mathbf{1}$ and $f_1 = f$. Choose a sufficiently small closed disc $D \subset S^n$ so that $X := \bigcup_t f_t(D) \neq S^n$. Pick a point $z \notin X$. Since $S^n \setminus \{z\}$ is diffeomorphic to $\mathbb{R}^n$, there exists a path $\{g_t\}$ of diffeomorphisms from $\mathrm{Diff}_0(S^n \setminus \{z\})$ such that $g_0 = \mathbf{1}, g_t|_D = f_t|_D$. Pick a point w in the interior of D. Note the path $\{g_t^{-1} f_t\}$ is compactly supported in $S^n \setminus \{w\}$. Thus the diffeomorphisms $g := g_1$ and $h := g^{-1} f$ are as required in the lemma. ■

3.3. Three-manifolds

Here we prove Theorem 1.11(iii). By a *graph* in a manifold we mean a piecewise smoothly embedded graph. By a smooth isotopy of a graph we mean an isotopy which extends to a smooth isotopy of its tubular neighborhood. We shall use without a special mentioning the following fact (see e.g. [20]): any smooth compactly supported diffeomorphism ϕ of an open handlebody U is isotopic to the identity through compactly supported diffeomorphisms, that is $f \in \mathrm{Diff}_0(U)$.

Lemma 3.4 (Fundamental Lemma). *Let Γ and K be two disjoint graphs and M. Let $f_t : \Gamma \to M$, $t \in [0;1]$ be a smooth isotopy with $f_0|_\Gamma = \mathbf{1}$ and $f_1(\Gamma) \cap K = \emptyset$. Then there exist a diffeomorphism h of M supported in a ball and a diffeomorphism $\phi \in \mathrm{Diff}_0(M \setminus K)$ so that*

$$f_1|_\Gamma = h \circ \phi|_\Gamma \,.$$

Let us prove the theorem assuming the lemma.

Proof of Theorem 1.11(iii): Take any norm ν on $\mathrm{Diff}_0(M)$. A graph is called *the Heegard graph* if its complement is diffeomorphic to an open handlebody. Every three-manifold contains a Heegard graph (for instance, a neighborhood of the 1-skeleton of a triangulation of M). Choose a pair of disjoint Heegard graphs L and K in M. Fix a sufficiently small tubular neighborhood U of L. Since $U, M \setminus K$ and $M \setminus L$

are open handlebodies and therefore are portable, Theorem 1.17 implies that the norm ν, when restricted to Diff_0 of these submanifolds, does not exceed some constant $C > 0$. We shall assume also that the same inequality holds for the restriction of ν to Diff_0 of any ball in M (we use here that all balls are pair-wise isotopic and portable).

We shall show that

$$\nu(f) \leq 5C \tag{9}$$

for every $f \in \mathrm{Diff}_0(M)$ with $f(U) \cap K = \emptyset$. Note that this yields the same inequality for *every* f. Indeed, perturbing K to K' by a small ambient isotopy of M and shrinking U to U' by an ambient isotopy of M we can always achieve that $f(U') \cap K' = \emptyset$. But the subgroups $\mathrm{Diff}_0(U')$ and $\mathrm{Diff}_0(M \setminus K')$ are conjugate in $\mathrm{Diff}_0(M)$ to $\mathrm{Diff}_0(U)$ and $\mathrm{Diff}_0(M \setminus K)$ respectively, and hence the restriction of the norm ν to these subgroups is bounded by the same constant C which yields inequality (9). From now on we assume that $f(U) \cap K = \emptyset$.

Let $N \subset U \setminus L$ be any embedded graph so that the induced homomorphism $\pi_1(N) \to \pi_1(U \setminus L)$ is a surjection. Put $\Gamma = L \cup N$, and apply the Fundamental Lemma. We get a diffeomorphism h supported in a ball, and a diffeomorphism $\phi \in \mathrm{Diff}_0(M \setminus K)$ so that $f|_\Gamma = h \circ \phi|_\Gamma$. Denote $\psi = (h\phi)^{-1} f$ and observe that $\psi|_\Gamma = \mathbf{1}$.

In particular, ψ fixes L. We wish to correct ψ and get a diffeomorphism fixing *a neighborhood* of L. This is the point where the graph N enters the play. More precisely, we claim that there exist diffeomorphisms $\xi, \theta \in \mathrm{Diff}_0(U)$ and $\eta \in \mathrm{Diff}_0(M \setminus L)$ so that $\psi = \xi\eta\theta$. Indeed, since ψ fixes L, there exists a sufficiently small tubular neighborhood $V \subset U$ of L and a diffeomorphism $\theta \in \mathrm{Diff}_0(U)$ so that $\psi\theta^{-1}(V) = V$. Put $\tau := \psi\theta^{-1}$. Since $U \setminus L$ retracts to ∂V and ψ fixes N we conclude that τ induces the identity isomorphism of $\pi_1(\partial V)$. It is well known (see e.g. [36, 19, 20]) that therefore $\tau|_V : V \to V$ is isotopic to the identity. Hence there exists a diffeomorphism $\xi \in \mathrm{Diff}_0(U)$ which coincides with τ on V, and so $\eta := \xi^{-1}\tau$ is supported in $M \setminus L$. The claim follows.

Finally, write

$$f = h\phi\psi = h\phi\xi\eta\theta \ .$$

Note that $h \in \mathrm{Diff}_0(B)$ where B is a ball, and hence $\nu(h) \leq C$ where the constant C was chosen in the beginning of the proof. Furthermore, $\phi \in \mathrm{Diff}_0(M \setminus K)$, $\xi, \theta \in \mathrm{Diff}_0(U)$ and $\eta \in \mathrm{Diff}_0(M \setminus L)$. Thus $\nu(f) \leq 5C$ which proves inequality (9). This completes the proof. ■

Proof of Theorem 1.13(ii): In the proof above we represented every diffeomorphism from Diff_0 of a closed connected three-manifold M as a product of 5 diffeomorphisms from Diff_0 of portable manifolds. Applying Theorem 1.18 we get the desired estimate $cld(M) \le 10$. ∎

Proof of Lemma 3.4: The proof is divided into several steps.

STEP 1: Let $\Gamma, L \subset M$ be disjoint embedded graphs, and $f_t : \Gamma \to M$ be a smooth isotopy. Put $\Gamma_t := f_t(\Gamma)$. We say that the crossing point $y = f_\tau(x) \in \Gamma_\tau \cap L$ is *generic* if the points x and y lie in smooth interior parts of Γ and L respectively and

$$f_{\tau *}(T_x\Gamma) \oplus T_y L \oplus \mathbb{R} \cdot \frac{\partial}{\partial t}\Big|_{t=\tau} f_t x = T_y M \ .$$

Introduce two modifications of the isotopy f_t at a generic crossing point.

TYPE I MODIFICATION (REMOVING THE CROSSING POINT): Here we assume that L is **a segment** with the endpoints A and B and $y = \Gamma_\tau \cap L$ is a generic crossing point. Choose $\epsilon > 0$ small enough so that y is the only crossing point on the time interval $I := [\tau - \epsilon; \tau + \epsilon]$. Choose a sufficiently small neighborhood U of L. Let $h_s, s \in I$ be a path in $\mathrm{Diff}_0(U)$ so that $h_s = \mathbf{1}$ outside a small neighborhood of $s = \tau$, $h_s(L) \subset L$ and $h_s(B) = B$ for all s, and h_τ shrinks L so that $y \notin h_\tau(L)$. Replace the piece $\{\Gamma_t\}_{t \in I}$ of the original isotopy by $\{\Gamma'_t\}_{t \in I}$ where $\Gamma'_t = h_t^{-1}\Gamma_t$. Note that $\Gamma_t \cap h_t(L) = \emptyset$, and hence $\Gamma'_t \cap L = \emptyset$, for all $t \in I$.

TYPE II MODIFICATION (DECOMPOSITION): Here Γ and L are arbitrary graphs, and $y = f_\tau(x) \in \Gamma_\tau \cap L$ is a generic crossing point. Choose $\epsilon > 0$ small enough so that y is the only crossing point on the time interval $I := [\tau - 2\epsilon; \tau + 2\epsilon]$. There exists a neighborhood E of y diffeomorphic to a Euclidean cube

$$Q = \{(u, v, w) \in \mathbb{R}^3 \ \Big| \ |u|, |v|, |w| < 2\epsilon\}$$

so that $L \cap Q$ is the vertical segment $\{u = v = 0, w \in [-2\epsilon; 2\epsilon]\}$ and $\Gamma_t \cap Q$ is the segment $c_{t-\tau} := \{u = t - \tau, v \in [-2\epsilon; 2\epsilon], w = 0\}$ for $t \in I$. Thus the isotopy Γ_t inside Q is given by the motion of the segment $c_{-2\epsilon}$ in the (u, v)-plane in the direction of the u-axis. In this picture, the crossing point y is the origin.

Let us agree on the following wording: Suppose that two curves α_0 and α_1 in the (u, v)-plane are given by the graphs $\{u = F_0(v)\}$ and $\{u = F_1(v)\}$ of smooth functions $F_0, F_1 : [-2\epsilon; 2\epsilon] \to \mathbb{R}$. The *linear isotopy* between α_0 and α_1 is formed by graphs of $(1 - s)F_0 + sF_1$, $s \in [0; 1]$.

The modification we are going to describe is local. Fix a smooth cut-off function $\rho : [-2\epsilon; 2\epsilon] \to [0; 3\epsilon/2]$ which is supported in a very small neighborhood of 0 and which satisfies $\rho(0) = 3\epsilon/2$. Denote $\beta^{\pm} = c_{\pm\epsilon}$. Consider the curve

$$\alpha = \{u = -\epsilon + \rho(v), v \in [-2\epsilon; 2\epsilon], w = 0\} .$$

Modify the original isotopy on the time interval $I' := [\tau - \epsilon; \tau + \epsilon]$ as follows: first make a linear isotopy from β^- to α, and then a linear isotopy from α to β^+. We extend the curves appearing in the process of this isotopy outside Q by appropriate Γ_t's and make an obvious change of time in order to fit into the time interval I'.

The following features of the modified isotopy are crucial for our further purposes. The isotopy from β^- to α can be realized by an isotopy of diffeomorphisms of M supported in a ball $B \subset Q$. The isotopy from α to β^+ does not hit L and hence can be extended to an ambient isotopy of M which is fixed near L.

STEP 2: After these preliminaries, we pass to the situation described in the formulation of the lemma: Let Γ, K be two disjoint graphs in M and let $f_t : \Gamma \to M$, $t \in [0; 1]$ be a smooth isotopy with $f_1(\Gamma) \cap K = \emptyset$. After a small perturbation of the isotopy with fixed end points we can assume that the following conditions hold:

(C1) The set

$$\{(x, t) \in \Gamma \times [0; 1] \;\Big|\; f_t(x) \in K\}$$

consists of N pairs (x_i, t_i), $i = 1, ..., N$ so that $\{x_i\}$ are distinct points of Γ, $0 < t_1 < ... < t_N < 1$ and $y_i = f_{t_i}(x)$ are distinct generic crossing points.

(C2) The curves $\gamma_i := \{f_t(x_i)\}_{t \in [0;1]}$ are pairwise disjoint embedded segments.

(C3) For each i, the isotopy $f_t : \Gamma \setminus \{x_i\} \to M$ crosses γ_i generically.

We shall remove the latter crossings using the Type I modification (see Step 1): Note that each such crossing occurs in the subsegment of γ_i which is either of the form $[x_i; f_{t_i - \delta} x_i]$ or $[f_{t_i + \delta} x_i; f_1 x_i]$, where $\delta > 0$ is small enough. We apply Type I modification to these segments keeping the end point $f_{t_i \pm \delta} x_i$ fixed (such an end point is denoted by B in the local description of a Type I modification above). Note that each such modification is localized near some γ_i and hence does not create new crossings, so the process stops after a finite number of modifications. Thus we replace assumption (C3) above by a stronger one:

(C3') For each i, the isotopy $f_t : \Gamma \setminus \{x_i\} \to M$ does not hit γ_i.

STEP 3: It would be convenient to make a change of time in our isotopy as follows. We assume that f_t is defined on the time interval $t \in [0; N+1]$ and the crossings times are consecutive integers $t_i = i$, $i = 1, ..., N$. Assumptions (C1) and (C2) of the previous step yield existence of embedded pair-wise disjoint parallelepipeds $P_i \subset M$, $i = 1, ..., N$ (each parallelepiped P_i is a neighborhood of the segment γ_i) equipped with local coordinates $u \in [-1; N+2], v \in [-1; 1], w \in [-1; 1]$ so that the following holds:

$$\gamma_i = \{(u, 0, 0) \mid u \in [0; N+1]\} ,$$

$$K \cap P_i = \{(i, 0, w) \mid w \in [-1; 1]\}, \quad \Gamma \cap P_i = \{(0, v, 0) \mid v \in [-1; 1]\},$$

and

$$f_t(0, v, 0) = (t, v, 0) \ \forall t \in [0; N+1], v \in [-1; 1] .$$

In addition, assumption (C3') of the previous step guarantees that P_i's can be chosen so thin that

$$f_t(\Gamma \setminus P_i) \cap P_i = \emptyset \ \forall t \in [0; N+1] . \tag{10}$$

STEP 4: Let $Q_i \subset P_i$ be a sufficiently small cube centered at the crossing $(i, 0, 0)$ whose edges have the length 4ϵ and are parallel to the coordinate axes. Perform a Type II modification of our isotopy inside Q_i: We keep notations $\alpha_i, \beta_i^{\pm}$ (with the extra sub-index i) for special curves appearing in the description of the modification presented in Step 1. The reader should have in mind that the current u-coordinate is shifted by i in comparison to the one of Step 1, and the crossing time τ equals i.

Thus we assume that

$$\beta_i^{\pm} = \{(i \pm \epsilon, v, 0) \mid v \in [-2\epsilon, 2\epsilon] \} .$$

Set

$$\Gamma_i^- = f_{i-\epsilon}(\Gamma) \ \text{ and } \ \Gamma_i^+ = (f_{i-\epsilon}(\Gamma) \setminus \beta_i^-) \cup \alpha_i \ , \ \ i = 1, ..., N .$$

Note that $\Gamma_i^+ = h_i(\Gamma_i^-)$, where $h_i \in \mathrm{Diff}_0(Q_i)$.

It will be convenient to put $\Gamma_0^+ = \Gamma$ and $\Gamma_{N+1}^- = f_1(\Gamma)$. Recall that we write $\Gamma_t = f_t(\Gamma)$.

STEP 5: Fix $i \in \{0; ...; N\}$. Let us focus on the following isotopy taking Γ_i^+ to Γ_{i+1}^- : we proceed according to the description of the Type II modification (see Step 1) until we reach the graph $\Gamma_{i+\epsilon}$ which extends β_i^+

(this move is empty when $i = 0$), and then move on with the original isotopy f_t until Γ_{i+1}^-. Note that this isotopy does not hit K. Furthermore, the (time-dependent) vector field $\zeta_t^{(i)}$ of this isotopy, which is defined along the image of Γ_i^+ at the time moment t, is parallel to the u-axis *in each of the parallelepipeds* P_j, $j = 1, ..., N$. Now we shall use property (10) of the original isotopy: It guarantees that one can cut off $\zeta_t^{(i)}$ near K and extend it to the whole M so that *it remains parallel to the u-axis in all P_j's.* After such an extension we get an isotopy supported in $M \setminus K$ so that its time-1-map ϕ_i sends Γ_i^+ to Γ_{i+1}^-.

The following property of maps ϕ_i, which readily follows from the above discussion on vector fields $\zeta_t^{(i)}$, is crucial for the final step of the proof:

$$\phi_N \circ ... \circ \phi_i(Q_i) \subset P_i \;\; \forall i = 1, ..., N \; . \tag{11}$$

STEP 6: We have

$$f_1|_\Gamma = \phi_N h_N \circ ... \circ \phi_1 h_1 \phi_0|_\Gamma \; , \tag{12}$$

where the diffeomorphisms $h_i \in \text{Diff}_0(Q_i)$ and the cubes Q_i appear in Step 4, and the diffeomorphisms $\phi_i \in \text{Diff}_0(M \setminus K)$ are constructed in the previous step. Put

$$g_i = (\phi_N \circ ... \circ \phi_i) h_i (\phi_N \circ ... \circ \phi_i)^{-1} \; , \; i = 1, ..., N \; .$$

Note that $g_i \in \text{Diff}_0(Q_i')$ where $Q_i' = \phi_N \circ ... \circ \phi_i(Q_i)$. By (11), the sets Q_i' are pair-wise disjoint. Since each of Q_i' is diffeomorphic to an open ball, the diffeomorphism $h := g_N \circ ... \circ g_1$ is supported in a ball. Finally, put

$$\phi = \phi_N \circ ... \circ \phi_0 \in \text{Diff}_0(M \setminus K)$$

and observe that in view of equation (12) $f_1|_\Gamma = h\phi|_\Gamma$. This finishes off the proof of the lemma. ■

Acknowledgments. We thank A. Furman, D. Calegari, E. Ghys, A. Iozzi, N. Ivanov, D. Kotschick, F. Laudenbach, D. McDuff, A. Muranov, Y. Shalom, T. Tsuboi and A. Wienhard for useful discussions. We are grateful to the referee for an extraordinary detailed and useful report, and to the anonymous editorial board member for fixing a mathematical error indicated by the referee.

References

[1] R. D. Anderson, On homeomorphisms as products of conjugates of a given homeomorphism and its inverse, In: Topology of 3-manifolds and related topics, Proc. The Univ. of Georgia Institute, 1961, Prentice-Hall, Englewood Cliffs, NJ, 1962, pp. 231–234.

[2] A. Banyaga, Sur la structure du groupe des difféomorphismes qui préservent une forme symplectique, Comm. Math. Helv., **53** (1978), 174–227.

[3] A. Banyaga, The structure of classical diffeomorphism groups, Math. Appl., **400**, Kluwer Academic Publishers Group, Dordrecht, 1997.

[4] Barge, J., Ghys, E., *Cocycles d'Euler et de Maslov,* Math. Ann. **294**:2 (1992), 235–265.

[5] C. Bavard, Longueur stable des commutateurs, Enseign. Math. (2), **37** (1991), 109–150.

[6] P. Biran, From Symplectic Packing to Algebraic Geometry and Back, In: Proceedings of the 3'rd European Congress of Mathematics, Barcelona, 2000, Vol II, Progr. Math., **202**, Birkhäuser, 2001, pp. 507–524.

[7] P. Biran, M. Entov and L. Polterovich, Calabi quasimorphisms for the symplectic ball, Commun. Contemp. Math., **6** (2004), 793–802.

[8] M. Burger, A. Iozzi and A. Wienhard, Surface group representations with maximal Toledo invariant, preprint, arXiv:math/0605656.

[9] D. Carter and G. Keller, Bounded elementary generation of $\mathrm{SL}_n(\mathcal{O})$, Amer. J. Math., **105** (1983), 673–687.

[10] Y. de Cornulier, Strong boundedness of $Homeo(S^n)$, Appendix to D. Calegari and M. Freedman, Distortion in transformation groups, Geom. Topol., **10** (2006), 267–293.

[11] Y. Eliashberg and L. Polterovich, Bi-invariant metrics on the group of Hamiltonian diffeomorphisms, Internat. J. Math., **4** (1993), 727–738.

[12] M. Entov, Commutator length of symplectomorphisms, Comment. Math. Helv., **79** (2004), 58–104.

[13] M. Entov and L. Polterovich, Calabi quasimorphism and quantum homology, Int. Math. Res. Not., **30** (2003), 1635–1676.

[14] M. Entov and L. Polterovich, Quasi-states and symplectic intersections, Comm. Math. Helv., **81** (2006), 75–99.

[15] J.-M. Gambaudo and E. Ghys, Commutators and diffeomorphisms of surfaces, Ergodic Theory Dynam. Systems, **24** (2004), 1591–1617.

[16] M. Herman, Simplicite du groupe des diffeomorphismes de classe $C\infty$, isotopes a l'identite, du tore de dimension n, C. R. Acad. Sci. Paris Ser. A-B, **273** (1971), A232–A234.

[17] H. Hofer, On the topological properties of symplectic maps, Proc. Roy. Soc. Edinburgh Sect. A, **115** (1990), 25–38.

[18] D. Kotschick, Stable length in stable groups, in this volume, pp. 401–413.

[19] F. Laudenbach, Topologie de la dimension trois: homotopie et isotopie, Astérisque, **12**, Société Mathématique de France, Paris, 1974.

[20] E. Luft, Actions of the homeotopy group of an orientable 3-dimensional handlebody, Math. Ann., **234** (1978), 279–292.

[21] J. Mather, Commutators of diffeomorphisms, Comment. Math. Helv., **49** (1974), 512–528.

[22] J. Mather, Commutators of diffeomorphisms II, Comment. Math. Helv., **50** (1975), 33–40.

[23] D. McDuff, Monodromy in Hamiltonian Floer theory, preprint, arXiv: math/0801.1328.

[24] D. McDuff and L. Polterovich, Symplectic packings and algebraic geometry, Invent. Math., **115** (1994), 405–434.

[25] D. McDuff and D. Salamon, J-holomorphic curves and symplectic topology, Amer. Math. Soc., Providence, RI, 2004.

[26] A. Muranov, Finitely generated infinite simple groups of infinite commutator width, Internat. J. Algebra Comput., **17** (2007), 607–659.

[27] Y.-G. Oh, Construction of spectral invariants of Hamiltonian diffeomorphisms on general symplectic manifolds, In: The breadth of symplectic and Poisson geometry, Birkhäuser, Boston, 2005, pp. 525–570.

[28] L. Polterovich, The geometry of the group of symplectic diffeomorphisms, Lectures Math. ETH Zürich, Birkhäuser, 2001.

[29] L. Polterovich, Floer homology, dynamics and groups, In: Morse theoretic methods in nonlinear analysis and in symplectic topology, Proceedings of the NATO Advanced Study Institute, Montreal, Canada, July 2004, (eds. P. Biran, O. Cornea and F. Lalonde), NATO Sci. Ser. II Math. Phys. Chem., **217**, Springer, Dordrecht, 2006.

[30] P. Py, Quasi-morphismes et invariant de Calabi, Ann. Sci. École Norm. Sup. (4), **39** (2006), 177–195.

[31] P. Py, Quasi-morphismes de Calabi et graphe de Reeb sur le tore, C. R. Math. Acad. Sci. Paris, **343** (2006), 323–328.

[32] M. Schwarz, On the action spectrum for closed symplectically aspherical manifolds, Pacific J. Math., **193** (2000), 419–461.

[33] J.-C. Sikorav, Systemes Hamiltoniens et topologie symplectique, ETS Editrice, Pisa, 1990.

[34] W. Thurston, Foliations and groups of diffeomorphisms, Bull. Amer. Math. Soc., **80** (1974), 304–307.

[35] T. Tsuboi, On the uniform perfectness of diffeomorphism groups, in this volume, pp. 505–524.

[36] F. Waldhausen, On irreducible 3-manifolds which are sufficiently large, Ann. of Math. (2), **87** (1968), 56–88.

[37] D. Witte Morris, Bounded generation of $SL(n, A)$ (after D. Carter, G. Keller and E. Paige), preprint, arXiv:math/0503083.

Dmitri Burago
Department of Mathematics
The Pennsylvania State University
University Park, PA 16802, U.S.A.
E-mail address: burago@math.psu.edu

Sergei Ivanov
Steklov Math. Institute
St. Petersburg, Russia
E-mail address: svivanov@pdmi.ras.ru

Leonid Polterovich
School of Mathematical Sciences
Tel Aviv University
Tel Aviv 69978, Israel
E-mail address: polterov@post.tau.ac.il

Advanced Studies in Pure Mathematics 52, 2008
Groups of Diffeomorphisms
pp. 251–282

A generalization of Chakiris' fibrations

Hisaaki Endo

Abstract.

Chakiris [5] constructed examples of holomorphic Lefschetz fibrations of genus 2 with separating singular fibers and proved a classification theorem for such fibrations in the late 1970's. We generalize some parts of his construction topologically to give new examples of hyperelliptic Lefschetz fibrations of arbitrary genus with separating singular fibers which include homeomorphic but non-diffeomorphic 4-manifolds.

§1. Introduction

According to remarkable works of Donaldson [6] and Gompf [14] (see also [1]), there is a Lefschetz fibration over the 2-sphere with prescribed fundamental group. The classification of all Lefschetz fibrations over the 2-sphere is not possible in nature. Many examples of Lefschetz fibrations are given in terms of positive relations in mapping class groups (see [14], [35], [4], [21], [15], [16], and [8]).

Hyperelliptic Lefschetz fibrations, which are (relative minimalizations of) double branched coverings of simple 4-manifolds (see [32], [12]), include all Lefschetz fibrations of genus 1 and 2 and many important examples. It would be rather hopeful to classify hyperelliptic Lefschetz fibrations over the 2-sphere. Siebert and Tian [32] conjectured that every hyperelliptic Lefschetz fibration over the 2-sphere without separating singular fibers is holomorphic. They solved it affirmatively in genus 2 case under assumption of monodromy transitivity [33]. Their conjecture is closely related to a smooth analogue of an earlier theorem of Chakiris [5] which asserts that every holomorphic fibration of genus 2 without virtual reducible singular fibers is a fiber sum of three typical fibrations. On the other hand, it does not seem to be known how many hyperelliptic

Received April 3, 2007.
Revised November 6, 2007.
This research is partially supported by Grant-in-Aid for Scientific Research (No. 18540083), Japan Society for the Promotion of Science.

Lefschetz fibrations with separating singular fibers exist. Matsumoto's genus 2 Lefschetz fibration [25] and its generalization in arbitrary even genus due to Cadavid [4] and Korkmaz [21] are well-known examples of such fibrations. Chakiris [5] also showed a mysterious classification theorem of Lefschetz fibrations of genus 2 with separating singular fibers, which we would like to call the '1/19-theorem'.

In this paper we generalize some parts of Chakiris' construction topologically to give new examples of hyperelliptic Lefschetz fibrations of arbitrary genus with separating singular fibers. In Section 2 we review definitions and basic properties of Lefschetz fibrations and relations in mapping class groups. In Section 3 we construct new positive relations in hyperelliptic mapping class groups and in Section 4 we investigate various properties of the corresponding hyperelliptic Lefschetz fibrations over the 2-sphere. In particular, we exhibit infinitely many pairs of homeomorphic but non-diffeomorphic hyperelliptic Lefschetz fibrations with separating singular fibers.

The author is grateful to Y. Matsumoto for helpful suggestions on Chakiris' work and to D. Kotschick, S. Hirose, and Y. Sato for helpful discussions and useful comments. He also thank the referee for various helpful comments and suggestions.

§2. Lefschetz fibrations and positive relations

In this section we briefly review Lefschetz fibrations and relations in mapping class groups.

2.1. Lefschetz fibrations and their monodromies

We first review the definition and basic properties of Lefschetz fibrations. More details can be found in Matsumoto [25] and Gompf and Stipsicz [14].

Let Σ_g be a closed oriented surface of genus g.

Definition 2.1. Let M be a closed oriented smooth 4-manifold. A smooth map $f : M \to S^2$ is called a *Lefschetz fibration* of genus g if it satisfies the following conditions:

(i) f has finitely many critical values $b_1, \dots, b_n \in S^2$ and f is a smooth fiber bundle over $S^2 - \{b_1, \dots, b_n\}$ with fiber Σ_g;

(ii) for each i ($i = 1, \dots, n$), there exists a unique critical point p_i in the *singular fiber* $F_i := f^{-1}(b_i)$ such that f is locally written as $f(z_1, z_2) = z_1^2 + z_2^2$ with respect to some local complex coordinates around p_i and b_i which are compatible with orientations of M and S^2;

(iii) no fiber contains a (-1)-sphere.

Let $\mathcal{M}_g$ be the mapping class group of Σ_g, namely the group of all isotopy classes of orientation-preserving diffeomorphisms of Σ_g. We follow the functional notation: for $\varphi, \psi \in \mathcal{M}_g$, the symbol $\psi\varphi$ means that we apply φ first and then ψ. We denote by $\mathcal{F}$ the free group generated by all isotopy classes $\mathcal{S}$ of simple closed curves on Σ_g. There is a natural epimorphism $\varpi : \mathcal{F} \to \mathcal{M}_g$ which sends (the isotopy class of) a simple closed curve a on Σ_g to the right-handed Dehn twist t_a along a. We often denote the image $\varpi(W)$ of a word W in the generators $\mathcal{S}$ by $\overline{W}$. We set $\mathcal{R} := \mathrm{Ker}\,\varpi$ and call each element of $\mathcal{R}$ a *relator* in the generators $\mathcal{S}$ of $\mathcal{M}_g$. We put ${}_W(c) := t_{a_r}^{\varepsilon_r} \cdots t_{a_1}^{\varepsilon_1}(c) \in \mathcal{S}$ for $c \in \mathcal{S}$ and $W = a_r^{\varepsilon_r} \cdots a_1^{\varepsilon_1} \in \mathcal{F}$ ($a_1, \ldots, a_r \in \mathcal{S}, \varepsilon_1, \ldots, \varepsilon_r \in \{\pm 1\}$) and ${}_W V := {}_W(c_1) \cdots {}_W(c_s) \in \mathcal{F}$ for $V = c_1 \cdots c_s \in \mathcal{F}$ ($c_1, \ldots c_s \in \mathcal{S}$).

Let $f : M \to S^2$ be a Lefschetz fibration of genus g as in the definition above. Since f restricted over $S^2 - \{b_1, \ldots, b_n\}$ is a smooth fiber bundle with fiber Σ_g, we consider the homomorphism

$$\chi : \pi_1(S^2 - \{b_1, \ldots, b_n\}) \to \pi_1(\mathrm{BDiff}_+\Sigma_g) \cong \pi_0(\mathrm{Diff}_+\Sigma_g) = \mathcal{M}_g$$

induced by the classifying map $S^2 - \{b_1, \ldots, b_n\} \to \mathrm{BDiff}_+\Sigma_g$, which is called the *holonomy homomorphism* (cf. Morita [28]) or the *monodromy representation* of f. Let γ_i $(i = 1, \ldots, n)$ be the loop consisting of the boundary circle of a small disk neighborhood of b_i oriented clockwise and a path connecting a point on the circle to the base point $b_0 \in S^2 - \{b_1, \ldots, b_n\}$. We choose these loops $\gamma_1, \ldots, \gamma_n$ so that the composition $\gamma_1 \cdots \gamma_n$ is null-homotopic on $S^2 - \{b_1, \ldots, b_n\}$ and any two of them intersect only at b_0. Thus we obtain a presentation

$$\pi_1(S^2 - \{b_1, \ldots, b_n\}, b_0) = \langle \gamma_1, \ldots, \gamma_n | \gamma_1 \cdots \gamma_n = 1 \rangle.$$

For each i $(i = 1, \ldots, n)$, $\chi(\gamma_i)$ is known to be a right-handed Dehn twist t_{c_i} along some essential simple closed curve c_i on Σ_g. Hence we have a *positive* relation

$$t_{c_1} \cdots t_{c_n} = \chi(\gamma_1 \cdots \gamma_n) = 1 \in \mathcal{M}_g$$

or a *positive* relator $c_1 \cdots c_n \in \mathcal{R}$ associated to the Lefschetz fibration $f : M \to S^2$. Each c_i is called the *vanishing cycle* of the singular fiber F_i. F_i is called *non-separating* (or *irreducible*, *type* I) if c_i does not separate Σ_g into two connected components and *separating* (or *reducible*, *type* II_h) if c_i separates Σ_g into subsurfaces of genus h and $g - h$.

Suppose that $g \geq 2$. Kas [19] and Matsumoto [25] proved that there exists a one-to-one correspondence between the isomorphism classes of Lefschetz fibrations $f : M \to S^2$ and the conjugacy classes of homomorphisms χ which sends each loop going around b_i to a right-handed

Dehn twist along an essential simple closed curve on Σ_g. Although the positive relator $c_1 \cdots c_n \in \mathcal{R}$ actually depends on a choice of a loop system $(\gamma_1, \ldots, \gamma_n)$ on $S^2 - \{b_1, \ldots, b_n\}$, its equivalence class modulo conjugations of all factors $c_1, \ldots, c_n$ by a fixed element W of $\mathcal{F}$:

$$c_1 \cdot \cdots \cdot c_n \sim {}_W(c_1) \cdot \cdots \cdot {}_W(c_n),$$

and *elementary transformations* :

$$c_1 \cdot \cdots \cdot c_i \cdot c_{i+1} \cdot \cdots \cdot c_n \sim c_1 \cdot \cdots \cdot c_{i+1} \cdot {}_{c_{i+1}^{-1}}(c_i) \cdot \cdots \cdot c_n,$$

$$c_1 \cdot \cdots \cdot c_i \cdot c_{i+1} \cdot \cdots \cdot c_n \sim c_1 \cdot \cdots \cdot {}_{c_i}(c_{i+1}) \cdot c_i \cdot \cdots \cdot c_n$$

$(i = 1, \ldots, n-1)$, is uniquely determined by the isomorphism class of the Lefschetz fibration $f : M \to S^2$. Conversely, any positive relator $\varrho \in \mathcal{R}$ can be realized as a relator associated to some Lefschetz fibration over S^2. We denote (the isomorphism class of) such a Lefschetz fibration by $M_\varrho \to S^2$.

Let $f : M \to S^2$ and $f' : M' \to S^2$ be Lefschetz fibrations of genus g and $\varrho, \varrho' \in \mathcal{R}$ corresponding positive relators. Take regular values $b_0, b_0' \in S^2$ of f, f' and consider the fiber $F := f^{-1}(b_0), F' := f'^{-1}(b_0')$ and their open fibered neighborhoods $\nu F \subset M, \nu F' \subset M'$, respectively. Using a fiber-preserving, orientation-reversing diffeomorphism $\varphi : \partial(M - \nu F) \to \partial(M' - \nu F')$, we can glue $M - \nu F$ and $M' - \nu F'$ together and construct a new manifold $M \#_F M'$ which will admit a Lefschetz fibration $f \# f' : M \#_F M' \to S^2$ of genus g. We call this fibration a *fiber sum* of $f : M \to S^2$ and $f' : M' \to S^2$. The diffeomorphism type of $M \#_F M'$ and the isomorphism type of $f \# f'$ might depend on the choice of the diffeomorphism φ. A positive relator corresponding to the fiber sum $f \# f'$ can be written as $\varrho \cdot {}_W\varrho'$ for some $W \in \mathcal{F}$ which depends on the choice of φ.

Let $\iota : \Sigma_g \to \Sigma_g$ be (the mapping class of) a hyperelliptic involution, an involution on Σ_g with $2g+2$ fixed points, and $\mathcal{H}_g$ the centralizer of ι in $\mathcal{M}_g$, which is called the *hyperelliptic mapping class group*. Note that $\mathcal{H}_1 = \mathcal{M}_1$ and $\mathcal{H}_2 = \mathcal{M}_2$, while $\mathcal{H}_g \neq \mathcal{M}_g$ for $g \geq 3$. We set $\mathcal{S}^H := \{a \in \mathcal{S} \mid t_a \in \mathcal{H}_g\}$. We denote by $\mathcal{F}^H$ the subgroup of $\mathcal{F}$ generated by $\mathcal{S}^H$ and put $\mathcal{R}^H := \mathcal{R} \cap \mathcal{F}^H$. A Lefschetz fibration $f : M \to S^2$ of genus g is said to be *hyperelliptic* if its holonomy homomorphism χ can be chosen in the conjugacy class so that the image $\operatorname{Im}\chi$ is included in $\mathcal{H}_g$. If the canonical projection $\mathcal{H}_g \to \mathcal{M}_{0,2g+2} \to S_{2g+2}$ maps $\operatorname{Im}\chi$ onto a transitive subgroup of S_{2g+2}, we say that the monodromy of f is *transitive*, otherwise *intransitive*, where $\mathcal{M}_{0,2g+2}$ is the mapping class group of the 2-sphere with $2g+2$ marked points and S_{2g+2} is the symmetric group of degree $2g+2$.

2.2. Basic relations in mapping class groups

We next review several relations in the mapping class group $\mathcal{M}_g$ (cf. [42], [21], and [8]).

The relator $A = A(a) := a \in \mathcal{R}$ is called an *identity relator*, where a is a null-homotopic simple closed curve on Σ_g.

For (isotopy classes of) simple closed curves a and b, we denote their geometric intersection number by $i(a, b)$. Let a and b be simple closed curves on Σ_g and c the simple closed curve $t_b(a)$. The relation

$$t_c = t_b t_a t_b^{-1}$$

in $\mathcal{M}_g$ is called the *braid relation.* (It follows from the braid relation $t_{t_b(a)} = t_b t_a t_b^{-1}$ that $\overline{{}_b(a)} = \overline{bab^{-1}}$, $\overline{{}_{b^{-1}}(a)} = \overline{b^{-1}ab}$, and elementary transformations keep positive relators being positive relators.) If $i(a, b) = n$, we put

$$T_n = T(a, b) := bab^{-1}c^{-1} \in \mathcal{R}.$$

If $i(a, b) = 1$, we have another braid relation $t_b = t_a t_c t_a^{-1}$. This relation together with the original relation $t_c = t_b t_a t_b^{-1}$ yields Artin's relation $t_a t_b t_a = t_b t_a t_b$.

An ordered n-tuple $(c_1, \ldots, c_n)$ of simple closed curves on Σ_g is called a *chain* of length n if c_i and c_{i+1} intersect transversely at one point $(i = 1, \ldots, n-1)$ and other c_i and c_j never intersect. When the length n is even (resp. odd), a regular neighbourhood of a chain $(c_1, \ldots, c_n)$ is a subsurface of Σ_g which is of genus $h = n/2$ (resp. $h = (n-1)/2$) and has one boundary component (resp. two boundary components). We denote simple closed curves parallel to the boundary by d (resp. d_1 and d_2). The relation

$$t_d = (t_{c_1} \cdots t_{c_{2h}})^{4h+2} \quad (\text{resp. } t_{d_1} t_{d_2} = (t_{c_1} \cdots t_{c_{2h+1}})^{2h+2})$$

is called the *chain relation* of length n, or the *even* (resp. *odd*) chain relation (see Wajnryb [42]). We put

$$C_{2h} = C(c_1, \ldots, c_{2h}) := (c_1 \cdots c_{2h})^{4h+2} d^{-1} \in \mathcal{R},$$
$$C_{2h+1} = C(c_1, \ldots, c_{2h+1}) := (c_1 \cdots c_{2h+1})^{2h+2} d_1^{-1} d_2^{-1} \in \mathcal{R}.$$

Remark 2.2. The even (resp. odd) chain relation above holds even if we permute the factors of the chain:

$$(c_{\sigma(1)} \cdots c_{\sigma(2h)})^{4h+2} d^{-1} \in \mathcal{R} \ (\text{resp. } (c_{\tau(1)} \cdots c_{\tau(2h+1)})^{2h+2} d_1^{-1} d_2^{-1} \in \mathcal{R}),$$

where $\sigma \in S_{2h}$ (resp. $\tau \in S_{2h+1}$) is an arbitrary permutation (cf. Matsumoto [24]). Hirose told the author an elementary proof of this fact.

We need the following definition and lemma for constructions of positive relators in the next section.

Definition 2.3 (Smith [35], cf. [8]). Let $\varrho = W_1^{-1}W_2 \in \mathcal{R}$ be a relator with W_1 and W_2 positive words in $\mathcal{F}$. Suppose that a positive relator $\varsigma \in \mathcal{R}$ includes W_1 as a subword: $\varsigma = UW_1V$, where U and V are positive words in $\mathcal{F}$. Then we can construct a new positive relator $\varsigma' = \varsigma V^{-1}\varrho V = UW_2V$ in $\mathcal{R}$. This operation $\varsigma \mapsto \varsigma'$ is called a ϱ-*substitution* to ς. If a positive relator $\hat{\varsigma}$ is obtained by applying a sequence of $\varrho^{\pm 1}$-substitutions to ς, we denote it by $\varsigma \equiv \hat{\varsigma} \pmod{\varrho}$.

Lemma 2.4. *Let $(c_1, \ldots, c_n)$ be a chain of length n on Σ_g. The following equivalence holds for $k = 1, \ldots, n-1$ and $i = 1, \ldots, k+1$.*

$$(c_1c_2\cdots c_n)^i \equiv (c_1c_2\cdots c_k)^i \cdot (c_{k+1}c_k\cdots c_{k-i+2}) \cdot (c_{k+2}c_{k+1}\cdots c_{k-i+3}) \cdot\cdots\cdot (c_nc_{n-1}\cdots c_{n-i+1}) \pmod{T_0, T_1}$$

Proof. Straightforward from the proof of Lemma 4.6 of [8]. Q.E.D.

§3. Hyperelliptic Chakiris relations

In this section we construct new examples of positive relators in the hyperelliptic mapping class group $\mathcal{H}_g$. We first review basic relations in $\mathcal{H}_g$ (cf. [3] and [8]).

Let $(c_1, \ldots, c_{2g+1})$ be a chain of length $2g+1$ on Σ_g. Suppose that each c_i is invariant under the hyperelliptic involution ι. Hence the right-handed Dehn twists $t_{c_1}, \ldots, t_{c_{2g+1}}$ belong to $\mathcal{H}_g$. The chain relator $C_{2g+1} = (c_1\cdots c_{2g+1})^{2g+2}d_1^{-1}d_2^{-1}$ of length $2g+1$ combined with two identity relators $A(d_1) = d_1$ and $A(d_2) = d_2$ is a positive relator

$$C_{\mathrm{I}} := C_{2g+1}A(d_2)A(d_1) = (c_1\cdots c_{2g+1})^{2g+2} \in \mathcal{R}^H.$$

The chain relator $C_{2g} = (c_1\cdots c_{2g})^{4g+2}d^{-1}$ of length $2g$ combined with an identity relator $A(d) = d$ is a positive relator

$$C_{\mathrm{II}} := C_{2g}A(d) = (c_1\cdots c_{2g})^{4g+2} \in \mathcal{R}^H.$$

It is well-known that the images of

$$I := c_1c_2\cdots c_{2g}c_{2g+1}^2c_{2g}\cdots c_2c_1 \in \mathcal{F}^H,$$
$$J := (c_1c_2\cdots c_{2g})^{2g+1} \in \mathcal{F}^H$$

under ϖ represent (the mapping class of) the hyperelliptic involution ι (see Birman and Hilden [3], p. 108, Equation (8), and [7], Lemma 4.13).

3.1. Even genus

Suppose that $g \geq 2$ and g is even. We first consider the following three elements of $\mathcal{F}^H$.

$$\begin{aligned}
P_0 &:= ((c_1c_2\cdots c_g)^{g+1} \\
&\quad \cdot (c_{g+1}\cdots c_3c_2)\cdot(c_{g+2}\cdots c_4c_3)\cdot\cdots\cdot(c_{2g}\cdots c_{g+2}c_{g+1}))^2, \\
Q_0 &:= (c_1c_2\cdots c_{2g+1})^{g+1}\cdot(c_g\cdots c_2c_1)^{2g+2} \\
&\quad \cdot (c_1^{-1}c_2^{-1}\cdots c_g^{-1}c_{g+1}c_g\cdots c_2c_1)\cdot(c_2^{-1}c_3^{-1}\cdots c_{g+1}^{-1}c_{g+2}c_{g+1}\cdots c_3c_2) \\
&\quad \cdot\cdots\cdot(c_{g+1}^{-1}c_{g+2}^{-1}\cdots c_{2g}^{-1}c_{2g+1}c_{2g}\cdots c_{g+2}c_{g+1}), \\
R_0 &:= (c_g\cdots c_2c_1)^{2g+2} \\
&\quad \cdot (c_1^{-1}c_2^{-1}\cdots c_g^{-1}c_{g+1}c_g\cdots c_2c_1)\cdot(c_2^{-1}c_3^{-1}\cdots c_{g+1}^{-1}c_{g+2}c_{g+1}\cdots c_3c_2) \\
&\quad \cdot\cdots\cdot(c_{g+1}^{-1}c_{g+2}^{-1}\cdots c_{2g}^{-1}c_{2g+1}c_{2g}\cdots c_{g+2}c_{g+1})\cdot(c_{2g+1}\cdots c_2c_1)^{g+1}.
\end{aligned}$$

Lemma 3.1. *The words P_0, Q_0, and R_0 represent $\iota, 1$, and ι in $\mathcal{H}_g$, respectively.*

Proof. Using Lemma 2.4, we have

$$\begin{aligned}
(c_1c_2\cdots c_{2g})^{g+1} \equiv{}& (c_1c_2\cdots c_g)^{g+1}\cdot(c_{g+1}\cdots c_2c_1)\cdot(c_{g+2}\cdots c_3c_2) \\
&\cdot\cdots\cdot(c_{2g}\cdots c_{g+1}c_g) \pmod{T_0, T_1}.
\end{aligned}$$

We rewrite the right-hand side by using braid relations to obtain

$$\begin{aligned}
(c_1c_2\cdots c_{2g})^{g+1} \equiv{}& (c_1c_2\cdots c_g)^{g+1}\cdot(c_{g+1}\cdots c_3c_2)\cdot(c_{g+2}\cdots c_4c_3) \\
&\cdot\cdots\cdot(c_{2g}\cdots c_{g+2}c_{g+1})\cdot c_1c_2\cdots c_g \pmod{T_0, T_1}.
\end{aligned}$$

Again from Lemma 2.4, we have

$$\begin{aligned}
(c_1c_2\cdots c_{2g})^{g} \equiv{}& (c_1c_2\cdots c_g)^{g}\cdot(c_{g+1}\cdots c_3c_2)\cdot(c_{g+2}\cdots c_4c_3) \\
&\cdot\cdots\cdot(c_{2g}\cdots c_{g+2}c_{g+1}) \pmod{T_0, T_1}.
\end{aligned}$$

We combine the last two equivalences to obtain

$$\begin{aligned}
J &= (c_1c_2\cdots c_{2g})^{2g+1} \\
&\equiv ((c_1c_2\cdots c_g)^{g+1}\cdot(c_{g+1}\cdots c_3c_2)\cdot(c_{g+2}\cdots c_4c_3) \\
&\quad \cdot\cdots\cdot(c_{2g}\cdots c_{g+2}c_{g+1}))^2 \pmod{T_0, T_1} \\
&= P_0.
\end{aligned}$$

Hence we have $\overline{P_0} = \overline{J} = \iota$.

As is already mentioned, the image of

$$C_{\mathrm{I}} = (c_1 c_2 \cdots c_{2g+1})^{2g+2} \in \mathcal{F}^H$$

under ϖ is equal to 1 in $\mathcal{H}_g$. By virtue of Lemma 2.4 and Corollary A.2, we have

$$\begin{aligned}(c_1 c_2 \cdots c_{2g+1})^{g+1} &\equiv (c_1 c_2 \cdots c_g)^{g+1} \cdot (c_{g+1} \cdots c_2 c_1) \cdot (c_{g+2} \cdots c_3 c_2) \\ &\quad \cdots\cdot (c_{2g+1} \cdots c_{g+2} c_{g+1}) \pmod{T_0, T_1} \\ &\equiv (c_g \cdots c_2 c_1)^{g+1} \cdot (c_{g+1} \cdots c_2 c_1) \cdot (c_{g+2} \cdots c_3 c_2) \\ &\quad \cdots\cdot (c_{2g+1} \cdots c_{g+2} c_{g+1}) \pmod{T_0, T_1}.\end{aligned}$$

It is easy to check by manipulating braid relations as Lemma 2.1 of [21] that the following equivalence holds.

$$\begin{aligned}&(c_1^{-1} c_2^{-1} \cdots c_g^{-1})^{g+1} \cdot (c_{g+1} \cdots c_2 c_1) \\ &\quad \cdot (c_{g+2} \cdots c_3 c_2) \cdots\cdot (c_{2g+1} \cdots c_{g+2} c_{g+1}) \\ &\equiv (c_1^{-1} c_2^{-1} \cdots c_g^{-1} c_{g+1} c_g \cdots c_2 c_1) \cdot (c_2^{-1} c_3^{-1} \cdots c_{g+1}^{-1} c_{g+2} c_{g+1} \cdots c_3 c_2) \\ &\quad \cdots\cdot (c_{g+1}^{-1} c_{g+2}^{-1} \cdots c_{2g}^{-1} c_{2g+1} c_{2g} \cdots c_{g+2} c_{g+1}) \qquad \pmod{T_0, T_1}\end{aligned}$$

Gathering these equivalences, we obtain

$$\begin{aligned}&(c_1 c_2 \cdots c_{2g+1})^{g+1} \\ &\equiv (c_g \cdots c_2 c_1)^{2g+2} \cdot (c_1^{-1} c_2^{-1} \cdots c_g^{-1})^{g+1} (c_{g+1} \cdots c_2 c_1) \cdot (c_{g+2} \cdots c_3 c_2) \\ &\quad \cdots\cdot (c_{2g+1} \cdots c_{g+2} c_{g+1}) \pmod{T_0, T_1} \\ &\equiv (c_g \cdots c_2 c_1)^{2g+2} \cdot (c_1^{-1} c_2^{-1} \cdots c_g^{-1} c_{g+1} c_g \cdots c_2 c_1) \\ &\quad \cdot (c_2^{-1} c_3^{-1} \cdots c_{g+1}^{-1} c_{g+2} c_{g+1} \cdots c_3 c_2) \\ &\quad \cdots\cdot (c_{g+1}^{-1} c_{g+2}^{-1} \cdots c_{2g}^{-1} c_{2g+1} c_{2g} \cdots c_{g+2} c_{g+1}) \pmod{T_0, T_1}.\end{aligned}$$

We multiply both sides of the equivalence by $(c_1 c_2 \cdots c_{2g+1})^{g+1}$ and conclude

$$Q_0 \equiv C_{\mathrm{I}} \pmod{T_0, T_1} \quad \text{and} \quad \overline{Q}_0 = \overline{C}_{\mathrm{I}} = 1 \in \mathcal{H}_g.$$

It follows from the equivalence above that

$$\begin{aligned}R_0 &= (c_1 c_2 \cdots c_{2g+1})^{-(g+1)} \cdot Q_0 \cdot (c_{2g+1} \cdots c_2 c_1)^{g+1} \\ &\equiv (c_1 c_2 \cdots c_{2g+1})^{-(g+1)} \cdot C_{\mathrm{I}} \cdot (c_{2g+1} \cdots c_2 c_1)^{g+1} \pmod{T_0, T_1} \\ &= (c_1 c_2 \cdots c_{2g+1})^{g+1} \cdot (c_{2g+1} \cdots c_2 c_1)^{g+1}.\end{aligned}$$

The element

$$D_i := [I, c_i] = I c_i I^{-1} c_i^{-1} \quad (i = 1, \ldots, 2g+1)$$

of $\mathcal{F}^H$ is a relator of $\mathcal{H}_g$: $D_i \in \mathcal{R}^H$ (see Birman and Hilden [3], Theorem 8). Applying D_i-substitutions repeatedly, we have

$$\begin{aligned}
&(c_1 c_2 \cdots c_{2g+1})^{g+1} \cdot (c_{2g+1} \cdots c_2 c_1)^{g+1} \\
\equiv\ & I \cdot (c_1 c_2 \cdots c_{2g+1})^{g} \cdot (c_{2g+1} \cdots c_2 c_1)^{g} \quad (\text{mod } D_1, \ldots, D_{2g+1}) \\
\equiv\ & \cdots\cdots \\
\equiv\ & I^{g+1} \quad (\text{mod } D_1, \ldots, D_{2g+1}).
\end{aligned}$$

We combine these equivalence to obtain

$$R_0 \equiv I^{g+1} \ (\text{mod } T_0, T_1, D_1, \ldots, D_{2g+1}) \quad \text{and} \quad \overline{R}_0 = \overline{I}^{g+1} = \iota$$

because g is even. This completes the proof of the lemma. Q.E.D.

Let $d \in \mathcal{S}^H$ be the boundary curve of a regular neighborhood of $c_1 \cup \cdots \cup c_g$. We now define three positive words in the generators $\mathcal{S}^H$:

$$\begin{aligned}
P :=\ & d \cdot {}_W(c_{g+1} \cdots c_3 c_2) \cdot \cdots \cdot {}_W(c_{2g} \cdots c_{g+2} c_{g+1}) \\
& \cdot (c_{g+1} \cdots c_3 c_2) \cdot \cdots \cdot (c_{2g} \cdots c_{g+2} c_{g+1}) \quad (W := (c_1 c_2 \cdots c_g)^{-(g+1)}); \\
Q :=\ & (c_1 c_2 \cdots c_{2g+1})^{g+1} \cdot d \cdot {}_{W_1}(c_{g+1}) \cdot {}_{W_2}(c_{g+2}) \cdot \cdots \cdot {}_{W_{g+1}}(c_{2g+1}); \\
R :=\ & d \cdot {}_{W_1}(c_{g+1}) \cdot {}_{W_2}(c_{g+2}) \cdot \cdots \cdot {}_{W_{g+1}}(c_{2g+1}) \cdot (c_{2g+1} \cdots c_2 c_1)^{g+1} \\
& (W_i := (c_{i+g-1} \cdots c_{i+1} c_i)^{-1} \ \ (i = 1, \ldots, g+1)).
\end{aligned}$$

Theorem 3.2. *The words P, Q, and R in $\mathcal{F}^H$ are positive words representing $\iota, 1$, and ι in $\mathcal{H}_g$, respectively.*

Proof. We apply C_g^{-1}-substitutions to P_0, Q_0, and R_0 and rewrite them as follows.

$$\begin{aligned}
P_0 =\ & (c_1 c_2 \cdots c_g)^{2g+2} \cdot (c_1 c_2 \cdots c_g)^{-(g+1)} \\
& \cdot (c_{g+1} \cdots c_3 c_2) \cdot \cdots \cdot (c_{2g} \cdots c_{g+2} c_{g+1}) \\
& \cdot (c_1 c_2 \cdots c_g)^{g+1} \cdot (c_{g+1} \cdots c_3 c_2) \cdot \cdots \cdot (c_{2g} \cdots c_{g+2} c_{g+1}) \\
\equiv\ & (c_1 c_2 \cdots c_g)^{2g+2} \cdot ({}_W(c_{g+1}) \cdots {}_W(c_3) {}_W(c_2)) \\
& \cdot \cdots \cdot ({}_W(c_{2g}) \cdots {}_W(c_{g+2}) {}_W(c_{g+1})) \\
& \cdot (c_{g+1} \cdots c_3 c_2) \cdot \cdots \cdot (c_{2g} \cdots c_{g+2} c_{g+1}) \quad (\text{mod } T_n) \\
\equiv\ & P \quad (\text{mod } C_g),
\end{aligned}$$

$$\begin{aligned}
Q_0 &\equiv (c_1c_2\cdots c_{2g+1})^{g+1}\cdot(c_g\cdots c_2c_1)^{2g+2}\\
&\quad\cdot w_1(c_{g+1})\cdot w_2(c_{g+2})\cdot\cdots\cdot w_{g+1}(c_{2g+1}) \qquad (\mathrm{mod}\ T_n)\\
&\equiv Q \quad (\mathrm{mod}\ C_g),\\
R_0 &\equiv (c_g\cdots c_2c_1)^{2g+2}\cdot w_1(c_{g+1})\cdot w_2(c_{g+2})\cdot\cdots\cdot w_{g+1}(c_{2g+1})\\
&\quad\cdot(c_{2g+1}\cdots c_2c_1)^{g+1} \qquad (\mathrm{mod}\ T_n)\\
&\equiv R \quad (\mathrm{mod}\ C_g).
\end{aligned}$$

Thus the proof is completed. Q.E.D.

The next corollary immediately follows from the theorem.

Corollary 3.3. *The words* $P^2, Q, R^2, PR, PI, PJ, RI, RJ$ *in* $\mathcal{F}^H$ *are positive relators for* $\mathcal{H}_g$.

3.2. Odd genus

Suppose that $g \geq 3$ and g is odd. We first consider the following two elements of $\mathcal{F}^H$.

$$\begin{aligned}
Q_0 &:= (c_1c_2\cdots c_{2g+1})^{g+2}\cdot(c_{g-1}\cdots c_2c_1)^{2g+2}\\
&\quad\cdot(c_1^{-1}c_2^{-1}\cdots c_{g-1}^{-1}c_gc_{g-1}\cdots c_2c_1)\cdot(c_2^{-1}c_3^{-1}\cdots c_g^{-1}c_{g+1}c_g\cdots c_3c_2)\\
&\quad\cdot\cdots\cdot(c_{g+2}^{-1}c_{g+3}^{-1}\cdots c_{2g}^{-1}c_{2g+1}c_{2g}\cdots c_{g+3}c_{g+2}),\\
R_0 &:= c_1c_2\cdots c_{2g+1}\cdot(c_{g-1}\cdots c_2c_1)^{2g+2}\\
&\quad\cdot(c_1^{-1}c_2^{-1}\cdots c_{g-1}^{-1}c_gc_{g-1}\cdots c_2c_1)\cdot(c_2^{-1}c_3^{-1}\cdots c_g^{-1}c_{g+1}c_g\cdots c_3c_2)\\
&\quad\cdot\cdots\cdot(c_{g+2}^{-1}c_{g+3}^{-1}\cdots c_{2g}^{-1}c_{2g+1}c_{2g}\cdots c_{g+3}c_{g+2})\cdot(c_{2g+1}\cdots c_2c_1)^{g+1}.
\end{aligned}$$

Lemma 3.4. *Both of the words* Q_0 *and* R_0 *are relators for* $\mathcal{H}_g$.

Proof. As is already mentioned, the image of

$$C_{\mathrm{I}} = (c_1c_2\cdots c_{2g+1})^{2g+2} \in \mathcal{F}^H$$

under ϖ is equal to 1 in $\mathcal{H}_g$. By virtue of Lemma 2.4 and Corollary A.2, we have

$$\begin{aligned}
(c_1c_2\cdots c_{2g+1})^g &\equiv (c_1c_2\cdots c_{g-1})^g\cdot(c_g\cdots c_2c_1)\cdot(c_{g+1}\cdots c_3c_2)\\
&\quad\cdot\cdots\cdot(c_{2g+1}\cdots c_{g+3}c_{g+2}) \quad (\mathrm{mod}\ T_0, T_1)\\
&\equiv (c_{g-1}\cdots c_2c_1)^g\cdot(c_g\cdots c_2c_1)\cdot(c_{g+1}\cdots c_3c_2)\\
&\quad\cdot\cdots\cdot(c_{2g+1}\cdots c_{g+3}c_{g+2}) \quad (\mathrm{mod}\ T_0, T_1).
\end{aligned}$$

It is easy to check by manipulating braid relations as Lemma 2.1 of [21] that the following equivalence holds.

$$\begin{aligned}
&(c_1^{-1}c_2^{-1}\cdots c_{g-1}^{-1})^{g+2}\cdot(c_g\cdots c_2c_1)\\
&\qquad\cdot(c_{g+1}\cdots c_3c_2)\cdot\cdots\cdot(c_{2g+1}\cdots c_{g+3}c_{g+2})\\
\equiv\ &(c_1^{-1}c_2^{-1}\cdots c_{g-1}^{-1}c_gc_{g-1}\cdots c_2c_1)\cdot(c_2^{-1}c_3^{-1}\cdots c_g^{-1}c_{g+1}c_g\cdots c_3c_2)\\
&\cdot\cdots\cdot(c_{g+2}^{-1}c_{g+3}^{-1}\cdots c_{2g}^{-1}c_{2g+1}c_{2g}\cdots c_{g+3}c_{g+2})\pmod{T_0,T_1}
\end{aligned}$$

Gathering these equivalences, we obtain

$$\begin{aligned}
&(c_1c_2\cdots c_{2g+1})^g\\
\equiv\ &(c_{g-1}\cdots c_2c_1)^{2g+2}\cdot(c_1^{-1}c_2^{-1}\cdots c_{g-1}^{-1})^{g+2}(c_g\cdots c_2c_1)\cdot(c_{g+1}\cdots c_3c_2)\\
&\cdot\cdots\cdot(c_{2g+1}\cdots c_{g+3}c_{g+2})\pmod{T_0,T_1}\\
\equiv\ &(c_{g-1}\cdots c_2c_1)^{2g+2}\cdot(c_1^{-1}c_2^{-1}\cdots c_{g-1}^{-1}c_gc_{g-1}\cdots c_2c_1)\\
&\cdot(c_2^{-1}c_3^{-1}\cdots c_g^{-1}c_{g+1}c_g\cdots c_3c_2)\\
&\cdot\cdots\cdot(c_{g+2}^{-1}c_{g+3}^{-1}\cdots c_{2g}^{-1}c_{2g+1}c_{2g}\cdots c_{g+3}c_{g+2})\pmod{T_0,T_1}.
\end{aligned}$$

We multiply both sides of the equivalence by $(c_1c_2\cdots c_{2g+1})^{g+2}$ and conclude

$$Q_0\equiv C_{\mathrm{I}}\pmod{T_0,T_1}\quad\text{and}\quad\overline{Q}_0=\overline{C}_{\mathrm{I}}=1\in\mathcal{H}_g.$$

It follows from the equivalence above that

$$\begin{aligned}
R_0&=(c_1c_2\cdots c_{2g+1})^{-(g+1)}\cdot Q_0\cdot(c_{2g+1}\cdots c_2c_1)^{g+1}\\
&\equiv(c_1c_2\cdots c_{2g+1})^{-(g+1)}\cdot C_{\mathrm{I}}\cdot(c_{2g+1}\cdots c_2c_1)^{g+1}\pmod{T_0,T_1}\\
&=(c_1c_2\cdots c_{2g+1})^{g+1}\cdot(c_{2g+1}\cdots c_2c_1)^{g+1}.
\end{aligned}$$

Applying D_i-substitutions repeatedly, we have

$$\begin{aligned}
&(c_1c_2\cdots c_{2g+1})^{g+1}\cdot(c_{2g+1}\cdots c_2c_1)^{g+1}\\
\equiv\ &I\cdot(c_1c_2\cdots c_{2g+1})^g\cdot(c_{2g+1}\cdots c_2c_1)^g\pmod{D_1,\ldots,D_{2g+1}}\\
\equiv\ &\cdots\cdots\\
\equiv\ &I^{g+1}\pmod{D_1,\ldots,D_{2g+1}}.
\end{aligned}$$

We combine these equivalence to obtain

$$R_0\equiv I^{g+1}\ (\mathrm{mod}\ T_0,T_1,D_1,\ldots,D_{2g+1})\quad\text{and}\quad\overline{R}_0=\overline{I}^{g+1}=1$$

because g is odd. This completes the proof of the lemma. Q.E.D.

Let $d \in \mathcal{S}^H$ be the boundary curve of a regular neighborhood of $c_1 \cup \cdots \cup c_{g-1}$. We now define two positive words in the generators $\mathcal{S}^H$:

$$\begin{aligned} Q :=& \ (c_1 c_2 \cdots c_{2g+1})^{g+2} \cdot d \cdot (c_{g-1} \cdots c_2 c_1)^2 \\ & \cdot {}_{W_1}(c_g) \cdot {}_{W_2}(c_{g+1}) \cdot \cdots \cdot {}_{W_{g+2}}(c_{2g+1}); \\ R :=& \ c_1 c_2 \cdots c_{2g+1} \cdot d \cdot (c_{g-1} \cdots c_2 c_1)^2 \\ & \cdot {}_{W_1}(c_g) \cdot {}_{W_2}(c_{g+1}) \cdot \cdots \cdot {}_{W_{g+2}}(c_{2g+1}) \cdot (c_{2g+1} \cdots c_2 c_1)^{g+1} \\ & (W_i := c_i^{-1} c_{i+1}^{-1} \cdots c_{i+g-2}^{-1} \ (i = 1, \ldots, g+2)). \end{aligned}$$

Theorem 3.5. *Both of the words Q and R in $\mathcal{F}^H$ are positive relators for $\mathcal{H}_g$.*

Proof. We apply C_g^{-1}-substitutions to Q_0 and R_0 and rewrite them as follows.

$$\begin{aligned} Q_0 &\equiv (c_1 c_2 \cdots c_{2g+1})^{g+2} \cdot (c_{g-1} \cdots c_2 c_1)^{2g} \cdot (c_{g-1} \cdots c_2 c_1)^2 \\ &\quad \cdot {}_{W_1}(c_g) \cdot {}_{W_2}(c_{g+1}) \cdot \cdots \cdot {}_{W_{g+2}}(c_{2g+1}) \pmod{T_n} \\ &\equiv Q \pmod{C_{g-1}}, \\ R_0 &\equiv c_1 c_2 \cdots c_{2g+1} \cdot (c_{g-1} \cdots c_2 c_1)^{2g} \cdot (c_{g-1} \cdots c_2 c_1)^2 \\ &\quad \cdot {}_{W_1}(c_g) \cdot {}_{W_2}(c_{g+1}) \cdot \cdots \cdot {}_{W_{g+2}}(c_{2g+1}) \cdot (c_{2g+1} \cdots c_2 c_1)^{g+1} \pmod{T_n} \\ &\equiv R \pmod{C_{g-1}}. \end{aligned}$$

Thus the proof is completed. Q.E.D.

Let $d_1, d_2 \in \mathcal{S}^H$ be the boundary curves of a regular neighborhood of $c_1 \cup \cdots \cup c_g$.

Remark 3.6. Substituting the same words as P_0, Q_0, and R_0 for even genus by the inverse C_g^{-1} of a chain relator C_g, we have the following positive words for odd genus:

$$\begin{aligned} P' :=& \ d_1^2 d_2^2 \cdot {}_W(c_{g+1} \cdots c_3 c_2) \cdot \cdots \cdot {}_W(c_{2g} \cdots c_{g+2} c_{g+1}) \\ & \cdot (c_{g+1} \cdots c_3 c_2) \cdot \cdots \cdot (c_{2g} \cdots c_{g+2} c_{g+1}) \quad (W := (c_1 c_2 \cdots c_g)^{-(g+1)}); \\ Q' :=& \ (c_1 c_2 \cdots c_{2g+1})^{g+1} \cdot d_1^2 d_2^2 \cdot {}_{W_1}(c_{g+1}) \cdot {}_{W_2}(c_{g+2}) \cdot \cdots \cdot {}_{W_{g+1}}(c_{2g+1}); \\ R' :=& \ d_1^2 d_2^2 \cdot {}_{W_1}(c_{g+1}) \cdot {}_{W_2}(c_{g+2}) \cdot \cdots \cdot {}_{W_{g+1}}(c_{2g+1}) \cdot (c_{2g+1} \cdots c_2 c_1)^{g+1} \\ & (W_i := (c_{i+g-1} \cdots c_{i+1} c_i)^{-1} \ (i = 1, \ldots, g+1)), \end{aligned}$$

which are not in $\mathcal{F}^H$. The words P, Q, and R in $\mathcal{F}$ are positive words representing $\iota, 1$, and ι in $\mathcal{M}_g$, respectively.

§4. Generalized Chakiris fibrations

In this section we study various properties of hyperelliptic Lefschetz fibrations arising from hyperelliptic Chakiris relations given in the previous section.

We denote the signature and the Euler characteristic of a compact oriented smooth 4-manifold M by $\sigma = \mathrm{Sign}(M)$ and e, respectively. It is easily seen that $e = -4(g-1) + n$ for a Lefschetz fibration $f : M \to S^2$ of genus g with n singular fibers. We often denote by n_0 (resp. n_+) the number of non-separating (resp. separating) singular fibers of f: $n = n_0 + n_+$.

Let $(c_1, \ldots, c_{2g+1})$ be a chain of length $2g+1$ on Σ_g. Suppose that each c_i is invariant under the hyperelliptic involution ι. Hence the right-handed Dehn twists $t_{c_1}, \ldots, t_{c_{2g+1}}$ belong to $\mathcal{H}_g$.

Three hyperelliptic Lefschetz fibrations $M_{I^2}, M_{C_{\mathrm{I}}}$, and $M_{C_{\mathrm{II}}} = M_{J^2}$ without separating singular fibers have been studied from various points of view (see [25], [26], [34], [14], [32], [33], [2], [30], [7], and [8]). For example, the number n of singular fibers, and the values of signature σ and the Euler characteristic e for these manifolds are calculated as in the following table.

LF	$n = n_0$	σ	e
M_{I^2}	$4(2g+1)$	$-4(g+1)$	$4(g+2)$
$M_{C_{\mathrm{I}}}$	$2(g+1)(2g+1)$	$-2(g+1)^2$	$2(2g^2+g+3)$
$M_{C_{\mathrm{II}}}$	$4g(2g+1)$	$-4g(g+1)$	$4(2g^2+1)$

$M_{I^2}, M_{C_{\mathrm{I}}}$, and $M_{C_{\mathrm{II}}} = M_{J^2}$ are known to be simply-connected, non-spin, and have a (-1)-section. M_{I^2} and $M_{C_{\mathrm{I}}}$ have transitive monodromy, whereas $M_{C_{\mathrm{II}}}$ has intransitive monodromy.

Lemma 4.1 (cf. Wajnryb [42], Lemma 21, Auroux [2], Lemma 3.4). *Three fiber sums* $M_{C_{\mathrm{I}}^2} = \#_F 2M_{C_{\mathrm{I}}}$, $M_{I^{2g+2}} = \#_F(g+1)M_{I^2}$, *and* $M_{C_{\mathrm{II}}I^2} = M_{C_{\mathrm{II}}}\#_F M_{I^2}$ *are isomorphic as Lefschetz fibrations.*

Proof. Using Corollary A.3 and applying elementary transformations repeatedly, we have

$$\begin{aligned} C_{\mathrm{I}}^2 &\sim (c_1c_2\cdots c_{2g+1})^{4g+4} \sim (c_1c_2\cdots c_{2g+1})^{2g+2}\cdot(c_{2g+1}\cdots c_2c_1)^{2g+2} \\ &= (c_1c_2\cdots c_{2g+1})^{2g+1}\cdot I\cdot(c_{2g+1}\cdots c_2c_1)^{2g+1} \\ &\sim (c_1c_2\cdots c_{2g+1})^{2g+1}\cdot {}_I(c_{2g+1}\cdots c_2c_1)^{2g+1}\cdot I \\ &= (c_1c_2\cdots c_{2g+1})^{2g+1}\cdot(c_{2g+1}\cdots c_2c_1)^{2g+1}\cdot I \\ &\sim \cdots\cdots \sim I^{2g+2}, \end{aligned}$$

where we use ${}_I(c_i) = \iota(c_i) = c_i$ $(i = 1, \ldots, 2g+1)$.

We apply elementary transformations and Lemma 2.4 to obtain $JI \sim C_{\mathrm{I}}$:

$$\begin{aligned}
&JI \\
&= J \cdot c_1 c_2 \cdots c_{2g} c_{2g+1}^2 c_{2g} \cdots c_2 c_1 \sim {}_J(c_1 c_2 \cdots c_{2g+1}) \cdot J \cdot c_{2g+1} \cdots c_2 c_1 \\
&= c_1 c_2 \cdots c_{2g+1} \cdot J \cdot c_{2g+1} \cdots c_2 c_1 \sim c_1 c_2 \cdots c_{2g+1} \cdot (c_1 c_2 \cdots c_{2g+1})^{2g+1} \\
&= C_{\mathrm{I}}.
\end{aligned}$$

From this equivalence, we have

$$C_{\mathrm{I}}^2 \sim (JI)^2 = JIJI \sim J \cdot {}_I J \cdot I^2 = J^2 I^2 = C_{\mathrm{II}} \cdot I^2$$

as claimed. Q.E.D.

4.1. Even genus

Suppose that $g \geq 2$ and g is even. Let $d \in \mathcal{S}^H$ be the boundary curve of a regular neighborhood of $c_1 \cup \cdots \cup c_g$. We obtain positive relators $P^2, Q, R^2, PR, PI, PJ, RI, RJ \in \mathcal{R}^H$ and corresponding hyperelliptic Lefschetz fibrations

$$M_{P^2},\ M_Q,\ M_{R^2},\ M_{PR},\ M_{PI},\ M_{PJ},\ M_{RI},\ M_{RJ}$$

of genus g over S^2 from Corollary 3.3. These are non-spin 4-manifolds because a component of a separating singular fiber represents a homology class of square -1. Each of M_{P^2}, M_Q, M_{PI}, and M_{PJ}, which does not include the word R in its monodromy, has a smooth (-1)-section which naturally comes from C_{II} or $C_{\mathrm{I}} \sim JI$ (cf. Smith [36], Lemma 2.3). $M_Q, M_{R^2}, M_{PR}, M_{PI}, M_{RI}$, and M_{RJ} have transitive monodromy, whereas M_{P^2} and M_{PJ} have intransitive monodromy.

We first examine the fundamental groups of these manifolds.

Proposition 4.2. *The fundamental group $\pi_1(M_{P^2})$ of M_{P^2} is isomorphic to $\mathbb{Z}_2$, while the manifolds $M_Q, M_{R^2}, M_{PR}, M_{PI}, M_{PJ}, M_{RI}, M_{RJ}$ are simply connected.*

Proof. We orient $c_1, c_2, \ldots, c_{2g+1}$ so that $c_i \cdot c_{i+1} = +1$ $(i = 1, \ldots, 2g)$ and take oriented simple closed curves $e_1, e_2, \ldots, e_g$ so that $\{c_2, c_4, \ldots, c_{2g}, e_1, e_2, \ldots, e_g\}$ is a symplectic basis of $H_1(\Sigma_g; \mathbb{Z})$ (i.e. $c_{2i} \cdot e_j = \delta_{ij}, c_{2i} \cdot c_{2j} = e_i \cdot e_j = 0$ $(i, j = 1, \ldots, g)$). Connecting these curves to a base point $*$ of Σ_g by appropriate arcs, we consider them also to be elements of $\pi_1(\Sigma_g, *)$ which satisfy $[c_2, e_1][c_4, e_2] \cdots [c_{2g}, e_g] = 1$. Namely,

$$\pi_1(\Sigma_g, *) = \langle c_2, c_4, \ldots, c_{2g}, e_1, e_2, \ldots, e_g \mid [c_2, e_1][c_4, e_2] \cdots [c_{2g}, e_g] \rangle.$$

Let $i : \Sigma_g \hookrightarrow M_\varrho$ be the inclusion map from a general fiber into the total space M_ϱ, where $\varrho = P^2, Q, R^2, PR, PI, PJ, RI, RJ$. The induced homomorphism $i_\# : \pi_1(\Sigma_g) \to \pi_1(M_\varrho)$ is surjective and the kernel of $i_\#$ includes the normal subgroup N of $\pi_1(M_\varrho)$ generated by the vanishing cycles of M_ϱ (cf. Amorós et al. [1], Lemma 3.2).

If $\varrho \neq P^2$, then M_ϱ has vanishing cycles $c_1, c_2, \ldots, c_{2g}$. We can choose arcs connecting $c_1, c_3, \ldots, c_{2g-1}$ to the base point $*$ such that $c_1 = e_1^{-1}$, $c_{2i-1} = e_{i-1}^{-1} c_{2i} e_i c_{2i}^{-1}$ $(i = 2, \ldots, g)$ as elements of $\pi_1(\Sigma_g, *)$. Thus we obtain a presentation

$$\begin{aligned}\pi_1(\Sigma_g, *)/N = \langle c_2, c_4, \ldots, c_{2g}, e_1, e_2, \ldots, e_g \mid [c_2, e_1][c_4, e_2] \cdots [c_{2g}, e_g], \\ c_2, c_4, \ldots, c_{2g}, e_1^{-1}, e_{i-1}^{-1} c_{2i} e_i c_{2i}^{-1}\ (i = 2, \ldots, g), \text{etc...}\rangle \\ = \{1\}.\end{aligned}$$

Hence we have $\pi_1(M_\varrho) = \{1\}$.

If $\varrho = P^2$, the kernel of $i_\#$ coincides with N and $\pi_1(M_{P^2})$ is isomorphic to $\pi_1(\Sigma_g, *)/N$ because M_{P^2} has a smooth (-1)-section which naturally comes from a chain relation C_{2g} of length $2g$ (cf. Smith [36], Lemma 2.3, Amorós et al. [1], Lemma 3.2). Since M_{P^2} has vanishing cycles $c_2, c_3, \ldots, c_{2g}$, we obtain a presentation

$$\begin{aligned}\pi_1(\Sigma_g, *)/N' = \langle c_2, c_4, \ldots, c_{2g}, e_1, e_2, \ldots, e_g \mid [c_2, e_1][c_4, e_2] \cdots [c_{2g}, e_g], \\ c_2, c_4, \ldots, c_{2g}, e_{i-1}^{-1} c_{2i} e_i c_{2i}^{-1}\ (i = 2, \ldots, g)\rangle \\ = \langle e_1 \rangle,\end{aligned}$$

where N' is the normal subgroup of $\pi_1(\Sigma_g, *)$ generated by $c_2, c_3, \ldots, c_{2g}$. Thus $\pi_1(M_\varrho)$ is cyclic because there is a natural surjective homomorphism $\pi_1(\Sigma_g, *)/N' \to \pi_1(\Sigma_g, *)/N \cong \pi_1(M_\varrho)$. The first homology group $H_1(M_{P^2}; \mathbb{Z})$ is isomorphic to $H_1(\Sigma_g; \mathbb{Z})/N_0$, where $N_0 := ND/N$ and $D := [\pi_1(\Sigma_g), \pi_1(\Sigma_g)]$. Since $c_{2i-1} = e_i - e_{i-1}\,(i = 2, \ldots, g)$, $w(c_i) = -c_i\,(i = 2, \ldots, g)$, $w(c_i) = c_i\,(i = g+2, \ldots, 2g)$, $w(c_{g+1}) = e_g + e_{g-1}$ in $H_1(\Sigma_g; \mathbb{Z})$, we have

$$H_1(M_{P^2}; \mathbb{Z}) \cong H_1(\Sigma_g; \mathbb{Z})/N_0 = \mathbb{Z}[e_1]/(2e_1) \cong \mathbb{Z}_2.$$

Hence $\pi_1(M_{P^2})$ is isomorphic to $\mathbb{Z}_2$. Q.E.D.

We next mention other invariants for our examples. The numbers n_0, n_+ of singular fibers, and the values of signature σ and the Euler characteristic e for the manifolds above are calculated as in the following table. (By virtue of Proposition 4.2, these manifolds satisfy $b_2^+ = (e + \sigma)/2 - 1$ and $b_2^- = (e - \sigma)/2 - 1$.)

LF	n_0	n_+	σ	e
M_{P^2}	$4g^2$	2	$-2(g^2+1)$	$2(2g^2-2g+3)$
M_Q	$2(g+1)^2$	1	$-(g^2+2g+3)$	$2g^2+7$
M_{R^2}	$4(g+1)^2$	2	$-2(g^2+2g+3)$	$2(2g^2+2g+5)$
M_{PR}	$2(2g^2+2g+1)$	2	$-2(g^2+g+2)$	$4(g^2+2)$
M_{PI}	$2(g+1)^2$	1	$-(g^2+2g+3)$	$2g^2+7$
M_{PJ}	$2g(3g+1)$	1	$-(3g^2+2g+1)$	$6g^2-2g+5$
M_{RI}	$2(g^2+4g+2)$	1	$-(g^2+4g+5)$	$2g^2+4g+9$
M_{RJ}	$2(3g^2+3g+1)$	1	$-(3g^2+4g+3)$	$6g^2+2g+7$

b_2^+ is always odd for these manifolds and it is greater than 1 except in the case of M_{P^2}, M_Q, and M_{PI} for $g=2$.

Remark 4.3. The values of signature of hyperelliptic Lefschetz fibrations in the table above are calculated by using the local signature formula [27], [7] or by computing signature contributions of relators in their monodromies [8]. Ozbagci's signature formula [29] and these methods are suited for explicit computation of signature of Lefschetz fibrations. For example, Hasegawa [17] and Yun [43] independently computed signatures of Gurtas' fibrations [15], [16] by using these formulae.

Remark 4.4. It is likely that M_Q and M_{PI} are isomorphic as Lefschetz fibrations although the author could not relate them by elementary transformations.

We recall a theorem of Chakiris [5].

Theorem 4.5 (Chakiris' 1/19-theorem [5], Theorem 4.9). *Let $M \to \mathbb{CP}^1$ be a holomorphic Lefschetz fibration of genus 2 with $n \geq 19n_+$.*

(1) *If the monodromy is transitive, then M is isomorphic to M_W, where $W = I^p R^q Q^r C_{\mathrm{I}}^s$ $(p,q,r,s \geq 0,\ p \equiv q \pmod 2)$. In particular, M is a fiber sum of copies of $M_{I^2}, M_{C_{\mathrm{I}}}, M_Q, M_{R^2}$, and M_{RI}.*

(2) *If the monodromy is intransitive, then M is isomorphic to M_W, where $W = P^k J^l$ $(k,l \geq 0,\ k \equiv l \pmod 2)$. In particular, M is a fiber sum of copies of $M_{C_{\mathrm{II}}}, M_{P^2}$, and M_{PJ}.*

Although we do not give any proof of this theorem, we generalize some lemmas in [5] which were used to prove the theorem.

We define an element of $\mathcal{F}^H$ by

$$K := (d \cdot {}_{W_1}(c_{g+1}) \cdot {}_{W_2}(c_{g+2}) \cdot \cdots \cdot {}_{W_{g+1}}(c_{2g+1}))^2.$$

It immediately follows from the proofs of Lemma 3.1 for $Q_0 \in \mathcal{R}^H$ and Theorem 3.2 for $Q \in \mathcal{R}^H$ that K is a positive relator for $\mathcal{H}_g$.

Proposition 4.6 (cf. Chakiris [5], 'the second' Lemma 4.8)**.** *Both of the hyperelliptic Lefschetz fibrations* $M_{Q^2} = M_Q \#_F M_Q$ *and* M_{R^2} *are isomorphic to a fiber sum* $M_K \#_F M_{C_{\mathrm{I}}} = M_{KC_{\mathrm{I}}}$ *of* M_K *and* $M_{C_{\mathrm{I}}}$.

Proof. We apply elementary transformations to Q^2 as follows.

$$
\begin{aligned}
Q^2 &\sim (d \cdot {}_{W_1}(c_{g+1}) \cdot {}_{W_2}(c_{g+2}) \cdot \cdots \cdot {}_{W_{g+1}}(c_{2g+1}) \cdot (c_1 c_2 \cdots c_{2g+1})^{g+1})^2 \\
&\sim (d \cdot {}_{W_1}(c_{g+1}) \cdot {}_{W_2}(c_{g+2}) \cdot \cdots \cdot {}_{W_{g+1}}(c_{2g+1}))^2 \\
&\qquad \cdot (c_1 c_2 \cdots c_{2g+1})^{g+1} \cdot ({}_U(c_1) \cdot {}_U(c_2) \cdot \cdots \cdot {}_U(c_{2g+1}))^{g+1} \\
&(U := (d \cdot {}_{W_1}(c_{g+1}) \cdot {}_{W_2}(c_{g+2}) \cdot \cdots \cdot {}_{W_{g+1}}(c_{2g+1}) \cdot (c_1 c_2 \cdots c_{2g+1})^{g+1})^{-1}) \\
&= (d \cdot {}_{W_1}(c_{g+1}) \cdot {}_{W_2}(c_{g+2}) \cdot \cdots \cdot {}_{W_{g+1}}(c_{2g+1}))^2 \cdot (c_1 c_2 \cdots c_{2g+1})^{2g+2} \\
&= K \cdot C_{\mathrm{I}},
\end{aligned}
$$

where we use ${}_U(c_i) = \overline{U}(c_i) = c_i$ $(i = g+1, \ldots, 2g+1)$ because U is conjugate to Q^{-1} and then $\overline{U} = 1$ in $\mathcal{H}_g$.

We apply similar elementary transformations to R^2 as follows.

$$
\begin{aligned}
R^2 &\sim (d \cdot {}_{W_1}(c_{g+1}) \cdot {}_{W_2}(c_{g+2}) \cdot \cdots \cdot {}_{W_{g+1}}(c_{2g+1}))^2 \\
&\qquad \cdot (c_{2g+1} \cdots c_2 c_1)^{g+1} \cdot ({}_{R^{-1}}(c_{2g+1}) \cdot \cdots \cdot {}_{R^{-1}}(c_2) \cdot {}_{R^{-1}}(c_1))^{g+1} \\
&= (d \cdot {}_{W_1}(c_{g+1}) \cdot {}_{W_2}(c_{g+2}) \cdot \cdots \cdot {}_{W_{g+1}}(c_{2g+1}))^2 \cdot (c_{2g+1} \cdots c_2 c_1)^{2g+2} \\
&\sim (d \cdot {}_{W_1}(c_{g+1}) \cdot {}_{W_2}(c_{g+2}) \cdot \cdots \cdot {}_{W_{g+1}}(c_{2g+1}))^2 \cdot (c_1 c_2 \cdots c_{2g+1})^{2g+2} \\
&= K \cdot C_{\mathrm{I}},
\end{aligned}
$$

where we use ${}_{R^{-1}}(c_i) = \overline{R}^{-1}(c_i) = \iota(c_i) = c_i$ $(i = g+1, \ldots, 2g+1)$ and Corollary A.3. Q.E.D.

Remark 4.7. It is not difficult to show that M_K is isomorphic to Cadavid-Korkmaz' generalization, which we denote by M_{CK} in [8], of Matsumoto's genus 2 Lefschetz fibration on $S^2 \times T^2 \# 4\overline{\mathbb{CP}}^2$ (see [25], Example B). Hirose told the author a combinatorial proof of this fact. It is known that $n_0 = 2g+2, n_+ = 2, \sigma = -4$, and $e = 8 - 2g$ for M_K (see [4], [21], and [8]).

Both of the hyperelliptic Lefschetz fibrations M_{PJ} and $M_{RI^{g-1}}$ have $6g^2+2g$ non-separating singular fibers and one separating singular fiber. From the local signature formula [27], [7], they have the same signature $\sigma = -(3g^2+2g+1)$ and the same Euler characteristic $e = 6g^2 - 2g + 5$. They are simply-connected and non-spin. It follows from Freedman's classification theorem that both M_{PJ} and $M_{RI^{g-1}}$ are homeomorphic to $(3g^2/2 - 2g + 1)\mathbb{CP}^2 \# (9g^2/2 + 2)\overline{\mathbb{CP}}^2$. However they are not isomorphic as Lefschetz fibrations because the monodromy of $M_{RI^{g-1}}$ is transitive while that of M_{PJ} is intransitive.

Theorem 4.8. *If $g \geq 4$ and g is even, then $M_{PJ}, M_{RI^{g-1}}$, and $(3g^2/2 - 2g + 1)\mathbb{CP}^2 \#(9g^2/2 + 2)\overline{\mathbb{CP}}^2$ are mutually non-diffeomorphic.*

Proof. We first note that $b_2^+ = 3g^2/2 - 2g + 1 > 1$ and odd for these manifolds. If $g \geq 4$, $M_{RI^{g-1}}$ is isomorphic to a non-trivial fiber sum $M_{RI} \#_F (g/2 - 1) M_{I^2}$. It follows from a theorem of Usher [41] that $M_{RI^{g-1}}$ is a minimal symplectic 4-manifold. Since $b_2^+ > 1$, $M_{RI^{g-1}}$ does not contain any smooth (-1)-sphere as a consequence of Seiberg-Witten theory [39], [40], [22] (cf. [14], Remark 10.2.4(a)). On the other hand, M_{PJ} has a smooth (-1)-section which naturally comes from a chain relation C_{2g} of length $2g$ (cf. Smith [36], Lemma 2.3). Hence M_{PJ} and $M_{RI^{g-1}}$ can not be diffeomorphic.

By Gompf's theorem ([14], Theorem 10.2.18) M_{PJ} (resp. $M_{RI^{g-1}}$) admits a symplectic structure ω_{PJ} (resp. $\omega_{RI^{g-1}}$). It follows from results of Taubes [38] (cf. [14], Theorem 10.1.11) that the classes $\pm c_1(M_{PJ}, \omega_{PJ})$ (resp. $\pm c_1(M_{RI^{g-1}}, \omega_{RI^{g-1}})$) are Seiberg-Witten basic classes. On the other hand, the 4-manifold $(3g^2/2 - 2g + 1)\mathbb{CP}^2 \#(9g^2/2 + 2)\overline{\mathbb{CP}}^2$ has vanishing Seiberg-Witten invariants because it decomposes as the connected sum of $(3g^2/2 - 2g)\mathbb{CP}^2$ and $\mathbb{CP}^2 \#(9g^2/2 + 2)\overline{\mathbb{CP}}^2$ [23] (cf. [14], Theorem 2.4.6). Hence M_{PJ} (resp. $M_{RI^{g-1}}$) is not diffeomorphic to $(3g^2/2 - 2g + 1)\mathbb{CP}^2 \#(9g^2/2 + 2)\overline{\mathbb{CP}}^2$. Q.E.D.

Remark 4.9. Theorem 4.8 is a variant of Fuller's theorem [11], which states that $\#_F g M_{I^2}, M_{C_{\rm II}}$, and $(2g^2 - 2g + 1)\mathbb{CP}^2 \#(6g^2 + 2g + 1)\overline{\mathbb{CP}}^2$ are homeomorphic but mutually non-diffeomorphic for every $g \geq 2$ (see also [25], [13], and [7]). Fuller's theorem can be reproved by the same method as the proof of Theorem 4.8 without using Kirby calculus.

In contrast to Theorem 4.8 we show the following theorem about fiber sums.

Theorem 4.10 (cf. Chakiris [5], 'the first' Lemma 4.8). *The fiber sum $M_{PJI^2} = M_{PJ} \#_F M_{I^2}$ is isomorphic to the fiber sum $M_{RI^{g+1}} = M_{RI^{g-1}} \#_F M_{I^2}$ as Lefschetz fibrations for every even $g \geq 2$. In particular, these manifolds are diffeomorphic to each other.*

We postpone the proof of this theorem to Appendix B.

Remark 4.11. Theorem 4.10 is a variant of Lemma 4.1. Such kinds of stability for hyperelliptic Lefschetz fibrations under taking fiber sums with copies of M_{I^2} were formulated by Auroux [2] and Kharlamov and Kulikov [20].

If $g = 2$, the manifolds M_{PJ}, M_{RI}, and $K3\#\overline{\mathbb{CP}}^2$ are homeomorphic to $3\mathbb{CP}^2\#20\overline{\mathbb{CP}}^2$ by Freedman's classification theorem. The next theorem was suggested by the referee.

Theorem 4.12. *If $g = 2$, then M_{PJ}, $K3\#\overline{\mathbb{CP}}^2$, and $3\mathbb{CP}^2\#20\overline{\mathbb{CP}}^2$ are mutually non-diffeomorphic.*

Proof. $0 \in H^2(K3;\mathbb{Z})$ is the only Seiberg-Witten basic class of $K3$ [10] (cf. [14], Corollary 3.1.15). The blowup formula for Seiberg-Witten invariants [9] (cf. [14], Theorem 2.4.9) tells us that the only basic classes of $K3\#\overline{\mathbb{CP}}^2$ are $\pm E$, where $E \in H^2(K3\#\overline{\mathbb{CP}}^2;\mathbb{Z})$ is the Poincaré dual of the homology class of the exceptional sphere. On the other hand, consider M_{PJ} and its symplectic structure ω_{PJ}. The classical adjunction formula implies that the canonical class $K_{PJ} = -c_1(M_{PJ}, \omega_{PJ})$ of the symplectic manifold (M_{PJ}, ω_{PJ}) satisfies $K_{PJ} \cdot F = 2$, where $F \in H^2(M_{PJ};\mathbb{Z})$ is the Poincaré dual of the homology class of the fiber. Moreover, as observed in the proof of Theorem 4.8, M_{PJ} has a smooth (-1)-section and we have $S \cdot F = 1$, where $S \in H^2(M_{PJ};\mathbb{Z})$ is the Poincaré dual of the homology class of the section. Taubes' result [38] (cf. [14], Theorem 10.1.11) shows that $\pm K_{PJ}$ are basic classes for M_{PJ}. But the blowup formula shows that there must be other basic classes $\pm(K_{PJ} - 2S)$ which are distinct from $\pm K_{PJ}$ since they pair trivially with F (see [14], Exercise 10.1.20). Therefore M_{PJ} is not diffeomorphic to $K3\#\overline{\mathbb{CP}}^2$.

The manifold $3\mathbb{CP}^2\#20\overline{\mathbb{CP}}^2$ has vanishing Seiberg-Witten invariants for the same reason as the proof of Theorem 4.8. Hence this manifold is diffeomorphic neither to M_{PJ} nor to $K3\#\overline{\mathbb{CP}}^2$. Q.E.D.

We notice that M_{RI} for $g = 2$ is not diffeomorphic to $3\mathbb{CP}^2\#20\overline{\mathbb{CP}}^2$, but we can not distinguish M_{RI} from other two manifolds.

Problem 4.13. *Determine whether M_{PJ} and M_{RI} are diffeomorphic or not when $g = 2$. Is M_{RI} diffeomorphic to $K3\#\overline{\mathbb{CP}}^2$?*

Remark 4.14. Sato [30] listed the pairs (n_0, n_+) of numbers of singular fibers possibly realized by some genus 2 Lefschetz fibration with (-1)-sphere. Hirose [18] has constructed examples of genus 2 Lefschetz fibrations with (-1)-sphere which actually realize the pairs $(16, 2), (18, 1)$, and $(28, 1)$. If $g = 2$, the pairs (n_0, n_+) of numbers of singular fibers of the Lefschetz fibrations $M_{P^2}, M_Q, M_{PR}, M_{PI}, M_{PJ}, M_{RI}$ are $(16, 2), (18, 1), (26, 2), (18, 1), (28, 1), (28, 1)$, respectively. $M_{P^2}, M_Q, M_{PI}, M_{PJ}$ also realize three pairs (n_0, n_+) in Sato's table ([30], Table 1) because they contain (-1)-spheres. M_{RI} for $g = 2$ turns out to be isomorphic to Auroux's fibration X_2 in [2] (see Appendix B). Sato told the author that X_2 admits no (-1)-section but contains a (-1)-sphere as a 'double section'. Hence M_{RI} also realizes the pair $(28, 1)$.

Remark 4.15. Usher's theorem [41], which is used in the proof of Theorem 4.8 and is an affirmative solution to a conjecture of Stipsicz [37], is proved also by Sato [31] when $g = 2$.

4.2. Odd genus

Suppose that $g \geq 3$ and g is odd. Let $d \in \mathcal{S}^H$ be the boundary curve of a regular neighborhood of $c_1 \cup \cdots \cup c_{g-1}$. We obtain positive relators $Q, R \in \mathcal{R}^H$ by Theorem 3.5. We also have positive relators for $\mathcal{H}_g$ defined by

$$\begin{aligned} K_1 := &(c_1 c_2 \cdots c_{2g+1} \cdot d \cdot (c_{g-1} \cdots c_2 c_1)^2 \\ &\cdot w_1(c_g) \cdot w_2(c_{g+1}) \cdots w_{g+2}(c_{2g+1}))^2, \\ K_2 := &(c_1 c_2 \cdots c_{2g+1})^2 (d \cdot (c_{g-1} \cdots c_2 c_1)^2 \\ &\cdot w_1(c_g) \cdot w_2(c_{g+1}) \cdots w_{g+2}(c_{2g+1}))^2, \end{aligned}$$

which are constructed in the same way as K for even genus (see the proofs of Lemma 3.4 for $Q_0 \in \mathcal{R}^H$ and Theorem 3.5 for $Q \in \mathcal{R}^H$).

Thus we obtain the corresponding hyperelliptic Lefschetz fibrations

$$M_Q, \ M_R, \ M_{K_1}, \ M_{K_2}$$

of genus g over S^2 with transitive monodromy. These are non-spin 4-manifolds because a component of a separating singular fiber represents a homology class of square -1. Each of M_Q, M_{K_1}, and M_{K_2} has a smooth (-1)-section which naturally comes from C_{I} (cf. Smith [36], Lemma 2.3).

Proposition 4.16. *The manifolds M_Q, M_R, M_{K_1}, and M_{K_2} are simply connected.*

Proof. Quite similar to the proof of Proposition 4.2. Q.E.D.

The numbers n_0, n_+ of singular fibers, and the values of signature σ and the Euler characteristic e for the manifolds above are calculated as in the following table. (By virtue of Proposition 4.16, these manifolds satisfy $b_2^+ = (e+\sigma)/2 - 1$ and $b_2^- = (e-\sigma)/2 - 1$.)

LF	n_0	n_+	σ	e
M_Q and M_R	$2(g^2+4g+1)$	1	$-(g+2)^2$	$2g^2+4g+7$
M_{K_1} and M_{K_2}	$2(5g+1)$	2	$-2(2g+3)$	$2(3g+4)$

The values of signature of hyperelliptic Lefschetz fibrations in the table above are calculated by using formulae in [27], [7], or [8]. b_2^+ is always odd for these manifolds and it is greater than 1.

Proposition 4.17. *The four fiber sums* $M_{Q^2} = M_Q \#_F M_Q$, $M_{R^2} = M_R \#_F M_R$, $M_{K_i C_{\mathrm{I}}} = M_{K_i} \#_F M_{C_{\mathrm{I}}}$ $(i = 1, 2)$ *of copies of hyperelliptic Lefschetz fibrations* M_Q, M_R, $M_{C_{\mathrm{I}}}$, M_{K_i} $(i = 1, 2)$ *are isomorphic to each other.*

Proof. We apply elementary transformations to Q^2 as follows.

$$
\begin{aligned}
Q^2 &\sim (c_1 c_2 \cdots c_{2g+1} \cdot d \cdot (c_{g-1} \cdots c_2 c_1)^2 \\
&\qquad \cdot {}_{W_1}(c_g) \cdot {}_{W_2}(c_{g+1}) \cdot \cdots \cdot {}_{W_{g+2}}(c_{2g+1}) \cdot (c_1 c_2 \cdots c_{2g+1})^{g+1})^2 \\
&\sim (c_1 c_2 \cdots c_{2g+1} \cdot d \cdot (c_{g-1} \cdots c_2 c_1)^2 \\
&\qquad \cdot {}_{W_1}(c_g) \cdot {}_{W_2}(c_{g+1}) \cdot \cdots \cdot {}_{W_{g+2}}(c_{2g+1}))^2 \\
&\qquad \cdot (c_1 c_2 \cdots c_{2g+1})^{g+1} \cdot ({}_U(c_1) \cdot {}_U(c_2) \cdot \cdots \cdot {}_U(c_{2g+1}))^{g+1} \\
(U &:= (c_1 c_2 \cdots c_{2g+1} \cdot d \cdot (c_{g-1} \cdots c_2 c_1)^2 \\
&\qquad \cdot {}_{W_1}(c_g) \cdot {}_{W_2}(c_{g+1}) \cdot \cdots \cdot {}_{W_{g+2}}(c_{2g+1}) \cdot (c_1 c_2 \cdots c_{2g+1})^{g+1})^{-1}) \\
&= (c_1 c_2 \cdots c_{2g+1} \cdot d \cdot (c_{g-1} \cdots c_2 c_1)^2 \cdot {}_{W_1}(c_g) \cdot {}_{W_2}(c_{g+1}) \cdot \cdots \\
&\qquad \cdot {}_{W_{g+2}}(c_{2g+1}))^2 \cdot (c_1 c_2 \cdots c_{2g+1})^{2g+2} \\
&= K_1 \cdot C_{\mathrm{I}},
\end{aligned}
$$

where we use ${}_U(c_i) = \overline{U}(c_i) = c_i$ $(i = 1, \ldots, 2g+1)$ because U is conjugate to Q^{-1} and then $\overline{U} = 1$ in $\mathcal{H}_g$.

$$
\begin{aligned}
Q^2 &\sim (d \cdot (c_{g-1} \cdots c_2 c_1)^2 \\
&\qquad \cdot {}_{W_1}(c_g) \cdot {}_{W_2}(c_{g+1}) \cdot \cdots \cdot {}_{W_{g+2}}(c_{2g+1}) \cdot (c_1 c_2 \cdots c_{2g+1})^{g+2})^2 \\
&\sim (d \cdot (c_{g-1} \cdots c_2 c_1)^2 \cdot {}_{W_1}(c_g) \cdot {}_{W_2}(c_{g+1}) \cdot \cdots \cdot {}_{W_{g+2}}(c_{2g+1}))^2 \\
&\qquad \cdot (c_1 c_2 \cdots c_{2g+1})^{g+2} \cdot ({}_V(c_1) \cdot {}_V(c_2) \cdot \cdots \cdot {}_V(c_{2g+1}))^{g+2} \\
(V &:= (d \cdot (c_{g-1} \cdots c_2 c_1)^2 \\
&\qquad \cdot {}_{W_1}(c_g) \cdot {}_{W_2}(c_{g+1}) \cdot \cdots \cdot {}_{W_{g+2}}(c_{2g+1}) \cdot (c_1 c_2 \cdots c_{2g+1})^{g+2})^{-1}) \\
&= (d \cdot (c_{g-1} \cdots c_2 c_1)^2 \cdot {}_{W_1}(c_g) \cdot {}_{W_2}(c_{g+1}) \cdot \cdots \cdot {}_{W_{g+2}}(c_{2g+1}))^2 \\
&\qquad \cdot (c_1 c_2 \cdots c_{2g+1})^{2g+4} \\
&\sim K_2 \cdot C_{\mathrm{I}},
\end{aligned}
$$

where we use ${}_V(c_i) = \overline{V}(c_i) = c_i$ $(i = 1, \ldots, 2g+1)$ because V is conjugate to Q^{-1} and then $\overline{V} = 1$ in $\mathcal{H}_g$.

We apply similar elementary transformations to R^2 as follows.

$$\begin{aligned}
R^2 &= (c_1c_2\cdots c_{2g+1}\cdot d\cdot(c_{g-1}\cdots c_2c_1)^2 \\
&\quad \cdot {}_{W_1}(c_g)\cdot {}_{W_2}(c_{g+1})\cdot\cdots\cdot {}_{W_{g+2}}(c_{2g+1})\cdot(c_{2g+1}\cdots c_2c_1)^{g+1})^2 \\
&\sim (c_1c_2\cdots c_{2g+1}\cdot d\cdot(c_{g-1}\cdots c_2c_1)^2 \\
&\quad \cdot {}_{W_1}(c_g)\cdot {}_{W_2}(c_{g+1}))\cdot\cdots\cdot {}_{W_{g+2}}(c_{2g+1}))^2 \\
&\quad \cdot(c_{2g+1}\cdots c_2c_1)^{g+1}\cdot({}_{R^{-1}}(c_{2g+1})\cdot\cdots\cdot {}_{R^{-1}}(c_2)\cdot {}_{R^{-1}}(c_1))^{g+1} \\
&= (c_1c_2\cdots c_{2g+1}\cdot d\cdot(c_{g-1}\cdots c_2c_1)^2 \\
&\quad \cdot {}_{W_1}(c_g)\cdot {}_{W_2}(c_{g+1}))\cdot\cdots\cdot {}_{W_{g+2}}(c_{2g+1}))^2\cdot(c_{2g+1}\cdots c_2c_1)^{2g+2} \\
&= (c_1c_2\cdots c_{2g+1}\cdot d\cdot(c_{g-1}\cdots c_2c_1)^2 \\
&\quad \cdot {}_{W_1}(c_g)\cdot {}_{W_2}(c_{g+1}))\cdot\cdots\cdot {}_{W_{g+2}}(c_{2g+1}))^2\cdot(c_1c_2\cdots c_{2g+1})^{2g+2} \\
&= K_1\cdot C_{\mathrm{I}},
\end{aligned}$$

where we use ${}_{R^{-1}}(c_i)=\overline{R}^{-1}(c_i)=c_i$ $(i=1,\ldots,2g+1)$ and Corollary A.3. Q.E.D.

Remark 4.18. The author does not know any explicit examples of hyperelliptic Lefschetz fibrations of odd genus with separating singular fibers other than $M_Q, M_R, M_{K_1}, M_{K_2}$, and fiber sums of their copies. M_Q (resp. M_{K_1}) might not be isomorphic to M_R (resp. M_{K_2}) although the author does not know any invariants which distinguish these fibrations.

Let $d_1, d_2 \in \mathcal{S}^H$ be the boundary curves of a regular neighborhood of $c_1\cup\cdots\cup c_g$.

Remark 4.19. Substituting the same word as Q_0 for even genus by the inverse C_g^{-1} of a chain relator C_g twice, we have the following positive word for odd genus:

$$K' := (d_1^2d_2^2\cdot {}_{W_1}(c_{g+1})\cdot {}_{W_2}(c_{g+2})\cdot\cdots\cdot {}_{W_{g+1}}(c_{2g+1}))^2$$

which is not in $\mathcal{F}^H$ (see Remark 3.6). This is a positive relator in $\mathcal{M}_g$ and the corresponding Lefschetz fibration $M_{K'}$ is nothing but Cadavid-Korkmaz' fibration for odd genus, which we denote by M_{CK} in [8]. It is known that $n_0=2g+10, n_+=0, \sigma=-8$, and $e=14-2g$ for M_K (see [4], [21], and [8]).

§5. Concluding remarks

Hyperelliptic Lefschetz fibrations form a very special and beautiful class of Lefschetz fibrations. But it seems that there is much room to be studied. For example, a famous conjecture of Siebert and Tian [32] for hyperelliptic Lefschetz fibrations without separating singular fibers is only partially solved in genus 2 case. Moreover it is not clear whether all hyperelliptic Lefschetz fibrations over the 2-sphere have been discovered.

On the other hand, hyperelliptic Lefschetz fibrations are rich enough to include many explicit examples with interesting properties. For example, generalizations of Matsumoto's fibrations [25] and Chakiris' fibrations [5] give examples of homeomorphic but non-diffeomorphic 4-manifolds which become diffeomorphic after taking fiber sums with only one copy of M_{I^2} (see Remark 4.9, Remark 4.11, Theorem 4.8, and Theorem 4.10). These examples together with stabilization theorems of Auroux [2] and Kharlamov and Kulikov [20] seem to be 'fiber sum analogues' of 4-manifolds which dissolve after taking connected sums with only one copy of $S^2 \times S^2$ together with Wall's stabilization theorem for connected sums of simply-connected 4-manifolds with copies of $S^2 \times S^2$.

If hyperelliptic Lefschetz fibrations are investigated very well, they would be recognized as new fundamental 4-manifolds and might play interesting roles such as elliptic surfaces in 4-manifold topology.

§Appendix A. A reversing lemma

Let $(c_1, c_2, \ldots, c_n)$ be a chain of length n on Σ_g and W_1, W_2 positive words in the generators $c_1, c_2, \ldots, c_n \in \mathcal{S}$. We write $W_1 \approx W_2$ if W_1 can be transformed into W_2 by replacing $c_i c_{i+1} c_i$ with $c_{i+1} c_i c_{i+1}$, $c_{i+1} c_i c_{i+1}$ with $c_i c_{i+1} c_i$ for $i = 1, \ldots, n-1$, and $c_i c_j$ with $c_j c_i$ for $|i - j| > 1$ repeatedly. This relation is an equivalence relation on the set of positive words in the generators $c_1, c_2, \ldots, c_n$. It is not difficult to verify that Lemma 2.4 is true even if we replace $\equiv \pmod{T_0, T_1}$ with $\approx$.

Lemma A.1 (Chakiris [5], Lemma 3.5). *The following equivalence holds.*

$$(c_1 c_2 \cdots c_n)^{n+1} \approx (c_n \cdots c_2 c_1)^{n+1} \quad (n = 1, \ldots, 2g).$$

Proof. We set $W_1 := (c_1 c_2 \cdots c_n)^{n+1}$ and $W_2 := (c_n \cdots c_2 c_1)^{n+1}$ for $n = 1, \ldots, 2g$. We prove $W_1 \approx W_2$ by induction on n.

If $n = 2$, then W_1 is transformed into W_2 as follows:

$$W_1 = (c_1 c_2)^3 = c_1 c_2 c_1 \cdot c_2 c_1 c_2 \approx c_2 c_1 c_2 \cdot c_1 c_2 c_1 = (c_2 c_1)^3 = W_2.$$

Suppose that $W_1 \approx W_2$ is valid for $n - 1$:

$$(c_1 c_2 \cdots c_{n-1})^n \approx (c_{n-1} \cdots c_2 c_1)^n.$$

We consider W_1 and W_2 for n. Applying Lemma 2.4 for $\approx$ and using the assumption, we have

$$\begin{aligned} W_1 = (c_1c_2\cdots c_n)^{n+1} &= (c_1c_2\cdots c_n)^n \cdot c_1c_2\cdots c_n \\ &\approx (c_1c_2\cdots c_{n-1})^n \cdot c_n\cdots c_2c_1 \cdot c_1c_2\cdots c_n \\ &\approx (c_{n-1}\cdots c_2c_1)^n \cdot c_n\cdots c_2c_1 \cdot c_1c_2\cdots c_n. \end{aligned}$$

Manipulating braid relations as Lemma 2.1 of [21], we transform the right-hand side as follows:

$$\begin{aligned} &(c_{n-1}\cdots c_2c_1)^n \cdot c_n\cdots c_2c_1 \cdot c_1c_2\cdots c_n \\ \approx\ & c_n\cdots c_2c_1 \cdot (c_n\cdots c_3c_2)^n \cdot c_1c_2\cdots c_n \\ =\ & c_n\cdots c_2c_1 \cdot (c_n\cdots c_3c_2)^{n-1} \cdot c_n\cdots c_2c_1 \cdot c_2\cdots c_n \\ \approx\ & c_n\cdots c_2c_1 \cdot (c_n\cdots c_3c_2)^{n-2} \cdot (c_n\cdots c_2c_1)^2 \cdot c_3\cdots c_n \\ \approx\ & \cdots\cdots \\ \approx\ & c_n\cdots c_2c_1 \cdot c_n\cdots c_3c_2 \cdot (c_n\cdots c_2c_1)^{n-1} \cdot c_n \\ \approx\ & (c_n\cdots c_2c_1)^{n+1} = W_2. \end{aligned}$$

We have thus shown $W_1 \approx W_2$ as claimed. Q.E.D.

Let $\varrho \in \mathcal{R}$ be a relator including $W_1 = (c_1c_2\cdots c_n)^{n+1}$ as a subword: $\varrho = UW_1V$ $(U, V \in \mathcal{F})$. We put $\varrho' := UW_2V \in \mathcal{F}$, where $W_2 = (c_n\cdots c_2c_1)^{n+1}$.

Corollary A.2. *The word ϱ' is also a relator in $\mathcal{R}$ and $\varrho \equiv \varrho'$* (mod T_0, T_1).

Proof. It immediately follows from the definition of $\approx$ that $W_1 \approx W_2$ implies $\varrho \equiv \varrho'$ (mod T_0, T_1) and then $\varrho' \in \mathcal{R}$. Q.E.D.

We set $W_1 = (c_1c_2\cdots c_n)^{n+1}$ and $W_2 = (c_n\cdots c_2c_1)^{n+1}$ for $n = 1, \ldots, 2g$ again. Let $\varrho \in \mathcal{R}$ be a positive relator including W_1 as a subword: $\varrho = UW_1V$ ($U, V \in \mathcal{F}$ and U, V are positive). We put $\varrho' := UW_2V \in \mathcal{F}$.

Corollary A.3. *The word ϱ' is also a positive relator in $\mathcal{R}$ and $\varrho \sim \varrho'$.*

Proof. The word ϱ' is obviously positive. From Corollary A.2, $\varrho \equiv \varrho'$ (mod T_0, T_1) and then $\varrho' \in \mathcal{R}$. We can show that $W_1 \approx W_2$ implies $\varrho \sim \varrho'$ because

$$\begin{aligned} &\cdots\cdot c_i \cdot c_{i+1} \cdot c_i \cdot\cdots \sim \cdots\cdot c_i \cdot {}_{c_{i+1}}(c_i) \cdot c_{i+1} \cdot\cdots \\ \sim\ & \cdots\cdot {}_{c_1c_{i+1}}(c_i) \cdot c_i \cdot c_{i+1} \cdot\cdots = \cdots\cdot c_{i+1} \cdot c_i \cdot c_{i+1} \cdot\cdots \end{aligned}$$

for $i = 1, \ldots, n-1$ and

$$\cdots \cdot c_i \cdot c_j \cdot \cdots \sim \cdots \cdot {}_{c_j}(c_i) \cdot c_i \cdot \cdots = \cdots \cdot c_j \cdot c_i \cdot \cdots$$

for $|i-j| > 1$ by braid relations. Q.E.D.

§Appendix B. Proof of Theorem 4.10

It is easy to see that the image of

$$H := c_{2g+1}c_{2g} \cdots c_2 c_1^2 c_2 \cdots c_{2g} c_{2g+1} \in \mathcal{F}^H$$

under ϖ represents the hyperelliptic involution ι.

We need the following lemma to show that $PJI^2 \sim RI^{g+1}$.

Lemma B.1 (cf. Chakiris [5], Lemma 4.7). *The equivalence*

$$P \cdot {}_{c_{2g+1}^{-1}} J \sim RH^{g-1}$$

holds for every even $g \geq 2$.

Proof. We first notice that $P \cdot {}_{c_{2g+1}^{-1}} J$ and RH^{g-1} belong to $\mathcal{R}^H$ because $\varpi(P \cdot {}_{c_{2g+1}^{-1}} J) = \iota t_{c_{2g+1}}^{-1} \iota t_{c_{2g+1}} = 1$ and $\varpi(RH^{g-1}) = \iota^g = 1$.

We use such elementary transformations as in the proof of Corollary A.3 and cyclic permutations, which are expressed as compositions of elementary transformations, repeatedly in this proof. Using Corollary A.3, Lemma 2.4, and manipulation of braid relations as Lemma 2.1 of [21], we obtain the following long sequence of elementary transformations.

$$\begin{aligned}
& P \cdot {}_{c_{2g+1}^{-1}} J \\
\sim\ & {}_{c_{2g+1}^{-1}} J \cdot P = {}_{c_{2g+1}^{-1}} (c_1 c_2 \cdots c_{2g})^{2g+1} \cdot P \sim {}_{c_{2g+1}^{-1}} (c_{2g} \cdots c_2 c_1)^{2g+1} \cdot P \\
\sim\ & {}_{c_{2g+1}^{-1}} ((c_{2g} \cdots c_2 c_1)^g \cdot (c_g c_{g-1} \cdots c_{2g}) \cdot \\
& \qquad \cdots \cdot (c_1 c_2 \cdots c_{g+1}) \cdot (c_g \cdots c_2 c_1)^{g+1}) \cdot P \\
=\ & {}_{c_{2g+1}^{-1}} ((c_{2g} \cdots c_2 c_1)^g \cdot (c_g c_{g-1} \cdots c_{2g}) \cdot \\
& \qquad \cdots \cdot (c_1 c_2 \cdots c_{g+1})) \cdot (c_g \cdots c_2 c_1)^{g+1} \cdot P \\
\sim\ & {}_{c_{2g+1}^{-1}} ((c_{2g} \cdots c_2 c_1)^g \cdot (c_g c_{g-1} \cdots c_{2g}) \cdot \cdots \cdot (c_1 c_2 \cdots c_{g+1})) \cdot W^{-1} \cdot P \\
=\ & {}_{c_{2g+1}^{-1}} ((c_{2g} \cdots c_2 c_1)^g \cdot (c_g c_{g-1} \cdots c_{2g}) \cdot \cdots \cdot (c_1 c_2 \cdots c_{g+1})) \cdot W^{-1} \\
& \cdot d \cdot {}_W(c_{g+1} \cdots c_3 c_2) \cdot \cdots \cdot {}_W(c_{2g} \cdots c_{g+2} c_{g+1}) \\
& \cdot (c_{g+1} \cdots c_3 c_2) \cdot \cdots \cdot (c_{2g} \cdots c_{g+2} c_{g+1}) \\
\sim\ & {}_{c_{2g+1}^{-1}} ((c_{2g} \cdots c_2 c_1)^g \cdot (c_g c_{g-1} \cdots c_{2g}) \cdot \cdots \cdot (c_1 c_2 \cdots c_{g+1})) \cdot W^{-1}
\end{aligned}$$

$$
\begin{aligned}
&\cdot {}_{W}(c_{g+1}\cdots c_3c_2)\cdot\cdots\cdot {}_{W}(c_{2g}\cdots c_{g+2}c_{g+1})\\
&\cdot (c_{g+1}\cdots c_3c_2)\cdot\cdots\cdot(c_{2g}\cdots c_{g+2}c_{g+1})\cdot {}_{(d^{-1}P)^{-1}}(d)\\
\sim\ & {}_{c_{2g+1}^{-1}}((c_{2g}\cdots c_2c_1)^g\cdot(c_gc_{g-1}\cdots c_{2g})\cdot\cdots\cdot(c_1c_2\cdots c_{g+1}))\\
&\cdot (c_{g+1}\cdots c_3c_2)\cdot\cdots\cdot(c_{2g}\cdots c_{g+2}c_{g+1})\cdot W^{-1}\\
&\cdot (c_{g+1}\cdots c_3c_2)\cdot\cdots\cdot(c_{2g}\cdots c_{g+2}c_{g+1})\cdot d\\
\sim\ & ({}_{c_{2g}}c_{2g+1}c_{2g-1}\cdots c_2c_1)^g\\
&\cdot(c_gc_{g-1}\cdots c_{2g-1}\,{}_{c_{2g}}c_{2g+1})\cdot(c_{g-1}c_{g-2}\cdots c_{2g})\cdot\cdots\cdot(c_1c_2\cdots c_{g+1})\\
&\cdot (c_{g+1}\cdots c_3c_2)\cdot\cdots\cdot(c_{2g}\cdots c_{g+2}c_{g+1})\cdot(c_1c_2\cdots c_g)^{g+1}\\
&\cdot (c_{g+1}\cdots c_3c_2)\cdot\cdots\cdot(c_{2g}\cdots c_{g+2}c_{g+1})\cdot d\\
&\qquad (\text{n.b. } {}_{c_{2g+1}^{-1}}c_{2g} = {}_{c_{2g}}c_{2g+1})\\
\sim\ & ({}_{c_{2g}}c_{2g+1}c_{2g-1}\cdots c_2c_1)^g\\
&\cdot(c_gc_{g-1}\cdots c_{2g-1}\,{}_{c_{2g}}c_{2g+1})\cdot(c_{g-1}c_{g-2}\cdots c_{2g})\cdot\cdots\cdot(c_1c_2\cdots c_{g+1})\\
&\cdot(c_{g+1}\cdots c_2c_1)\cdot(c_{g+2}\cdots c_2c_1)\cdot\cdots\cdot(c_{2g}\cdots c_2c_1)\\
&\cdot c_g\cdot(c_{g-1}c_g)\cdot(c_{g-2}c_{g-1}c_g)\cdot\cdots\cdot(c_1c_2\cdots c_g)\\
&\cdot (c_{g+1}\cdots c_3c_2)\cdot\cdots\cdot(c_{2g}\cdots c_{g+2}c_{g+1})\cdot d\\
\sim\ & ({}_{c_{2g}}c_{2g+1}c_{2g-1}\cdots c_2c_1)^g\\
&\cdot(c_gc_{g-1}\cdots c_{2g-1}\,{}_{c_{2g}}c_{2g+1})\cdot(c_{g-1}c_{g-2}\cdots c_{2g})\cdot\cdots\cdot(c_1c_2\cdots c_{g+1})\\
&\cdot(c_{g+1}\cdots c_2c_1)\cdot(c_{g+2}\cdots c_2c_1)\cdot\cdots\cdot(c_{2g}\cdots c_2c_1)\\
&\cdot c_g\cdot(c_{g-1}c_g)\cdot\cdots\cdot(c_2c_3\cdots c_g)\cdot(c_1c_2\cdots c_g)\\
&\cdot(c_{g+1}c_{g+2}\cdots c_{2g})\cdot\cdots\cdot(c_3c_4\cdots c_{g+2})\cdot(c_2c_3\cdots c_{g+1})\cdot d\\
\sim\ & ({}_{c_{2g}}c_{2g+1}c_{2g-1}\cdots c_2c_1)^g\\
&\cdot(c_gc_{g-1}\cdots c_{2g-1}\,{}_{c_{2g}}c_{2g+1})\cdot(c_{g-1}c_{g-2}\cdots c_{2g})\cdot\cdots\cdot(c_1c_2\cdots c_{g+1})\\
&\cdot(c_{g+1}\cdots c_2c_1)\cdot(c_{g+2}\cdots c_2c_1)\cdot\cdots\cdot(c_{2g}\cdots c_2c_1)\\
&\cdot(c_1c_2\cdots c_{2g})\cdot\cdots\cdot(c_1c_2\cdots c_{g+2})\cdot(c_1c_2\cdots c_{g+1})\cdot d\\
\sim\ & ({}_{c_{2g}}c_{2g+1}c_{2g-1}\cdots c_2c_1)^g\\
&\cdot(c_gc_{g-1}\cdots c_{2g-1}\,{}_{c_{2g}}c_{2g+1})\cdot(c_{g-1}c_{g-2}\cdots c_{2g})\cdot\cdots\cdot(c_1c_2\cdots c_{g+1})\\
&\cdot(c_{2g}\cdots c_2c_1^2c_2\cdots c_{2g})\cdot\cdots\cdot(c_{g+2}\cdots c_2c_1^2c_1\cdots c_{g+2})\\
&\cdot(c_{g+1}\cdots c_2c_1^2c_1\cdots c_{g+1})\cdot d\\
\sim\ & ({}_{c_{2g}}c_{2g+1}c_{2g-1}\cdots c_2c_1)^g\\
&\cdot(c_gc_{g-1}\cdots c_{2g-1}\,{}_{c_{2g}}c_{2g+1})\cdot(c_{g-1}c_{g-2}\cdots c_{2g})\cdot\cdots\cdot(c_1c_2\cdots c_{g+2})\\
&\cdot c_{g+1}\cdots c_2c_1^2c_2\cdots c_{2g}\cdot(c_{2g-1}\cdots c_2c_1^2c_2\cdots c_{2g-1})
\end{aligned}
$$

$$
\begin{aligned}
&\cdots\cdot(c_{g+2}\cdots c_2c_1^2c_1\cdots c_{g+2})\cdot(c_{g+1}\cdots c_2c_1^2c_1\cdots c_{g+1})\cdot d\\
\sim\ &({}_{c_{2g}}c_{2g+1}c_{2g-1}\cdots c_2c_1)^g\\
&\cdot(c_gc_{g-1}\cdots c_{2g-1\,c_{2g}}c_{2g+1}c_{2g})\cdot(c_{g-1}c_{g-2}\cdots c_{2g})\cdot\cdots\cdot(c_1c_2\cdots c_{g+2})\\
&\cdot c_g\cdots c_2c_1^2c_2\cdots c_{2g}\cdot(c_{2g-1}\cdots c_2c_1^2c_2\cdots c_{2g-1})\\
&\cdots\cdot(c_{g+2}\cdots c_2c_1^2c_1\cdots c_{g+2})\cdot(c_{g+1}\cdots c_2c_1^2c_1\cdots c_{g+1})\cdot d\\
\sim\ &({}_{c_{2g}}c_{2g+1}c_{2g-1}\cdots c_2c_1)^g\cdot(c_gc_{g-1}\cdots c_{2g}c_{2g+1})\cdot(c_{g-1}c_{g-2}\cdots c_{2g})\\
&\cdots\cdot(c_1c_2\cdots c_{g+2})\cdot c_g\cdots c_2c_1^2c_2\cdots c_{2g}\cdot(c_{2g-1}\cdots c_2c_1^2c_2\cdots c_{2g-1})\\
&\cdots\cdot(c_{g+2}\cdots c_2c_1^2c_1\cdots c_{g+2})\cdot(c_{g+1}\cdots c_2c_1^2c_1\cdots c_{g+1})\cdot d\\
\sim\ &({}_{c_{2g}}c_{2g+1}c_{2g-1}\cdots c_2c_1)^g\cdot c_{2g}\cdots c_{g+2}c_{g+1}\cdot(c_gc_{g-1}\cdots c_{2g+1})\\
&\cdot(c_{g-1}c_{g-2}\cdots c_{2g})\cdot\cdots\cdot(c_1c_2\cdots c_{g+2})\cdot c_1c_2\cdots c_{2g}\\
&\cdot(c_{2g-1}\cdots c_2c_1^2c_2\cdots c_{2g-1})\cdot\cdots\cdot(c_{g+1}\cdots c_2c_1^2c_1\cdots c_{g+1})\cdot d\\
\sim\ &({}_{c_{2g}}c_{2g+1}c_{2g-1}\cdots c_2c_1)^g\cdot c_{2g}\cdots c_2c_1\cdot(c_{g+1}c_{g+2}\cdots c_{2g+1})\\
&\cdot(c_gc_{g-1}\cdots c_{2g})\cdot\cdots\cdot(c_2c_3\cdots c_{g+2})\cdot c_1c_2\cdots c_{2g}\\
&\cdot(c_{2g-1}\cdots c_2c_1^2c_2\cdots c_{2g-1})\cdot\cdots\cdot(c_{g+1}\cdots c_2c_1^2c_1\cdots c_{g+1})\cdot d\\
\sim\ &c_{2g}\cdots c_2c_1\cdot(c_{2g+1}c_{2g}\cdots c_3c_2)^g\cdot(c_{g+1}c_{g+2}\cdots c_{2g+1})\\
&\cdot(c_gc_{g-1}\cdots c_{2g})\cdot\cdots\cdot(c_2c_3\cdots c_{g+2})\cdot c_1c_2\cdots c_{2g}\\
&\cdot(c_{2g-1}\cdots c_2c_1^2c_2\cdots c_{2g-1})\cdot\cdots\cdot(c_{g+1}\cdots c_2c_1^2c_1\cdots c_{g+1})\cdot d\\
\sim\ &c_{g+1}\cdots c_2c_1\cdot(c_{2g+1}c_{2g}\cdots c_3c_2)^g\cdot(c_{g+1}c_{g+2}\cdots c_{2g+1})\\
&\cdot(c_gc_{g-1}\cdots c_{2g})\cdot\cdots\cdot(c_2c_3\cdots c_{g+2})\cdot c_1c_2\cdots c_{2g}\\
&\cdot(c_{2g-1}\cdots c_2c_1^2c_2\cdots c_{2g-1})\cdot\cdots\cdot(c_{g+1}\cdots c_2c_1^2c_1\cdots c_{g+1})\\
&\cdot c_{2g}\cdot\cdots\cdot c_{g+3}c_{g+2}\cdot d\\
\sim\ &c_{g+1}\cdots c_2c_1\cdot(c_{2g+1}c_{2g}\cdots c_3c_2)^g\cdot(c_{g+1}c_{g+2}\cdots c_{2g+1})\\
&\cdot(c_gc_{g-1}\cdots c_{2g})\cdot\cdots\cdot(c_2c_3\cdots c_{g+2})\cdot(c_1c_2\cdots c_{g+1})\\
&\cdot(c_{2g}\cdots c_2c_1^2c_2\cdots c_{2g})\cdot\cdots\cdot(c_{g+2}\cdots c_2c_1^2c_1\cdots c_{g+2})\cdot d\\
\sim\ &(c_{2g+1}\cdots c_{g+4}c_{g+3})\cdot\cdots\cdot(c_{2g+1}c_{2g})\cdot c_{2g+1}\\
&\cdot(c_{g+1}\cdots c_2c_1)\cdot(c_{g+2}\cdots c_3c_2)\cdot\cdots\cdot(c_{2g+1}\cdots c_3c_2)\\
&\cdot(c_{g+1}c_{g+2}\cdots c_{2g+1})\cdot\cdots\cdot(c_2c_3\cdots c_{g+2})\cdot(c_1c_2\cdots c_{g+1})\\
&\cdot(c_{2g}\cdots c_2c_1^2c_2\cdots c_{2g})\cdot\cdots\cdot(c_{g+2}\cdots c_2c_1^2c_1\cdots c_{g+2})\cdot d\\
\sim\ &(c_{g+1}\cdots c_2c_1)\cdot(c_{g+2}\cdots c_3c_2)\cdot\cdots\cdot(c_{2g+1}\cdots c_3c_2)\\
&\cdot(c_{g+1}c_{g+2}\cdots c_{2g+1})\cdot\cdots\cdot(c_2c_3\cdots c_{g+2})\cdot(c_1c_2\cdots c_{g+1})\\
&\cdot(c_{2g}\cdots c_2c_1^2c_2\cdots c_{2g})\cdot\cdots\cdot(c_{g+2}\cdots c_2c_1^2c_1\cdots c_{g+2})
\end{aligned}
$$

$$
\begin{aligned}
&\cdot (c_{2g+1} \cdots c_{g+4}c_{g+3}) \cdot \cdots \cdot (c_{2g+1}c_{2g}) \cdot c_{2g+1} \cdot d \\
\sim\ & (c_{g+1} \cdots c_2c_1) \cdot (c_{g+2} \cdots c_3c_2) \cdot \cdots \cdot (c_{2g+1} \cdots c_{g+2}c_{g+1}) \\
&\cdot c_2 \cdot (c_3c_2) \cdot \cdots \cdot (c_g \cdots c_3c_2) \\
&\cdot (c_{g+1}c_{g+2} \cdots c_{2g+1}) \cdot \cdots \cdot (c_2c_3 \cdots c_{g+2}) \cdot (c_1c_2 \cdots c_{g+1}) \\
&\cdot (c_{2g} \cdots c_2c_1^2c_2 \cdots c_{2g}) \cdot \cdots \cdot (c_{g+2} \cdots c_2c_1^2c_1 \cdots c_{g+2}) \\
&\cdot (c_{2g+1} \cdots c_{g+4}c_{g+3}) \cdot \cdots \cdot (c_{2g+1}c_{2g}) \cdot c_{2g+1} \cdot d \\
\sim\ & (c_{g+1} \cdots c_2c_1) \cdot (c_{g+2} \cdots c_3c_2) \cdot \cdots \cdot (c_{2g+1} \cdots c_{g+2}c_{g+1}) \\
&\cdot c_g \cdot (c_{g-1}c_g) \cdot \cdots \cdot (c_2c_3 \cdots c_g) \\
&\cdot (c_{g+1}c_{g+2} \cdots c_{2g+1}) \cdot \cdots \cdot (c_2c_3 \cdots c_{g+2}) \cdot (c_1c_2 \cdots c_{g+1}) \\
&\cdot (c_{2g} \cdots c_2c_1^2c_2 \cdots c_{2g}) \cdot \cdots \cdot (c_{g+2} \cdots c_2c_1^2c_1 \cdots c_{g+2}) \\
&\cdot (c_{2g+1} \cdots c_{g+4}c_{g+3}) \cdot \cdots \cdot (c_{2g+1}c_{2g}) \cdot c_{2g+1} \cdot d \\
\sim\ & (c_{g+1} \cdots c_2c_1) \cdot (c_{g+2} \cdots c_3c_2) \cdot \cdots \cdot (c_{2g+1} \cdots c_{g+2}c_{g+1}) \\
&\cdot c_g \cdot (c_{g-1}c_g) \cdot \cdots \cdot (c_2c_3 \cdots c_g) \\
&\cdot (c_{g+1} \cdots c_2c_1) \cdot (c_{g+2} \cdots c_3c_2) \cdot \cdots \cdot (c_{2g+1} \cdots c_{g+2}c_{g+1}) \\
&\cdot (c_{2g} \cdots c_2c_1^2c_2 \cdots c_{2g}) \cdot \cdots \cdot (c_{g+2} \cdots c_2c_1^2c_1 \cdots c_{g+2}) \\
&\cdot (c_{2g+1} \cdots c_{g+4}c_{g+3}) \cdot \cdots \cdot (c_{2g+1}c_{2g}) \cdot c_{2g+1} \cdot d \\
\sim\ & (c_{g+1} \cdots c_2c_1) \cdot (c_{g+2} \cdots c_3c_2) \cdot \cdots \cdot (c_{2g+1} \cdots c_{g+2}c_{g+1}) \\
&\cdot (c_{g+1} \cdots c_2c_1) \cdot (c_{g+2} \cdots c_3c_2) \cdot \cdots \cdot (c_{2g+1} \cdots c_{g+2}c_{g+1}) \\
&\cdot c_{2g+1} \cdot (c_{2g}c_{2g+1}) \cdot \cdots \cdot (c_{g+3}c_{g+4} \cdots c_{2g+1}) \\
&\cdot (c_{2g} \cdots c_2c_1^2c_2 \cdots c_{2g}) \cdot \cdots \cdot (c_{g+2} \cdots c_2c_1^2c_1 \cdots c_{g+2}) \\
&\cdot (c_{2g+1} \cdots c_{g+4}c_{g+3}) \cdot \cdots \cdot (c_{2g+1}c_{2g}) \cdot c_{2g+1} \cdot d \\
\sim\ & (c_{g+1} \cdots c_2c_1) \cdot (c_{g+2} \cdots c_3c_2) \cdot \cdots \cdot (c_{2g+1} \cdots c_{g+2}c_{g+1}) \\
&\cdot (c_{g+1} \cdots c_2c_1) \cdot (c_{g+2} \cdots c_3c_2) \cdot \cdots \cdot (c_{2g+1} \cdots c_{g+2}c_{g+1}) \\
&\cdot (c_{2g+1} \cdots c_2c_1^2c_2 \cdots c_{2g+1})^{g-1} \cdot d \\
\sim\ & (c_{g+1} \cdots c_2c_1) \cdot (c_{g+2} \cdots c_3c_2) \cdot \cdots \cdot (c_{2g+1} \cdots c_{g+2}c_{g+1}) \\
&\cdot (c_{g+1}c_{g+2} \cdots c_{2g+1}) \cdot \cdots \cdot (c_2c_3 \cdots c_g) \cdot (c_1c_2 \cdots c_{g+1}) \cdot H^{g-1} \cdot d \\
\sim\ & (c_g \cdots c_2c_1)^{g+1} \cdot {}_{W_1}(c_{g+1}) \cdot {}_{W_2}(c_{g+2}) \cdot \cdots \cdot {}_{W_{g+1}}(c_{2g+1}) \\
&\cdot (c_{g+1}c_{g+2} \cdots c_{2g+1}) \cdot \cdots \cdot (c_2c_3 \cdots c_g) \cdot (c_1c_2 \cdots c_{g+1}) \cdot H^{g-1} \cdot d \\
\sim\ & {}_{W_1}(c_{g+1}) \cdot {}_{W_2}(c_{g+2}) \cdot \cdots \cdot {}_{W_{g+1}}(c_{2g+1}) \\
&\cdot (c_{g+1}c_{g+2} \cdots c_{2g+1}) \cdot \cdots \cdot (c_2c_3 \cdots c_g) \cdot (c_1c_2 \cdots c_{g+1}) \\
&\cdot (c_g \cdots c_2c_1)^{g+1} \cdot H^{g-1} \cdot d \quad (\text{n.b. } {}_{H^{g-1}}(c_i) = c_i \ (i = 1, \ldots, g))
\end{aligned}
$$

$$\begin{aligned}
&\sim {}_{W_1}(c_{g+1}) \cdot {}_{W_2}(c_{g+2}) \cdot \cdots \cdot {}_{W_{g+1}}(c_{2g+1}) \cdot (c_{2g+1} \cdots c_2 c_1)^{g+1} \cdot H^{g-1} \cdot d \\
&\sim d \cdot {}_{W_1}(c_{g+1}) \cdot {}_{W_2}(c_{g+2}) \cdot \cdots \cdot {}_{W_{g+1}}(c_{2g+1}) \cdot (c_{2g+1} \cdots c_2 c_1)^{g+1} \cdot H^{g-1} \\
&= R \cdot H^{g-1}
\end{aligned}$$

Thus the proof is completed. Q.E.D.

By virtue of Lemma 4.1, Corollary A.3, and Lemma B.1, we can show the equivalence $PJI^2 \sim RI^{g+1}$ as follows.

$$\begin{aligned}
PJI^2 &= P \cdot JI \cdot I = P \cdot C_{\mathrm{I}} \cdot c_1 c_2 \cdots c_{2g} c_{2g+1}^2 c_{2g} \cdots c_2 c_1 \\
&\sim {}_{PC_{\mathrm{I}}}(c_1 c_2 \cdots c_{2g+1}) \cdot PC_{\mathrm{I}} \cdot c_{2g+1} \cdots c_2 c_1 \sim H \cdot PC_{\mathrm{I}} \\
&\sim H \cdot P \cdot (c_{2g+1} \cdots c_2 c_1)^{2g+2} \\
&\sim c_{2g} \cdots c_2 c_1 \cdot HP \cdot c_{2g+1} \cdot (c_{2g} \cdots c_2 c_1 c_{2g+1})^{2g+1} \\
&\sim HP \cdot {}_{(HP)^{-1}}(c_{2g} \cdots c_2 c_1) \cdot c_{2g+1} \cdot (c_{2g} \cdots c_2 c_1 c_{2g+1})^{2g+1} \\
&\sim HP \cdot (c_{2g} c_{2g+1} c_{2g-1} \cdots c_2 c_1)^{2g+2} \\
&\sim HP \cdot (c_{2g+1} \cdot {}_{c_{2g+1}^{-1}}(c_{2g}) \cdot c_{2g-1} \cdots c_2 c_1)^{2g+2} \\
&\sim HP \cdot {}_{c_{2g+1}^{-1}}(c_{2g+1} \cdots c_2 c_1)^{2g+2} \sim HP \cdot {}_{c_{2g+1}^{-1}} C_{\mathrm{I}} \\
&\sim HP \cdot {}_{c_{2g+1}^{-1}}(JH) \sim HP \cdot {}_{c_{2g+1}^{-1}} J \cdot {}_{c_{2g+1}^{-1}} H \\
&\sim H \cdot RH^{g-1} \cdot {}_{c_{2g+1}^{-1}} H \sim RH^{g-1} \cdot {}_{c_{2g+1}^{-1}} H \cdot H \\
&= RH^{g-1} \cdot c_{2g+1} \cdot {}_{c_{2g+1}^{-1}}(c_{2g}) \cdot c_{2g-1} \cdots c_2 c_1 \\
&\qquad \cdot c_1 c_2 \cdots c_{2g-1} \cdot {}_{c_{2g+1}^{-1}}(c_{2g}) \cdot c_{2g+1} \cdot H \\
&\sim RH^{g-1} \cdot c_{2g} c_{2g-1} \cdots c_2 c_1^2 c_2 \cdots c_{2g-1} c_{2g} c_{2g+1}^2 \cdot H \\
&\sim c_{2g+1} \cdot H \cdot RH^{g-1} \cdot c_{2g} \cdots c_2 c_1^2 c_2 \cdots c_{2g} c_{2g+1} \\
&\sim H \cdot RH^{g-1} \cdot {}_{(HRH^{g-1})^{-1}}(c_{2g+1}) \cdot c_{2g} \cdots c_2 c_1^2 c_2 \cdots c_{2g} c_{2g+1} \\
&= H \cdot RH^{g-1} \cdot H \sim RH^{g+1} \sim c_1 c_2 \cdots c_{2g+1} \cdot RH^g \cdot c_{2g+1} \cdots c_2 c_1 \\
&\sim R \cdot {}_{R^{-1}}(c_1 c_2 \cdots c_{2g+1}) \cdot H^g \cdot c_{2g+1} \cdots c_2 c_1 \\
&= RI^{g+1}
\end{aligned}$$

This completes the proof of Theorem 4.10.

References

[1] J. Amorós, F. Bogomolov, L. Katzarkov and T. Pantev, Symplectic Lefschetz fibrations with arbitrary fundamental groups, J. Diff. Geom., **54** (2000), 489–545.

[2] D. Auroux, Fiber sums of genus 2 Lefschetz fibrations, Turkish J. Math., **25** (2001), 1–10.

[3] J. Birman and H. Hilden, On mapping class groups of closed surfaces as covering spaces, Advances in the Theory of Riemann surfaces, Ann. Math. Stud., **66**, Princeton Univ. Press, 1971, pp. 81–115.

[4] C. Cadavid, A remarkable set of words in the mapping class group, Dissertation, Univ. of Texas, Austin, 1998.

[5] K. N. Chakiris, The monodromy of genus two pencils, Dissertation, Colombia Univ., 1978.

[6] S. K. Donaldson, Lefschetz pencils on symplectic manifolds, J. Differential Geom., **53** (1999), 205–236.

[7] H. Endo, Meyer's signature cocycle and hyperelliptic fibrations (with Appendix written by T. Terasoma), Math. Ann., **316** (2000), 237–257.

[8] H. Endo and S. Nagami, Signature of relations in mapping class groups and non-holomorphic Lefschetz fibrations, Trans. Amer. Math. Soc., **357** (2005), 3179–3199.

[9] R. Fintushel and R. J. Stern, Immersed spheres in 4-manifolds and the immersed Thom conjecture, Turkish J. Math., **19** (1995), 27–40.

[10] R. Fintushel and R. J. Stern, Rational blowdowns of smooth 4-manifolds, J. Differential Geom., **46** (1997), 181–235.

[11] T. Fuller, Diffeomorphism types of genus 2 Lefschetz fibrations, Math. Ann., **311** (1998), 163–176.

[12] T. Fuller, Hyperelliptic Lefschetz fibrations and branched covering spaces, Pacific J. Math., **196** (2000), 369–393.

[13] T. Fuller, Lefschetz fibrations of 4-dimensional manifolds, Cubo A Math. J., **5** (2003), 275–294.

[14] R. E. Gompf and A. I. Stipsicz, 4-manifolds and Kirby calculus, Grad. Stud. Math., **20**, Amer. Math. Soc., 1999.

[15] Y. Z. Gurtas, Positive Dehn twist expressions for some new involutions in mapping class group I, II, preprint, arXiv:math.GT/0404310, math.GT/0404311.

[16] Y. Z. Gurtas, Positive Dehn twist expressions for some elements of finite order in mapping class group, preprint, arXiv:math.GT/0501385.

[17] I. Hasegawa, A combinatorial proof for the signature of the Lefschetz fibration with Gurtas' monodromy, preprint, 2005.

[18] S. Hirose, Examples of non-minimal Lefschetz fibrations of genus 2, preprint, 2006.

[19] A. Kas, On the handlebody decomposition associated to a Lefschetz fibration, Pacific J. Math., **89** (1980), 89–104.

[20] V. M. Kharlamov and V. S. Kulikov, On braid monodromy factorizations, Izv. Math., **67** (2003), 499–534.

[21] M. Korkmaz, Noncomplex smooth 4-manifolds with Lefschetz fibrations, Internat. Math. Res. Notices, 2001, no. 3, 115–128.
[22] D. Kotschick, The Seiberg-Witten invariants of symplectic four-manifolds (after C. H. Taubes), Séminaire Bourbaki, 48ème annèe, 1995-96, n° 812, Astérisque, **241** (1997), 195–220.
[23] D. Kotschick, J. W. Morgan and C. H. Taubes, Four-manifolds without symplectic structures but with nontrivial Seiberg-Witten invariants, Math. Res. Letters, **2** (1995), 119–124.
[24] M. Matsumoto, A presentation of mapping class groups in terms of Artin groups and geometric monodromy of singularities, Math. Ann., **316** (2000), 401–418.
[25] Y. Matsumoto, Lefschetz fibrations of genus two – a topological approach, In: Topology and Teichmüller Spaces, Proceedings of the 37th Taniguchi Symposium, World Scientific, Singapore, 1996, pp. 123–148.
[26] Y. Matsumoto, Splitting of certain singular fibers of genus 2, Bol. Soc. Nat. Mexicana, **10** (2004), 331–355.
[27] T. Morifuji, On Meyer's function of hyperelliptic mapping class groups, J. Math. Soc. Japan, **55** (2003), 117–129.
[28] S. Morita, Characteristic classes of surface bundles, Invent. Math., **90** (1987), 551–577.
[29] B. Ozbagci, Signatures of Lefschetz fibrations, Pacific J. Math., **202** (2002), 99–118.
[30] Y. Sato, 2-spheres of square -1 and the geography of genus-2 Lefschetz fibrations, preprint, 2003.
[31] Y. Sato, A conjecture of Stipsicz – minimality of fiber sums of genus 2 Lefschetz fibrations, preprint (in Japanese), 2006.
[32] B. Siebert and G. Tian, On hyperelliptic C^∞-Lefschetz fibrations of four-manifolds, Commun. Contemp. Math., **1** (1999), 466–488.
[33] B. Siebert and G. Tian, On the holomorphicity of genus two Lefschetz fibrations, Ann. of Math. (2), **161** (2005), 959–1020.
[34] I. Smith, Lefschetz fibrations and the Hodge bundle, Geom. Topol., **3** (1999), 211–233.
[35] I. Smith, Lefschetz pencils and divisors in moduli space, Geom. Topol., **5** (2001), 579–608.
[36] I. Smith, Geometric monodromy and the hyperbolic disc, Quart. J. Math., **52** (2001), 217–228.
[37] A. I. Stipsicz, Indecomposability of certain Lefschetz fibrations, Proc. Amer. Math. Soc., **129** (2000), 1499–1502.
[38] C. H. Taubes, The Seiberg-Witten invariants and symplectic forms, Math. Res. Letters, **1** (1994), 809–822.
[39] C. H. Taubes, Counting pseudo-holomorphic submanifolds in dimension 4, J. Differential Geom., **44** (1996), 818–893.
[40] C. H. Taubes, The Seiberg-Witten and Gromov invariants, Math. Res. Letters, **2** (1995), 221–238.

[41] M. Usher, Minimality and symplectic sums, Internat. Math. Res. Notices, 2006, Article ID 49857, 1–17.
[42] B. Wajnryb, An elementary approach to the mapping class group of a surface, Geom. Topol., **3** (1999), 405–466.
[43] K.-H. Yun, On the signature of a Lefschetz fibration coming from an involution, Topology Appl., **153** (2006), 1994–2012.

Department of Mathematics
Graduate School of Science
Osaka University
Toyonaka, Osaka 560-0043
Japan

Advanced Studies in Pure Mathematics 52, 2008
Groups of Diffeomorphisms
pp. 283–296

Subgroups generated by two pseudo-Anosov elements in a mapping class group. I. Uniform exponential growth

Koji Fujiwara

Abstract.

Suppose G acts acylindrically by isometries on a δ-hyperbolic graph Γ. We discuss subgroups generated by two hyperbolic elements in G and give sufficient conditions for them to be free of rank two.

We apply our results to the mapping class group $\mathrm{Mod}(S)$ of a compact orientable surface S and its action on the curve graph such that S is non-sporadic. There exists a constant Q, depending only on S, with the following property. If $a, b \in \mathrm{Mod}(S)$ are pseudo-Anosovs such that $\langle a, b\rangle$ is not virtually cyclic, then there exists $M > 0$, which depends on a, b, such that either $\langle a^n, b^m\rangle$ is free of rank two for all $n \geq Q, m \geq M$, or $\langle a^m, b^n\rangle$ is free of rank two for all $n \geq Q, m \geq M$ (Theorem 3.1).

At the end we ask a question in connection to the uniformly exponential growth of subgroups in a mapping class group (Question 3.4).

§1. δ-hyperbolic geometry and the Nielsen condition

In the paper, we expect that the readers are familiar to δ-hyperbolic geometry. We give definitions and references, and describe the idea of the argument without all details for standard facts and techniques. We recommend [BrHa, III, H] as a good reference book.

1.1. Nielsen condition for free generators

A geodesic space is called *δ-hyperbolic* for $\delta \geq 0$ if for any geodesics α, β, γ which form a triangle, α is contained in the δ-neighborhood of $\beta \cup \gamma$ ([Gr]). Let Γ be a δ-hyperbolic graph. Let a be an isometry of Γ. If there exist a point $x \in \Gamma$ and a constant $C > 0$ such that $d(x, a^n(x)) \geq Cn$ for any $n > 0$, then a is called *hyperbolic*.

Received July 13, 2007.
Revised April 3, 2008.

Suppose a is a hyperbolic isometry. If there exists a bi-infinite geodesic α such that $a(\alpha)$ is contained in the C-neighborhood of α for some $C \geq 0$, α is called a *quasi-axis* of a. By δ-hyperbolicity of Γ, it then follows that $a(\alpha)$ is in the 2δ-neighborhood of α (use [BrHa, III.H.3.3 Lemma]). If α and β are quasi-axes of a (they are geodesics by definition), then they are contained in the 2δ-neighborhood of each other. If $C = 0$ we say α is an *axis* of a.

For two points $x, y \in \Gamma$, we may denote a (non-unique) geodesic joining them by $[x, y]$. We may write the distance between the two points as $|x - y|$. For an isometry a, we define its *translation length*, (or *stable length*), $\mathrm{tr}(a)$, by

$$\mathrm{tr}(a) = \lim_{n \to \infty} \frac{|x - a^n(x)|}{n} \geq 0$$

for a point x. It is easy to see $\mathrm{tr}(a)$ does not depend on the choice of x. Also, $\mathrm{tr}(a^n) = |n|\mathrm{tr}(a)$. The isometry a is hyperbolic iff $\mathrm{tr}(a) > 0$.

For any point $z \in \Gamma$, we have $|z - a(z)| \geq \mathrm{tr}(a)$ by the triangle inequality. If a has an axis α, then $|z - a(z)| = \mathrm{tr}(a)$ for any point $z \in \alpha$. If α is a quasi-axis of a, then $|z - a(z)| \leq \mathrm{tr}(a) + 10\delta$ (otherwise, use that $a^n(\alpha)$ is in the 2δ-neighborhood of α for any $n > 0$ and show $|z - a^n(z)| \geq n(\mathrm{tr}(a) + \delta)$, which gives a contradiction if $\delta > 0$). This last inequality will be used, sometimes implicitly since we do not always spell out all details, in the rest of the paper. We recommend interested readers who want to know all details, in particular estimates regarding δ, first to imagine $\delta = 0$ and/or quasi-axes are axes, then try to modify the estimates and the arguments.

Let $C \geq 0$ be a constant. For geodesics α and β, we define *the C-overlap*, denoted by $\alpha \cap_C \beta$, by

$$\alpha \cap_C \beta = (\alpha \cap N_C(\beta)) \cup (\beta \cap N_C(\alpha)),$$

where $N_C(\alpha)$ is the C-neighborhood of α. Let $|\alpha \cap_C \beta|$ denote the diameter of this set.

In the following argument, we often take $C = 10\delta$. By δ-hyperbolicity, if $|\alpha \cap_{10\delta} \beta|$ is finite, then the longest segment of α, the longest segment of β and the longest geodesics which are contained in $\alpha \cap_{10\delta} \beta$ all have length between $|\alpha \cap_{10\delta} \beta| - 20\delta$ and $|\alpha \cap_{10\delta} \beta| + 20\delta$, and those segments are in the 20δ-neighborhood of each other.

The following result is fundamental and classical. It has its origin in combinatorial group theory and was generalized in [Gr] to the setting of δ-hyperbolic geometry. It was used in [De] and [Ko] as an essential tool. For a proof, see [Ko, Lemma 2.4] or [Fu], where the convexity (see [FaMo]) of $\langle a^n, b^m \rangle$ is also discussed.

Proposition 1.1 (Nielsen condition). *Suppose a, b act as hyperbolic isometries on a δ-hyperbolic graph Γ with quasi-axes α, β, respectively. Suppose $|\alpha \cap_{10\delta} \beta| < \infty$. If $1 \le n, m \in \mathbb{Z}$ are such that*

$$\mathrm{tr}(a^n) \ge |\alpha \cap_{10\delta} \beta| + 100(\delta + 1),\ \mathrm{tr}(b^m) \ge |\alpha \cap_{10\delta} \beta| + 100(\delta + 1),$$

then a^n and b^m freely generate a rank-two free subgroup in $\mathrm{Isom}(\Gamma)$.

1.2. Quasi-geodesic axis

The existence of quasi-axes, which are geodesics by our definition, is not strictly necessary for Proposition 1.1 or for some other results in this paper. Indeed, we can use certain quasi-geodesics as discussed in section 2.2. As we do not use this for our main application to mapping class groups, uninterested readers may skip this section. First, a path α, parametrized by arclength, is called a (K, ε)-*quasi-geodesic* for $0 < K \le 1$ and $0 \le \varepsilon$ if for all t, s we have $K|t-s|-\varepsilon \le d(\alpha(t), \alpha(s))$. (A standard definition of quasi-geodesics also requires $d(\alpha(t), \alpha(s)) \le |t-s|/K + \varepsilon$ but this is trivially satisfied since α is parametrized by arclength.)

If a is a hyperbolic isometry of a δ-hyperbolic graph Γ, there exists a (K, ε)-quasi-geodesic α for some K, ε such that

(1) $a^n(\alpha)$ and α are in the 30δ-neighborhood of each other for any n. (Namely, α is almost invariant by a.)
(2) Let $p, q \in \alpha$. Then the subpath of α between p, q and any geodesic $[p, q]$ are in the 10δ-neighborhood of each other.

We call such a path α a *quasi-geodesic axis* of a in this paper. To be precise, we should use the term *quasi-geodesic quasi-axis*, but we make it shorter. One can easily show from (1) and (2) that any two quasi-geodesic axes of a are in the 30δ-neighborhood of each other. Note that (2) concerns only the path and not the element a. Also, the quasi-geodesic constants of α are not important for our purpose. What is useful for us is (2).

We briefly review how to find a quasi-geodesic axis for a. Fix a point $x \in \Gamma$. Choose $I > 0$ such that $\mathrm{tr}(a^I) \ge 1000\delta$. Set $y = a^I(x)$. Then, $|x - y| \ge 1000\delta$. Choose $N > 0$ such that $|a^N(x) - x| \ge 100|a^I(x) - x|$. Let m be the mid point of a geodesic $[x, a^N(x)]$. Define a path, which is invariant by a^I, by

$$\alpha = \cup_{n \in \mathbb{Z}} a^{nI}([m, a^I(m)]).$$

Since a is hyperbolic, α is a quasi-geodesic. Observe that α trivially satisfies (1) for the element a^I (not for the element a yet). We claim that α also satisfies (2). Firstly, by the way we chose I and N, most parts of the geodesics, except for a small portion near each end, $[x, a^N(x)]$

and $[y, a^N(y)] = a^I([x, a^N(x)])$ are in the 2δ-neighborhood of each other. This is because $|x - a^N(x)|$ is much larger than $|x - y|$. (Draw a geodesic rectangle joining $x, a^N(x), a^N(y)$ and y in this order. Then the rectangle is narrow.) Also, $|m - a^I(m)| \geq \mathrm{tr}(a^I) \geq 1000\delta$. Those two estimates imply (2) for α by δ-hyperbolic geometry. Note that $|m - a^I(m)|$ is much smaller than $|x - m| = |m - a^N(x)|$. (See Figure 1.)

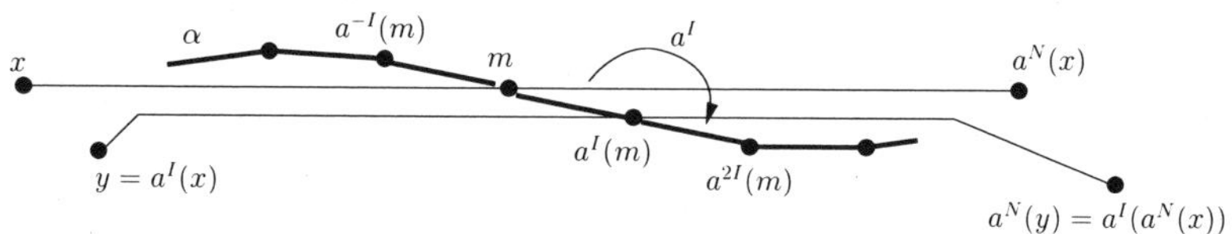

Fig. 1. Two geodesics $[x, a^N(x)]$ and $[y, a^N(y)]$ stay 2δ- close for the most part. The geodesic $[m, a^I(m)]$ is "pinched" between those two geodesics.

Now we claim that α satisfies (1) for the element a, namely, it is a quasi-geodesic axis for a. Let $L > 0$ be a large integer, which we decide later. By (2), the geodesic $[m, a^{LI}(m)]$ and the subpath α' of α between the two points m and $a^{LI}(m)$ are in the 10δ-neighborhood of each other. Therefore, the geodesic $a([m, a^{LI}(m)])$ and $a(\alpha')$ are also in the 10δ-neighborhood of each other. On the other hand, by δ-hyperbolic geometry, most parts of $[m, a^{LI}(m)]$ and $a([m, a^{LI}(m)])$ are in the 2δ-neighborhood of each other if $|m - a^{LI}(m)|$ is much larger than $|m - a(m)|$. (This follows from the same argument using a narrow geodesic rectangle as before.) We choose L this way. As a consequence, most parts of α' and $a(\alpha')$, except for a small segment at each end, are in the 30δ-neighborhood of each other. Replacing L by a larger integer, we find that α and $a(\alpha)$ are in the 30δ-neighborhood of each other. A similar argument works for $a^n(\alpha)$ for all n. This proves that α satisfies (1) for a.

§2. Acylindricity and free subgroups

2.1. Free subgroups

Suppose G acts on Γ. Bowditch [Bo] defined that the action of G is *acylindrical* if for any $R > 0$, there exist $K(R), L(R) \geq 1$ such that for any vertices $x, y \in \Gamma$ with $d(x, y) \geq L$, the following set has at most K elements.

$$\{g \in G | d(x, g(x)) \leq R, d(y, g(y)) \leq R\}.$$

We show one lemma (see Lemma 2.5).

Lemma 2.1. *Suppose G acts on a δ-hyperbolic graph Γ. If the action is acylindrical, then there exists an integer $P \geq 1$ such that $\mathrm{tr}(a^P) \geq 1$ for any element $a \in G$ which acts hyperbolically on Γ with a quasi-axis.*

Proof. If $\delta = 0$, then Γ is a tree. Therefore $\mathrm{tr}(a) \geq 1$. Set $P = 1$. Suppose $\delta > 0$. Fix a constant $R \geq 100\delta$. Let α be a (geodesic) quasi-axis of a. Take a point $x \in \alpha$. Let $y \in \alpha$ be a point with $|x - y| \geq L(2R)$. If $|a^i(x) - x| \leq R$ for some i, then $|a^i(y) - y| \leq 2R$. This is because $a^i(\alpha)$ is in the 2δ-neighborhood of α. Therefore, by the acylindricity, there is some I, with $1 \leq I \leq K(2R)$, such that $|a^I(x) - x| > R$. Since x lies on a quasi-axis of a, it then follows that $|a^{In}(x) - x| > n(R - 10\delta)$ for any $n \geq 1$. To verify this estimate, imagine first that the geodesic α is exactly invariant by a, namely, an axis. Then, clearly, $|a^{In}(x) - x| > nR$. Now, try to estimate the error terms using that α is only a quasi-axis. We leave the details to readers.

This implies that

$$\mathrm{tr}(a) \geq \frac{R - 10\delta}{I} \geq \frac{R - 10\delta}{K(2R)}.$$

Take P such that $P \geq \frac{K(2R)}{R-10\delta}$. Q.E.D.

The following lemma is a generalization of a result by Koubi ([Ko, Lemma 5.4]). He discusses the case such that the action of G is (uniformly) proper and $\mathrm{tr}(a) = \mathrm{tr}(b)$. The commutator of two elements is defined by

$$[f, g] = f^{-1}g^{-1}fg.$$

Lemma 2.2. *Let Γ be a δ-hyperbolic graph and G a group acting acylindrically on Γ with constants $K(R), L(R)$. Suppose $a, b \in G$ act hyperbolically with quasi-axes $\alpha, \beta \subset \Gamma$, respectively.*

(1) *If $a^n b \neq b a^n$ for all $n \neq 0$ or $b^n a \neq a b^n$ for all $n \neq 0$, then*

$$|\alpha \cap_{10\delta} \beta| < 4PK(20\delta)L(20\delta)\max(\mathrm{tr}(a), \mathrm{tr}(b)) + 100\delta,$$

where P is the constant from Lemma 2.1.

(2) *If $\mathrm{tr}(a) = \mathrm{tr}(b)$ and for all $n \neq 0$, $a^n \neq b^{\pm n}$, then we have the same inequality as above.*

Proof. 1. To argue by contradiction, assume that the inequality is false. It suffices to show that $a^n b = b a^n$ for some $n \neq 0$ and also $b^n a = a b^n$ for some $n \neq 0$. Set $K = K(20\delta), L = L(20\delta)$. For concreteness, suppose $\mathrm{tr}(b) \leq \mathrm{tr}(a)$. By our assumption, since $|\alpha \cap_{10\delta} \beta|$ is much larger than 2δ, the set $\alpha \cap_{10\delta} \beta$ looks like a narrow tube.

Let $\ell \subset \alpha$ be the longest segment which is contained in $\alpha \cap_{10\delta} \beta$. Then, by our assumption, $|\ell| \geq 4PKL\mathrm{tr}(a) + 80\delta$. Take a point $p \in \ell$ such that the following points are in $N_{2\delta}(\ell)$:

$$p, a(p), a^2(p), \cdots, a^{4PKL}(p).$$

(See Figure 2. For simplicity, we put those points on ℓ in the figure.) To see that we can take such a point p, as usual, first imagine that α is invariant by a. Then most parts of ℓ and $a(\ell)$ coincide, therefore, one can take p such that all the above points are on ℓ. Now, in general, most parts of ℓ and $a(\ell)$ are in the 2δ-neighborhood of each other by δ-hyperbolic geometry, hence a required point p exists.

Set

$$x = a^{PKL}(p),\ y = a^{2PKL}(p).$$

Since $y = a^{PKL}(x)$, it follows from Lemma 2.1 that $d(x, y) \geq PKL\mathrm{tr}(a) \geq KL \geq L$.

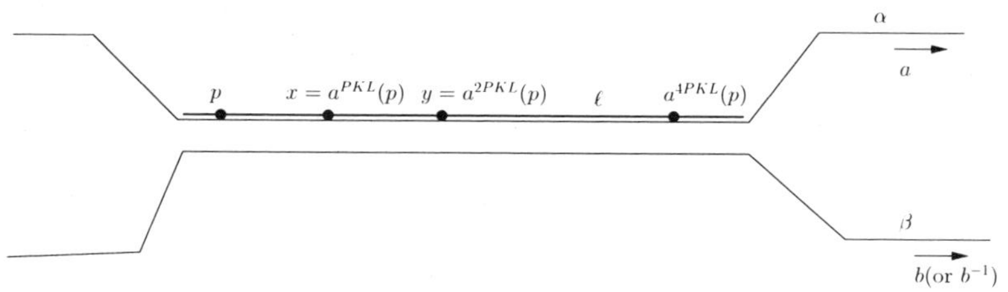

Fig. 2. Apply the acylindricity to the pair x, y.

Claim. For each $i, (1 \leq i \leq PKL)$,

$$d(x, [b, a^i](x)) \leq 20\delta,\ d(y, [b, a^i](y)) \leq 20\delta.$$

We first consider the special case that $\delta = 0$, namely, Γ is a tree. Then, $\alpha \cap_{10\delta} \beta$ coincides the segment $\alpha \cap \beta$, and also the segment ℓ, therefore, all above points $a^n(p), 1 \leq n \leq 4PKL$, are in $\alpha \cap \beta$. We want to show $x = [b, a^i](x)$, but this is obvious since when we apply a^i, b, a^{-i}, then b^{-1} to x, the point moves within ℓ. Thus, $[b, a^i](x) = x$. If $\delta > 0$, we can show that the point moves in the 10δ-neighborhood of ℓ when we apply a^i, b, a^{-i} followed by b^{-1} to x. Therefore, we get $d(x, [b, a^i](x)) \leq 20\delta$ by estimating the error terms from the tree case using triangle inequality. We leave the details to readers. (See Figure 3.) We can show $d(y, [b, a^i](y)) \leq 20\delta$ in the same way. This proves the claim.

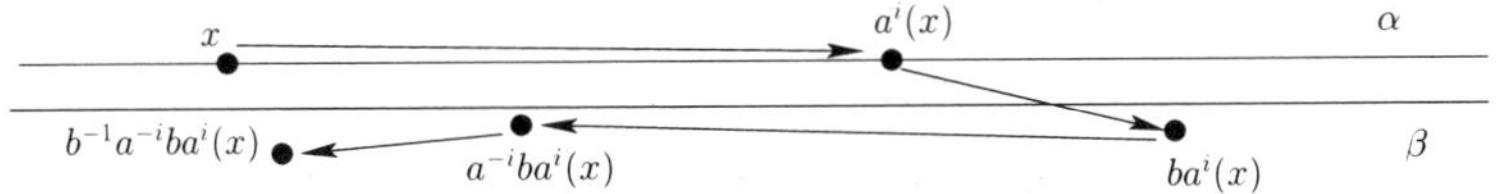

Fig. 3. How commutators act near $\alpha \cap_{10\delta} \beta$.

Since $|x-y| \geq L = L(20\delta)$, by the acylindricity of the action, it follows from the claim that there are at most K distinct elements in the set $[b,a^i], (1 \leq i \leq PKL)$. By the pigeon-hole principle, $[b,a^i]=[b,a^j]$ for some $i \neq j, (1 \leq i,j \leq PKL)$. It follows that $a^i b^{-1} a^{-i} = a^j b^{-1} a^{-j}$, therefore, $a^{i-j}b^{-1} = b^{-1}a^{i-j}$. We get $[b,a^n]=1$ for some $n \neq 0$. The same argument applies to the elements $[a,b^i]$ since $\mathrm{tr}(b) \leq \mathrm{tr}(a)$, therefore we also get $[a,b^n]=1$ for some $n \neq 0$ as well.

2. This is similar to 1. Assume that the inequality is false. Take a segment $\ell \subset \alpha$ as before. Then, as we said, the set $\alpha \cap_{10\delta} \beta$ looks like a narrow tube. Therefore, it makes sense to talk about the direction of the action by a and b along this tube, and furthermore, the direction of a coincides with the direction of one of b or b^{-1}. We did not need this consideration in 1 since we used commutators. Now, take points $p,x,y \in \ell$ as before and apply the same argument to the set of elements $\{b^n a^n : 1 \leq n \leq PKL\}$ (if the action of a,b along ℓ have the opposite direction) or $\{b^{-n}a^n : 1 \leq n \leq PKL\}$ (if the actions have the same direction). Then we conclude that there must be $n \neq m$ with $b^n a^n = b^m a^m$, or $b^{-n}a^n = b^{-m}a^m$, respectively. This is a contradiction. Q.E.D.

As a consequence, a generalization of [Ko, Proposition 5.5] follows.

Proposition 2.3. *Let Γ be a δ-hyperbolic graph, and G a group acting acylindrically by isometries on Γ. Then there exists a constant $N>0$, which depends on δ and the set of the acylindricity constants $K(R), L(R)$, with the following property.*

(1) *Suppose $a,b \in G$ act by hyperbolic isometries with quasi-axes such that $a^n b \neq ba^n$ for all $n \neq 0$, or $b^n a \neq ab^n$ for all $n \neq 0$. Then there exists a constant $M>0$, which depends on a,b, such that either $\langle a^n, b^m \rangle$ is free of rank two for all $n \geq N, m \geq M$, or $\langle a^m, b^n \rangle$ is a free group of rank two for all $n \geq N, m \geq M$.*

(2) *Suppose $a \in G$ acts by a hyperbolic isometry with a quasi-axis. Let $c \in G$ be such that for any $n>0$, $ca^n c^{-1} \neq a^{\pm n}$. Then the subgroup $\langle a^n, ca^m c^{-1} \rangle$ is free of rank two for all $n,m \geq N$.*

Proof. Let $K = K(20\delta), L = L(20\delta)$ be the acylindricity constants evaluated at 20δ. Let P be the constant from Lemma 2.1. Set

$$N = 4PKL + 200P(\delta + 1).$$

1. Let α, β be quasi-axes of a, b, respectively. Suppose $\mathrm{tr}(b) \leq \mathrm{tr}(a)$. Set $M = N\frac{\mathrm{tr}(a)}{\mathrm{tr}(b)}$. We will show that a^n, b^m generate a free group of rank two if $n \geq N, m \geq M$.

By Lemma 2.2 (1), we have

$$|\alpha \cap_{10\delta} \beta| < 4PKL\mathrm{tr}(a) + 100\delta.$$

It follows that if $n \geq N, m \geq M$ then

$$\mathrm{tr}(a^n) \geq |\alpha \cap_{10\delta} \beta| + 100(\delta + 1),\ \mathrm{tr}(b^m) \geq |\alpha \cap_{10\delta} \beta| + 100(\delta + 1)$$

by the way we chose N, M, P. Note that the first inequality follows from our assumption and

$$\begin{aligned} \mathrm{tr}(a^n) \geq N\mathrm{tr}(a) &= 4PKL\mathrm{tr}(a) + 200P(\delta + 1)\mathrm{tr}(a) \\ &\geq 4PKL\mathrm{tr}(a) + 200(\delta + 1). \end{aligned}$$

The second inequality also holds by the way we chose M. Now we can apply Proposition 1.1 to a^n, b^m. If $\mathrm{tr}(b) \geq \mathrm{tr}(a)$, then set $M = N\frac{\mathrm{tr}(b)}{\mathrm{tr}(a)}$ and argue in the same way switching the roles of a and b.
2. Let α be a quasi-axis of a. Put $b = cac^{-1}$. Then $\beta = c\alpha$ is a quasi-axis of b and $\mathrm{tr}(a) = \mathrm{tr}(b)$. By the hypothesis, for all $n \neq 0$, $a^n \neq b^{\pm n}$. By Lemma 2.2 (2), we have

$$|\alpha \cap_{10\delta} \beta| < 4PKL\mathrm{tr}(a) + 100\delta.$$

We then argue in the same way as the previous case. Since $\mathrm{tr}(a) = \mathrm{tr}(b)$, $M = N$ in the previous argument, therefore, the subgroup generated by a^n, b^m is free of rank two for all $n, m \geq N$. Q.E.D.

It is more difficult to analyze a subgroup normally generated by two, or even one, elements. See an interesting paper by T. Delzant [De] which contains positive results and a warning example.

2.2. Hyperbolic isometries without quasi-axes

Although we have been assuming that hyperbolic isometries have quasi-axes, this assumption is not necessary for our purpose if we only assume the acylindricity of the action. In this section, we state results without assuming quasi-axes for potential application in the future since the existence of quasi-axes is a strong assumption. Uninterested readers may skip this section as it is unnecessary for our application to mapping class groups.

Our main goal is to drop the assumption on quasi-axes from Proposition 2.3 as follows.

Proposition 2.4. *Suppose that G acts acylindrically by isometries on a δ-hyperbolic graph Γ. Then there exists a constant $N > 0$, which depends on δ and the set of the acylindricity constants $K(R), L(R)$, such that the conclusion of Proposition* 2.3 (1), (2) *holds without the assumption that a, b have quasi-axes.* (*This constant N is maybe larger than the one which is obtained in Proposition* 2.3).

Proof. A hyperbolic isometry always has a quasi-geodesic axis (see Section 1.2). Basically, we use quasi-geodesic axes instead of quasi-axes for hyperbolic isometries. We then adapt the original argument to the new setting. We may first need to modify the statements since some of the original ones concern quasi-axes. Namely, we restate and prove Proposition 1.1, prove Lemma 2.1, restate and prove Lemma 2.2, and finally modify the proof of Proposition 2.3.

We only outline the arguments. As for Proposition 1.1, replace quasi-axes α, β by quasi-geodesic axes α, β and also $\alpha \cap_{10\delta} \beta$ by $\alpha \cap_{1000\delta} \beta$ in all places in the statement. We did not give a proof of Proposition 1.1 and only referred to [Ko] and [Fu]. The arguments there work with minor modification using the properties (1) and (2) of the quasi-geodesic axes.

We reprove Lemma 2.1 without using quasi-axes. This result says that the existence of such P is a consequence of the acylindricity.

Lemma 2.5. *Suppose G acts by isometries on a δ-hyperbolic graph Γ. If the action is acylindrical, then there exists an integer $P \geq 1$ such that for any element $a \in G$ which acts hyperbolically on Γ, we have* $\mathrm{tr}(a^P) \geq 1$.

The proof is essentially same as for Lemma 2.1. We use a quasi-geodesic axis α instead of a quasi-axis. The key property is (2), which says that a sub-path of α is at most 10δ-close to a geodesic with the common end points with the sub-path.

Proof. We may assume $\delta > 0$. Let α be a quasi-geodesic axis for a. Set $R = 1000\delta$. Take a point $x \in \alpha$, and let $y \in \alpha$ be a point with $|x - y| \geq L(1500\delta)$. Since α is a quasi-geodesic axis, it follows that if $|a^i(x) - x| \leq R$, then $|a^i(y) - y| \leq R + 500\delta = 1500\delta$. Therefore, by acylindricity (look at the set of elements $a^i, 1 \leq i \leq K(1500\delta)$, and a pair of points x, y), there exists some I, with $1 \leq I \leq K(1500\delta)$, such that $|a^I(x) - x| > 1000\delta = R$. From this, since again α is a quasi-geodesic axis, for any $n \geq 1$, $|a^{In}(x) - x| > 500\delta n$. Now, as before, we get

$$\mathrm{tr}(a) \geq \frac{500\delta}{K(1500\delta)}.$$

Take $P \geq \frac{K(1500\delta)}{500\delta}$. Q.E.D.

Next, we modify Lemma 2.2. Namely, we replace quasi-axes α, β for a, b by quasi-geodesic axes in the assumption. The inequality in the conclusion should be replaced, for example, by

$$|\alpha \cap_{1000\delta} \beta| < 4PK(20\delta)L(20\delta)\max(\mathrm{tr}(a), \mathrm{tr}(b)) + 10000\delta,$$

where P is the constant from Lemma 2.5. The proof is same after an appropriate modification regarding constants.

Having done all this, the argument for the proposition is same as for Proposition 2.3 with minor modifications. For example, replace all $\alpha \cap_{10\delta} \beta$ by $\alpha \cap_{1000\delta} \beta$ and define the constant $N = 4PKL + 20000P(\delta+1)$, where, P is the constant from Lemma 2.5. We omit details. Q.E.D.

§3. Application to mapping class groups

3.1. Two pseudo-Anosov maps

Let S be a compact orientable surface. Following [MM], we call S *sporadic* when S is a sphere with $p \leq 4$ punctures or a torus with $p \leq 1$ puncture. Mapping class groups $\mathrm{Mod}(S)$ are already well-understood in this case. Namely (see [Iv, 9.2]), $\mathrm{Mod}(S)$ is isomorphic to $\mathrm{SL}(2, \mathbb{Z})$ if S is a torus with ≤ 1 puncture; $\mathrm{Mod}(S)$ is commensurable with $\mathrm{PSL}(2, \mathbb{Z})$ when S is a sphere with 4 punctures; and $\mathrm{Mod}(S)$ is finite when S is a sphere with ≤ 3 punctures.

The following result is well-known (for example, see [Iv1]) except for a uniform bound on one of the exponents by Q.

Theorem 3.1. [1]

Let S be a compact orientable surface and $\mathrm{Mod}(S)$ *its mapping class group such that S is non-sporadic. Then there exists a constant $Q(S)$ with the following property.*

(1) *Suppose $a, b \in \mathrm{Mod}(S)$ are pseudo-Anosovs such that for all $n, m \neq 0$, $a^n b^m \neq b^m a^n$. Then the subgroup $\langle a^n, b^m \rangle$ is free of rank two*

[1] In [Fu], this result will be improved such that in (1), $\langle a^n, b^m \rangle$ is free of rank two for all $n, m \geq Q(S)$, where we may need to take a larger constant for $Q(S)$ than in Theorem 3.1.

for all sufficiently large $m > 0$ and all $n \geq Q$, or $\langle a^m, b^n \rangle$ is free of rank two for all sufficiently large $m > 0$ and all $n \geq Q$.

(2) *Suppose $a \in \mathrm{Mod}(S)$ is pseudo-Anosov and $c \in \mathrm{Mod}(S)$ is such that for any $n > 0$, $ca^nc^{-1} \neq a^{\pm n}$. Then the subgroup $\langle a^n, ca^mc^{-1} \rangle$ is free of rank two for all $n, m \geq Q$.*

Regarding the assumption on two pseudo-Anosov elements a, b in (1), it is equivalent to requiring that the subgroup generated by a, b is not *virtually* cyclic, namely, it does not contain a cyclic subgroup of finite index (see, for example, [Iv, Theorem 7.4.I]).

Proof. Let $C(S)$ be the curve graph of S (see [Iv]). $\mathrm{Mod}(S)$ acts on $C(S)$ by isometries. It is known that $C(S)$ is δ-hyperbolic, and an element $a \in \mathrm{Mod}(S)$ acts as a hyperbolic isometry if and only if it is pseudo-Anosov [MM, Theorem 1.1 and Proposition 4.6]. The action of $\mathrm{Mod}(S)$ is acylindrical, and there is a quasi-axis for a pseudo-Anosov element [Bo, Theorem 1.3 and 1.4].

Apply Proposition 2.3 to the action of $\mathrm{Mod}(S)$ on $C(S)$ and let N be the constant from the proposition. Set $Q(S) = N$.

1. We apply Proposition 2.3 (1) to a, b and obtain M. Then the claim holds for all $m \geq M$ and $n \geq Q$.

2. Apply Proposition 2.3 (2) to a and c. The claim holds for all $n, m \geq Q$. Q.E.D.

We remark that the existence of P in Lemma 2.1 for this setting was already known in [MM] before [Bo].

3.2. Exponential growth rate

Definition 3.2 (Growth rate (see [Har])). Let G be a finitely generated group. For a finite generating set A, and for an integer $n \geq 0$, let $b(G, A; n)$ be the number of elements in G whose word length in terms of A are at most n.

The *exponential growth rate* of (G, A), $\omega(G, A)$, is defined as

$$\omega(G, A) = \limsup_{n \to \infty} (b(G, A; n))^{\frac{1}{n}}.$$

The group G is said to be of *exponential growth* if $\omega(G, A) > 1$ for some (and consequently all) A.

The *minimal growth rate* of a finitely generated group G of exponential growth is defined to be

$$\omega(G) = \inf \omega(G, A),$$

where the infimum is taken over all finite generating sets A. G is said to be of *uniformly exponential growth* if $\omega(G) > 1$.

Let F_k be a free group of rank $k > 1$. By computation, $\omega(F_k, A_k) = 2k - 1$ for a free generating set A_k. It is a non-trivial fact ([Har, Proposition 13]) that

$$\omega(F_k) = 2k - 1.$$

If G is a non-elementary word-hyperbolic group, then G contains a free group of rank two as a subgroup, therefore clearly G has exponential growth. Moreover, Koubi [Ko] showed that G has uniformly exponential growth. The key result in his argument is Proposition 5.5 [Ko], which we generalized in Proposition 2.3. In his case, Γ is a Cayley graph of G, therefore the action is proper, in particular, acylindrical.

Corollary 3.3. *Let S be a compact orientable surface and* $\mathrm{Mod}(S)$ *its mapping class group such that S is non-sporadic. Let $Q(S) > 0$ be the constant from Theorem 3.1. Suppose $G <$ $\mathrm{Mod}(S)$ is a finitely generated subgroup which is not virtually cyclic.*

Let Σ be a finite generating set of G. If Σ contains a pseudo-Anosov element a, then there is an element $c \in \Sigma$ such that the subgroup $\langle a^n, ca^m c^{-1}\rangle$ is free of rank two for all $n, m \geq Q$. In particular,

$$\omega(G, \Sigma) \geq 3^{\frac{1}{Q+2}}.$$

Proof. There must be an element $c \in \Sigma$ such that $ca^n c^{-1} \neq a^{\pm n}$ for all $n > 0$, since otherwise, Σ would generate a virtually cyclic group. Apply Theorem 3.1 (2) to a and c. Then, $\langle a^Q, ca^Q c^{-1}\rangle$ is free of rank two. The word length of the two elements $a^Q, ca^Q c^{-1}$ in terms of Σ is at most $Q+2$. Thus, the inequality on $\omega(G, \Sigma)$ follows from $\omega(F_2, A_2) = 3$, where A_2 is a free generating set. Q.E.D.

It is not clear if every finitely generated, non-virtually-abelian subgroup $G <$ $\mathrm{Mod}(S)$ has uniformly exponential growth. [2] We ask the following question.

Question 3.4 (Short pseudo-Anosov elements)**.** Let $G <$ $\mathrm{Mod}(S)$ be a finitely generated, non-virtually-cyclic subgroup which contains a pseudo-Anosov element. Does there exist a constant U which depends only on G, or even just on the surface S, such that if Σ is a finite generating set of G, then there is a pseudo-Anosov element $a \in G$ whose word length in terms of Σ is at most U ?

A positive answer to this question would imply $\omega(G) \geq 3^{\frac{1}{UQ+2}}$ by Corollary 3.3. It seems the answer is not known even for $G =$ $\mathrm{Mod}(S)$,

[2]After this paper, an affirmative answer is announced by J. Mangahas in December 2007, using Theorem 3.1(2) combined with her result concerning non pseudo-Anosov generators. The question 3.4 seems to be still open.

although Mod(S) has uniformly exponential growth if Mod(S) is not virtually abelian [AAS]. This is because Mod(S) has a surjective homomorphism to $\mathrm{Aut}(H_1(\pi_1(S)), \mathbb{Z})$), which has uniformly exponential growth as it is linear and not virtually nilpotent [EMO]. The kernel of this homomorphism is called the *Torelli* group. We do not know if this subgroup has uniformly exponential growth (see the footnote 2).

Acknowledgment This work was done while the author was visiting H. Short at Université de Provence during March 2007 and attending the conference "Groups 007" at CIRM in Marseille. He appreciates their hospitality. He would like to thank G. Besson, G. Courtois and B. Farb for discussion during the conference. He is grateful to the referee for the quick and professional job, who read first manuscripts very carefully and made numbers of useful comments and suggestions, which significantly improved the paper.

References

[AAS] J. W. Anderson, J. Aramayona and K. J. Shackleton, Uniformly exponential growth and mapping class groups of surfaces, In the tradition of Ahlfors-Bers. IV, 1–6, Contemp. Math., **432**, Amer. Math. Soc., 2007.

[Bo] B. Bowditch, Tight geodesics in the curve complex, Invent. Math., **171** (2008), 281–300.

[BrHa] M. R. Bridson and A. Haefliger, Metric spaces of non-positive curvature, Grundlehren Math. Wiss., **319**, Springer-Verlag, 1999.

[De] T. Delzant, Sous-groupes distingués et quotients des groupes hyperboliques, Duke Math. J., **83** (1996), 661–682.

[EMO] A. Eskin, S. Mozes and H. Oh, On uniform exponential growth for linear groups, Invent. Math., **160** (2005), 1–30.

[FaMo] B. Farb and L. Mosher, Convex cocompact subgroups of mapping class groups, Geom. Topol., **6** (2002), 91–152.

[Fu] K. Fujiwara, Subgroups generated by two pseudo-Anosov elements in a mapping class group. II. Uniform bound on exponents, in preparation.

[Gr] M. Gromov, Hyperbolic groups, In: Essays in group theory, Math. Sci. Res. Inst. Publ., **8**, Springer, New York, 1987, pp. 75–263.

[Har] P. de la Harpe, Topics in geometric group theory. Chicago Lectures in Math., Univ. of Chicago Press, 2000.

[Har2] P. de la Harpe, Uniform growth in groups of exponential growth, Geom. Dedicata, **95** (2002), 1–17.

[Iv1] N. V. Ivanov, Subgroups of Teichmüller modular groups, Transl. Math. Monogr., **115**, Amer. Math. Soc., 1992.

[Iv] N. V. Ivanov, Mapping class groups, In: Handbook of Geometric Topology, (eds. R. Daverman and R. Sher), North-Holland, 2002, pp. 523–633.

[Ko] M. Koubi, Croissance uniforme dans les groupes hyperboliques. Ann. Inst. Fourier (Grenoble), **48** (1998), 1441–1453.

[MM] H. Masur and Y. Minsky, Geometry of the complex of curves. I. Hyperbolicity, Invent. Math., **138** (1999), 103–149.

Graduate School of Information Science
Tohoku University
Sendai 980-8579
Japan
E-mail address: `fujiwara@math.is.tohoku.ac.jp`

Advanced Studies in Pure Mathematics 52, 2008
Groups of Diffeomorphisms
pp. 297–308

Differential characters and the Steenrod squares

Kiyonori Gomi

Abstract.

The groups of differential characters of Cheeger and Simons admit a natural multiplicative structure. The map given by the squares of degree $2k$ differential characters reduces to a homomorphism of ordinary cohomology groups. We prove that the homomorphism factors through the Steenrod squaring operation of degree $2k$. A simple application shows that five-dimensional Chern-Simons theory for pairs of B-fields is $SL(2, \mathbb{Z})$-invariant on spin manifolds.

§1. Introduction

The group of *Differential characters*, introduced by Cheeger and Simons [1], is a certain refinement of ordinary cohomology involving information of differential forms. From the beginning, differential characters enjoy numerous applications to geometry, topology and mathematical physics.

We recall the definition here: let X be a smooth manifold. We denote the group of singular p-chains with coefficients in $\mathbb{Z}$ by $C_p(X) = C_p(X; \mathbb{Z})$, and singular p-cycles by $Z_p(X) \subset C_p(X)$. A *differential character* of degree ℓ is defined to be a homomorphism $\chi : Z_\ell(X) \to \mathbb{R}/\mathbb{Z}$ such that there exists a differential $(\ell+1)$-form ω satisfying $\chi(\partial\tau) = \int_\tau \omega$ mod $\mathbb{Z}$ for all $\tau \in C_{\ell+1}(X)$. The group of differential characters of degree ℓ is denoted by $\widehat{H}^\ell(X, \mathbb{R}/\mathbb{Z})$.

It is known [1] that there is a bilinear map on differential characters:

$$\cup : \widehat{H}^{\ell_1}(X, \mathbb{R}/\mathbb{Z}) \times \widehat{H}^{\ell_2}(X, \mathbb{R}/\mathbb{Z}) \longrightarrow \widehat{H}^{\ell_1+\ell_2+1}(X, \mathbb{R}/\mathbb{Z}).$$

The bilinear map $\cup$, which we call the *cup product*, is associative. The cup product is also graded commutative in the sense that $\chi_1 \cup \chi_2 = (-1)^{(\ell_1+1)(\ell_2+1)} \chi_2 \cup \chi_1$ for $\chi_i \in \widehat{H}^{\ell_i}(X, \mathbb{R}/\mathbb{Z})$.

Received March 31, 2007.
Revised February 25, 2008.

Now we consider the quadratic map

$$\hat{q}^\ell : \widehat{H}^\ell(X, \mathbb{R}/\mathbb{Z}) \longrightarrow \widehat{H}^{2\ell+1}(X, \mathbb{R}/\mathbb{Z})$$

defined by $\hat{q}^\ell(\chi) = \chi \cup \chi$. If we take an even integer $\ell = 2k$, then $\hat{q}^{2k}$ gives rise to a homomorphism by the graded commutativity. Moreover, $\hat{q}^{2k}$ descends to give the following homomorphism of ordinary cohomology (see Lemma 3.8):

$$q^{2k} : H^{2k+1}(X; \mathbb{Z}) \otimes \mathbb{Z}_2 \longrightarrow \mathrm{Hom}(H_{4k+1}(X; \mathbb{Z}), \mathbb{Z}_2).$$

In the case of $k = 0$ and $X = S^1$, an explicit formula of the cup product in [1] leads to $q^0 = 1$, so that q^{2k} is non-trivial in general.

The homomorphism q^2 has a relationship with the 5-dimensional topological field theory with Chern-Simons action studied by Witten in [10]. To be more precise, we assume for a moment that X is a compact oriented 5-dimensional manifold. We take $\mathcal{B}_{\mathrm{RR}} \times \mathcal{B}_{\mathrm{NS}} = \widehat{H}^2(X, \mathbb{R}/\mathbb{Z}) \times \widehat{H}^2(X, \mathbb{R}/\mathbb{Z})$ to be the space of fields (modulo gauge transformation), and consider the action functional $I_{\mathrm{CS}} : \mathcal{B}_{\mathrm{RR}} \times \mathcal{B}_{\mathrm{NS}} \to \mathbb{R}/\mathbb{Z}$ given by $I_{\mathrm{CS}}(B_{\mathrm{RR}}, B_{\mathrm{NS}}) = -(B_{\mathrm{RR}} \cup B_{\mathrm{NS}})(X)$. Notice that $SL(2, \mathbb{Z})$ acts on $\mathcal{B}_{\mathrm{RR}} \times \mathcal{B}_{\mathrm{NS}}$ in the standard manner, however, I_{CS} is not generally invariant under the action. This is the point q^2 appears: I_{CS} has the $SL(2, \mathbb{Z})$-invariance if and only if $q^2 = 0$. In the case where X is the direct product of a 4-dimensional spin manifold and S^1, it is shown [10] that I_{CS} acquires the $SL(2, \mathbb{Z})$-invariance on a certain subgroup in $\mathcal{B}_{\mathrm{RR}} \times \mathcal{B}_{\mathrm{NS}}$.

In the present paper, we prove generally that the homomorphism q^{2k} factors through the *Steenrod squaring operation* ([9], see [7, 8]):

$$\mathrm{Sq}^{2k} : H^{2k+1}(X; \mathbb{Z}_2) \longrightarrow H^{4k+1}(X; \mathbb{Z}_2).$$

Let ι and π be the homomorphisms in the universal coefficients theorems:

$$0 \to H^{2k+1}(X; \mathbb{Z}) \otimes \mathbb{Z}_2 \xrightarrow{\iota} H^{2k+1}(X; \mathbb{Z}_2) \to \mathrm{Tor}(H^{2k+2}(X; \mathbb{Z}), \mathbb{Z}_2) \to 0,$$

$$0 \to \mathrm{Ext}(H_{4k}(X; \mathbb{Z}), \mathbb{Z}_2) \to H^{4k+1}(X; \mathbb{Z}_2) \xrightarrow{\pi} \mathrm{Hom}(H_{4k+1}(X; \mathbb{Z}), \mathbb{Z}_2) \to 0.$$

The main result of this paper is:

Theorem 1. $q^{2k} = \pi \circ \mathrm{Sq}^{2k} \circ \iota$.

As a consequence, we can immediately see that the Chern-Simons action I_{CS} has the $SL(2, \mathbb{Z})$-invariance for a compact 5-dimensional spin manifold X.

As the group of differential characters refines ordinary cohomology theory, the notion of *generalized differential cohomology* [6] refines generalized cohomology theory involving information of differential forms.

For differential (twisted) K-theory, a quantity similar to q^{2k} is shown to be generally non-trivial [5]. It may be interesting to establish counterparts of Theorem 1 in generalized differential cohomology theories, which should be answered in a future work.

The present paper is organized as follows. In Section 2, we review the cup product on ordinary cohomology and the Steenrod squaring operations. We also review homotopies between the cup product on cohomology with coefficients in $\mathbb{R}$ and the wedge product of differential forms. Such a homotopy is used in defining the cup product of differential characters. In Section 3, we describe the group of differential characters as a cohomology of a cochain complex, and introduce the cup product. After a study of the graded commutativity of the cup product, we prove Theorem 1.

§2. The cup product on ordinary cohomology

2.1. The cup product

Let X be a topological space. We denote by $C_p(X) = C_p(X; \mathbb{Z})$ the group of singular p-chains on X, and by $\partial : C_p(X) \to C_{p-1}(X)$ the boundary operator. The subgroup of cycles in $C_p(X)$ is denoted by $Z_p(X)$ as usual.

We write $C_*(X)$ for the singular chain complex, and $C_*(X) \otimes_{\mathbb{Z}} C_*(X)$ for the chain complex obtained by the tensor product. Both chain complexes are augmented over the $\mathbb{Z}$-module $\mathbb{Z}$. We introduce a chain map

$$T: C_*(X) \otimes_{\mathbb{Z}} C_*(X) \to C_*(X) \otimes_{\mathbb{Z}} C_*(X)$$

by $T(\sigma_1 \otimes \sigma_2) = (-1)^{|\sigma_1||\sigma_2|} \sigma_2 \otimes \sigma_1$.

By the *method of acyclic models* ([3], see [4, 8]), we have:

Lemma 2.1. *There exists a sequence* $\{D_i\}_{i \geq 0}$ *of functorial homomorphisms* $D_i : C_*(X) \to C_*(X) \otimes_{\mathbb{Z}} C_*(X)$ *raising the degree by* i *such that:*

(a) D_0 *is a chain map preserving the augmentations;*
(b) $\partial D_i - (-1)^i D_i \partial = D_{i-1} + (-1)^i T D_{i-1}$ *for* $i \geq 1$.

If $\{D_i\}$ *and* $\{D'_i\}$ *are as above, then there exists a sequence* $\{E_i\}_{i \geq 0}$ *of functorial homomorphisms* $E_i : C_*(X) \to C_*(X) \otimes_{\mathbb{Z}} C_*(X)$ *raising the degree by* i *such that:*

(c) $E_0 = 0$;
(d) $D'_i - D_i = E_i + (-1)^i T E_i + \partial E_{i+1} + (-1)^i E_{i+1} \partial$ *for* $i \geq 0$.

Proof. We follow [8] (Chapter 5, Section 9). Let $R = \mathbb{Z}[t]/(t^2 - 1)$ be the group ring of $\mathbb{Z}_2 = \mathbb{Z}/2\mathbb{Z}$. We define a chain complex (F_*, ∂)

over R as follows. For $i \geq 0$, $F_i = R\langle e_i \rangle$ is the free R-module of rank 1 generated by e_i. For $i < 0$, we put $F_i = 0$. The boundary operator $\partial : F_i \to F_{i-1}$ is $\partial(e_i) = (1 + (-1)^i t)e_{i-1}$. We make $\mathbb{Z}$ into an R-module by letting $t \in R$ act as the identity. Note that F_* is augmented over the R-module $\mathbb{Z}$. The R-module structure on F_* makes $F_* \otimes_{\mathbb{Z}} C_*(X)$ into a chain complex over R. We also make $C_*(X) \otimes_{\mathbb{Z}} C_*(X)$ into a chain complex over R by letting $t \in R$ act as T. Both of these chain complexes are augmented over the R-module $\mathbb{Z}$. Now, by means of the method of acyclic models, there exists a functorial chain map $D : F_* \otimes_{\mathbb{Z}} C_*(X) \to C_*(X) \otimes_{\mathbb{Z}} C_*(X)$ preserving the augmentations, and such chain maps D and D' are naturally chain homotopic. Then D gives the sequence $\{D_i\}_{i \geq 0}$ of functorial homomorphisms $D_i : C_*(X) \to C_*(X) \otimes_{\mathbb{Z}} C_*(X)$ stated in the lemma by setting $D(e_i \otimes \sigma) = D_i(\sigma)$. Similarly, a natural chain homotopy E between D and D' gives $\{E_i\}_{i \geq 0}$ by setting $E(e_i \otimes \sigma) = E_{i+1}(\sigma)$. Q.E.D.

Let Λ be either $\mathbb{Z}, \mathbb{Z}_2$ or $\mathbb{R}$. We denote by $(C^p(X; \Lambda), \delta)$ the singular cochain complex with coefficients in Λ. Let $\{D_i\}$ be a sequence of functorial homomorphisms in Lemma 2.1. Using the natural multiplicative structure on Λ, we define a homomorphism of Λ-modules

$$\cup : \ C^p(X; \Lambda) \otimes_\Lambda C^p(X; \Lambda) \longrightarrow C^{p+q}(X; \Lambda)$$

by $f \cup g = D_0^*(f \otimes g)$. Since $D_0 : C_*(X) \to C_*(X) \otimes_{\mathbb{Z}} C_*(X)$ gives a *diagonal approximation* ([8]), $\cup$ induces the cup product on $H^*(X; \Lambda)$.

For later convenience, we let $\{D^i\}_{i \geq 0}$ be the sequence of functorial homomorphisms $D^i : C^*(X; \Lambda) \otimes_\Lambda C^*(X; \Lambda) \to C^*(X; \Lambda)$ given by $D^i(f \otimes g) = D_i^*(f \otimes g)$. By definition, D^i lowers the degree by i, and satisfies

$$D^i \delta - (-1)^i \delta D^i = D^{i-1} + (-1)^i D^{i-1} T,$$

where $T : C^*(X; \Lambda) \otimes_\Lambda C^*(X; \Lambda) \to C^*(X; \Lambda) \otimes_\Lambda C^*(X; \Lambda)$ is the cochain map given by $T(f \otimes g) = (-1)^{|f||g|} g \otimes f$.

2.2. The Steenrod squaring operations

Recall that the *Steenrod squaring operations* [7, 8, 9] are the additive cohomology operations

$$\mathrm{Sq}^i : \ H^*(X; \mathbb{Z}_2) \longrightarrow H^{*+i}(X; \mathbb{Z}_2)$$

characterized by the following axioms:

(a) $\mathrm{Sq}^0 = 1$,
(b) $\mathrm{Sq}^p(c) = c \cup c$ for $c \in H^p(X; \mathbb{Z}_2)$,
(c) $\mathrm{Sq}^i(c) = 0$ for $c \in H^p(X; \mathbb{Z}_2)$ with $i > p$,

(d) $\mathrm{Sq}^i(c \cup c') = \sum_{i=j+j'} \mathrm{Sq}^j(c) \cup \mathrm{Sq}^{j'}(c')$ for $c, c' \in H^*(X; \mathbb{Z}_2)$.

We follow [8] to realize the Steenrod squaring operations: let $\{D_i\}_{i\geq 0}$ be as in Lemma 2.1, and $\{D^i\}_{i\geq 0}$ the associated sequence of functorial homomorphisms $D^i : C^p(X;\mathbb{Z}_2) \otimes_{\mathbb{Z}_2} C^q(X;\mathbb{Z}_2) \to C^{p+q-i}(X;\mathbb{Z}_2)$. For $i \geq 0$ we define homomorphisms $\mathrm{Sq}^i : C^p(X;\mathbb{Z}_2) \to C^{p+i}(X;\mathbb{Z}_2)$ by

$$\mathrm{Sq}^i(c) = \begin{cases} 0 & i > p, \\ D^{p-i}(c \otimes c) & i \leq p. \end{cases}$$

These homomorphisms induce the Steenrod squaring operations $\mathrm{Sq}^i : H^p(X;\mathbb{Z}_2) \to H^{p+i}(X;\mathbb{Z}_2)$.

2.3. The cup product and the wedge product

Let X be a smooth manifold. The integration on singular simplices gives a functorial cochain map from the de Rham complex $(\Omega^*(X), d)$ to $(C^*(X;\mathbb{R}), \delta)$. As is well-known, there exists a homotopy between the wedge product $\omega_1 \wedge \omega_2$ and the cup product $\omega_1 \cup \omega_2$. In other words, the diagram:

$$\begin{array}{ccc} \Omega^*(X) \otimes_{\mathbb{R}} \Omega^*(X) & \xrightarrow{\wedge} & \Omega^*(X) \\ \downarrow & & \downarrow \\ C^*(X;\mathbb{R}) \otimes_{\mathbb{R}} C^*(X;\mathbb{R}) & \xrightarrow[\cup]{} & C^*(X;\mathbb{R}) \end{array}$$

is commutative up to a homotopy. In this subsection, we review such a homotopy, since it will be used in defining the cup product on the groups of differential characters (Definition 3.4).

Lemma 2.2. *There exists a sequence $\{B^i\}_{i\geq 0}$ of functorial homomorphisms $B^i : \Omega^*(X) \otimes_{\mathbb{R}} \Omega^*(X) \to C^*(X;\mathbb{R})$ lowering the degree by i such that:*

(a) $B^0 = 0$;
(b) $\alpha \wedge \beta - \alpha \cup \beta = B^1 d(\alpha \otimes \beta) + \delta B^1(\alpha \otimes \beta)$ *for* $\alpha, \beta \in \Omega^*(X)$;
(c) $-D^i = B^i + (-1)^i B^i T + B^{i+1} d + (-1)^i \delta B^{i+1}$ *for* $i \geq 1$.

Proof. The proof is almost the same as that of Lemma 2.1: we put $R' = \mathbb{R}[t]/(t^2 - 1)$, and regard $\mathbb{R}$ as an R'-module by letting $t \in R'$ act as the identity. We define a cochain complex (F^*, δ) over R' as follows. For $i \geq 0$, $F^i = R'\langle e^i \rangle$ is the free R'-module of rank 1. For $i < 0$, we put $F^i = 0$. The coboundary operator $\delta : F^i \to F^{i+1}$ is $\delta(e^i) = (1 - (-1)^i t)e^{i+1}$. Then we have two cochain complexes $F^* \otimes_{\mathbb{R}} C^*(X;\mathbb{R})$ and $\Omega^*(X) \otimes_{\mathbb{R}} \Omega^*(X)$ over R'. We define functorial cochain maps $W, U : \Omega^*(X) \otimes_{\mathbb{R}} \Omega^*(X) \to F^* \otimes_{\mathbb{R}} C^*(X;\mathbb{R})$ by

$$W(\alpha \otimes \beta) = (1+t)e^0 \otimes (\alpha \wedge \beta),$$
$$U(\alpha \otimes \beta) = \sum_{i \geq 0} \left(e^i \otimes D^i(\alpha \otimes \beta) + te^i \otimes D^i T(\alpha \otimes \beta)\right).$$

We now appeal to the method of acyclic models. (In particular, Theorem 7B in [4] suits the present case.) Then there exists a natural cochain homotopy B between W and U. Because B is an R'-module homomorphism, we have the following expression:

$$B(\alpha \otimes \beta) = \sum_{i \geq 0} \left(e^i \otimes B^{i+1}(\alpha \otimes \beta) + te^i \otimes B^{i+1} T(\alpha \otimes \beta)\right).$$

We can easily verify that the sequence of homomorphisms $\{B^i\}$ above has the properties stated in the present lemma. Q.E.D.

§3. Differential characters

3.1. Cohomology presentation

We realize the group of differential characters as a cohomology group ([2, 6]).

Definition 3.1. Let X be a smooth manifold. For a positive integer p, we define $(\check{C}(p)^*, \check{d})$ to be the following cochain complex:

$$\check{C}(p)^q = \begin{cases} C^q(X;\mathbb{Z}) \times C^{q-1}(X;\mathbb{R}), & q < p, \\ C^q(X;\mathbb{Z}) \times C^{q-1}(X;\mathbb{R}) \times \Omega^q(X), & q \geq p, \end{cases}$$
$$\check{d}(b,f) = \begin{cases} (\delta b, -b - \delta f), & (b,f) \in \check{C}(p)^q,\ q < p-1, \\ (\delta b, -b - \delta f, 0), & (b,f) \in \check{C}(p)^{p-1}, \end{cases}$$
$$\check{d}(c,h,\omega) = (\delta c, \omega - c - \delta h, d\omega), \qquad (c,h,\omega) \in \check{C}(p)^q,\ q \geq p.$$

We denote the cohomology group of this complex by $\check{H}(p)^q = \check{H}(p)^q(X)$.

The following lemma is easily shown.

Lemma 3.2. *For a positive integer p, we have the exact sequences:*

$$0 \to H^{p-1}(X;\mathbb{R}/\mathbb{Z}) \to \check{H}(p)^p(X) \xrightarrow{\delta_1} \Omega^p(X)_{\mathbb{Z}} \to 0,$$
$$0 \to \Omega^{p-1}(X)/\Omega^{p-1}(X)_{\mathbb{Z}} \to \check{H}(p)^p(X) \xrightarrow{\delta_2} H^p(X;\mathbb{Z}) \to 0,$$

where $\Omega^q(X)_{\mathbb{Z}}$ is the group of closed integral q-forms.

As in Section 1, we denote by $\widehat{H}^{\ell}(X,\mathbb{R}/\mathbb{Z})$ the group of differential characters of degree ℓ.

Lemma 3.3. *There is an isomorphism:*

$$\check{H}(\ell+1)^{\ell+1}(X) \to \widehat{H}^{\ell}(X,\mathbb{R}/\mathbb{Z}).$$

Proof. For $x \in \check{H}(\ell+1)^{\ell+1}(X)$, let $(c,h,\omega) \in \check{Z}(\ell+1)^{\ell+1}$ be a representative of x. Note that $h \in C^{\ell}(X;\mathbb{R})$. We define a homomorphism $\chi : Z_{\ell}(X) \to \mathbb{R}/\mathbb{Z}$ by $\chi(\sigma) = \langle h, \sigma\rangle \mod \mathbb{Z}$. Because (c,h,ω) is a cocycle, we have $\omega = c + \delta h$. Hence $\chi(\partial\tau) = \int_{\tau}\omega \mod \mathbb{Z}$ for $\tau \in C_{\ell+1}(X)$, so that $\chi \in \widehat{H}^{\ell}(X,\mathbb{R}/\mathbb{Z})$. The assignment $x \mapsto \chi$ gives rise to a well-defined homomorphism $\check{H}(\ell+1)^{\ell+1}(X) \to \widehat{H}^{\ell}(X,\mathbb{R}/\mathbb{Z})$. It is known [1] that $\widehat{H}^{\ell}(X,\mathbb{R}/\mathbb{Z})$ fits into the exact sequences:

$$0 \to H^{\ell}(X;\mathbb{R}/\mathbb{Z}) \to \widehat{H}^{\ell}(X,\mathbb{R}/\mathbb{Z}) \xrightarrow{\delta_1} \Omega^{\ell+1}(X)_{\mathbb{Z}} \to 0,$$
$$0 \to \Omega^{\ell}(X)/\Omega^{\ell}(X)_{\mathbb{Z}} \to \widehat{H}^{\ell}(X,\mathbb{R}/\mathbb{Z}) \xrightarrow{\delta_2} H^{\ell+1}(X;\mathbb{Z}) \to 0.$$

Comparing the exact sequences above with those in Lemma 3.2, we can see that the homomorphism is bijective. Q.E.D.

In the remainder of this paper, we will mean by $\check{H}(\ell+1)^{\ell+1}(X)$ the group of differential characters $\widehat{H}^{\ell}(X,\mathbb{R}/\mathbb{Z})$.

3.2. The cup product

Now we introduce the cup product on $\check{H}(p)^{p}(X) \cong \widehat{H}^{p-1}(X,\mathbb{R}/\mathbb{Z})$. Recall that, by Lemma 2.2, we have a homotopy $B^1 : \Omega^p(X)\otimes_{\mathbb{R}}\Omega^q(X) \to C^{p+q-1}(X;\mathbb{R})$ such that $\omega_1\wedge\omega_2 - \omega_1\cup\omega_2 = B^1 d(\omega_1\otimes\omega_2) + \delta B^1(\omega_1\otimes\omega_2)$.

Definition 3.4. We define a homomorphism

$$\cup : \check{C}(p)^p \otimes_{\mathbb{Z}} \check{C}(q)^q \longrightarrow \check{C}(p+q)^{p+q}$$

by setting

$$\begin{aligned}&(c_1,h_1,\omega_1)\cup(c_2,h_2,\omega_2)\\ &\quad = (c_1\cup c_2,\ (-1)^p c_1\cup h_2 + h_1\cup\omega_2 + B^1(\omega_1\otimes\omega_2),\ \omega_1\wedge\omega_2).\end{aligned}$$

The *cup product on differential characters* is defined to be the induced homomorphism $\cup : \check{H}(p)^p(X)\otimes_{\mathbb{Z}}\check{H}(q)^q(X) \to \check{H}(p+q)^{p+q}(X)$.

It is known [1] that the cup product is associative and graded commutative:

$$(x_1\cup x_2)\cup x_3 = x_1\cup(x_2\cup x_3),$$
$$x_1\cup x_2 = (-1)^{p_1p_2}x_2\cup x_1,$$

where $x_i \in \check{H}(p_i)^{p_i}(X)$. The graded commutativity will be shown in the next subsection.

It would be worth while to describe an example due to Cheeger and Simons [1]. Notice that $\check{H}(1)^1(X) \cong \widehat{H}^0(X, \mathbb{R}/\mathbb{Z}) \cong C^\infty(X, U(1))$. We consider the case of $X = S^1$. For $f : S^1 \to U(1)$ we can find a map $F : \mathbb{R} \to \mathbb{R}$ such that $f(\theta) = \exp(2\pi\sqrt{-1}F(\theta))$. We define $\Delta_f \in \mathbb{Z}$ by $F(\theta + 2\pi) = F(\theta) + \Delta_f$. Similarly, we introduce $G : \mathbb{R} \to \mathbb{R}$ to $g : S^1 \to U(1)$. Then the cup product $f \cup g \in \check{H}(2)^2(X) \cong \widehat{H}^1(S^1, \mathbb{R}/\mathbb{Z})$ is expressed as

$$(f \cup g)(S^1) = \Delta_f G(0) - \int_0^{2\pi} F\frac{dG}{d\theta} d\theta \mod \mathbb{Z}.$$

3.3. The graded commutativity of the cup product

We here introduce an analogy of the sequence of homomorphisms $\{D^i\}_{i \geq 0}$ to the cochain complex $(\check{C}(p)^*, \check{d})$.

Definition 3.5. Let $p, q, \bar{p}, \bar{q}$ be positive integers.
(a) For a non-negative integer i, we define a homomorphism

$$F^i : \check{C}(p)^{\bar{p}} \otimes_{\mathbb{Z}} \check{C}(q)^{\bar{q}} \longrightarrow C^{\bar{p}+\bar{q}-i-1}(X; \mathbb{R})$$

as follows:

$$\begin{aligned} &F^{2j}((c_1, h_1, \omega_1) \otimes (c_2, h_2, \omega_2)) = \\ &\qquad (-1)^{\bar{p}} D^{2j-1}(h_1 \otimes h_2) + (-1)^{\bar{p}} D^{2j}(c_1 \otimes h_2) \\ &\qquad + D^{2j}(h_1 \otimes \omega_2) + B^{2j+1}(\omega_1 \otimes \omega_2), \\ &F^{2j+1}((c_1, h_1, \omega_1) \otimes (c_2, h_2, \omega_2)) = \\ &\qquad (-1)^{\bar{p}} D^{2j}(h_1 \otimes h_2) - D^{2j+1}(h_1 \otimes c_2) \\ &\qquad - (-1)^{\bar{p}} D^{2j+1}(\omega_1 \otimes h_2) - B^{2j+2}(\omega_1 \otimes \omega_2), \end{aligned}$$

where $D^{-1}(h_1 \otimes h_2) = 0$. When ω_1 or ω_2 is irrelevant, we substitute 0 for it.
(b) We define a sequence $\{G^i\}_{i \geq 0}$ of homomorphisms

$$G^i : \check{C}(p)^{\bar{p}} \otimes_{\mathbb{Z}} \check{C}(q)^{\bar{q}} \longrightarrow \check{C}(p+q)^{\bar{p}+\bar{q}-i}$$

by setting

$$\begin{aligned} &G^i((c_1, h_1, \omega_1) \otimes (c_2, h_2, \omega_2)) = \\ &\qquad (D^i(c_1 \otimes c_2),\ F^i((c_1, h_1, \omega_1) \otimes (c_1, h_2, \omega_2)),\ W^i(\omega_1 \otimes \omega_2)), \end{aligned}$$

where $W^i(\omega_1 \otimes \omega_2) = 0$ for $i > 0$ and $W^0(\omega_1 \otimes \omega_2) = \omega_1 \wedge \omega_2$. When ω_1 or ω_2 is irrelevant, we substitute 0 for it.

Proposition 3.6. *The sequence of homomorphisms* $\{G^i\}_{i\geq 0}$ *obeys*

(a) $G^0\check{d} = \check{d}G^0$;
(b) $G^i\check{d} - (-1)^i\check{d}G^i = G^{i-1} + (-1)^i G^{i-1}T$ *for* $i \geq 1$,

where the cochain map $T : \check{C}(p)^{\bar{p}} \otimes_{\mathbb{Z}} \check{C}(q)^{\bar{q}} \to \check{C}(q)^{\bar{q}} \otimes_{\mathbb{Z}} \check{C}(p)^{\bar{p}}$ *is defined by* $T(x_1 \otimes x_2) = (-1)^{\bar{p}\bar{q}} x_2 \otimes x_1$.

Proof. We can directly prove (a) by Lemma 2.2. The proof of (b) amounts to showing the following formulae:

$$F^{2j} - F^{2j}T = -D^{2j+1} - \delta F^{2j+1} + F^{2j+1}\check{d},$$
$$F^{2j+1} + F^{2j+1}T = D^{2j+2} + \delta F^{2j+2} + F^{2j+2}\check{d},$$

where $j \geq 0$. We can show them by Lemma 2.1 and Lemma 2.2. Q.E.D.

Since $G^0 : \check{C}(p)^p \otimes_{\mathbb{Z}} \check{C}(q)^q \to \check{C}(p+q)^{p+q}$ induces the cup product, we have:

Corollary 3.7. *The cup product is graded commutative.*

3.4. The main theorem

Recall that we defined $\hat{q}^\ell : \check{H}(\ell+1)^{\ell+1}(X) \longrightarrow \check{H}(2\ell+2)^{2\ell+2}(X)$ by setting $\hat{q}^\ell(x) = x \cup x$.

Lemma 3.8. *For* $k \geq 0$, *the map* $\hat{q}^{2k}$ *induces the homomorphism:*

$$q^{2k} : \; H^{2k+1}(X;\mathbb{Z}) \otimes \mathbb{Z}_2 \longrightarrow \mathrm{Hom}(H_{4k+1}(X;\mathbb{Z}), \mathbb{Z}_2).$$

Proof. Since the cup product is graded commutative, we have

$$\hat{q}^{2k}(x+y) = x\cup x + x\cup y + y\cup x + y\cup y = x\cup x + y\cup y = \hat{q}^{2k}(x) + \hat{q}^{2k}(y).$$

We focus on the exact sequences in Lemma 3.2. Clearly, we have $\delta_1(\hat{q}^{2k}(x)) = \delta_1(x) \wedge \delta_1(x) = 0$. For $\alpha \in \Omega^{2k}(X)/\Omega^{2k}(X)_{\mathbb{Z}}$, we also have $\hat{q}^{2k}(\alpha) = \frac{1}{2}d(\alpha \wedge \alpha) = 0$ in $\Omega^{4k+1}(X)/\Omega^{4k+1}(X)_{\mathbb{Z}}$. Hence the homomorphism $\hat{q}^{2k}$ descends to

$$\hat{q}^{2k} : \; H^{2k+1}(M;\mathbb{Z}) \longrightarrow H^{4k+1}(M;\mathbb{R}/\mathbb{Z}) \cong \mathrm{Hom}(H_{4k+1}(M;\mathbb{Z}), \mathbb{R}/\mathbb{Z}).$$

Using again the graded commutativity, we have $2\hat{q}^{2k}(x) = 0$, so that $(\hat{q}^{2k}(x))(\sigma)$ belongs to $(\frac{1}{2}\mathbb{Z})/\mathbb{Z} \subset \mathbb{R}/\mathbb{Z}$. The identification $(\frac{1}{2}\mathbb{Z})/\mathbb{Z} \cong \mathbb{Z}/2\mathbb{Z}$ gives

$$\hat{q}^{2k} : \; H^{2k+1}(M;\mathbb{Z}) \longrightarrow \mathrm{Hom}(H_{4k+1}(M;\mathbb{Z}), \mathbb{Z}_2).$$

Because $(\hat{q}^{2k}(2x))(\sigma) = 2(\hat{q}^{2k}(x))(\sigma) = 0$, the $\hat{q}^{2k}$ descends to give q^{2k}. Q.E.D.

We are in the position of proving the main result:

Theorem 3.9. $q^{2k} = \pi \circ \mathrm{Sq}^{2k} \circ \iota$.

Proof. For any abelian group A, we have the natural injection

$$\mathrm{Hom}(H^{2k+1}(M;\mathbb{Z}) \otimes \mathbb{Z}_2, A) \longrightarrow \mathrm{Hom}(H^{2k+1}(M;\mathbb{Z}), A).$$

So we will verify the coincidence of q^{2k} and $\pi \circ \mathrm{Sq}^{2k} \circ \iota$ regarding them as homomorphisms $H^{2k+1}(M;\mathbb{Z}) \to \mathrm{Hom}(H_{4k+1}(X;\mathbb{Z}), \mathbb{Z}_2)$. Let $x = (c, h, \omega)$ be a cocycle in $\check{Z}(2k+1)^{2k+1}$. Because any cochain with coefficients in $\mathbb{R}$ is divisible by 2, the following formula is derived from Proposition 3.6:

$$x \cup x = \left(c \cup c, \ -\frac{1}{2} D^1(c \otimes c) - \delta\left(\frac{1}{2} F^1(x \otimes x)\right), \ 0 \right).$$

Therefore we have the following expression of q^{2k}:

$$\left(q^{2k}([c])\right)(\sigma) = \langle D^1(c \otimes c), \sigma \rangle \mod 2\mathbb{Z},$$

where $\sigma \in Z_{4k+1}(X)$. This expression of q^{2k} coincides with that of $\pi \circ \mathrm{Sq}^{2k} \circ \iota$, by means of the realization of Sq^{2k} in Section 2. Q.E.D.

An example of q^{2k} is given by taking $X = S^1$. Then $\mathrm{Sq}^0 = 1$ implies that $q^0 : \mathbb{Z}_2 \to \mathbb{Z}_2$ is the identity map. We can verify this example directly by using the formula at the end of Subsection 3.2.

In general, for a compact oriented $(4k+1)$-dimensional smooth manifold X without boundary, the Steenrod squaring operation Sq^{2k} is expressed as $\mathrm{Sq}^{2k}(c) = v_{2k}(X) \cup c$ for $c \in H^{2k+1}(X;\mathbb{Z}_2)$, where $v_{2k}(X) \in H^{2k}(X;\mathbb{Z}_2)$ is the $2k$-th *Wu class* of X, ([7, 8]). Note that, in this case, the evaluation of the fundamental class $[X] \in H_{4k+1}(X;\mathbb{Z})$ of X simplifies q^{2k} as

$$q^{2k} : H^{2k+1}(X;\mathbb{Z}) \otimes \mathbb{Z}_2 \longrightarrow \mathbb{Z}_2.$$

Corollary 3.10. *If X is a compact oriented $(4k+1)$-dimensional smooth manifold without boundary, then q^{2k} has the expression:*

$$q^{2k}(c) = \langle v_{2k}(X) \cup \iota(c), [X] \rangle.$$

For example, when X is 5-dimensional and spin, we have $v_2(X) = w_2(X) = 0$, so that $q^2 = 0$. As is mentioned in Introduction, q^2 is the obstruction to the $SL(2,\mathbb{Z})$-invariance of the Chern-Simons action functional $I_{\mathrm{CS}}(B_{\mathrm{RR}}, B_{\mathrm{NS}})$ for pairs of "B-fields" $B_{\mathrm{RR}}, B_{\mathrm{NS}} \in \check{H}(3)^3(X)$.

Thus, if X is spin, then I_{CS} has the $SL(2,\mathbb{Z})$-invariance. In the case that X is the direct product of a compact oriented 4-dimensional manifold M and S^1, the $SL(2,\mathbb{Z})$-invariance was known by Witten [10] for B-fields corresponding to closed paths in $H^2(M;\mathbb{R})/H^2(M;\mathbb{Z})$. This fact follows from our result, since $X = M \times S^1$ is spin.

Acknowledgments. The relation between q^{2k} and Sq^{2k} was originally suggested by M. Furuta in replying to my question. I would like to express my gratitude to him. I would also like to thank T. Kohno, Y. Terashima, Y. Kametani and H. Sasahira for valuable discussions. The author's research is supported by Research Fellowship of the Japan Society for the Promotion of Science for Young Scientists.

References

[1] J. Cheeger and J. Simons, Differential characters and geometric invariants, Lecture Notes in Math., **1167** (1985), Springer-Verlag, 50–80.

[2] P. Deligne and D. S. Freed, Classical field theory, Quantum fields and strings: a course for mathematicians, Vol. 1, 2, (Princeton, NJ, 1996/1997), Amer. Math. Soc., Providence, RI, 1999, pp. 137–225.

[3] S. Eilenberg and S. MacLane, Acyclic models, Amer. J. Math., **75** (1953), 189–199.

[4] A. Forrester, Acyclic models and de Rham's theorem, Trans. Amer. Math. Soc., **85** (1957), 307–326.

[5] D. S. Freed, G. W. Moore and G. Segal, Heisenberg groups and noncommutative fluxes, Ann. Phys., **322** (2007), 236–285, arXiv:hep-th/0605200.

[6] M. J. Hopkins and I. M. Singer, Quadratic functions in geometry, topology, and M-theory, J. Differential Geom., **70** (2005), 329–452, arXiv:math/0211216.

[7] J. W. Milnor and J. D. Stasheff, Characteristic classes, Ann. of Math. Stud., **76**, Princeton Univ. Press, Princeton, NJ; Univ. of Tokyo Press, Tokyo, 1974.

[8] E. H. Spanier, Algebraic topology, corrected reprint of the 1966 original, Springer-Verlag, New York.

[9] N. E. Steenrod, Cohomology operations, Lectures by N. E. Steenrod written and revised by D. B. A. Epstein, Ann. of Math. Stud., **50**, Princeton Univ. Press, Princeton, NJ, 1962.

[10] E. Witten, AdS/CFT correspondence and topological field theory, J. High Energy Phys., 1998, no. 12, Paper 12, 31 pp. (electronic), arXiv:hep-th/9812012.

Graduate school of Mathematical Sciences
The University of Tokyo
Komaba 3-8-1, Meguro-Ku, Tokyo
153-8914 Japan

Current address:
The Department of Mathematics
The University of Texas at Austin
1 University Station C1200, Austin
TX 78712-0257 USA
E-mail address: kgomi@math.utexas.edu

Advanced Studies in Pure Mathematics 52, 2008
Groups of Diffeomorphisms
pp. 309–368

Relative weight filtrations on completions of mapping class groups

Richard Hain

CONTENTS

§1. Introduction

One of the primary themes of Morita's work over the last 20 years has been the study of the structure of mapping class groups via their actions on nilpotent quotients of surface groups. A secondary theme has been the relation of this work to the study of invariants of 3-manifolds and homology spheres. The goal of this paper is to introduce to topologists a new tool that may be useful in these pursuits.

The tool is the relative weight filtration of the relative completion of a mapping class group of a surface that is associated to a system of simple closed curves on the surface. Establishing the existence of relative

Received May 11, 2007.
Revised May 21, 2008.
Supported in part by the National Science Foundation through grants DMS-0405440 and DMS-0706955.

weight filtrations on completions of mapping class groups is non-trivial and was established with Makoto Matsumoto using Galois actions in [20], and with Gregory Pearlstein and Tomohide Terasoma using Hodge theory in [22]. The existence of relative weight filtrations on completions of mapping class groups follows from general results about fundamental groups of smooth varieties and because all mapping class groups occur as fundamental groups of moduli spaces of curves. The general theories of the Hodge and Galois theory of fundamental groups of algebraic varieties imply that relative weight filtrations have strong exactness properties.

The paper begins with an exposition for topologists of the theory of relative completion of discrete groups. It is illustrated by the examples of mapping class groups and automorphism groups of free groups. The paper continues with an exposition of the weight filtration associated to a nilpotent endomorphism of a vector space, and its generalization, the relative weight filtration associated to a nilpotent endomorphism of a *filtered* vector space, which is due to Deligne [11] and was further developed by Steenbrink and Zucker [48], and Kashiwara [31]. Since the generic nilpotent endomorphism of a filtered vector space does not have a relative weight filtration, the existence of a relative weight filtration of a nilpotent endomorphism of a filtered vector space imposes non-trivial restrictions on the endomorphism. (Cf. [48].)

Relative weight filtrations appear in the study of mapping class groups in the following context. Suppose that S is a compact oriented surface of genus g which, for simplicity, we suppose to be ≥ 3. Denote the relative completion of its mapping class group Γ_S by $\mathcal{G}_S$. This is a proalgebraic group (defined over $\mathbb{Q}$) that is an extension

$$1 \to \mathcal{U}_S \to \mathcal{G}_S \to \mathrm{Sp}(H_1(S)) \to 1$$

of the symplectic group that is associated to the first homology of S and its intersection form by a prounipotent group $\mathcal{U}_S$, which is essentially (but not quite) the unipotent completion of the Torelli group T_S of S. (Cf. [18].) There is a natural Zariski dense homomorphism $\Gamma_S \to \mathcal{G}_S$; the image of the Torelli group T_S is Zariski dense in $\mathcal{U}_S$. The Lie algebra of $\mathcal{G}_S$ is an extension

$$0 \to \mathfrak{u}_S \to \mathfrak{g}_S \to \mathfrak{sp}(H_1(S)) \to 0$$

of the symplectic Lie algebra associated to $H_1(S)$ by $\mathfrak{u}_S$, which is a pronilpotent Lie algebra. It has a natural weight filtration which is defined by

$$W_0\mathfrak{g}_S = \mathfrak{g}_S,\ W_{-1}\mathfrak{g}_S = \mathfrak{u}_S,\ W_{-m}\mathfrak{g}_S = L^m\mathfrak{u}_S\ (m \geq 1),$$

where $L^m \mathfrak{u}_S$ denotes the mth term of the lower central series of $\mathfrak{u}_S$. A system $\gamma = \{c_0, \dots, c_m\}$ of disjoint simple closed curves on S determines commuting Dehn twists $\tau_0, \dots, \tau_m$. Their product τ_γ lies in a prounipotent subgroup of $\mathcal{G}_S$ and has a unique logarithm $N_\gamma := \log \tau_\gamma \in \mathfrak{g}_S$ whose adjoint action

$$\operatorname{ad}(N_\gamma) : \mathfrak{g}_S \to \mathfrak{g}_S$$

preserves the weight filtration $W_\bullet$. It therefore induces an inverse system of nilpotent endomorphisms

$$\operatorname{ad}(N_\gamma) : \mathfrak{g}_S / W_{-n}\mathfrak{g}_S \to \mathfrak{g}_S / W_{-n}\mathfrak{g}_S$$

each of which preserves the induced weight filtration $W_\bullet$. General results in Hodge and Galois theory imply the existence of the relative weight filtration $M_\bullet^\gamma$ of $\mathfrak{g}_S$ associated to the curve system γ. This filtration is compatible with the bracket in the sense that the bracket induces a map

$$M_r^\gamma \mathfrak{g}_S \otimes M_s^\gamma \mathfrak{g}_S \to M_{r+s}^\gamma \mathfrak{g}_S.$$

This filtration and the weight filtration $W_\bullet$ have very strong exactness and naturality properties. For example, there is a natural (though not canonical) isomorphism of pronilpotent Lie algebras

$$\mathfrak{g}_S \cong \prod_{m,k} \operatorname{Gr}_k^{M^\gamma} \operatorname{Gr}_m^W \mathfrak{g}_S.$$

There are similar results when S has decorations, such as points and boundary components. The Lie algebra $\mathfrak{p}(S,x)$ of the unipotent completion of $\pi_1(S,x)$ also has a relative weight filtration associated to each curve system γ of (S,x). The natural action $\mathfrak{g}_S \to \operatorname{Der} \mathfrak{p}(S,x)$ preserves the relative weight filtration $M_\bullet^\gamma$ as well as the weight filtration $W_\bullet$. This map has strong exactness properties with respect to both of these filtrations. In particular, it is compatible with their identifications with their associated bigraded objects.

Since the bracket preserves the relative weight filtration, $M_k^\gamma \mathfrak{g}_S$ is a subalgebra of $\mathfrak{g}_S$ whenever $k \le 0$. These correspond to subgroups $M_k^\gamma \mathcal{G}_S$ ($k \le 0$) of $\mathcal{G}_S$. The subalgebras $M_0^\gamma \mathfrak{g}_S$, parametrized by curve systems γ on S are natural analogues of parabolic subalgebras of semi-simple Lie algebras as they are parametrized by the boundary components of the corresponding moduli space of curves and they equal their normalizers in $\mathfrak{g}_S$, as we establish in Proposition 8.5.

A particularly interesting case occurs when the curve system γ is maximal. Maximal curve systems correspond to pants decompositions. Since each oriented "pair of pants" naturally bounds a ball, a pants decomposition of S determines a handlebody U with boundary S in which each $c_j \in \gamma$ bounds an imbedded disk in U. The handlebody U is unique in the sense that if V is another such handlebody, then there is a diffeomorphism $U \to V$ which restricts to the identity on S. In Section 9 we use Morse Theory to show that the relative weight filtration corresponding to a pants decomposition depends only on the handlebody U that it determines. We denote it by $M_\bullet^U$. In Section 10 we use the exactness properties of the relative weight filtration and results of Griffiths, Luft and Pitsch [14, 35, 43] to show that the relative weight filtrations $M_\bullet^U$ and $M_\bullet^V$ of $\mathfrak{g}_S$ associated to two handlebodies U and V are equal if and only if there is a diffeomorphism $U \to V$ whose restriction to their boundaries is the identity $S \to S$.

Each handlebody U with boundary S determines a subgroup Λ_U of Γ_S which consists of those elements of Γ_S that extend to an isotopy class of diffeomorphisms of U. In the pointed case, we combine results of Griffiths, Luft and Pitsch [14, 35, 43] with the exactness properties of $M_\bullet^U$ to prove that

$$\Lambda_{U,x} = \Gamma_{S,x} \cap M_0^U \mathcal{G}_{S,x}, \quad \Lambda_{U,x} \cap M_{-2}^U \mathcal{G}_{S,x} = \ker\{\Lambda_{U,x} \to \operatorname{Aut} \pi_1(U,x)\}.$$

These results provide an upper bound on the size of $\Lambda_{U,x}$ in $\Gamma_{S,x}$. They also imply that there is an injection

$$\operatorname{Aut} \pi_1(U,x) \to \operatorname{Gr}_0^{M_U} \mathcal{G}_{S,x}.$$

This induces a homomorphism $\mathcal{A}_g \to \operatorname{Gr}_0^{M_U} \mathcal{G}_{S,x}$ from the relative completion $\mathcal{A}_g$ of $\operatorname{Aut} \pi_1(U,x)$. In Section 10 we show that this homomorphism is not surjective, which implies the unexpected result that Λ_U is not Zariski dense in $M_0^U \mathcal{G}_S$. This implies that the relative weight filtration of $\mathcal{G}_S$ is not obtained simply by taking the Zariski closure in $\mathcal{G}_S$ of a filtration of Γ_S.

We give an application to the problem of determining which elements of Γ_S extend to a handlebody. Very similar results have been obtained independently by Jamie Jorgensen [30] by different methods. If S bounds the handlebody U, then the elements of Γ_S that extend to some handlebody is

$$C = \bigcup_{\phi \in \Gamma_S} \phi \Lambda_U \phi^{-1}.$$

In Section 11, we use properties of $M_\bullet^U$ to show that the Zariski closure of the intersection of C with the mth term of the lower central series of $\mathcal{U}_S$ is a proper (i.e., $\subsetneq$) closed subvariety of the mth term of the lower central series of $\mathcal{U}_S$ for all $m \geq 1$ when $g \geq 7$ and slightly restricted ranges when $3 \leq g < 7$.

A more substantial potential application should be to finite type invariants of 3-manifolds and homology 3-spheres. The set of all genus g Heegaard decompositions of 3-manifolds and homology 3-spheres are the double coset spaces

$$\Lambda_U \backslash \Gamma_S / \Lambda_U \text{ and } T\Lambda_V \backslash T_S / T\Lambda_U,$$

where $S^3 = U \cup V$ is a Heegaard decomposition of the 3-sphere, $\partial U = \partial V = S$, and $T\Lambda_U := T_S \cap \Lambda_U$. The sets of all 3-manifolds and homology 3-manifolds are obtained from these by a suitable stabilization where the genus of S goes to infinity.[1] It is thus natural to consider the double coset spaces

$$\mathcal{L}_U \backslash \mathcal{G}_S / \mathcal{L}_U \text{ and } W_{-1}\mathcal{L}_V \backslash \mathcal{U}_S / W_{-1}\mathcal{L}_U,$$

where $\mathcal{L}_U$ denotes the Zariski closure of Λ_U in $\mathcal{G}_S$, as well as their stabilizations as $g(S) \to \infty$. There are natural mappings

$$\Lambda_U \backslash \Gamma_S / \Lambda_U \to \mathcal{L}_U \backslash \mathcal{G}_S / \mathcal{L}_U$$

and

$$T\Lambda_V \backslash T_S / T\Lambda_U \to W_{-1}\mathcal{L}_V \backslash \mathcal{U}_S / W_{-1}\mathcal{L}_U.$$

and their stabilizations as $g(S) \to \infty$. Functions on $\mathcal{L}_U \backslash \mathcal{G}_S / \mathcal{L}_U$ should yield finite type invariants of Heegaard decompositions and functions on the stabilization should yield finite type invariants of 3-manifolds — and similarly for homology 3-spheres. However, there is one major difficulty in carrying out this program. One needs to take the quotients

$$\mathcal{L}_U \backslash \mathcal{G}_S / \mathcal{L}_U$$

using geometric invariant theory (GIT). But since the group $\mathcal{L}_U$ is not reductive, the GIT problems are more difficult and less likely to be well behaved. (Cf. [12].) Amassa Fauntleroy and I are attempting to use the strictness properties of $M_\bullet^U$ to construct and study the GIT quotients above.

The reader should be aware that, in an attempt to make this material more accessible to non-experts, the Hodge and Galois theoretic aspects of the theory have been suppressed. This choice comes at the expense of giving the basic properties of relative weight filtrations a false

[1]The stabilization is by taking the connected sum with the standard genus one Heegaard decomposition of the 3-sphere, cf. [6]. To construct the stabilization maps, one needs to use the mapping class group $\Gamma_{S,D}$ associated to a surface with one boundary component and the corresponding handlebody subgroup $\Lambda_{S,D}$.

aura of mystery. Readers wanting more background should consult the papers of Deligne [11], Steenbrink-Zucker [48] and Kashiwara [31].

Convention 1.1. Throughout the paper the default coefficient group in all homology and cohomology groups is $\mathbb{Q}$, the rational numbers. So, for example, $H_\bullet(S)$ denotes the rational homology of S and $H^\bullet(\Gamma)$ denotes the rational cohomology of the group Γ. All other coefficients will be made explicit.

Acknowledgments: First and foremost, it is a great pleasure to thank Shigeyuki Morita for his decades of inspiring work. His work complements my own and has had significant impact upon it. I would also like to thank my collaborators Gregory Pearlstein, Tomohide Terasoma and especially Makoto Matsumoto for joint work which makes the current work possible. I would also like to thank Alan Hatcher for communicating a proof of the folk result, Proposition 10.1 and Amassa Fauntleroy for discussions of geometric quotients by non-reductive groups. Finally, I thank the anonymous referee for his very careful and reading of the manuscript and for his suggested improvements to Proposition 3.15.

§2. Filtrations

Since filtrations play a central role in this paper, it is wise to first lay out the general conventions used in this paper. Deligne's conventions on filtrations [9] are used systematically as they work well and as they are used in Hodge theory and the study of Galois actions.

Suppose that V is a vector space over a field F of characteristic zero. An increasing filtration $G_\bullet$ of V is a sequence of subspaces

$$\cdots \subseteq F_{m-1}V \subseteq F_m V \subseteq F_{m+1}V \subseteq \cdots$$

where $m \in \mathbb{Z}$. When V is finite dimensional, we require that the intersection of the $F_m V$ be trivial and that their union be all of V. The infinite dimensional case is more subtle and is discussed below.

The mth graded quotient $F_m V/F_{m-1}V$ of $F_\bullet$ will be denoted by $\mathrm{Gr}^F_m V$. The associated graded vector space will be denoted by $\mathrm{Gr}^F_\bullet V$.

Decreasing filtrations of V will be denoted with an upper index:

$$\cdots \supseteq F^{m-1}V \supseteq F^m V \supseteq F^{m+1}V \supseteq \cdots$$

The mth graded quotient $F^m V/F^{m+1}V$ will be denoted by $\mathrm{Gr}_F^\bullet V$. We require that the intersection of the $F^m V$ be trivial and that their union be all of V.

An increasing filtration $F_\bullet$ of V can be regarded as a decreasing filtration by "raising indices": $F^m V := F_{-m}V$. For this reason, we will discuss the remaining properties only for increasing filtrations.

If $(V, F_\bullet)$ and $(W, F_\bullet)$ are filtered vector spaces, then $V \otimes W$ and $\operatorname{Hom}(V, W)$ inherit natural filtrations:

$$F_m(V \otimes W) := \sum_{j+k=m} F_j V \otimes F_k W$$

and

$$F_m \operatorname{Hom}(V, W) := \{\phi : V \to W : \phi(F_k V) \subseteq F_{m+k} W \text{ for all } k \in \mathbb{Z}\}.$$

In particular, the dual V^* of V has a natural filtration

$$F_m V^* = \{\phi \in V^* : \phi(F_{-m-1}V) = 0\}.$$

With these definitions, there are natural isomorphisms

$$\begin{aligned} \operatorname{Gr}^F_m(V \otimes W) &\cong \bigoplus_{j+k=m} \operatorname{Gr}^F_j V \otimes \operatorname{Gr}^F_k V \\ \operatorname{Gr}^F_m \operatorname{Hom}(V, W) &\cong \bigoplus_k \operatorname{Hom}(\operatorname{Gr}^F_k V, \operatorname{Gr}^F_{m+k} W) \\ \operatorname{Gr}^F_m (V^*) &\cong (\operatorname{Gr}^F_{-m} V)^*. \end{aligned}$$

A filtration $F_\bullet$ of V naturally induces one on every subspace W and every quotient $p : V \to V/U$ by

$$F_m W := W \cap F_m V \text{ and } F_m(V/U) := p(F_m V).$$

There are thus two ways of inducing a filtration on a subquotient $q : W \to p(W)$. One way is to restrict the quotient filtration on V/U to the subspace $p(W)$; the other is to give $p(W)$ the image of the filtration induced by $F_\bullet$ on W. These are easily seen to agree. (Cf. [9]).

In particular, if a vector space V has two filtrations $F_\bullet$ and $G_\bullet$, then the filtration $F_\bullet$ induces a natural filtration (also denoted by $F_\bullet$) on each $G_\bullet$-graded quotient $\operatorname{Gr}^G_m V$ of V. There are then natural isomorphisms

$$\operatorname{Gr}^F_m \operatorname{Gr}^G_n V \cong \operatorname{Gr}^G_n \operatorname{Gr}^F_m V.$$

2.1. The infinite dimensional case

Unless otherwise noted, all infinite dimensional vector spaces considered will be either ind- or pro- objects of the category $\mathsf{Vec}_F^{\text{fin}}$ of finite dimensional F-vector spaces. The dual of a pro-object of $\mathsf{Vec}_F^{\text{fin}}$ is an ind-object of $\mathsf{Vec}_F^{\mathrm{fin}}$, and vice-versa. If V is an ind-object of $\mathsf{Vec}_F^{\mathrm{fin}}$, then it is naturally isomorphic to its double dual. Similarly, an ind-object of $\mathsf{Vec}_F^{\mathrm{fin}}$ is naturally isomorphic to its double dual. A filtration of a pro-object (resp. ind-object) of $\mathsf{Vec}_F^{\mathrm{fin}}$ is simply a filtration of it by pro-objects (resp. ind-objects) of $\mathsf{Vec}_F^{\mathrm{fin}}$.

§3. Relative Completion of Discrete Groups

Here we summarize the theory of relative unipotent completion of discrete groups. Some of the statements are stronger than results in the literature. Full proofs will appear in [21]. Versions of many of these results for the related notion of weighted completion can be found in [19].

3.1. Unipotent and Prounipotent Groups

Suppose that F is a field of characteristic zero. Recall that a unipotent algebraic group over F is a subgroup U, for some n, of the group of the $n \times n$ unipotent upper triangular matrices

$$\{X \in \mathrm{GL}_n(F) : X - I \text{ is strictly upper triangular}\}$$

that is defined by polynomial equations. The Baker-Campbell-Hausdorff formula implies that

$$\mathfrak{u} = \{\log u \in \mathfrak{gl}_n(F) : u \in U\}$$

is a Lie algebra with bracket $[x, y] = xy - yx$. The exponential mapping $\exp : \mathfrak{u} \to U$ is a polynomial bijection. The algebraic subgroups of U correspond bijectively to the Lie subalgebras of $\mathfrak{u}$ via the exponential mapping.

A *pronilpotent* Lie algebra is, by definition, the inverse limit of finite dimensional nilpotent Lie algebras:

$$\mathfrak{u} = \varprojlim_{\alpha} \mathfrak{u}_\alpha$$

It has a natural topology; a base of neighbourhoods of 0 consists of the kernels of the projections of $\mathfrak{u}$ to each of the $\mathfrak{u}_\alpha$.

A *prounipotent group* $\mathcal{U}$ is the inverse limit

$$\mathcal{U} = \varprojlim_{\alpha} U_\alpha$$

of an inverse system of unipotent groups. The Lie algebra $\mathfrak{u}$ of $\mathcal{U}$ is the inverse limit of the Lie algebras of the U_α. It is a pronilpotent Lie algebra The exponential mapping $\exp : \mathcal{U} \to \mathfrak{u}$ is an isomorphism of proalgebraic varieties.

Pronilpotent Lie algebras have nice presentations. Suppose that $\mathfrak{u}$ is a pronilpotent Lie algebra. Define $H_1(\mathfrak{u})$ to be the abelianization of $\mathfrak{u}$. It is a topological vector space — if $\mathfrak{u}$ is the inverse limit of the finite dimensional nilpotent Lie algebras $\mathfrak{u}_\alpha$, then $H_1(\mathfrak{u})$ is the inverse limit of the $H_1(\mathfrak{u}_\alpha)$. A continuous section $s : H_1(\mathfrak{u}) \to \mathfrak{u}$ of the natural projection $\mathfrak{u} \to H_1(\mathfrak{u})$ induces a continuous Lie algebra homomorphism

$$s_* : \mathbb{L}(H_1(\mathfrak{u}))^\wedge \to \mathfrak{u}$$

from the free completed Lie algebra generated by $H_1(\mathfrak{u})$ to $\mathfrak{u}$. This induces an isomorphism on abelianizations. Since $\mathfrak{u}$ is pronilpotent, s_* is surjective. It follows that $\mathfrak{u}$ has a presentation of the form

$$\mathfrak{u} \cong \mathbb{L}(H_1(\mathfrak{u}))^\wedge/\mathfrak{r}$$

where $\mathfrak{r} = \ker s_*$ is a closed ideal contained in the commutator subalgebra of $\mathbb{L}(H_1(\mathfrak{u}))^\wedge$. Such presentations are said to be *minimal*.

Define the continuous cohomology $H^\bullet(\mathfrak{u})$ of a pronilpotent Lie algebra $\mathfrak{u}$ that is the inverse limit of the finite dimensional nilpotent Lie algebras $\mathfrak{u}_\alpha$ by

$$H^\bullet(\mathfrak{u}) := \varinjlim_\alpha H^\bullet(\mathfrak{u}_\alpha).$$

This will be regarded an ind-object of the category of finite dimensional F-vector spaces. It is easy to check that for all $k \geq 0$,

$$H^k(\mathfrak{u}) = \mathrm{Hom}_{\mathrm{cts}}(H_k(\mathfrak{u}), F) \text{ and } H_k(\mathfrak{u}) = \mathrm{Hom}(H^k(\mathfrak{u}), F).$$

This generalizes to pronilpotent coefficients (i.e., projective systems of nilpotent coefficients): If $V = \varprojlim V_\alpha$ where V_α is a nilpotent $\mathfrak{u}_\alpha$-module, then

$$H_k(\mathfrak{u}, V) := \varprojlim_\alpha H_k(\mathfrak{u}_\alpha, V_\alpha) \text{ and } H^k(\mathfrak{u}, V^*) := \varinjlim_\alpha H^k(\mathfrak{u}_\alpha, V_\alpha^*).$$

There are natural isomorphisms

$$H^k(\mathfrak{u}, V^*) = \mathrm{Hom}_{\mathrm{cts}}(H_k(\mathfrak{u}, V), F) \text{ and } H_k(\mathfrak{u}, V) = \mathrm{Hom}(H^k(\mathfrak{u}, V^*), F)$$

If $\mathfrak{f}$ is a free pronilpotent Lie algebra, then $H^k(\mathfrak{f}; V) = 0$ for all $k > 1$ and all nilpotent $\mathfrak{f}$-modules V.

A complete proof of the following analogue of Hopf's Theorem will appear in [21].

Proposition 3.1. *If $\mathfrak{r}$ is a closed ideal in the free pronilpotent Lie algebra $\mathfrak{f}$ that is contained in $[\mathfrak{f},\mathfrak{f}]$, then there is a natural isomorphism*

$$H^2(\mathfrak{f}/\mathfrak{r}) \cong \mathrm{Hom}_{\mathrm{cts}}(\mathfrak{r}/[\mathfrak{r},\mathfrak{f}], F)$$

of ind-vector spaces. Moreover, if $\theta : \mathfrak{r}/[\mathfrak{r},\mathfrak{f}] \to \mathfrak{r}$ is a continuous section of the quotient mapping, then $\mathfrak{r}$ is generated as a closed ideal by $\mathrm{im}\,\theta$.

Proof. The fact that every subalgebra of a free Lie algebra is free [45], implies that every subalgebra of a free pronilpotent Lie algebra is also a free pronilpotent Lie algebra. Consequently, $\mathfrak{r}$ is a free pronilpotent Lie algebra and has vanishing cohomology in degrees > 1. Using standard cochains, one can show that there is a spectral sequence

$$E_2^{s,t} = H^s(\mathfrak{f}/\mathfrak{r}, H^t(\mathfrak{r})) \Rightarrow H^{s+t}(\mathfrak{f}).$$

Since $\mathfrak{r} \subseteq [\mathfrak{f},\mathfrak{f}]$, $H^1(\mathfrak{f}) = H^1(\mathfrak{f}/\mathfrak{r})$. Since $H^1(\mathfrak{r}) = \mathrm{Hom}_{\mathrm{cts}}(H_1(\mathfrak{r}), F)$ and since

$$H_0(\mathfrak{f}/\mathfrak{r}, H_1(\mathfrak{r})) = H_0(\mathfrak{f}, H_1(\mathfrak{r})) \cong \mathfrak{r}/[\mathfrak{r},\mathfrak{f}],$$

the vanishing of the higher cohomology of $\mathfrak{f}$ and $\mathfrak{r}$ imply (when plugged into the spectral sequence) that

$$H^2(\mathfrak{f}/\mathfrak{r}) = H^0(\mathfrak{f}, H^1(\mathfrak{r})) = \mathrm{Hom}_{\mathrm{cts}}(H_0(\mathfrak{f}, H_1(\mathfrak{r})), F) = \mathrm{Hom}_{\mathrm{cts}}(\mathfrak{f}/[\mathfrak{r},\mathfrak{f}], F).$$

Q.E.D.

An immediate corollary is an analogue of Stallings' result [47]. A detailed proof will appear in [21].

Corollary 3.2. *A homomorphism $\phi : \mathfrak{u}_1 \to \mathfrak{u}_2$ of pronilpotent Lie algebras is an isomorphism if and only if it induces an isomorphism $H^1(\mathfrak{u}_2) \to H^1(\mathfrak{u}_1)$ and a monomorphism $H^2(\mathfrak{u}_2) \to H^2(\mathfrak{u}_1)$ of ind-vector spaces.*

Sketch of Proof. The only if assertion is trivially true. Suppose that $H^k(\mathfrak{u}_2) \to H^k(\mathfrak{u}_1)$ is an isomorphism when $k = 1$ and a monomorphism when $k = 2$. Since $\mathfrak{u}_1$ and $\mathfrak{u}_2$ are pronilpotent, the isomorphism on H^1 implies that ϕ is a quotient map in the category of pronilpotent Lie algebras. Choose a minimal presentation $\mathfrak{u}_1 = \mathfrak{f}/\mathfrak{r}_1$. Let $\mathfrak{r}_2$ be the kernel of $\mathfrak{f} \to \mathfrak{u}_1 \to \mathfrak{u}_2$. Then $\mathfrak{u}_2 = \mathfrak{f}/\mathfrak{r}_2$ and ϕ is an isomorphism if and only if the inclusion $\phi : \mathfrak{r}_1 \to \mathfrak{r}_2$ is a quotient mapping. But this holds if and only if

$$H^2(\mathfrak{u}_2) = \mathrm{Hom}_{\mathrm{cts}}(\mathfrak{r}_2/[\mathfrak{r}_2,\mathfrak{f}], F) \to \mathrm{Hom}_{\mathrm{cts}}(\mathfrak{r}_1/[\mathfrak{r}_1,\mathfrak{f}], F) = H^2(\mathfrak{u}_1)$$

is a monomorphism. Q.E.D.

Corollary 3.3. *A pronilpotent Lie algebra* $\mathfrak{u}$ *is trivial if and only if* $H^1(\mathfrak{u}) = 0$ *and free if and only if* $H^2(\mathfrak{u}) = 0$. □

3.2. Relative Unipotent Completion

The data for relative completion are:

(i) a discrete group Γ;
(ii) a field F of characteristic zero;
(iii) a reductive algebraic group R over F, such as $\mathrm{GL}_n(F)$, $\mathrm{SL}_n(F)$ or $\mathrm{Sp}_n(F)$;
(iv) a Zariski dense homomorphism $\rho : \Gamma \to R$.[2]

The completion of Γ with respect to ρ consists of a proalgebraic group (i.e., an inverse limit of algebraic groups) $\mathcal{G}$ over F that is an extension

$$1 \to \mathcal{U} \to \mathcal{G} \to R \to 1 \tag{1}$$

where $\mathcal{U}$ is prounipotent and a homomorphism $\hat{\rho} : \Gamma \to \mathcal{G}$ whose composition with $\mathcal{G} \to R$ is ρ. It is characterized by the following universal mapping property:

If G is an affine (pro)algebraic group over F that is an extension

$$1 \to U \to G \to R \to 1$$

of R by a (pro)unipotent group U, and if $\tilde{\rho} : \Gamma \to G$ is a homomorphism whose composition with $G \to R$ is ρ, then there is a unique homomorphism $\phi : \mathcal{G} \to G$ of (pro)algebraic F-groups such that

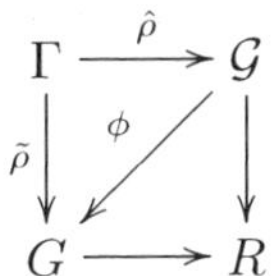

commutes.

The universal mapping property implies that the homomorphism $\hat{\rho} : \Gamma \to \mathcal{G}$ is Zariski dense — that is, if $\mathcal{G}'$ is a proalgebraic subgroup of $\mathcal{G}$ defined over F that contains the image of $\hat{\rho}$, then $\mathcal{G}' = \mathcal{G}$. The point being that the Zariski closure of $\operatorname{im} \hat{\rho}$ in $\mathcal{G}$ has the same universal mapping property as $\mathcal{G}$.

Suppose that K is an extension field of F. Every (pro)algebraic group G over F gives rise to a (pro)algebraic group $G \otimes_F K$ over K

[2]In this paper I will not distinguish between an algebraic group G over F and its group of F- rational points $G(F)$.

by extension of scalars. The universal mapping property of the relative completion $\mathcal{G}_K$ of Γ over K with respect to $\rho : \Gamma \to R \otimes_F K$ implies that $\Gamma \to \mathcal{G} \otimes_F K$ induces a homomorphism $\mathcal{G}_K \to \mathcal{G} \otimes_F K$.

Theorem 3.4 (Hain-Matsumoto [21]). *The homomorphism $\mathcal{G}_K \to \mathcal{G} \otimes_F K$ is an isomorphism.*

In general we will work over the smallest field F possible, which is the smallest field over which both R and ρ are defined. In all principal examples in this paper, the field will be $\mathbb{Q}$.

3.3. Unipotent Completion

When R is the trivial group, relative completion reduces to classical unipotent completion, which is also known as Malcev completion and which can be computed by the methods of rational homotopy theory due to Quillen, Chen and Sullivan. We shall denote the unipotent completion of Γ over F by $\Gamma^{\mathrm{un}}_{/F}$. The default field will be $\mathbb{Q}$. We shall abbreviate $\Gamma^{\mathrm{un}}_{/\mathbb{Q}}$ to Γ^{un}. The unipotent completion of Γ over F is obtained from Γ^{un} by extension of scalars:

$$\Gamma^{\mathrm{un}}_{/F} = \Gamma^{\mathrm{un}} \otimes_{\mathbb{Q}} F.$$

If Γ is a free group $F_n = \langle x_1, \dots, x_n \rangle$ on n-generators, the Lie algebra of $F^{\mathrm{un}}_{n/F}$ is the completed free Lie algebra

$$\mathfrak{f}_n := \mathbb{L}(X_1, \dots, X_n)^{\wedge}$$

which is the closure of the free Lie algebra $\mathbb{L}(X_1, \dots, X_n)$ in the noncommutative power series ring $F\langle\langle X_1, \dots, X_n \rangle\rangle$. The prounipotent group $F^{\mathrm{un}}_{n/F}$ is

$$F^{\mathrm{un}}_{n/F} = \{\exp u \in F\langle\langle X_1, \dots, X_n \rangle\rangle : u \in \mathfrak{f}_n\}.$$

The natural homomorphism $\hat{\rho} : F_n \to F^{\mathrm{un}}_{n/F}$ is defined by $\hat{\rho}(x_j) = \exp X_j$. This can be proved using universal mapping properties. A theorem of Magnus [36] implies that the homomorphism $F_n \to F^{\mathrm{un}}_n$ is injective.

3.4. Completions of $\operatorname{Aut} F_n$ and $\operatorname{Out} F_n$

Denote the automorphism group of the free group F_n by $\operatorname{Aut} F_n$ and its quotient by inner automorphisms of F_n by $\operatorname{Out} F_n$. There are natural surjections

$$\operatorname{Aut} F_n \to \mathrm{GL}_n(\mathbb{Z}) \text{ and } \operatorname{Out} F_n \to \mathrm{GL}_n(\mathbb{Z}).$$

Denote their kernels by IA_n and IO_n, respectively. Let $\operatorname{Aut}^+ F_n$ be the index 2 subgroup of $\operatorname{Aut} F_n$ whose image in $\mathrm{GL}_n(\mathbb{Z})$ is $\mathrm{SL}_n(\mathbb{Z})$. Let $\operatorname{Out}^+ F_n$ be its quotient by the group of inner automorphisms.

Take $F = \mathbb{Q}$, $R = \mathrm{SL}_n(\mathbb{Q})$ and let

$$\rho : \mathrm{Aut}^+ F_n \to \mathrm{SL}_n(\mathbb{Q})$$

be the natural representation. This is Zariski dense. The completion of $\mathrm{Aut}^+ F_n$ with respect to ρ is an extension

$$1 \to \mathcal{IA}_n \to \mathcal{A}_n \to \mathrm{SL}_n(\mathbb{Q}) \to 1.$$

Similarly, we have the completion of $\mathrm{Out}^+ F_n$. It is an extension

$$1 \to \mathcal{IO}_n \to \mathcal{O}_n \to \mathrm{SL}_n(\mathbb{Q}) \to 1.$$

The corresponding sequences of Lie algebras

$$0 \to \mathfrak{ia}_n \to \mathfrak{a}_n \to \mathfrak{sl}_n \to 0 \text{ and } 0 \to \mathfrak{io}_n \to \mathfrak{o}_n \to \mathfrak{sl}_n \to 0$$

are exact. There are natural homomorphisms

$$\mathfrak{a}_n \to \mathrm{Der}\,\mathfrak{f}_n \text{ and } \mathfrak{o}_n \to \mathrm{OutDer}\,\mathfrak{f}_n,$$

where $\mathrm{OutDer}\,\mathfrak{f}_n$ denotes the Lie algebra of outer derivations of $\mathfrak{f}_n$.

The natural homomorphisms $IA_n \to \mathcal{IA}_n$ and $IO_n \to \mathcal{IO}_n$ induce homomorphisms $IA_n^{\mathrm{un}} \to \mathcal{IA}_n$ and $IO_n^{\mathrm{un}} \to \mathcal{IO}_n$. We shall see later (cf. Cor. 3.14) that these homomorphisms are isomorphisms when $n \geq 4$, surjective when $n = 3$ and are far from surjective when $n = 2$.

Proposition 3.5. *The natural homomorphism* $\hat{\rho} : \mathrm{Aut}^+ F_n \to \mathcal{A}_n$ *is injective.*

Proof. Since the unipotent completion $F_n \to F_n^{\mathrm{un}}$ is injective, it follows that the natural representation $\theta : \mathrm{Aut}\, F_n \to \mathrm{Aut}\, F_n^{\mathrm{un}}$ is injective. The Zariski closure of the image of $\mathrm{Aut}^+ F_n$ under θ is easily seen to be an extension of $\mathrm{SL}_n(\mathbb{Q})$ by a prounipotent group. The universal mapping property of relative completion induces a homomorphism $\psi : \mathcal{A}_n \to \mathrm{Aut}\, F_n^{\mathrm{un}}$ such that the diagram

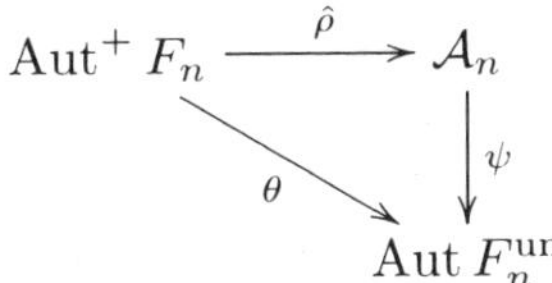

The injectivity of $\hat{\rho}$ follows from the injectivity of θ. Q.E.D.

The injectivity of $\mathrm{Out}^+ F_n \to \mathrm{Out}\, F_n^{\mathrm{un}}$ would follow if one could prove that $F_n^{\mathrm{un}} \cap \mathrm{Aut}\, F_n = F_n$ in $\mathrm{Aut}\, F_n^{\mathrm{un}}$, where F_n and F_n^{un} are regarded as subgroups of $\mathrm{Aut}\, F_n^{\mathrm{un}}$ via the inner action. This is not clear.

3.5. Properties of Relative Completion

Here we list some of the basic properties of relative completion that we shall need. The proofs of these are sprinkled throughout the literature and are sometimes proved for the related notion of *weighted completion* [19]. The notes [21] will give an efficient and uniform presentation of the theory relative and related completions of discrete and profinite groups.

Proposition 3.6 (Naturality). *Suppose that $\rho_j : \Gamma_j \to R_j$, $j = 1, 2$ are Zariski dense homomorphisms from discrete groups to reductive groups over F. Let $\Gamma_j \to \mathcal{G}_j$ be the completion of Γ_j with respect to ρ_j over F. If the diagram*

$$\begin{array}{ccc} \Gamma_1 & \xrightarrow{\rho_1} & R_1 \\ {\scriptstyle \phi_\Gamma}\downarrow & & \downarrow{\scriptstyle \phi_R} \\ \Gamma_2 & \xrightarrow{\rho_2} & R_2 \end{array}$$

commutes where ϕ_R is a homomorphism of algebraic groups, then there is a unique homomorphism $\phi_{\mathcal{G}} : \mathcal{G}_1 \to \mathcal{G}_2$ such that the diagram

$$\begin{array}{ccccc} \Gamma_1 & \xrightarrow{\hat{\rho}_1} & \mathcal{G}_1 & \longrightarrow & R_1 \\ {\scriptstyle \phi_\Gamma}\downarrow & & \downarrow{\scriptstyle \phi_{\mathcal{G}}} & & \downarrow{\scriptstyle \phi_R} \\ \Gamma_2 & \xrightarrow{\hat{\rho}_2} & \mathcal{G}_2 & \longrightarrow & R_2 \end{array}$$

Completions are, in general, right exact. Here we state a useful special case.

Proposition 3.7 (Right exactness). *Suppose that $\rho : \Gamma \to R$ is a Zariski dense homomorphism from a discrete group to a reductive F-group. Denote the completion Γ with respect to ρ by $\mathcal{G}$ and the completion of $\operatorname{im}\rho$ with respect to the inclusion $\operatorname{im}\rho \hookrightarrow R$ by $\mathcal{R}$. Then the sequence*

$$\left(\ker\rho\right)^{\mathrm{un}}_{/F} \to \mathcal{G} \to \mathcal{R} \to 1$$

is exact.

A generalization of Levi's Theorem implies that, when Γ is finitely generated, the extension (1) is split, and that any two splitting are conjugate by an element of $\mathcal{U}$. It follows that $\mathfrak{u}$ is a Lie algebra in the category of pro-representations[3] of R and that there is an isomorphism

$$\mathcal{G} \cong R \ltimes \exp \mathfrak{u}$$

that is unique up to conjugation by an element of $\exp \mathfrak{u}$.

Relative completions are manageable and somewhat computable as they are quite tightly controlled by cohomology.

Suppose that $\overline{F}$ is an algebraic closure of F. An irreducible representation V of R is *absolutely irreducible* if $V \otimes_F \overline{F}$ is an irreducible representation of $R \otimes_F \overline{F}$.

Theorem 3.8. *For all finite dimensional R-modules V, there is a homomorphism*

$$\mathrm{Hom}_R(H_k(\mathfrak{u}), V) \cong \left(H^k(\mathfrak{u}) \otimes V\right)^R \to H^k(\Gamma; V)$$

that is natural with respect to the maps described in Proposition 3.6. It is an isomorphism when $k = 1$ and injective when $k = 2$. If every irreducible finite dimensional representation of R is absolutely irreducible, then there is a natural R-module isomorphism

$$H^1(\mathfrak{u}) \cong \bigoplus_{\alpha} H^1(\Gamma; V_\alpha) \otimes V_\alpha^*$$

and a natural R-module injection

$$H^2(\mathfrak{u}) \hookrightarrow \bigoplus_{\alpha} H^2(\Gamma; V_\alpha) \otimes V_\alpha^*$$

where $\{V_\alpha\}$ is a set of representatives of the isomorphism classes of irreducible finite dimensional R-modules and where each $H^1(\Gamma; V_\alpha)$ is regarded as a trivial R-module.

This theorem alone and in combination with the Base Change Theorem 3.4 can often be used to compute $\mathfrak{u}$. Combined with Corollary 3.3, it gives the following criterion for the vanishing of $\mathfrak{u}$.

Corollary 3.9. *The prounipotent radical of $\mathcal{G}$ vanishes if and only if $H^1(\Gamma; V) = 0$ for all irreducible finite dimensional R-modules.* □

Combining this with right exactness (Prop. 3.7) yields:

[3]That is, an inverse limit of finite dimensional R-modules. Since R is reductive, the pro-representations of R are direct products of finite dimensional R-modules.

Corollary 3.10. *If $H^1(\operatorname{im}\rho; V) = 0$ for all finite dimensional R-modules V, then*

$$\big(\ker\rho\big)^{\mathrm{un}}_{/F} \to \mathcal{G} \to R \to 1$$

is exact. That is, $\mathcal{U}$ is a quotient of $\big(\ker\rho\big)^{\mathrm{un}}_{/F}$.

Additional hypotheses give an upper bound on the kernel. The following result is a refined version of [16, Prop. 4.13].

Theorem 3.11. *Suppose that $\rho : \Gamma \to R$ is a Zariski dense homomorphism. Denote $\ker\rho$ by T. If the $\operatorname{im}\rho$ module $H_1(T; F)$ is finite dimensional and the restriction (via ρ) of a finite dimensional R-module, and if $H^1(\operatorname{im}\rho; V) = 0$ for all irreducible finite dimensional representations of R, then there is a natural exact sequence*

$$1 \to K \to T^{\mathrm{un}}_{/F} \to \mathcal{G} \to R \to 1$$

where K is contained in the center of $T^{\mathrm{un}}_{/F}$. Moreover, if $H^2(\operatorname{im}\rho; V)$ is finite dimensional for all irreducible R-modules V, then K is an R-submodule of the abelian prounipotent group

$$\prod_\alpha H^2(\operatorname{im}\rho; V_\alpha)^* \otimes V_\alpha.$$

where V_α ranges over representatives of the isomorphism classes of finite dimensional R-modules.

3.6. Examples

Equipped with the results of the previous section, we can approach the problem of computing the relative completions in natural examples.

Example 3.12 (Lattices)**.** If Γ is an irreducible lattice in a semi-simple real Lie group G of rank ≥ 2, then Raghunathan's vanishing theorem [44] states that

$$H^1(\Gamma; V) = 0$$

for all irreducible representations V of G. Corollary 3.9 implies that the completion of Γ with respect to the inclusion $\Gamma \to G$ over $\mathbb{R}$ is G.

In particular, when $n \geq 3$, the completion of any finite index subgroup Γ of $\mathrm{SL}_n(\mathbb{Z})$ with respect to the inclusion $\Gamma \to \mathrm{SL}_n(\mathbb{Q})$ is $\mathrm{SL}_n(\mathbb{Q})$. When $g \geq 2$, the completion of any finite index subgroup of $\mathrm{Sp}_g(\mathbb{Z})$ with respect to the inclusion $\Gamma \to \mathrm{Sp}_g(\mathbb{Q})$ is $\mathrm{Sp}_g(\mathbb{Q})$.

The rank condition is necessary. The groups $\mathrm{SL}_2(\mathbb{R})$ and $\mathrm{Sp}_1(\mathbb{R})$ are isomorphic and have real rank 1. If we take Γ to be one of the isomorphic groups $\mathrm{SL}_2(\mathbb{Z})$, $\mathrm{Aut}^+ F_2$, $\Gamma_{S,x}$, where S is a genus 1 surface, then the prounipotent radical of the completion of Γ with respect to the inclusion $\Gamma \to \mathrm{SL}_2(\mathbb{Q})$ is a free prounipotent group whose abelianization is infinite dimensional. (Cf. [18, Rem. 3.9].) It is closely connected with classical modular forms and elliptic motives. (Cf. [22].)

Results of Borel [2] imply that if $\operatorname{im}\rho$ is arithmetic of sufficiently high rank, then $H^2(\operatorname{im}\rho; V)$ vanishes for all non-trivial R-modules and $H^2(\operatorname{im}\rho;\mathbb{Q})$ is isomorphic to the corresponding cohomology group of the compact dual of the symmetric space of $R \otimes \mathbb{R}$. In particular, Borel's formula implies the vanishing of $H^2(\mathrm{SL}_2(\mathbb{Z}); V)$ for *all* SL_n-modules V when $n \geq 4$. It also implies the vanishing of $H^2(\mathrm{Sp}_g(\mathbb{Z}), V)$ for all non-trivial irreducible Sp_g-modules when $g \geq 3$.

Example 3.13 (Universal Central Extensions). Suppose that Γ is a non-zero multiple of the universal central extension of $\mathrm{Sp}_g(\mathbb{Z})$, where $g \geq 2$:

$$0 \to \mathbb{Z} \to \Gamma \to \mathrm{Sp}_g(\mathbb{Z}) \to 1.$$

Let $R = \mathrm{Sp}_g(\mathbb{Q})$ and $\rho : \Gamma \to \mathrm{Sp}_g(\mathbb{Q})$ be the obvious homomorphism. Denote the relative completion of Γ with respect to ρ by $\mathcal{G}$. By Example 3.12, the completion of $\mathrm{Sp}_g(\mathbb{Z})$ with respect to the inclusion $\mathrm{Sp}_g(\mathbb{Z}) \to \mathrm{Sp}_g(\mathbb{Q})$ is $\mathrm{Sp}_g(\mathbb{Q})$. Raghunathan's Theorem implies that $H^1(\mathrm{Sp}_g(\mathbb{Z}); V)$ vanishes for all finite dimensional representations V of Sp_g. An elementary spectral sequence argument implies that $H^1(\Gamma, V)$ also vanishes for all finite dimensional Sp_g-modules V. Cor. 3.9 then implies that $\mathcal{G} \to \mathrm{Sp}_g(\mathbb{Q})$ is an isomorphism.

This provides an interesting example of Theorem 3.11. Borel's vanishing theorem implies that, when $g \geq 3$, $H^2(\mathrm{Sp}_g(\mathbb{Z}), V)$ vanishes for all non-trivial irreducible $\mathrm{Sp}_g(\mathbb{Q})$-modules and that $H^2(\mathrm{Sp}_g(\mathbb{Z}), \mathbb{Q})$ is 1-dimensional. Since the unipotent completion of $\mathbb{Z}$ is $\mathbb{Q}$, Theorem 3.11 implies that we have an exact sequence

$$H^2(\mathrm{Sp}_g(\mathbb{Q});\mathbb{Q})^* \to \mathbb{Q} \to \mathcal{G} \to \mathrm{Sp}_g(\mathbb{Q}) \to 1.$$

Since $\mathcal{G} \to \mathrm{Sp}_g(\mathbb{Q})$ is an isomorphism, it follows that $H^2(\mathrm{Sp}_g(\mathbb{Q});\mathbb{Q})^* \to \mathbb{Q}$ is an isomorphism and that $\mathbb{Q} \to \mathcal{G}$ is trivial. □

As remarked in Example 3.12, $IA_2^{\mathrm{un}} \to \mathcal{IA}_2$ and $IO_2^{\mathrm{un}} \to \mathcal{IO}_2$ are far from surjective. However, when $n \geq 3$, the situation improves.

Corollary 3.14. *If $n \geq 3$, then the natural homomorphisms $IA_n^{\mathrm{un}} \to \mathcal{IA}_n$ and $IO_n^{\mathrm{un}} \to \mathcal{IO}_n$ are surjective. If $n \geq 4$, they are isomorphisms.*

Proof. By results of Magnus [36] and Kawazumi [33], there are natural $\mathrm{GL}_n(\mathbb{Z})$-equivariant isomorphisms

$$H_1(IA_n) \cong \mathrm{Hom}(V, \Lambda^2 V) \text{ and } H_1(IO_n) \cong \mathrm{Hom}(V, \Lambda^2 V)/V,$$

where $V = H_1(F_n)$, from which it follows that the $\mathrm{SL}_n(\mathbb{Z})$-modules $H_1(IA_n)$ and $H_1(IO_n)$ are the restrictions of $\mathrm{SL}(V)$-modules. Surjectivity when $n \geq 3$ follows from Corollary 3.10 and Raghunathan's vanishing result. When $n \geq 4$, the result follows from Theorem 3.11 and Borel's vanishing result, stated above. Q.E.D.

Another situation in which left exactness holds, that we shall need later, is the following. Suppose that

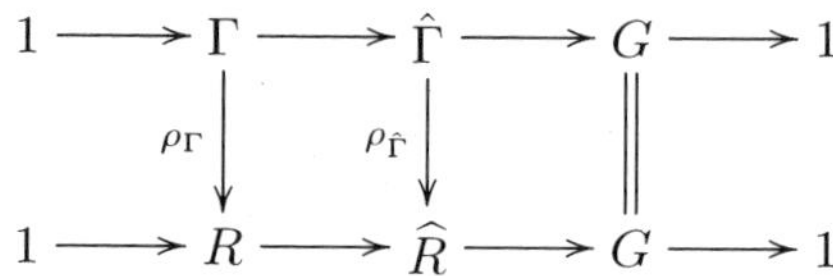

is a commutative diagram of groups with exact rows where:

(i) Γ and $\hat{\Gamma}$ are discrete groups;
(ii) R and $\widehat{R}$ are reductive F-groups;
(iii) G is a finite group;
(iv) ρ_Γ is Zariski dense (which implies that $\rho_{\hat{\Gamma}}$ is also Zariski dense).

Denote the completion of Γ with respect to ρ_Γ by $\mathcal{G}$ and the completion of $\hat{\Gamma}$ with respect to $\rho_{\hat{\Gamma}}$ by $\widehat{\mathcal{G}}$. Naturality implies that there is a homomorphism $\mathcal{G} \to \widehat{\mathcal{G}}$ such that the diagram

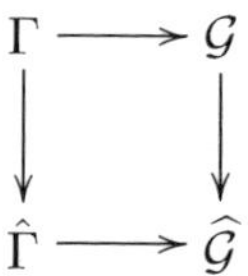

commutes. Right exactness implies that the sequence $\mathcal{G} \to \widehat{\mathcal{G}} \to G \to 1$ is exact. Denote the prounipotent radicals of $\mathcal{G}$ and $\widehat{\mathcal{G}}$ by $\mathcal{U}$ and $\widehat{\mathcal{U}}$, respectively.

Proposition 3.15. *The natural homomorphism $\mathcal{G} \to \widehat{\mathcal{G}}$ is injective. Consequently, the induced mapping $\mathcal{U} \to \widehat{\mathcal{U}}$ of prounipotent radicals is an isomorphism.*

Proof. Stallings' criterion (Cor. 3.2) will be used to show that $\mathfrak{u} \to \widehat{\mathfrak{u}}$ is an isomorphism. To prove this we need the notion of an induced module. (This is sometimes called a co-induced module, cf. [3, p. 67].)

For an R-module V, define the representation induced from V to $\widehat{R}$ by

$$\mathrm{Ind}_R^{\widehat{R}} V = \mathrm{Fun}_R(\widehat{R}, V),$$

where Fun_R denotes the set of left R-invariant functions $\phi : \widehat{R} \to V$. This is a left $\widehat{R}$-module with respect to the action $(r\phi)(x) = \phi(xr)$, where $r, x \in \widehat{R}$. Since R has finite index in $\widehat{R}$, the induced representation is a rational representation of $\widehat{R}$ whenever V is a rational representation of R.

For all R-modules U and $\widehat{R}$-modules V, there is a natural isomorphism

$$\mathrm{Hom}_R(\mathrm{Res}_R^{\widehat{R}} U, V) \cong \mathrm{Hom}_{\widehat{R}}(U, \mathrm{Ind}_R^{\widehat{R}} V),$$

where $\mathrm{Res}_R^{\widehat{R}} U$ denotes the restriction of U to R.

Likewise, for any Γ module V, we can define $\mathrm{Ind}_\Gamma^{\hat{\Gamma}} V = \mathrm{Fun}_\Gamma(\hat{\Gamma}, V)$. If V is an R-module, viewed as a Γ-module via ρ_Γ, then the restriction mapping

$$\mathrm{Ind}_R^{\widehat{R}} V \xrightarrow{\simeq} \mathrm{Ind}_\Gamma^{\hat{\Gamma}} V,$$

is an isomorphism of $\hat{\Gamma}$-modules.

To apply Stallings' criterion, we need to show that $H^k(\widehat{\mathfrak{u}}) \to H^k(\mathfrak{u})$ is an isomorphism (resp., injection) when $k = 1$ (resp., $k = 2$). Since R is reductive, it suffices to show that the natural mapping

$$\phi_k : \mathrm{Hom}_R(\mathrm{Res}_R^{\widehat{R}} H_k(\widehat{\mathfrak{u}}), V) \to \mathrm{Hom}_R(H_k(\mathfrak{u}), V)$$

is an isomorphism (resp., injection) for all finite dimensional R-modules V when $k = 1$ (resp., $k = 2$). Consider the commutative diagram

$$\begin{array}{ccccc}
\mathrm{Hom}_R(\mathrm{Res}_R^{\widehat{R}} H_k(\widehat{\mathfrak{u}}), V) & \xrightarrow{\simeq} & \mathrm{Hom}_{\widehat{R}}(H_k(\widehat{\mathfrak{u}}), \mathrm{Ind}_R^{\widehat{R}} V) & \longrightarrow & H^k(\hat{\Gamma}; \mathrm{Ind}_\Gamma^{\hat{\Gamma}} V) \\
\downarrow{\scriptstyle \phi_k} & & & & \Vert \\
\mathrm{Hom}_R(H_k(\mathfrak{u}), V) & & \longrightarrow & & H^k(\Gamma; V)
\end{array}$$

The right hand vertical map is an isomorphism by Shapiro's Lemma [3, p. 73]. We apply Theorem 3.8. When $k = 1$, all horizontal mappings are isomorphisms, which implies that ϕ_1 is an isomorphism. When $k = 2$, all horizontal mappings are injective, which implies that ϕ_2 is injective.
Q.E.D.

Example 3.16. Suppose that $n \geq 1$. Denote the subgroup of $\mathrm{GL}_n(R)$ that consists of matrices with determinant ± 1 by $\widehat{\mathrm{SL}}(R)$. Then

Proposition 3.15 implies that the commutative diagram

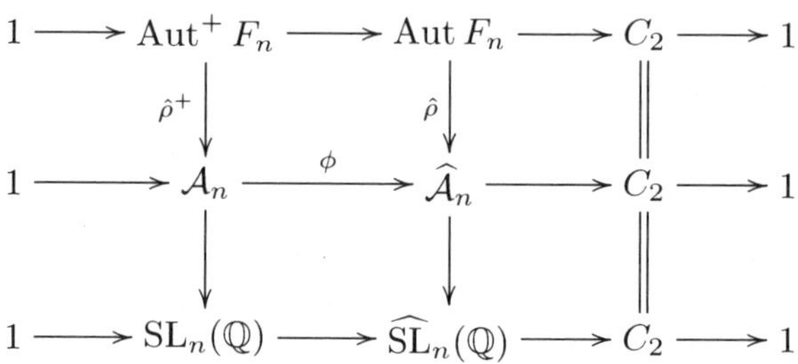

has exact rows. It follows that $\mathcal{A}_n$ is the identity component of $\widehat{\mathcal{A}}_n$. There is a similar story for $\operatorname{Out} F_n$. It is for this reason that in Section 3.4 we considered only the completions of $\operatorname{Aut}^+ F_n$ and $\operatorname{Out}^+ F_n$.

§4. Mapping Class Groups and their Completions

Suppose that g, n, r are non-negative integers. A decorated surface of type (g, n, r) is a pair (S, D) where S is a compact oriented surface of genus g and $D = P \cup V$ is a set of decorations, where $P = \{x_1, \dots, x_n\}$ is a set of n points of S and $V = \{v_1, \dots, v_r\}$ is a set of r non-zero tangent vectors of S. If $v_j \in T_{y_j} S$, the points $x_1, \dots, x_n, y_1, \dots, y_r$ are required to be distinct. The decorated surface (S, D) is *stable* if the punctured surface $S'_D := S - \{x_1, \dots, x_n, y_1, \dots, y_r\}$ has negative Euler characteristic:

$$\chi(S'_D) = \chi(S) - |P| - |V| = 2 - 2g - (r + n) < 0.$$

The mapping class group $\hat{\Gamma}_{S,D}$ of a stable decorated surface (S, D) is the group of connected components of the group of orientation preserving diffeomorphisms of S that fix P and V set wise. There is a natural surjection

$$\hat{\Gamma}_{S,D} \to \operatorname{Aut} D := \operatorname{Aut} P \times \operatorname{Aut} V.$$

For a subgroup G of $\operatorname{Aut} D$ define $\Gamma^G_{S,D}$ to be the inverse image of G under this homomorphism. The classical mapping group of (S, D) corresponds to the trivial group $\mathbf{1}$:

$$\Gamma_{S,D} := \Gamma^{\mathbf{1}}_{S,D} = \pi_0 \operatorname{Diff}^+(S, D).$$

There is a natural extension

$$1 \to \Gamma_{S,D} \to \Gamma^G_{S,D} \to G \to 1.$$

The classification of surfaces implies that $\Gamma_{S,D}^G$ depends only on (g, n, r) and the subgroup G of $S_n \times S_r$.

For a commutative ring R, set $H_R = H_1(S; R)$. The group of automorphisms of H that preserve the intersection pairing is an algebraic group over $\mathbb{Q}$ that we shall denote by $\mathrm{Sp}(H)$. There is a natural surjective homomorphism

$$\rho : \Gamma_{S,D}^G \to G \times \mathrm{Sp}(H_{\mathbb{Z}}).$$

Its kernel is, by definition, the Torelli group $T_{S,D}$.

4.1. Boundary Components versus Tangent Vectors

Tangent vectors are essentially interchangeable with marked boundary components. Because boundary components are less natural in algebraic geometry, we prefer to work with tangent vectors. A marked boundary component of a surface is a boundary component of the surface together with a point on the boundary component. Marked boundary components can be exchanged with tangent vectors as follows:

If $v \in T_y S$ is a non-zero tangent vector of a surface S, then one can replace (S, v) by a surface $\hat{S}$ with a marked boundary component. Here $\hat{S}$ is the real oriented blowup of S at y. This is the surface obtained from S by replacing y by the circle of rays in $T_y S$. The marked point on the boundary of $\hat{S}$ corresponds to the ray $\mathbb{R}^+ v$ in $T_y S$ determined by v. It will be denoted by $[v]$.

This process may be reversed by collapsing the boundary component to a point y and choosing any non-zero tangent vector at y that lies in the ray in $T_y S$ determined by the marked point. These identifications are well defined and mutually inverse up to isotopy.

The corresponding mapping class groups are isomorphic. For example, if S is compact, then the natural homomorphisms

$$\pi_0 \,\mathrm{Diff}^+(S, v) \xrightarrow{\simeq} \pi_0 \,\mathrm{Diff}^+(\hat{S}, [v]) \xleftarrow{\simeq} \pi_0 \,\mathrm{Diff}^+(\hat{S}, \partial \hat{S})$$

are isomorphisms.

4.2. Completions of Mapping Class Groups

The ground field F will be $\mathbb{Q}$ unless otherwise stated. Suppose that (S, D) is a stable decorated surface and that G is a subgroup of $\mathrm{Aut}\, D$, where $D = P \cup V$. The group $G \times \mathrm{Sp}(H)$ is a reductive algebraic group over $\mathbb{Q}$ and the representation $\rho : \Gamma_{S,D}^G \to G \times \mathrm{Sp}(H)$ is Zariski dense. Denote the completion of $\Gamma_{S,D}^G$ relative to ρ by $\mathcal{G}_{S,D}^G$. It is an extension

$$1 \to \mathcal{U}_{S,D}^G \to \mathcal{G}_{S,D}^G \to G \times \mathrm{Sp}(H) \to 1.$$

The next result follows directly from Proposition 3.15.

Proposition 4.1. *For all subgroups G of* $\operatorname{Aut} D$, *the sequence*

$$1 \to \mathcal{G}_{S,D} \to \mathcal{G}_{S,D}^G \to G \to 1$$

is exact.

The proposition implies that $\mathcal{G}_{S,D}$ is the connected component of the identity of $\mathcal{G}_{S,D}^G$.

Corollary 4.2. *For all subgroups G of* $\operatorname{Aut} D$, *the Lie algebra of $\mathcal{G}_{S,D}^G$ is $\mathfrak{g}_{S,D}$.*

Theorem 4.3. *If (S, D) is a stable decorated surface where $g(S) \geq 3$, then*

$$0 \to \mathbb{Q} \to T_{S,D}^{\mathrm{un}} \to \mathcal{G}_{S,D} \to \mathrm{Sp}(H) \to 1$$

is exact. When $g = 2$, the homomorphism $T_{S,D}^{\mathrm{un}} \to \mathcal{U}_{S,D}$ is surjective.

This result deserves some comment. Corollary 3.10 implies that $T_{S,D}^{\mathrm{un}} \to \mathcal{U}_{S,D}$ is surjective when $g \geq 2$. Theorem 3.11 implies that

$$\mathbb{Q} \to T_{S,D}^{\mathrm{un}} \to \mathcal{G}_{S,D} \to \mathrm{Sp}(H) \to 1$$

is exact when $g \geq 3$. The injectivity of $\mathbb{Q} \to T_{S,D}^{\mathrm{un}}$ is equivalent to the non-vanishing of a Chern class. A clumsy proof of the non-vanishing is given in [16]. However, the non-vanishing follows directly from an earlier result of Morita [39], as explained in [23].

4.3. Tautological Homomorphisms

Suppose that (S, D) is a decorated surface. A decoration $\widetilde{D} = \widetilde{P} \cup \widetilde{V}$ of S is a *refinement* of D if

$$D \subseteq \widetilde{D},\ P \subseteq \widetilde{P} \cup \widetilde{V} \text{ and } V \subseteq \widetilde{V},$$

where $D = P \cup V$ and $\widetilde{D} = \widetilde{P} \cup \widetilde{V}$. Thus, in passing from $\widetilde{D}$ to D, tangent vectors can become points, and points and tangent vectors can be forgotten.

Suppose that (S, D) is stable. This implies that $(S, \widetilde{D})$ is also stable. For each $G \subseteq \operatorname{Aut} D \cap \operatorname{Aut} \widetilde{D}$, there is a natural homomorphism $\Gamma_{S,\widetilde{D}}^G \to \Gamma_{S,D}^G$.

4.4. Natural Actions

The natural actions of mapping class groups on the fundamental groups of associated surfaces can be completed.

Suppose that (S, D) is a stable decorated surface where $D = P \cup V$. Recall that S'_D is the surface obtained from S by removing the support of D.

Definition 4.4. An *admissible base point* x of S'_D is either (1) a point x of S'_D or (2) a tangent vector $x \in V$. Let $\widetilde{D} = D \cup \{x\}$. This equals D when $x \in V$.

If x is an admissible base point of S'_D, then $\pi_1(S'_D, x)$ is defined.

Suppose that G is a subgroup of $\operatorname{Aut}\widetilde{D}$ that fixes x. It can also be viewed as a subgroup of $\operatorname{Aut} D$. Denote the Lie algebra of $\pi_1(S'_D, x)^{\mathrm{un}}$ by $\mathfrak{p}(S'_D, x)$. There are natural actions

$$\tilde{\theta}_x : \Gamma^G_{S,\widetilde{D}} \to \operatorname{Aut}\mathfrak{p}(S'_D, x) \text{ and } \theta : \Gamma^G_{S,D} \to \operatorname{Out}\mathfrak{p}(S'_D).$$

The Zariski closure of the image of each of these is an extension of $G \times \mathrm{Sp}(H)$ by a prounipotent group. The universal mapping property of relative completion implies that $\tilde{\theta}_x$ and θ induce homomorphisms

$$\tilde{\phi}_x : \mathcal{G}^G_{S,\widetilde{D}} \to \operatorname{Aut}\mathfrak{p}(S'_D, x) \text{ and } \phi : \mathcal{G}^G_{S,D} \to \operatorname{Out}\mathfrak{p}(S'_D).$$

These, in turn, induce Lie algebra homomorphisms

$$d\tilde{\phi}_x : \mathfrak{g}_{S,\widetilde{D}} \to \operatorname{Der}\mathfrak{p}(S'_D, x) \text{ and } d\phi : \mathfrak{g}_{S,D} \to \operatorname{OutDer}\mathfrak{p}(S'_D).$$

Proposition 4.5. *If D is non-empty, then $\hat{\rho} : \Gamma^G_{S,D} \to \mathcal{G}^G_{S,D}$ is injective.*

Proof. It suffices to prove that $T_{S,D}$ injects into $\mathcal{U}_{S,D}$. It also suffices to prove the case where D consists only of points. Write $D = D' \cup \{x\}$. Set $S' = S - D'$ and $\pi = \pi_1(S', x)$. Then the natural homomorphism $\Gamma_{S,D} \to \operatorname{Aut}\pi$ is injective. Since π is resdidually torsion free nilpotent, $\pi \to \pi^{\mathrm{un}}$ is injective. It follows that $\Gamma_{S,D} \to \operatorname{Aut}\mathfrak{p}$ is injective, where $\mathfrak{p}$ is the Lie algebra of π^{un}. The result follows as this homomorphism factors $\Gamma_{S,D} \to \mathcal{G}_{S,D} \to \operatorname{Aut}\mathfrak{p}$, which forces $\Gamma_{S,D} \to \mathcal{G}_{S,D}$ to be injective. Q.E.D.

Denote the Lie algebra of $T^{\mathrm{un}}_{S,D}$ by $\mathfrak{t}_{S,D}$. Since the natural representations $T^{\mathrm{un}}_{S,x} \to \operatorname{Aut}\mathfrak{p}(S, x)$ and $T^{\mathrm{un}}_S \to \operatorname{Out}\mathfrak{p}(S)$ factor through $\mathcal{G}_{S,x} \to \operatorname{Aut}\mathfrak{p}(S, x)$ and $\mathcal{G}_S \to \operatorname{Out}\mathfrak{p}(S)$, Theorem 4.3 implies:

Proposition 4.6. *When $g \geq 3$, the natural representations $T_{S,x}^{\mathrm{un}} \to \operatorname{Aut} \mathfrak{p}(S,x)$ and $T_S^{\mathrm{un}} \to \operatorname{Out} \mathfrak{p}(S)$ have non-trivial kernel. Equivalently, both $\mathfrak{t}_{S,x} \to \operatorname{Der} \mathfrak{p}(S,x)$ and $\mathfrak{t}_S \to \operatorname{OutDer} \mathfrak{p}(S)$ have non-trivial kernel.* □

When $g \geq 3$, the only known elements of the kernel of $\mathfrak{t}_{S,x} \to \operatorname{Der} \mathfrak{p}(S,x)$ are those in $\ker\{\mathfrak{t}_{S,x} \to \mathfrak{u}_{S,x}\}$. So it is natural (and interesting) to ask whether $\mathfrak{u}_{S,x} \to \operatorname{Der} \mathfrak{p}(S,x)$ is injective when $g \geq 3$. (This homomorphism fails to be injective when $g = 1, 2$.)

§5. Weight Filtrations on Homology and Cohomology

The rational cohomology[4] of a complex algebraic variety X carries a natural filtration

$$0 = W_0 H^m(X) \subseteq W_1 H^m(X) \subseteq \cdots$$
$$\cdots \subseteq W_{2m-1} H^m(X) \subseteq W_{2m} H^m(X) = H^m(X)$$

called the *weight filtration*, which was constructed by Deligne using Hodge theory in [9, 10]. Weight filtrations can be constructed Galois actions as well [11]. Algebraic maps between complex algebraic varieties induce weight filtration preserving maps of their cohomology [10]. In particular, $(H^\bullet(X), W_\bullet)$ is a filtered algebra. The weight filtration is a powerful tool for studying the topology of complex algebraic varieties due to its strong exactness properties. In this section we give a brief introduction to weight filtrations directed at topologists. Deligne's paper [7] provides a more complete exposition of the yoga of weight filtrations. Full details can be found in [9].

An integer k is a (non-trivial) weight of $H^m(X)$ if its kth weight graded quotient

$$\operatorname{Gr}_k^W H^m(X) := W_k H^m(X)/W_{k-1} H^m(X)$$

is non-zero. We say that $H^m(X)$ is *pure* of weight k if k is the only non-trivial weight of $H^m(X)$. The weights on $H^m(X)$ are $\geq m$ when X is smooth and $\leq m$ when X is compact. So if X is smooth and projective, then $H^m(X)$ is pure of weight m.

Since we are working with fundamental groups, it is more natural to work with weight filtrations on homology than on cohomology. The

[4] Recall Convention 1.1: all (co)homology is with rational coefficients unless otherwise noted.

weight filtration on $H_m(X)$ is defined by

$$W_{-k}H_m(X) = \mathrm{Hom}(H^m(X)/W_{k-1}, \mathbb{Q}).$$

When X is smooth, the weights on $H_m(X)$ lie in $\{-2m, \ldots, -m\}$.

Example 5.1. The weight filtration on the homology of a smooth complex algebraic curve is determined by the topology of the underlying surface. Suppose that S is a compact oriented surface and that D is a finite subset. Set $S' = S - D$. Then one has the exact sequence (the dual of the Gysin sequence):

$$0 \to \widetilde{H}_0(D) \to H_1(S') \to H_1(S) \to 0.$$

The weight filtration on $H_1(S')$ is given by

$$W_{-k}H_1(S') = \begin{cases} 0 & k \geq 3 \\ \widetilde{H}_0(D) & k = 2 \\ H_1(S') & k \leq 1. \end{cases}$$

Note that $\mathrm{Gr}^W_{-1} H_1(S') = H_1(S)$. The weight filtration on $H_0(S')$ and $H_2(S')$ are uninteresting.

Higher dimensional examples with non-trivial weight filtrations can be constructed by taking products of curves. The weight filtration on the product of two varieties is the tensor product of the two weight filtrations:

$$W_k H^n(X \times Y) = \bigoplus_{\ell+m=n} \sum_{i+j=k} W_i H^\ell(X) \otimes W_j H^m(Y).$$

This induces an isomorphism

$$\mathrm{Gr}^W_k H^n(X \times Y) \cong \bigoplus_{\ell+m=n} \sum_{i+j=k} Gr^W_i H^\ell(X) \otimes \mathrm{Gr}^W_j H^m(Y).$$

5.1. Strictness and Exactness Properties

Weight filtrations that arise from Hodge and/or Galois theory have strong exactness properties which make them a powerful tool in studying the topology of algebraic varieties and algebraic maps.

Definition 5.2. A morphism $f(V_1, W_\bullet) \to (V_2, W_\bullet)$ of filtered vector spaces is *strict* with respect to $W_\bullet$ if for all $m \in \mathbb{Z}$

$$W_m V_2 \cap f(V_1) = f(W_m V_1).$$

Suppose that V is a vector space and that A is a subspace of V and $q : V \to B$ is a quotient. A filtration $W_\bullet$ of V induces one on A and B by restriction and projection, respectively:

$$W_m A := A \cap W_m V \text{ and } W_m B = q(W_m V).$$

In particular, we can induce filtrations on the kernel and cokernel of a filtration preserving map $f : (V_1, W_\bullet) \to (V_2, W_\bullet)$.

It is easy to check that f is strict with respect to $W_\bullet$ if and only if

$$0 \to \operatorname{Gr}_m^W \ker f \to \operatorname{Gr}_m^W V_1 \to \operatorname{Gr}_m^W V_2 \to \operatorname{Gr}_m^W \operatorname{coker} f \to 0$$

is exact for all $m \in \mathbb{Z}$.

Another important property of weight filtrations on cohomology groups of algebraic varieties, established in [9], is that there are natural (but not canonical) isomorphisms

$$H^m(X;\mathbb{C}) \cong \bigoplus \operatorname{Gr}_k^W H^m(X;\mathbb{C})$$

that are preserved by algebraic maps and which are compatible with tensor products, cup products, the Künneth isomorphism, etc. Establishing the existence of natural splittings of the weight filtration is the essential ingredient in establishing the strictness and exactness properties stated above.

Many other invariants of algebraic varieties and maps (such as the Leray spectral sequence, Gysin sequences, long exact sequences of a pair) carry natural weight filtrations, and all of their internal maps (differentials, Gysin maps, connecting homomorphisms) and all maps induced between them by algebraic maps preserve the weight filtration (sometimes with a shift) and are strict. The following example of Deligne [10] illustrates the basic yoga of weights and how it can be used to prove a topological result.

Example 5.3 (Deligne). Suppose that G is a connected linear algebraic group over $\mathbb{C}$ and that X is a smooth complex projective variety. Suppose that $\mu : G \times X \to X$ is an algebraic action. Deligne [10] shows that the weights on $H^k(G)$ are strictly larger than k except when $k = 0$. Since X is smooth and projective, $H^k(X)$ is pure of weight k. The mapping

$$\mu^* : H^n(X) \to H^n(G \times X) \cong \bigoplus_{\ell+m=n} H^\ell(G) \otimes H^m(X)$$

is thus filtration preserving. Since $H^n(X) = W_n H^n(X)$, strictness implies that

$$\operatorname{im} \mu^* = \operatorname{im} \mu^* \cap W_n H^n(G \times X) = H^0(G) \otimes H^n(X)$$

from which it follows that μ^* is the inclusion

$$H^n(X) \cong H^0(G) \otimes H^n(X) \hookrightarrow H^n(G \times X).$$

That is, rational cohomology cannot distinguish μ from the trivial action.

§6. Weight Filtrations on Completed Mapping Class Groups

Completions of mapping class groups have natural weight filtrations that are preserved by the natural homomorphisms between them. They arise because mapping class groups occur as fundamental groups of smooth stacks (moduli spaces of curves) and are constructed using either Hodge theory [18] or Galois theory [20].

Denote the lower central series of a Lie algebra $\mathfrak{h}$ by

$$\mathfrak{h} = L^1\mathfrak{h} \supseteq L^2\mathfrak{h} \supseteq \mathfrak{h} \supseteq \cdots$$

where $L^{m+1}\mathfrak{h} := [\mathfrak{h}, L^m\mathfrak{h}]$.

Theorem 6.1 (Hain [18]). *If (S, D) is a stable decorated surface and G is a subgroup of $\operatorname{Aut} D$, then $\mathcal{O}(\mathcal{G}^G_{S,D})$ has a natural weight filtration with which the product, antipode and coproduct are strictly compatible. This corresponds to a weight filtration*

$$\cdots \subseteq W_{-2}\mathcal{G}^G_{S,D} \subseteq W_{-1}\mathcal{G}^G_{S,D} \subseteq W_0\mathcal{G}^G_{S,D} = \mathcal{G}^G_{S,D}$$

by subgroups, where $W_{-1}\mathcal{G}^G_{S,D} = \mathcal{U}_{S,D}$. It also induces a filtration of the Lie algebra $\mathfrak{g}_{S,D}$ of the identity component. It has the property that $\mathfrak{g}_{S,D} = W_0\mathfrak{g}_{S,D}$ and $\mathfrak{u}_{S,D} = W_{-1}\mathfrak{g}_{S,D}$. The adjoint action

$$\mathfrak{g}_{S,D} \otimes \mathcal{O}(\mathcal{G}^G_{S,D}) \to \mathcal{O}(\mathcal{G}^G_{S,D}),$$

the bracket $\mathfrak{g}_{S,D} \otimes \mathfrak{g}_{S,D} \to \mathfrak{g}_{S,D}$ and the natural homomorphisms $\mathfrak{g}_{S,\tilde{D}} \to \mathfrak{g}_{S,D}$ are all strictly compatible with the weight filtration. When $g \geq 3$ and $\#D = 1$, the weight filtration is related to the lower central series of $\mathfrak{u}_{S,D}$ by

$$W_{-m}\mathfrak{g}_{S,D} = L^m\mathfrak{u}_{S,D}.$$

When $g = 0$ and $m \geq 1$,

$$W_{-2m+1}\mathfrak{g}_{S,D} = W_{-2m}\mathfrak{g}_{S,D} = L^m\mathfrak{u}_{S,D}.$$

In particular, when $g = 0$, $\mathfrak{g}_{S,D} = W_{-2}\mathfrak{g}_{S,D}$ and all odd weight graded quotients of $\mathfrak{g}_{S,D}$ are trivial.

For each subgroup G of $\operatorname{Aut} D$, conjugation induces infinitesimal actions

$$\operatorname{ad} : \mathfrak{g}_{S,D} \to \operatorname{Der} \mathcal{O}(\mathcal{G}^G_{S,D}) \text{ and } \operatorname{ad} : \mathfrak{g}_{S,D} \to \operatorname{Der} \mathfrak{g}_{S,D}.$$

Since $\operatorname{Gr}^W_0 \mathfrak{g}_{S,D} = \mathfrak{g}_{S,D}/\mathfrak{u}_{S,D} \cong \mathfrak{sp}(H)$, we have:

Corollary 6.2. *Each* $\operatorname{Gr}^W_m \mathcal{O}(\mathcal{G}^G_{S,D})$ *is a direct sum of finite dimensional* $\mathfrak{sp}(H)$*-modules and each* $\operatorname{Gr}^W_m \mathfrak{g}_{S,D}$ *is a direct product of finite dimensional* $\mathfrak{sp}(H)$*-modules.* □

These weight filtrations are compatible with those constructed (in [38, 15]) on fundamental groups of algebraic curves and their configuration spaces:

Theorem 6.3 (Morgan, Hain). *If* (S, D) *is a stable decorated surface, then the Lie algebra* $\mathfrak{p}$ *of the unipotent completion of the fundamental group of the configuration space of* m *ordered points in* S'_D *has a natural weight filtration that satisfies* $\mathfrak{p} = W_{-1}\mathfrak{p}$. *In particular,* $\mathfrak{p}(S'_D, x)$ *has a natural weight filtration that satisfies* $\mathfrak{p}(S'_D, x) = W_{-1}\mathfrak{p}(S'_D, x)$. *The bracket and the surjection* $\mathfrak{p}(S'_D, x) \to H_1(S'_D)$ *are strictly compatible with the weight filtration. When* $\#D \leq 1$, *the weight filtration of* $\mathfrak{p}(S'_D)$ *is given by its lower central series:*

$$W_{-m}\mathfrak{p}(S'_D, x) = L^m\mathfrak{p}(S'_D, x)$$

when $m \geq 1$.

The natural action of the $\mathfrak{g}_{S,D}$ on the $\mathfrak{p}(S'_D)$ is compatible with these weight filtrations.

Theorem 6.4 (Hain [18]). *If* (S, D) *is a stable decorated surface and* x *is an admissible base point of* S'_D, *then the natural homomorphisms*

$$\mathfrak{g}_{S,D\cup\{x\}} \to \operatorname{Der} \mathfrak{p}(S'_D, x) \text{ and } \mathfrak{g}_{S,D} \to \operatorname{OutDer} \mathfrak{p}(S'_D)$$

are strictly compatible with the natural weight filtrations.

For all stable decorated surfaces (S, D), the weight filtrations on

$$\mathfrak{g}_{S,D},\ \mathfrak{p}(S'_D, x),\ \operatorname{Der} \mathfrak{p}(S'_D, x),\ \operatorname{OutDer} \mathfrak{p}(S'_D)$$

all have natural splittings.[5] That is, if $\mathfrak{g}$ is such a Lie algebra, then there is a natural isomorphism of complete Lie algebras

$$\mathfrak{g} \cong \prod_m \operatorname{Gr}^W_m \mathfrak{g},$$

and if $\phi : \mathfrak{g} \to \mathfrak{h}$ is a natural homomorphism between two such Lie algebras, then the diagram

$$\begin{array}{ccc} \mathfrak{g} & \xrightarrow{\simeq} & \prod_m \operatorname{Gr}^W_m \mathfrak{g} \\ {\scriptstyle\phi}\downarrow & & \downarrow{\scriptstyle \operatorname{Gr}^W_\bullet \phi} \\ \mathfrak{h} & \xrightarrow{\simeq} & \prod_m \operatorname{Gr}^W_m \mathfrak{h} \end{array}$$

commutes.[6]

The existence of natural splittings allows one to study, *without loss of information*, the infinitesimal actions

$$d\tilde{\phi}_x : \mathfrak{u}_{S,\tilde{D}} \to \operatorname{Der} \mathfrak{p}(S'_D, x) \text{ and } d\phi : \mathfrak{u}_{S,D} \to \operatorname{OutDer} \mathfrak{p}(S'_D)$$

using the associated graded actions

$$\operatorname{Gr}^W_\bullet \mathfrak{u}_{S,\tilde{D}} \to \operatorname{Der} \operatorname{Gr}^W_\bullet \mathfrak{p}(S'_D, x) \text{ and } \operatorname{Gr}^W_\bullet \mathfrak{u}_{S,D} \to \operatorname{OutDer} \operatorname{Gr}^W_\bullet \mathfrak{p}(S'_D).$$

It also allows us to construct presentations of $\mathfrak{u}_{S,D}$ by giving presentations to their associated weight graded quotients as was done in [18] for the $\mathfrak{u}_S$ when $g \geq 6$.

Remark 6.5. One might hope that there are natural weight filtrations on the Lie algebras $\mathfrak{f}_n$, $\mathfrak{a}_n$ and $\mathfrak{o}_n$ associated to $\operatorname{Aut}^+ F_n$ and $\operatorname{Out}^+ F_n$ with respect to which the natural actions $\mathfrak{a}_n \to \operatorname{Der} \mathfrak{f}_n$ and $\mathfrak{o}_n \to \operatorname{OutDer} \mathfrak{f}_n$ are strict and for which each Lie algebra is naturally isomorphic to its associated graded. This would simplify the problem of finding presentations of $\mathfrak{ia}_n$ and $\mathfrak{io}_n$.

---

[5] In Hodge theory, one usually tensors with $\mathbb{C}$ first to construct these splittings. However, the machinery of tannakian categories implies the existence of such splittings over $\mathbb{Q}$. Cf. [9, 38]

[6] Examples of morphisms $\phi : \mathfrak{g} \to \mathfrak{h}$ that are strictly compatible with the weight filtration are those which are induced by morphisms of moduli spaces of curves or are associated with monodromy representations of fundamental groups of moduli spaces of curves associated to natural local systems over moduli spaces such as those associated to families of unipotent completions of fundamental groups of universal curves and other tautological bundles.

Such weight filtrations would probably exist if $\mathrm{Aut}\, F_n$ or $\mathrm{Out}\, F_n$ were the fundamental group of an algebraic variety or stack defined over a number field, or if one could construct an action of the Galois group of (say) $\mathbb{Q}(\boldsymbol{\mu}_n)$, where $\boldsymbol{\mu}_n$ denotes the nth roots of unity, on their profinite completions that was compatible with the action of the Galois group on the profinite completion of $\pi_1(\mathbb{A}^1 - \boldsymbol{\mu}_n, 0)$. It is not clear how to proceed, or if this could ever be true.

§7. The Relative Weight Filtration of a Nilpotent Endomorphism

This section is an exposition of the linear algebra of nilpotent endomorphisms of filtered vector spaces, which arises naturally in the study of degenerations of complex algebraic varieties. For example, suppose that

$$f : X \to \Delta$$

is a family of complex algebraic varieties over the unit disk that is topologically locally trivial over the punctured disk Δ^*. Denote the fiber of f over $t \in \Delta$ by X_t. Fix a base point $t_o \in \Delta^*$. Since the family is locally topologically trivial over Δ^*, there is a monodromy operator[7]

$$h : H^m(X_{t_o}) \to H^m(X_{t_o})$$

for each $m \in \mathbb{N}$. A general result of Griffiths-Landman-Grothendieck (Cf. [32, 34]) implies that the eigenvalues of h are roots of unity. So, by replacing the family by its pullback along a finite covering $\Delta \to \Delta, s \mapsto s^c$ if necessary, we may assume that h is unipotent (i.e., all of its eigenvalues are 1). In this case it is the exponential of a nilpotent matrix $N = \log h$.

Example 7.1. A classical and relevant example occurs when the fiber X_t over $t \in \Delta^*$ is a compact Riemann surface and the central fiber X_0 is obtained from $S = X_{t_o}$ by contracting a a finite set of disjoint simple closed curves (the *vanishing cycles*) $\{c_1, c_2, \dots, c_r\}$ in S. The geometric monodromy τ is the product of the Dehn twists about the c_j. The induced mapping

$$h : H_1(S) \to H_1(S)$$

[7]Recall Convention 1.1: all (co)homology is with rational coefficients unless otherwise noted.

is given by the *Picard-Lefschetz formula*:

$$h(x) = x + \sum_{j=1}^{r} \langle c_j, x \rangle c_j.$$

This is clearly unipotent. Its logarithm $N : H_1(S) \to H_1(S)$ is the operator

$$N : x \mapsto \sum_{j=1}^{r} \langle c_j, x \rangle c_j$$

which satisfies $N^2 = 0$. (This formula is independent of the orientations assigned to the c_j.) $\square$

7.1. The Weight Filtration of a Nilpotent Endomorphism

There is a natural weight filtration of a vector space associated to a nilpotent endomorphism N of it.

Proposition 7.2. *If N is a nilpotent endomorphism of a finite dimensional vector space V, then there is a unique filtration*

$$0 = W(N)_{-m-1} \subseteq W(N)_{-m} \subseteq W(N)_{-m+1} \subseteq \cdots$$
$$\cdots \subseteq W(N)_{m-1} \subseteq W(N)_m = V$$

of V such that

(i) *for all $n \in \mathbb{Z}$, $NW(N)_n \subseteq W(N)_{n-2}$;*
(ii) *for each $k \in \mathbb{Z}$,*

$$N^k : \mathrm{Gr}_k^{W(N)} V \to \mathrm{Gr}_{-k}^{W(N)} V$$

is an isomorphism.

The filtration $W(N)_\bullet$ of V is called the *weight filtration of N*.

Proof. To prove existence , it is enough to consider the case where N has a single Jordan block. There is a basis

$$e_{-m}, e_{-m+2}, e_{-m+4}, \ldots, e_{m-2}, e_m$$

of V such that $Ne_j = e_{j-2}$. Define

$$W(N)_j = \mathrm{span}\{e_k : k \leq j\}.$$

Uniqueness is proved by induction on the exponent of nilpotency of N. If $N = 0$, then uniqueness is clear. Suppose that $m > 0$ and that $N^{m+1} = 0$, but that $N^m \neq 0$. The vanishing of N^{m+1} implies that

$$W(N)_k = V \text{ and } W(N)_{-k-1} = 0 \text{ for all } k \geq m.$$

Since $N^m : \mathrm{Gr}_m V \to \mathrm{Gr}_{-m} V$ is an isomorphism, we must have

$$W(N)_{m-1} = \ker N^m \text{ and } W(N)_{-m} = \operatorname{im} N^m.$$

Since $N^m \neq 0$,

$$0 \neq W(N)_{-m} \subseteq W(N)_{m-1} \neq V.$$

The induced endomorphism $\bar{N}$ of $V' := W(N)_{m-1}/W(N)_{-m}$ satisfies $\bar{N}^m = 0$. By induction, the weight filtration of $W(\bar{N})$ of $\bar{N}$ is unique. All weight filtrations of V must satisfy

$$W(N)_k = \text{ inverse image of } W(\bar{N})_k \text{ whenever } -m < k < m.$$

Uniqueness follows. Q.E.D.

Note that $W(N)_\bullet$ is centered at 0. If $V = H^m(X)$, where X is a smooth projective variety, it is natural to reindex the weight filtration of a nilpotent endomorphism N of V so that it is centered at the weight m of V. The shifted filtration

$$M_k V := W(N)_{k-m}$$

is centered at m. The reindexed filtration $M_\bullet$ satisfies $NM_k \subseteq M_{k-2}$ and

$$N^k : \mathrm{Gr}^M_{m+k} V \xrightarrow{\simeq} \mathrm{Gr}^M_{m-k} V$$

is an isomorphism for all $k \in \mathbb{Z}$. We will call the shifted weight filtration $M_\bullet$ the monodromy weight filtration of $N : V \to V$.

Example 7.3. The monodromy weight filtration for nilpotent endomorphism N of the weight -1 vector space $H_1(S)$ in Example 7.1 is:

$$\begin{aligned} M_0 &= W(N)_1 = H_1(S) \\ M_{-1} &= W(N)_0 = \ker N \\ M_{-2} &= W(N)_{-1} = \operatorname{im} N = \operatorname{span}\{c_1, \dots, c_r\} \\ M_{-3} &= W(N)_{-2} = 0. \end{aligned} \tag{2}$$

The existence of a weight filtration extends to arbitrary direct sums and direct products of nilpotent N-modules.

7.2. Curve Systems

A *curve system* on a stable decorated surface (S, D) is a set

$$\gamma = \{c_0, \ldots, c_r\}$$

of disjoint simple closed curves such that each connected component of

$$S'_D - |\gamma| := S'_D - \bigcup_{j=1}^{r} c_j$$

has negative Euler characteristic. Equivalently, no two c_j are isotopic in S'_D and no c_j bounds a disk or punctured disk in S'_D. Two curve systems are considered to be equal if they are isotopic in S.

Denote the Dehn twist about c_j by τ_j. Since the c_j are disjoint, these commute. Set $\tau = \prod_j \tau_j$. The action

$$\tau_* : H_1(S'_D) \to H_1(S'_D)$$

of τ on $H_1(S'_D)$ is given by the *Picard-Lefschetz formula:*

$$\tau_*(x) = x + \sum_{j=0}^{r} \langle c_j, x \rangle c_j.$$

This is clearly unipotent. Its logarithm $N_\gamma : H_1(S'_D) \to H_1(S'_D)$ is the operator $\tau_* - \mathrm{id}$ which is given by

$$N_\gamma : x \mapsto \sum_{j=1}^{r} \langle c_j, x \rangle c_j.$$

It satisfies $N_\gamma^2 = 0$. Note that it preserves the weight filtration $W_\bullet$ defined on $H_1(S'_D)$ in Example 5.1 and acts trivially on $W_{-2} H_1(S'_D)$.

The following example will be used later in the paper.

Example 7.4. Suppose that $H = H_1(S)$ and $N = N_\gamma$, where H has weight -1. Denote the corresponding monodromy weight filtration by $M_\bullet$. Set

$$A = \mathrm{Gr}_0^M H, \ H_0 = \mathrm{Gr}_{-1}^M H, \ B = \mathrm{Gr}_{-2}^M H.$$

There is a natural isomorphism $\mathfrak{sp}(H) \cong S^2 H$, which we consider to have weight 0 as it is a subspace of $\mathrm{End}(H)$, which has weight 0. Note that

$$\mathrm{Gr}_\bullet^M \mathfrak{sp}(H) = \mathfrak{sp}(\mathrm{Gr}_\bullet^M H)$$

and that there is a natural Lie algebra isomorphism

$$\operatorname{Gr}_0^M \mathfrak{sp}(H) \cong \mathfrak{gl}(A) \oplus \mathfrak{sp}(H_0).$$

Denote by ξ the element of $\mathfrak{sp}(\operatorname{Gr}_\bullet^M H)$ that corresponds to the identity element of $\mathfrak{gl}(A)$. Note that ξ acts as the identity on A, minus the identity on B and trivially on H_0. It follows that if we consider $H^{\otimes n}$ to have weight $-n$, then ξ acts on $\operatorname{Gr}_k^M H^{\otimes n}$ as multiplication by $k-n$. It follows that if V is an $\mathfrak{sp}(H)$-submodule of $H^{\otimes n}$, then ξ acts on $\operatorname{Gr}_k^M V$ as multiplication by $k-n$.

7.3. The Weight Filtration of a Nilpotent Endomorphism of a Filtered Vector Space

Now suppose that N is a nilpotent endomorphism of a filtered finite dimensional vector space V. That is, V has a filtration

$$0 \subseteq \cdots \subseteq W_{m-1}V \subseteq W_m V \subseteq W_{m+1}V \subseteq \cdots \subseteq V$$

which is stable under N. This is extended to the infinite dimensional case using the conventions of Section 2.1. Namely, infinite dimensional examples are either ind- or pro-objects of the category of finite dimensional filtered vector spaces; the nilpotent endomorphism N is replaced by a locally nilpotent endomorphism (i.e., a direct limit of nilpotent endomorphisms) in the ind case and a pronilpotent endomorphism (i.e., an inverse limit of nilpotent endomorphisms) in the pro case.

We will often call the filtration $W_\bullet$ of V the *weight filtration* of V and $\operatorname{Gr}_m^W V$ the mth weight graded quotient of V.

Natural examples of a filtered vector space $(V, W_\bullet)$ with a nilpotent endomorphism arise from degenerations of smooth (not necessarily compact) varieties. In this case $(V, W_\bullet)$ is $H^m(X_t)$ endowed with its natural weight filtration, and N is the logarithm of the unipotent part of the monodromy operator.

Since N preserves the weight filtration, it induces an endomorphism

$$N_m := \operatorname{Gr}_m^W N : \operatorname{Gr}_m^W V \to \operatorname{Gr}_m^W V.$$

of the mth weight graded quotient of V. Since, by assumption, $\operatorname{Gr}_m^W V$ is the product or sum of nilpotent N-modules, Proposition 7.2 implies that each graded quotient has a weight filtration $W(N_m)$. The reindexed filtration $W(N_m)[m]_\bullet$ is centered at m. Denote it by $M_\bullet^{(m)}$

Definition 7.5. A filtration $M_\bullet$ of V is called a *relative weight filtration* of $N : (V, W_\bullet) \to (V, W_\bullet)$ if

(i) for each $k \in \mathbb{Z}$, $NM_k \subseteq M_{k-2}$;

(ii) the filtration induced by $M_\bullet$ on $\mathrm{Gr}^W_m V$ is the reindexed weight filtration $M^{(m)}_\bullet$.

Relative weight filtrations, if they exist, are unique. (Cf. [48]).

Example 7.6. If $N : (V, W_\bullet) \to (V, W_\bullet)$ satisfies $N(W_m V) \subseteq W_{m-2}V$ for all $m \in \mathbb{Z}$, then each $N_m = 0$ and the relative weight flirtation $M_\bullet$ of N exists and equals the original weight filtration $W_\bullet$.

Example 7.7. Suppose that γ is a curve system on a stable decorated surface (S, D). Take $V = H_1(S'_D)$ with the weight filtration defined in Example 5.1 and N to be the nilpotent endomorphism N_γ associated to γ defined in Section 7.2. The non-trivial weight graded quotients of $H_1(S'_D)$ are

$$\mathrm{Gr}^W_{-1} H_1(S'_D) \cong H_1(S) \text{ and } \mathrm{Gr}^W_{-2} H_1(S'_D) \cong \widetilde{H}_0(D).$$

Note that $N_{-1} : H_1(S) \to H_1(S)$ is the operator given in Example 7.1. Consequently $M^{(-1)}_\bullet$ is given by Example 7.3:

$$M^{(-1)}_{-2}H_1(S) = \operatorname{im} N_{-1}, M^{(-1)}_{-1}H_1(S) = \ker N_{-1}, M^{(-1)}_0 H_1(S) = H_1(S).$$

Since $N_{-2} = 0$,

$$0 = M^{(-2)}_{-3} \subseteq M^{(-2)}_{-2} = \widetilde{H}_0(D).$$

The relative weight filtration of $N_\gamma : H_1(S'_D) \to H_1(S'_D)$ exists. It is defined by

$$M_{-3} = 0, M_{-2} = \operatorname{im} N_\gamma + \widetilde{H}_0(D), M_{-1} = \ker N_\gamma + \widetilde{H}_0(D), M_0 = H_1(S'_D).$$

Even though the weight filtration of a nilpotent endomorphism of a finite dimensional vector space always exists, the relative weight filtration of a nilpotent endomorphism of a *filtered* vector space $(V, W_\bullet)$ usually does not exist. Necessary and sufficient conditions for the existence of a relative weight filtration are given in [48].

The existence of a relative weight filtration on the rational cohomology of the general fiber of a degeneration of complex algebraic varieties was first established for degenerations of varieties by Deligne [11, (1.8)] using ℓ-adic methods, and for smooth varieties over the complex numbers using Hodge theory by Steenbrink and Zucker [48]. It provides non-trivial restrictions on the possible monodromy operators of degenerations of algebraic varieties. For example, the existence relative weight filtration is a strong enough invariant to show that a bounding pair (BP) map cannot be the geometric monodromy of a degeneration of complex algebraic curves:[8]

Example 7.8. Suppose that $g(S) \geq 1$ and that the curve system $\{c_0, c_1\}$ is a bounding pair of simple closed curves in S. (That is, $S - |\gamma|$ has two connected components.) Suppose that $P = \{x_0, x_1\}$ is a pair of points in $S - |\gamma|$, one in each component. Denote the Dehn twist about c_j by τ_j. The associated bounding pair map is $\tau = \tau_1 \tau_0^{-1}$. It acts non-trivially and unipotently on $H_1(S'_D)$. Its logarithm $N = \tau_* - \mathrm{id}$ acts trivially on both weight graded quotients of $H_1(S'_D)$. Because of this, the relative weight filtration $M_\bullet$, if it exists, must agree with the weight filtration $W_\bullet$. But since these satisfy

$$M_{-3}H_1(S'_D) = W_{-3}H_1(S'_D) = 0 \text{ and}$$
$$M_{-1}H_1(S'_D) = W_{-1}H_1(S'_D) = H_1(S'_D),$$

the condition $NM_{-1} \subseteq M_{-3}$ implies that $N = 0$. But this contradicts the non-triviality of τ_*. Consequently, the endomorphism N of $(H_1(S'_D), W_\bullet)$ has no relative weight filtration.

§8. Relative Weight Filtrations on Mapping Class Groups

An element σ of a proalgebraic group $\mathcal{G}$ is *prounipotent* if it lies in a prounipotent subgroup. An element N of a pro-Lie algebra $\mathfrak{g}$ is *pronilpotent* if it lies in a pronilpotent subalgebra.

Lemma 8.1. *Each prounipotent element of a proalgebraic group $\mathcal{G}$ can be written uniquely as the exponential of a pronilpotent element of $\mathfrak{g}$, the Lie algebra of $\mathcal{G}$.*

Proof. Suppose that τ is a prounipotent element of $\mathcal{G}$. The existence of a pronilpotent logarithm of τ is clear as it lies in a prounipotent subgroup. Since every algebraic subgroup of a prounipotent group is prounipotent, the intersection of two prounipotent subgroups of $\mathcal{G}$ is also prounipotent. If $\tau = \exp N_1 = \exp N_2$, then lies in the intersection of the two unipotent 1-parameter subgroups $\{\exp tN_j : t \in F\}$, $j = 1, 2$. If $\tau \neq 1$, this forces $N_1 = N_2$. If $\tau = 1$, the unique logarithm is $N = 0$. Q.E.D.

---

[8]This can be proved by elementary and direct arguments. However, using the non-existence of the relative weight filtration to prove that a BP map cannot be the geometric monodromy of degeneration of curves illustrates the kinds of restrictions that the existence of relative weight filtrations places on the monodromy of degenerations of varieties in general.

Proposition 8.2. *If $\tau \in \Gamma_{S,D}^G$ is a Dehn twist, then $\hat{\rho}(\tau)$ is a prounipotent element of $\mathcal{G}_{S,D}$.*

Proof. The Picard-Lefschetz formula implies that $\rho(\tau)$ is a unipotent element of $\mathrm{Sp}(H_1(S))$ and that it is the exponential of $N = \rho(\tau) - \mathrm{id}$. The inverse image $\mathcal{H}$ of the unipotent subgroup $L := \{\exp tN : t \in \mathbb{Q}\}$ of $\mathrm{Sp}(H)$ in $\mathcal{G}_{S,D}$ is prounipotent as it is an extension

$$1 \to \mathcal{U}_{S,D} \to \mathcal{H} \to L \to 1$$

of a unipotent group by a prounipotent group. Since $\hat{\rho}(\tau) \in \mathcal{H}$, it is prounipotent and thus has a unique pronilpotent logarithm. Q.E.D.

Suppose that (S, D) is a stable decorated surface and that $\gamma = \{c_0, \dots, c_m\}$ is a curve system on (S, D). Denote the Dehn twist on c_j by τ_j. By Proposition 8.2 $\hat{\rho}(\tau_j)$ has a canonical logarithm $N_j \in \mathfrak{g}_{S,D}$.

Define the closed cone in $\mathfrak{g}_{S,D}$ associated to γ by

$$C(\gamma) = \{\sum_{j=0}^{m} r_j N_j : r_j \in \mathbb{Q} \text{ and } r_j \geq 0\}$$

and the open cone by

$$C^o(\gamma) = \{\sum_{j=0}^{m} r_j N_j : r_j \in \mathbb{Q} \text{ and } r_j > 0\}$$

Then $C(\gamma)$ is a simplicial cone in $\mathbb{Q}^\gamma$ whose faces correspond to the subsets σ of γ:

$$C(\gamma) = \coprod_{\sigma \subseteq \gamma} C^o(\sigma).$$

Suppose that $N \in C(\gamma)$ and that G is a subgroup of $\mathrm{Aut}\, D$. The infinitesimal actions

$$\mathrm{ad} : \mathfrak{g}_{S,D} \to \mathrm{Der}\, \mathcal{O}(\mathcal{G}_{S,D}^G) \text{ and } \mathrm{ad} : \mathfrak{g}_{S,D} \to \mathrm{Der}\, \mathfrak{g}_{S,D}.$$

induce actions of N on each weight graded quotient of $\mathcal{O}(\mathcal{G}_{S,D}^G)$ and $\mathfrak{g}_{S,D}$. By Corollary 6.2, each weight graded quotient of $\mathcal{O}(\mathcal{G}_{S,D}^G)$ is a direct sum of finite dimensional $\mathfrak{sp}(H)$-modules and each weight graded quotient of $\mathfrak{g}_{S,D}$ is a direct product of finite dimensional $\mathfrak{sp}(H)$-modules. Consequently, each weight graded quotient of $\mathcal{O}(\mathcal{G}_{S,D}^G)$ is a direct sum of finite dimensional nilpotent N-modules and and each weight graded quotient of $\mathfrak{g}_{S,D}$ is a direct product of finite dimensional nilpotent N-modules.

The following theorem is a special case of more general results proved in [20] using Galois theory and in [22] using Hodge theory.

Theorem 8.3. *For all curve systems γ of a stable decorated surface (S, D), all subgroups G of* $\operatorname{Aut} D$, *and all $N \in C^o(\gamma)$, there is a (necessarily unique) relative weight filtration $M_\bullet^\gamma$ (denoted $M_\bullet$) of $\mathcal{O}(\mathcal{G}_{S,D})$ of the endomorphism*

$$\operatorname{ad}(N) \in \operatorname{Der}(\mathcal{O}(\mathcal{G}_{S,D}^G), W_\bullet)$$

that is compatible with the product, antipode and coproduct of $\mathcal{O}(\mathcal{G}_{S,D})$. It induces relative weight filtrations on $\mathfrak{g}_{S,D}$ and $\mathfrak{p}(S'_D, x)$, where $x \in S'_D - |\gamma|$ is an admissible base point. The bracket of each of these Lie algebras is strictly compatible with $M_\bullet$. These relative weight filtrations depends only on γ and will be denoted by $M_\bullet^\gamma$. Each $N_j \in C(\gamma)$ lies in $M_{-2}\mathfrak{g}_{S,D}$. Moreover, for each curve system γ there is a natural (though not canonical) isomorphism

$$\mathfrak{g}_{S,D} \cong \prod_{k,m} \operatorname{Gr}_k^M \operatorname{Gr}_m^W \mathfrak{g}_{S,D}$$

of completed Lie algebras. In addition, if $\widetilde{D}$ is a refinement of D that contains the base point x and $\tilde{\gamma}$ is a curve system on $(S, \widetilde{D})$ whose image in (S, D) is γ, then the natural actions

$$d\tilde{\phi}_x : \mathfrak{g}_{S,\widetilde{D}} \to \operatorname{Der} \mathfrak{p}(S'_D, x) \text{ and } d\phi : \mathfrak{g}_{S,D} \to \operatorname{OutDer} \mathfrak{p}(S'_D).$$

are strictly compatible with $W_\bullet$ and the relative weight filtrations $M_\bullet^\gamma$ and $M_\bullet^{\tilde{\gamma}}$. The filtrations $M_\bullet$ and $W_\bullet$ can be simultaneously split. That is, they can be chosen so that the diagram

$$\begin{array}{ccc} \mathfrak{g}_{S,\widetilde{D}} & \xrightarrow{\simeq} & \prod_{k,m} \operatorname{Gr}_k^M \operatorname{Gr}_m^W \mathfrak{g}_{S,\widetilde{D}} \\ \downarrow & & \downarrow \\ \operatorname{Der} \mathfrak{p}(S'_D, x) & \xrightarrow{\simeq} & \operatorname{Der} \operatorname{Gr}_\bullet^M \operatorname{Gr}_\bullet^W \mathfrak{p}(S'_D, x) \end{array}$$

commutes. There is a similar diagram for $d\phi : \mathfrak{g}_{S,D} \to \operatorname{OutDer} \mathfrak{p}(S'_D)$.

Since the diagonal $\Delta : \mathcal{O}(\mathcal{G}_{S,D}^G) \to \mathcal{O}(\mathcal{G}_{S,D}^G) \otimes \mathcal{O}(\mathcal{G}_{S,D}^G)$ preserves $M_\bullet$, the image of $M_{-1}\mathcal{O}(\mathcal{G}_{S,D}^G)$ under the diagonal is contained in

$$M_{-1}\mathcal{O}(\mathcal{G}_{S,D}^G) \otimes \mathcal{O}(\mathcal{G}_{S,D}^G) + \mathcal{O}(\mathcal{G}_{S,D}^G) \otimes M_{-1}\mathcal{O}(\mathcal{G}_{S,D}^G).$$

This implies that $M_{-1}\mathcal{O}(\mathcal{G}_{S,D}^G)$ is a Hopf ideal of $\mathcal{O}(\mathcal{G}_{S,D}^G)$. Define

$$M_0\mathcal{G}_{S,D}^G = \operatorname{Spec}\left(\mathcal{O}(\mathcal{G}_{S,D}^G)/M_{-1}\mathcal{O}(\mathcal{G}_{S,D}^G)\right).$$

This is a subgroup of $\mathcal{G}^G_{S,D}$. Since $M_\bullet$ is preserved by the bracket, $M_k\,\mathfrak{g}_{S,D}$ is a pronilpotent Lie subalgebra of $\mathfrak{g}_{S,D}$ whenever $k \le 0$. The Lie algebra of $M_0\mathcal{G}^G_{S,D}$ is $M_0\mathfrak{g}_{S,D}$. When $k < 0$, $M_k\mathfrak{g}_{S,D}$ is pronilpotent. Denote the corresponding prounipotent subgroup of $\mathcal{G}^G_{S,D}$ by $M_k\mathcal{G}_{S,D}$.

The uniqueness of relative weight filtrations is a strong condition which implies that $M^\gamma_\bullet$ has many nice properties. Some are established in the following paragraphs.

Proposition 8.4. *Suppose that γ is a curve system on a stable decorated surface (S, D). If $\phi \in \Gamma^G_{S,D}$, then the relative weight filtrations of $\mathcal{O}(\mathcal{G}^G_{S,D})$ and $\mathfrak{g}_{S,D}$ satisfy $M^{\phi(\gamma)}_\bullet = \mathrm{Ad}(\phi)M^\gamma_\bullet$.*

Proof. For a curve system σ, let $\tau_\sigma = \prod_{c\in\sigma}\tau_c$ where τ_c denotes the Dehn twist about c. Denote the logarithm of τ_σ by N_σ. Since $\tau_{\phi(c)} = \phi\tau_c\phi^{-1}$, it follows that $\tau_{\phi(\gamma)} = \phi\tau_\gamma\phi^{-1}$, which implies $N_{\phi(\gamma)} = \mathrm{Ad}(\phi)N_\gamma$. Since $N_\gamma(M^\gamma_k) \subseteq M^\gamma_{k-2}$, it follows that

$$\begin{aligned} N_{\phi(\gamma)}\big(\mathrm{Ad}(\phi)M^\gamma_k\big) &= \big(\mathrm{Ad}(\phi)N_\gamma\big)\big(\mathrm{Ad}(\phi)M^\gamma_k\big) \\ &= \mathrm{Ad}(\phi)\big(N_\gamma\big(M^\gamma_k\big)\big) \subseteq \mathrm{Ad}(\phi)M^\gamma_{k-2}. \end{aligned}$$

The result now follows from the uniqueness of the relative weight filtration. Q.E.D.

The subalgebra $M^\gamma_0\,\mathfrak{g}_{S,D}$ of $\mathfrak{g}_{S,D}$ behaves like a parabolic subalgebra of a semi-simple Lie algebra:

Parabolic subalgebras of semi-simple and Kac-Moody Lie algebras are self normalizing, and correspond to boundary strata in the semi-simple case. The following result suggests that when $g \ge 2$, the subalgebras $M^\gamma_0\,\mathfrak{g}_{S,D}$ of $\mathfrak{g}_{S,D}$ might provide a good notion of parabolic subalgebra of $\mathfrak{g}_{S,D}$.

Proposition 8.5. *If γ is a curve system on a stable decorated surface (S, D) where $g(S) \ge 3$, then the normalizer of $M^\gamma_0\mathfrak{g}_{S,D}$ in $\mathfrak{g}_{S,D}$ is $M^\gamma_0\,\mathfrak{g}_{S,D}$.*

Proof. Since the functor $\mathrm{Gr}^M_\bullet\,\mathrm{Gr}^W_\bullet$ is exact, it suffices to prove the result for the associated bigraded Lie algebra $\mathrm{Gr}^M_\bullet\,\mathrm{Gr}^W_\bullet\,\mathfrak{g}_{S,D}$. Set $M_\bullet = M^\gamma_\bullet$ and $H = H_1(S)$. Let $\xi \in \mathrm{Gr}^M_0\,\mathfrak{sp}(H)$ be the element defined in Example 7.4. It lies in $\mathrm{Gr}^M_0\,\mathrm{Gr}^W_0\,\mathfrak{g}_{S,D}$. Johnson's theorem [29] implies that $\mathrm{Gr}^W_m\,\mathfrak{g}_{S,D}$ is an $\mathfrak{sp}(H)$-quotient of $(H^n \oplus \Lambda^3 H)^{\otimes m}$, where $n = \#D - 1$. It follows from Example 7.4 that if $k > 0$ and $X \in \mathrm{Gr}^M_k\,\mathrm{Gr}^W_m\,\mathfrak{g}_{S,D}$, then $[\xi, X] = (k - m)X$, which is non-zero whenever $k > 0$. Thus, if $X \notin M_0\,\mathrm{Gr}^M_\bullet\,\mathrm{Gr}^W_\bullet\,\mathfrak{g}_{S,D}$, then X does not normalize $M_0\,\mathrm{Gr}^M_\bullet\,\mathrm{Gr}^W_\bullet\,\mathfrak{g}_{S,D}$. Q.E.D.

This result also holds in genus 2, but not in genus 1.

8.1. Dependence on γ

For a curve system γ of a stable decorated surface (S, D), denote the subspace of $H_1(S)$ spanned by the homology classes of the $c \in \gamma$ by $\langle\gamma\rangle$.

Proposition 8.6. *If γ is a curve system of a stable decorated surface (S, D) and $\sigma \subseteq \gamma$, then the relative weight filtrations $M_\bullet^\gamma$ and $M_\bullet^\sigma$ of $\mathcal{O}(\mathcal{G}_{S,D}^G)$, $\mathfrak{g}_{S,D}$ and $\mathfrak{p}(S'_D)$ are equal if and only if $\langle\sigma\rangle = \langle\gamma\rangle$.*

Proof. The condition that $\langle\gamma\rangle = \langle\sigma\rangle$ implies that the monodromy weight filtrations on $H_1(S)$ are equal. Since each weight graded quotient of $\mathcal{O}(\mathcal{G}_{S,D}^G)$, $\mathfrak{g}_{S,D}$ and $\mathfrak{p}(S'_D)$ is a subquotient of a tensor power of $H_1(S)$, it follows that the monodromy weight filtrations associated to γ and σ on $\mathrm{Gr}_\bullet^W \mathfrak{g}_{S,D}$ are equal if and only if $\langle\gamma\rangle = \langle\sigma\rangle$. Similarly for $\mathfrak{p}(S'_D)$ and $\mathcal{O}(\mathcal{G}_{S,D}^G)$.

To complete the proof, we now show that if the monodromy weight filtrations on $\mathrm{Gr}_\bullet^W V$ agree, where $V = \mathcal{O}(\mathcal{G}_{S,D}^G)$, $\mathfrak{g}_{S,D}$ or $\mathfrak{p}(S'_D)$, then the relative weight filtrations on V agree. Since $\sigma \subseteq \gamma$, for each $c \in \sigma$, $N_c(M_k^\gamma) \subseteq M_{k-2}^\gamma$. That is $M_\bullet^\gamma$ is a relative weight filtration on V for each $N \in C^o(\sigma)$. Uniqueness implies that $M_\bullet^\sigma = M_\bullet^\gamma$. Q.E.D.

When $g(S) = 0$, $H_1(S) = 0$ and the hypotheses of Proposition 8.6 are satisfied when σ is empty. This implies that the relative weight filtration is the existing weight filtration:

Corollary 8.7. *If $\gamma = \{c_0, \ldots, c_m\}$ is a curve system on a stable decorated surface (S, D) of genus 0 and G is a subgroup of* Aut D, *then the relative weight filtrations of $\mathcal{O}(\mathcal{G}_{S,D}^G)$, $\mathfrak{p}(S'_D, x)$ and $\mathfrak{g}_{S,D}$ equal their natural weight filtrations $W_\bullet$. Consequently, $\mathcal{G}_{S,D}^G = M_0^\gamma \mathcal{G}_{S,D}^G$.* □

Suppose that (S, D) is a stable decorated surface, where $D = P \cup V$. For the purposes of this definition, we will consider V as a set of boundary components. Following Hatcher-Thurston [27] we say that two curve systems γ' and γ'' of (S, D) differ by an *A-move* if

$$\{c_1, c_2, c_3, c_4\} \subseteq (\gamma' \cap \gamma'') \cup V$$

that bound a genus 0 subsurface T of S and there are $c'_0 \in \gamma'$ and $c''_0 \in \gamma''$ that lie in T and

$$\gamma' = \sigma \cup \{c'_0\} \text{ and } \gamma' = \sigma \cup \{c'_0\}.$$

Figure 1 illustrates an A-move.

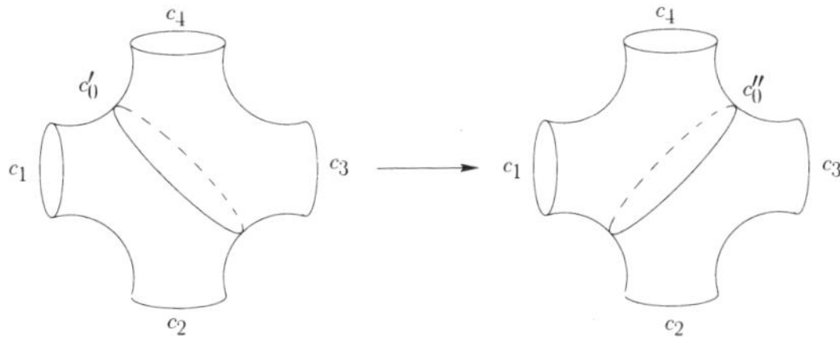

Fig. 1. An A-move

Suppose that σ and γ are curve systems on the stable decorated surface (S, D). We say that γ *is obtained from* σ *by homology neutral insertions* (or that σ is obtained from γ by homology neutral deletions) if $\sigma \subseteq \gamma$ and the subspaces $\langle\sigma\rangle$ and $\langle\gamma\rangle$ of $H_1(S)$ are equal.

The invariance of the $M_\bullet^\gamma$ under homology neutral insertions and deletions follows directly from Proposition 8.6.

Proposition 8.8. *Suppose that* (S, D) *is a stable decorated surface. If* γ_1 *and* γ_2 *are two curve systems on* (S, D) *that differ by a sequence of* A*-moves and by homology neutral insertions and deletions, then the relative weight filtrations* $M_\bullet^{\gamma_1}$ *and* $M_\bullet^{\gamma_2}$ *of* $\mathfrak{g}_{S,D}$ *are equal.* □

8.2. Glueing Lemma

Suppose that $\gamma = \{c_0, \ldots, c_m\}$ is a curve system on the stable decorated surface (S, D). For each connected component T' of $S'_D - |\gamma|$ there is a compact oriented surface T and decorations $D_T = P_T \cup V_T$ such that

$$P_T = T' \cap P$$

and V_T is the union of $T' \cap V$ with the new tangent vectors obtained by collapsing the boundary components created by removing the c_j.[9]

There is a natural homomorphism (the *glueing map*)

$$\prod_T \Gamma_{T,D_T} \to \Gamma_{S,D} \tag{3}$$

whose image is the subgroup of $\Gamma_{S,D}$ that are represented by diffeomorphisms that restrict to the identity on each c_j. The kernel is isomorphic to $\mathbb{Z}^\gamma$.

[9]It is convenient and natural to choose a point on each c_j, as it will provide a marking on each boundary component of $S'_D - |\gamma|$.

Proposition 8.9 (Glueing Lemma). *The glueing map induces a homomorphism on relative completions such that the diagram*

$$\begin{array}{ccc} \prod_T \Gamma_{T,D_T} & \longrightarrow & \Gamma_{S,D} \\ \downarrow & & \downarrow \\ \prod_T \mathcal{G}_{T,D_T} & \longrightarrow & \mathcal{G}_{S,D} \end{array}$$

commutes. The induced Lie algebra homomorphism

$$\bigoplus_T \mathfrak{g}_{T,D_T} \to \mathfrak{g}_{S,D}$$

is strict with respect to the weight filtration $W_\bullet$ on the $\mathfrak{g}_{T,D_T}$ and the relative weight filtration $M_\bullet^\gamma$ on $\mathfrak{g}_{S,D}$.

Proof. Here we prove the existence of the induced homomorphism $\prod_T \mathcal{G}_{T,D_T} \to \mathcal{G}_{S,D}$. Its compatibility with the filtrations follows from the existence of limit mixed Hodge structures [22] or, alternatively, the Galois equivariance of the corresponding map of profinite mapping class groups.

Set $H = H_1(S)$ and $H_T = H_1(T)$. The relative weight filtration $M_\bullet := M_\bullet^\gamma$ of H satisfies

$$\mathrm{Gr}^M_{-1} H = \oplus_T H_T \text{ and } M_{-2}H = \langle \gamma \rangle.$$

Set

$$M_m \mathrm{Sp}(H) := \{\phi \in \mathrm{Sp}(H) : (\phi - \mathrm{id})(M_k H) \subseteq M_{k+m} H\}.$$

This is the subgroup of $\mathrm{Sp}(H)$ with Lie algebra $M_m\mathfrak{sp}(H)$. The Zariski closure of the image of $\prod_T \Gamma_{T,D_T}$ in $\mathrm{Sp}(H)$ is contained in $M_0\mathrm{Sp}(H)$ and is the extension of the subgroup $\prod_T \mathrm{Sp}(H_T)$ of $\mathrm{Gr}^M_0 \mathrm{Sp}(H)$ by a unipotent subgroup of $M_{-2}\mathrm{Sp}(H)$. Denote by $\mathcal{H}$ the inverse image of $\prod_T \Gamma_{T,D_T}$ under the surjection $M_0\mathcal{G}_{S,D} \to \mathrm{Gr}^M_0 \mathrm{Sp}(H)$. It is an extension of $\prod_T \Gamma_{T,D_T}$ by the prounipotent group $M_{-2}\mathcal{G}_{S,D}$.

The completion of $\prod_T \Gamma_{T,D_T}$ with respect to the natural homomorphism $\prod_T \Gamma_{T,D_T} \to \prod_T \mathrm{Sp}(H_T)$ is $\prod_T \mathcal{G}_{T,D_T}$. The universal mapping property of relative completion implies that the homomorphism $\prod_T \Gamma_{T,D_T} \to \mathcal{H}$ induces a homomorphism $\prod_T \mathcal{G}_{T,D_T} \to \mathcal{H}$ such that the diagram

$$\begin{array}{ccccc} \prod_T \Gamma_{T,D_T} & & \longrightarrow & & \Gamma_{S,D} \\ \downarrow & & & & \downarrow \\ \prod_T \mathcal{G}_{T,D_T} & \longrightarrow & \mathcal{H} & \longrightarrow & \mathcal{G}_{S,D} \end{array}$$

commutes. Q.E.D.

There is a more elaborate version of the Glueing Lemma that applies to groups that contain

$$\prod_T \Gamma_{T,D_T}^{G_T}$$

as a finite index subgroup, where $G_T \subseteq \operatorname{Aut} D_T$, and which map to $\Gamma_{S,D}^G$ for certain $G \subseteq \operatorname{Aut} D$. Rather than formulate such a result in general, we now state and prove a very special case that we shall need when investigating handlebody subgroups of $\Gamma_{S,D}$.

Consider the decorated surface (S, D) of genus $h-1$ with one marked boundary component that is constructed as the double covering of the 2-sphere branched at the $2h$th roots of unity, $\boldsymbol{\mu}_{2h}$, and with the disk of radius $1/2$ removed from one of the branches. This surface can be described as the Riemann surface of the algebraic function

$$y^2 = x^{2h} - 1$$

with the disk $|x| < 1/2$ removed from one of the two branches. The marked boundary point is chosen to be $x = 1/2$.

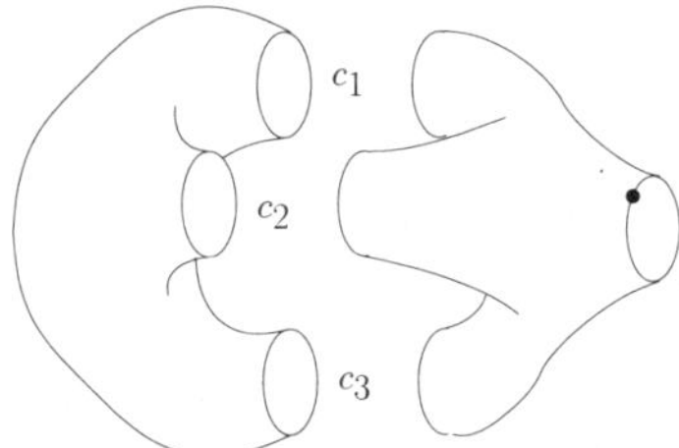

Fig. 2. The surface $S - |\gamma|$ when $h = 3$

For $j = 1, \ldots, h$, let c_j be the circle in S that is the inverse image in S of the interval $[\zeta^{2j-1}, \zeta^{2j}]$ in $\mathbb{C}$, where $\zeta = \exp 2\pi i/2h$. Then $\gamma := \{c_1, \ldots, c_h\}$ is a curve system that separates S into two genus 0 subsurfaces (T_0, D_0) and (T_1, D_1), where

$$D_0 = \{c_1, \ldots, c_h\} \text{ and } D_1 = \{\partial S\} \cup D_0.$$

The case $h = 3$ is illustrated in Figure 2.

The group $\boldsymbol{\mu}_h$ of hth roots of unity acts on S: $\zeta^{2j} : (x, y) \mapsto (\zeta^{2j}x, y)$. It acts on D_0 and D_1 by taking c_j to c_{j+1}, where the indices are considered mod h. Denote the natural homomorphism $\Gamma_{T_j,D_j}^G \to \boldsymbol{\mu}_h$ by p_j.

Set

$$[\Gamma_{T_0,D_0} \times \Gamma_{T_1,D_1}]^{\boldsymbol{\mu}_h} = \{(\phi,\phi_2) \in \Gamma^{\boldsymbol{\mu}_h}_{T_0,D_0} \times \Gamma^{\boldsymbol{\mu}_h}_{T_1,D_1} : p_1(\phi_1) = p_2(\phi_2)\}.$$

There is an obvious glueing homomorphism

$$[\Gamma_{T_0,D_0} \times \Gamma_{T_1,D_1}]^{\boldsymbol{\mu}_h} \to \Gamma_{S,D}. \tag{4}$$

The completion of $[\Gamma_{T_0,D_0} \times \Gamma_{T_1,D_1}]^{\boldsymbol{\mu}_h}$ with respect to the natural homomorphism to $\boldsymbol{\mu}_h$ is easily seen to be the restriction of $\mathcal{G}^{\boldsymbol{\mu}_h}_{T_0,D_0} \times \mathcal{G}^{\boldsymbol{\mu}_h}_{T_1,D_1}$ to the diagonal of $\boldsymbol{\mu}_h \times \boldsymbol{\mu}_h$. Denote it by

$$[\mathcal{G}^{\boldsymbol{\mu}_h}_{T_0,D_0} \times \mathcal{G}^{\boldsymbol{\mu}_h}_{T_1,D_1}]^{\boldsymbol{\mu}_h}.$$

Proposition 8.10. *The homomorphism* (4) *induces a homomorphism*

$$[\mathcal{G}^{\boldsymbol{\mu}_h}_{T_0,D_0} \times \mathcal{G}^{\boldsymbol{\mu}_h}_{T_1,D_1}]^{\boldsymbol{\mu}_h} \to \mathcal{G}_{S,D}$$

that is strictly compatible with the induced mapping

$$(\mathcal{O}(\mathcal{G}_{S,D}), M^{\gamma}_{\bullet}) \to (\mathcal{O}([\mathcal{G}^{\boldsymbol{\mu}_h}_{T_0,D_0} \times \mathcal{G}^{\boldsymbol{\mu}_h}_{T_1,D_1}]^{\boldsymbol{\mu}_h}), W_{\bullet}).$$

The existence of the induced homomorphism is proved by an argument similar to the proof of Proposition 8.9. The strictness with respect to the filtrations follows from either Hodge theory or Galois theory.

Since T_0 and T_1 are spheres, Corollary 8.7 implies that $\mathcal{G}^{\boldsymbol{\mu}_h}_{T_j,D_j} = W_0\mathcal{G}^{\boldsymbol{\mu}_h}_{T_j,D_j}$. This gives the following important consequence of strictness.

Corollary 8.11. *The image of* $[\mathcal{G}^{\boldsymbol{\mu}_h}_{T_0,D_0} \times \mathcal{G}^{\boldsymbol{\mu}_h}_{T_1,D_1}]^{\boldsymbol{\mu}_h} \to \mathcal{G}_{S,D}$ *lies in* $M_0\mathcal{G}_{S,D}$. □

Define $\phi_h \in \Gamma_{S,D}$ to be the mapping class of the diffeomorphism

$$(x,y) \mapsto (e^{2\pi i/h}x, y)$$

of S composed with $1/h$th of the inverse of the Dehn twist about ∂T. This fixes the boundary point $1/2$. Observe that ϕ_h^h is the inverse of the Dehn twist about ∂T. Since ϕ_h lies in the image of (4), we have:

Proposition 8.12. *The image of* ϕ_h *in* $\mathcal{G}_{S,D}$ *lies in* $M_0\mathcal{G}_{S,D}$.

We shall also need a certain diffeomorphism ψ of a surface S of genus 2 with one boundary component. It is convenient to take S to the Riemann surface

$$y^2 = (x^2-1)(x^2-4)(x^2-9)$$

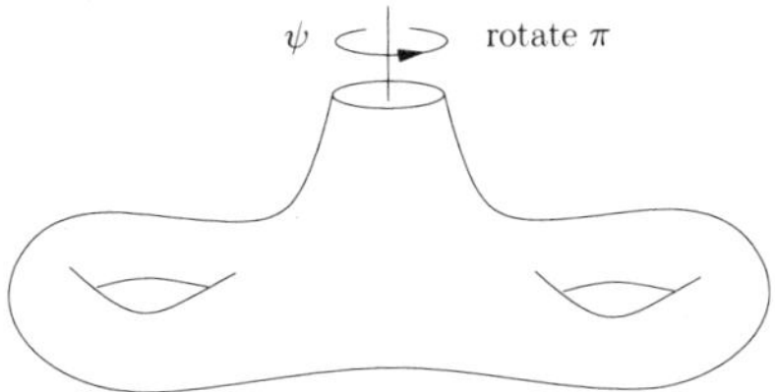

Fig. 3. The diffeomorphism ψ

with one of the two preimages of the disk $|x| < 1/2$ removed. It is illustrated in Figure 3. The element ψ of $\Gamma_{S,\partial S}$ is the composition of the diffeomorphism $(x,y) \mapsto (-x,y)$ of S with the square root of the inverse of the Dehn twist about the boundary of S. In terms of the illustration in Figure 3, it is obtained by rotating S by π about the vertical axis of the boundary circle and composing with the square root of the inverse of the Dehn twist about the boundary of S. An argument similar to the one used to prove Proposition 8.12 can be used to prove:

Proposition 8.13. *The image of ψ in $\mathcal{G}_{S,\partial S}$ lies in $M_0\mathcal{G}_{S,\partial S}$.*

§9. Handlebodies and Relative Weight Filtrations

A curve system $\gamma = \{c_0, \dots, c_m\}$ on a stable decorated surface (S, D) is said to be *rational* if each connected component of $S - |\gamma|$ is a genus 0 surface.[10]

Lemma 9.1. *If γ is a rational curve system on a stable decorated surface (S, D), then there is a unique handlebody U_γ such that $S = \partial U_\gamma$ and each curve $c \in \gamma$ bounds a disk in U.*[11]

Proof. This follows directly from the elementary fact that an oriented 2-sphere bounds a handlebody (necessarily a 3-ball) in a unique way, as the handle body is the cone over S. Q.E.D.

A maximal curve system on a stable decorated surface (S, D) is called a *pants decomposition* of S'_D. If γ is a pants decomposition of S'_D,

[10]This is equivalent to the condition that the nodal surface S/γ obtained from S by collapsing each $c \in \gamma$ to point has the topological type of a stable rational curve. It is also equivalent to the condition that $\mathrm{Gr}^M_{-1} H_1(S) = 0$.

[11]That is, if U_1 and U_2 are handlebodies with $\partial U_1 = \partial U_2 = S$ where each $c \in \gamma$ bounds in each U_j, then there is a homeomorphism $f : U_1 \to U_2$ that is the identity on S.

then each component of $S'_D - |\gamma|$ is a sphere with r boundary components and n punctures, where $r + n = 3$. Thus pants decompositions are rational and determine a handlebody U_γ. It is proved in [26]that any two pants decompositions of S can be joined by A-moves and S-moves. It is clear that A-moves leave the handlebody U_γ unchanged while S-moves change the handlebody.

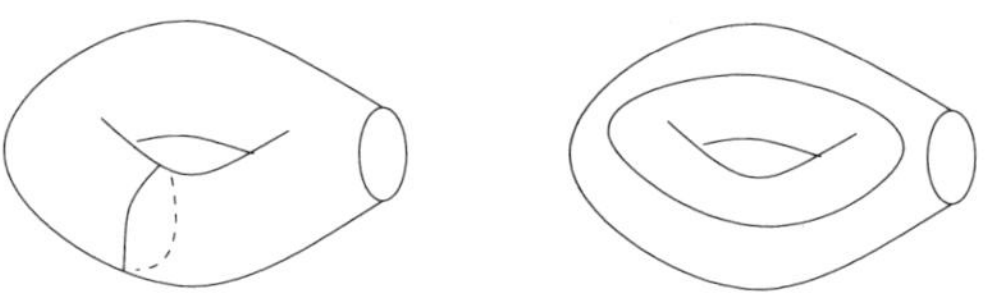

Fig. 4. An S-move

Theorem 9.2. *Two pants decompositions of a decorated stable surface determine the same handlebody if and only if they can be joined by A-moves.*

Sketch of Proof. It is clear that A-moves do not change the handlebody. We use ideas from Morse theory to prove the converse. They are an elaboration of ideas used by Hatcher and Thurston [27] (see also [26]). We will use boundary components instead of tangent vectors and will assume, without loss of generality, that the decorations D consist entirely of boundary components.

If γ is a pants decomposition of (S, D), then the boundary of the associated handlebody U equals

$$\partial U = S \cup \bigcup_{v \in \pi_0(\partial S)} \mathbb{D}_v,$$

where $\mathbb{D}_v$ is a disk that corresponds to the boundary component v of S. We regard this as a manifold with corners at the submanifold ∂S of ∂U.

To a pants decomposition γ of S, we associate a graph G_γ. This has one white vertex for each $c \in \gamma$ and one black vertex for each pair of pants. Edges connect a black and a white vertex; each black vertex is joined to the white vertex corresponding to its boundary components. All black vertices have valence 3; white vertices have valence 1 if they are boundary components of S, otherwise they have valence 2. The handlebody U is a regular neighbourhood of G_γ. The circles retract onto the white vertices.

A height function $f : G_\gamma \to [0, 1]$ on such a graph is a continuous function whose restriction to each edge has no critical points and whose local extrema occur only at white vertices. We also require that f vanish on each 1-valent white vertex. Every G_γ has a height function.

We will say that a smooth function $F : U \to [0,1]$ is *Morse* if it vanishes identically on each $\mathbb{D}_v$, it has no critical points in U, and if its restriction $F|_S$ to S is a Morse function. A Morse function $F : U \to [0,1]$ is *convex* if $F^{-1}(a)$ is a disjoint union of contractible sets for each $a \in [0,1]$.

A height function f extends to a convex Morse function $F : U \to [0,1]$ where $F|_S$ has one critical point for each black vertex and two for each 2-valent white vertex that is a local extremum of f. The critical values of $F|_S$ equal the values of f on the black vertices and the values of $f \pm \epsilon$ on the 2-valent white vertices that are local extrema. The stable or unstable manifold in S of the critical point of $F|_S$ corresponding to a 2-valent white vertex is the isotopy class of the corresponding $c \in \gamma$. The white vertices that are not local extrema of f correspond to components of the level sets of F. The isotopy class of the vertex corresponding to $c \in \gamma$ is c.

Suppose now that γ_0 and γ_1 are two pants decompositions of (S, D) that determine the same handlebody U. Choose height functions $f_j : \gamma_j \to [0,1]$. Extend these to convex Morse functions $F_j : U \to [0,1]$ using the construction in [27]. It follows from [27] that the result will follow if we can show that there is a smooth function $F : U \times [0,1] \to [0,1]$ that vanishes identically on each $\mathbb{D}_v \times [0,1]$, whose restriction to $S \times [0,1] \to [0,1]$ is a generic 1-parameter family of Morse functions, and where the restriction $F_t : U \to [0,1]$ to each $U \times \{t\}$ has no critical points and is a convex Morse functions for all but finitely many $t \in [0,1]$.

To construct such an F, first extend F_0 and F_1 to Morse functions $H_j : (M, \partial M) \to ([0,1], 0)$, where

$$M = S^3 - \cup_{v \in \pi_0(\partial S)} B^3$$

and where $\mathbb{D}_v$ is a hemisphere of the boundary of the corresponding 3-ball. Join these by a generic 1-parameter family of functions $H_t : (M, \partial M) \to ([0,1], 0)$ that have no critical points in a neighbourhood of ∂M and where $H : (M, \partial M) \times [0,1] \to ([0,1], 0) \times [0,1]$ that takes (x,t) to $(H_t(x), t)$ is smooth. The critical set $\Sigma \subset M \times [0,1]$ of H is 1-dimensional and has relative dimension 0 over $[0,1]$. On the other hand, we can choose U so that it is a regular neighbourhood of a graph Γ in M. Then $U \times [0,1]$ is a regular neighbourhood of $\Gamma \times [0,1]$ in $M \times [0,1]$. Since $\Gamma \times [0,1]$ has relative dimension 1 over $[0,1]$ and is disjoint from $\Sigma_t := \Sigma \cap M \times \{t\}$ when $t = 0, 1$, there is a vector field on $M \times [0,1]$ that is tangent to the t-slices and whose flow moves $\Gamma \times [0,1]$ in $M \times [0,1]$ to a subset J that is disjoint from Σ and intersects each fiber in a graph J_t homeomorphic to Γ. Then we may identify a regular neighbourhood W of J in M with $U \times [0,1]$ where $W_t := W \cap (M \times \{t\})$ is homeomorphic with U and does not intersect Σ. We may then deform the restriction of H to W so that it is a generic 1-parameter family of functions $\partial S \times [0,1] \subset \partial W$ and has no critical points in the interior of W.

The issue now is that the restriction $H_t : U \to [0,1]$ of H_t to $U \to [0,1]$ may not be convex as there may be $a, t \in [0,1]$ where $H_t^{-1}(a)$ is not a disjoint union of contractible sets. There are parameter values $0 < t_1 < t_2 < \cdots < t_m < 1$ where $H_t|_U$ is Morse when $t \notin \{t_1, \ldots, t_m\}$. On each interval (t_{j-1}, t_j) of $[0,1] - \{t_1, \ldots, t_m\}$, the Morse function $H_t|_S$ is represented by a graph G_j and a height function $h_j : G_j \to [0,1]$. As t moves through t_j, two vertices of G_j reverse height and G_j changes into G_{j+1} by one of the elementary moves described in [27]. The extra data of the function $H_t : U \to [0,1]$, when $t \in (t_{j-1}, t_j)$, is determined by an equivalence relation on imbedded intervals in G_j that correspond to non-critical values of H_t: two "strands" are equivalent if the corresponding circles are boundary components of the same component of $H_j^{-1}(a)$ for some non-critical value $a \in [0,1]$.

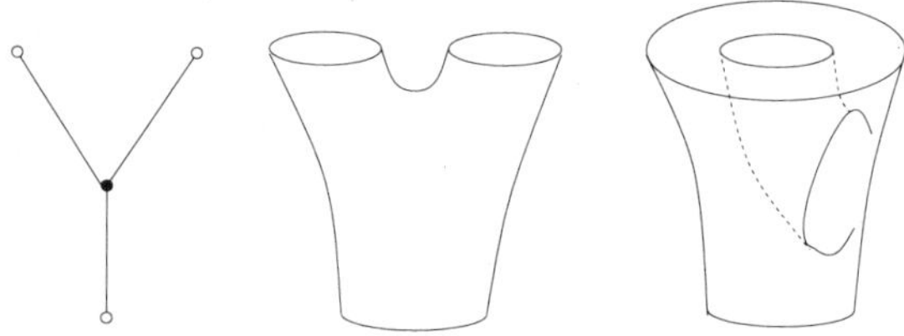

Fig. 5. Interior versus Exterior

Even if the Morse functions H_t are not convex, we can use the sequence of graphs $(G_0, h_0), \ldots, (G_m, h_m)$ to construct a sequence of A-moves that join γ_0 to γ_1. Note that at a black vertex, the level set can change, for example, from a disk to two disks or to an annulus, as illustrated in Figure 5. However, we can apply the Hatcher-Thurston construction to each graph to obtain a sequence of convex Morse functions $K_j : U \to [0,1]$ where $0 \leq j \leq m$ where K_0 and K_1 correspond to the pants decompositions γ_0 and γ_1. Denote by μ_j the pants decomposition of S that corresponds to the restriction of K_j to S. Then it follows by examinining the list of elementary moves in [27] that the pants decompositions determined by two convex Morse functions that differ by an elementary move, themselves differ by an A-move. In particular, μ_{j-1} and μ_j differ by an A-move. This shows that $\gamma_0 = \mu_0$ and $\gamma_1 = \mu_m$ are connected by a sequence of A-moves. Q.E.D.

This result has the important consequence that each way of writing S as the boundary of a handlebody U defines a relative weight filtration on invariants of (S, D), such as $\mathfrak{g}_{S,D}$.

Corollary 9.3. *Suppose that (S, D) is a stable decorated surface and that x is an admissible base point of S'_D. If U is a handlebody and $S = \partial U$, then U determines relative weight filtrations $M^U_\bullet$ on $\mathfrak{g}_{S,D}$, $\mathcal{O}(\mathcal{G}_{S,D})$, $\mathfrak{g}_{S,D\cup\{x\}}$ and $\mathfrak{p}(S'_D, x)$. The actions*

$$\mathfrak{g}_{S,D} \to \operatorname{OutDer}\mathfrak{p}(S'_D) \text{ and } \mathfrak{g}_{S,D\cup\{x\}} \to \operatorname{Der}\mathfrak{p}_{S'_D,x}$$

are strict with respect to the weight filtrations $W_\bullet$ and the relative weight filtrations $M^U_\bullet$.

Proof. Choose a pants decomposition γ of $(S, D \cup \{x\})$ such that each $c \in \gamma$ bounds a disk in U. Define $M^U_\bullet = M^\gamma_\bullet$. Theorem 9.2 and Proposition 8.8 implies that this is independent of the choice of γ. The strictness properties follow from Theorem 8.3. Q.E.D.

For a handlebody U and $x \in U$, denote $\pi_1(U, x)^{\mathrm{un}}$ by $\mathcal{F}(U, x)$ and its Lie algebra by $\mathfrak{f}(U, x)$.

Proposition 9.4. *If S bounds the handlebody U and $x \in S$, then*

$$M^U_0\mathfrak{p}(S, x) = \mathfrak{p}(S, x) \text{ and } M^U_{-2}\mathfrak{p}(S, x) = \ker\{\mathfrak{p}(S, x) \to \mathfrak{f}(U, x)\}.$$

Consequently, $\mathfrak{f}(U, x) \cong \operatorname{Gr}^{M^U}_0 \mathfrak{p}(S, x)$.

Proof. This proof can be made in either the Galois category or the Hodge category, according to taste. Choose a pants decomposition γ of (S, x) as in the proof of Corollary 9.3 such that $M^\gamma_\bullet = M^U_\bullet$. The relative weight filtration on $\mathfrak{p}(S, x)$ arises from a degeneration of (S, x) to the stable rational curve C_0 whose underlying surface is $(S/\gamma, x)$. The degeneration[12] can be chosen to be defined over $\mathbb{Q}$ as each of its components is a 3-pointed $\mathbb{P}^1$. The inclusion of a nearby fiber C_t (the fiber over a first order smoothing of C_0) into the total space of the local deformation $X \to \mathbb{D}$ induces the homomorphism $\pi_1(S, x) \cong \pi_1(C_t, x) \to \pi_1(C_0, x) \cong \pi_1(U, x)$. This implies that the homomorphism $\mathfrak{p}(S, x) \to \mathfrak{f}(U, x)$ is Galois equivariant and a morphism of mixed Hodge structures.

[12]We shall denote the degeneration by $X \to \mathbb{D}$, where $\mathbb{D}$ is an analytic disk in the Hodge case and a formal disk $\operatorname{Spec}\mathbb{Q}[[t]]$ in the Galois case.

Since the Galois action on the algebraic fundamental group of (C_0, x) is trivial (resp., the MHS on $\mathfrak{f}(U, x)$ is pure of weight 0 and type $(0,0)$), it follows that $\mathfrak{f}(U, x)$ is a trivial Galois module (resp., is also pure of weight 0 and type $(0,0)$). Since $\mathfrak{p}(S, x) \to \mathfrak{f}(U, x)$ is a morphism (Galois, Hodge), it is strict with respect to the weight filtration $M^U_\bullet$. This and the strictness of the bracket of $\mathfrak{p}(S, x)$ with respect to both $W_\bullet$ and $M^U_\bullet$ imply that

$$M^U_{-2} H_1(\mathfrak{p}(S, x)) = \ker\{H_1(\mathfrak{p}(S, x)) \to H_1(\mathfrak{f}(U, x))\}.$$

Q.E.D.

§10. Handlebody Groups

Suppose that (S, D) is a stable decorated surface and that S is the boundary of a handlebody U. Define the *handlebody group* of (U, D) by

$$\Lambda_{U,D} = \pi_0 \operatorname{Diff}^+(U, D),$$

where each diffeomorphism acts trivially on D.[13] Griffiths [14], Suzuki [49], Luft [35], and Pitsch [43] have proved fundamental results about handlebody groups and found generating sets of Λ_U. For example, Griffiths [14] proved that homomorphisms

$$\Lambda_{U,x} \to \operatorname{Aut} \pi_1(U, x) \text{ and } \Lambda_U \to \operatorname{Out} \pi_1(U)$$

are surjective. Luft proved that the kernels of each of these is generated by twists on imbedded disks $(\mathbb{D}, \partial\mathbb{D}) \hookrightarrow (U, S)$.[14]

Restriction to the boundary defines a homomorphism

$$r_{U,D} : \Lambda_{U,D} \to \Gamma_{S,D}$$

It is straightforward to show that if $\widetilde{D}$ is a refinement of D, then $r_{U,\widetilde{D}}$ and $r_{U,D}$ induce an isomorphism

$$r : \ker\{\Lambda_{U,\widetilde{D}} \to \Lambda_{U,D}\} \xrightarrow{\simeq} \ker\{\Gamma_{U,\widetilde{D}} \to \Gamma_{U,D}\}. \tag{5}$$

[13]For a subgroup G of $\operatorname{Aut} D$, one can also define $\Lambda^G_{U,D}$ as we did for mapping class groups. However, we shall not need such groups.

[14]The twist on an imbedded disk is the isotopy class of a smoothing of the homeomorphism $(re^{i\theta}, t) \mapsto (re^{i(\theta+2\pi t)}, t)$ of a tubular neighbourhood $\mathbb{D} \times [0,1]$ of $\mathbb{D}$ in U. Its restriction to S is the Dehn twist about the loop $\partial\mathbb{D}$ in S.

For this reason, we will mainly restrict our attention to the cases where $\#D \leq 1$.

The following appears to be well-known to the experts. I am grateful to Alan Hatcher for communicating a proof.

Proposition 10.1. *If (S, D) is a stable surface, then the homomorphism $r_{U,D}$ is injective.*

Sketch of Proof. By the isomorphism (5), it suffices to prove the result when D is empty and $g \geq 2$ and when (g, n) is $(0, 3)$ and $(1, 1)$.

In genus 0 (with any number of points), the result follows directly from a result of Cerf [4]. The general case is proved by induction on g. Suppose $g \geq 1$. Suppose that $\phi \in \mathrm{Diff}^+ U$ is a diffeomorphism whose restriction to the boundary is isotopic to the identity. Then ϕ is isotopic to a diffeomorphism whose restriction to S is the identity. We will assume that this is the case. Now choose an imbedded disk $(\mathbb{D}, \partial\mathbb{D}) \subset (U, S)$. Let $(\mathbb{D}', \partial\mathbb{D}')$ be a disk imbedded in (U, S) parallel to and disjoint from $\mathbb{D}$. After altering ϕ by an isotopy fixing S if necessary, we may assume that the restriction of ϕ to $\mathbb{D}$ is transverse to $\mathbb{D}'$. Since U is an irreducible 3-manifold [28], $\mathbb{D}' \cup \phi(\mathbb{D}) \cup A$, where $A \subset S$ is the annulus between $\partial\mathbb{D}$ and $\partial\mathbb{D}'$, bounds a 3-ball. We can then deform ϕ by an isotopy fixing S so that the number of connected components of the complement in U of $\mathbb{D}' \cup \phi(\mathbb{D}) \cup A$ is reduced by one. We may therefore assume that $\mathbb{D}' \cup \phi(\mathbb{D}) \cup A$ bounds a ball. By further modifying ϕ by an isotopy fixing S, we may assume that ϕ fixes $\mathbb{D}$ pointwise. Now cut U apart along $\mathbb{D}$ to obtain a diffeomorphism ϕ' of a handlebody U' of genus one less whose restriction to $\partial U'$ is the identity. The result now follows by induction. Q.E.D.

The handlebody group Λ_U is bounded by the 0th term M_0^U of the relative weight filtration.

Lemma 10.2. *If (S, D) is a stable decorated surface and S bounds the handlebody U, then the image of $\Lambda_{U,D} \to \mathcal{G}_{S,D}$ lies in $M_0^U \mathcal{G}_{S,D}$.*

Proof. The proof would be straightforward if $\mathfrak{u}_{S,x} \to \mathrm{Der}\, \mathfrak{p}(S, x)$ were injective. Since this is not known, we need a direct proof. As noted above, Luft [35] proved that $\ker\{\Lambda_{U,x} \to \mathrm{Aut}\, \pi_1(U, x)\}$ is generated by twists on imbedded disks. If c is a simple closed curve in S that bounds an imbedded disk in U, then there is a pants decomposition γ of (S, D) that contains c where each $c' \in \gamma$ bounds in U. It follows from Theorem 8.3 that that the Dehn twist on c lies in $M_{-2}^U \mathcal{G}_{S,D} := M_{-2}^\gamma \mathcal{G}_{S,D}$.

Thus, to prove the result, it suffices to show that there are elements of $M_0 \Lambda_{U,x} := \Lambda_{U,x} \cap M_0^U \mathcal{G}_{S,x}$ whose images in $\mathrm{Aut}\, \pi_1(U, x)$ generate $\mathrm{Aut}\, \pi_1(U, x)$. To do this, we use elements of $\Lambda_{U,x}$ closely related to those used by Luft in [35]. That these generators lie in $M_0^U \mathcal{G}_{S,x}$ follows from Propositions 8.12 and 8.13.

Represent $\pi_1(U,x)$ as a free group $\langle a_1, \dots, a_g \rangle$, where each a_j is a simple closed curve on the boundary of S. Note that the automorphisms conjugate to $\phi_2 \in M_0 \Lambda_{U,x}$ constructed in Section 8.2 can be used to invert any generators a_j of $\pi_1(U,x)$ while leaving the remaining generators fixed. The automorphism $\psi \in M_0 \Lambda_{U,x}$ defined there can be used to define an automorphism that fixes all but two of the generators a_j and a_k and acts on them via

$$a_j \mapsto a_k^{-1} \text{ and } a_k \mapsto a_j^{-1}.$$

Composing this with the first kind of automorphism, we see that there are elements of $M_0 \Lambda_{U,x}$ that transpose any two of the generators a_j. We can therefore realize all permutations of the generators a_j by elements of $M_0 \Lambda_{U,x}$. Finally, the elements $\phi_3 \in M_0 \Lambda_{U,x}$ realize the automorphism that fixes a_j when $j > 2$ and satisfies

$$a_1 \mapsto a_2 \text{ and } a_2 \mapsto (a_1 a_2)^{-1}.$$

By a Theorem of Nielsen [41] (cf. [35]) these automorphisms of $\pi_1(U,x)$ generate $\operatorname{Aut} \pi_1(U,x)$. This completes the proof. Q.E.D.

When $\#D \geq 1$, the homomorphism $\Gamma_{S,D} \to \mathcal{G}_{S,D}$ is injective.

Theorem 10.3. *If (S,D) is a stable decorated surface that bounds the handlebody U, where $\#D = 1$, then*

(i) $\Lambda_{U,D} = \Gamma_{S,D} \cap M_0^U \mathcal{G}_{S,D}$;
(ii) $\ker\{\Lambda_{U,D} \to \operatorname{Aut} \pi_1(U,x)\} = \Lambda_{U,D} \cap M_{-2}^U \mathcal{G}_{S,D}$;
(iii) $\Lambda_{U,D} \cap M_{-2}^U \mathcal{U}_{S,D}$ *is generated by opposite twists on disjoint bounding pairs of imbedded disks $(\mathbb{D}_j, \partial \mathbb{D}_j) \hookrightarrow (U,S)$ $(j = 1,2)$ whose images avoid D.*

The second assertion is a consequence of a result of Griffith [14] and the third a restatement of a result of Pitsch [43, Prop. 6].

Proof. We prove the result when x is a point of S. The case where D is a non-zero tangent vector $v \in T_x S$ follows as $\mathfrak{g}_{S,v} \to \mathfrak{g}_{S,x}$ is strict with respect to $M_\bullet^U$ and the kernel is central and generated by a Dehn twist on a curve that bounds a disk in U, and therefore lies in M_{-2}^U.

Proposition 9.4 combined with the fact that the homomorphisms $\pi_1(S,x) \to \mathcal{P}(S,x)$ and $\pi_1(U,x) \to \mathcal{F}(U,x)$ are injective implies that the commutative diagram

$$\begin{array}{ccccccccc} 1 & \longrightarrow & \pi_1(S,x)\cap M_{-2}^U\mathcal{P}(S,x) & \longrightarrow & \pi_1(S,x) & \longrightarrow & \pi_1(U,x) & \longrightarrow & 1 \\ & & \downarrow & & \downarrow & & \downarrow & & \\ 1 & \longrightarrow & M_{-2}^U\mathcal{P}(S,x) & \longrightarrow & \mathcal{P}(S,x) & \longrightarrow & \mathcal{F}(U,x) & \longrightarrow & 1 \end{array}$$

has exact rows. Since $\Gamma_{S,x}$ is a subgroup of $\operatorname{Aut}\pi_1(S,x)$, it follows that $\Gamma_{S,x}$ is a subgroup of $\operatorname{Aut}\mathcal{P}(S,x)$. Consequently $\Gamma_{S,x}\cap M_0^U\operatorname{Aut}\mathcal{P}(S,x)$ consists of those automorphisms of $\pi_1(S,x)$ that preserve $\ker\{\pi_1(S,x)\to\pi_1(U,x)\}$. By a result of Griffiths [14] this is $\Lambda_{U,x}$, so that

$$\Gamma_{S,x}\cap M_0^U\operatorname{Aut}\mathcal{P}(S,x)=\Lambda_{U,x}. \tag{6}$$

Since $\mathcal{P}(S,x)=M_0^U\mathcal{P}(S,x)$, and since $\pi_1(U,x)\to\mathcal{F}(U,x)$ is injective, the commutativity of the diagram above implies that

$$\Gamma_{S,x}\cap M_{-2}^U\operatorname{Aut}\mathcal{P}(S,x)\subseteq\ker\{\Lambda_{U,x}\to\operatorname{Aut}\pi_1(U,x)\}. \tag{7}$$

Since the homomorphism $\mathfrak{g}_{S,x}\to\operatorname{Der}\mathfrak{p}(S,x)$ preserves the filtration $M_\bullet^U$, it follows that for all k we have

$$\Gamma_{S,x}\cap M_k^U\mathcal{G}_{S,x}\subseteq\Gamma_{S,x}\cap M_k^U\operatorname{Aut}\mathcal{P}(S,x). \tag{8}$$

Lemma 10.2 implies that $\Lambda_{U,x}\subseteq\Gamma_{S,x}\cap M_0^U\mathcal{G}_{S,x}$. The first assertion follows by combining this with the inclusions (6) and (8) with $k=0$.

By a result of Luft [35], the kernel of $\Lambda_{U,x}\to\operatorname{Aut}\pi_1(U,x)$ is generated by Dehn twists on simple closed curves in S that bound a disk imbedded in U. But these lie in $M_{-2}^U\mathcal{G}_{S,x}$ by Theorem 8.3. Therefore

$$\ker\{\Lambda_{U,x}\to\operatorname{Aut}\pi_1(U,x)\}\subseteq\Gamma_{S,x}\cap M_{-2}^U\mathcal{G}_{S,x}.$$

The second assertion follows by combining this with the inclusions (7) and (8) with $k=-2$.

The final assertion follows from this and a result of Pitsch [43, Prop. 6]. Q.E.D.

Corollary 10.4. *There is a natural injective homomorphism*

$$\operatorname{Aut}\pi_1(U,x)\hookrightarrow\operatorname{Gr}_0^{M^U}\mathcal{G}_{S,D}.$$

This homomorphism induces homomorphisms on the relative completion of $\operatorname{Aut}^+\pi_1(U,x)$ and $\operatorname{Out}^+\pi_1(U,x)$. Surprisingly, these are not surjective. Equivalently, the injection in the previous corollary is not Zariski dense.

Proposition 10.5. *If $g \geq 3$, then the induced homomorphisms*

$$\mathfrak{a}_n \to \mathrm{Gr}_0^{M^U} \mathfrak{g}_{S,x} \text{ and } \mathfrak{ia}_n \to \mathrm{Gr}_0^{M^U} \mathfrak{g}_S$$

are not surjective.

Sketch of Proof. Denote $H_1(S)$ by H and $H_1(U)$ by A. Denote the relative weight filtration $M_\bullet^U$ by $M_\bullet$. Denote the kernel of $H \to A$ by B. Then $A = \mathrm{Gr}_0^M H$ and $B = \mathrm{Gr}_{-2}^M H$. Since $\mathrm{Gr}_0^W \mathfrak{g}_{S,x} = \mathfrak{sp}(H) \cong S^2 H$, it follows that

$$\mathrm{Gr}_0^M \mathrm{Gr}_0^W \mathfrak{g}_{S,x} \cong A \otimes B \cong \mathrm{End}(A) \cong \mathfrak{gl}(A).$$

There are natural $\mathfrak{sp}(H)$-equivariant isomorphisms

$$H_1(\mathfrak{u}_{S,x}) \cong \mathrm{Gr}_{-1}^W \mathfrak{g}_{S,x} \cong H_1(T_{S,x}) \cong \Lambda^3 H$$

given by the Johnson homomorphism and general results in [18]. The exactness properties of $\mathrm{Gr}_\bullet^M$ and $\mathrm{Gr}_\bullet^W$ imply that there are $\mathfrak{gl}(A)$-equivariant isomorphisms

$$\mathrm{Gr}_0^M \mathrm{Gr}_{-1}^W \mathfrak{g}_{S,x} \cong B \otimes \Lambda^2 A \cong \mathrm{Hom}(A, \mathbb{L}_2(A)),$$

where $\mathbb{L}_m(A)$ denotes the mth graded quotient of the free Lie algebra generated by A. Moreover, the mapping $IA_n \to \mathrm{Gr}_0^M \mathcal{U}_{S,x}$ induces Magnus' isomorphism

$$H_1(IA_n) \to \mathrm{Hom}(A, \mathbb{L}_2(A)).$$

By [18, (10.1),§11], the second weight graded quotient of $\mathfrak{g}_{S,x}$ is the sum of the $\mathrm{Sp}(H)$-modules that corresponds to the partitions $[2, 2]$ and $[1, 1]$. A straightforward linear algebra computation shows that, as $\mathfrak{gl}(A)$-modules,

$$\mathrm{Gr}_0^M \mathrm{Gr}_{-2}^W \mathfrak{g}_{S,x} \cong B \otimes \mathbb{L}_3(A) \cong \mathrm{Hom}(A, \mathbb{L}_3(A)).$$

Alternatively, it is isomorphic to the kernel of the natural surjection $S^2\Lambda^2 H \to \Lambda^4 H$ minus a copy of the trivial representation. The image of $[\mathfrak{ia}_n, \mathfrak{ia}_n]$ in this group is a quotient of

$$\Lambda^2 H_1(IA_n) = \Lambda^2 \mathrm{Hom}(A, \mathbb{L}_2(A)).$$

Since $S^2 A$ is a summand of $\mathrm{Hom}(A, \mathbb{L}_3(A))$ but not of this group, the homomorphism $\mathfrak{ia}_n \to \mathrm{Gr}_0^M \mathfrak{u}_{S,x}/W_{-3}$ is not surjective. The result follows. Q.E.D.

This result shows that the relative weight filtration of $\mathcal{G}_{S,x}$ is not simply obtained by taking the Zariski closure of a filtration of $\Gamma_{S,x}$.

At first glance, this result appears to contradict Theorem 10.3. and the fact that the image of $T_{S,x} \to \mathcal{U}_{S,x}$ is Zariski dense. However, these results simply say that given $n \geq 1$ and a $\mathbb{Q}$-rational element ϕ of $M_0\mathcal{U}_{S,x}/W_{-n}$, there exists $\psi \in T_{S,x}$ and a positive integer m such that

$$\phi^m \equiv \psi \bmod W_{-n}\mathcal{U}_{S,x}.$$

The previous two results imply that when $n > 2$, it is not always possible to choose ψ to lie in $\Lambda_{U,x} = \Gamma_{S,x} \cap M_0\mathcal{G}_{S,x}$.

Theorem 10.3 and Proposition 8.5 yield the following strengthening of Theorem 9.2. It says that the different ways of writing (S,x) as the boundary of a handlebody is faithfully represented in the set of relative weight filtrations of $\mathfrak{g}_{S,x}$.

Corollary 10.6. *For two pants decompositions γ_1 and γ_2 of a stable decorated surface (S,D), where $\#D = 1$, the following are equivalent:*

(i) *γ_1 and γ_2 are connected by A-moves;*
(ii) *$U^{\gamma_1} = U^{\gamma_2}$;*
(iii) *the associated relative weight filtrations $M_\bullet^{\gamma_1}$ and $M_\bullet^{\gamma_2}$ of $\mathfrak{g}_{S,D}$ are equal.*

Proof. Theorem 9.2 gives the equivalence of (i) and (ii). Proposition 8.8 established that (i) implies (iii). It remains to prove that (iii) implies (ii). We will show that not (ii) implies not (iii).

Set $U_j = U^{\gamma_j}$ and $M_\bullet^{\gamma_j} = M_\bullet^{U_j}$. There exists $\phi \in \Gamma_{S,D}$ that extends to a diffeomorphism $\tilde{\phi} : U_1 \to U_2$. Then

$$\Lambda_{U_2} = \phi \Lambda_{U_1} \phi^{-1} \text{ and } M_\bullet^{U_2} = \mathrm{Ad}(\phi) M_\bullet^{U_1}.$$

If $U_1 \neq U_2$, then $\phi \notin \Lambda_{U_1}$. Theorem 10.3 implies that $\phi \notin M_0^{U_1}\mathcal{G}_{S,D}$. We will prove the result by showing that ϕ does not normalize $M_0^{U_1}\mathcal{G}_{S,D}$.

Since $\mathcal{G}_{S,D}$ is connected, it suffices to prove the Lie algebra version: if $X \in \mathfrak{g}_{S,D}$ and $X \notin M_0^{U_1}\mathfrak{g}_{S,D}$, then X does not normalize $M_0^{U_1}\mathfrak{g}_{S,D}$. But this follows directly from Proposition 8.5. Q.E.D.

§11. Extending Diffeomorphisms to Handlebodies

In this section we give an application to the problem of bounding the subset of elements of $\Gamma_{S,D}$ consisting of mapping classes that extend to some handlebody. Similar results have been obtained independently by Jamie Jorgensen [30].

View $\mathcal{G}_{S,D}$ as a proalgebraic variety over $\mathbb{Q}$. It is filtered by its weight filtration

$$\mathcal{G}_{S,D} = W_0\mathcal{G}_{S,D} \supseteq W_{-1}\mathcal{G}_{S,D} \supseteq W_{-2}\mathcal{G}_{S,D} \supseteq \cdots$$

where $W_{-1}\mathcal{G}_{S,D} = \mathcal{U}_{S,D}$ and $W_{-m}\mathcal{G}_{S,D}$ is the mth term of the lower central series of $\mathcal{U}_{S,D}$. Recall from [18] that when $g \geq 3$ and $m \neq 2$, $\mathrm{Gr}^W_{-m}\mathcal{U}_{S,D}$ is isomorphic to the mth graded quotient of the lower central series of the Torelli group $T_{S,D}$ tensored with $\mathbb{Q}$.

Write S as the boundary of a handlebody U. Then the set of elements of $\Gamma_{S,D}$ that extend across some handlebody with boundary S is

$$C := \bigcup_{\phi\in\Gamma_{S,D}} \phi\Lambda_{U,D}\phi^{-1}.$$

For all $m \geq 1$, set $C_m = C \cap W_{-m}\mathcal{G}_{S,D}$. Denote the Zariski closure of C_m in $\mathcal{G}_{S,D}$ by X_m.

Theorem 11.1. *If (S, D) is a stable decorated surface, then X_m is a proper subvariety of $W_{-m}\mathcal{G}_{S,D}$ for all*

$$m \geq \begin{cases} 4 & \text{when } g = 3; \\ 2 & \text{when } g = 4, 5, 6; \\ 1 & \text{when } g \geq 7. \end{cases}$$

In some sense, this theorem says that most elements of $W_{-m}\Gamma_{S,D}$ do not extend to any handle body.

Denote the Zariski closure of $\Lambda_{U,D}$ in $\mathcal{G}_{S,D}$ by $\mathcal{L}_{U,D}$ and its intersection with $W_{-m}\mathcal{U}_{S,D}$ by $W_{-m}\mathcal{L}_{U,D}$. Since $\Gamma_{S,D}$ is Zariski dense in $\mathcal{G}_{S,D}$, C_m is contained in the Zariski closure of the image of the map

$$F : \mathcal{G}_{S,D} \times W_{-m}\mathcal{L}_{U,D} \to \mathcal{G}_{S,D}$$

defined by $F(g, \lambda) = g\lambda g^{-1}$.

To prove the result we show that the image of C_m in $\mathrm{Gr}^W_{-m}\mathcal{G}_{S,D}$ is contained in a proper subvariety. Since $\Lambda_{U,D}$ is contained in $M_0^U\mathcal{G}_{S,D}$, $W_{-m}\mathcal{L}_{U,D}$ is a subgroup of $M_0^U W_{-m}\mathcal{G}_{S,D}$. Consequently, the image of C_m in $\mathrm{Gr}^W_{-m}\mathcal{G}_{S,D}$ is contained in the Zariski closure of the image of the map

$$\mathrm{Gr}^W_0 \mathcal{G}_{S,D} \times M_0^U \mathrm{Gr}^W_{-m}\mathcal{G}_{S,D} \to \mathrm{Gr}^W_{-m}\mathcal{G}_{S,D}.$$

induced by conjugation.

The following is an immediate consequence of a theorem of Chevalley [5], which can be found in exercises 3.18 and 3.19 of [25, Chap. II, sect. 3].

Lemma 11.2. *Suppose that X is a quasi-projective variety over a field and that Y is a closed subvariety. If $G \times X \to X$ is the action of an algebraic group on X, then the image $G \cdot Y$ of the restricted action $G \times Y \to X$ is a constructable subset of X whose Zariski closure in X has dimension*

$$\dim \overline{G \cdot Y} = \dim Y + \dim G - \dim G_Y$$

where $G_Y = \{g \in G : g(Y) \subseteq Y\}$. $\square$

Recall that $\mathrm{Gr}_0^W \mathcal{G}_{S,D} \cong \mathrm{Sp}(H)$ where $H = H_1(S)$. We apply the Lemma to the adjoint action of $G = \mathrm{Sp}(H)$ on

$$X = \mathrm{Gr}_{-m}^W \mathcal{G}_{S,D} \cong \mathrm{Gr}_{-m}^W \mathfrak{g}_{S,D} \text{ where } Y = M_0^U \mathrm{Gr}_{-m}^W \mathfrak{g}_{S,D}.$$

Proposition 8.5 implies that $G_Y = M_0^U \mathrm{Sp}(H)$. Applying the Lemma, and using the fact that $\mathfrak{sp}(H) = M_2^U \mathfrak{sp}(H)$, we see that the codimension of the closure of the $\mathrm{Sp}(H)$ orbit of $M_0^U \mathrm{Gr}_{-m}^W \mathfrak{g}_{S,D}$ in $\mathrm{Gr}_{-m}^W \mathfrak{g}_{S,D}$ satisfies

$$\begin{aligned} \operatorname{codim} & \overline{\mathrm{Sp}(H) \cdot M_0^U \mathrm{Gr}_{-m}^W \mathfrak{g}_{S,D}} \\ &= \dim \mathrm{Gr}_{-m}^W \mathfrak{g}_{S,D}/M_0^U - \dim \mathrm{Sp}(H)/M_0 \mathrm{Sp}(H) \\ &\qquad \geq \dim \mathrm{Gr}_2^M \mathrm{Gr}_{-m}^W \mathfrak{g}_{S,D} - \dim \mathrm{Gr}_2^M \mathfrak{sp}(H). \end{aligned}$$

It remains to show this is positive for all m in the statement of the theorem. First, since $\mathrm{Gr}_2^M \mathfrak{sp}(H)$ is the symmetric square of a maximal isotropic subspace of H, it has dimension $g(g+1)/2$.

We use representation theory to find a lower bound for the other term. Each $\mathrm{Gr}_k^M \mathrm{Gr}_m^W \mathfrak{g}_{S,D}$ is a $\mathrm{Gr}_0^M \mathrm{Gr}_0^W \mathfrak{g}_{S,D}$-module. Recall from the proof of Corollary 10.6 that $\mathrm{Gr}_0^M \mathrm{Gr}_0^W \mathfrak{g}_{S,D}$ is isomorphic to $\mathfrak{gl}_g$, so that its irreducible representations are given by Young diagrams with $\leq g$ rows. These are the same Young diagrams that parametrize the irreducible $\mathfrak{sp}(H)$-modules, where $H = H_1(S)$.

Proposition 11.3. *If $g \geq 3$ and $m > 1$, then $\mathrm{Gr}_2^M \mathrm{Gr}_{-m}^W \mathfrak{g}_{S,D}$ contains the $\mathfrak{gl}_g$-module corresponding to the partition $[k,k]$ when $m = 2k$ and $[k,k,1]$ when $m = 2k-1$.*

Proof. Results of Oda [42] and Asada-Nakamura [1] imply that if $m > 0$, then the $\mathrm{Sp}(H)$-module $\mathrm{Gr}_{-m}^W \mathfrak{u}_{S,D}$ contains the representation $[k,k]$ when $m = 2k$ and $[k,k,1]$ when $m = 2k-1$. If we take $\mathrm{Gr}_2^M \mathfrak{sp}(H)$ to be positive roots of $\mathfrak{sp}(H)$, then the highest weight vectors of each of these representations lies in $\mathrm{Gr}_2^M \mathrm{Gr}_{-m}^W \mathfrak{g}_{S,D}$. Since $\mathfrak{gl}_g$ is a subalgebra of $\mathfrak{sp}(H)$ with the same Cartan subalgebra, the $\mathfrak{gl}_g$-submodule

of $\mathrm{Gr}_2^M \mathrm{Gr}_{-m}^W \mathfrak{g}_{S,D}$ generated by v will correspond to the same partition.
Q.E.D.

Using the formula [13, (6.4)], when $k \geq 2$ we have:

$$\dim V_{[k,k]} = \frac{(g-1)(g+k-1)\prod_{j=0}^{k-2}(g+j)^2}{k!(k+1)!}$$

$$\dim V_{[k,k,1]} = \frac{(g-1)(g-2)(g+k-1)\prod_{j=0}^{k-2}(g+j)^2}{(k-1)!(k+2)!}$$

where V_λ denotes the $\mathfrak{gl}_g$-module corresponding to the partition λ. These dimensions increase monotonically with k. The proof is completed by an elementary computation, which is left to the reader.

References

[1] M. Asada and H. Nakamura, On graded quotient modules of mapping class groups of surfaces, Israel J. Math., **90** (1995), 93–113.
[2] A. Borel, Stable real cohomology of arithmetic groups II, Manifolds and Lie groups, Notre Dame, Ind., 1980, Progr. Math. **14**, Birkhäuser, 1981, pp. 21–55.
[3] K. Brown, Cohomology of groups, Grad. Texts in Math., **87**, Springer-Verlag, 1982.
[4] J. Cerf, Sur les difféomorphismes de la sphère de dimension trois ($\Gamma_4 = 0$), Lecture Notes in Math., **53**, Springer-Verlag, Berlin-New York, 1968.
[5] C. Chevalley, Fondements de la géométrie algébrique, Secrétariat Mathématique, Paris, 1958.
[6] R. Craggs, A new proof of the Reidemeister-Singer theorem on stable equivalence of Heegaard splittings, Proc. Amer. Math. Soc. **57** (1976), 143–147.
[7] P. Deligne, Poids dans la cohomologie des variétés algébriques, In: Proceedings of the International Congress of Mathematicians, Vancouver, B. C., 1974, Vol. 1, Canad. Math. Congress, Montreal, Que., 1975, pp. 79–85.
[8] P. Deligne, Théorie de Hodge, I, Actes du Congrès International des Mathématiciens, Nice, 1970, Tome 1, Gauthier-Villars, Paris, 1971, pp. 425–430,
[9] P. Deligne, Théorie de Hodge, II, Inst. Hautes Études Sci. Publ. Math., **40** (1971), 5–57.
[10] P. Deligne, Théorie de Hodge, III, Inst. Hautes Études Sci. Publ. Math., **44** (1974), 5–77.
[11] P. Deligne, La conjecture de Weil, II, Inst. Hautes Études Sci. Publ. Math., **52** (1980), 137–252.
[12] B. Doran and F. Kirwan, Towards non-reductive geometric invariant theory, Pure Appl. Math. Q., **3** (2007), 61–105.

[13] W. Fulton and J. Harris, Representation Theory, Grad. Texts in Math., **129**, Springer-Verlag, 1991.
[14] H. Griffiths, Automorphisms of a 3-dimensional handlebody, Abh. Math. Sem. Univ. Hamburg, **26** (1963/1964), 191–210.
[15] R. Hain, The de Rham homotopy theory of complex algebraic varieties, I & II, K-Theory, **1** (1987), 271–324 and 481–497.
[16] R. Hain, Completions of mapping class groups and the cycle $C - C^{-}$, Mapping class groups and moduli spaces of Riemann surfaces, Contemp. Math., **150**, Amer. Math. Soc., pp.75–105.
[17] R. Hain, Hodge-de Rham theory of relative Malcev completion, Ann. Sci. École Norm. Sup., **t. 31** (1998), 47–92.
[18] R. Hain, Infinitesimal presentations of Torelli groups, J. Amer. Math. Soc., **10** (1997), 597–651.
[19] R. Hain and M. Matsumoto, Weighted completion of Galois groups and Galois actions on the fundamental group of $\mathbb{P}^1 - \{0, 1, \infty\}$, Compositio Math., **139** (2003), 119–167.
[20] R. Hain and M. Matsumoto, Completions of Arithmetic Mapping Class Groups, in preparation, 2007.
[21] R. Hain and M. Matsumoto, Lectures on Completions of Fundamental Groups, in preparation, 2007.
[22] R. Hain, G. Pearlstein and T. Terasoma, Variation of Mixed Hodge Structure and Manin's Periods of Iterated Integrals of Modular Forms, in preparation, 2007.
[23] R. Hain and D. Reed, Geometric proofs of some results of Morita, J. Algebraic Geom., **10** (2001), 199–217.
[24] R. Hain and S. Zucker, Unipotent variations of mixed Hodge structure, Invent. Math., **88** (1987), 83–124.
[25] R. Hartshorne, Algebraic geometry, Grad. Texts in Math., **52**, Springer-Verlag, 1977.
[26] A. Hatcher, P. Lochak and L. Schneps, On the Teichmüller tower of mapping class groups, J. Reine Angew. Math., **521** (2000), 1–24.
[27] A. Hatcher and W. Thurston, A presentation for the mapping class group of a closed orientable surface, Topology, **19** (1980), 221–237.
[28] W. Jaco, Lectures on three-manifold topology, CBMS Regional Conf. Ser. in Math., **43**, Amer. Math. Soc., 1980.
[29] D. Johnson, The structure of the Torelli group. III. The abelianization of $\mathcal{T}$, Topology, **24** (1985), 127–144.
[30] J. Jorgensen, Surface homeomorphisms that do not extend to any handlebody and the Johnson filtration, preprint, 2008, arXiv:0805.4253.
[31] M. Kashiwara, A study of variation of mixed Hodge structure, Publ. Res. Inst. Math. Sci., **22** (1986), 991–1024.
[32] N. Katz, Nilpotent connections and the monodromy theorem: Applications of a result of Turrittin, Inst. Hautes Études Sci. Publ. Math., **39** (1970), 175–232.

[33] N. Kawazumi, Cohomological Aspects of Magnus Expansions, preprint, math.GT/0505497.

[34] A. Landman, On the Picard-Lefschetz transformation for algebraic manifolds acquiring general singularities, Trans. Amer. Math. Soc., **181** (1973), 89–126.

[35] E. Luft, Actions of the homeotopy group of an orientable 3-dimensional handlebody, Math. Ann., **234** (1978), 279–292.

[36] W. Magnus, Über n-dimensionale Gittertransformationen, Acta Math., **64** (1935), 353–367.

[37] G. Margulis, Discrete subgroups of semisimple Lie groups, Ergeb. Math. Grenzgeb. (3), **17**, Springer-Verlag, 1991.

[38] J. Morgan, The algebraic topology of smooth algebraic varieties, Inst. Hautes Études Sci. Publ. Math., **48** (1978), 137–204.

[39] S. Morita, A linear representation of the mapping class group of orientable surfaces and characteristic classes of surface bundles, Topology and Teichmüller spaces, Katinkulta, 1995, World Sci. Publ., 1996, pp. 159–186.

[40] S. Morita, Structure of the mapping class groups of surfaces: a survey and a prospect, In: Proceedings of the Kirbyfest, Berkeley, CA, 1998, Geom. Topol. Monogr., **2**, Geom. Topol. Publ., Coventry, 1999, pp. 349–406.

[41] J. Nielsen, Die Isomorphismengruppe der freien Gruppen. Math. Ann., **91** (1924), 169–209.

[42] T. Oda, A lower bound for the weight graded quotients associated with the relative weight filtration on the Teichmüller group, preprint, 1992.

[43] W. Pitsch, Trivial Cocycles and Invariants of Homology 3-Spheres, preprint, math.GT/0605725.

[44] M. S. Raghunathan, Cohomology of arithmetic subgroups of algebraic groups, I & II, Ann. of Math. (2),**86** (1967), 409–424; ibid. (2), **87** (1967), 279–304.

[45] C. Reutenauer, Free Lie algebras, London Math. Soc. Monogr. (N.S.), **7**, Oxford Univ. Press, 1993.

[46] J. Singer, Three-dimensional manifolds and their Heegaard diagrams, Trans. Amer. Math. Soc., **35** (1933), 88–111.

[47] J. Stallings, Homology and central series of groups, J. Algebra, **2** (1965), 170–181.

[48] J. Steenbrink and S. Zucker, Variation of mixed Hodge structure, I, Invent. Math., **80** (1985), 489–542.

[49] S. Suzuki, On homeomorphisms of a 3-dimensional handlebody, Canad. J. Math., **29** (1977), 111–124.

Department of Mathematics
Duke University
Durham, NC 27708-0320
E-mail address: hain@math.duke.edu

Advanced Studies in Pure Mathematics 52, 2008
Groups of Diffeomorphisms
pp. 369–381

Remarks on the faithfulness of the Jones representations

Yasushi Kasahara

Abstract.

We consider the linear representations of the mapping class group of an n–punctured 2–sphere constructed by V. F. R. Jones using Iwahori–Hecke algebras of type A. We show that their faithfulness is equivalent to that of certain related Iwahori–Hecke algebra representation of Artin's braid group of $n-1$ strands. In the case of $n=6$, we provide a further restriction for the kernel using our previous result, as well as a certain relation to the Burau representation of degree 4.

§1. Introduction

A linear representation of a group is said to be faithful if it is injective as a homomorphism of the group into the corresponding group of linear transformations. In the seminal paper [7], V. F. R. Jones constructed a family of linear representations of $\mathcal{M}_0^n$, the mapping class group of an n–punctured 2–sphere with one parameter. Each of the family is obtained as a modification of the Iwahori–Hecke algebra representation of B_n, Artin's braid group of n strands, provided with a *rectangular* Young diagram with n boxes. Except for the trivial cases, which correspond to the Young diagrams $[1^n]$ or $[n]$, it remains open whether these Jones representations of $\mathcal{M}_0^n$ are faithful or not. It might be therefore possible that some of these Jones representations gives a *naturally defined faithful* linear representation of $\mathcal{M}_0^n$, which seems missing even after the work of Korkmaz [11], and Bigelow–Budney [3], who independently constructed a faithful linear representation of $\mathcal{M}_0^n$ as an *induced* representation of a faithful representation of a certain subgroup of finite index defined by modifying the Lawrence–Krammer representation of B_{n-1}, the faithfulness of which was established by the celebrated works of Bigelow [2] and Krammer [12]. On the other hand, if there exists a nontrivial element in

Received April 30, 2007.
Revised October 23, 2007.

the kernel of a Jones representation of $\mathcal{M}_0^n$, probably of infinite order, it might be expected to give a nontrivial knot with the same value of the Jones polynomial as the unknot.

In this note, we show, for arbitrary $n \geq 6$, that the faithfulness of the Jones representation of $\mathcal{M}_0^n$ corresponding to the rectangular Young diagram Y is equivalent to that of the Iwahori–Hecke algebra representation of the braid group B_{n-1} which corresponds to the *unique* Young diagram obtained from Y by removing a single box. We note that the faithfulness problem for the latter representation also remains open. Furthermore, in the case of $n = 6$, we apply our previous result [8] to obtain a restriction for the kernel of the Jones representation of $\mathcal{M}_0^6$ in terms of the mapping class group of genus 2 via the Birman–Hilden theory. The case $n < 6$ is mentioned in Remark 3.2 (1), and its details are discussed elsewhere [10].

§2. Preliminaries

We fix some notation and briefly recall necessary material to describe Jones' construction.

2.1. Mapping class groups of genus 0

Let D^2 be a 2–disk, P the set of distinct n points $p_1, p_2, \ldots, p_n$ in $\operatorname{Int} D^2$. We call P the set of "punctures". We denote by D_n the pair of spaces (D^2, P). The n–strand braid group B_n is defined as $\pi_0 \operatorname{Homeo}^+(D_n; \partial D_n)$, the mapping class group of D_n. Namely, it is the group of the orientation preserving homeomorphisms of D_n which restrict to the identity on the boundary, modulo the isotopy in the same class of homeomorphisms.

We choose a simple closed curve in $\operatorname{Int} D^2$ so that the disk component of its complement intersects with P at $P_0 = \{p_1, \ldots, p_{n-1}\}$. Identifying the bounding 2–disk with D^2 again, the inclusion defines $D_{n-1} \hookrightarrow D_n$, which induces an injective homomorphism

$$i : B_{n-1} \to B_n$$

by extending the homeomorphisms with the identity on $D_n \setminus D_{n-1}$.

Next, capping off a 2–disk on ∂D^2, we obtain a 2–sphere S^2. We choose a single point p_{n+1} in the interior of the added disk, and set $P_+ = P \cup \{p_{n+1}\}$. As in the case of D^2, we set $S_n = (S^2, P)$ and $S_{n+1} = (S^2, P_+)$, and define their mapping class groups as $\mathcal{M}_0^n = \pi_0 \operatorname{Homeo}^+(S_n)$ and $\mathcal{M}_0^{n+1} = \pi_0 \operatorname{Homeo}^+(S_{n+1})$, respectively. We denote by $\mathcal{M}_0^{n+1}(p_{n+1})$ the subgroup of $\mathcal{M}_0^{n+1}$ consisting of those mapping classes which fix p_{n+1}. By "forgetting" p_{n+1}, we obtain the natural surjective homomorphism

$$p : \mathcal{M}_0^{n+1}(p_{n+1}) \to \mathcal{M}_0^n$$

The inclusion $D_n \hookrightarrow S_{n+1}$ induces the homomorphism $j_n : B_n \to \mathcal{M}_0^{n+1}$ by again extending the homeomorphism with the identity.

Proposition 2.1. *For $n \geq 2$, there exists the following short exact sequence:*

$$0 \to \mathbb{Z} \to B_n \xrightarrow{j_n} \mathcal{M}_0^{n+1}(p_{n+1}) \to 1 \tag{1}$$

Here, the image of $\mathbb{Z}$ in B_n is generated by the single element D_∂, the Dehn twist about the simple closed curve parallel to the boundary, and coincides with $\mathrm{Center}(B_n)$ *if $n \geq 3$. If $n = 2$,* $\mathrm{Center}(B_2)$ *coincides with B_2 and the image of $\mathbb{Z}$ is an index 2 subgroup of* $\mathrm{Center}(B_2)$.

Remark 2.2. In the case of $n \geq 3$, this proposition is nothing but [3, Lemma 2.2]. In the case of $n = 2$, the proposition follows from the fact that $\mathcal{M}_0^3$ is isomorphic to the symmetric group of three letters.

Finally, we denote by $k : B_n \to \mathcal{M}_0^n$ the homomorphism induced by the inclusion $D_n \hookrightarrow S_n$, and obtain the commutative diagram:

$$\begin{array}{ccccc} & & B_n & & \\ & & \downarrow j & \searrow k & \\ B_n/\langle D_\partial \rangle & = & \mathcal{M}_0^{n+1}(p_{n+1}) & \xrightarrow[p]{} & \mathcal{M}_0^n \end{array}$$

where $\langle D_\partial \rangle$ denotes the subgroup generated by D_∂.

2.2. Iwahori–Hecke algebra representations of B_n

We denote by $H(q,n)$ the Iwahori–Hecke algebra of type A_{n-1}. As its ground ring, we take $\mathbb{Q}(q)$, the quotient field of the polynomial ring $\mathbb{Q}[q]$ with q an indeterminate. Let σ be the right-handed half twist about an arbitrary embedded arc in $\mathrm{Int}D^2$ joining two distinct points of P such that no interior points of the arc intersect with P. Such σ is unique up to conjugation in B_n. Then $H(q,n)$ can be defined as the quotient of the group ring $\mathbb{Q}(q)[B_n]$ of B_n by the two–sided ideal generated by the single element $(\sigma - 1)(\sigma + q)$. Hence the projection defines a natural multiplication–preserving mapping $B_n \to H(q,n)$. From any representation of $H(q,n)$, this mapping gives rise to a representation of B_n.

Here, we summarize the facts necessary in this note about $H(q,n)$. For details, we refer to [15].

Proposition 2.3. *As a $\mathbb{Q}(q)$–algebra, $H(q,n)$ is semisimple. Furthermore, all the irreducible representations of $H(q,n)$ are in one-to-one correspondence with all the Young diagrams with n boxes.*

Let Y be a Young diagram with n boxes. We denote by V_Y the representation space of the corresponding representation of $H(q,n)$, and by $\pi_Y : B_n \to \mathrm{GL}(V_Y)$ the representation obtained from V_Y as above. The identity element of $\mathrm{GL}(V_Y)$ will be denoted by I. As usual, we adopt the notation that $\pi_{[n]}$ denotes the one–dimensional scalar representation $\sigma \mapsto q \cdot I$, which coincides with the notation of [7]. Then the representation $\pi_{[1^n]}$ is the one–dimensional representation defined by $\sigma \mapsto -I$. We call this representation sgn.

Proposition 2.4. *Let Y be an arbitrary Young diagram with n boxes. We denote all the distinct Young diagrams obtained from Y by removing a single box by Y_0^1, Y_0^2, $\ldots$, Y_0^s. Then the representation of B_{n-1} obtained as the composition of π_Y with the injection $i : B_{n-1} \to B_n$ is equivalent to the direct sum $\pi_{Y_0^1} \oplus \pi_{Y_0^2} \oplus \cdots \oplus \pi_{Y_0^s}$ as representations over $\mathbb{Q}(q)$.*

Next, let $\mathfrak{S}_n$ denote the permutation group of n letters, and $\nu : B_n \to \mathfrak{S}_n$ the homomorphism induced by the permutation of the puncture set P.

Proposition 2.5. *Let Y be a Young diagram with n boxes. Then the specialization of the representation π_Y of B_n at $q = 1$ descends via ν to the irreducible representation of $\mathfrak{S}_n$ over $\mathbb{Q}$ which corresponds to the same Young diagram Y.*

2.3. Jones' construction.

Let Y be an arbitrary Young diagram with n boxes, and π_Y the corresponding irreducible representation of B_n with the representation space V_Y. We denote by d $(= d_Y)$ the dimension of V_Y over $\mathbb{Q}(q)$. According to the analysis in [7], the Dehn twist D_∂ along the boundary curve is mapped under π_Y to the scalar $q^{rn(n-1)/d}$. Here, r is a certain non-negative integer defined as $\mathrm{rank}\,(I + \pi_Y(\sigma))$. A precise combinatorial description of r can be found in [7]. Important here is the fact that r is equal to 0 if and only if $Y = [1^n]$. We also remark that $rn(n-1)/d$ is always an integer.

We now adjust π_Y by rescaling so that the image of D_∂ becomes trivial, appealing to the well-known fact that the abelianization of B_n is $\mathbb{Z}$. For that purpose, we need the formal power $q^{1/d}$. To avoid confusion, .we introduce another indeterminate corresponding to $q^{1/d}$. Let t be another indeterminate. We consider the $\mathbb{Q}$–algebra homomorphism

$\mathbb{Q}(q) \to \mathbb{Q}(t)$ defined by $q \mapsto t^d$. This homomorphism naturally gives rise to a structure of a $\mathbb{Q}(q)$–algebra on $\mathbb{Q}(t)$. From now on, we consider every representation over $\mathbb{Q}(q)$ also as that over $\mathbb{Q}(t)$ by coefficient extension.

We denote the representation obtained as the composition of the abelianization of B_n with the mapping $m \in \mathbb{Z} \mapsto (t^{-r})^m$ by

$$\alpha : B_n \to \mathrm{GL}(\mathbb{Q}(t)).$$

Clearly, α is trivial if and only if $r = 0$, *i.e.*, if and only if $Y = [1^n]$. We now consider the representation $\alpha \otimes_{\mathbb{Q}(q)} \pi_Y$. Its representation space is $M_Y = \mathbb{Q}(t) \otimes_{\mathbb{Q}(q)} V_Y$ which is a d-dimensional vector space over $\mathbb{Q}(t)$. Since the abelianization maps D_∂ to $n(n-1) \in \mathbb{Z}$, we have $\alpha \otimes_{\mathbb{Q}(q)} \pi_Y(D_\partial) = I$. Therefore, by Proposition 2.1, the representation $\alpha \otimes_{\mathbb{Q}(q)} \pi_Y$ descends via j_n to that of $\mathcal{M}_0^{n+1}(p_{n+1})$. The condition that this representation further descends to that of $\mathcal{M}_0^n$ is given in [7]:

Proposition 2.6. *Via the homomorphism $k : B_n \to \mathcal{M}_0^n$, the representation $\alpha \otimes_{\mathbb{Q}(q)} \pi_Y$ descends to that of $\mathcal{M}_0^n$ if and only if Y is* rectangular.

For a rectangular Young diagram Y with n boxes, we call the representation of $\mathcal{M}_0^n$ given by Proposition 2.6 the Jones representation of the n–punctured sphere corresponding to Y, and denote it by $\bar{\pi}_Y : \mathcal{M}_0^n \to \mathrm{GL}(M_Y)$, with $M_Y = V_Y \otimes \mathbb{Q}(t)$.

§3. The faithfulness of $\bar{\pi}_Y$

With the preparation above, we can now state our main result.

Theorem 3.1. *Suppose that Y is a rectangular Young diagram with n boxes, $\neq [n]$, $[1^n]$. Let Y_0 be the* unique *Young diagram obtained from Y by removing a single box. Assume further that $n \geq 6$. Then the representation*

$$\bar{\pi}_Y : \mathcal{M}_0^n \to \mathrm{GL}(M_Y)$$

is faithful if and only if the representation

$$\pi_{Y_0} : B_{n-1} \to \mathrm{GL}(V_{Y_0})$$

is faithful.

Remark 3.2. (1) For $n < 6$, there exists only a single case where the other assumptions of Theorem 3.1 on Y hold: $n = 4$, $Y = [2, 2]$, and $Y_0 = [2, 1]$. In this case, the theorem is not true. In fact, on the one hand, π_{Y_0} is the reduced Burau representation of B_3, and is classically known to be faithful [14]. On the other hand, we can verify that the kernel of $\bar{\pi}_Y$ is isomorphic to $\mathbb{Z}/2\mathbb{Z} \oplus \mathbb{Z}/2\mathbb{Z}$, and coincides with the kernel of the natural homomorphism $\mathcal{M}_0^4 \to \mathrm{PSL}(2, \mathbb{Z})$ described in [1]. In another word, $\bar{\pi}_{[2,2]}$ descends to a faithful representation of $\mathrm{PSL}(2, \mathbb{Z})$. We also note $\mathrm{PSL}(2, \mathbb{Z}) \cong B_3/\,\mathrm{Center}(B_3)$. We discuss the details in [10].

(2) If n is prime, there exist no Young diagrams satisfying the assumptions of the theorem. Therefore, we cannot expect that the Jones representations here would provide faithful representations of B_n for *all* n, different from the representations obtained from the Lawrence–Krammer representations.

Proof of Theorem 3.1. Via the injective homomorphism $i : B_{n-1} \to B_n$, we consider B_{n-1} as a subgroup of B_n. Since Y_0 is the unique Young diagram obtained from Y by removing a single box, the restriction of π_Y to B_{n-1} coincides with π_{Y_0} by Proposition 2.4. Hence we have the following commutative diagram:

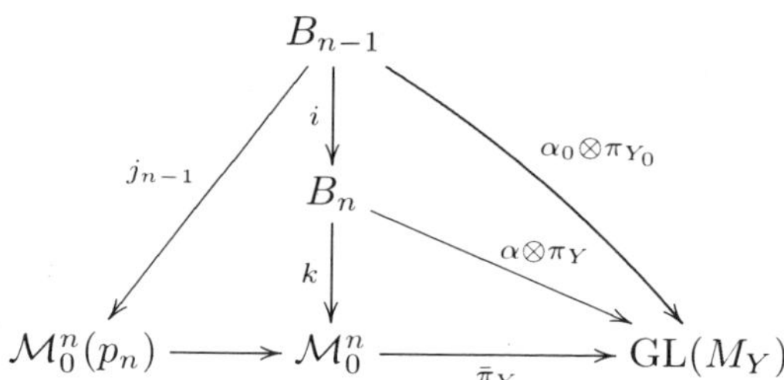

where α_0 denotes the restriction of α to B_{n-1}.

Suppose now that $\bar{\pi}_Y$ is faithful. For $x \in B_{n-1}$, $\bar{x}$ denotes the corresponding element $j_{n-1}(x)$ in $\mathcal{M}_0^n(p_n)$. If $\pi_{Y_0}(x) = 1$, then

$$\bar{\pi}_Y(\bar{x}) = \pi_{Y_0}(x) \cdot \alpha_0(x) = \alpha(x) \cdot I.$$

Therefore, by the faithfulness of $\bar{\pi}_Y$, $\bar{x}$ lies in the center of $\mathcal{M}_0^n$. On the other hand, by Gillette–Buskirk [5], it is known that the center of $\mathcal{M}_0^n$ is trivial. Hence, we have $\bar{x} = 1$, that is, $x \in \mathrm{Ker}\, j_{n-1}$. Now, by Proposition 2.1 for $n-1$, $x \in \mathrm{Center}(B_{n-1})$. Then the faithfulness of π_{Y_0} reduces to its faithfulness on $\mathrm{Center}(B_{n-1})$, which can be seen as follows. Since $Y \neq [1^n]$, we can see that α, and therefore α_0, is non-trivial, as noted in Section 2.3. Since α_0 factors through the abelianization of B_{n-1}, it is faithful on $\mathrm{Center}(B_{n-1})$. Together with the fact that $\alpha_0 \otimes \pi_{Y_0}$ is trivial on $\mathrm{Ker}\, j_{n-1} = \mathrm{Center}(B_{n-1})$, this implies that π_{Y_0} is faithful on $\mathrm{Center}(B_{n-1})$. This completes the proof that π_{Y_0} is faithful.

Suppose next that π_{Y_0} is faithful. Assume that $f \in \mathcal{M}_0^n$ and $\bar{\pi}_Y(f) = 1$. By Proposition 2.5, the specialization of $\bar{\pi}_Y$ at $t = 1$ descends to the irreducible representation of the symmetric group $\mathfrak{S}_n$ corresponding to Y. By the assumption $Y \neq [1^n]$, $[n]$, we can see that this specialization is a faithful representation of $\mathfrak{S}_n$. In fact, since $n \geq 5$, it is a classical fact that the alternating group $\mathfrak{A}_n$ is a simple finite group. Therefore, the kernel of the specialization is either $\mathfrak{S}_n$, $\mathfrak{A}_n$, or $\{1\}$. But since $Y \neq [1^n]$, $[n]$, we can conclude that the kernel is not $\mathfrak{S}_n$ nor $\mathfrak{A}_n$. Hence the specialization is faithful on $\mathfrak{S}_n$. Therefore, the permutation of the set of punctures P induced by f is trivial. In particular, we have $f \in \mathcal{M}_0^n(p_n)$. Hence, there exists some $x \in B_{n-1}$ such that $\bar{x} = f$. Then we have

$$\alpha(x) \cdot \pi_{Y_0}(x) = \bar{\pi}_Y(f) = I.$$

So, we have $\pi_{Y_0}(x) = \alpha(x)^{-1} \cdot I$, and then by the assumption that π_{Y_0} is faithful, $x \in \mathrm{Center}(B_{n-1})$. Hence, $f = \bar{x} = 1 \in \mathcal{M}_0^n$. Therefore, $\mathrm{Ker}\,\bar{\pi}_Y = \{1\}$. This completes the proof of Theorem 3.1. Q.E.D.

Remark 3.3. In the course of the proof above, the use of the result in [5], to show $x \in \mathrm{Center}(B_{n-1})$ under the assumption that $\bar{\pi}_Y$ be faithful and $\pi_{Y_0}(x) = 1$, is not necessary as follows. Since $\bar{x} \in \mathrm{Center}(\mathcal{M}_0^n)$, it holds for any $\tau \in B_{n-1}$ that $[\bar{x}, \bar{\tau}] = 1$ in $\mathcal{M}_0^n$. Hence $[x, \tau] \in \mathrm{Ker}\, j_{n-1} = \mathrm{Center}(B_{n-1})$. On the other hand, we see $\pi_{Y_0}[x, \tau] = [\pi_{Y_0}(x), \pi_{Y_0}(\tau)] = [1, \pi_{Y_0}(\tau)] = 1$. Since we have already seen the faithfulness of π_{Y_0} on $\mathrm{Center}(B_{n-1})$, we have $[x, \tau] = 1$ for any $\tau \in B_{n-1}$, *i.e.*, $x \in \mathrm{Center}(B_{n-1})$.

§4. Hyperelliptic mapping class group

In order to consider the case $n = 6$ further, we need to consider the Jones representation of $\mathcal{M}_0^n$, with n even, as a representation of hyperelliptic mapping class group, as follows. Let Σ_g be a closed oriented surface of genus $g \geq 2$ and let $\mathcal{M}_g$ be its mapping class group. The hyperelliptic mapping class group $\mathcal{H}_g$ of genus g is defined as the subgroup of $\mathcal{M}_g$ consisting of those elements which commute with the class of a fixed hyperelliptic involution $\iota : \Sigma_g \to \Sigma_g$. We identify the quotient orbifold of Σ_g by the action of ι with S_{2g+2} where the singular locus of the orbifold corresponds to P, the set of punctures. Due to Birman–Hilden [4], the natural projection $\Sigma_g \to S_{2g+2}$ induces the following short exact sequence:

$$0 \to \mathbb{Z}/2\mathbb{Z} \to \mathcal{H}_g \xrightarrow{q} \mathcal{M}_0^{2g+2} \to 1$$

where the image of $\mathbb{Z}/2\mathbb{Z}$ in $\mathcal{H}_g$ is generated by the class of ι. Provided with an arbitrarily *rectangular* Young diagram Y with $(2g + 2)$ boxes, we obtain, following Jones [7], the representation of $\mathcal{H}_g$

$$\rho_Y : \mathcal{H}_g \to \mathrm{GL}(M_Y)$$

as the composition of $\bar{\pi}_Y$ with the above homomorphism q. As a corollary to Theorem 3.1, we have:

Corollary 4.1. $\mathrm{Ker}\,\rho_Y \cong \mathbb{Z}/2\mathbb{Z}$ *if and only if* π_{Y_0} *is faithful.*

§5. The case of $g = 2$

In the case of $g = 2$, it is classically known that $\mathcal{H}_2 = \mathcal{M}_2$. Also, because of the "column-row symmetry", the representation of $\mathcal{M}_2$ obtained as above is essentially unique (c.f. [8, Section 2.2]). So, we take $Y = [2^3]$, and consider the representation

$$\rho = \rho_{[2^3]} : \mathcal{M}_2 \to \mathrm{GL}(M_{[2^3]}).$$

We can now say a little bit about $\mathrm{Ker}\,\rho$, and hence about $\mathrm{Ker}\,\bar{\pi}_{[2^3]}$. Let

$$\rho_0 : \mathcal{M}_2 \to \mathrm{Sp}(4, \mathbb{Z})$$

be the symplectic representation induced by the action of $\mathcal{M}_2$ on the homology group $H_1(\Sigma_2; \mathbb{Z})$. The Torelli group $\mathcal{I}_2$ is defined as $\mathrm{Ker}\,\rho_0$.

Theorem 5.1. $\mathrm{Ker}\,\rho = \mathbb{Z}/2\mathbb{Z} \oplus (\mathrm{Ker}\,\rho \cap \mathcal{I}_2)$. *Here,* $\mathbb{Z}/2\mathbb{Z}$ *is generated by the class of the hyperelliptic involution* ι. *In particular,* ρ *is* faithful on $\mathcal{I}_2$ *if and only if* $\pi_{[2^2,1]}$ *is a faithful representation of* B_5.

By the definition $\rho = \bar{\pi}_{[2^3]} \circ q$, we have $\mathrm{Ker}\,\bar{\pi}_{[2^3]} = q(\mathrm{Ker}\,\rho)$. Therefore, Theorem 5.1 implies immediately a restriction of the kernel for the case $n = 6$:

Corollary 5.2. $\mathrm{Ker}\,\bar{\pi}_{[2^3]} \subset q(\mathcal{I}_2) \cong \mathcal{I}_2$.

Proof of Theorem 5.1. It suffices to prove the isomorphism in the first part of the theorem since the rest of the theorem is then a direct consequence of Corollary 4.1. We construct the isomorphism directly. Set $K = \mathrm{Ker}\,\rho$, and $K_0 = K \cap \mathcal{I}_2$. Note that ι lies in K by the definition of ρ. We then define a mapping $h_0 : \mathbb{Z} \oplus K_0 \to K$ by $h_0(a, x) = \iota^a \cdot x$. Since $\iota^2 = 1$, h_0 descends to a well-defined mapping

$$h : \mathbb{Z}/2\mathbb{Z} \oplus K_0 \to K.$$

Hereafter, the multiplication in $\mathbb{Z}/2\mathbb{Z}$ is written additively, and each element of $\mathbb{Z}/2\mathbb{Z}$ will be denoted by its representative in $\mathbb{Z}$, so that $1 + 1 = 0$ in $\mathbb{Z}/2\mathbb{Z}$. Since ι lies in $\mathrm{Center}\,(\mathcal{M}_2)$, the mapping h is actually a homomorphism of group. Suppose next that $h(a,x)=1$, *i.e.*, $\iota^a \cdot x = 1$. Then, by taking the image under ρ_0, we have $(-I)^a \cdot \rho_0(x) = I$. Since $\rho_0(x) = I$, we have $(-I)^a = 1$. Therefore, we have $a = 0$ in $\mathbb{Z}/2\mathbb{Z}$, and hence $x = 1$. This shows that the homomorphism h is injective.

Next, to prove that h is surjective, we consider the specialization of ρ at $t = -1$. By our previous result [8], this specialization is trivial on $\mathcal{I}_2$ and can be described as follows. Let sgn denote the one-dimensional representation of $\mathcal{M}_2$ which sends the Dehn twist along every non-separating simple closed curve to -1. It is easy to see that sgn is trivial on $\mathcal{I}_2$, and therefore descends to a representation of $\mathrm{Sp}(4,\mathbb{Z})$, denoted by $\overline{\mathrm{sgn}}$. Now, let λ denote the linear representation of $\mathrm{Sp}(4,\mathbb{Z})$ which is induced by the natural action on $\Lambda^2 H_1(\Sigma_2;\mathbb{Z})/\omega\mathbb{Z}$ where ω denotes the symplectic class in $\Lambda^2 H_1(\Sigma_2;\mathbb{Z})$. Then the specialization of ρ at $t=-1$ is equivalent to $(\overline{\mathrm{sgn}} \otimes \lambda) \circ \rho_0$ ([8, Lemma 2.1]).

Claim 5.3. As a subgroup of $\mathrm{Sp}(4,\mathbb{Z})$, the kernel of $\overline{\mathrm{sgn}} \otimes \lambda$ coincides with $\{\pm I\}$.

This claim implies the surjectivity of h as follows. Note that we have an obvious relation

$$K \subset \mathrm{Ker}\,(\text{the specialization of } \rho \text{ at } t=-1) = \mathrm{Ker}\,((\overline{\mathrm{sgn}} \otimes \lambda) \circ \rho_0).$$

Taking the images of the both ends under ρ_0, we have $\rho_0(K) \subset \{\pm I\}$ by the claim. On the other hand, it is easy to see that $\rho_0(\iota) = -I$. Therefore, recalling that $\iota \in K$ again, we have the equality $\rho_0(\mathrm{Ker}\,\rho) = \{\pm I\}$.

Now, let z be an arbitrary element of K. Then, we have either $\rho_0(z) = I$, or $\rho_0(z) = -I$. In the case of $\rho_0(z) = I$, we have $z \in H_0$ and hence $z = h(0,z)$. In the case of $\rho_0(z) = -I$, take $x = \iota z$. We then have $\rho_0(x) = \rho_0(\iota) \cdot \rho_0(x) = (-I)^2 = I$, and hence $x \in H_0$. Since $z = \iota x$, we have $z = h(1,x)$. This shows that h is surjective. Therefore, we have proven that h is an isomorphism. This finishes the proof of Theorem 5.1 Q.E.D.

We now prove Claim 5.3 to complete the proof of Theorem 5.1. Most essential is Lemma 5.4 below, the proof of which is postponed until Appendix. It is clear that $\{\pm I\} \subset \mathrm{Ker}\,(\overline{\mathrm{sgn}} \otimes \lambda)$. To show the converse, suppose $X \in \mathrm{Ker}\,(\overline{\mathrm{sgn}} \otimes \lambda)$. By the definition of $\overline{\mathrm{sgn}}$, we see that $\lambda(X)$ is equal to either I or $-I$. On the other hand, it is easy to see that $\lambda(-I) = I$, and thus the representation λ descends to a representation of $\mathrm{PSp}(4,\mathbb{Z}) = \mathrm{Sp}(4,\mathbb{Z})/\{\pm I\}$, denoted by

$$\bar{\lambda} : \mathrm{PSp}(4,\mathbb{Z}) \to \mathrm{GL}(V).$$

Here, V denotes $\Lambda^2 H_1(\Sigma_2;\mathbb{Z})/\omega\mathbb{Z}$. Then, for the element $\overline{X} \in \mathrm{PSp}(4,\mathbb{Z})$ corresponding to X, we have $\bar{\lambda}(\overline{X}) \in \mathrm{Center}\,(\mathrm{GL}(V))$. Now the lemma is in order.

Lemma 5.4. (1) $\mathrm{PSp}(4,\mathbb{Z})$ *has no nontrivial centers.*
(2) *As a representation of* $\mathrm{PSp}(4,\mathbb{Z})$, $\bar{\lambda}$ *is faithful.*

Then the part (2) of the lemma implies that $\overline{X}$ lies in Center $(\mathrm{PSp}(4,\mathbb{Z}))$. Next, by the part (1), we have $\overline{X} = I$ in $\mathrm{PSp}(4,\mathbb{Z})$. Hence, we have $X \in \{\pm I\}$. This completes the proof of Claim 5.3. Q.E.D.

Remark 5.5. By the celebrated theorem of Mess [16], the Torelli group $\mathcal{I}_2$ is an infinitely generated free group. Therefore, we can see, by Corollary 5.2, that $\mathrm{Ker}\,\bar{\pi}_{[2^3]}$ is, if non-trivial, a free group. On the other hand, our previous results on the non-triviality of $\rho_{[2^3]}$ [8, 9] could be considered as showing the non-triviality of $\bar{\pi}_{[2^3]}$, and hence of $\pi_{[2^2,1]}$.

Remark 5.6. It seems quite difficult to determine whether or not the representation $\pi_{[2^2,1]}$, and therefore $\rho|_{\mathcal{I}_2}$, is faithful. As an example, let us consider the restriction of $\pi_{[2^2,1]}$ to B_4. By Proposition 2.4, this restriction decomposes as $\pi_{[2,1^2]} \oplus \pi_{[2,2]}$. Then, by a result of Long [13], this direct sum is faithful if and only if either one of the summands is faithful. On the other hand, it is easy to see that the representation $\pi_{[2,2]}$ is *not faithful* since $\pi_{[2,2]}$ can be expressed as the composition of $\pi_{[2,1]}$ with a certain homomorphism $B_4 \to B_3$ with non-trivial kernel. Therefore, we can see that $\pi_{[2^2,1]}$ is faithful on B_4 if and only if $\pi_{[2,1^2]}$ is faithful. Now, it is well-known that $\pi_{[2,1^2]}$ is equivalent to the tensor product of sgn and the reduced Burau representation of B_4 (see [7, Note 5.7]). One can easily check that tensoring sgn does not affect the kernel, and so we can conclude that $\pi_{[2^2,1]}$ is faithful on B_4 if and only if the reduced Burau representation of B_4 is faithful. In particular, the unfaithfulness of the reduced Burau representation of B_4 implies the unfaithfulness of $\pi_{[2^2,1]}$ and $\rho|_{\mathcal{I}_2}$. We note that the reduced Burau representation of B_4 is the only representation among all the reduced Burau representations the faithfulness of which remains open.

§ Appendix—the proof of Lemma 5.4

It seems that Lemma 5.4 is well-known to experts, but because we are unable to provide any suitable reference in the literature, we include a proof here. We will use some well-known properties of $\mathrm{PSp}(4,\mathbb{Z}/p\mathbb{Z}) = \mathrm{Sp}(4,\mathbb{Z}/p\mathbb{Z})/\{\pm I\}$, for the details of which we refer to [6]. Let p be a prime number. Taking the mod p reduction $\mathbb{Z} \to \mathbb{Z}/p\mathbb{Z}$ for each matrix entry, we obtain a homomorphism of group

$$k_p : \mathrm{Sp}(4, \mathbb{Z}) \to \mathrm{Sp}(4, \mathbb{Z}/p\mathbb{Z}).$$

Since $k_p(-I) = -I$, k_p induces a homomorphism $\bar{k}_p : \mathrm{PSp}(4, \mathbb{Z}) \to \mathrm{PSp}(4, \mathbb{Z}/p\mathbb{Z})$. Note that $\mathrm{Ker}\, k_p$ consists of those matrices in $\mathrm{Sp}(4, \mathbb{Z})$ for which every diagonal entry is equal to 1 mod p and every off-diagonal entry is equal to 0 mod p. Therefore, we can observe that the intersection of $\mathrm{Ker}\, k_p$'s, when p varies among infinitely many arbitrary primes, consists of the single matrix I. It is also well-known that $\mathrm{Sp}(4, \mathbb{Z}/p\mathbb{Z})$ is generated by transvections, and every transvection in $\mathrm{PSp}(4, \mathbb{Z}/p\mathbb{Z})$ comes from the one in $\mathrm{Sp}(4, \mathbb{Z})$ via k_p. Therefore, k_p and hence $\bar{k}_p$ are surjective.

Now we prove the first part of the lemma. For each $X \in \mathrm{Sp}(4, \mathbb{Z})$, we denote by $\overline{X}$ the corresponding element of $\mathrm{PSp}(4, \mathbb{Z})$. Clearly, every element of $\mathrm{PSp}(4, \mathbb{Z})$ has an expression of this form. Let us choose an arbitrary element in $\mathrm{Center}\,(\mathrm{PSp}(4, \mathbb{Z}))$ and denote it by $\overline{Z}$ with $Z \in \mathrm{Sp}(4, \mathbb{Z})$. Since the homomorphism $\bar{k}_p$ is surjective, $\bar{k}_p(\overline{Z})$ lies in $\mathrm{Center}\,(\mathrm{PSp}(4, \mathbb{Z}/p\mathbb{Z}))$. We appeal to the following classical theorem.

Proposition A. *The finite group* $\mathrm{PSp}(4, \mathbb{Z}/p\mathbb{Z})$ *is simple for every prime* $p \geq 3$.

In particular, for $p \geq 3$, $\mathrm{PSp}(4, \mathbb{Z}/p\mathbb{Z})$ has no nontrivial centers. Therefore, we have $\bar{k}_p(\overline{Z}) = I$ in $\mathrm{PSp}(4, \mathbb{Z}/p\mathbb{Z})$. In other words, we have $k_p(Z) = I$, or $-I$ for $p \geq 3$. Therefore, either one of Z or $-Z$ is contained in infinitely many $(\mathrm{Ker}\, k_p)$ s'. Then the observation above implies that $Z = I$, or $-I$. In either case, we have $\overline{Z} = I$ in $\mathrm{PSp}(4, \mathbb{Z})$. This proves the first part of the lemma.

Next, we proceed to (2). Recall that the representation $\bar{\lambda}$ of $\mathrm{PSp}(4, \mathbb{Z})$ is induced by the natural action λ of $\mathrm{Sp}(4, \mathbb{Z})$ on the free abelian group $V = \Lambda^2 H_1(\Sigma_2; \mathbb{Z})/\omega\mathbb{Z}$, via the symplectic representation ρ_0. Let V_p denote $V \otimes \mathbb{Z}/p\mathbb{Z}$. We consider the representation $\lambda \otimes 1_{\mathbb{Z}/p\mathbb{Z}} : \mathrm{Sp}(4, \mathbb{Z}) \to \mathrm{GL}(V_p)$. Here, $1_{\mathbb{Z}/p\mathbb{Z}}$ denotes the identity on $\mathbb{Z}/p\mathbb{Z}$. Clearly, we have $\mathrm{Ker}\, k_p \subset \mathrm{Ker}\,(\lambda \otimes 1_{\mathbb{Z}/p\mathbb{Z}})$, and thus the representation $\lambda \otimes 1_{\mathbb{Z}/p\mathbb{Z}}$ descends to that of $\mathrm{Sp}(4, \mathbb{Z}/p\mathbb{Z})$, denoted by

$$\lambda_p : \mathrm{Sp}(4, \mathbb{Z}/p\mathbb{Z}) \to \mathrm{GL}(V_p).$$

Furthermore, recalling that $\lambda(-I) = I$ in $\mathrm{GL}(V)$, we see that λ_p descends to a representation of $\mathrm{PSp}(4, \mathbb{Z}/p\mathbb{Z})$ denoted by $\bar{\lambda}_p$. It is easy to check that $\bar{\lambda}_p$ is nontrivial for arbitrary prime p, for instance by computing the image under $\lambda_p \circ k_p \circ \rho_0$ of the Dehn twist along any non-separating simple closed curve. Then, by Proposition A again, $\bar{\lambda}_p$

is a faithful representation of $\mathrm{PSp}(4, \mathbb{Z}/p\mathbb{Z})$ for $p \geq 3$. Therefore, for arbitrary $X \in \operatorname{Ker} \lambda$, we have $k_p(X) = I$, or $-I$. Now the same argument as in (1) implies $\overline{X} = I$ in $\mathrm{PSp}(4, \mathbb{Z})$. This proves (2), and hence the lemma. Q.E.D.

We remark that the same proof as above works for $\mathrm{PSp}(2g, \mathbb{Z})$ with general $g \geq 2$, and its nontrivial linear representation defined naturally on an arbitrary free abelian subquotient of the tensor product of copies of the fundamental representation.

References

[1] J. E. Andersen, G. Masbaum and K. Ueno, Topological quantum field theory and the Nielsen-Thurston classification of $M(0,4)$, Math. Proc. Cambridge Philos. Soc., **141** (2006), 477–488.

[2] S. Bigelow, Braid groups are linear, J. Amer. Math. Soc., **14** (2001), 471–486.

[3] S. Bigelow and R. Budney, The mapping class group of a genus two surface is linear, Algebr. Geom. Topol., **1** (2001), 699–708.

[4] J. S. Birman and H. Hilden, Isotopies of homeomorphisms of Riemann surfaces and a theorem about Artin's braid group, Ann. of Math. (2), **97** (1973), 424–439.

[5] R. Gillette and J. van Buskirk, The word problem and consequences for the braid groups and mapping class groups of the 2-sphere, Trans. Amer. Math. Soc., **131** (1968), 277–296.

[6] L. C. Grove, Classical groups and geometric algebra, Grad. Stud. in Math., **39**, Amer. Math. Soc., Providence, RI, 2002.

[7] V. F. R. Jones, Hecke algebra representations of braid groups and link polynomials, Ann. of Math. (2), **126** (1987), 335–388.

[8] Y. Kasahara, An expansion of the Jones representation of genus 2 and the Torelli group, Algebr. Geom. Topol., **1** (2001), 39–55.

[9] ———, An expansion of the Jones representation of genus 2 and the Torelli group II, J. Knot Theory Ramifications, **13** (2004), 297–306.

[10] ———, The Jones representation of genus 1, to appear in: Intelligence of Low Dimensional Topology 2006, Series on Knot and Everything, **40**, World Sci. Publ., River Edge, NJ, 2007.

[11] M. Korkmaz, On the linearity of certain mapping class groups, Turkish J. Math., **24** (2000), 367–371.

[12] D. Krammer, Braid groups are linear, Ann. of Math. (2), **155** (2002), 131–156.

[13] D. Long, A note on the normal subgroups of mapping class groups, Math. Proc. Camb. Philos. Soc., **99** (1986), 79–87.

[14] W. Magnus and A. Peluso, On a theorem of V. I. Arnol'd, Comm. Pure Appl. Math., **22** (1969), 683–692.

[15] A. Mathas, Iwahori-Hecke algebras and Schur algebras of the symmetric group, Univ. Lecture Ser., **15**, Amer. Math. Soc., 1999.
[16] G. Mess, The Torelli groups for genus 2 and 3 surfaces, Topology, **31** (1992), 775–790.

Department of Mathematics
Kochi University of Technology
Tosayamada, Kami City, Kochi
782-8502 Japan
E-mail address: kasahara.yasushi@kochi-tech.ac.jp

Advanced Studies in Pure Mathematics 52, 2008
Groups of Diffeomorphisms
pp. 383–400

On the stable cohomology algebra of extended mapping class groups for surfaces

Nariya Kawazumi

Abstract.

Let $\Sigma_{g,1}$ be an oriented compact surface of genus g with 1 boundary component, and $\Gamma_{g,1}$ the mapping class group of $\Sigma_{g,1}$. We determine the stable cohomology group of $\Gamma_{g,1}$ with coefficients in $H^1(\Sigma_{g,1};\mathbb{Z})^{\otimes n}$, $n \geq 1$, explicitly modulo the stable cohomology group with trivial coefficients. As a corollary the rational stable cohomology algebra of the semi-direct product $\Gamma_{g,1} \ltimes H_1(\Sigma_{g,1};\mathbb{Z})$ (which we call the *extended mapping class group*) is proved to be freely generated by the generalized Morita-Mumford classes $\widetilde{m_{i,j}}$'s ($i \geq 0$, $j \geq 1$, $i+j \geq 2$) [11] over the rational stable cohomology algebra of the group $\Gamma_{g,1}$.

Introduction

Let $g \geq 2$ be an integer, $\Sigma_{g,1}$ an oriented compact surface of genus g with 1 boundary component, and $\Gamma_{g,1}$ the mapping class group of $\Sigma_{g,1}$. Similarly let Σ_g be an oriented closed surface of genus g, and Γ_g the mapping class group of Σ_g. We denote by H the first integral homology group of the surface $\Sigma_{g,1}$, $H_1(\Sigma_{g,1};\mathbb{Z})$, which can be regarded as that of Σ_g. The mapping class groups Γ_g and $\Gamma_{g,1}$ act on it in an obvious symplectic way.

One of the earliest works in the study of twisted (co)homology of the mapping class group with symplectic coefficients is Morita's paper [18] computing the first homology group with coefficients in H. Here he proved $H_1(\Gamma_{g,1};H) = \mathbb{Z}$ and $H_1(\Gamma_g;H) = \mathbb{Z}/(2-2g)$. In particular, the integral homology $H_1(\Gamma_g;H)$ depends on the genus g. This suggests to us the situation in the symplectic twisted (co)homology for $\Gamma_{g,1}$, which

---

Received May 5, 2007.
Revised October 28, 2007.
2000 *Mathematics Subject Classification.* Primary 57R20. Secondary 14H15, 32G15, 57M20, 57M50.
The research is supported in part by the Grant-in-Aid for Scientific Research (No. 18204002), the Ministry of Education, Culture, Sports, Science and Technology, Japan.

we would like to call the 'bordered' case, is different from that in the 'closed' case Γ_g.

The Harer stability theorem [3] and the theorem of Madsen and Weiss [16] are the most important facts on (co)homology of the mapping class groups. The former states the cohomology group of the mapping class group with trivial coefficients is independent of the genus g and the number of boundary components of the surface, provided that the degree is smaller than $g/3$ [3] or $g/2$ [8]. Moreover Ivanov [7] has generalized this theorem to those with twisted coefficients in the 'bordered' case. These theorems enable us to consider the stable cohomology group of the mapping class groups for surfaces. When we consider trivial coefficients $\mathbb{Z}$ (resp. $\mathbb{Q}$), we denote it by $H^*(\Gamma_\infty;\mathbb{Z})$ (resp. $H^*(\Gamma_\infty;\mathbb{Q})$). The latter theorem established in 2002 [16] gave a loop-space description for $H^*(\Gamma_\infty;\mathbb{Z})$. As a corollary it is proved the rational stable cohomology algebra $H^*(\Gamma_\infty;\mathbb{Q})$ is generated by the Morita-Mumford classes $e_i = (-1)^{i+1}\kappa_i$ [17] [21].

Looijenga [15] proved that the rational stable cohomology group of the mapping class group Γ_g, the 'closed' one, with coefficients in any irreducible representation of the rational symplectic group was a free module over $H^*(\Gamma_\infty;\mathbb{Q})$, and described its free basis. His computation is involved with geometric considerations on the moduli orbifold of complex algebraic curves including a theorem on Hodge theory [2]. Here it is remarkable that his results are based on the Harer stability theorem with trivial coefficients.

In [11] we independently constructed a bigraded series of cohomology classes of the mapping class group $\Gamma_{g,1}$, the 'bordered' one, with coefficients in the exterior algebra $\bigwedge^* H$. These series are twisted generalizations of the Morita-Mumford classes [17] [21], and are easily modified to those with coefficients in the n-fold tensor product $H^{\otimes n}$, $n \geq 1$.

The present paper is a revised version of the author's preprint [12]. In §§1-3 we will consider only the 'bordered' case, i.e., the case when the surface has boundaries. Our purpose is to prove the stable cohomology group of the mapping class group with coefficients in the n-fold tensor product $H^{\otimes n}$, $n \geq 1$, is a free module over the algebra $H^*(\Gamma_\infty;\mathbb{Z})$ and some combinations of the (modified) twisted Morita-Mumford classes give its free basis (Theorems 1.A and 1.B). This implies the Ivanov stability theorem with coefficients in any finite dimensional rational symplectic coefficients. Following Looijenga [15] we deduce them from the Harer stability theorem with trivial coefficients, but use the Lyndon-Hochschild-Serre spectral sequence for a pair of groups introduced in [11] instead of geometric considerations including Hodge theory. As a corollary the rational cohomology algebra of the semi-direct product

$H \rtimes \Gamma_{g,1}$ (which we call *the extended mapping class group*) is proved to be stabilized and to be freely generated by the generalized Morita-Mumford classes over the rational stable cohomology algebra of the mapping class group in the stable range (Theorem 1.C). After the original version was completed, the twisted Morita-Mumford classes turned out to lift to the mapping class group with punctures [13]. This will be discussed in §4, where we will also discuss briefly what the theorem of Madsen and Weiss [16] brought to our results.

The author would like to express his gratitude to Shigeyuki Morita and Youichi Shibukawa for helpful discussions, and to the organizing committee of the 37th Taniguchi Symposium, especially to Mika Seppälä, for part of the original version of this paper was achieved during the symposium.

CONTENTS

§1. Results

Let $g \geq 2$, $r, s \geq 0$ be integers, and $\Sigma^s_{g,r}$ a 2-dimensional oriented connected C^∞ manifold (i.e., oriented surface) of genus g with r ordered boundary components and s ordered punctures. The group of path-components $\pi_0(\mathrm{Diff}^+(\Sigma^s_{g,r}))$ is denoted by $\Gamma^s_{g,r}$ (or $\mathcal{M}^s_{g,r}$) and called the mapping class group of genus g with r ordered boundary components and s ordered punctures. Here $\mathrm{Diff}^+(\Sigma^s_{g,r})$ denotes the topological group (endowed with C^∞ topology) consisting of all orientation preserving diffeomorphisms of $\Sigma^s_{g,r}$ which fix all the boundary points and the punctures pointwise. When $s = 0$, we drop the indices: $\Sigma_{g,r} = \Sigma^0_{g,r}$, $\Gamma_{g,r} = \Gamma^0_{g,r}$ and similarly $\Sigma_g = \Sigma^0_{g,0}$, $\Gamma_g = \Gamma^0_{g,0}$. Throughout this paper we often denote by $H^1(\Sigma^s_{g,r})$ the first integral singular cohomology group of the space $\Sigma^s_{g,r}$. The group $\Gamma^t_{g,q}$ acts on it in an obvious way provided that $q \geq r$ and $t \geq s$. When $s = 0$ and $r = 1$, we often write simply

$$H := H_1(\Sigma_{g,1};\mathbb{Z}) = H_1(\Sigma_g;\mathbb{Z}) = H^1(\Sigma_g;\mathbb{Z}) = H^1(\Sigma_{g,1};\mathbb{Z}). \tag{1.1}$$

The isomorphism $H_1(\Sigma_g;\mathbb{Z}) = H^1(\Sigma_g;\mathbb{Z})$ is the Poincaré duality map, which is invariant under the action of the mapping class group Γ_g.

In view of the Harer stability theorem [3] there exists an integer $N(g)$ depending only on the genus g such that the forgetful map $\Gamma^s_{g,r+1} \to \Gamma^s_{g,r}$ given by forgetting the $(r+1)$-st boundary component induces an isomorphism

$$H^*(\Gamma^s_{g,r+1};\mathbb{Z}) = H^*(\Gamma^s_{g,r};\mathbb{Z})$$

for any $* \leq N(g)$ and $s, r \geq 0$. Harer [3] proved $N(g) \geq g/3$, and later Ivanov [8] improved the inequality; $N(g) \geq g/2$. Now consider a natural central extension

$$0 \to \mathbb{Z} \to \Gamma^s_{g,r+1} \to \Gamma^{s+1}_{g,r} \to 1 \tag{1.2}$$

given by collapsing the $(r+1)$-st boundary component to the $(s+1)$-st puncture. Let $e \in H^2(\Gamma^{s+1}_{g,r};\mathbb{Z})$ be the Euler class of the central extension. Then, substituting Harer's isomorphism into the Gysin sequence of the extension (1.2), we obtain a natural decomposition

$$H^*(\Gamma^{s+1}_{g,r};\mathbb{Z}) = H^*(\Gamma^s_{g,r};\mathbb{Z}) \oplus eH^{*-2}(\Gamma^{s+1}_{g,r};\mathbb{Z}) = H^*(\Gamma^s_{g,r};\mathbb{Z})[e] \tag{1.3}$$

for $* \leq N(g)$ (cf. [17] [4] [15]).

Our first theorem in the present paper is

Theorem 1.A. *If $s \geq 0$, $r \geq 1$ and $n \geq 0$, we have*

$$H^*(\Gamma^s_{g,r}; H^1(\Sigma^s_{g,r};\mathbb{Z})^{\otimes n}) = H^*(\Gamma_{g,1}; H^{\otimes n}) \otimes_{H^*(\Gamma_{g,1};\mathbb{Z})} H^*(\Gamma^s_{g,r};\mathbb{Z})$$

for degrees $\leq N(g) - n$. Here the RHS is a tensor product over the graded algebra $H^*(\Gamma_{g,1};\mathbb{Z})$.

If we denote by $e_{(a)}$ the Euler class corresponding to the a-th puncture, $1 \leq a \leq s$, then $H^*(\Gamma^s_{g,r};\mathbb{Z}) = H^*(\Gamma_{g,1};\mathbb{Z})[e_{(1)}, \ldots, e_{(s)}]$ for $* \leq N(g)$ from (1.3). Hence Theorem 1.A means

$$H^*(\Gamma^s_{g,r}; H^1(\Sigma^s_{g,r};\mathbb{Z})^{\otimes n}) = H^*(\Gamma_{g,1}; H^{\otimes n}) \otimes_{\mathbb{Z}} \mathbb{Z}[e_{(1)}, \ldots, e_{(s)}]$$

for degrees $\leq N(g) - n$.

As a consequence one deduces the Ivanov stability theorem [7] for the $\Gamma^s_{g,r}$-module $H^1(\Sigma^s_{g,r};\mathbb{Z})^{\otimes n}$ and any finite dimensional rational Sp-modules.

To describe the cohomology group $H^*(\Gamma_{g,1}; H^{\otimes n})$ we need to introduce some notions related to the mapping class groups. Observe the surface $\Sigma^1_{g,1}$ is obtained by glueing the surfaces $\Sigma_{g,1}$ and $\Sigma^1_{0,2}$ along boundary components. The infinite cyclic group $\mathbb{Z}$ acts on the surface

$\Sigma^1_{0,2}$ by rotating the puncture and fixing all the boundary pointwise. So the group $\Gamma_{g,1} \times \mathbb{Z}$ is embedded into the group $\Gamma^1_{g,1}$ (cf. e.g.,[6]). The Lyndon-Hochschild-Serre spectral sequence of the pair of groups $(\Gamma^1_{g,1}, \Gamma_{g,1} \times \mathbb{Z})$ introduced in [11]§1 induces the fiber integral

$$\pi_! : H^q(\Gamma^1_{g,1}, \Gamma_{g,1} \times \mathbb{Z}; M) \to H^{q-2}(\Gamma_{g,1}; M) \tag{1.4}$$

for any $\Gamma_{g,1}$-module M. Here we mean by $H^q(\Gamma^1_{g,1}, \Gamma_{g,1} \times \mathbb{Z}; M)$ the q-th cohomology group of the kernel of the restriction map

$$C^*(\Gamma^1_{g,1}, \Gamma_{g,1} \times \mathbb{Z}; M) := \operatorname{Ker}(C^*(\Gamma^1_{g,1}; M) \to C^*(\Gamma_{g,1} \times \mathbb{Z}; M))$$

of the normalized standard cochain complexes $C^*(\cdot;\cdot)$ [5].

The cohomology class ω defined below plays an important role throughout this paper. Regard the surface $\Sigma^1_{g,1}$ as a subsurface obtained by deleting one interior point from the surface $\Sigma_{g,1}$. The cohomology exact sequence of the pair of spaces $(\Sigma_{g,1}, \Sigma^1_{g,1})$ gives a $\Gamma^1_{g,1}$-exact sequence

$$0 \to H^1(\Sigma_{g,1}) = H \to H^1(\Sigma^1_{g,1}) \to H^2(\Sigma_{g,1}, \Sigma^1_{g,1}) = \mathbb{Z} \to 0. \tag{1.5}$$

We denote by ω the image of $1 \in \mathbb{Z} = H^0(\Gamma^1_{g,1}; \mathbb{Z})$ under the connecting homomorphism δ^* induced by (1.5):

$$\omega := \delta^*(1) \in H^1(\Gamma^1_{g,1}; H). \tag{1.6}$$

The restriction of ω to the subgroup $\Gamma_{g,1} \times \mathbb{Z} (\subset \Gamma^1_{g,1})$ is null cohomologous. In fact, choose a simple curve l inside the subsurface $\Sigma^1_{0,2} (\subset \Sigma^1_{g,1})$ connecting the puncture to a point on the boundary of $\Sigma^1_{g,1}$. The 1-cocycle $\omega_l \in Z^1(\Gamma^1_{g,1}; H)$ given by

$$\omega_l(\gamma) = \gamma l - l \in H, \quad \gamma \in \Gamma^1_{g,1}, \tag{1.7}$$

represents the cohomology class $\omega \in H^1(\Gamma^1_{g,1}; H)$. Clearly we have $\omega_l(\gamma) = 0$ for any $\gamma \in \Gamma_{g,1} \times \mathbb{Z}$.

Thus, in view of the cohomology exact sequence

$$0 \to H^1(\Gamma^1_{g,1}, \Gamma_{g,1} \times \mathbb{Z}; H) \to H^1(\Gamma^1_{g,1}; H) \to H^1(\Gamma_{g,1} \times \mathbb{Z}; H),$$

there exists a unique element of $H^1(\Gamma^1_{g,1}, \Gamma_{g,1} \times \mathbb{Z}; H)$ mapping to ω. We also denote it by

$$\omega \in H^1(\Gamma^1_{g,1}, \Gamma_{g,1} \times \mathbb{Z}; H).$$

For a finite subset S of the natural numbers $\mathbb{N}$ we form the power of ω

$$\omega^S \in H^{\sharp S}(\Gamma^1_{g,1}, \Gamma_{g,1} \times \mathbb{Z}; H^{\otimes S}),$$

which we multiply in numerical order. Let $i \geq 0$ be an integer satisfying the condition

$$i + \sharp S \geq 2 \tag{1.8}$$

Then we define *the twisted*[1] *Morita-Mumford class* $m_{i,S}$ by

$$m_{i,S} := \pi_!(e^i \omega^S) \in H^{2i+\sharp S-2}(\Gamma_{g,1}; H^{\otimes S}), \tag{1.9}$$

where $\pi_!$ is the fiber integral given in (1.4) and $e \in H^2(\Gamma_{g,1}; \mathbb{Z})$ is the Euler class of the central extension (1.2) for $r = 1$, $s = 0$.

Definition 1.1. A set $\widehat{P} = \{(S_1, i_1), (S_2, i_2), \ldots, (S_\nu, i_\nu)\}$ is a *weighted partition* [2] *of the index set* $\{1, 2, \ldots, n\}$ if

(1) The set $\{S_1, S_2, \ldots, S_\nu\}$ is a partition of the set $\{1, 2, \ldots, n\}$

$$\{1, 2, \ldots, n\} = \coprod_{a=1}^{\nu} S_a, \quad S_a \neq \emptyset \quad (1 \leq \forall a \leq \nu).$$

(2) $i_1, i_2, \ldots, i_\nu \geq 0$ are non-negative integers.
(3) Each (S_a, i_a) satisfies the condition (1.8): $i_a + \sharp S_a \geq 2$.

We denote by $\mathcal{P}_n$ the set consisting of all weighted partitions of the index set $\{1, 2, \ldots, n\}$. For each weighted partition $\widehat{P} = \{(S_1, i_1), (S_2, i_2), \ldots, (S_\nu, i_\nu)\} \in \mathcal{P}_n$ we define *the twisted Morita-Mumford class*

$$m_{\widehat{P}} := m_{i_1,S_1} m_{i_2,S_2} \cdots m_{i_\nu,S_\nu} \in H^{2(\sum i_a)+n-2\nu}(\Gamma_{g,1}; H^{\otimes n}). \tag{1.10}$$

Theorem 1.B. *For degrees* $\leq N(g) - n$

$$H^*(\Gamma_{g,1}; H^{\otimes n}) = \bigoplus_{\widehat{P} \in \mathcal{P}_n} H^*(\Gamma_{g,1}; \mathbb{Z}) m_{\widehat{P}} = \bigoplus_{\widehat{P} \in \mathcal{P}_n} H^{*-\deg m_{\widehat{P}}}(\Gamma_{g,1}; \mathbb{Z}).$$

By the *extended mapping class group* we mean the semi-direct product

$$\widetilde{\Gamma^s_{g,r}} := H \rtimes \Gamma^s_{g,r} = H_1(\Sigma_{g,1}; \mathbb{Z}) \rtimes \Gamma^s_{g,r}.$$

This group was studied in [1]. The generalized Morita-Mumford classes $\widetilde{m_{i,j}} \in H^*(\widetilde{\Gamma_{g,1}}; \mathbb{Z})$ are constructed as follows [11]. In a similar way to $\Gamma_{g,1} \times \mathbb{Z} \subset \Gamma^1_{g,1}$ the group $\widetilde{\Gamma_{g,1}} \times \mathbb{Z}$ is embedded into the group $\widetilde{\Gamma^1_{g,1}}$. Using

[1] In [11] and [12] the author called it the *generalized* Morita-Mumford class.
[2] We use the term 'partition of a set' following Stanley [22] p.33.

the simple curve l in (1.7), we define a 2-cocycle $\widetilde{\omega_l} \in Z^2(\widetilde{\Gamma^1_{g,1}}, \widetilde{\Gamma_{g,1}} \times \mathbb{Z}; \mathbb{Z})$ by

$$(1.11) \quad \widetilde{\omega_l}(u_1\gamma_1, u_2\gamma_2) := \gamma_1(\gamma_2 l - l) \cdot u_1, \quad u_1, u_2 \in H,\ \gamma_1, \gamma_2 \in \Gamma^1_{g,1},$$

where $\cdot$ denotes the intersection product on $H = H_1(\Sigma_g; \mathbb{Z})$. Its image $\widetilde{\omega}$ in $H^2(\widetilde{\Gamma^1_{g,1}}; \mathbb{Z})$ is equal to the Euler class of the central extension

$$0 \to \mathbb{Z} \to H_1(\Sigma^1_{g,1}; \mathbb{Z}) \rtimes \Gamma^1_{g,1} \to H_1(\Sigma_{g,1}; \mathbb{Z}) \rtimes \Gamma^1_{g,1} = \widetilde{\Gamma^1_{g,1}} \to 1.$$

The forgetful map $\widetilde{\pi} : \widetilde{\Gamma^1_{g,1}} \to \widetilde{\Gamma_{g,1}}$ induces the fiber integral

$$(1.12) \qquad \widetilde{\pi}_! : H^q(\widetilde{\Gamma^1_{g,1}}, \widetilde{\Gamma_{g,1}} \times \mathbb{Z}; \mathbb{Z}) \to H^{q-2}(\widetilde{\Gamma_{g,1}}; \mathbb{Z}).$$

Thus we can define *the generalized Morita-Mumford class*

$$\widetilde{m_{i,j}} := \widetilde{\pi}_!(e^i \widetilde{\omega}^j) \in H^{2i+2j-2}(\widetilde{\Gamma_{g,1}}; \mathbb{Z})$$

for $i \geq 0$, $j \geq 0$ with $i + j \geq 2$. Clearly $\widetilde{m_{i+1,0}}$ is equal to (the image of) the i-th Morita-Mumford class e_i $(= (-1)^{i+1}\kappa_i) \in H^{2i}(\Gamma_g; \mathbb{Z})$ [17] [21]:

$$\widetilde{m_{i+1,0}} = e_i \in H^{2i}(\widetilde{\Gamma_{g,1}}; \mathbb{Z}).$$

Theorem 1.C.

$$H^*(\widetilde{\Gamma_{g,1}}; \mathbb{Q}) = H^*(\Gamma_{g,1} \ltimes H_1(\Sigma_{g,1}; \mathbb{Z}); \mathbb{Q}) = H^*(\Gamma_{g,1}; \mathbb{Q}) \otimes \mathbb{Q}[\widetilde{m_{i,j}}]$$

for degrees $\leq N(g)$, where integers i and j run over the domain

$$\{(i, j) \in \mathbb{Z} \times \mathbb{Z}; \quad i \geq 0,\ j \geq 1 \text{ and } i + j \geq 2\}.$$

§2. Stable Cohomology with Coefficients in $H^1(\Sigma_{g,1}; \mathbb{Z})^{\otimes n}$

This section is devoted to the proof of Theorems 1.A and 1.B.

Suppose $r, s \geq 1$. We have a natural commutative diagram of forgetful maps

$$(2.1) \qquad \begin{array}{ccc} \Gamma^s_{g,r} & \xrightarrow{\pi} & \Gamma^{s-1}_{g,r} \\ \varpi \downarrow & & \varpi \downarrow \\ \Gamma^1_{g,1} & \xrightarrow{\pi} & \Gamma_{g,1}. \end{array}$$

Here the upper and the lower π's are given by forgetting the s-th and the first punctures, respectively, and the left and the right ϖ's by forgetting the punctures from the first to the $(s-1)$-st and the boundary components except the first.

We regard the surface $\Sigma^s_{g,r}$ as a subsurface obtained by deleting one interior point from the surface $\Sigma^{s-1}_{g,r}$ and numbering the resulting puncture the s-th. The inclusion homomorphism $H^1(\Sigma^{s-1}_{g,r}) \to H^1(\Sigma^s_{g,r})$ is equivariant under the forgetful map $\pi : \Gamma^s_{g,r} \to \Gamma^{s-1}_{g,r}$, and so induces a $\Gamma^s_{g,r}$-exact sequence

$$0 \to H^1(\Sigma^{s-1}_{g,r}) \to H^1(\Sigma^s_{g,r}) \to H^2(\Sigma^{s-1}_{g,r}, \Sigma^s_{g,r}) = \mathbb{Z} \to 0. \tag{2.2}$$

We denote by $\omega = \omega_{(s-1)}$ the image of $1 \in \mathbb{Z} = H^0(\Gamma^s_{g,r}; \mathbb{Z})$ under the connecting homomorphism δ^* induced by (2.2):

$$\omega = \omega_{(s-1)} := \delta^*(1) \in H^1(\Gamma^s_{g,r}; H^1(\Sigma^{s-1}_{g,r})). \tag{2.3}$$

From the commutative diagram (2.1), the homomorphism induced by the forgetful map ϖ

$$H^1(\Gamma^1_{g,1}; H) \to H^1(\Gamma^s_{g,r}; H) \to H^1(\Gamma^s_{g,r}; H^1(\Sigma^{s-1}_{g,r}))$$

maps ω defined in (1.6) to $\omega = \omega_{(s-1)}$ defined in (2.3).

The kernel of the forgetful map $\pi : \Gamma^{s+1}_{g,r} \to \Gamma^s_{g,r}$ forgetting the $(s+1)$-st puncture is naturally isomorphic to $\pi_1(\Sigma^s_{g,r})$ $(s \geq 0)$, and so we have a Gysin exact sequence

$$\begin{aligned} &\cdots \to H^q(\Gamma^s_{g,r}; M) \xrightarrow{\pi^*} H^q(\Gamma^{s+1}_{g,r}; M) \\ &\xrightarrow{\pi_\sharp} H^{q-1}(\Gamma^s_{g,r}; H^1(\Sigma^s_{g,r}) \otimes M) \to H^{q+1}(\Gamma^s_{g,r}; M) \to \cdots \end{aligned} \tag{2.4}$$

for any $\Gamma^s_{g,r}$-module M. Here we denote the Gysin map (the fiber integral) by $\pi_\sharp$ to distinguish it from the fiber integral $\pi_!$ introduced in (1.4).

Theorem 2.1. *Let I and $J \subset \mathbb{N}$ be mutually disjoint finite index sets. Suppose $s \geq 0$ and $r \geq 1$. Moreover if $I \neq \emptyset$, assume $s \geq 1$. Then the forgetful map ϖ induces an isomorphism*

$$\begin{aligned} &H^*(\Gamma^s_{g,r}; H^1(\Sigma^{s-1}_{g,r})^{\otimes I} \otimes H^1(\Sigma^s_{g,r})^{\otimes J}) \\ = \ &(\bigoplus_{S \subset I} \omega^S \otimes H^*(\Gamma_{g,1}; H^{\otimes(J \cup I - S)})) \otimes_{H^*(\Gamma_{g,1};\mathbb{Z})} H^*(\Gamma^s_{g,r}; \mathbb{Z}) \end{aligned} \tag{2.5}$$

for degrees $\leq N(g) - \sharp(I \cup J)$, where $\omega^S \in H^{\sharp S}(\Gamma^1_{g,1}; H^{\otimes S})$ is the power of the class ω defined in (1.6). *The RHS is a tensor product over the graded algebra $H^*(\Gamma_{g,1}; \mathbb{Z})$.*

In particular, if $I = \emptyset$ and $J = \{1, 2, \ldots, n\}$, we obtain Theorem 1.A stated in §1.

Proof. We write simply $H_{(s)} := H^1(\Sigma^s_{g,r}) = H^1(\Sigma^s_{g,r};\mathbb{Z})$. The $\Gamma^s_{g,r}$-exact sequence (2.2) is rewritten to

$$0 \to H_{(s-1)} \to H_{(s)} \to \mathbb{Z} \to 0. \tag{2.6}$$

We prove the theorem by double induction on $\sharp(I \cup J)$ and $\sharp I$. When $I \cup J = \emptyset$, the theorem is trivial. Suppose $\sharp(I \cup J) \geq 1$.

(A). The case $I = \emptyset$: Let j_0 be the minimum of J and set $J_- := J - \{j_0\}$. From the inductive assumption applied to J_- the forgetful homomorphism

$$\pi^* : H^*(\Gamma^s_{g,r}; {H_{(s)}}^{\otimes J_-}) \to H^*(\Gamma^{s+1}_{g,r}; {H_{(s)}}^{\otimes J_-})$$

can be identified with $\pi^* : H^*(\Gamma^s_{g,r};\mathbb{Z}) \to H^*(\Gamma^{s+1}_{g,r};\mathbb{Z})$ tensored by $H^*(\Gamma_{g,1}; H^{\otimes J_-}))\otimes_{H^*(\Gamma_{g,1};\mathbb{Z})}$, and so has a left inverse over $H^*(\Gamma^s_{g,r})$. Hence the Gysin sequence (2.4) splits into the $H^*(\Gamma^s_{g,r})$-split exact sequences

$$\begin{aligned} &0 \to H^*(\Gamma^s_{g,r}; {H_{(s)}}^{\otimes J_-}) \to \\ &H^*(\Gamma^{s+1}_{g,r}; {H_{(s)}}^{\otimes J_-}) \overset{\pi_\sharp}{\to} H^{*-1}(\Gamma^s_{g,r}; H_{(s)} \otimes {H_{(s)}}^{\otimes J_-}) \to 0 \end{aligned} \tag{2.7}$$

for $* \leq N(g) - \sharp J_-$. When $s = 0$ and $r = 1$, we have

$$\begin{aligned} &0 \to H^*(\Gamma_{g,1}; H^{\otimes J_-}) \to \\ &H^*(\Gamma^1_{g,1}; H^{\otimes J_-}) \overset{\pi_\sharp}{\to} H^{*-1}(\Gamma_{g,1}; H \otimes H^{\otimes J_-}) \to 0 \end{aligned} \tag{2.8}$$

for $* \leq N(g) - \sharp J_-$. Compare the exact sequence (2.7) with the sequence (2.8) tensored by $\otimes_{H^*(\Gamma_{g,1})} H^*(\Gamma^s_{g,r})$. Then the forgetful map ϖ induces an isomorphism

$$\varpi^* : H^*(\Gamma^s_{g,r}; H_{(s)} \otimes {H_{(s)}}^{\otimes J_-}) \overset{\cong}{\to} H^*(\Gamma_{g,1}; H \otimes H^{\otimes J_-}) \otimes_{H^*(\Gamma_{g,1})} H^*(\Gamma^s_{g,r})$$

for $* \leq N(g) - \sharp J$. Here we use the fact the map ϖ induces an isomorphism

$$\varpi^* : H^*(\Gamma^1_{g,1}) \otimes_{H^*(\Gamma_{g,1})} H^*(\Gamma^s_{g,r}) \cong H^*(\Gamma^{s+1}_{g,r})$$

deduced from (1.3) and (2.1). Finally label the first H and $H_{(s)}$ the index j_0. Thus the induction proceeds.

(B). The case $I \neq \emptyset$: Then we suppose $s \geq 1$. Choose an index $i_0 \in I$. Set $I_- := I - \{i_0\}$ and $J_0 := J \cup \{i_0\}$. The sequence (2.6) induces a $\Gamma^s_{g,r}$-exact sequence

$$\begin{aligned} &0 \to {H_{(s-1)}}^{\otimes I} \otimes {H_{(s)}}^{\otimes J} \to \\ &{H_{(s-1)}}^{\otimes I_-} \otimes {H_{(s)}}^{\otimes J_0} \to {H_{(s-1)}}^{\otimes I_-} \otimes \mathbb{Z}^{\otimes \{i_0\}} \otimes {H_{(s)}}^{\otimes J} \to 0. \end{aligned} \tag{2.9}$$

By the inductive assumption applied to the index sets I_- and J_0

$$\begin{aligned} & H^*(\Gamma^s_{g,r}; H_{(s-1)}{}^{\otimes I_-} \otimes H_{(s)}{}^{\otimes J_0}) \\ = \ & (\bigoplus_{S \subset I_-} \omega^S \otimes H^*(\Gamma_{g,1}; H^{\otimes (J_0 \cup I_- - S)})) \otimes_{H^*(\Gamma_{g,1})} H^*(\Gamma^s_{g,r}). \end{aligned}$$

Each element of the RHS lifts to a (uniquely determined) element of $H^*(\Gamma^s_{g,r}; H^{\otimes(I_- \cup J_0)})$. Hence the map

$$H^*(\Gamma^s_{g,r}; H^{\otimes(I_- \cup J_0)}) \to H^*(\Gamma^s_{g,r}; H_{(s-1)}{}^{\otimes I_-} \otimes H_{(s)}{}^{\otimes J_0})$$

has a right inverse for $* \leq N(g) - \sharp(I \cup J)$. Therefore the cohomology exact sequence induced by the sequence (2.9) splits into a split exact sequence

$$\begin{aligned} 0 \to H^{*-1}(\Gamma^s_{g,r}; H_{(s-1)}{}^{\otimes I_-} \otimes H_{(s)}{}^{\otimes J}) \xrightarrow{\omega^{\{i_0\}} \otimes} \\ H^*(\Gamma^s_{g,r}; H_{(s-1)}{}^{\otimes I} \otimes H_{(s)}{}^{\otimes J}) \to H^*(\Gamma^s_{g,r}; H_{(s-1)}{}^{\otimes I_-} \otimes H_{(s)}{}^{\otimes J_0}) \to 0 \end{aligned}$$

for $* \leq N(g) - \sharp(I \cup J)$. Thus we have

$$\begin{aligned} & H^*(\Gamma^s_{g,r}; H_{(s-1)}{}^{\otimes I} \otimes H_{(s)}{}^{\otimes J}) \\ = \ & H^*(\Gamma^s_{g,r}; H_{(s-1)}{}^{\otimes I_-} \otimes H_{(s)}{}^{\otimes J_0}) \\ & \oplus (\omega^{\{i_0\}} \otimes H^*(\Gamma^s_{g,r}; H_{(s-1)}{}^{\otimes I_-} \otimes H_{(s)}{}^{\otimes J})) \\ = \ & (\bigoplus_{S \subset I_-} \omega^S \otimes H^*(\Gamma_{g,1}; H^{\otimes (J_0 \cup I_- - S)})) \otimes_{H^*(\Gamma_{g,1})} H^*(\Gamma^s_{g,r}) \\ & \oplus (\bigoplus_{S \subset I_-} \omega^{S \cup \{i_0\}} \otimes H^*(\Gamma_{g,1}; H^{\otimes (J \cup I_- - S)})) \otimes_{H^*(\Gamma_{g,1})} H^*(\Gamma^s_{g,r}) \\ = \ & (\bigoplus_{S \subset I} \omega^S \otimes H^*(\Gamma_{g,1}; H^{\otimes (J \cup I - S)})) \otimes_{H^*(\Gamma_{g,1})} H^*(\Gamma^s_{g,r}) \end{aligned}$$

for $* \leq N(g) - \sharp(I \cup J)$. This completes the induction. Q.E.D.

We have introduced two sorts of fiber integrals or Gysin maps induced by the forgetful map $\pi : \Gamma^1_{g,1} \to \Gamma_{g,1}$ in (1.4) and (2.4). These two Gysin maps are related to each other in the following way.

Lemma 2.2. *For any $\Gamma_{g,1}$-module M we have*

$$\begin{array}{ccc} H^p(\Gamma^1_{g,1}; M) & = & H^p(\Gamma^1_{g,1}; M) \\ {\scriptstyle \omega\cup}\downarrow & & \downarrow{\scriptstyle \pi_\sharp} \\ H^{p+1}(\Gamma^1_{g,1}, \Gamma_{g,1} \times \mathbb{Z}; H \otimes M) & \xrightarrow{\pi_!} & H^{p-1}(\Gamma_{g,1}; H \otimes M). \end{array}$$

Here we write simply $H = H^1(\Sigma_{g,1}; \mathbb{Z})$ *as in* (1.1).

Proof. Let G be a group, K a subgroup of G and M a G-module. We define the cohomology group $H^*(G, K; M)$ by that of the kernel of the restriction map

$$C^*(G, K; M) := \mathrm{Ker}(C^*(G; M) \to C^*(K; M))$$

of the <u>normalized</u> standard cochain complexes $C^*(\cdot;\cdot)$ [11]§1. Consider a normal subgroup N of G satisfying the condition: $KN = G$. In [5] p.118, l.27ff and p.119, l.6ff two mutually equivalent filtrations (A_j) and (A_j^*) are introduced on the normalized standard cochain complex, and induce the (ordinary) Lyndon-Hochschild-Serre spectral sequence. The filtration (A_j^*) (or equivalently (A_j)) restricted to $C^*(G, K; M)$ induces the Lyndon-Hochschild-Serre spectral sequence of pairs of groups [11]:

$$E_2^{p,q} = H^p(G/N; H^q(N, N \cap K; M)) \Rightarrow H^{p+q}(G, K; H).$$

In our situation we consider the case $G = \Gamma^1_{g,1}$, $K = \Gamma_{g,1} \times \mathbb{Z}$ and $N = \pi_1(\Sigma_{g,1}) \subset \Gamma^1_{g,1}$. Since $H^{p-i}(\Gamma_{g,1}; H^i(\pi_1(\Sigma_{g,1}); M)) = 0$ for $i \geq 2$, any $u \in H^p(\Gamma^1_{g,1}; M)$ is represented by a cocycle z whose value $z(\gamma_1, \gamma_2, \ldots, \gamma_p)$, $\gamma_1, \gamma_2, \ldots, \gamma_p \in \Gamma^1_{g,1}$, depends only on γ_1 and the cosets $\gamma_2\pi_1(\Sigma_{g,1}), \ldots, \gamma_p\pi_1(\Sigma_{g,1})$. We denote by $r_{p-1}z$ the cocycle given by restricting γ_1 into $\pi_1(\Sigma_{g,1})$ and regarding $\gamma_2, \ldots, \gamma_p$ as elements of $\Gamma_{g,1} = \Gamma^1_{g,1}/\pi_1(\Sigma_{g,1})$. By definition we have $\pi_\sharp u = [r_{p-1}z] \in H^{p-1}(\Gamma_{g,1}; H \otimes M)$.

On the other hand the cocycle $\omega \cup z$ defined by

$$(\omega \cup z)(\gamma_0, \gamma_1, \ldots, \gamma_p) = \omega(\gamma_0) \otimes \gamma_0(z(\gamma_1, \gamma_2, \ldots, \gamma_p)),$$

for $\gamma_0, \gamma_1, \ldots, \gamma_p \in \Gamma^1_{g,1}$, represents the cup product $\omega \cup u \in H^{p+1}(\Gamma^1_{g,1}, \Gamma_{g,1} \times \mathbb{Z}; H \otimes M)$. The value $(\omega \cup z)(\gamma_0, \gamma_1, \ldots, \gamma_p)$ depends only on γ_0, γ_1 and the cosets $\gamma_2\pi_1(\Sigma_{g,1}), \ldots, \gamma_p\pi_1(\Sigma_{g,1})$.

Hence, from a computation involved with [11] Lemma 2.3,

$$\pi_!(\omega \cup z)(\gamma_2, \ldots, \gamma_p) = -\sum_{i=1}^{g} a_i \otimes z(b_i, \gamma_2 \ldots, \gamma_p) + b_i \otimes z({a_i}^{-1}, \gamma_2 \ldots, \gamma_p),$$

where $\{a_1, a_2, \ldots, a_g, b_1, b_2, \ldots, b_g\}$ is a usual symplectic generating system of the fundamental group $\pi_1(\Sigma_{g,1})$. This implies $\pi_!(\omega \cup z) = r_{p-1}z$ and so $\pi_!(\omega \cup u) = [\pi_!(\omega \cup z)] = [r_{p-1}z] = \pi_\sharp u$, as was to be shown.

Q.E.D.

Proof of Theorem 1.B. We prove the theorem by induction on n. When $n = 0$, the theorem is trivial, so we assume $n \geq 1$. Set $J = \{1, 2, \ldots, n\}$, $j_0 = 1$ and $J_- = \{2, \ldots, n\}$. Recall the exact sequence (2.8) in the proof of Theorem 2.1. From Theorem 2.1 and (1.3) the Gysin map $\pi_\sharp$ restricted to

$$(H^*(\Gamma_{g,1}; H^{\otimes J_-}) \otimes e\mathbb{Z}[e]) \oplus \bigoplus_{\emptyset \neq S \subset J_-} \omega^S \otimes H^*(\Gamma_{g,1}; H^{\otimes (J_- - S)}) \otimes \mathbb{Z}[e]$$

is an isomorphism onto $H^{*-1}(\Gamma_{g,1}; H^{\otimes J})$ for $* \leq N(g) - n + 1$. We denote $S_+ := S \cup \{1\}$ for $S \subset J_-$. Lemma 2.2 implies

$$\pi_\sharp(e^i \omega^S) = \pi_!(e^i \omega^{S_+}) = m_{i,S_+} \in H^*(\Gamma_{g,1}; H^{\otimes S_+}).$$

Therefore from the inductive assumption we obtain

$$\begin{aligned}
(2.10) \qquad H^*(\Gamma_{g,1}; H^{\otimes n}) &= H^*(\Gamma_{g,1}; H^{\otimes J}) \\
&= \bigoplus_{\emptyset \neq S \subset J_-} \bigoplus_{i=0}^{\infty} m_{i,S_+} H^*(\Gamma_{g,1}; H^{\otimes (J_- - S)}) \\
&\qquad \oplus \bigoplus_{i=1}^{\infty} m_{i,\{1\}} H^*(\Gamma_{g,1}; H^{\otimes J_-}) \\
&= \bigoplus_{\widehat{P} \in \mathcal{P}_n} m_{\widehat{P}} H^*(\Gamma_{g,1}; \mathbb{Z})
\end{aligned}$$

for $* \leq N(g) - n$, which completes the induction. Q.E.D.

§3. Stable Cohomology Algebra of Extended Mapping Class Groups

Let i and j be integers with $i \geq 0$, $j \geq 1$ and $i + j \geq 2$. As in [11], we define

$$m_{i,j} := \pi_!(e^i \omega^j) \in H^{2i+j-2}(\Gamma_{g,1}; \textstyle\bigwedge^j H),$$

where $\pi_!$ is the fiber integral (1.4), and $\omega^j \in H^j(\Gamma^1_{g,1}, \Gamma_{g,1} \times \mathbb{Z}; \bigwedge^j H)$ is the j-th power of ω.

The n-th symmetric group $\mathfrak{S}_n$ acts on the set $\mathcal{P}_n$ of the weighted partitions of $\{1, 2, \ldots, n\}$ by

$$\sigma \widehat{P} := \{(\sigma(S_1), i_1), (\sigma(S_2), i_2), \ldots, (\sigma(S_\nu), i_\nu)\},$$

where $\sigma \in \mathfrak{S}_n$ and $\widehat{P} = \{(S_1, i_1), (S_2, i_2), \ldots, (S_\nu, i_\nu)\} \in \mathcal{P}_n$. The $\mathfrak{S}_n$-orbits in $\mathcal{P}_n$ are parametrized by the set $\mathcal{Q}_n$ defined as follows.

Definition 3.1. A sequence $\widehat{Q} = ((j_1, i_1), (j_2, i_2), \ldots, (j_\nu, i_\nu))$ is a *weighted partition of the number* n if

(1) The sequence $(j_1, j_2, \ldots, j_\nu)$ is a partition of the number n

$$j_1 + j_2 + \cdots + j_\nu = n, \quad j_1 \geq j_2 \geq \cdots \geq j_\nu \geq 1.$$

(2) $i_1, i_2, \ldots, i_\nu \geq 0$ are non-negative integers.
(3) $i_a \geq i_{a+1}$ if $j_a = j_{a+1}$.
(4) Each (j_a, i_a) satisfies the condition: $i_a + j_a \geq 2$.

We denote by $\mathcal{Q}_n$ the set consisting of all weighted partitions of the number n. Define

$$\lambda\widehat{P} := ((\sharp S_1, i_1), (\sharp S_2, i_2), \ldots, (\sharp S_\nu, i_\nu)) \in \mathcal{Q}_n$$

for $\widehat{P} = \{(S_1, i_1), (S_2, i_2), \ldots, (S_\nu, i_\nu)\} \in \mathcal{P}_n$ provided that $\sharp S_1 \geq \sharp S_2 \geq \cdots \geq \sharp S_\nu$ and $\sharp S_a = \sharp S_{a+1} \Rightarrow i_a \geq i_{a+1}$, $(1 \leq a < \nu)$. Then the map $\lambda : \mathcal{P}_n \to \mathcal{Q}_n$, $\widehat{P} \mapsto \lambda\widehat{P}$ induces a bijection $\lambda : \mathcal{P}_n/\mathfrak{S}_n = \mathcal{Q}_n$. Define

$$m_{\widehat{Q}} := m_{i_1,j_1} m_{i_2,j_2} \cdots m_{i_\nu,j_\nu} \in H^{\sum(2i_a+j_a-1)}(\Gamma_{g,1}; \bigwedge^n H)$$

for $\widehat{Q} = ((j_1, i_1), (j_2, i_2), \ldots, (j_\nu, i_\nu)) \in \mathcal{Q}_n$. The canonical projection $\lambda : H^{\otimes n} \to \bigwedge^n H$ maps $m_{\widehat{P}}$ to $\pm m_{\lambda\widehat{P}}$ for any $\widehat{P} \in \mathcal{P}_n$

$$\lambda_*(m_{\widehat{P}}) = \pm m_{\lambda\widehat{P}} \in H^*(\Gamma_{g,1}; \bigwedge^n H). \tag{3.1}$$

Theorem 3.2. *Let* $\mathbf{k}$ *be a field with* $\mathrm{ch}\,\mathbf{k} > n$ *or* $= 0$. *Then we have*

$$\begin{aligned} H^*(\Gamma_{g,1}; \bigwedge^n H^1(\Sigma_{g,1}; \mathbf{k})) &= \bigoplus_{\widehat{Q} \in \mathcal{Q}_n} H^*(\Gamma_{g,1}; \mathbf{k}) m_{\widehat{Q}} \\ &= \bigoplus_{\widehat{Q} \in \mathcal{Q}_n} H^{*-\deg m_{\widehat{Q}}}(\Gamma_{g,1}; \mathbf{k}) \end{aligned}$$

for degrees $\leq N(g) - n$.

As a corollary we obtain Theorem 1.C. In fact, let $h : H \rtimes \Gamma_{g,1} = \widetilde{\Gamma_{g,1}} \to H$ denote the twisted 1-cocycle $u\gamma \mapsto u$, $u \in H$, $\gamma \in \Gamma_{g,1}$. Then we have

$$\widetilde{\omega_l} = -\mu(h \cup \omega_l) \in Z^2(\widetilde{\Gamma^1_{g,1}}, \widetilde{\Gamma_{g,1}} \times \mathbb{Z}),$$

where $\mu : H^{\otimes 2} \to \mathbb{Z}$ is the intersection pairing. If we define $M_j : H^{\otimes 2j} \to \mathbb{Z}$ by $M_j(u_1 \otimes \cdots \otimes u_{2j}) := \prod_{k=1}^j \mu(u_k \otimes u_{j+k})$, then

$$e^i \widetilde{\omega_l}^j = \pm M_j(h^j e^i \omega_l{}^j) \in H^{2i+2j}(\widetilde{\Gamma^1_{g,1}}, \widetilde{\Gamma_{g,1}} \times \mathbb{Z}).$$

Since h^j comes from the group $\widetilde{\Gamma_{g,1}}$, we obtain

$$\widetilde{m_{i,j}} = \pm M_j(h^j m_{i,j}) \in H^{2i+2j-2}(\widetilde{\Gamma_{g,1}}). \tag{3.2}$$

The Lyndon-Hochschild-Serre spectral sequence of the group extension $H \to \widetilde{\Gamma_{g,1}} = H \rtimes \Gamma_{g,1} \to \Gamma_{g,1}$ is given by

$$E_2^{p,q} = H^p(\Gamma_{g,1}; \bigwedge^q H) \Rightarrow H^{p+q}(\widetilde{\Gamma_{g,1}}).$$

From (3.2) the class $m_{i,j} \in E_2^{2i+j-2,j}$ lifts to the class $\pm\widetilde{m_{i,j}} \in H^{2i+2j-2}(\widetilde{\Gamma_{g,1}})$. This proves Theorem 1.C.

Proof of Theorem 3.2. We define the order in each $\widehat{P} = \{(S_1, i_1), (S_2, i_2), \ldots, (S_\nu, i_\nu)\} \in \mathcal{P}_n$ by

(1) $\sharp S_1 \geq \sharp S_2 \geq \cdots \geq \sharp S_\nu$.
(2) If $\sharp S_a = \sharp S_{a+1}$, $i_a \geq i_{a+1}$.
(3) If $\sharp S_a = \sharp S_{a+1}$ and $i_a = i_{a+1}$, then the minimum of S_a is smaller than that of S_{a+1}.

Furthermore we denote

$$\tau_{\widehat{P}} = \begin{pmatrix} 1\ 2 & \cdots & n \\ S_1 & \cdots & S_\nu \end{pmatrix} \in \mathfrak{S}_n,$$

where the indices are set in numerical order inside each subset S_a. In other words, if $\psi_a : \{1, 2, \ldots, \sharp S_a\} \to S_a$ is the unique order-preserving bijection, then we have $\tau_{\widehat{P}}(i) = \psi_a(i - \sum_{b=1}^{a-1} \sharp S_b)$ for $\sum_{b=1}^{a-1} \sharp S_b < i \leq \sum_{b=1}^{a} \sharp S_b$.

Let the n-th symmetric group $\mathfrak{S}_n$ act on $H^{\otimes n}$ by

$$\sigma(u_1 \otimes u_2 \otimes \cdots \otimes u_n) = (\operatorname{sign}\sigma) u_{\sigma(1)} \otimes u_{\sigma(2)} \otimes \cdots \otimes u_{\sigma(n)}$$

for $\sigma \in \mathfrak{S}_n$, $u_i \in H$ $(1 \leq i \leq n)$. Then we have

$$\tau_* m_{\widehat{P}} = m_{\widehat{P}} \in H^*(\Gamma_{g,1}; H^{\otimes n}). \tag{3.3}$$

for $\tau \in \mathfrak{S}_n$ and $\widehat{P} \in \mathcal{P}_n$ with $\tau(\widehat{P}) = \widehat{P}$. In fact, from the (anti-)commutativity of cup products we have $\tau_* \omega^S = \omega^S$ if $S \subset \{1, 2, \ldots, n\}$, $\tau \in \mathfrak{S}_n$ and $\tau(S) = S$. Hence $\tau_* m_{i,S} = m_{i,S}$. Furthermore, since $\deg m_{i,S} \equiv \sharp S \mod 2$, we have

$$\begin{pmatrix} S & T \\ T & S \end{pmatrix}_* m_{i,S} m_{i,T} = m_{i,S} m_{i,T}$$

for $S,T \subset \{1,2,\ldots,n\}$ with $\sharp S = \sharp T$ and $S \cap T = \emptyset$. Here if $\varphi : S \to T$ is the unique order-preserving bijection, then $\begin{pmatrix} S & T \\ T & S \end{pmatrix}$ maps $s \in S$ to $\varphi(s)$, and $t \in T$ to $\varphi^{-1}(t)$.

Now, for any $\sigma \in \mathfrak{S}_n$, the permutation $\sigma^{-1}\tau_{\sigma(\widehat{P})}\tau_{\widehat{P}}{}^{-1}$ fixes $\widehat{P}$. From (3.3)

$$\sigma_* m_{\widehat{P}} = \tau_{\sigma(\widehat{P})}{}_* \tau_{\widehat{P}}{}^{-1}{}_* m_{\widehat{P}} = (\operatorname{sign} \tau_{\sigma(\widehat{P})}\tau_{\widehat{P}}{}^{-1}) m_{\sigma(\widehat{P})}.$$

Therefore for any $\widehat{P_0} \in \mathcal{P}_n$ the sum $\sum_{\widehat{P} \in \lambda^{-1}\lambda(\widehat{P_0})} (\operatorname{sign} \tau_{\widehat{P}}) m_{\widehat{P}}$ is invariant under the $\mathfrak{S}_n$-action. This means $m_{\lambda\widehat{P_0}} \neq 0$ in $H^*(\Gamma_{g,1}; \bigwedge^n H \otimes \mathbf{k})$.

The group $H^*(\Gamma_{g,1}; H^{\otimes n})$ is decomposed into a direct sum of $\mathfrak{S}_n$-submodules parametrized by $\mathcal{Q}_n = \mathcal{P}_n/\mathfrak{S}_n$, which implies $m_{\widehat{Q}}$'s, $\widehat{Q} \in \mathcal{Q}_n$, are linearly independent over $H^*(\Gamma_{g,1}; \mathbf{k})$.

From the assumption on the characteristic of the field $\mathbf{k}$ the map $\lambda_* : H^*(\Gamma_{g,1}; (H \otimes \mathbf{k})^{\otimes n}) \to H^*(\Gamma_{g,1}; \bigwedge^n H \otimes \mathbf{k})$ is surjective. By Theorem 1.B the $H^*(\Gamma_{g,1}; \mathbf{k})$-module $H^*(\Gamma_{g,1}; (H \otimes \mathbf{k})^{\otimes n})$ is generated by $m_{\widehat{P}}$, $\widehat{P} \in \mathcal{P}_n$ for $* \leq N(g) - n$. Hence, from (3.1), the $H^*(\Gamma_{g,1}; \mathbf{k})$-module $H^*(\Gamma_{g,1}; \bigwedge^n H \otimes \mathbf{k})$ is generated by $m_{\widehat{Q}}$, $\widehat{Q} \in \mathcal{Q}_n$. Q.E.D.

§4. Concluding Remarks

After the original version was completed in 1995, the twisted Morita-Mumford classes turned out to lift to the mapping class group $\mathcal{M}_{g,*} := \Gamma^1_{g,0}$ [13] [14]. The cohomology class $\omega \in H^1(\Gamma^1_{g,1}; H)$ in the previous sections has already appeared in Morita's work [19]. To explain it following [14], we introduce the fiber product $\overline{\mathcal{M}_{g,*}} := \mathcal{M}_{g,*} \times_{\Gamma_g} \mathcal{M}_{g,*}$ of the group $\mathcal{M}_{g,*}$ with respect to the forgetful map $\pi : \mathcal{M}_{g,*} = \Gamma^1_{g,0} \to \Gamma_g$. Since the kernel of π is naturally isomorphic to the fundamental group $\pi_1\Sigma_g$, we have an isomorphism

$$\overline{\mathcal{M}_{g,*}} \cong \pi_1\Sigma_g \rtimes \mathcal{M}_{g,*}, \quad (\varphi, \psi) \mapsto (\psi\varphi^{-1}, \varphi).$$

In [19] Morita introduced a twisted 1-cocycle

$$k_0 : \overline{\mathcal{M}_{g,*}} \to H, \quad (\varphi, \psi) \mapsto [\psi\varphi^{-1}]$$

taking the homology class of $\psi\varphi^{-1} \in \pi_1\Sigma_g$. As in §1 we choose a simple curve ℓ on $\Sigma^1_{g,1}$ connecting the puncture p_1 to a point p_0 on the boundary. Take a diffeomorphism $\psi_\ell : (\Sigma_g, p_1) \to (\Sigma_g, p_0)$ sliding the point p_1 along the curve ℓ. Define a homomorphism $\varpi' : \Gamma^1_{g,1} \to \mathcal{M}_{g,*}$ by forgetting the puncture p_1 and collapsing the boundary to a new puncture p_0, and $\varpi'' : \Gamma^1_{g,1} \to \mathcal{M}_{g,*}$ by forgetting the boundary. A homomorphism

$\alpha_\ell : \Gamma^1_{g,1} \to \mathcal{M}_{g,*}$ is defined by $\varphi \mapsto (\varpi'(\varphi), \psi_\ell \varpi''(\varphi){\psi_\ell}^{-1})$. Then we obtain

$$\alpha_\ell{}^*(k_0) = -\omega_\ell \in Z^1(\Gamma^1_{g,1}; H) \tag{4.1}$$

by a straight forward computation ([14] Lemma 4.1).

Now let e be the Euler class of the forgetful central extension $\mathbb{Z} \to \Gamma_{g,1} \to \mathcal{M}_{g,*}$, and $\overline{e}$ the pull-back of e by the second projection $\overline{\pi} : \overline{\mathcal{M}_{g,*}} \to \mathcal{M}_{g,*}$, $(\varphi, \psi) \mapsto \psi$. Then the twisted Morita-Mumford class

$$m_{i,S} := \pi_!(\overline{e}^i(-k_0)^{\otimes S}) \in H^{2i+\sharp S-2}(\mathcal{M}_{g,*}; H^{\otimes S}) \tag{4.2}$$

is defined for any i and $S \subset \mathbb{N}$ with $i + \sharp S \geq 2$. Here $\pi_!$ is the Gysin map of the first projection $\pi : \overline{\mathcal{M}_{g,*}} \to \mathcal{M}_{g,*}$, $(\varphi, \psi) \mapsto \varphi$. As was proved in [13] [14] by (4.1), this is exactly a lift of $m_{i,S} \in H^*(\Gamma_{g,1}; H^{\otimes S})$ defined in §1. We define $m_{\widehat{P}} \in H^*(\mathcal{M}_{g,*}; H^{\otimes n})$ for $\widehat{P} \in \mathcal{P}_n$ in a similar way to §1. From Theorem 1.B we have

Theorem 4.1. *For degrees $\leq N(g) - n$*

$$H^*(\mathcal{M}_{g,*}; H^{\otimes n}) = \bigoplus_{\widehat{P} \in \mathcal{P}_n} H^*(\mathcal{M}_{g,*}; \mathbb{Z}) m_{\widehat{P}}.$$

Consider the semi-direct products $\widetilde{\mathcal{M}_{g,*}} := H \rtimes \mathcal{M}_{g,*}$ and $H \rtimes \overline{\mathcal{M}_{g,*}}$. Making use of k_0 instead of ω, we can define a 2-cocycle $\tilde{k}_0 \in Z^2(H \rtimes \overline{\mathcal{M}_{g,*}}; \mathbb{Z})$ and the generalized Morita-Mumford class $\widetilde{m_{i,j}}$ for $i, j \geq 0$ with $i + j \geq 2$ in a similar way to (1.11) and (1.12). Then from Theorem 1.C we obtain

Theorem 4.2. *For degrees $\leq N(g)$*

$$H^*(\widetilde{\mathcal{M}_{g,*}}; \mathbb{Q}) = H^*(\mathcal{M}_{g,*}; \mathbb{Q}) \otimes \mathbb{Q}[\widetilde{m_{i,j}};\ i \geq 0,\ j \geq 1 \text{ and } i + j \geq 2].$$

In 2002 Madsen and Weiss [16] gave a loop-space description of $H^*(\Gamma_\infty; \mathbb{Z})$ and proved the algebra $H^*(\Gamma_\infty; \mathbb{Q})$ is generated by the Morita-Mumford classes. This result improves ours in the present paper in a surprising way. Theorem 1.C and Theorem 1.B are improved to the following.

Theorem 4.3. *For degrees $\leq N(g)$*

$$H^*(\widetilde{\Gamma_{g,1}}; \mathbb{Q}) = \mathbb{Q}[\widetilde{m_{i,j}};\ i, j \geq 0, \text{ and } i + j \geq 2].$$

Theorem 4.4. *For degrees $\leq N(g) - n$*

$$H^*(\Gamma_{g,1}; H^{\otimes n}) = \bigoplus_{\widehat{P} \in \mathcal{P}_n} \mathbb{Q}[e_j;\ j \geq 1] m_{\widehat{P}}.$$

Theorems 4.1 and 4.2 also have appropriate improvements.

Finally we give a brief remark on a relation to the Johnson homomorphism [9]. Let $\mathcal{I}_{g,1}$ be the Torelli group, i.e., the kernel of the action of the mapping class group $\Gamma_{g,1}$ on the integral homology group H. Johnson [10] proved the first Johnson homomorphism $\tau_1 : \mathcal{I}_{g,1} \to \bigwedge^3 H$ induces the isomorphism $H_1(\mathcal{I}_{g,1};\mathbb{Q}) \cong \bigwedge^3 H \otimes \mathbb{Q}$. The twisted Morita-Mumford classes are related to the extended Johnson homomorphism $\tilde{k} : \Gamma_{g,1} \to \frac{1}{2}\bigwedge^3 H$ introduced by Morita [20]. It extends the homomorphism τ_1 and is equal to $\frac{1}{6}m_{0,3}$. Let $Sp(H)$ denote the symplectic group of H. The extended homomorphism $\tilde{k}$ induces a natural homomorphism

$$\tilde{k}_M : ((\bigwedge^* H^1(\mathcal{I}_{g,1};\mathbb{Q})) \otimes M)^{Sp(H)} \to H^*(\Gamma_{g,1};M) \tag{4.3}$$

for any finite dimensional $\mathbb{Q}[Sp(H)]$-module M. It was proved in [13] [14] the image of the map is exactly the submodule generated by the Morita-Mumford classes e_i's and the twisted ones. The Madsen-Weiss theorem implies the map $\tilde{k}_M$ is stably surjective. In fact, Theorem 4.4 is applicable, because M is regarded as a $Sp(H)$-submodule of some $H^{\otimes n} \otimes \mathbb{Q}$.

References

[1] E. Arbarello, C. DeContini, V. G. Kac and C. Procesi, Moduli spaces of curves and representation theory, Commun. Math. Phys., **117** (1988), 1–36.

[2] P. Deligne, Théorème de Lefschetz et critères de dégénerescence de suites spectrales, Inst. Hautes Études Sci. Publ. Math., **35** (1968), 259–278.

[3] J. L. Harer, Stability of the homology of the mapping class group of orientable surfaces, Ann. of Math. (2), **121** (1985), 215–249.

[4] J. L. Harer, The third homology group of the moduli space of curves, Duke Math. J., **63** (1991), 25–55.

[5] G. Hochschild and J.-P. Serre, Cohomology of group extensions, Trans. Amer. Math. Soc., **74** (1953), 110–134.

[6] A. Ishida, Master thesis (in Japanese), Univ. of Tokyo, 1994.

[7] N. Ivanov, On the homology stability for Teichmüller modular groups: closed surfaces and twisted coefficients, Contemp. Math., **150** (1993), 149–194.

[8] N. Ivanov, Complexes of curves and the Teichmüller modular group, Russian Math. Survey, **42** (1987), 55–107.

[9] D. Johnson, A survey of the Torelli group, Contemp. Math., **20** (1983), 165–179

[10] D. Johnson, The structure of the Torelli group, II, III, Topology, **24** (1985), 113–144
[11] N. Kawazumi, A generalization of the Morita-Mumford classes to extended mapping class groups for surfaces, Invent. Math., **131** (1998), 137–149.
[12] N. Kawazumi, On the stable cohomology algebra of extended mapping class groups for surfaces, preprint, Hokkaido Univ. Preprint Series in Mathematics, **311** (1995).
[13] N. Kawazumi and S. Morita, The primary approximation to the cohomology of the moduli space of curves and cocycles for the stable characteristic classes, Math. Research Lett., **3** (1996), 629–641.
[14] N. Kawazumi and S. Morita, The primary approximation to the cohomology of the moduli space of curves and cocycles for the Mumford-Morita-Miller classes, preprint, Univ. of Tokyo, UTMS 2001-13 (2001).
[15] E. Looijenga, Stable cohomology of the mapping class group with symplectic coefficients and of the universal Abel-Jacobi map, J. Algebraic Geom., **5** (1996), 135–150.
[16] I. Madsen and M. Weiss, The stable moduli space of Riemann surfaces: Mumford's conjecture, preprint, arXiv:math.AT/0212321.
[17] S. Morita, Characteristic classes of surface bundles, Invent. Math., **90** (1987), 551–577.
[18] S. Morita, Families of Jacobian manifolds and characteristic classes of surface bundles, I, Ann. Inst. Fourier (Grenoble), **39** (1989), 777–810.
[19] S. Morita, Families of Jacobian manifolds and characteristic classes of surface bundles, II, Math. Proc. Camb. Phil. Soc., **105** (1989), 79–101.
[20] S. Morita, The extension of Johnson's homomorphism from the Torelli group to the mapping class group, Invent. Math., **111** (1993), 197–224.
[21] D. Mumford, Towards an enumerative geometry of the moduli space of curves, In: Arithmetic and Geometry, Progr. Math., **36**, 1983, pp. 271–328.
[22] R. P. Stanley, *Enumerative Combinatorics, I*, Wadsworth & Brooks/Cole, Monterey, California, 1986.

Department of Mathematical Sciences
University of Tokyo
Tokyo, 153-8914
Japan
E-mail address: kawazumi@ms.u-tokyo.ac.jp

Advanced Studies in Pure Mathematics 52, 2008
Groups of Diffeomorphisms
pp. 401–413

Stable length in stable groups

Dieter Kotschick

Abstract.

We show that the stable commutator length vanishes for certain groups defined as infinite unions of smaller groups. The argument uses a group-theoretic analogue of the Mazur swindle, and goes back to the works of Anderson, Fisher, and Mather on homeomorphism groups.

§1. Introduction

In this paper we show that the stable commutator length vanishes for certain groups defined as unions of subgroups that have many conjugate embeddings that commute element-wise. The argument used is a group-theoretic analogue of the Mazur swindle. Informally, it can be paraphrased by saying that as the size of a table grows, it becomes easier to sort objects on its surface, so that commutation can be done very efficiently on a table of infinite size[1]. In the words of Poénaru [18]: "The infinite comes with magic and power."

Let Γ be a group. The commutator length $c(g)$ of an element g in the commutator subgroup $[\Gamma, \Gamma] \subset \Gamma$ is the minimal number of factors needed to write g as a product of elements that can be expressed as single commutators. The stable commutator length of g is defined to be

$$||g|| = \lim_{n \to \infty} \frac{c(g^n)}{n} .$$

Direct proofs of the vanishing of the stable commutator length sometimes proceed by showing a lot more, namely that the commutator length itself is bounded. A prototypical argument for this in the context of homeomorphism groups goes back to Anderson [2] and Fisher [11], and was later used and refined by Mather [16] and by Matsumoto–Morita [17].

Received August 24, 2007.
Revised June 16, 2008.

[1] I owe this description to N. A'Campo.

This argument uses an infinite iteration that leads to complicated behaviour near one point, and is therefore not suitable for the study of diffeomorphism groups. The argument we employ here is a variation on this classical one. It works for diffeomorphism groups because instead of an infinite iteration we do only a finite iteration, however it is important that the finite number of iterations can be taken to be arbitrarily large. This argument does not prove that the commutator length is bounded, it only proves the weaker conclusion that the stable commutator length vanishes.

Bavard [5], using a Hahn–Banach argument influenced by that of Matsumoto and Morita [17], showed that the vanishing of the stable commutator length on the commutator subgroup is equivalent to the injectivity of the comparison map $H_b^2(\Gamma; \mathbb{R}) \to H^2(\Gamma; \mathbb{R})$, where $H_b^*(\Gamma)$ denotes bounded group cohomology in the sense of Gromov. Another way to express this condition is to say that every homogeneous quasi-homomorphism $\Gamma \to \mathbb{R}$ is in fact a homomorphism. If Γ is a perfect group, then the stable commutator length is defined on the whole of Γ, and vanishes identically if and only if there are no non-trivial homogeneous quasi-homomorphisms from Γ to $\mathbb{R}$.

Bavard's result suggests that the commutator calculus arguments discussed above should have a dual formulation in terms of quasi-homomorphisms. It is this dual formulation that we shall discuss, although we could equally well argue directly on the commutator side. Thus, what we prove for certain groups is that every homogeneous quasi-homomorphism is a homomorphism. Using Bavard's result this is equivalent to the vanishing of the stable commutator length.

In Section 2 we give the algebraic mechanism behind the vanishing results we shall prove. Section 3 applies this mechanism to the stable mapping class group, the braid group on infinitely many strands, and the stable automorphism groups of free groups. In Section 4 we give applications to diffeomorphism groups, and in Section 5 we compare our methods and results to those of Burago, Ivanov and Polterovich [8].

We refer the reader to [5, 15] for background on quasi-homomorphisms. In fact, the proofs of the vanishing results below are somewhat reminiscent of the discussion of weak bounded generation in [15].

§2. The algebraic vanishing result

A map $f\colon \Gamma \to \mathbb{R}$ is called a quasi-homomorphism if its deviation from being a homomorphism is bounded; in other words, there exists a constant $D(f)$, called the defect of f, such that

$$|f(xy) - f(x) - f(y)| \leq D(f)$$

for all $x, y \in \Gamma$. We will always take $D(f)$ to be the smallest number with this property, i.e. it is the supremum of the left hand sides over all x and $y \in \Gamma$.

Every quasi-homomorphism can be homogenized by defining

$$\varphi(g) = \lim_{n\to\infty} \frac{f(g^n)}{n} .$$

Then φ is again a quasi-homomorphism, is homogeneous in the sense that $\varphi(g^n) = n\varphi(g)$, and is constant on conjugacy classes. (Compare [5], Proposition 3.3.1.) Throughout this paper we shall only consider homogeneous quasi-homomorphisms.

We begin with the following preliminary result.

Lemma 2.1. *Let $\varphi\colon \Gamma \to \mathbb{R}$ be a homogeneous quasi-homomorphism. Then the following two properties hold:*

(1) *If $x,\ y \in \Gamma$ commute, then $\varphi(xy) = \varphi(x) + \varphi(y)$.*
(2) *If φ vanishes on every single commutator, then φ is a homomorphism.*

Proof. By homogeneity we have the following:

$$\begin{aligned} |\varphi(xy) - \varphi(x) - \varphi(y)| &= \lim_{n\to\infty} \frac{1}{n} |\varphi((xy)^n) + \varphi(x^{-n}) + \varphi(y^{-n})| \\ &\leq \lim_{n\to\infty} \frac{1}{n} \left(|\varphi((xy)^n x^{-n} y^{-n})| + 2D(\varphi) \right) \\ &= \lim_{n\to\infty} \frac{1}{n} |\varphi((xy)^n x^{-n} y^{-n})| . \end{aligned}$$

If x and y commute, then $\varphi((xy)^n x^{-n} y^{-n}) = 0$ for every n, giving the first statement.

Assume now that φ vanishes on single commutators. As $(xy)^n x^{-n} y^{-n}$ can be expressed as the product of $\frac{n}{2} + c$ commutators, see [5], the right hand side of the formula is bounded above by $\frac{1}{2}D(\varphi)$. However, taking the supremum of the left hand side over all x and y, we get the defect $D(\varphi)$. Thus $D(\varphi) \leq \frac{1}{2}D(\varphi)$, showing that the defect vanishes, and φ is a homomorphism. Q.E.D.

Here is the main mechanism for the vanishing theorems.

Proposition 2.2. *Let $\Lambda \subset \Gamma$ be a subgroup with the property that there is an arbitrarily large number of conjugate embeddings $\Lambda_i \subset \Gamma$*

of Λ in Γ with the property that elements of Λ_i and of Λ_j commute with each other in Γ whenever $i \neq j$. Then every homogeneous quasi-homomorphism on Γ restricts to Λ as a homomorphism.

Note that homogeneous quasi-homomorphisms are constant on conjugacy classes, so that the restriction of φ to Λ_i is independent of i.

Proof. By the second part of Lemma 2.1 we only have to prove that $\varphi([x,y]) = 0$ for any $x, y \in \Lambda$. Let x_i, $y_i \in \Gamma$ be the images of x and y under the embedding $\Lambda_i \subset \Gamma$. We have the following equalities:

$$\begin{aligned} n\varphi([x,y]) &= \varphi([x_1,y_1]) + \ldots + \varphi([x_n,y_n]) \\ &= \varphi([x_1,y_1]\ldots[x_n,y_n]) \\ &= \varphi([x_1\ldots x_n, y_1\ldots y_n]) \ , \end{aligned}$$

where the first one comes from the constancy of φ on conjugacy classes, the second is the first part of Lemma 2.1 applied to the commuting embeddings, and the third follows directly from the commuting property of the embeddings $\Lambda_i \subset \Gamma$. On the right hand side φ is applied to a single commutator in Γ, and therefore the right hand side is bounded in absolute value by the defect of φ on Γ. However, if $\varphi([x,y]) \neq 0$, then the left hand side is unbounded because by assumption we can make n arbitrarily large. Thus $\varphi([x,y]) = 0$ for all $x, y \in \Lambda$. Q.E.D.

There are several ways in which this mechanism can be applied to obtain the vanishing of the stable commutator length in various groups. We try to give a general statement, in the hope of unifying several different applications of the argument.

Theorem 2.3. *Let Γ be a group in which every element can be decomposed as a product of some fixed number k of elements contained in distinguished subgroups $\Lambda \subset \Gamma$. If each Λ is perfect and has the property in Proposition 2.2, then the stable commutator length of Γ vanishes.*

Proof. As the subgroups Λ are assumed perfect, the restrictions of quasi-homomorphisms on Γ to Λ vanish, because by Proposition 2.2 they are homomorphisms. Therefore the value of a quasi-homomorphism on every element of Γ is bounded by $k-1$ times the defect. But every bounded homogeneous quasi-homomorphism is trivial. Q.E.D.

§3. Applications to discrete groups

3.1. The stable mapping class group

Let Γ_g^1 be the group of isotopy classes of diffeomorphisms with compact support in the interior of a compact surface Σ_g^1 of genus g with one boundary component. Attaching a two-holed torus along the boundary defines the stabilization homomorphism $\Gamma_g^1 \to \Gamma_{g+1}^1$. The stable mapping class group Γ_∞ is defined as the limit

$$\Gamma_\infty = \varinjlim_g \Gamma_g^1 .$$

For $g \geq 3$ the groups Γ_g^1 are perfect, see [19], hence Γ_∞ is also perfect. Recall that in [10] it was proved that the stable commutator length is non-trivial on every Γ_g^1 with $g \geq 2$, see also [6, 7, 9, 15]. In contrast with this we have:

Theorem 3.1. *The stable commutator length for Γ_∞ vanishes identically.*

Proof. We apply Theorem 2.3 with $k = 1$. The subgroups Λ are the images of the Γ_g^1. In detail, every element of Γ_∞ is in the image of some $\Gamma_g^1 = \Lambda$. This has arbitrarily large numbers of commuting conjugate embeddings in Γ_∞ given by taking the boundary connected sum of an arbitrarily large number of copies of the surface of genus g with one boundary component. Any homogeneous quasi-homomorphism on Γ_∞ restricts to (the image of) Γ_g^1 as a homomorphism. However, for $g \geq 3$ this group is perfect, and so the homomorphism vanishes. Q.E.D.

Remark 3.2. Theorem 3.1 shows that the second bounded cohomology of mapping class groups does not stabilize. This contrasts sharply with the Harer stability theorem [13] for the ordinary group cohomology.

Remark 3.3. Theorem 3.1 fits in nicely with the form of the estimates for the stable commutator length obtained in [10, 7, 15]. The lower bounds given there for the stable commutator length of specific elements in Γ_g^1 go to zero for $g \to \infty$. A similar phenomenon seems to appear in the work of Calegari and Fujiwara [9].

Remark 3.4. The discussion in this subsection also applies to the stable mapping class groups for surfaces with several boundary components.

3.2. The braid group on infinitely many strands

Let B_n be the Artin braid group on n strands. Adding strands defines injective stabilization homomorphisms $B_n \longrightarrow B_{n+1}$. The braid group on infinitely many strands is

$$B_\infty = \bigcup_n B_n .$$

Theorem 3.5. *Any homogeneous quasi-homomorphism on the infinite braid group B_∞ is a homomorphism.*

Proof. We apply Proposition 2.2. The subgroups Λ are the B_n. Any two x, $y \in B_\infty$ are contained in some finite $B_n = \Lambda$. This has arbitrarily large numbers of commuting conjugate embeddings in B_∞ given by considering braids on n strands placed side by side. Any homogeneous quasi-homomorphism φ on B_∞ therefore restricts to B_n as a homomorphism. Thus $\varphi(xy) = \varphi(x) + \varphi(y)$, showing that φ is a homomorphism on B_∞. Q.E.D.

We have chosen this formulation of the result because the braid groups have infinite Abelianizations, so that it does not make sense to speak of the stable commutator lengths of elements. The stabilization is compatible with Abelianization, so that B_∞, like B_n for finite n, has infinite cyclic Abelianization. The conclusion of Theorem 3.5 is that every homogeneous quasi-homomorphism is a constant multiple of the Abelianization homomorphism.

For finite n, the braid groups do admit non-trivial homogeneous quasi-homomorphisms that are not proportional to the Abelianization map $B_n \longrightarrow \mathbb{Z}$, see [3, 12]. Thus Theorem 3.5 shows that the bounded cohomology of braid groups does not stabilize. The quotients B_n/C of finite braid groups modulo their centers have finite Abelianizations, so that all elements have powers that are products of commutators. Baader [3] proves some lower bounds for the stable commutator length of certain elements in B_n/C. These lower bounds tend to zero as $n \to \infty$. This of course fits with Theorem 3.5. The corresponding homogeneous quasi-homomorphisms defined on B_n/C, and, by composition, on B_n, do not extend to the braid group on infinitely many strands.

3.3. Automorphism groups of free groups

The canonical homomorphisms $F_n \longrightarrow F_n \star \mathbb{Z} = F_{n+1}$ give rise to injective homomorphisms $\mathrm{Aut}(F_n) \longrightarrow \mathrm{Aut}(F_{n+1})$. Thus we define

$$\mathrm{Aut}_\infty(F) = \bigcup_n \mathrm{Aut}(F_n) \ .$$

Note that this is smaller than $\mathrm{Aut}(F_\infty)$.

Theorem 3.6. *The stable commutator length for* $\mathrm{Aut}_\infty(F)$ *vanishes identically.*

Proof. The Abelianization of $\mathrm{Aut}(F_n)$ is of order 2, and is stable. Thus, up to taking squares, all elements in $\mathrm{Aut}(F_n)$ and in $\mathrm{Aut}_\infty(F)$ are products of commutators. We again apply Proposition 2.2. The subgroups Λ are the $\mathrm{Aut}(F_n)$. In detail, every element of $\mathrm{Aut}_\infty(F)$ is contained in some $\mathrm{Aut}(F_n) = \Lambda$. This has arbitrarily large numbers of commuting conjugate embeddings in $\mathrm{Aut}_\infty(F)$ given by embedding $F_n \star F_n \star \ldots \star F_n$ in some large F_N. Although the elements in the different copies of F_n do not commute, the induced embeddings of $\mathrm{Aut}(F_n)$ in $\mathrm{Aut}_\infty(F)$ do commute and are conjugate to each other. Any homogeneous quasi-homomorphism on $\mathrm{Aut}_\infty(F)$ restricts to $\mathrm{Aut}(F_n)$ as a homomorphism. As the Abelianization of $\mathrm{Aut}(F_n)$ is finite, the homomorphism is trivial. Q.E.D.

One could also consider the outer automorphism groups $\mathrm{Out}(F_n)$, and of course quasi-homomorphisms on these induce quasi-homomorphisms on $\mathrm{Aut}(F_n)$ by composition with the projection. However, there is no natural way of stabilizing the outer automorphism groups.

Unlike for mapping class groups, no unbounded quasi-homomorphisms are known on the automorphism groups of free groups. The groups $\mathrm{Aut}(F_n)$ are analogous to the so-called extended mapping class groups, consisting of the isotopy classes of all diffeomorphisms of surfaces. The usual mapping class groups we considered in Theorem 3.1 are index 2 subgroups of the extended mapping class groups. A single Dehn twist has non-zero stable commutator length in a mapping class group [10, 7, 15], but is conjugate to its inverse in the extended mapping class group [1], so that in this latter group its stable commutator length must vanish. By analogy with this phenomenon, it may be more promising to look for quasi-homomorphisms on the special automorphism group $S\,\mathrm{Aut}(F_n)$, rather than on $\mathrm{Aut}(F_n)$, where by $S\,\mathrm{Aut}(F_n)$ we denote the automorphisms preserving an orientation on the Abelianization $\mathbb{Z}^n$ of F_n. Theorem 3.6 also applies to the stabilized special automorphism group $S\,\mathrm{Aut}_\infty(F)$.

§4. Applications to diffeomorphism groups

In this section we apply the algebraic vanishing result to some groups of diffeomorphisms. If M is a smooth closed manifold, we consider the identity component $G = \mathrm{Diff}_0(M)$ of the full diffeomorphism group. If M is open, we consider the identity component $G = \mathrm{Diff}_0^c(M)$ of the group of compactly supported diffeomorphisms. (Sometimes this notation is redundant because the group of compactly supported diffeomorphisms may be connected.) In both cases we assume that the diffeomorphisms are of class C^r with $1 \leq r \leq \infty$ and $r \neq 1 + \dim(M)$. The classical results of Herman, Thurston, Epstein and Mather then ensure that G is a perfect group; see [4] and the references quoted there.

4.1. The full diffeomorphism groups

We would like to prove that the stable commutator length vanishes on the diffeomorphism groups. Unfortunately, we can only achieve this goal in the following special cases:

Theorem 4.1. *The stable commutator length vanishes for the following diffeomorphism groups:*

(1) $\mathrm{Diff}_0^c(D)$ *for a ball* D,
(2) $\mathrm{Diff}_0^c(D \times M)$ *for any* M *and a ball* D *of positive dimension,*
(3) $\mathrm{Diff}_0(S^n)$ *for any sphere,*
(4) $\mathrm{Diff}_0(\Sigma^n)$ *for any exotic sphere of dimension* $\neq 4$.

Proof. Let D be an n-dimensional ball, and $G = \mathrm{Diff}_0^c(D)$ the group of diffeomorphisms of D with compact support in the interior of D. Fix an exhaustion of D by smaller nested balls D_i so that each D_i has closure contained in the interior of D_{i+1}. If G_i is the group of diffeomorphisms of D_i with compact support in the interior of D_i, then we have injective homomorphisms $G_i \to G_{i+1}$ induced by the inclusion $D_i \subset D_{i+1}$, and

$$G = \bigcup_i G_i \ .$$

We now apply Theorem 2.3 with $k = 1$. The subgroups Λ are the G_i. In detail, every element of G is contained in some G_i. This has arbitrarily large numbers of commuting conjugate embeddings in G, because we can embed arbitrarily large numbers of disjoint (smaller) balls in the annulus $D \setminus D_i$. Any homogeneous quasi-homomorphism on G then restricts to G_i as a homomorphism. But this group is perfect, and so the homomorphism vanishes. This completes the proof of the first claim. The same argument works for $D \times M$ using the exhaustion by $D_i \times M$.

Now let $G = \mathrm{Diff}_0(S^n)$. We can compose any $g \in G$ with a suitable $f_1^{-1} \in G$ supported in an open ball to achieve that $f_1^{-1}g$ has a fixed point. It follows that we can write $f_1^{-1}g = f_2 f_3$, with f_3 having support in a ball around the fixed point, and f_2 having support in the complement of a (smaller) ball around the fixed point. Because the manifold is a sphere, this complement is again a ball. Using $g = f_1 f_2 f_3$, we apply Theorem 2.3 with $k = 3$. Let $\Lambda = \mathrm{Diff}^c(D)$ be the group of compactly supported diffeomorphisms of a ball. This is perfect and admits infinitely many conjugate commuting embeddings in G.

The argument given for spheres also works for exotic spheres in dimensions $n \neq 4$, because they are all twisted spheres obtained from two standard balls by gluing along the boundary [14]. Q.E.D.

After this paper had been submitted, Tsuboi [20] proved a much stronger result. He proved that $\mathrm{Diff}_0(M)$ is uniformly perfect for every closed manifold, with the possible exception of even-dimensional manifolds having no handle decomposition without handles of middle dimension. The simplest M for which the question is unresolved, is T^2. Because of Tsuboi's results, I have removed some material from the original version of this paper, which was concerned with the continuity, with respect to the C^0 topology, of potential homogeneous quasi-homomorphisms on diffeomorphism groups. That discussion is rather delicate technically, and I hope to return to it elsewhere, unless it is rendered completely empty by a generalization of Tsuboi's work.

4.2. Diffeomorphism groups preserving a symplectic or volume form

The first statement in Theorem 4.1 can be phrased for the compactly supported diffeomorphism group of Euclidean space $\mathbb{R}^n$ instead of a ball D. However, the two cases are rather different if we consider diffeomorphism groups preserving a symplectic or volume form ω, and we fix our conventions so that $\mathbb{R}^n$ has infinite (symplectic) volume and a ball has finite volume.

Let $G = \mathrm{Diff}_0^c(\mathbb{R}^n, \omega)$ be the identity component of the group of compactly supported diffeomorphisms preserving ω. This is not perfect because the Calabi invariant

$$\mathrm{Cal}\colon \mathrm{Diff}_0^c(\mathbb{R}^n, \omega) \longrightarrow \mathbb{R} \tag{1}$$

is a non-trivial homomorphism. (See for example [4] for the definition of Cal.) By results of Thurston and Banyaga, see [4], its kernel is a perfect group. It follows that Cal is the Abelianization homomorphism. So far there is no difference between the finite and infinite volume cases, and we have the same statements if we replace $(\mathbb{R}^n, \omega)$ by a ball of finite volume.

Theorem 4.2. *The stable commutator length vanishes on the kernel of the Calabi homomorphism in* $\mathrm{Diff}_0^c(\mathbb{R}^n, \omega)$. *On* $\mathrm{Diff}_0^c(\mathbb{R}^n, \omega)$ *every homogeneous quasi-homomorphism is a constant multiple of the Calabi invariant.*

Proof. As in the previous proof we can write the kernel of Cal as

$$G = \bigcup_i G_i \ ,$$

with each G_i consisting of those elements of G supported in a ball of radius i, say. Then each element of G is contained in some G_i, but this

G_i has arbitrarily large numbers of commuting conjugate embeddings given by disjoint Hamiltonian displacements of D_i in $\mathbb{R}^n$. Therefore we can apply Theorem 2.3 to conclude that every homogeneous quasi-homomorphism vanishes on Ker Cal.

Applying the same argument to $G = \mathrm{Diff}_0^c(\mathbb{R}^n, \omega)$ with each G_i equal to $\mathrm{Diff}_0^c(D_i, \omega)$, we conclude that the restriction to G_i of any homogeneous quasi-homomorphism φ on G is a homomorphism, and is therefore a constant multiple of the Calabi invariant. But any two balls in $\mathbb{R}^n$ are both contained in some larger ball, and therefore the restriction of φ is the same multiple of Cal on all balls. Q.E.D.

This argument clearly does not work in the finite volume case because the number of commuting conjugate embeddings one can construct is bounded in terms of the available volume. The abstract isomorphism type of the group of symplectic or volume-preserving diffeomorphisms of a ball is always the same. In particular it does not depend on the size or volume of the ball. Therefore one can always find arbitrarily large numbers of commuting embeddings of such a group in the corresponding group of a larger ball, or in itself, but these embeddings are conjugate only in the full diffeomorphism group, and not in the diffeomorphism group preserving ω.

§5. Comparison with results on quasi-norms

In this section we compare our results on the non-existence of homogeneous quasi-homomorphisms on various stable groups to results about existence and non-existence of quasi-norms. The following definitions are due to Burago, Ivanov and Polterovich [8].

Definition 5.1. A function $q\colon G \longrightarrow \mathbb{R}$ on a group is called a quasi-norm if it is almost subadditive, almost invariant under conjugacy, and unbounded.

A group G is called unbounded if it admits a quasi-norm, and is called bounded otherwise.

Here almost subadditive and almost invariant under conjugation means that there is a constant c such that

$$q(xy) \leq q(x) + q(y) + c$$

and

$$|q(xyx^{-1}) - q(y)| \leq c$$

for all x and $y \in G$.

If φ is a non-trivial homogeneous quasi-homomorphism, then its absolute value is a quasi-norm. Therefore boundedness of a group is an even stronger property than the non-existence of homogeneous quasi-homomorphisms. A bounded group has finite Abelianization and vanishing stable commutator length. If the Abelianization is trivial, then the group is uniformly perfect [8].

Burago, Ivanov and Polterovich [8] prove boundedness of many diffeomorphism groups by an argument that is rather stronger than ours. In particular, they prove boundedness for the groups we considered in Theorem 4.1. An easy example that illustrates the difference between the vanishing mechanism in [8] and the one considered here is the following.

Example 5.2. Consider the group $G = \mathrm{Diff}_0^c(\mathbb{R}^n, \omega)$ from Theorem 4.2. Then the function q which assigns to every $g \in G$ the (symplectic) volume of its support is a quasi-norm which is non-trivial on the kernel of the Calabi homomorphism. Therefore this kernel is an unbounded group which nevertheless has vanishing stable commutator length.

This also brings out another aspect of the difference between the finite and infinite volume cases mentioned earlier. If we take $G = \mathrm{Diff}_0^c(D, \omega)$ with D of finite (symplectic) volume, then q is bounded and therefore not a quasi-norm.

This idea of a quasi-norm defined by support size actually applies to all the discrete groups we considered in Section 3:

Theorem 5.3. *The following groups are unbounded:*

(1) *the stable mapping class group* Γ_∞,
(2) *the braid group* B_∞ *on infinitely many strands,*
(3) *the stable automorphism group of free groups* $Aut_\infty(F)$, *and*
(4) *the stable special automorphism group of free groups* $S\,\mathrm{Aut}_\infty(F)$.

Proof. Perhaps the case of the braid group is easiest to visualize. Define $q\colon B_\infty \longrightarrow \mathbb{Z}$ by sending a braid x to the smallest number k of strands needed to express it. These need not be the first k strands, but can be any k strands. The function q is clearly invariant under conjugation and is unbounded. It is also subadditive, because a set of strands used to express x and a set of strands used to express y can together be used to express xy. Thus q is a quasi-norm.

On the stable mapping class group define $q\colon \Gamma_\infty \longrightarrow \mathbb{Z}$ by mapping an element x to the smallest genus g of a compact surface with one boundary component on which a diffeomorphism representing the isotopy class x can be supported. Again this does not have to be the embedding of Σ_g^1 from the definition of the stabilization procedure, but can be any such subsurface of the infinite genus surface. By definition, this function is conjugacy-invariant. It is also not hard to see that it is unbounded, for example by considering the action of mapping classes on homology. To check (almost) subadditivity, one has to consider several cases depending on how compact surfaces with one boundary component supporting representatives for x and y sit in the infinite genus surface. In all cases the union of these two subsurfaces can be enlarged slightly, without increasing the genus, to obtain a compact surface with one boundary component supporting a representative for xy. The genus of this surface is at most $q(x) + q(y)$. Therefore q is in fact subadditive and a quasi-norm.

The argument for $Aut_\infty(F)$ and $S\,\mathrm{Aut}_\infty(F)$ is completely analogous. Q.E.D.

Acknowledgements: I am very grateful to all the colleagues with whom I discussed the subject matter of this paper, including in particular N. A'Campo, D. Burago, D. Calegari, K. Fujiwara, V. Markovic, S. Matsumoto, S. Morita, L. Polterovich, P. Py and T. Tsuboi. Special thanks are due to P. Py for an inspiring conversation about the work of Burago, Ivanov and Polterovich, which got me started in the first place.

References

[1] N. A'Campo, Monodromy of real isolated singularities, Topology, **42** (2003), 1229–1240.

[2] R. D. Anderson, On homeomorphisms as products of conjugates of a given homeomorphism and its inverse, In: Topology of 3-manifolds, (ed. M. K. Fort), Prentice-Hall, Englewood Cliffs, NJ, 1962.

[3] S. Baader, Asymptotic Rasmussen invariant, preprint, arXiv:math.GT/0702335v1, 12 Feb 2007.

[4] A. Banyaga, The structure of classical diffeomorphism groups, Math. Appl., **400**, Kluwer Academic Publishers Group, Dordrecht, 1997.

[5] C. Bavard, Longeur stable des commutateurs, Enseign. Math., **37** (1991), 109–150.

[6] M. Bestvina and K. Fujiwara, Bounded cohomology of subgroups of mapping class groups, Geom. Topol., **6** (2002), 69–89.

[7] V. Braungardt and D. Kotschick, Clustering of critical points in Lefschetz fibrations and the symplectic Szpiro inequality, Trans. Amer. Math. Soc., **355** (2003), 3217–3226.

[8] D. Burago, S. Ivanov and L. Polterovich, Conjugation-invariant norms on groups of geometric origin, in this volume, pp. 221–250.

[9] D. Calegari and K. Fujiwara, Stable commutator length in word-hyperbolic groups, preprint, arXiv:math.GR/0611889v3, 9 Feb 2007.

[10] H. Endo and D. Kotschick, Bounded cohomology and non-uniform perfection of mapping class groups, Invent. Math., **144** (2001), 169–175.

[11] G. M. Fisher, On the group of all homeomorphisms of a manifold, Trans. Amer. Math. Soc., **97** (1960), 193–212.

[12] J.-M. Gambaudo and É. Ghys, Braids and signatures, Bull. Soc. Math. France, **133** (2005), 541–579.

[13] J. Harer, Stability of the homology of the mapping class group of an orientable surface, Ann. of Math. (2), **121** (1985), 215–249.

[14] A. Kosinski, Differential Manifolds, Academic Press, Boston, MA, 1993.

[15] D. Kotschick, Quasi-homomorphisms and stable lengths in mapping class groups, Proc. Amer. Math. Soc., **132** (2004), 3167–3175.

[16] J. N. Mather, The vanishing of the homology of certain groups of homeomorphisms, Topology, **10** (1971), 297–298.

[17] S. Matsumoto and S. Morita, Bounded cohomology of certain groups of homeomorphisms, Trans. Amer. Math. Soc., **94** (1985), 539–544.

[18] V. Poénaru, What is an infinite swindle?, Notices Amer. Math. Soc., **54** (2007), 619–622.

[19] J. Powell, Two theorems on the mapping class group of a surface, Proc. Amer. Math. Soc., **68** (1978), 347–350.

[20] T. Tsuboi, On the uniform perfectness of diffeomorphism groups, in this volume, pp. 505–524.

Mathematisches Institut
Ludwig-Maximilians-Universität München
Theresienstr. 39
80333 München
Germany
E-mail address: dieter@member.ams.org

Advanced Studies in Pure Mathematics 52, 2008
Groups of Diffeomorphisms
pp. 415–442

Foliations and compact leaves on 4-manifolds I. Realization and self-intersection of compact leaves

Yoshihiko Mitsumatsu and Elmar Vogt[1]

Abstract.

We introduce an easily tractable cohomological criterion for the existence of 2-dimensional foliations with a prescribed compact leaf on a 4-manifold relying on standard methods, Milnor's inequality for the existence of a flat connection on an $\mathbb{R}^2$-bundle over a surface, and Thurston's h-principle. This is used to investigate the self-intersection numbers of compact leaves of foliations on the product of two surfaces, in particular the question whether these numbers are bounded on a given 4-manifold.

§1. Introduction

This article is the first in a series of papers to study 2-dimensional foliations with compact leaves on 4-manifolds, in particular the self-intersection of these compact leaves. By [HM] the self-intersection homology class of a Ruelle-Sullivan foliation cycle without compact leaves in its support is zero. Thus only the self-intersection classes of compact leaves in the support of a foliation cycle contribute to the self-intersection class of this cycle. From this fact results our interest in self-intersection numbers of compact leaves of n-dimensional foliations on $2n$- manifolds, with 2-dimensional foliations on 4-manifolds being the first case of interest.

This article treats two themes. First we give a general criterion in (co)homological terms for the existence of a foliation on a given closed

Received May 7, 2007.
Revised April 4, 2008.
This research is supported in part by the Grant-in-Aid for Scientific Research (No. 18340020), the Ministry of Education, Culture, Sports, Science and Technology, Japan.
[1] The research for this paper was carried out during three stays of the second author as a Visiting Professor in Tokyo, two of these in 1999 and 2003 at Chuo University, the third one in 2006-2007 at the University of Tokyo.

4-manifold with a prescribed embedded closed surface as a compact leaf. This will be explained in §4. The basic tools for solving this problem are Thurston's h-principle for foliations of codimension greater than one ([Th]), Milnor's inequality for the existence of a flat connection on an $\mathbb{R}^2$-bundle over a surface ([M]), and well known relations for characteristic classes concerning the existence of plane fields on 4-manifolds due to Hirzebruch-Hopf ([HH]).

To deal with this question was prompted by the second theme. Here we ask whether for any given closed oriented 4-manifold M there exists an upper bound for the self-intersection numbers of compact leaves of 2-dimensional foliations on M. This question originated in the work of the first author in his thesis [Mi1] under the direction of Professor Shigeyuki Morita. See also [Mi2].

In 1999 the authors showed that there exists for each $g \geq 2$ and any even number k a foliation on $M = \Sigma_g \times T^2$ with a compact leaf with self-intersection number k. The method to produce these foliations is a good illustration in the use of Thurston's h-principle, so that we feel justified to include it. This we do in §3. The arguments we use in our treatment of the first theme can be regarded as cohomological improvements of this method.

While the results concerning our first theme apply to all closed orientable 4-manifolds, we only consider products of two orientable closed surfaces for the second. We show in §6 that the self-intersection numbers of compact leaves are unbounded for 2-dimensional foliations on products of closed orientable surfaces where either one factor is a torus or no factor is a sphere. In all other cases, *i.e.*, at least one factor is a 2-sphere and no factor is a torus, the set of intersection numbers is bounded. This is proved in §7. In fact, the set of homology classes in $H_2(\Sigma_g \times S^2) \cong \mathbb{Z} \oplus \mathbb{Z}$ which can be realized as a compact leaf of some foliation is essentially bounded.

Apart from the sections mentioned above there are three more. We begin in Section 2 by recalling some basic facts about compact leaves of foliations. Section 5 contains the proof of a Theorem of Section 4 which says that two splittings of an $\mathbb{R}^4$-bundle over a surface into a sum of two plane bundles are homotopic as splittings if their Euler classes agree. In the final section, apart from raising several questions related to this article, we supplement the information of Section 7 on the foliations with compact leaves on $\Sigma_g \times S^2$. As will become clear, there are many interesting foliations with compact leaves representing a surprising number of elements in $H_2(\Sigma_g \times S^2)$.

We only deal with smooth oriented foliations on smooth oriented manifolds in this article, but everything will work also in the C^r-category, $r \geq 2$.

The authors are grateful to Dieter Kotschick for suggesting to them to introduce cohomological considerations for these problems. As was mentioned earlier, the cohomological method gives rise to the existence of some very interesting foliations. In the second article of the present series we will describe some of them explicitly.

§2. Fundamentals on compact leaves and basic examples

In this section and in the next, we explain the constraints on a prescribed surface to be a compact leaf of some foliation. The first one is due to the integrability of the plane field.

Before we describe this we fix some notation which we will use throughout this paper. M will always denote an oriented closed 4-manifold, $\mathcal{F}$ a 2-dimensional smooth foliation on a 4-manifold, τ [*resp.* ν] the tangent [*resp.* normal] bundle to a foliation, e the Euler class of bundles, and Σ_g will denote the closed orientable surface of genus g with some fixed orientation. We deal only with smooth objects.

The normal bundle $\nu\mathcal{F}$ of any foliation $\mathcal{F}$ admits a so-called Bott connection, that is, a connection which is flat along the leaves of $\mathcal{F}$([B]). (This is true in all dimensions.)

On the other hand, Milnor's seminal inequality ([M]) asserts that an oriented vector bundle of rank 2 over $\Sigma_g, g \geq 1$, admits a flat connection if and only if the Euler number of the bundle is bounded in absolute value by one half of the absolute value of the Euler characteristic of the surface. Combining these two theorems, we have the following.

Proposition 2.1. *A compact leaf $L \cong \Sigma_g, g \geq 1$, of a foliation $\mathcal{F}$ on M satisfies the following inequality between its homological self-intersection $[L]^2 = [L] \cdot [L] \in \mathbb{Z}$ and its Euler characteristic.*

$$|\langle e(\nu\mathcal{F}), [L]\rangle| = |[L]^2| \leq \frac{1}{2}|\langle e(\tau\mathcal{F}), [L]\rangle| = \frac{1}{2}|\chi(L)| = g - 1$$

Conversely, if an embedded surface $L \cong \Sigma_g \subset M$ $(g \geq 1)$ satisfies $|[L]^2| \leq g - 1$, then L is a leaf of some foliation of some open neighborhood of L.

The following example is standard and fundamental. It also shows how to compactify and modify a foliated tubular neighbourhood of a closed surface into a foliated closed manifold.

Example 2.2. Take an oriented flat $\mathbb{R}^2$-vector bundle E_ρ over a closed oriented surface Σ_g of genus $g \geq 1$ with holonomy $\rho : \pi(\Sigma_g) \to SL(2;\mathbb{R})$. Its Euler number $k = \langle e(E_\rho), [\Sigma_g] \rangle$ satisfies Milnor's inequality $|k| \leq g-1$. Take the product $\overline{\rho}$ of the one-dimensional trivial representation and ρ, namely, embed $SL(2;\mathbb{R})$ into the second and the third rows and columns of $SL(3;\mathbb{R})$. By taking the associated action of $SL(3;\mathbb{R})$ on the space of oriented lines, we obtain an oriented flat S^2-bundle $\overline{E}_\rho$ and consequently a foliation $\mathcal{F}_\rho$ on $\overline{E}_\rho$.

The points $P_\pm \in S^2$ which correspond to $(\pm 1, 0, 0) \in \mathbb{R}^3$ are fixed points of the action (in fact, the only fixed points, if $k \neq 0$), and the tangential representation at P_+ exactly coincides with ρ, while at P_- we obtain ρ with the orientation reversed. Therefore the foliation $\mathcal{F}_\rho$ has two compact leaves $L_\pm$ correponding to $P_\pm$ with $[L_\pm]^2 = \pm k$. The hemispheres $H_\pm\{(x,y,z) \in S^2 \,;\, \pm x > 0\}$ and the equator $S^1 = \{x = 0\}$ are also invariant. The flat H_+-bundle is diffeomorphic as an oriented foliated manifold to the original flat vector bundle E_ρ. The same holds for the H_--bundle with the reversed orientation.

Also remark that if k is even, the S^2-bundle is diffeomorphic to the product bundle once the flat structure is forgotten. The identification via this diffeomorphism does not preserve the equator nor $P_\pm$. In the product bundle the compact leaf $L_\pm$ is identified with the graph of a map $f_\pm : \Sigma_g \to S^2$ with $\deg f_\pm = \pm k/2$. For k odd, the bundle is twisted. See also Proposition 5.1. □ 2.2.

§3. Unbounded self-intersection on $\Sigma_g \times T^2$. A geometric construction

In this section, we construct a family of foliations with prescribed compact leaves on $M = \Sigma_g \times T^2$ $(g \geq 2)$, relying on Thurston's h-principle, and we also explain how the h-principle is used in our context.

Theorem 3.1. (Thurston [Th]) *Let ξ be a smooth 2-plane field on a smooth n-manifold N, $n \geq 4$, let $K \subset N$ be closed, and assume that ξ is completely integrable (i.e., it defines a 2-dimensional foliation) in a neighborhood of K. Then ξ is homotopic to a completely integrable plane field via a homotopy which is constant on K.*

There is an unpublished somewhat simplified proof of this theorem, which is due to A. Haefliger. It makes the resulting foliations almost visible. We will present this proof in the forthcoming paper [MV].

Example 3.2. Fix $g \geq 2$. Then for each pair of positive integers a and b satisfying $1 \leq b \leq g-1$, there exists a foliation $\{\mathcal{F}_{a,b}\}$ on $M = \Sigma_g \times T^2$ with a compact leaf $L_{a,b}$, which has self-intersection number $[L_{a,b}]^2 = 2ab$. Especially, on each such M, the self-intersection numbers of closed orientable surfaces which are leaves of foliations on M are not bounded.

General strategy: The following is the basic strategy for the construction, which will again be used in the next section.

First we fix an embedded surface L satisfying Milnor's inequality, *i.e.*, $|[L]^2| \leq |\chi(L)|/2$. Then as we have seen in 2.1, we can foliate a neighbourhood of L. Next we extend the tangent plane field of this foliation to a plane field on all of M. In this section this process is done by a geometric argument, while a cohomological argument replaces this in the next section. Then Thurston's h-principle modifies the plane field into a foliation keeping L as a compact leaf.

Geometric Construction: To begin the construction, let us take a distinct points $\{P_1, ..., P_a\}$ of T^2 and b distinct points $\{Q_1, ..., Q_b\}$ of Σ_g. The singular oriented surface

$$L'_{a,b} = (\bigcup_{i=1}^{a} \Sigma_g \times P_i) \cup (\bigcup_{j=1}^{b} Q_j \times T^2)$$

has ab double points. Resolving the double points by rounding off, *i.e.*, by removing small disk neighborhoods of the points from the horizontal and vertical branches of $L'_{a,b}$, and reconnecting each pair of the resulting boundary circles by an annulus respecting the orientation, we obtain a connected closed surface $L_{a,b}$ with the desired self-intersection number. However, we do this slightly more carefully so that we can easily extend the tangent plane field of this surface to all of M.

Step 1 (resolving the double points): Take a small holomorphic coordinate neighbourhood (U_j, z) around Q_j in Σ_g so that at Q_j we have $z = 0$. Also fix a complex structure on T^2 as $\mathbb{C}/\mathbb{Z} \oplus \sqrt{-1}\mathbb{Z}$ and place P_1 at 0. Let w denote the standard holomorphic local coordinate on T^2. Then around the double point $(Q_j, P_1) = (0, 0)$ consider the graph of $zw = \varepsilon^4$ for a small constant $0 < \varepsilon \ll 1$. Inside the polydisk $\{(z, w)\,;\, |z|, |w| \leq \varepsilon^2\}$ of radius ε^2 we adopt the graph as part of the connecting smooth annulus.

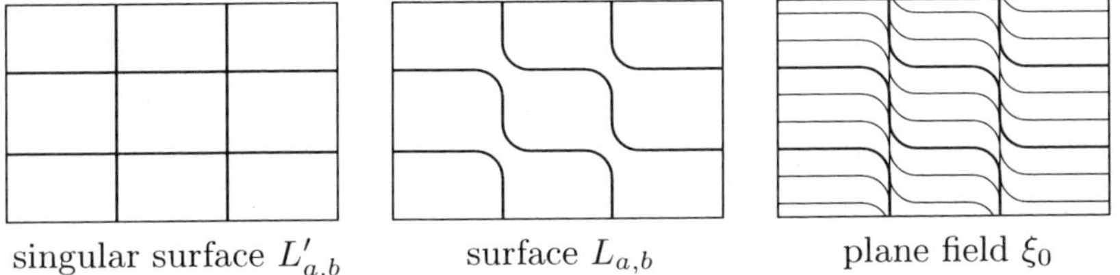

singular surface $L'_{a,b}$ surface $L_{a,b}$ plane field ξ_0

Outside the polydisk, we deform the graph as follows. In $\{(z,w)\,;\,|z| \le \varepsilon^2 \le |w|\}$ we deform the graph horizontally, *i.e.*, in z-direction, so that outside $\{(z,w)\,;\,\varepsilon \le |w|\}$ it is vertical, *i.e.*, it coincides with $\{z=0\}$. In $\{(z,w)\,;\,|w| \le \varepsilon^2 \le |z|\}$ it is deformed vertically so that outside $\{(z,w)\,;\,\varepsilon \le |z|\}$ it is horizontal. This can be done so that the resulting surface is smooth and is away from $\{z=0\}$ the graph of a function. We can achieve this by a C^1-small deformation if we take ε small enough.

The connecting annulus is then the part of the deformed graph inside $\{(z,w)\,;\,|z|,|w| \le \varepsilon\}$ and the disks $\{(z,0)\,;\,|z| \le \varepsilon\}$ and $\{(0,w)\,;\,|w| \le \varepsilon\}$ have been removed from $F'_{a,b}$.

Step 2 (coherent resolutions and the plane field): The remaining double points get resolved in exactly the same way using exactly the same graphs and deformations as above where at (Q_j, P_i) we choose the same coordinate system (U_j, z) around Q_j as before and around P_i the translate of the coordinate system of P_1.

The resulting smooth oriented embedded surface $L_{a,b}$ has the property that any translation of M in the T^2-direction which maps a point x of $L_{a,b}$ to another point y of $L_{a,b}$ also maps a neighborhood of x in $L_{a,b}$ diffeomorphically onto a neighborhood of y. Therefore, by translations in the T^2 direction, we can extend the tangent plane field of $L_{a,b}$ to an oriented plane field ξ_0 on M. This plane field is smooth in the complement of $\bigcup_{j=1}^{b} Q_j \times T^2$, but globally only C^0.

Let ξ_1 be a smooth approximation of ξ_0, close enough so that the arguments in the next step are valid.

Step 3 (application of 2.1 and 3.1): The surface $L_{a,b}$ represents the homology class $a[\Sigma_g] + b[T^2]$ and thus has self-intersection number $2ab$. Also it is easy to see that $\chi(L_{a,b}) = 2a(1-g-b)$. Therefore, our condition $1 \le b \le g-1$ is nothing but Milnor's inequality in 2.1 provided that a and b are positive. It follows that there is a foliation of some open neighbourhood V of $L_{a,b}$ which has $L_{a,b}$ as a compact leaf.

If V is small enough (and ξ_1 close enough to ξ_0), the plane field of this foliation is smoothly homotopic to ξ_1 on V. Therefore, by taking a closed neighborhood $U \subset V$ of $L_{a,b}$ and a smooth partition of unity subordinate to $\{M \setminus U, V\}$ we find an oriented plane field ξ on M which coincides with the tangent plane field of the foliation on U (and with ξ_1 outside of V).

Now, Thurston's h-principle 3.1 enables us to deform the plane field ξ into the tangent plane field of a foliation on M which admits $L_{a,b}$ as a compact leaf. This completes the construction of Example 3.2. □ 3.2.

Remark 3.3. The surface $L_{a,b}$ is genus-minimizing in its homology class, *i.e.*, the genus of any connected embedded surface which is homologous to $L_{a,b}$ is not smaller than the genus of $L_{a,b}$.

This is so, because $L_{a,b}$ is close to an almost holomorphic submanifold and so is a symplectic submanifold for the product symplectic structure on $M = \Sigma_g \times T^2$. Now, the validity of the symplectic Thom conjecture ([OS]) assures that $L_{a,b}$ is genus-minimizing in its homology class. We can also realize $L_{a,b}$ as a holomorphic submanifold of M.

§4. A cohomological criterion

We introduce a (co)homological method to decide when the tangent plane field of a given surface can be extended to the ambient manifold. The following result is well known, and it is an immediate corollary of a theorem accredited in [DW] to Pontrjagin [P]. This theorem states that for $n \geq 4$ isomorphism classes of principal $SO(n)$-bundles over a 4-complex K without 2-torsion in $H^4(K;\mathbb{Z})$ are completely classified by w_2, p_1, and W_4 . Here, $w_2 \in H^2(K;\mathbb{Z}/2)$ is the second Stiefel-Whitney class, $p_1 \in H^4(K;\mathbb{Z})$ the first Pontrjagin class, and W_4 the fourth Stiefel-Whitney class in $H^4(K;\mathbb{Z}), n = 4$, or in $H^4(K;\mathbb{Z}/2), n \geq 5$. (See also the corollary in Section 3 of [DW]). Note that for $SO(4)$-bundles, W_4 is just the Euler class.

Theorem 4.1. *Let M be a closed oriented connected 4-manifold and assume that the two cohomology classes e_1 and $e_2 \in H^2(M;\mathbb{Z})$ satisfy the follwoing three conditions.*

$$(1)\ e_1^2 + e_2^2 = p_1(M) \in H^2(M;\mathbb{Z})$$
$$(2)\ \langle e_1 \cup e_2, [M]\rangle = \chi(M)$$
$$(3)\ e_1 + e_2 \equiv w_2(M) \in H^2(M;\mathbb{Z}/2)$$

Then, there exists a pair of oriented $\mathbb{R}^2$-bundles $\mathcal{L}_1$ and $\mathcal{L}_2$ over M such that $\mathcal{L}_1 \oplus \mathcal{L}_2$ is isomorphic to the tangent bundle TM and $e(\mathcal{L}_i) = e_i$ for $i = 1, 2$. In other words, e_1 and e_2 are the Euler classes of a transverse pair of oriented plane fields on M.

Assume now that we have a pair of transverse plane fields of M. The next theorem will provide us with a simple check involving the Euler classes of these fields to decide when the first one of them can be homotoped into one which is tangent to a given embedded closed surface. A similar statement for splittings of $\mathbb{R}^2$-bundles over the circle is not true.

Theorem 4.2. *Let E be an oriented $\mathbb{R}^4$-bundle over a closed oriented surface L and $\mathcal{L}_1 \oplus \mathcal{L}_2 = E$ and $\mathcal{L}_1' \oplus \mathcal{L}_2' = E$ be two splittings of E*

into sums of oriented $\mathbb{R}^2$-subbundles. Then the following two statements are equivalent.

(4) $e(\mathcal{L}_1) = e(\mathcal{L}_1')$ *and* $e(\mathcal{L}_2) = e(\mathcal{L}_2')$.

(5) $\mathcal{L}_1 \oplus \mathcal{L}_2$ *and* $\mathcal{L}_1' \oplus \mathcal{L}_2'$ *are homotopic as splittings of* E.

The proof will be given in the next section.

Corollary 4.3. *Let L be an embedded closed oriented connected surface in M, e_1 and $e_2 \in H^2(M;\mathbb{Z})$ be two cohomology classes satisfying the conditions* (1) - (3) *of Theorem* 4.1 *and the following condition.*

(4') $\langle e_1, [L]\rangle = \chi(L)$ *and* $\langle e_2, [L]\rangle = [L]^2$.

Then the tangent plane field $TL \subset TM$ of L extends to an oriented plane field $\mathcal{L}$ on M.

Proof of Corollary 4.3.

This follows from Theorems 4.1 and 4.2 in a straight forward way.

Assuming the hypotheses of Corollary 4.3, Theorem 4.1 assures the existence of a transverse pair of plane fields $\mathcal{L}_1$ and $\mathcal{L}_2$ on M whose Euler classes are e_1 and e_2. Theorem 4.2 asserts that $\mathcal{L}_1|_L$ and TL are homotopic as subbundles of $E := TM|_L$. Let L_t', $0 \le t \le 1$, be a continuous family of 2-dimensional oriented subbundles of E with $\mathrm{L}_t' = TL, 0 \le t \le 1/2$, and $\mathrm{L}_1' = \mathcal{L}_1|_L$.

Next take a tubular neighbourhood $p : U_3 \to L$ of L in M which we view as a normal disk bundle of radius 3 and let $S_1 \subset U_3$ be the corresponding unit sphere bundle. Every point of U_3 can and will be written in the form $r \cdot x$, $0 \le r \le 3$, $x \in S_1$, and we identify $0 \cdot x$ with $p(x) \in L$. Also for $s < 3$, U_s denotes $\{r \cdot x \in U_3 \,|\, 0 \le r \le s\}$.

The bundle $TM|_{U_3}$ is isomorphic to

$$p^*E = \{(r \cdot x, v) \in U_3 \times E \,|\, v \in E_{0\cdot x}\},$$

where as usual E_b denotes the fibre of E over $b \in L$. We identify $TM|_{U_3}$ with p^*E.

The restriction $\mathcal{L}|_{U_3}$ of any oriented 2-dimensional subbundle $\mathcal{L}$ of TM is described by a map L which assigns to each $r \cdot x \in U_3$ an oriented 2-dimensional subspace $\mathrm{L}(r \cdot x)$ of $E_{0\cdot x}$. With this notation we define a continuous family $\mathcal{L}_t$, $0 \le t \le 1$, of oriented plane fields of M, where $\mathcal{L}_1$ is our original $\mathcal{L}_1$ from above, as follows.

(a) $\mathcal{L}_t|_{M\setminus U_2} = \mathcal{L}_1|_{M\setminus U_2}$

(b) $$\mathrm{L}_t(r \cdot x) = \begin{cases} \mathrm{L}_1\left(\frac{2(r-1+t)}{1+t} \cdot x\right), & 1-t \le r \le 2\,, \\ \mathrm{L}_{t+r}'(0 \cdot x)\,, & 0 \le r \le 1-t\,. \end{cases}$$

The plane field $\mathcal{L}_0$ restricted to $U_{1/2}$ is the pull back of TL by $p : U_{1/2} \to L$. Therefore it is tangent to L and smooth in the interior of $U_{1/2}$, and we may appproximate it by a smooth field, still called $\mathcal{L}_0$, which is tangent to L and homotopic to $\mathcal{L}_1$ as an oriented subbundle of TM. □ 4.3.

Applying our general strategy stated in Example 3.2, we obtain by combining Milnor's inequality 2.1, Thurston's h-principle 3.1, and Corollary 4.3 the following theorem, which is one of the main results of the present article.

Theorem 4.4. *Let L be a closed oriented connected embedded surface of genus greater than 0 in a closed oriented 4-manifold M which satisfies Milnor's inequality 2.1. Assume that two cohomology classes e_1 and $e_2 \in H^2(M;\mathbb{Z})$ satisfy the following five conditions.*

(1) $e_1^2 + e_2^2 = p_1(M) \in H^2(M;\mathbb{Z})$,
(2) $\langle e_1 \cup e_2, [M]\rangle = \chi(M)$,
(3) $e_1 + e_2 \equiv w_2(M) \in H^2(M;\mathbb{Z}/2)$,
(4) $\langle e_1, [L]\rangle = \chi(L)$,
(5) $\langle e_2, [L]\rangle = [L]^2$.

Then there exists a foliaton $\mathcal{F}$ on M which has L as a compact leaf.

This theorem translates the existence problem of foliations with prescribed compact leaves into a purely cohomological problem, *i.e.*, looking for solutions of a system of quadratic equations in the integral cohomology ring of the ambient 4-manifold. For 4-manifolds with accessible cohomology rings these equations can be dealt with. We will see examples of this in Sections 6, 7, and 8.

Remark 4.5. The above theorem implies that the question of whether an embedded surface L is a leaf of a foliation depends only on its homology class and its genus. Obviously, given a homology class and genus satisfying the equations of the theorem, we can realize it by an embedded surface if and only if the genus is not smaller than the minimal genus of a connected embedded surface in this homology class. So it is important to know this minimal genus. Of course, quite often the solution of the symplectic Thom conjecture [OS] provides the answer.

Here is an immediate corollary to the theorem, which is already interesting.

Corollary 4.6. *Let $\mathcal{F}$ be a 2 dimensional oriented foliation on a 4-manifold and let L be a homologically trivially embedded 2-torus in M, i.e., $[L] = 0 \in H_2(M;\mathbb{Z})$. Then we can modify the foliation in a* *neighbourhood of L into new foliations $\mathcal{F}_1$ and $\mathcal{F}_2$ such that the following holds.*

(1) *$\mathcal{F}_1$ has L as a compact leaf.*

(2) *$\mathcal{F}_2$ is transverse to L.*

Of course one is tempted to construct these foliations explicitly. Some of these constructions will be presented in the forthcoming paper [MV]. Scorpan showed the existence of foliations like $\mathcal{F}_2$ in the framework of singular foliations [S].

§5. Proof of Theorem 4.2

In this section we give a proof of Theorem 4.2, though it might be a folk theorem. It can be explained in a more general setting, as a general story about homotopy classes of subplane bundle of an oriented $\mathbb{R}^4$-bundle. (See for example the two excellent articles of Dold and Whitney [DW] and Hirzebruch and Hopf [HH]). However, we deal with the material in a rather down-to-earth manner which works in our particular range of dimensions. In this section, Σ refers to the oriented connected closed surface L in the statement of Theorem 4.2.

Let us start with reviewing the following well known facts, one of which was already mentioned in Example 2.2. Let $\mathcal{L}(k)$ denote an oriented $\mathbb{R}^2$-bundle over a closed oriented connected surface Σ with the Euler number $\langle e(\mathcal{L}(k)), [\Sigma]\rangle = k$. We also regard $\mathcal{L}(k)$ as a complex line bundle with $c_1 = k$. Let ε^r denote the product $\mathbb{R}^r$-bundle and $\varepsilon^r_{\mathbb{C}}$ the product $\mathbb{C}^r$-bundle. The isomorphism classes of complex line bundles or oriented $\mathbb{R}^2$-bundles over surfaces are determined by $c_1 = e$.

The statements in the following proposition are also fairly elementary and well known. Nevertheless, we review proofs of them , because this will aid in understanding the arguments leading up to the proof of Theorem 4.2.

Proposition 5.1.

1) *Isomorphism classes of oriented $\mathbb{R}^3$-bundles over Σ are determined by $w_2 \in H^2(\Sigma; \mathbb{Z}/2)$. In other words, any oriented $\mathbb{R}^3$-bundle over Σ is isomorphic to $\mathcal{L}(k) \oplus \varepsilon^1$ for some $k \in \mathbb{Z}$ and $\mathcal{L}(k_1) \oplus \varepsilon^1$ and $\mathcal{L}(k_2) \oplus \varepsilon^1$ are isomorophic if and only if $k_1 \equiv k_2 (\mathrm{mod} 2)$.*
2) *For any oriented $\mathbb{R}^3$-bundle over Σ with $w_2 = 0$ [resp. $w_2 \neq 0$], the self-intersection number of any cross section of the associated unit S^2-bundle is an even [resp. odd] number.*
3) *Isomorphism classes of oriented $\mathbb{R}^4$-bundles over Σ are also determined by $w_2 \in H^2(\Sigma; \mathbb{Z}/2)$. Any oriented $\mathbb{R}^4$-bundle over*

Σ is isomorphic to some $\mathcal{L}(k) \oplus \varepsilon^2$ and its isomorphism class is determined by the parity of k.

4) *$\mathbb{C}^2$-bundles over Σ are determined by $c_1 \in H^2(\Sigma; \mathbb{Z})$. In other words, any $\mathbb{C}^2$-bundle over Σ is isomorphic to $\mathcal{L}(k_1) \oplus \mathcal{L}(k_2)$ for some k_1 and $k_2 \in \mathbb{Z}$, and $\mathcal{L}(k_1) \oplus \mathcal{L}(k_2)$ and $\mathcal{L}(m_2) \oplus \mathcal{L}(m_2)$ are isomorophic as $\mathbb{C}^2$-bundles if and only if $k_1 + k_2 = m_1 + m_2$.*

Proof. 1) Reversing the steps in the construction for Example 2.2, where a trivial $\mathbb{R}$-bundle was added to an oriented $\mathbb{R}^2$-bundle, gives rise to the proof. Let $\overline{E}$ be an oriented $\mathbb{R}^3$-bundle over Σ and $S^2\overline{E}$ be the associated S^2-bundle. Because the fibre of $S^2\overline{E}$ is simply connected, it is easy to find a section of $S^2\overline{E}$. This section defines a trivial 1- dimensional subbundle of $\overline{E}$ which gives rise to an isomorphism $\overline{E} \cong \mathcal{L}(k) \oplus \varepsilon^1$ for some $k \in \mathbb{Z}$. Also notice here that the normal bundle of this section in $S^2\overline{E}$ is canonically isomorphic to the complementary subbundle $\mathcal{L}(k)$ and thus $k = |[L]^2|$, if L denotes the image of this section.

We can change the homotopy class of the cross section locally over a disc $D \subset \Sigma$. There the bundle looks like a product and the section is a graph of a smooth map $f : D \to S^2$. Up to homotopy we may assume that the map is constant on the boundary ∂D. So $f : (D, \partial D) \to (S^2, f(\partial D))$ has a degree. Changing the degree of this map by l changes the self-intersection of the cross section by $2l$.

The statement 2) also follows from the argument above. Remark also that if $w_2 = 0$, there is no homology class whose self-intersection is an odd number.

3) It is also easy to find a cross section to the associated S^3-bundle of a given oriented $\mathbb{R}^4$-bundle, which splits off a trivial line bundle. Then what we have to show reduces to statement 1) for the complementary $\mathbb{R}^3$-bundle.

Of course, an analogous statement is true for oriented vector bundles over Σ of any rank greater than 2.

4) The proof is again as in 1). Consider the $\mathbb{C}\mathrm{P}^1$-bundle associated to the given $\mathbb{C}^2$-bundle E. Because the fibre is $\mathbb{C}\mathrm{P}^1 \cong S^2$, we can again find a cross section, which gives rise to some complex line bundle $\mathcal{L} \subset E$. Therefore we obtain a splitting $E = \mathcal{L}_1 \oplus \mathcal{L}_2$ of E into two complex line bundles $\mathcal{L}_1 \cong \mathcal{L}(k_1)$ and $\mathcal{L}_2 \cong \mathcal{L}(k_2)$ for some k_1, $k_2 \in \mathbb{Z}$.

As $c_1(E) = k_1 + k_2$ is an invariant, any splitting satisfies this relation. On the other hand, assume $m_1 + m_2 = k_1 + k_2$. Then we can locally change the splitting so as to have $E \cong \mathcal{L}(m_1) \oplus \mathcal{L}(m_2)$ by changing the degree of the cross section to the associated $\mathbb{C}\mathrm{P}^1$-bundle, as the following famous example illustrates. Take $E = \varepsilon^2_{\mathbb{C}}$ over $\Sigma = \mathbb{C}\mathrm{P}^1$. Then we have two canonical splittings $\varepsilon^2_{\mathbb{C}} = \varepsilon^1_{\mathbb{C}} \oplus \varepsilon^1_{\mathbb{C}}$ and $\varepsilon^2_{\mathbb{C}} = \gamma^1 \oplus \overline{\gamma}^1$ where γ^1 denotes the tautological line bundle and $\overline{\gamma}^1$ denotes its complement. Of course we have $\gamma^1 \cong \mathcal{L}(-1)$ and $\overline{\gamma}^1 \cong \mathcal{L}(1)$. $\square$5.1.

We will also use the following proposition.

Proposition 5.2. *For an oriented S^2-bundle over Σ, two cross sections are homotopic if and only if their self-intersections coincide.*

Next, we review some basic facts about the oriented Grassmannian. Let us fix the standard orientation and the inner product on $\mathbb{R}^4$. We denote by $\widetilde{Gr}(4,2)$ the set of oriented planes through the origin in $\mathbb{R}^4$ with its natural structure as a smooth 4-manifold. Let L be an oriented plane in $\mathbb{R}^4$ and $L^\perp$ be its oriented orthogonal complement. Then we obtain a pair (J_L^+, J_L^-) of complex structures on $\mathbb{R}^4$ as follows.

$$\begin{array}{llll} J_L^+|_L & \text{rotates } L & \text{by} & \pi/2. \\ J_L^+|_{L^\perp} & \text{rotates } L^\perp & \text{by} & \pi/2. \\ J_L^-|_L & \text{rotates } L & \text{by} & \pi/2. \\ J_L^-|_{L^\perp} & \text{rotates } L^\perp & \text{by} & -\pi/2. \end{array}$$

Notice that J_+ defines the standard orientation on $\mathbb{R}^4$ while J_- defines the opposite one. We call a complex structure on $\mathbb{R}^4$ positive or negative depending on whether it induces the standard orientation or not. Also, we assume that any complex structure on $\mathbb{R}^n$ is an orthogonal transformation of $\mathbb{R}^4$. Let $\mathcal{J}_\pm$ denote the set of positive/negative complex structures considered as subspaces of $O(4)$. We have obtained the map $\Phi : \widetilde{Gr}(4,2) \to \mathcal{J}_+ \times \mathcal{J}_-$.

Proposition 5.3.

1) *Both of $\mathcal{J}_\pm$ are diffeomorphic to S^2.*
2) *The map $\Phi : \widetilde{Gr}(4,2) \to \mathcal{J}_+ \times \mathcal{J}_-$ is a diffeomorphism.*

Proof. 1) Let $\mathbf{e}_1, \ldots, \mathbf{e}_4$ be the standard basis of $\mathbb{R}^4$. A positive (or negative) almost complex structure J is uniquely determined by $J(\mathbf{e}_1)$ which lies in the unit two sphere $S^2(<\mathbf{e}_2, \mathbf{e}_3, \mathbf{e}_4>)$ in the span $\mathbf{e}_1^\perp = <\mathbf{e}_2, \mathbf{e}_3, \mathbf{e}_4>$ of $\mathbf{e}_2$, $\mathbf{e}_3$, and $\mathbf{e}_4$, because J rotates $L = <\mathbf{e}_1, J(\mathbf{e}_1)>$ positively and $L^\perp$ positively (or negatively). Here $<\ ,\ >$ denotes the linear span.

Conversely a free choice of the image of $\mathbf{e}_1$ from $S^2(\mathbf{e}_1^\perp)$ determines a positive (or negative) almost complex structure.

We orient $S^2(\mathbf{e}_1^\perp)$ as the unit sphere in $<\mathbf{e}_2, \mathbf{e}_3, \mathbf{e}_4>$ and orient $\mathcal{J}_\pm$ accordingly.

2) $SO(4)$ naturally acts transitively on $\widetilde{Gr}(4,2)$. It is not difficult to check at one point of $\widetilde{Gr}(4,2)$ that Φ is a local diffeomorphism. Therefore Φ is a covering map over a simply connected space and hence is a diffeomorphism. □ 5.3.

For a given oriented $\mathbb{R}^4$-bundle E over Σ, we always assume that a fibre-wise inner product has been fixed. Then denote by $\widetilde{Gr}(E)$ the associated Grassmannian bundle and by $\mathcal{J}_\pm(E)$ the associated bundles of positive and negative complex structures. The S^2-bundles $\mathcal{J}_\pm(E)$ over Σ carry natural orientations.

With regard to the next lemma, the proof of Proposition 5.3 is more important than the statement.

Lemma 5.4. *If $w_2(E) \neq 0$, both bundles $\mathcal{J}_\pm E$ are twisted as S^2-bundles, i.e., $w_2(\mathcal{J}_\pm E) \neq 0$. If $w_2(E) = 0$, of course, both bundles $\mathcal{J}_\pm(E)$ are trivial.*

Proof. By Proposition 5.1 3), we may assume that $E \cong \varepsilon^1 \oplus E'$ with $E' = \varepsilon^1 \oplus \mathcal{L}(k)$ for some $k \in \mathbb{Z}$, with $k \equiv \langle w_2(E), [\Sigma] \rangle \pmod 2$. It follows then from the proof of Proposition 5.3 that both $\mathcal{J}_\pm(E)$ are oriented isomorphic to $S^2(E')$. Therefore we have $w_2(\mathcal{J}_\pm(E)) = w_2(E') = w_2(E)$. □ 5.4.

The next Theorem is the key to prove Theorem 4.2. Before stating it, we fix some notations. For an oriented $\mathbb{R}^2$-subbundle $\mathcal{L} \subset E$, let $J_\pm(\mathcal{L})$ denote the complex structure of E determined by $E = \mathcal{L} \oplus \mathcal{L}^\perp$. $J_\pm(\mathcal{L})$ are also considered as sections of $\mathcal{J}_\pm(E)$.

Theorem 5.5. *For a splitting $E = \mathcal{L} \oplus \mathcal{L}^\perp$ with $\mathcal{L} \cong \mathcal{L}(k_1)$ and $\mathcal{L}^\perp \cong \mathcal{L}(k_2)$, the self-intersection of $J_\pm(\mathcal{L})$ in $\mathcal{J}_\pm(E)$ is given as follows.*

$$[J_+(\mathcal{L})]^2 = k_1 + k_2\,, \quad [J_-(\mathcal{L})]^2 = -k_1 + k_2\,.$$

Proof. First we prove this for the case $\mathcal{L} \cong \varepsilon^2 \cong \varepsilon^1_1 \oplus \varepsilon^1_2$ and $\mathcal{L}^\perp \cong \mathcal{L}(k)$. Let us follow the notations in the proof of 5.4. In this case we can identify $\mathcal{J}_+(E)$ with $S^2(E')$ and through this identification, $J_\pm(\mathcal{L})$ is identified with the cross section of $S^2(E')$ which corresponds to the canonical frame of $\varepsilon^1_2 \subset E'$, because $J_\pm(\mathcal{L})(\varepsilon^1_1) = \varepsilon^1_2$. As $E' = \varepsilon^1_2 \oplus \mathcal{L}(k)$, the proof of Proposition 5.1 1) tells us that the normal bundle of $J_\pm(\mathcal{L})$ in $S^2(E')$ is isomorphic to $\mathcal{L}(k)$. This implies

$$[J_+(\mathcal{L})]^2 = [J_-(\mathcal{L})]^2 = k\,.$$

Now let us prove the general case. By Proposition 5.1 4), we have an isomorphim $\mathcal{L}(k_1) \oplus \mathcal{L}(k_2) \cong \mathcal{L}(0) \oplus \mathcal{L}(k)$ for $k = k_1 + k_2$ as complex vector bundles, where $\mathcal{L}(k_1)$, $\mathcal{L}(k_2)$, $\mathcal{L}(0)$, and $\mathcal{L}(k)$ are considered as complex line bundles. This implies that the complex structures on $E = \mathcal{L}(k_1) \oplus \mathcal{L}(k_2)$ given by the two splittings $\mathcal{L}(k_1) \oplus \mathcal{L}(k_2)$ and $\mathcal{L}(0) \oplus \mathcal{L}(k)$ coincide. On the other hand, the complex structure on $\mathcal{L}(k_1) \oplus \mathcal{L}(k_2)$ [*resp.* $\mathcal{L}(0) \oplus \mathcal{L}(k)$] is nothing but $J_+(\mathcal{L}(k_1))$ [*resp.* $J_+(\mathcal{L}(0))$]. Hence their self-intersections also coincide. Therefore, computing the self-intersection of the first complex structure reduces to that of the second one and we obtain $[J_+(\mathcal{L}(k_1))]^2 = k = k_1 + k_2$.

The computation of $[J_-(\mathcal{L})]^2$ reduces to that of $[J_+(\mathcal{L})]^2$ by reversing the orientations. Let $\check{E}$ be the $\mathbb{R}^4$-bundle which is identical to E but with the opposite orientation. We can realize this by only reversing the orientation of $\mathcal{L}^\perp$ and leaving $\mathcal{L}$ as it is. Therefore $\check{E} \cong \mathcal{L} \oplus \check{\mathcal{L}}^\perp$ where $\check{\mathcal{L}}^\perp$ the same as $\mathcal{L}^\perp$ with the opposite orientation, and thus $\check{\mathcal{L}}^\perp \cong \mathcal{L}(-k_2)$. Therefore $J_-(\mathcal{L} \subset E)$ for $E = \mathcal{L} \oplus \mathcal{L}^\perp$ is nothing but $J_+(\mathcal{L} \subset \check{E})$ for $\check{E} = \mathcal{L} \oplus \check{\mathcal{L}}^\perp$. However, the orientation of $S^2(\check{E}')$ is the opposite of that of $S^2(E')$ where $\check{E}' = \varepsilon_2^1 \oplus \check{\mathcal{L}}^\perp$. Therefore we obtain

$$[J_-(\mathcal{L} \subset E)]^2 = -[J_+(\mathcal{L} \subset \check{E})]^2 = -(k_1 - k_2)\,.$$

This completes the proof. □5.1.

Proof of Theorem 4.2.

We only have to show (4) ⇒ (5) because the converse is trivial. Now (4) implies

$$[J_+(\mathcal{L}_1)]^2 = [J_+(\mathcal{L}_1')]^2 \quad \text{and} \quad [J_-(\mathcal{L}_1)]^2 = [J_-(\mathcal{L}_1')]^2\,.$$

By Proposition 5.2 $J_+(\mathcal{L}_1)$ and $J_+(\mathcal{L}_1')$ are homotopic to each other as sections of $\mathcal{J}_+(E)$ and so are $J_-(\mathcal{L}_1)$ and $J_-(\mathcal{L}_1')$. Therefore the two sections $(J_+(\mathcal{L}_1),\ J_-(\mathcal{L}_2))$ and $(J_+(\mathcal{L}_1'),\ J_-(\mathcal{L}_2'))$ of $\widetilde{Gr}(E)$ are homotopic. This completes the proof of Theorem 4.2. □4.2.

§6. Unboundedness for $\Sigma_g \times \Sigma_h$

As an application of Theorem 4.4, we show that for most products $\Sigma_g \times \Sigma_h$ of two closed oriented surfaces there is no bound on the set of self-intersection numbers of surfaces in $\Sigma_g \times \Sigma_h$ which can be realized as leaves of a foliation on $\Sigma_g \times \Sigma_h$. Precisely, we show the following.

Theorem 6.1. *Let M be one of the following products.*

(a) $M = \Sigma_g \times \Sigma_h$, *where* $g, h \geq 1$,
(b) $M = T^2 \times S^2$.

Then, there exists a family of 2-dimensional oriented foliations on M with compact leaves such that the set of self-intersection numbers of these compact leaves is unbounded.

This result is a substantial generalization of Example 3.2. It gives an indication that there should be many more manifolds which exhibit the same phenomenon. We will see in the proof of this theorem the power and ease of use of Theorem 4.4.

Proof of (a).

First let us fix some notation, which will be used throughout this section and also in the next section. In the cohomology ring

$$H^*(M;\mathbb{Z}) \cong H^*(\Sigma_g;\mathbb{Z}) \otimes H^*(\Sigma_h;\mathbb{Z})$$

we identify elements $x \in H^*(\Sigma_g;\mathbb{Z})$ with $x \otimes 1 \in H^*(M;\mathbb{Z})$. Similarly $y \in H^*(\Sigma_h;\mathbb{Z})$ is identified with $1 \otimes y$, so that xy is identified with $x \otimes y \in H^*(M;\mathbb{Z})$. If N is one of M, Σ_g, or Σ_h, $\{N\}$ denotes the cofundamental class. With this understanding, we have

$$H^2(M;Z) = H^2(\Sigma_g;\mathbb{Z}) \oplus H^1(\Sigma_g;\mathbb{Z}) \otimes H^1(\Sigma_h;\mathbb{Z}) \oplus H^2(\Sigma_h;\mathbb{Z}).$$

We use similar notations for homology as well.

Because both surfaces Σ_g and Σ_h have positive genus, there exists pairs of classes

$$c_i \in H^1(\Sigma_g;\mathbb{Z}) \quad \text{and} \quad d_j \in H^1(\Sigma_h;\mathbb{Z}) \quad \text{for} \quad i,j = 1,\, 2$$

which satisfy

$$c_1c_2 = \{\Sigma_g\} \quad \text{and} \quad d_1d_2 = \{\Sigma_h\}.$$

For two positive integers a and b, take an embedded surface $L = L_{a,b}$ in M which represents the homology class $a[\Sigma_g]+b[\Sigma_h] \in H_2(M;Z)$ constructed in the same way as the surface with the same name in Example 3.2. It has Euler characteristic

$$\chi(L) = 2a(1-g) + 2b(1-h) - 2ab$$

and is genus-minimizing in its homology class, as remarked in 3.3.

Now let us take two cohomology classes

$$\begin{aligned} e_1 &= \alpha\{\Sigma_g\} + \beta\{\Sigma_h\} + \sum_{i,j=1}^{2} \eta_{ij} c_i d_j \\ e_2 &= \gamma\{\Sigma_g\} + \delta\{\Sigma_h\} + \sum_{i',j'=1}^{2} \zeta_{ij} c_i d_j \end{aligned}$$

in $H^2(M;\mathbb{Z})$ with indeterminate integers α, β, γ, δ, η_{ij}, and ζ_{ij} (i, $j = 1$, 2). These are candidates for the Euler classes of the tangent and normal bundles of the foliations that we are looking for. We show that for an unbounded family of (a, b)'s, there exist solutions e_1 and e_2, *i.e.*, α, β, γ, δ, η_{ij}, and ζ_{ij} $(i, j = 1, 2)$, for the equations (1) - (5) in Theorem 4.4 which also satisfy Milnor's inequality. If $g \geq 2$ or $h \geq 2$ we have more room in $H^2(M; \mathbb{Z})$ and therefore more freedom to choose e_1 or e_2, but, as we will see, even with this limited choice for the Euler classes of tangent and normal bundles of foliations there is no bound on the set of self-intersection numbers of compact leaves.

To prove the theorem, it is enough to present the solutions. However, here we demonstrate a procedure to solve the equations briefly to appreciate the power of the cohomological criterion.

Let us first assume that $h \geq 2$. Then, for the surface $L_{a,b}$, Milnor's inequality $|[L_{a,b}]^2| \leq \frac{1}{2}|\chi(_{a,b})|$ becomes

$$a(g-1) + b(h-1) \geq ab\,.$$

Therefore, by putting $a = h - 1 \geq 1$, the inequality is satisfied for any b. The other equations (1) - (5) which have to be fulfilled are as follows in terms of the integral indeterminants.

$$\begin{array}{lll}
(1)\ p_1(M) & : & \alpha\delta + \beta\gamma + \eta_{12}\eta_{21} - \eta_{11}\eta_{22} + \zeta_{12}\zeta_{21} - \zeta_{11}\zeta_{22} = 0, \\
(2)\ \chi(M) & : & \alpha\beta + \gamma\delta - \eta_{11}\zeta_{22} + \eta_{12}\zeta_{21} + \eta_{21}\zeta_{12} - \eta_{22}\zeta_{11} \\
 & & \qquad = 4(1-g)(1-h), \\
(3)\ w_2(M) & : & \alpha \equiv \gamma,\ \beta \equiv \delta,\ \eta_{ij} \equiv \zeta_{ij}\ (i, j = 1, 2) \pmod 2, \\
(4)\ \chi(L_{a,b}) & : & a\alpha + b\beta = 2a(1-g) + 2b(1-h) - 2ab, \\
(5)\ [L_{a,b}]^2 & : & a\gamma + b\delta = 2ab.
\end{array}$$

Now (4) and (5) are also expressed as

$$\begin{aligned}
\alpha &= -\frac{b}{a}\beta + 2(1-g) + 2\frac{b}{a}(1-h) - 2b \\
\gamma &= -\frac{b}{a}\delta + 2b\,.
\end{aligned}$$

Therefore take b to be a multiple of a. Then any integral choice for β and δ determines integral α and γ. Also we assume here that β and δ are even so that α and γ are also even. To fulfill (3), we also assume that all η_{ij}'s and ζ_{ij}'s are even. Already (3) - (5) are fulfilled. Now put

$$\eta_{22} = \zeta_{11} = \zeta_{12} = \zeta_{21} = 0, \quad \eta_{21} = \zeta_{22} = 2\,.$$

Then for any choice for a, b and any even choice of β and δ, it is easy to find even integers η_{11} and η_{12} for which (1) and (2) are satisfied. Therefore the existence of a solution is proved for $a = h - 1$ and $b = k(h-1)$ for any $k \in \mathbb{N}$.

In the case $g = h = 1$, the argument has to be slightly modified. In particular, we need extra handles on L to achieve Milnor's inequality. So add local tiny handles to $L_{a,b}$ l-times to obtain a new surface L' which belongs to the same homology class. Then we have $\chi(L') = -2ab - 2l$. Putting $l = ab$, we can achieve the inequality as an extremal case, *i.e.*, with equality. Then the five equations are unchanged except for (4), which now takes the following form;

$$\alpha = -\frac{b}{a}\beta - 4b\,.$$

For the rest, the same argument works as above. □ 6.1 (a).

Proof of the case (b).

Here we only present some families of the solutions. We use the previous notations. In this case, we have $H_2(M;\mathbb{Z}) \cong H_2(T^2;\mathbb{Z}) \oplus H_2(S^2;\mathbb{Z})$ so that none of c_i, d_j, η_{ij}, or ζ_{ij} appear in the equations. However we need extra handles, whose number is denoted by l.

It is easy to verify that the requirements (1)-(5) and Milnor's inequality are satisfied in the families of solutions presented below. Here again, Milnor's inequality is satisfied as an extremal case.

Claim 6.2. *For any a, $b \geq 1$, the following families satisfiy all the requirements of Theorem* 4.4.

Family 1 : $\alpha = 0$, $\beta = -4a$, $\gamma = 0$, $\delta = 2a$, $l = ab + b$,
Family 2 : $\alpha = -4b$, $\beta = 0$, $\gamma = 2b$, $\delta = 0$, $l = ab + b$,
Family 3 : $\alpha = -3b$, $\beta = -a$, $\gamma = 3b$, $\delta = -a$, $l = ab + b$,
Family 4 : $\alpha = -b$, $\beta = -3a$, $\gamma = -b$, $\delta = 3a$, $l = ab + b$.

This completes the proof for case (b). □ 6.1 (b).

In case (b), one family would have been enough. There might exist many more.

§7. Boundedness for $\Sigma_g \times S^2$

Contrary to the case in the previous section, for $M = \Sigma_g \times S^2$, $H^2(M;\mathbb{Z})$ is small enough and we can conclude that the set of homology classes which can be realized as a compact leaf of some foliation is essentially bounded. More precisely the following holds.

Theorem 7.1. *For $M = \Sigma_g \times S^2$ with $g \neq 1$, there exists a number B_g so that any homology class $a[\Sigma_g] + b[S^2] \in H_2(M;\mathbb{Z}) \cong H_2(\Sigma_g;\mathbb{Z}) \oplus H_2(S^2;\mathbb{Z})$ which is represented by a compact leaf of some oriented foliation satisfies one of the following three conditions; $a = 0$, $b = 0$, or $a^2 + b^2 \leq B_g$.*

Corollary 7.2. *For $M = \Sigma_g \times S^2$ with $g \neq 1$, for any compact leaf $[L]$ of any foliation on M, we have $|[L]^2| \leq B_g$.*

Remark 7.3. Our simple proof below shows that we get a bound $|a|, |b| \leq 8(g-1)^2 + 2|g-1|$, so that $B_g \leq 2(8(g-1)^2 + 2|g-1|)^2$. But this bound is far from optimal, and can be improved without much effort.

Proof of Theorem 7.1.

Assume that there exists a foliation $\mathcal{F}$ with $e(\tau\mathcal{F}) = \alpha\{\Sigma_g\} + \beta\{S^2\}$, $e(\nu\mathcal{F}) = \gamma\{\Sigma_g\} + \delta\{S^2\} \in H^2(M;\mathbb{Z}) \cong H^2(\Sigma_g;\mathbb{Z}) \oplus H^2(S^2;\mathbb{Z})$ which has a compact leaf L representing $a[\Sigma_g] + b[S^2]$. Here $\{\Sigma_g\}$ [*resp.* $\{S^2\}$] denotes the cofundamental class of Σ_g [*resp.* S^2] pulled back to M. Without loss of generality we may assume that a and b are non-negative integers. The Euler classes of the plane fields and the homology class of L must satisfy the following equations.

$$\begin{array}{llll}
p_1(M): & e(\tau\mathcal{F})^2 + e(\nu\mathcal{F})^2 = p_1(TM) & \textit{i.e.}, & \alpha\beta + \gamma\delta = 0 \\
\chi(M): & e(\tau\mathcal{F})^2 \cup e(\nu\mathcal{F})^2 = e(TM) & \textit{i.e.}, & \alpha\delta + \beta\gamma = 4(1-g) \\
\chi(L): & \langle e(\tau\mathcal{F})^2, [L]\rangle = \chi(L) & \textit{i.e.}, & a\alpha + b\beta = \chi(L) \\
[L]^2: & \langle e(\nu\mathcal{F})^2, [L]\rangle = [L]^2 & \textit{i.e.}, & a\gamma + b\delta = 2ab \\
\text{Milnor:} & |[L]^2| \leq |\chi(L)|/2 & \textit{i.e.}, & 4ab \leq |a\alpha + b\beta|
\end{array}$$

The proof is broken down into several steps. In the first step we show that $e(\tau\mathcal{F})$ and $e(\nu\mathcal{F})$ are bounded by using the first two equations above. In the second, we prove that $[L]$ is bounded in the case $\alpha\beta\gamma\delta \neq 0$ by using equation ($[L]^2$) only. In the third step, we deal with the case $\alpha\beta\gamma\delta = 0$. Milnor's inequlity is used only in this step.

Step 1 (bound for the Euler classes): From the equations above we have the following:

$$\begin{array}{ll}
p_1(M) + \chi(M): & (\alpha+\gamma)(\beta+\delta) = -4(g-1) \\
p_1(M) - \chi(M): & (\alpha-\gamma)(\beta-\delta) = 4(g-1)
\end{array}$$

This implies that

$$1 \leqq |\alpha| + |\gamma|,\ |\beta| + |\delta| \leqq 4|g-1|,$$

and therefore

$$|\alpha|,\ |\beta|,\ |\gamma|,\ |\delta| < 4|g-1|.$$

This completes step 1.

Of course this estimate is far from being optimal. The arguments used in this step do not apply to $M = T^2 \times S^2$ because then $g - 1 = 0$.

Step 2 (bound for $[L]$, when $\alpha\beta\gamma\delta \neq 0$): Assuming $\alpha\beta\gamma\delta \neq 0$, we show that a and b are bounded. The equation ($[L]^2$) reads

$$(a - \frac{\delta}{2})(b - \frac{\gamma}{2}) = \frac{\gamma\delta}{4}$$

and we have assumed $\gamma\delta \neq 0$. Therefore, in the ab-plane, (a, b) lies on a hyperbola. Put $\overline{a} = a - \delta/2$ and $\overline{b} = b - \frac{\gamma}{2}$. Then the hyperbola is given by $\overline{a}\overline{b} = \frac{\gamma\delta}{4}$. To our integral point (a, b) corresponds an integral or a half-integral point $(\overline{a}, \overline{b})$. Therefore we have $|\overline{a}|, |\overline{b}| \geqq \frac{1}{2}$. This immediately implies $|\overline{a}|, |\overline{b}| \leqq 8(g-1)^2$, because we have seen that $|\gamma|, |\delta| \leqq 4|g-1|$. Thus we obtain

$$|a|, |b| \leqq 8(g-1)^2 + 2|g-1|.$$

Step 3 (the case $\alpha\beta\gamma\delta = 0$): The equations ($p_1(M)$) and ($\chi(M)$) imply that "$\alpha\beta\gamma\delta = 0$" is equivalent to

$$\text{"}\alpha = \delta = 0,\ \beta\gamma = 4(1-g)\text{"} \quad \text{or} \quad \text{"}\alpha\delta = 4(1-g),\ \beta = \gamma = 0\text{"}.$$

Therefore this step further splits into two cases $\alpha = \delta = 0$ and $\beta = \gamma = 0$.

If $a = 0$ or $b = 0$ there is nothing to prove. So we may assume $a, b \geq 1$.

Assume $\alpha = \delta = 0$ and $\beta\gamma = 4(1-g)$. Then ($[L]^2$) implies $\gamma = 2b$ and (Milnor) implies $4a \leq |\beta|$. Therefore we have $a \leq |g-1|, b \leq 2|g-1|$. (In fact, in this case, $g > 2, a = 1, b \leq \frac{g-1}{2}, \beta b = 2(1-g)), \gamma = 2b, \beta \equiv 0 \pmod 2$, see Remark 7.5 below).

If $\alpha\delta = 4(1-g)$ and $\beta = \gamma = 0$, we obtain $\delta = 2a$ and $4b \leq |\alpha|$. So $b \leq |g-1|, a \leq 2|g-1|$. (In fact, one gets $g > 2, a = 1, \alpha = 2(1-g), \delta = 2, b \leq \frac{g-1}{2}$, see Remark 7.5 below). □ 7.1.

The next proposition deals with the case $a \cdot b = 0$ excluded in Theorem 7.1.

Proposition 7.4. *Let $M = \Sigma_g \times S^2, g \neq 1$. Then*

1) *for any $b \in \mathbb{Z}$ there is a 2-dimensional oriented foliation on M which has a compact leaf representing $b[S^2] \in H_2(\Sigma_g \times S^2)$;*
2) *for any $a \neq 0$ there is a 2-dimensional oriented foliation on M which has a compact leaf L representing $a[\Sigma_g] \in H_2(\Sigma_g \times S^2)$, and if $g > 1$, then for any such foliation the leaf L is genus-minimizing in its homology class.*

Proof of Proposition 7.4.

We first deal with the case $a = 0$, $b \in \mathbb{Z}$. We may assume that $b \geq 0$ by changing the orientation of S^2 if necessary. We also know that any homologically trivial embedded torus is a leaf of a foliation in any homotopy class of foliations. Therefore, we may assume $b > 0$. Now, pick b disjoint vertical copies of S^2, punch out a disk in each copy and join the resulting disks by attaching $(b-1)$ annuli in such a way that we obtain an embedded 2-sphere L' representing $b[S^2]$.

If $\alpha\{\Sigma_g\}+\beta\{S^2\}, \gamma\{\Sigma_g\}+\delta\{S^2\}$ are the Euler classes of the tangent and normal bundle of a foliation having a compact leaf homologous to L', then $([L]^2)$, $(\chi(M))$, and $(p_1(M))$ imply that $\alpha = \delta = 0$ and $\beta\gamma = 4(1-g)$. Now choose any even negative β which divides $2(1-g)$ and choose l so as to have

$$b\beta = 2(1-l).$$

Then attach to L' l homologically trivial handles to obtain a connected orientable surface L homologous to L' with $\chi(L) = b\beta$. Finally, put

$$\gamma = \frac{4(1-g)}{\beta}.$$

Then $(\alpha = 0, \beta, \gamma, \delta = 0, a = 0, b, g)$ satisfy all requirements of Theorem 4.4.

If $g > 1, a > 0$, and $b = 0$, we obtain $\gamma = \beta = 0, \alpha \equiv \delta \equiv 0 \bmod 2$, and $\alpha\delta = 4(1-g)$. It is easy to embed a surface L' in M such that the projection to Σ_g is a covering of degree a. Then L' is genus minimizing in its homology class. The easiest way to see this is to use the Gromov volume for surfaces, and the fact that the projection onto Σ_g is a map of degree a for any connected surface representing $[L']$. Therefore, for any surface L homologous to L' we have $\chi(L) = 2a(1-g) - 2l$ with $l \geq 0$. Then $(\chi(L))$ reads $a\alpha = 2a(1-g) - 2l$, and we obtain $\delta = 2, \alpha = 2(1-g), l = 0$. Again, all conditions of Theorem 4.4 are satisfied for these choices for $(\alpha, \beta = 0, \gamma = 0, \delta = 0, a, b = 0, g)$.

Of course we can also construct explicitly such a foliation very easily in the following way. Take a surjective homomorphism $\phi : \pi_1(\Sigma_g) \to \mathbb{Z}/a\mathbb{Z}$ and let $\mathbb{Z}/a\mathbb{Z}$ act on S^2 by mapping a generator of $\mathbb{Z}/a\mathbb{Z}$ onto a rotation of angle $2\pi/a$. Composing with ϕ we obtain an action of $\pi_1(\Sigma_g)$ on S^2, and the corresponding suspension foliation is a foliation with all leaves compact. The leaves corresponding to the two fixed points of the rotations are projected to Σ_g bijectively, but any other leaf covers Σ_g with degree a. □ 7.4.

While Proposition 7.4 says that a or b are unbounded, if $a \cdot b = 0$, it only deals with compact leaves with vanishing self-intersection.

In Section 2 we have seen examples of foliations on $\Sigma_g \times S^2$ with compact leaves representing $[\Sigma_g] + b[S^2]$, *i.e.*, with leaves of self-intersection number $2b$, as long as $|2b| \leq |g-1|$. The corresponding foliations were foliated S^2-bundles $\mathcal{F}$ with $e(\tau\mathcal{F}) = 2(1-g)\{\Sigma_g\}$ and $e(\nu\mathcal{F}) = 2\{S^2\}$. This corresponds to $\alpha = 2(1-g), \beta = \gamma = 0, \delta = 2$ in our notation above.

In fact, whenever $\beta\gamma = 0$ there are no other solutions to our equations. More generally, we have

Remark 7.5. Let $\mathcal{F}$ be a 2-dimensional oriented foliation on $M = \Sigma_g \times S^2, g \neq 1$, with $e(\tau\mathcal{F}) = \alpha\{\Sigma_g\} + \beta\{S^2\}$, $e(\nu\mathcal{F}) = \gamma\{\Sigma_g\} + \delta\{S^2\} \in H^2(M;\mathbb{Z}) \cong H^2(\Sigma_g;\mathbb{Z}) \oplus H^2(S^2;\mathbb{Z})$ which has a compact leaf L representing $a[\Sigma_g] + b[S^2]$ with $a, b \geq 1$.

1) If $\beta\gamma = 0$, then $g \geq 3, \alpha = 2(1-g), \beta = \gamma = 0, \delta = 2, a = 1, 2b \leq g-1$, and L is genus minimizing in its homology class. As we have seen in Section 2 these data are realized by a foliation.
2) If $\alpha\delta = 0$, then $\alpha = \delta = 0, \gamma = 2b, \beta = 2(1-g)/b, a = 1, b$ divides $g-1, 2b \leq (g-1)$, and L is genus minimizing in its homology class. Conversely, these data are realized by a foliation. We may choose this foliation as a pullback of a foliation on $\Sigma_{\overline{g}} \times S^2$ with $\overline{g} = (g-1)/b + 1$ via a covering map $\Sigma_g \longrightarrow \Sigma_{\overline{g}}$ where the compact leaf represents $[\Sigma_{\overline{g}}] + [S^2]$.

The proofs of these statements follow easily from the equations introduced in the proof of Theorem 7.1 and are left to the reader.

So far, with regard to a foliation on $\Sigma_g \times S^2$ $(g > 1)$ and its compact leaf representing $a[\Sigma_g] + b[S^2]$ with $a, b \geq 1$, if we assume $\alpha\beta\gamma\delta = 0$ we have $a = 1$. We will see in the next section that there are many foliations on the spaces $\Sigma_g \times S^2$, $g > 1$, having compact leaves representing $a[\Sigma_g] + b[S^2]$ where both $|a|$ and $|b|$ are large. Obviously, by Theorem 7.1, then g has to be large also.

§8. Problems and further discussions

To conclude this article, we present comments, discussions, and problems grouped together under three headings. The first is about geometric constructions of foliations guaranteed by our cohomological criteria. The second one contains comments and questions about foliations on $\Sigma_g \times S^2$ and their compact leaves, some of which resulted from computer calculations we conducted. The final one is concerned with our original motivation, the self-intersection of compact leaves of a foliation.

A. Geometric constructions

We have shown the existence of certain foliations with some specific compact leaves. We are naturally tempted to construct such foliations in more explicit ways. In the forthcoming paper, we are goint to introduce such a construction of a foliation on $\mathbb{R}^4$ which has trivial T^2-knot as a leaf and generalize it to those who have spun T^2-knots as their leaves.

Problem 8.1. *Give explicit constructions for a wider class of T^2-knots in $\mathbb{R}^4$.*

Of course we first need a geometric and convenient presentation of the knot.

B. Foliations on $\Sigma_g \times S^2$

Here, we comment on the existence of foliations on $M = \Sigma_g \times S^2$ with compact leaves representing $a[\Sigma_g] + b[S^2]$ with both a and b, and therefore also g large. Also the fact that for many homotopy classes of foliations these leaves are minimal genus representatives in their homology classes, independent of the choice of foliation in its homotopy class, seems to be noteworthy. We observed some of these phenomena by running computer experiments.

As before, let $\alpha\{\Sigma_g\} + \beta\{S^2\}$ be the Euler class of the tangent bundle and $\gamma\{\Sigma_g\}+\delta\{S^2\}$ be the Euler class of the normal bundle of our foliation. By Remark 7.5 we have to turn to foliations with $\alpha\beta\gamma\delta \neq 0$, if we are looking for foliations with compact leaves representing $a[\Sigma_g] + b[S^2]$ with $a > 1, b > 0$.

Initially, we were uncertain about the existence of such foliations. But searching for solutions of the four equations, Milnor's inequality, and the congruences coming from Theorem 4.4 with the help of a computer, we saw that solutions abound. Below we present two 3-parameter families of foliations with associated homology classes of compact leaves by listing the associated values of the eight variables $g, \alpha, \beta, \gamma, \delta, a, b, l$. Without loss of generality we may assume that a and b are positive. The non-negative integer l is the number of homologically trivial handles added to the surface of minimal genus in the homology class $a[\Sigma_g]+b[S^2]$. So $l = 0$ is equivalent to the statement that the compact leaf is genus-minimizing in its homology class.

Also notice the following. If

$$\overline{g},\ \overline{\alpha},\ \overline{\beta},\ \overline{\gamma},\ \overline{\delta},\ \overline{a},\ \overline{b},\ \overline{l}$$

are the homological data for a foliation with compact leaf on $\Sigma_{\overline{g}} \times S^2$, then pulling back the bundle, foliation and compact leaf by a d-fold covering map $\Sigma_g \longrightarrow \Sigma_{\overline{g}}$ we get a foliation with compact leaf on $\Sigma_g \times S^2$ with homological data

$$g = d(\overline{g} - 1) + 1,\ \alpha = d\overline{\alpha},\ \beta = \overline{\beta},\ \gamma = d\overline{\gamma},\ \delta = \overline{\delta},\ a = \overline{a}, b = d\overline{b},\ l = d\overline{l}.$$

Conversely, if $g - 1, \alpha, \gamma, b, l$ are divisible by d, then these data are obtained by pulling back a foliation with compact leaf via a d-fold cover.

Therefore, below we only list data not coming from coverings.

Example 8.2. (Family of solutions with trivial handles)

(F1) For integers $0 \leq x < y, 0 \leq z$ set

$$\begin{aligned}
&F_1(x, y, z):\\
&g = (2z + 1)((2y + 1)^2 - (2x + 1)^2)/4 + 1\\
&\alpha = 2x + 1,\ \beta = -((2y + 1)(2z + 1),\\
&\gamma = 2y + 1,\ \delta = (2x + 1)(2z + 1),\\
&a = (\delta + 1)/2,\ b = a\gamma,\\
&l = (z + 1)(y - x)(1 + (2z + 1)(y - x))
\end{aligned}$$

It is easy to check that all equations are satisfied and that the congruence holds. Milnor's inequality is a little messy to write down, but it will hold if y is large when compared with x. For example, if $z = 0$, then $y \geq 2x + 2$, and if $z > 0$, then $y \geq 2x + 1$ will suffice. □ 8.2.

Obviously, $F_1(x, y, z)$ comes from a covering, if and only if the the greatest common divisor of $g - 1, \alpha, \gamma, l$ is greater than 1.

Note that $l > 0$, so that L is not genus-minimizing in all these examples. Also the genus g of the base surface is always odd. In fact, we have observed in our computer calculations the following phenomenon: if $a, b > 0$ and L is not genus-minimizing, then g is odd, $\alpha > 0$, and α is small when compared with g. Furthermore, if the data do not come from a covering, then α is odd and δ is an odd multiple of α.

Therefore, in this case, α, β, γ, and δ must have the above form. Up to $g = 90$ all examples with $l > 0$ are in the family (F1) or cover an element of this family.

Problem 8.3. *Do all foliations with a compact leaf which is not genus-minimizing and representing a class with $a, b \geq 1$ belong to* (F1) *or cover a foliation from* (F1)*?*

The description of the next family is slightly less direct than that of (F1).

Example 8.4. (Family of solutions without trivial handles)

(F2) This family is parametrized by a rational number x with $0 < x < 1/3$, and for each such x by an arithmetic progression

for the genus g of the base surface Σ_g. The denominator and numerator of x make it into a 3-parameter family. Specifically, for given x, set
$a = xg + 1 - 2x$, $b = xg/(1-2x) + 1$, and $l = 0$.

Choose $g > 1$, so that a and b are positive integers, and then set
$\alpha = -g$, $\beta = -g + 2$, $\gamma = -g + 2$, and $\delta = g$.

Obviously, all equations and the congruence hold. In order that Milnor's inequality holds, g has to be sufficiently large.

The arithmetic progression for g is obtained as follows:
Let $x = p/q, (p,q) = 1, q > 3p$. Then for any $k \geq 0$ set

$$g = q'q + 2 + q(q-2p)k,$$

where $0 < q' < q - 2p$ solves $q'q + 2 \equiv 0 \bmod (q-2p)$. Then a and b are positive integers. In order that Milnor's inequality holds we have to choose k large enough.

Notice that in family (F2) α is always negative, and the compact leaves are genus-minimizing. Furthermore, since α and $g-1$ are coprime, no member is the result of a pull-back via a covering map of the base surfaces. □ 8.4.

For (F2), there are no restrictions on the parity of g. In fact, in our computer calculations, solutions with $l = 0$ occur for every $g > 2$, and for odd g their number exceeds the number of solutions with $l > 0$ by a factor of at least 2. This is probably due to the fact that the foliations with genus-minimizing compact leaves occur more often as coverings. In fact, we do not know whether for large odd g the ratio of the number of solutions with $l = 0$ to the number with $l > 0$ has a limsup greater than 0, once we discard foliations which are coverings.

Problem 8.5. *What can be said about this ratio?*

With regard to the number of classes $\alpha\{\Sigma_g\}+\beta\{S^2\}$, which occur as the Euler class of a foliation with a compact leaf representing $a[\Sigma_g]+b[S^2]$ with $a, b > 0$, there are many more classes with $\alpha < 0$ than classes with $\alpha > 0$.

Problem 8.6. *What is the reason for this?*

With the usual meanings of a, b, α, we have mentioned above that for $a, b > 0$, as far as we know, the compact leaves of all foliations with $\alpha > 0$ are not genus-minimizing, while the one's with $\alpha < 0$ are genus-minimizing.

Problem 8.7. *Do the Euler classes of a foliation determine whether a compact leaf of this foliation is genus-minimizing in its homology class?*

Looking at Remark 7.5 we see that a given compact surface which is genus-minimizing in its homology class can be a leaf of foliations with different Euler classes. For example, if $g > 2$ and the surface represents $[\Sigma_g]+[S^2]$, there exist foliations with Euler class equal to any of $2(1-g)\{\Sigma_g\}$, $2(1-g)\{S^2\}$, $-g\{\Sigma_g\}+(2-g)\{S^2\}$, $(2-g)\{\Sigma_g\}-g\{S^2\}$ having this surface as a compact leaf.

There is also an occurence where leaves in the same homology class, one genus-minimizing, the other not, are leaves of foliations. These necessarily have distinct Euler classes. But up to coverings we have only one example for this: $g = 19, a = 2, b = 10$. One foliation is given by $\alpha = -21$, $\beta = -5$, $\gamma = -15$, $\delta = 7$, and $l = 0$, *i.e.*, the leaf is genus-minimizing. The other is given by $\alpha = 1$, $\beta = -15$, $\gamma = 5$, $\delta = 3$, and $l = 28$.

So one might pose the following

Problem 8.8. *In general, does each homology class know its foliated genus, i.e., the genus of a representing surface which is a leaf of a foliation on the manifold?*

In all our examples on $M = \Sigma_g \times \Sigma_h$ we have seen that genus-minimizing leaves representing $a[\Sigma_g] + b[\Sigma_h]$ can be chosen to be symplectic submanifolds of M with its standard symplectic structure.

Problem 8.9. *If a surface L in $M = \Sigma_g \times S^2$ is genus-minimizing in its homology class and a leaf of a foliation, does there exist a symplectic foliation, i.e., a foliation such that all leaves are symplectic submanifolds of M with respect to some symplectic structure on M, with L as a leaf? If it is not true, then, look for a condition which guarantees that L is a leaf of a symplectic foliation.*

This question might make sense for more general closed symplectic 4-manifolds, but so far, we have not looked into this.

C. Bounds of self-intersection numbers of compact leaves

In Sections 6 and 7 we settled the question for which products of two surfaces there is a bound on the self-intersection numbers of compact surfaces which occur as leaves of a foliation.

Problem 8.10. *For which 4-manifolds M is the set of self-intersection numbers of compact leaves of foliations on M bounded?*

For which 4-manifolds with positive second Betti number is the set of homology classes which are represented by some compact leaf of some foliation essentially bounded?

Here, "essentially bounded" should be interpreted in some reasonable way. For example, like in our context: bounded when restricted to classes of non-zero self-intersection.

It is premature to venture a guess whether among closed 4-manifolds admitting a 2-dimensional foliation with a compact leaf those with a bound on the self-intersection numbers of these leaves are more prevalent or not. In the small set of 4-manifolds that we considered they did occur less often.

However, in the case of foliated bundles, the situation should be different.

Problem 8.11. *Prove that for a given Σ_h-bundle over Σ_g, there is a bound for the set of self-intersection numbers of compact leaves of foliations transverse to the fibres.*

More strongly, for any given g and h, prove that there is a bound for the set of self-intersection numbers of compact leaves of any foliated Σ_h-bundle over Σ_g.

This problem is strengthened further by dropping the flatness condition of the bundle. Then of course the bound should be larger.

Problem 8.12. *For given h and g, does there exist an upper bound for the self-intersection number of any multi-section of any Σ_h-bundle over Σ_g?*

This is no longer a problem of foliations.

Example 8.13. There exists a multi-section of $\Sigma_2 \times \Sigma_2$, which covers both the base and the fibre twice. Since its normal and tangent bundles are isomorphic, its self-intersection number is -4.

To prove the existence of such a multi-section, it is enough to find a pair of orientation preserving free involutions σ and τ on Σ_3 such that $\sigma \circ \tau$ has no fixed point. This is done as follows.

In $\mathbb{R}^3 = \{(x, y, z)\}$, take a spatial graph $\Gamma = S^2 \cap \{(x+y)(x-y) = 0\}$ consisting of four longuitudes connecting the north and south poles. We realize Σ_3 as the smooth boundary of a thin regular neighbourhood of Γ in $\mathbb{R}^3$. Then the involutions σ and τ on Σ_3 are defined as follows. σ is the rotation $(x, y, z) \mapsto (x, -y, -z)$ around the x-axis by π restricted to Σ_3. τ is the rotation around the great circle $S^2 \cap \{(x + y) = 0\}$ by π. The two 1-handles around this great circle are invariant and rotated while the other two 1-handles connecting the regions close to the poles are exchanged by τ. In other words τ is the composition of the inversion of $\mathbb{R}^3$ with respect to S^2 and the reflection at the plane $x + y = 0$.

It is easy to verify that this pair of free involutions fulfills our requirements.

Then, the involutions σ and τ define two double coverings

$$\pi_\sigma, \pi_\tau : \Sigma_3 \longrightarrow \Sigma_2$$

so that each involution is the non-trivial covering transformation.

Now we define the multi-section as the image of the following map.

$$\varphi : \Sigma_3 \ni p \mapsto (\pi_\sigma(p),\ \pi_\tau(p)) \in \Sigma_2 \times \Sigma_2.$$

Since $\sigma \circ \tau$ has no fixed point φ is an embedding. Therefore it defines a multi-section with the required properties. □ 8.13.

We believe that this multi-section is the one with the largest self-intersection number among all multi-sections of this product bundle. Also notice that Milnor's inequality prohibits this surface to be a leaf of a foliation on $\Sigma_2 \times \Sigma_2$.

Are there methods to prove the statement about the maximality of the self-intersection number?

Of course, questions about the existence of multi-sections of surface bundles with certain properties can be interpreted as questions about the pointed mapping class groups of these surfaces.

Many of the problems above have obvious generalizations to foliations of higher dimension or codimension. But our methods are very specific for 2-dimensional foliations on 4-manifolds and so one needs some new ideas to proceed. Independent of this, we think it is important and worthwhile to pursue the study of the subjects dealt with in this paper in other dimensions. On the other hand, it is true that there is still an abundance of problems that remain to be settled in dimension 4.

References

[B] R. Bott, On a topological obstruction to integrability, Proc. Symp. Pure Math. Amer. Math. Soc., **10** (1970), 127–131.

[DW] A. Dold and H. Whitney, Classification of oriented sphere bundles over a 4-complex, Ann. of Math. (2), **69** (1959), 667–677.

[HH] F. Hirzebruch and H. Hopf, Felder von Flächenelementen in 4-dimensionalen Mannigfaltigkeiten, Math. Ann., **136** (1958), 156–172.

[HM] S. Hurder and Y. Mitsumatsu, The Intersection Product of Transverse Invariant Measures, Indiana Univ. Math. J., **40** (1991), 1169–1183.

[M] J. W. Milnor, On the existence of a connection with curvature zero, Comment. Math. Helv., **32** (1958), 215–223.

[Mi1] Y. Mitsumatsu, Self-intersections and transverce Euler numbers of foliation cycles, Thesis for doctor of science, Univ. of Tokyo, 1985.

[Mi2] Y. Mitsumatsu, On the self-intersections of foliation cycles, Trans. Amer. Math. Soc., **334** (1992), 851–860.

[MV] Y. Mitsumatsu and E. Vogt, Foliations and compact leaves on 4-manifolds II, Constructions, preprint, in preparation.

[OS] P. Ozsváth and Z. Szabó, The symplectic Thom conjecture, Ann. of Math. (2), **151** (2000), 93–124.

[P] L. Pontrjagin, Classification of some skew products, Dokl. Acad. Nauk SSSR NS, **47** (1945), 322–325.

[S] A. Scorpan, Existence of foliations on 4-manifolds, Algebr. Geom. Topol., **3** (2003), 1225–1256.

[Th] W. Thurston, The theory of foliations of codimension greater than one, Comment. Math. Helv., **49** (1974), 214–231.

Yoshihiko Mitsumatsu
Department of Mathematics
Chuo University
1-13-27 Kasuga, Bunkyo-ku,
Tokyo, 113-8551
Japan
E-mail address: yoshi@math.chuo-u.ac.jp

Elmar Vogt
Mathematisches Institut
Freie Universität Berlin
Arnimallee 3
14195 Berlin
Germany
E-mail address: vogt@math.fu-berlin.de

Advanced Studies in Pure Mathematics 52, 2008
Groups of Diffeomorphisms
pp. 443–468

Symplectic automorphism groups of nilpotent quotients of fundamental groups of surfaces

Shigeyuki Morita

Abstract.

We describe the *group version* of the trace maps given in [29]. This gives rise to abelian quotients of *symplectic* IA-automorphism groups of nilpotent quotients of the fundamental groups of compact surfaces. By making use of them, we construct a representation of the group $\mathcal{H}_{g,1}$ of homology cobordism classes of homology cylinders introduced by Garoufalidis and Levine [6]. We define various cohomology classes of $\mathcal{H}_{g,1}$ and propose a few problems concerning them. In particular, we mention a possible relation to additive invariants for the group $\Theta^3_{\mathbb{Z}}$ of homology cobordism classes of homology 3-spheres.

§1. Introduction

As is well known, the mapping class group $\mathcal{M}(\Sigma_g)$ of a closed oriented surface Σ_g of genus $g \geq 2$ can be identified with the orientation preserving outer automorphism group of $\pi_1\Sigma_g$. This is the classical theorem of Dehn and Nielsen. In the case of a compact oriented surface Σ_g^0 with one boundary component, the corresponding mapping class group $\mathcal{M}(\Sigma_g^0, \mathrm{rel}\ \partial\Sigma_g^0)$ relative to the boundary is canonically isomorphic to the subgroup of the automorphism group of $\pi_1\Sigma_g^0$ consisting of elements which preserve a particular element representing the boundary curve. Henceforth, we simply denote this group $\mathcal{M}(\Sigma_g^0, \mathrm{rel}\ \partial\Sigma_g^0)$ by $\mathcal{M}_{g,1}$.

Since the action of $\mathcal{M}_{g,1}$ on $\pi_1\Sigma_g^0$ preserves its lower central series, it induces a series of representations ρ_d of $\mathcal{M}_{g,1}$ into the automorphism groups of free nilpotent quotients of $\pi_1\Sigma_g^0$ with nilpotency class $d = 1, 2, \cdots$.

Received October 15, 2007.
Revised January 9, 2008.
Partially supported by JSPS Grant 19204003.

On the other hand, the kernels of these representations induce a filtration on the Torelli subgroup $\mathcal{I}_{g,1} \subset \mathcal{M}_{g,1}$ and the associated graded Lie algebra can be mapped injectively into the graded Lie algebra consisting of all the derivations of the free graded Lie algebra generated by the first homology group $H = H_1(\Sigma_g^0; \mathbb{Z})$. This embedding is called the Johnson homomorphism, which we denote by τ. It can be said that τ gathers the "differences" between the successive representations ρ_d and ρ_{d+1} for all d.

Many works have been done concerning the Johnson homomorphism τ and through them the image of τ has been shown to capture a smaller and smaller part of the target. First it was shown to satisfy a certain symplecticity condition and then it was proved to be in the kernel of certain abelian quotients of the relevant Lie algebra which we called the *traces* (see [29]).

The purpose of the present paper is to *lift* the traces as homomorphisms with abelian targets defined on certain symplectic automorphism groups of nilpotent quotients of $\pi_1\Sigma_g^0$ rather than in the context of Lie algebra homomorphisms.

We expect that these lifts will be useful in the study of the group $\mathcal{H}_{g,1}$ of homology cobordism classes of homology cylinders over Σ_g^0 which was introduced by Garoufalidis and Levine [6]. This is because of their result that the above series of representations ρ_d can be extended to their group $\mathcal{H}_{g,1}$ and these are all *surjective* onto some symplectic automorphism groups of nilpotent quotients of $\pi_1\Sigma_g^0$.

Part of the results of this paper were sketched in section 11 of [33].

§2. Symplectic automorphism groups

Let Σ_g denote a closed oriented surface of genus g and we denote by $\Sigma_g^0 = \Sigma_g \setminus \mathrm{Int} D^2$ a compact oriented surface of genus g with one boundary component. Consider the mapping class group

$$\mathcal{M}_{g,1} = \pi_0 \mathrm{Diff}(\Sigma_g^0, \mathrm{rel}\ \partial\Sigma_g^0)$$

of Σ_g^0 *relative* to the boundary. Let $\Gamma = \pi_1\Sigma_g^0$ denote the fundamental group of Σ_g^0 which is a free group of rank $2g$. We have a distinguished element

$$\zeta = [\alpha_1, \beta_1] \cdots [\alpha_g, \beta_g] \in \Gamma$$

which corresponds to the boundary curve of Σ_g^0 so that it descends to the single defining relation for the fundamental group $\pi_1\Sigma_g$ of the closed surface, where α_i, β_i denotes a standard system of generators of Γ. Then by the classical theorem of Dehn-Nielsen adapted to the case of one boundary component due to Zieschang, we can write

$$\mathcal{M}_{g,1} \cong \mathrm{Aut}_0\Gamma = \{\varphi \in \mathrm{Aut}\,\Gamma; \varphi(\zeta) = \zeta\}. \tag{1}$$

Now we consider the lower central series

$$\Gamma_0 = \Gamma,\ \Gamma_1 = [\Gamma, \Gamma],\ \Gamma_2 = [\Gamma_1, \Gamma], \cdots,\ \Gamma_d = [\Gamma_{d-1}, \Gamma], \cdots$$

of Γ and also consider the quotient groups

$$N_d = \Gamma/\Gamma_d \quad (d = 1, 2, \cdots)$$

which we call the d-th nilpotent quotient of Γ. The first one N_1 is nothing but the abelianization of Γ so that we have an identification

$$N_1 = H = H_1(\Sigma_g^0; \mathbb{Z}).$$

We also denote by ζ_d the image of ζ in N_d under the natural projection $\Gamma \to N_d$. The first one is trivial, namely $\zeta_1 = 0 \in N_1 = H$. However the second one can be identified as

$$\zeta_2 = \omega_0 = \sum_{i=1}^{g} x_i \wedge y_i \in \Lambda^2 H \cong \Gamma_1/\Gamma_2 \subset N_2$$

where ω_0 is the symplectic class and x_i, y_i denote the homology classes of α_i, β_i respectively.

Definition 2.1. We define two subgroups of $\mathrm{Aut}\,N_d$ as follows.

$$\mathrm{Aut}_0' N_d = \{\varphi \in \mathrm{Aut}\,N_d; \varphi(\zeta_d) = \zeta_d\}$$
$$\mathrm{Aut}_0 N_d = p(\mathrm{Aut}_0' N_{d+1})$$

where $p : \mathrm{Aut}\,N_{d+1} \to \mathrm{Aut}\,N_d$ denotes the natural projection.

The subgroup $\mathrm{Aut}_0 N_d$ was introduced by Garoufalidis and Levine [6] which plays a fundamental role in their theory of the group, denoted by $\mathcal{H}_{g,1}$, of homology cobordism classes of homology cylinders over Σ_g^0. The concept of homology cylinders in turn was introduced by Habiro [9] and Goussarov [7] independently. The above definition is a slight modification of their original one and the reason why we consider another subgroup $\mathrm{Aut}_0' N_d$ will be explained in Remark 2.2 below.

It is easy to see that the first group $\mathrm{Aut}_0 N_1 = \mathrm{Aut}_0 H$ can be identified with the symplectic group $\mathrm{Sp}(2g, \mathbb{Z})$ (whereas notice that $\mathrm{Aut}_0' N_1 \cong$

$\mathrm{GL}(2g,\mathbb{Z})$). By definition, these subgroups $\mathrm{Aut}_0 N_d$ ($d = 1, 2, \cdots$) make a projective system of groups

$$\cdots \longrightarrow \mathrm{Aut}_0 N_d \longrightarrow \mathrm{Aut}_0 N_{d-1} \longrightarrow \cdots \longrightarrow \mathrm{Aut}_0 N_2 \longrightarrow \mathrm{Aut}_0 N_1.$$

The expression (1) induces representations

$$\rho_d : \mathcal{M}_{g,1} \longrightarrow \mathrm{Aut}_0 N_d \tag{2}$$

and the totality of these representations gives rise to a homomorphism

$$\rho_\infty : \mathcal{M}_{g,1} \cong \mathrm{Aut}_0 \Gamma \longrightarrow \varprojlim_{d\to\infty} \mathrm{Aut}_0 N_d$$

which is known to be *injective.*

The induced filtration

$$\{\mathcal{M}_{g,1}(d)\}_d, \quad \mathcal{M}_{g,1}(d) = \mathrm{Ker}(\rho_d : \mathcal{M}_{g,1} \to \mathrm{Aut}_0 N_d)$$

of the mapping class group was considered by Johnson [14] and it is called the *Johnson filtration.* The first group $\mathcal{M}_{g,1}(1)$ is nothing other than the Torelli group denoted by $\mathcal{I}_{g,1}$.

To describe the structure of $\mathrm{Aut}_0 N_d$ more precisely, let

$$\mathcal{L}_{g,1} = \bigoplus_{d=1}^{\infty} \mathcal{L}_{g,1}(d)$$

be the free graded Lie algebra generated by $H = H_1(\Sigma_g;\mathbb{Z})$ so that $\mathcal{L}_{g,1}(1) = H, \mathcal{L}_{g,1}(2) \cong \Lambda^2 H$, and $\mathcal{L}_{g,1}(d) \cong \Gamma_{d-1}/\Gamma_d$ in general (see [27]). Next let

$$\mathfrak{h}_{g,1} = \bigoplus_{d=0}^{\infty} \mathfrak{h}_{g,1}(d)$$

be the graded Lie algebra consisting of *symplectic* derivations of the free Lie algebra $\mathcal{L}_{g,1}$, namely those derivations which kill the symplectic class $\omega_0 \in \mathcal{L}_{g,1}(2)$. Here the submodule

$$\mathfrak{h}_{g,1}(d) = \{D \in \mathrm{Hom}(H, \mathcal{L}_{g,1}(d+1)); D(\omega_0) = 0\}$$

of $\mathfrak{h}_{g,1}$ is the one consisting of symplectic derivations with degree d. As was shown in [29], the Poincaré duality $H^* \cong H$ induces a canonical isomorphism

$$\mathfrak{h}_{g,1}(d) \cong \mathrm{Ker}([\ ,\ ] : H \otimes \mathcal{L}_{g,1}(d+1) \to \mathcal{L}_{g,1}(d+2)).$$

Now Garoufalidis and Levine proved in the above cited paper [6] that there is a short exact sequence

$$1 \longrightarrow \mathfrak{h}_{g,1}(d) \longrightarrow \mathrm{Aut}_0 N_{d+1} \longrightarrow \mathrm{Aut}_0 N_d \longrightarrow 1. \tag{3}$$

Remark 2.2. It may appear that the definition of the subgroup $\mathrm{Aut}_0' N_d$ given in Definition 2.1 is more natural than that of $\mathrm{Aut}_0 N_d$. However this is not the case because of the following reason. Since the group N_d is defined as the quotient of Γ by its d-th commutator subgroup Γ_d and since the element ζ belongs to Γ_1, the condition of preserving the element ζ_d becomes vacuous on

$$\mathrm{Ker}(\mathrm{Aut}\, N_d \rightarrow \mathrm{Aut}\, N_{d-1}) \cong \mathrm{Hom}(H, \mathcal{L}_{g,1}(d))$$

(see [30] for the above isomorphism). The subgroup $\mathrm{Aut}_0 N_d$ is defined so as to eliminate this point. In particular, we have an isomorphism

$$\mathrm{Aut}_0' N_d / \mathrm{Aut}_0 N_d \cong \mathrm{Hom}(H, \mathcal{L}_{g,1}(d)) / \mathfrak{h}_{g,1}(d-1)$$

for any $d \geq 2$.

If we restrict the homomorphism ρ_{d+1} given in (2) to $\mathcal{M}_{g,1}(d)$, then we obtain a homomorphism

$$\tau_d = \rho_{d+1}|_{\mathcal{M}_{g,1}(d)} : \mathcal{M}_{g,1}(d) \longrightarrow \mathfrak{h}_{g,1}(d) \subset \mathrm{Hom}(H, \mathcal{L}_{g,1}(d+1))$$

which was introduced by Johnson [12][14] and is now called the d-th Johnson homomorphism (but with the target narrowed by [29]). See also Kawazumi's work [17] for his theory of Johnson *maps* which are certain extensions of Johnson homomorphism.

In our paper [29], we constructed an Sp-homomorphism

$$\mathrm{trace}(2d+1) : \mathfrak{h}_{g,1}(2d+1) \longrightarrow S^{2d+1} H$$

for any $d = 0, 1, 2, \cdots$ where $S^* H$ denotes the symmetric algebra of H. We proved that $\mathrm{trace}(2d+1)$ vanishes identically on $\mathrm{Image}\, \tau_{2d+1}$ while $\mathrm{trace}(2d+1) \otimes \mathbb{Q}$ is surjective for any $d \geq 1$. It follows that the cokernel of the homomorphism ρ_d becomes larger and larger as d tends to infinity.

On the other hand, there exists a natural homomorphism

$$\iota : \mathcal{M}_{g,1} \longrightarrow \mathcal{H}_{g,1} \tag{4}$$

from the mapping class group to the group $\mathcal{H}_{g,1}$ of Garoufalidis and Levine, already mentioned above. They proved the following remarkable theorem.

Theorem 2.3 (Garoufalidis-Levine [6], see also Habegger [8]). *For any d, there exists a homomorphism*

$$\tilde{\rho}_d : \mathcal{H}_{g,1} \longrightarrow \mathrm{Aut}_0 N_d$$

such that $\rho_d = \tilde{\rho}_d \circ \iota$. It follows that the natural homomorphism $\iota : \mathcal{M}_{g,1} \to \mathcal{H}_{g,1}$ is injective because ρ_∞ is injective. Furthermore all the homomorphisms $\tilde{\rho}_d$ are surjective.

They apply a theorem of Stallings in [38] to show the existence of $\tilde{\rho}_d$ and for the surjectivity of it, they use an argument of Kervaire-Milnor in [20]. They call their group $\mathcal{H}_{g,1}$ an *enlargement* of the mapping class group (see [25]).

In contrast to the case of the mapping class group, they point out that the induced homomorphism

$$\tilde{\rho}_\infty : \mathcal{H}_{g,1} \longrightarrow \varprojlim_{d\to\infty} \mathrm{Aut}_0 N_d$$

is *not injective* because the group $\Theta^3_{\mathbb{Z}}$ of homology cobordism classes of homology 3-spheres is contained in the center of the group $\mathcal{H}_{g,1}$ while all the homomorphisms $\tilde{\rho}_d$ vanish on it.

In relation to the problem of determining the image of $\tilde{\rho}_\infty$, Sakasai [36] considered the *acyclic closure* Γ^{acy} of Γ due to Levine (see [24]) and proved the following theorem (we refer to the above cited papers for the definition of the acyclic closure of a group because it is rather complicated and we do not use it in this paper).

Theorem 2.4 (Sakasai [36]). *There exists a natural homomorphism*

$$\rho^{acy} : \mathcal{H}_{g,1} \to \mathrm{Aut}\, \Gamma^{acy}$$

and its image can be identified as

$$\mathrm{Image}\, \rho^{acy} = \mathrm{Aut}_0 \Gamma^{acy} = \{\varphi \in \mathrm{Aut}\, \Gamma^{acy}; \varphi(\zeta) = \zeta\}$$

where $\zeta \in \Gamma^{acy}$ denotes the image of $\zeta \in \Gamma$ under the natural injection $\Gamma \to \Gamma^{acy}$.

We refer to the above cited paper of Sakasai as well as [37] for further results concerning the structure of $\mathcal{H}_{g,1}$.

In view of the above results, it is a very important problem to analyze the structure of the groups $\mathrm{Aut}_0 N_d$.

§3. Rational forms of symplectic automorphism groups

In this section, we embed $\mathrm{Aut}_0 N_d$ into a linear algebraic group over $\mathbb{Q}$, which we denote by $\mathrm{Aut}_0(N_d \otimes \mathbb{Q})$, as a discrete and Zariski dense subgroup. We may call the latter group the *rational form* of the former group. We also describe the Lie algebra of $\mathrm{Aut}_0(N_d \otimes \mathbb{Q})$ explicitly.

For this, we begin by recalling a few results from section 2 of [30] where we analyzed the structure of $\mathrm{Aut}\, N_d$. We showed that we can embed $\mathrm{Aut}\, N_d$ into the automorphism group $\mathrm{Aut}(N_d \otimes \mathbb{Q})$ of the Mal'cev completion $N_d \otimes \mathbb{Q}$ of N_d and also that the group $\mathrm{Aut}(N_d \otimes \mathbb{Q})$ has a natural structure of a linear algebraic group over $\mathbb{Q}$.

We shall show that almost the same argument applies to the subgroup $\mathrm{Aut}_0 N_d \subset \mathrm{Aut}\, N_d$ which consists of elements satisfying the symplectic constraint (see Definition 2.1).

First let $\mathrm{IAut}_0 N_d$ denote the kernel of the projection

$$\mathrm{Aut}_0 N_d \longrightarrow \mathrm{Aut}_0 N_1 \cong \mathrm{Sp}(2g, \mathbb{Z}).$$

It is the subgroup of $\mathrm{Aut}_0 N_d$ consisting of elements which act on the abelianization $H_1(N_d) \cong H$ trivially and we have a short exact sequence

$$1 \longrightarrow \mathrm{IAut}_0 N_d \longrightarrow \mathrm{Aut}_0 N_d \longrightarrow \mathrm{Sp}(2g, \mathbb{Z}) \longrightarrow 1.$$

The short exact sequence (3) restricts to

$$0 \longrightarrow \mathfrak{h}_{g,1}(d) \longrightarrow \mathrm{IAut}_0 N_{d+1} \longrightarrow \mathrm{IAut}_0 N_d \longrightarrow 1 \tag{5}$$

which turns out to be a *central* extension. It follows that the group $\mathrm{IAut}_0 N_d$ is a *nilpotent* group for any d. Let $(\mathrm{IAut}_0 N_d) \otimes \mathbb{Q}$ denote its Mal'cev completion.

Now it was shown in [30] (Proposition 2.5) that there exists an embedding

$$i : N_d \rightarrow N_d \otimes \mathbb{Q} \tag{6}$$

of N_d into its Mal'cev completion $N_d \otimes \mathbb{Q}$ and this induces an injective homomorphism

$$i_* : \mathrm{Aut}\, N_d \longrightarrow \mathrm{Aut}(N_d \otimes \mathbb{Q})$$

whose image is Zariski dense. Furthermore $\mathrm{Aut}(N_d \otimes \mathbb{Q})$ is a linear algebraic group over $\mathbb{Q}$ so that the extension

$$1 \longrightarrow \mathrm{IAut}(N_d \otimes \mathbb{Q}) \longrightarrow \mathrm{Aut}(N_d \otimes \mathbb{Q}) \longrightarrow \mathrm{GL}(2g, \mathbb{Q}) \longrightarrow 1 \tag{7}$$

splits, where $\mathrm{IAut}(N_d \otimes \mathbb{Q})$ denotes the kernel of the natural projection $\mathrm{Aut}(N_d \otimes \mathbb{Q}) \rightarrow \mathrm{GL}(2g, \mathbb{Q})$ and it is the maximal normal unipotent subgroup.

We have a distinguished element $i(\zeta_d) \in N_d \otimes \mathbb{Q}$ and the subgroup

$$\mathrm{Aut}'_0(N_d \otimes \mathbb{Q}) = \{\varphi \in \mathrm{Aut}(N_d \otimes \mathbb{Q}); \varphi(i(\zeta_d)) = i(\zeta_d)\}$$

of $\mathrm{Aut}(N_d \otimes \mathbb{Q})$ is clearly an algebraic subgroup because the condition for an automorphism to preserve a particular element is an algebraic one. Furthermore $i_*(\mathrm{Aut}_0' N_d)$ is contained in $\mathrm{Aut}_0'(N_d \otimes \mathbb{Q})$ as a Zariski dense subgroup.

Finally, we consider the natural projection

$$p : \mathrm{Aut}(N_{d+1} \otimes \mathbb{Q}) \longrightarrow \mathrm{Aut}(N_d \otimes \mathbb{Q})$$

and define $\mathrm{Aut}_0(N_d \otimes \mathbb{Q})$ to be the image of $\mathrm{Aut}_0'(N_{d+1} \otimes \mathbb{Q})$ under p, namely we set

$$\mathrm{Aut}_0(N_d \otimes \mathbb{Q}) = p(\mathrm{Aut}_0'(N_{d+1} \otimes \mathbb{Q})).$$

Clearly the above projection p is a rational morphism. Hence we can conclude that $\mathrm{Aut}_0(N_d \otimes \mathbb{Q})$ is algebraic because it is the image of the algebraic group $\mathrm{Aut}_0'(N_{d+1} \otimes \mathbb{Q})$ by a rational morphism. It is then easy to see that $i(\mathrm{Aut}_0 N_d)$ is Zariski dense in $\mathrm{Aut}_0(N_d \otimes \mathbb{Q})$.

Summarizing the above argument, we have the following proposition.

Proposition 3.1. *The group* $\mathrm{Aut}_0 N_d$ *can be embedded into a linear algebraic group* $\mathrm{Aut}_0(N_d \otimes \mathbb{Q})$ *as a Zariski dense subgroup and we have a split short exact sequence*

$$1 \longrightarrow \mathrm{IAut}_0(N_d \otimes \mathbb{Q}) \longrightarrow \mathrm{Aut}_0(N_d \otimes \mathbb{Q}) \longrightarrow \mathrm{Sp}(2g, \mathbb{Q}) \longrightarrow 1$$

where $\mathrm{IAut}_0(N_d \otimes \mathbb{Q})$ *is the maximal normal unipotent subgroup. Furthermore, the subgroup* $\mathrm{IAut}_0 N_d$ *is mapped to* $\mathrm{IAut}_0(N_d \otimes \mathbb{Q})$ *as a Zariski dense subgroup and this gives an explicit construction of the Mal'cev completion* $(\mathrm{IAut}_0 N_d) \otimes \mathbb{Q}$ *of the nilpotent group* $\mathrm{IAut}_0 N_d$.

Next we describe the Lie algebra of $\mathrm{Aut}_0(N_d \otimes \mathbb{Q})$. As in the previous section, let $\mathfrak{h}_{g,1}$ denote the graded Lie algebra over $\mathbb{Z}$ consisting of all the symplectic derivations of the free Lie algebra $\mathcal{L}_{g,1}$. Also let

$$\mathfrak{h}_{g,1}^+ = \bigoplus_{k=1}^{\infty} \mathfrak{h}_{g,1}(k)$$

be the ideal of $\mathfrak{h}_{g,1}$ consisting of all the derivations with *positive* degrees. We consider the rational forms

$$\mathfrak{h}_{g,1}^{\mathbb{Q}} = \mathfrak{h}_{g,1} \otimes \mathbb{Q} = \bigoplus_{k=0}^{\infty} \mathfrak{h}_{g,1}^{\mathbb{Q}}(k)$$

where $\mathfrak{h}_{g,1}^{\mathbb{Q}}(k) = \mathfrak{h}_{g,1}(k) \otimes \mathbb{Q}$ and

$$\mathfrak{h}_{g,1}^{\mathbb{Q}+} = \bigoplus_{k=1}^{\infty} \mathfrak{h}_{g,1}^{\mathbb{Q}}(k).$$

In particular, we have

$$\mathfrak{h}_{g,1}^{\mathbb{Q}}(0) = \mathfrak{sp}(2g, \mathbb{Q}), \quad \mathfrak{h}_{g,1}^{\mathbb{Q}}(1) = \Lambda^3 H_{\mathbb{Q}}.$$

Definition 3.2. For each $d = 1, 2, \cdots$, we define the truncated Lie algebras, denoted by $\mathfrak{h}_{g,1}^{\mathbb{Q}}[d]$ and $\mathfrak{h}_{g,1}^{\mathbb{Q}+}[d]$, by setting

$$\mathfrak{h}_{g,1}^{\mathbb{Q}}[d] = \mathfrak{h}_{g,1}^{\mathbb{Q}}/I(d)$$
$$\mathfrak{h}_{g,1}^{\mathbb{Q}+}[d] = \mathfrak{h}_{g,1}^{\mathbb{Q}+}/I(d)$$

where $I(d)$ denotes the ideal consisting of derivations with degree $\geq d$. Additively, we can write

$$\mathfrak{h}_{g,1}^{\mathbb{Q}}[d] = \bigoplus_{k=0}^{d-1} \mathfrak{h}_{g,1}^{\mathbb{Q}}(k)$$
$$\mathfrak{h}_{g,1}^{\mathbb{Q}+}[d] = \bigoplus_{k=1}^{d-1} \mathfrak{h}_{g,1}^{\mathbb{Q}}(k).$$

Theorem 3.3. *The Lie algebras of the algebraic groups* $\mathrm{Aut}_0(N_d \otimes \mathbb{Q})$ *and* $\mathrm{IAut}_0(N_d \otimes \mathbb{Q})$ *are isomorphic to the truncated Lie algebras*

$$\mathfrak{h}_{g,1}^{\mathbb{Q}}[d], \quad \mathfrak{h}_{g,1}^{\mathbb{Q}+}[d]$$

respectively.

Before proving the above result, we prepare a few facts. First of all, let $\mathfrak{n}_d^{\mathbb{Q}}$ denote the Lie algebra of the Mal'cev completion $N_d \otimes \mathbb{Q}$ of the free nilpotent group N_d. Then, as is well known, the exponential map induces a natural bijection

$$\exp : \mathfrak{n}_d^{\mathbb{Q}} \cong N_d \otimes \mathbb{Q}.$$

This then gives rise to an isomorphism

$$\mathrm{Aut}\, \mathfrak{n}_d^{\mathbb{Q}} \cong \mathrm{Aut}(N_d \otimes \mathbb{Q}) \tag{8}$$

between the corresponding automorphism groups. Also the restriction of (8) gives rise to an isomorphism

$$\mathrm{IAut}\, \mathfrak{n}_d^{\mathbb{Q}} \cong \mathrm{IAut}(N_d \otimes \mathbb{Q}) \tag{9}$$

where $\mathrm{IAut}\, \mathfrak{n}_d^{\mathbb{Q}}$ is defined by the following exact sequence

$$1 \longrightarrow \mathrm{IAut}\, \mathfrak{n}_d^{\mathbb{Q}} \longrightarrow \mathrm{Aut}\, \mathfrak{n}_d^{\mathbb{Q}} \longrightarrow \mathrm{GL}(2g, \mathbb{Q}) \longrightarrow 1$$

which is essentially the same as (7). These bijections as well as isomorphisms are compatible with respect to the projection $N_d \otimes \mathbb{Q} \to N_{d-1} \otimes \mathbb{Q}$ so that the following diagram is commutative

(10)

$$\begin{array}{ccccc}
\mathrm{Hom}(H_{\mathbb{Q}}, \mathcal{L}^{\mathbb{Q}}_{g,1}(d)) & \xrightarrow{\text{inj}} & \mathrm{Aut}\,\mathfrak{n}^{\mathbb{Q}}_d & \xrightarrow{\text{surj}} & \mathrm{Aut}\,\mathfrak{n}^{\mathbb{Q}}_{d-1} \\
\| & & \exp\big\downarrow\cong & & \exp\big\downarrow\cong \\
\mathrm{Hom}(H_{\mathbb{Q}}, \mathcal{L}^{\mathbb{Q}}_{g,1}(d)) & \xrightarrow{\text{inj}} & \mathrm{Aut}(N_d \otimes \mathbb{Q}) & \xrightarrow{\text{surj}} & \mathrm{Aut}(N_{d-1} \otimes \mathbb{Q}).
\end{array}$$

Now let us consider the particular element

$$\tilde{\omega}_d := \exp^{-1}(i(\zeta_d)) \in \mathfrak{n}^{\mathbb{Q}}_d$$

and make the following definition.

Definition 3.4. We set

$$\begin{aligned}
\mathrm{Aut}'_0\,\mathfrak{n}^{\mathbb{Q}}_d &= \{\varphi \in \mathrm{Aut}\,\mathfrak{n}^{\mathbb{Q}}_d : \varphi(\tilde{\omega}_d) = \tilde{\omega}_d\} \\
\mathrm{Aut}_0\,\mathfrak{n}^{\mathbb{Q}}_d &= p(\mathrm{Aut}'_0\,\mathfrak{n}^{\mathbb{Q}}_{d+1})
\end{aligned}$$

where

$$p : \mathrm{Aut}\,\mathfrak{n}^{\mathbb{Q}}_{d+1} \longrightarrow \mathrm{Aut}\,\mathfrak{n}^{\mathbb{Q}}_d$$

denotes the natural projection.

Then, all the above argument can be adapted to the cases of Aut'_0 and Aut_0. In particular, the restrictions of (8) and (9) induce canonical isomorphisms

$$\mathrm{Aut}_0\,\mathfrak{n}^{\mathbb{Q}}_d \cong \mathrm{Aut}_0(N_d \otimes \mathbb{Q}) \tag{11}$$

$$\mathrm{IAut}_0\,\mathfrak{n}^{\mathbb{Q}}_d \cong \mathrm{IAut}_0(N_d \otimes \mathbb{Q}), \tag{12}$$

respectively.

As a final step towards the proof of Theorem 3.3, we recall a classical result (see [27]) that the Lie algebra of the Mal'cev completion $N_d \otimes \mathbb{Q}$ of the free nilpotent group N_d is isomorphic (though not canonically) to the free nilpotent Lie algebra which is defined as the truncated Lie algebra

$$\mathcal{L}^{\mathbb{Q}}_{g,1}[d] = \mathcal{L}^{\mathbb{Q}}_{g,1}/J_d$$

where J_d denotes the ideal of the free Lie algebra $\mathcal{L}^{\mathbb{Q}}_{g,1}$ consisting of elements with degree $> d$. Thus additively we can write

$$\mathcal{L}^{\mathbb{Q}}_{g,1}[d] = \bigoplus_{k=1}^{d} \mathcal{L}^{\mathbb{Q}}_{g,1}(k).$$

In this Lie algebra, we have a particular element, namely the symplectic class

$$\omega_0 \in \mathcal{L}^{\mathbb{Q}}_{g,1}(2) \cong \Lambda^2 H_{\mathbb{Q}} \subset \mathcal{L}^{\mathbb{Q}}_{g,1}[d] \quad (d \geq 2).$$

Lemma 3.5. *For any $d \geq 2$, the natural homomorphism*

$$\mathrm{Aut}\,\mathcal{L}^{\mathbb{Q}}_{g,1}[d] \longrightarrow \mathrm{Aut}\,\mathcal{L}^{\mathbb{Q}}_{g,1}[d-1]$$

is surjective and we have the following exact sequence

$$1 \longrightarrow \mathrm{Hom}(H_{\mathbb{Q}}, \mathcal{L}^{\mathbb{Q}}_{g,1}(d)) \longrightarrow \mathrm{Aut}\,\mathcal{L}^{\mathbb{Q}}_{g,1}[d] \longrightarrow \mathrm{Aut}\,\mathcal{L}^{\mathbb{Q}}_{g,1}[d-1] \longrightarrow 1.$$

Proof. If $d = 1$, $\mathcal{L}^{\mathbb{Q}}_{g,1}[1] = H_{\mathbb{Q}}$ so that

$$\mathrm{Aut}\,\mathcal{L}^{\mathbb{Q}}_{g,1}[1] = \mathrm{GL}(2g, \mathbb{Q}).$$

Since the abelianization of the Lie algebra $\mathcal{L}^{\mathbb{Q}}_{g,1}[d]$ can be identified with $H_{\mathbb{Q}}$, we have a representation

$$r : \mathrm{Aut}\,\mathcal{L}^{\mathbb{Q}}_{g,1}[d] \longrightarrow \mathrm{GL}(2g, \mathbb{Q}).$$

On the other hand, clearly the action of $\mathrm{GL}(2g, \mathbb{Q})$ on $H_{\mathbb{Q}}$ extends canonically to that on $\mathcal{L}^{\mathbb{Q}}_{g,1}[d]$ for any d so that we have a canonically *split extension*

$$1 \longrightarrow \mathrm{IAut}\,\mathcal{L}^{\mathbb{Q}}_{g,1}[d] \longrightarrow \mathrm{Aut}\,\mathcal{L}^{\mathbb{Q}}_{g,1}[d] \stackrel{r}{\longrightarrow} \mathrm{GL}(2g, \mathbb{Q}) \longrightarrow 1$$

where $\mathrm{IAut}\,\mathcal{L}^{\mathbb{Q}}_{g,1}[d]$ denotes the kernel of the representation r. Hence, for the surjectivity, it is enough to show that the natural homomorphism

$$\mathrm{IAut}\,\mathcal{L}^{\mathbb{Q}}_{g,1}[d] \longrightarrow \mathrm{IAut}\,\mathcal{L}^{\mathbb{Q}}_{g,1}[d-1]$$

is surjective. Any element $\varphi \in \mathrm{IAut}\,\mathcal{L}^{\mathbb{Q}}_{g,1}[d-1]$ gives rise to an *endomorphism* of $\mathcal{L}^{\mathbb{Q}}_{g,1}[d]$ which we denote by $\tilde{\varphi}$. Therefore it suffices to prove that this extended endomorphism $\tilde{\varphi}$ is in fact an isomorphism. For this, it is enough to show that for any element $u \in \mathcal{L}^{\mathbb{Q}}_{g,1}(1) = H_{\mathbb{Q}}$, there exists an element $\xi \in \mathcal{L}^{\mathbb{Q}}_{g,1}[d]$ such that

$$\tilde{\varphi}(\xi) = u.$$

Since φ is an isomorphism by the assumption, there exists an element $\xi' \in \mathcal{L}^{\mathbb{Q}}_{g,1}[d-1]$ such that

$$\varphi(\xi') = u.$$

Then, for some element $\eta \in \mathcal{L}_{g,1}^{\mathbb{Q}}(d)$, we have

$$\tilde{\varphi}(\xi') = u + \eta.$$

It is easy to see that the action of any element in $\mathrm{IAut}\,\mathcal{L}_{g,1}^{\mathbb{Q}}[d]$ on $\mathcal{L}_{g,1}^{\mathbb{Q}}(d)$ is trivial. If we set $\xi = \xi' - \eta$, then

$$\tilde{\varphi}(\xi) = u + \eta - \eta = u.$$

Therefore $\tilde{\omega}$ is an isomorphism as required.

Finally, the fact that the kernel of the homomorphism

$$\mathrm{Aut}\,\mathcal{L}_{g,1}^{\mathbb{Q}}[d] \rightarrow \mathrm{Aut}\,\mathcal{L}_{g,1}^{\mathbb{Q}}[d-1]$$

is isomorphic to $\mathrm{Hom}(H_{\mathbb{Q}}, \mathcal{L}_{g,1}^{\mathbb{Q}}(d))$ can be seen easily, where any element f in this module acts on $\mathcal{L}_{g,1}^{\mathbb{Q}}[d]$ by

$$H_{\mathbb{Q}} \ni u \longmapsto u + f(u) \in \mathcal{L}_{g,1}^{\mathbb{Q}}[d].$$

This completes the proof. Q.E.D.

The following is the main technical fact which will be used in the proof of Theorem 3.3.

Lemma 3.6. *Let*

$$\xi = \omega_0 + \xi_3 + \cdots + \xi_d \in \mathcal{L}_{g,1}^{\mathbb{Q}}[d]$$

be any element with $\xi_k \in \mathcal{L}_{g,1}^{\mathbb{Q}}(k)$ $(k = 3, \cdots, d)$. *Then there exists an element*

$$\varphi \in \mathrm{IAut}\,\mathcal{L}_{g,1}^{\mathbb{Q}}[d-1]$$

such that

$$\tilde{\varphi}(\omega_0) = \xi$$

where $\tilde{\varphi} \in \mathrm{IAut}\,\mathcal{L}_{g,1}^{\mathbb{Q}}[d]$ *is the extension of* φ *as in the proof of Lemma* 3.5.

Proof. We use the induction on d. The claim clearly holds for $d = 2$, because $\xi = \omega_0$ in this case so that we can set $\varphi = \mathrm{id}$. Suppose that it holds for the cases $2, \cdots, d-1$, and we prove the case d. By the induction assumption, there exists an element $\psi \in \mathrm{IAut}\,\mathcal{L}_{g,1}[d-2]$ such that

$$\tilde{\psi}(\omega_0) = \bar{\xi} \in \mathcal{L}_{g,1}^{\mathbb{Q}}[d-1]$$

where $\tilde{\psi} \in \mathrm{IAut}\,\mathcal{L}_{g,1}[d-1]$ denotes the extension of ψ as in the proof of Lemma 3.5 and

$$\bar{\xi} = \omega_0 + \xi_3 + \cdots + \xi_{d-1}.$$

Now consider the element

$$\tilde{\tilde{\psi}} \in \mathrm{IAut}\,\mathcal{L}_{g,1}[d]$$

which denotes further extension of $\tilde{\psi}$. Then we see that

$$\eta := \tilde{\tilde{\psi}}(\omega_0) - \xi \in \mathcal{L}^{\mathbb{Q}}_{g,1}(d)$$

because

$$\tilde{\tilde{\psi}}(\omega_0) \equiv \bar{\xi} \bmod \mathcal{L}^{\mathbb{Q}}_{g,1}(d).$$

Since the bracket operation

$$H_{\mathbb{Q}} \otimes \mathcal{L}^{\mathbb{Q}}_{g,1}(d-1) \longrightarrow \mathcal{L}^{\mathbb{Q}}_{g,1}(d)$$

is surjective, we can write

$$\eta = \sum_{i=1}^{g} ([x_i, \gamma_i] + [y_i, \delta_i]) \quad (\gamma_i, \delta_i \in \mathcal{L}^{\mathbb{Q}}_{g,1}(d-1))$$

where $x_1, \cdots, x_g, y_1, \cdots, y_g$ is a symplectic basis of H. Now define an element $\varphi \in \mathrm{IAut}\,\mathcal{L}^{\mathbb{Q}}_{g,1}[d-1]$ by setting

$$\varphi(x_i) = \tilde{\psi}(x_i) + \delta_i, \quad \varphi(y_i) = \tilde{\psi}(y_i) - \gamma_i$$

and let $\tilde{\varphi} \in \mathrm{IAut}\,\mathcal{L}^{\mathbb{Q}}_{g,1}[d]$ denotes its extension. Then we have

$$\begin{aligned}
\tilde{\varphi}(\omega_0) &= \sum_{i=1}^{g} \left\{ [\tilde{\tilde{\psi}}(x_i) + \delta_i, \tilde{\tilde{\psi}}(y_i) - \gamma_i] \right\} \\
&= \tilde{\tilde{\psi}}(\omega_0) - \sum_{i=1}^{g} \{[x_i, \gamma_i] - [\delta_i, y_i]\} \\
&= \xi + \eta - \eta = \xi
\end{aligned}$$

because

$$\tilde{\tilde{\psi}}(x_i) \equiv x_i \bmod J_1, \quad \tilde{\tilde{\psi}}(y_i) \equiv y_i \bmod J_1.$$

Thus the claim holds for the case of d and this completes the proof. Q.E.D.

Definition 3.7. We set

$$\operatorname{Aut}'_0 \mathcal{L}^{\mathbb{Q}}_{g,1}[d] = \{\varphi \in \operatorname{Aut} \mathcal{L}^{\mathbb{Q}}_{g,1}[d]; \varphi(\omega_0) = \omega_0\}$$
$$\operatorname{Aut}_0 \mathcal{L}^{\mathbb{Q}}_{g,1}[d] = p(\operatorname{Aut}'_0 \mathcal{L}^{\mathbb{Q}}_{g,1}[d+1])$$

where

$$p : \operatorname{Aut} \mathcal{L}^{\mathbb{Q}}_{g,1}[d+1] \longrightarrow \operatorname{Aut} \mathcal{L}^{\mathbb{Q}}_{g,1}[d]$$

denotes the natural projection.

Proposition 3.8. *The group* $\operatorname{Aut}_0 \mathcal{L}^{\mathbb{Q}}_{g,1}[d]$ *is a linear algebraic group over* $\mathbb{Q}$ *which can be expressed as the following canonically split extension*

$$1 \longrightarrow \operatorname{IAut}_0 \mathcal{L}^{\mathbb{Q}}_{g,1}[d] \longrightarrow \operatorname{Aut}_0 \mathcal{L}^{\mathbb{Q}}_{g,1}[d] \stackrel{r}{\longrightarrow} \operatorname{Sp}(2g,\mathbb{Q}) \longrightarrow 1.$$

Furthermore the Lie algebras of $\operatorname{Aut}_0 \mathcal{L}^{\mathbb{Q}}_{g,1}[d]$ *and* $\operatorname{IAut}_0 \mathcal{L}^{\mathbb{Q}}_{g,1}[d]$ *are canonically isomorphic to* $\mathfrak{h}^{\mathbb{Q}}_{g,1}[d]$ *and* $\mathfrak{h}^{\mathbb{Q}+}_{g,1}[d]$, *respectively.*

Proof. For the former half of the claim, almost the same arguments as in the case of $\operatorname{Aut}_0(N_d \otimes \mathbb{Q})$ (see those arguments given before the statement of Proposition 3.1) apply to the present case as well. In fact, this Lie algebra case is even simpler than the group case. In fact, it is easy to see first that $\operatorname{Aut} \mathcal{L}^{\mathbb{Q}}_{g,1}[d]$ is a linear algebraic group over $\mathbb{Q}$. Then $\operatorname{Aut}'_0 \mathcal{L}^{\mathbb{Q}}_{g,1}[d]$ is shown to be an algebraic subgroup because it consists of elements which fix a particular element ω_0. Then, since the natural projection p is rational, its image $\operatorname{Aut}_0 \mathcal{L}^{\mathbb{Q}}_{g,1}[d]$ is also algebraic. The splitting is canonical because $\operatorname{Sp}(2g,\mathbb{Q})$ acts naturally on $\mathcal{L}^{\mathbb{Q}}_{g,1}[d]$ as an automorphism group preserving the symplectic class ω_0.

Now we prove the latter half of the claim. By a well known fact, the Lie algebra of the Lie group $\operatorname{Aut} \mathcal{L}^{\mathbb{Q}}_{g,1}[d]$ is isomorphic to that of derivations of the Lie algebra $\mathcal{L}^{\mathbb{Q}}_{g,1}[d]$ which we denote by $\operatorname{Der} \mathcal{L}^{\mathbb{Q}}_{g,1}[d]$. Since $\mathcal{L}^{\mathbb{Q}}_{g,1}[d]$ is a graded Lie algebra, the above derivations are also graded and we can write

$$\operatorname{Der} \mathcal{L}^{\mathbb{Q}}_{g,1}[d] = \bigoplus_{k=0}^{d} \operatorname{Hom}(H_{\mathbb{Q}}, \mathcal{L}_{g,1}(k)).$$

Furthermore we have the following canonically split exact sequence of Lie algebras

$$0 \longrightarrow \operatorname{IDer} \mathcal{L}^{\mathbb{Q}}_{g,1}[d] \longrightarrow \operatorname{Der} \mathcal{L}^{\mathbb{Q}}_{g,1}[d] \longrightarrow \mathfrak{gl}(2g,\mathbb{Q}) \longrightarrow 0$$

where $\operatorname{IDer} \mathcal{L}^{\mathbb{Q}}_{g,1}[d]$ denotes the kernel of the natural homomorphism

$$\operatorname{Der} \mathcal{L}^{\mathbb{Q}}_{g,1}[d] \longrightarrow \mathfrak{gl}(2g,\mathbb{Q})$$

and it is the Lie algebra of $\mathrm{IAut}\,\mathcal{L}^{\mathbb{Q}}_{g,1}[d]$.

Keeping in mind Definition 3.7, we define

$$\mathrm{Der}'_0\,\mathcal{L}^{\mathbb{Q}}_{g,1}[d] = \{D \in \mathrm{IDer}\,\mathcal{L}^{\mathbb{Q}}_{g,1}[d] : D(\omega_0) = 0\}$$
$$\mathrm{Der}_0\,\mathcal{L}^{\mathbb{Q}}_{g,1}[d] = p(\mathrm{IDer}'_0\,\mathcal{L}^{\mathbb{Q}}_{g,1}[d+1])$$

where

$$p : \mathrm{IDer}\,\mathcal{L}^{\mathbb{Q}}_{g,1}[d+1] \longrightarrow \mathcal{L}^{\mathbb{Q}}_{g,1}[d]$$

denotes the natural projection. Then it is easy to see that $\mathrm{Der}_0\,\mathcal{L}^{\mathbb{Q}}_{g,1}[d]$ is the Lie algebra of $\mathrm{Aut}_0\,\mathcal{L}^{\mathbb{Q}}_{g,1}[d]$ and similarly $\mathrm{IDer}_0\,\mathcal{L}^{\mathbb{Q}}_{g,1}[d]$ is the Lie algebra of $\mathrm{IAut}_0\,\mathcal{L}^{\mathbb{Q}}_{g,1}[d]$. Since

$$\mathrm{Der}_0\,\mathcal{L}^{\mathbb{Q}}_{g,1}[d] = \mathfrak{h}^{\mathbb{Q}}_{g,1}[d], \quad \mathrm{IDer}_0\,\mathcal{L}^{\mathbb{Q}}_{g,1}[d] = \mathfrak{h}^{\mathbb{Q}+}_{g,1}[d]$$

by definition, the proof is finished.

Q.E.D.

Proof of Theorem 3.3. First of all, $\mathrm{Aut}_0(N_d \otimes \mathbb{Q})$ and $\mathrm{IAut}_0(N_d \otimes \mathbb{Q})$ are isomorphic to $\mathrm{Aut}_0\,\mathfrak{n}^{\mathbb{Q}}_d$ and $\mathrm{IAut}_0\,\mathfrak{n}^{\mathbb{Q}}_d$ by (11) and (12), respectively.

On the other hand, it is a classical result that the Lie algebra $\mathfrak{n}^{\mathbb{Q}}_d$ is isomorphic to the truncated Lie algebra $\mathcal{L}^{\mathbb{Q}}_{g,1}[d]$ (see [27]). Although there is no canonical isomorphism between the two Lie algebras, it is easy to see that any isomorphism

$$\iota : \mathfrak{n}^{\mathbb{Q}}_d \overset{\cong}{\longrightarrow} \mathcal{L}^{\mathbb{Q}}_{g,1}[d]$$

satisfies the condition

$$\iota(\tilde{\omega}_d) = \omega_0 + \xi_3 + \cdots + \xi_d$$

for some $\xi_k \in \mathcal{L}_{g,1}(k)$ $(k = 3, \cdots, d)$. By Lemma 3.6, there exists an isomorphism $\varphi \in \mathrm{Aut}\,\mathcal{L}^{\mathbb{Q}}_{g,1}[d]$ such that

$$\varphi(\omega_0) = \omega_0 + \xi_3 + \cdots + \xi_d.$$

Hence, replacing ι by $\varphi^{-1} \circ \iota$, we may assume

$$\iota(\tilde{\omega}_d) = \omega_0.$$

It follows that

$$\mathrm{Aut}_0\,\mathfrak{n}^{\mathbb{Q}}_d \cong \mathrm{Aut}_0\,\mathcal{L}^{\mathbb{Q}}_{g,1}[d].$$

Finally, we know by Proposition 3.8 that the Lie algebras of $\mathrm{Aut}_0\,\mathcal{L}^{\mathbb{Q}}_{g,1}[d]$ and $\mathrm{IAut}_0\,\mathcal{L}^{\mathbb{Q}}_{g,1}[d]$ are canonically isomorphic to $\mathfrak{h}^{\mathbb{Q}}_{g,1}[d]$ and

$\mathfrak{h}_{g,1}^{\mathbb{Q}+}[d]$, respectively. Summarizing the above, we can now conclude that the Lie algebras of $\mathrm{Aut}_0(N_d \otimes \mathbb{Q})$ and $\mathrm{IAut}_0(N_d \otimes \mathbb{Q})$ are isomorphic to $\mathfrak{h}_{g,1}^{\mathbb{Q}}[d]$ and $\mathfrak{h}_{g,1}^{\mathbb{Q}+}[d]$, respectively, as required.

This completes the proof.

Q.E.D.

Remark 3.9. As mentioned above, there is no canonical isomorphism between the two Lie algebras $\mathfrak{n}_d^{\mathbb{Q}}$ and $\mathcal{L}_{g,1}^{\mathbb{Q}}[d]$. The latter Lie algebra is canonically isomorphic to the graded Lie algebra associated to the lower central series of the Lie group $N_d \otimes \mathbb{Q}$. However if we consider the analogues of the above Lie algebras corresponding to a *closed surface*, then according to Hain [10] each complex structure on the surface induces a canonical isomorphism between the two Lie algebras tensored with $\mathbb{C}$, namely $\bar{\mathfrak{n}}_d^{\mathbb{C}}$ and $\mathcal{L}_g^{\mathbb{C}}[d]$. In fact, Hain showed this at the level of the Torelli group by making use of Hodge theory.

§4. Group version of the trace maps

In this section, we give a definition of the *group version* of the trace maps given in [29].

As was already mentioned in section 2, the trace maps consist of a series of certain homomorphisms

$$\mathrm{trace}(2k+1) : \mathfrak{h}_{g,1}(2k+1) \longrightarrow S^{2k+1}H \quad (k = 0, 1, 2, \cdots)$$

where $S^{2k+1}H$ denotes the $(2k+1)$-st symmetric power of H. As before, if we denote by $H_{\mathbb{Q}}$ the rational homology group $H \otimes \mathbb{Q} = H_1(\Sigma_g^0; \mathbb{Q})$ and also by $\mathfrak{h}_{g,1}^{\mathbb{Q}}$ the corresponding rational form of $\mathfrak{h}_{g,1}$, then we have the rational trace maps

$$\mathrm{trace}(2k+1) : \mathfrak{h}_{g,1}^{\mathbb{Q}}(2k+1) \longrightarrow S^{2k+1}H_{\mathbb{Q}} \quad (k = 0, 1, 2, \cdots)$$

which were proved to be surjective for all k. The first trace, namely trace(1) is nothing but the contraction

$$\mathfrak{h}_{g,1}^{\mathbb{Q}}(1) \cong \Lambda^3 H_{\mathbb{Q}} \rightarrow H_{\mathbb{Q}}$$

which is a unique (up to scalars) Sp-equivariant non-trivial morphism. The totality of these trace maps gives rise to an abelian quotient

$$\mathrm{trace} : \mathfrak{h}_{g,1}^{\mathbb{Q}+} \longrightarrow \Lambda^3 H_{\mathbb{Q}} \oplus \bigoplus_{k=1}^{\infty} S^{2k+1} H_{\mathbb{Q}}$$

of the ideal $\mathfrak{h}_{g,1}^{\mathbb{Q}+}$.

If we consider the truncated Lie algebra $\mathfrak{h}_{g,1}^{\mathbb{Q}+}[d]$, then we obtain an abelian quotient

$$\text{trace}: \mathfrak{h}_{g,1}^{\mathbb{Q}+}[d] \longrightarrow \Lambda^3 H_{\mathbb{Q}} \oplus \bigoplus_{k=1}^{\ell} S^{2k+1} H_{\mathbb{Q}} \tag{13}$$

where ℓ denotes the largest integer such that $2\ell+1 \leq d-1$.

On the other hand, we know by Theorem 3.3 that the Lie algebra of $\mathrm{IAut}_0(N_d \otimes \mathbb{Q})$ is isomorphic to $\mathfrak{h}_{g,1}^{\mathbb{Q}+}[d]$. Since the commutator subgroup of any Lie group corresponds to the commutator ideal of the associated Lie algebra, and since the real form $N_d \otimes \mathbb{R}$ of $N_d \otimes \mathbb{Q}$ is simply connected, we have a canonical isomorphism

$$H_1(\mathrm{IAut}_0(N_d \otimes \mathbb{Q})) \cong H_1(\mathfrak{h}_{g,1}^{\mathbb{Q}+}[d]). \tag{14}$$

By combining (13) with (14), we obtain a homomorphism

$$\widetilde{\text{trace}}: \mathrm{IAut}_0 N_d \overset{i}{\subset} \mathrm{IAut}_0(N_d \otimes \mathbb{Q}) \longrightarrow \Lambda^3 H_{\mathbb{Q}} \oplus \bigoplus_{k=1}^{\ell} S^{2k+1} H_{\mathbb{Q}}.$$

In particular, we call the homomorphism

$$\widetilde{\text{trace}}(2k+1): \mathrm{IAut}_0 N_d \longrightarrow S^{2k+1} H_{\mathbb{Q}} \quad (2k+1 \leq d-1)$$

the group version of the trace map. If we further use Proposition 3.1, then we can conclude that the above homomorphism can be extended to a crossed homomorphism

$$\widetilde{\text{trace}}(2k+1): \mathrm{Aut}_0 N_d \longrightarrow S^{2k+1} H_{\mathbb{Q}} \quad (2k+1 \leq d-1)$$

on the whole group $\mathrm{Aut}_0 N_d$. Thus we obtain the following theorem.

Theorem 4.1. *For any $d \geq 2$, let ℓ denote the largest integer such that $2\ell+1 \leq d-1$. Then the group version of traces $\widetilde{\text{trace}}(2k+1)$ gives rise to a homomorphism*

$$\mathrm{IAut}_0 N_d \longrightarrow \Lambda^3 H_{\mathbb{Q}} \oplus \bigoplus_{k=1}^{\ell} S^{2k+1} H_{\mathbb{Q}}$$

whose induced homomorphism on the Mal'cev completion $(\mathrm{IAut}_0 N_d) \otimes \mathbb{Q} \cong \mathrm{IAut}_0(N_d \otimes \mathbb{Q})$ is surjective. Furthermore the above homomorphism can be extended to the whole group $\mathrm{Aut}_0 N_d$ as a crossed homomorphism so that we obtain a homomorphism

$$\mathrm{Aut}_0 N_d \longrightarrow \left(\Lambda^3 H_{\mathbb{Q}} \oplus \bigoplus_{k=1}^{\ell} S^{2k+1} H_{\mathbb{Q}} \right) \rtimes \mathrm{Sp}(2g, \mathbb{Q}).$$

Conjecture 4.2. *The above theorem gives the abelianization of the nilpotent group* $\mathrm{IAut}_0 N_d$ *modulo torsion. More precisely*

$$H_1(\mathrm{IAut}_0 N_d; \mathbb{Q}) \cong \Lambda^3 H_{\mathbb{Q}} \oplus \bigoplus_{k=1}^{\ell} S^{2k+1} H_{\mathbb{Q}}.$$

As an evidence for the above conjecture, we would like to mention Kassabov's work in [15] which treated the case $\mathrm{IAut}\, N_d$ where there is no symplectic constraint. Here we also remark that suitably modified version of the results of this section are valid for $\mathrm{Aut}\, N_d$ and $\mathrm{IAut}\, N_d$ without the symplectic constraint.

If the above conjecture were true, then we could conclude that

$$H_1(\mathrm{Aut}_0 N_d; \mathbb{Q}) = 0.$$

This follows from the following exact sequence

$$\cdots \longrightarrow H_1(\mathrm{IAut}_0 N_d; \mathbb{Q})_{\mathrm{Sp}} \longrightarrow H_1(\mathrm{Aut}_0 N_d; \mathbb{Q}) \longrightarrow H_1(\mathrm{Sp}(2g; \mathbb{Q}); \mathbb{Q}) = 0$$

because clearly we have

$$H_1\left(\Lambda^3 H_{\mathbb{Q}} \oplus \bigoplus_{k=1}^{\ell} S^{2k+1} H_{\mathbb{Q}}\right)_{\mathrm{Sp}} = 0.$$

Remark 4.3. The results of this section are valid for the closed surface case as well. More precisely, let $\bar{\Gamma} = \pi_1 \Sigma_g$ be the fundamental group of Σ_g and let $\bar{N}_d$ denote its d-th nilpotent quotient. Then according to Labute [23], the graded Lie algebra associated to the lower central series of $\bar{\Gamma}$ is given by

$$\mathcal{L}_g := \mathcal{L}_{g,1}/(\omega_0)$$

where (ω_0) denotes the ideal generated by the symplectic class ω_0. Let

$$\mathrm{Out}^+ \bar{N}_d$$

denote the orientation preserving outer automorphism group of $\bar{N}_d$ and we define a subgroup of it by setting

$$\mathrm{Out}_0^+ \bar{N}_d = \mathrm{Image}\ (p : \mathrm{Out}^+ \bar{N}_{d+1} \rightarrow \mathrm{Out}^+ \bar{N}_d)$$

where p denotes the natural homomorphism. Then we have a short exact sequence

$$1 \longrightarrow \mathrm{IOut}_0^+ \bar{N}_d \longrightarrow \mathrm{Out}_0^+ \bar{N}_d \longrightarrow \mathrm{Sp}(2g, \mathbb{Z}) \longrightarrow 1.$$

On the other hand, there is a surjective homomorphism

$$\mathfrak{h}_{g,1} \longrightarrow \mathfrak{h}_g := \operatorname{Der} \mathcal{L}_g/\operatorname{Inn} \tag{15}$$

of Lie algebras, where Inn denotes the Lie subalgebra consisting of all the inner derivations. We can also consider truncated versions of these Lie algebras. Now it can be shown that the traces vanish on the kernel of the above homomorphism (15). Then suitably modified versions of the arguments in this paper imply the following result.

Namely, the group version of traces $\widetilde{\operatorname{trace}}(2k+1)$ give rise to a homomorphism

$$\operatorname{Out}_0^+ \bar{N}_d \longrightarrow \left((\Lambda^3 H_{\mathbb{Q}}/H_{\mathbb{Q}}) \oplus \bigoplus_{k=1}^{\ell} S^{2k+1} H_{\mathbb{Q}} \right) \rtimes \operatorname{Sp}(2g, \mathbb{Q})$$

whose induced homomorphism on $\operatorname{Out}_0^+(\bar{N}_d \otimes \mathbb{Q})$ is surjective.

In this way, we can avoid rather technical arguments related to the condition for automorphisms of free nilpotent groups or Lie algebras to preserve certain particular elements related to the symplectic class ω_0. However we treated the case of compact surfaces with boundary because of the following merits. One is that it is more suited to the group $\mathcal{H}_{g,1}$ and the other is that it is much more convenient for explicit computations.

§5. A representation of $\mathcal{H}_{g,1}$

As was already mentioned in section 2, Garoufalidis and Levine [6] (see also Habegger [8]) proved that the homomorphism

$$\tilde{\rho}_d : \mathcal{H}_{g,1} \longrightarrow \operatorname{Aut}_0 N_d$$

is *surjective* for all d. On the other hand, if we combine Proposition 3.1 with Theorem 4.1, we can conclude that there exists a homomorphism

$$\operatorname{Aut}_0 N_d \longrightarrow \left(\Lambda^3 H_{\mathbb{Q}} \oplus \bigoplus_{k=1}^{\ell} S^{2k+1} H_{\mathbb{Q}} \right) \rtimes \operatorname{Sp}(2g, \mathbb{Q})$$

whose image is Zariski dense. Since these homomorphisms are compatible with respect to d, by letting d go to the infinity, we obtain the following result.

Theorem 5.1. *There exists a homomorphism*

$$\tilde{\rho} : \mathcal{H}_{g,1} \longrightarrow \left(\Lambda^3 H_{\mathbb{Q}} \oplus \bigoplus_{k=1}^{\infty} S^{2k+1} H_{\mathbb{Q}} \right) \rtimes \operatorname{Sp}(2g, \mathbb{Q}) \tag{16}$$

whose image is Zariski dense.

Let $\mathcal{IH}_{g,1}$ denote the kernel of the homomorphism

$$\mathcal{H}_{g,1} \longrightarrow \mathrm{Sp}(2g, \mathbb{Q})$$

which is the analogue of the Torelli group in the context of the group of homology cylinders. Then the restriction of the representation $\tilde{\rho}$ to the subgroup $\mathcal{IH}_{g,1}$ gives rise to a homomorphism

$$\tilde{\rho} : \mathcal{IH}_{g,1} \longrightarrow \Lambda^3 H_{\mathbb{Q}} \bigoplus_{k=1}^{\infty} S^{2k+1} H_{\mathbb{Q}} \tag{17}$$

whose image is Zariski dense. Hence we obtain the following corollary.

Corollary 5.2. *The subgroup $\mathcal{IH}_{g,1}$ of $\mathcal{H}_{g,1}$ is not finitely generated because the rank of its abelianization is already infinite.*

Here we mention that Sakasai [37] proved that the abelianization of the IA automorphism group IAut F_n^{acy} of the acyclic closure F_n^{acy} of a free group F_n of rank $n \geq 2$ has infinite rank.

The above result shows a sharp contrast with the case of the Torelli group $\mathcal{I}_{g,1}$ which is known to be *finitely generated* by Johnson [13] for any $g \geq 3$.

Question 5.3. *Does the homomorphism $\tilde{\rho}$ in* (17) *give the abelianization of the group $\mathcal{IH}_{g,1}$ modulo torsion?*

This question should be very difficult to answer. If it would be affirmatively answered, then it would follow that

$$H_1(\mathcal{H}_{g,1}; \mathbb{Q}) = 0$$

because of the following exact sequence

$$\cdots \longrightarrow H_1(\mathcal{IH}_{g,1}; \mathbb{Q})_{\mathrm{Sp}} \longrightarrow H_1(\mathcal{H}_{g,1}; \mathbb{Q}) \longrightarrow H_1(\mathrm{Sp}(2g; \mathbb{Z}); \mathbb{Q}) = 0.$$

§6. Cohomology classes of $\mathcal{H}_{g,1}$

The homomorphism

$$\tilde{\rho} : \mathcal{H}_{g,1} \longrightarrow \left(\Lambda^3 H_{\mathbb{Q}} \bigoplus_{k=1}^{\infty} S^{2k+1} H_{\mathbb{Q}} \right) \rtimes \mathrm{Sp}(2g, \mathbb{Q})$$

given in Theorem 5.1 induces the following homomorphism in cohomology

$$(18) \quad \mathbb{Q}[c_1, c_3, \cdots] \otimes H^*\left(\Lambda^3 H_{\mathbb{Q}} \oplus \bigoplus_{k=1}^{\infty} S^{2k+1} H_{\mathbb{Q}}\right)^{\mathrm{Sp}} \longrightarrow H^*(\mathcal{H}_{g,1}; \mathbb{Q})$$

where

$$\mathbb{Q}[c_1, c_3, \cdots] = \lim_{g \to \infty} H^*(\mathrm{Sp}(2g, \mathbb{Q}); \mathbb{Q})$$

corresponds to the stable cohomology of $\mathrm{Sp}(2g, \mathbb{Q})$ determined by Borel. Thus we obtain many cohomology classes of the group $\mathcal{H}_{g,1}$ and it is an important problem to determine how non-trivial classes they are.

The mapping class group $\mathcal{M}_{g,1}$ is a subgroup of $\mathcal{H}_{g,1}$ by [6] so that we can consider the restriction of the above cohomology classes to $H^*(\mathcal{M}_{g,1}; \mathbb{Q})$. Since we know that the traces vanish identically on $\mathcal{M}_{g,1}$, it is enough to consider the following composition of homomorphisms.

$$(19) \quad \mathbb{Q}[c_1, c_3, \cdots] \otimes H^*(\Lambda^3 H_{\mathbb{Q}})^{\mathrm{Sp}} \longrightarrow H^*(\mathcal{H}_{g,1}; \mathbb{Q}) \longrightarrow H^*(\mathcal{M}_{g,1}; \mathbb{Q}).$$

It was proved in [31] that the factor $H^*(\Lambda^3 H_{\mathbb{Q}})^{\mathrm{Sp}}$ gives rise to all the Mumford-Morita-Miller classes in $H^*(\mathcal{M}_{g,1}; \mathbb{Q})$. Then by making use of Kawazumi's result in [16], an explicit formula for the above homomorphism was given in [18][19] thereby it was proved that its image is precisely the subalgebra generated by the above classes. Thus we can conclude that the Mumford-Morita-Miller classes are already defined as cohomology classes in $H^*(\mathcal{H}_{g,1}; \mathbb{Q})$.

On the other hand, the Grothendieck-Riemann-Roch theorem (or the Atiyah-Singer index theorem for families), applied to the Hodge bundle over the moduli space of curves, implies that there are close relations between the Chern classes of the Hodge bundle, which come from $\mathbb{Q}[c_1, c_3, \cdots]$ above, and the Mumford-Morita-Miller classes of *odd* indices (see [34][28]). Thus we have the following question.

Question 6.1. *Do the aforementioned relations between the Chern classes of the Hodge bundle and the odd Mumford-Morita-Miller classes, which hold at the level of $H^*(\mathcal{M}_{g,1}; \mathbb{Q})$, continue to hold in $H^*(\mathcal{H}_{g,1}; \mathbb{Q})$?*

There is a natural homomorphism

$$\mathcal{H}_{g,1} \longrightarrow \mathcal{H}_{g+1,1}$$

and all the cohomology classes defined above are *stable* with respect to g. Thus, keeping in mind Harer's stability theorem [11] as well as recent remarkable result of Madsen and Weiss [26] for the cohomology of the mapping class groups, we can ask the following.

Question 6.2. *Does the (rational) cohomology groups of $\mathcal{H}_{g,1}$ stabilize with respect to g? If so, what is the stable cohomology of $\mathcal{H}_{g,1}$?*

§7. Group of homology cobordism classes of homology 3-spheres

In this section, we consider a series of certain elements in the second cohomology of $\mathrm{Aut}_0 N_d$ and $\mathcal{H}_{g,1}$. These elements for the latter group are the simplest special cases of the construction of elements of $H^*(\mathcal{H}_{g,1};\mathbb{Q})$ given in the previous section. They are also the group version of previously defined classes given in the context of Lie algebras which we now recall.

We defined in [32][33] a series of cohomology classes

$$t_{2k+1} \in H^2(\mathfrak{h}_{g,1}^{\mathbb{Q}})$$

by setting

$$t_{2k+1} = \mathrm{trace}(2k+1)^*(\iota_{2k+1})$$

where $\mathrm{trace}(2k+1) : \mathfrak{h}_{g,1}^{\mathbb{Q}} \to S^{2k+1}H_{\mathbb{Q}}$ is the $(2k+1)$-st trace map and $\iota_{2k+1} \in H^2(S^{2k+1}H_{\mathbb{Q}})^{\mathrm{Sp}} \cong \mathbb{Q}$ is the generator.

These cohomology classes t_{2k+1} correspond to certain elements

$$\mu_k \in H_{4k}(\mathrm{Out}\, F_{2k+2};\mathbb{Q})$$

via a theorem of Kontsevich [21][22] (see also Conant and Vogtmann [1]) which relates $H^*(\mathfrak{h}_{g,1}^{\mathbb{Q}})$ with the rational homology groups of the outer automorphism groups $\mathrm{Out}\, F_n$ of free groups F_n with $n \geq 2$, or the quotient space of the Outer Space of Culler and Vogtmann [3] by $\mathrm{Out}\, F_n$. In particular, the non-triviality of t_{2k+1} is equivalent to that of μ_k. The non-triviality of t_3 was proved in [32] and that of both elements μ_1 and μ_2 was proved by Conant and Vogtmann [1] where they interpreted and extended these classes in the context of the geometry of the Outer Space.

However the non-triviality of higher classes t_{2k+1}, μ_k for $k \geq 3$ is still unknown at present.

Now we define the group version of the above classes as follows. By Theorem 4.1, we have a crossed homomorphism

$$\widetilde{\mathrm{trace}}(2k+1) : \mathrm{Aut}_0 N_d \longrightarrow S^{2k+1}H_{\mathbb{Q}} \quad (2k+1 \leq d-1).$$

Definition 7.1. We define

$$\tilde{t}_{2k+1} \in H^2(\mathrm{Aut}_0\, N_d; \mathbb{Q}) \quad (2k+1 \leq d-1)$$

by setting

$$\tilde{t}_{2k+1} = \widetilde{\mathrm{trace}}(2k+1)^*(\iota_{2k+1}).$$

Since these classes are compatible with respect to d, we may consider

$$\tilde{t}_{2k+1} \in H^2(\varprojlim_{d\to\infty} \mathrm{Aut}_0 N_d; \mathbb{Q})$$

As for the non-triviality of $\tilde{t}_{2k+1}$, we have the following result.

Proposition 7.2. *Suppose that $t_{2k+1} \in H^2(\mathfrak{h}^{\mathbb{Q}}_{g,1})$ is non-trivial. Then $\tilde{t}_{2k+1} \in H^2(\mathrm{Aut}_0\, N_d; \mathbb{Q})$ is also non-trivial. Conversely, the non-triviality of the latter class implies that of the former class.*

Proof. The class $t_{2k+1} \in H^2(\mathfrak{h}^{\mathbb{Q}}_{g,1})$ is already defined as a second cohomology class of the truncated Lie algebra $\mathfrak{h}^{\mathbb{Q}}_{g,1}[d]$ for any $d \geq 2k+2$. Hence if t_{2k+1} is non-trivial, it is so at the level of $H^2(\mathfrak{h}^{\mathbb{Q}}_{g,1}[d])$. Since $\mathfrak{h}^{\mathbb{Q}}_{g,1}[d]$ is the semi-direct product of $\mathfrak{h}^{\mathbb{Q}+}_{g,1}[d]$ with $\mathfrak{sp}(2g, \mathbb{Q})$, the above non-triviality condition is equivalent to that of the restricted element

$$t_{2k+1} \in H^2(\mathfrak{h}^{\mathbb{Q}+}_{g,1}[d])^{\mathrm{Sp}}$$

to the ideal $\mathfrak{h}^{\mathbb{Q}+}_{g,1}[d]$. On the other hand, by Theorem 3.3, $\mathfrak{h}^{\mathbb{Q}+}_{g,1}[d]$ is isomorphic to the Lie algebra of $\mathrm{IAut}_0(N_d \otimes \mathbb{Q})$ which is a nilpotent Lie group. Therefore, by a theorem of Nomizu [35], the non-triviality of the above restricted element is equivalent to that of the restricted element

$$\tilde{t}_{2k+1} \in H^2(\mathrm{IAut}_0(N_d \otimes \mathbb{Q}))^{\mathrm{Sp}}$$

which is the same as the non-triviality of the class $\tilde{t}_{2k+1} \in H^2(\mathrm{Aut}_0(N_d \otimes \mathbb{Q}))$ because $\mathrm{Aut}_0(N_d \otimes \mathbb{Q}))$ is the semi-direct product of $\mathrm{IAut}_0(N_d \otimes \mathbb{Q})$ with $\mathrm{Sp}(2g, \mathbb{Q})$. The result follows from this. Q.E.D.

Thus we know only the non-triviality of $\tilde{t}_3$ and $\tilde{t}_5$. However it seems to be reasonable to expect that all the classes would be non-trivial.

As was already mentioned in section 2, we know by Garoufalidis and Levine that the group $\Theta^3_{\mathbb{Z}}$ of homology cobordism classes of oriented homology 3-spheres is contained in their group $\mathcal{H}_{g,1}$ as a *central* subgroup. Hence we can consider the quotient group

$$\overline{\mathcal{H}}_{g,1} = \mathcal{H}_{g,1}/\Theta^3_{\mathbb{Z}}.$$

Since the representations $\tilde{\rho}_d$ are all trivial on $\Theta^3_{\mathbb{Z}}$, they induce a homomorphism

$$\bar{\rho}_\infty : \overline{\mathcal{H}}_{g,1} \longrightarrow \varprojlim_{d\to\infty} \mathrm{Aut}_0 N_d.$$

Conjecture 7.3. *All the classes*

$$\bar{\rho}^*_\infty(\tilde{t}_{2k+1}) \in H^2(\overline{\mathcal{H}}_{g,1};\mathbb{Q})$$

are non-trivial while all the classes

$$\tilde{\rho}^*_\infty(\tilde{t}_{2k+1}) \in H^2(\mathcal{H}_{g,1};\mathbb{Q})$$

are trivial.

If this conjecture would have an affirmative solution, then the 2-cocycles representing $\tilde{\rho}^*_\infty(\tilde{t}_{2k+1})$ would be coboundaries of certain 1-cochains which are functions $\mathcal{H}_{g,1}\to\mathbb{Q}$. The restriction of these functions would yield certain homomorphisms $\Theta^3_{\mathbb{Z}}\to\mathbb{Q}$. We expect that the above argument would be useful to analyze the structure of $\Theta^3_{\mathbb{Z}}$.

Recall here that Furuta [5] proved that the rank of $\Theta^3_{\mathbb{Z}}$ is infinite. See also Fintushel and Stern [4] for another proof.

Acknowledgment The author would like to express his hearty thanks to R. Hain, N. Kawazumi and T. Sakasai for useful information.

References

[1] J. Conant and K. Vogtmann, On a theorem of Kontsevich, Algebr. Geom. Topol., **3** (2003), 1167–1224.

[2] J. Conant and K. Vogtmann, Morita classes in the homology of automorphism group of free groups, Geom. Topol., **8** (2004), 1471–1499.

[3] M. Culler and K. Vogtmann, Moduli of graphs and automorphisms of free groups, Invent. Math., **84** (1986), 91–119.

[4] R. Fintushel and R. Stern, Instanton homology of Seifert fibred homology three spheres, Proc. London Math. Soc., **61** (1990), 109–137.

[5] M. Furuta, Homology cobordism group of homology 3-spheres, Invent. Math., **100** (1990), 339–355.

[6] S. Garoufalidis and J. Levine, Tree-level invariants of three-manifolds, Massey products and the Johnson homomorphism, In: Graphs and Patterns in Mathematics and Theoretical Physics, Proc. Sympos. Pure Math., **73**, Amer. Math. Soc., 2005, pp. 173–205.

[7] M. Goussarov, Finite type invariants and n-equivalence of 3-manifolds, C. R. Math. Acad. Sci. Paris, **329** (1999), 517–522.

[8] N. Habegger, Milnor, Johnson, and tree level perturbative invariants, preprint.
[9] K. Habiro, Claspers and finite type invariants of links, Geom. Topol., **4** (2000), 34–43.
[10] R. Hain, Infinitesimal presentations of the Torelli groups, J. Amer. Math. Soc., **10** (1997), 597–651.
[11] J. Harer, Stability of the homology of the mapping class group of orientable surfaces, Ann. of Math. (2), **121** (1985), 215–249.
[12] D. Johnson, An abelian quotient of the mapping class group $\mathcal{I}_g$, Math. Ann., **249** (1980), 225–242.
[13] D. Johnson, The structure of the Torelli group I: A finite set of generators for $\mathcal{I}_g$, Ann. of Math. (2), **118** (1983), 423–442.
[14] D. Johnson, A survey of the Torelli group, Contemp. Math., **20** (1983), 165–179.
[15] M. Kassabov, On the automorphism tower of free nilpotent groups, thesis, Yale Univ., 2003.
[16] N. Kawazumi, A generalization of the Morita-Mumford classes to extended mapping class groups for surfaces, Invent. Math., **131** (1998), 137–149.
[17] N. Kawazumi, Cohomological aspects of Magnus expansions, preprint, math.GT/0505497.
[18] N. Kawazumi and S. Morita, The primary approximation to the cohomology of the moduli space of curves and cocycles for the stable characteristic classes, Math. Res. Letters, **3** (1996), 629–641.
[19] N. Kawazumi and S. Morita, The primary approximation to the cohomology of the moduli space of curves and cocycles for the Mumford-Morita-Miller classes, preprint.
[20] M. Kervaire and J. Milnor, Groups of homotopy spheres, I, Ann. of Math. (2), **77** (1963), 504–537.
[21] M. Kontsevich, Formal (non-)commutative symplectic geometry, In: The Gelfand Mathematical Seminars 1990-1992, Birkhäuser Verlag, 1993, pp. 173–188.
[22] M. Kontsevich, Feynman diagrams and low-dimensional topology, In: Proceedings of the First European Congress of Mathematicians, Vol II, Paris, 1992, Progr. Math., **120**, Birkhäuser Verlag, 1994, pp. 97–121.
[23] J. Labute, On the descending central series of groups with a single defining relation, J. Algebra, **14** (1970), 16–23.
[24] J. Levine, Algebraic closure of groups, Contemp. Math., **109** (1990), 99–105.
[25] J. Levine, Homology cylinders: an enlargement of the mapping class group, Algebr. Geom. Topol., **1** (2001), 243–270.
[26] I. Madsen and M. Weiss, The stable moduli space of Riemann surfaces: Mumford's conjecture, Ann. of Math. (2), **165** (2007), 843–941.
[27] W. Magnus, A. Karrass and D. Solitar, Combinatorial Group Theory, Wiley, New York, 1966.

[28] S. Morita, Characteristic classes of surface bundles, Bull. Amer. Math. Soc., **11** (1984), 386–388.

[29] S. Morita, Abelian quotients of subgroups of the mapping class group of surfaces, Duke Math. J., **70** (1993), 699–726.

[30] S. Morita, The extension of Johnson's homomorphism from the Torelli group to the mapping class group, Invent. Math., **111** (1993), 197–224.

[31] S. Morita, A linear representation of the mapping class group of orientable surfaces and characteristic classes of surface bundles, In: Proceedings of the Taniguchi Symposium on Topology and Teichmülller Spaces, Finland, July 1995, World Scientific, 1996, pp. 159–186.

[32] S. Morita, Structure of the mapping class groups of surfaces: a survey and a prospect, In: Proceedings of the Kirbyfest, Geom. Topol. Monog., **2**, Geom. Topol. Publ., 1999, pp. 349–406.

[33] S. Morita, Cohomological structure of the mapping class group and beyond, In: Problems on Mapping Class Groups and Related Topics, (ed. Benson Farb), Proc. Sympos. Pure Math., **74**, Amer. Math. Soc., 2006, pp. 329–354.

[34] D. Mumford, Towards an enumerative geometry of the moduli space of curves, In: Arithmetic and Geometry, Progr. Math., **36**, Birkhäuser, 1983, pp. 271–328.

[35] K. Nomizu, On the cohomology of compact homogeneous spaces of nilpotent Lie groups, Ann. of Math. (2), **59** (1954), 167–189.

[36] T. Sakasai, Homology cylinders and the acyclic closure of a free group, Algebr. Geom. Topol., **6** (2006), 603–631.

[37] T. Sakasai, The Magnus representation and higher-order Alexander invariants for homology cobordisms of surfaces, preprint, math.GT/0507266.

[38] J. Stallings, Homology and central series of groups, J. Algebra, **2** (1965), 170–181.

Graduate School of Mathematical Sciences
University of Tokyo
Komaba, Tokyo 153-8914
Japan

Advanced Studies in Pure Mathematics 52, 2008
Groups of Diffeomorphisms
pp. 469–477

Mapping configuration spaces to moduli spaces

Graeme Segal and Ulrike Tillmann

§1. Introduction

Let $\mathcal{C}_n$ be the space of unordered n-tuples of distinct points in the interior of the unit disc $D = \{z \in \mathbb{C} : |z| \leq 1\}$, and let $\mathcal{M}_{g,2}$ denote the moduli space of connected Riemann surfaces of genus g with two ordered and parametrised boundary components. There is a natural map

$$\Phi : \mathcal{C}_{2g+2} \longrightarrow \mathcal{M}_{g,2};$$

it takes a subset $\underline{a} = \{a_1, \ldots, a_{2g+2}\} \subset D$ to the part $\Sigma_{\underline{a}}$ of the Riemann surface associated to the function

$$f_{\underline{a}}(z) = ((z - a_1) \ldots (z - a_{2g+2}))^{1/2}$$

which lies over the disc D.

The purpose of this paper is to describe this map in topological terms. On passing to the fundamental groups it gives a geometric definition of a well-known algebraic homomorphism $\phi : \mathfrak{Br}_{2g+2} \to \Gamma_{g,2}$ from the braid group on $2g+2$ strings to the mapping class group which is described below. The main result is that Φ is compatible with naturally defined actions of the framed little 2-discs operad on configuration spaces and moduli spaces. As immediate consequences we deduce that Φ is trivial in homology with field coefficients, and trivial on stable homology with any constant coefficient system. In particular, this proves an unstable version of a conjecture by Harer, and simplifies the proof of the triviality of the map in stable homology given in [ST], to which we refer for more background on this problem.

Received July 29, 2007.
Revised October 23, 2007.

In an appendix we discuss the map in homology given by the inclusion of the hyper-elliptic involution into the mapping class group.

§2. The induced map on fundamental groups

Let $F_{g,2}$ be a surface of genus g with two boundary circles. The moduli space $\mathcal{M}_{g,2}$ is homotopic to the classifying space of the group of homeomorphisms of $F_{g,2}$ which leave the boundary circles pointwise fixed:

$$\mathcal{M}_{g,2} \simeq B\mathrm{Homeo}^+(F_{g,2};\partial) \simeq B\Gamma_{g,2};$$

here $\Gamma_{g,2}$ is the associated mapping class group. Similarly, the configuration space $\mathcal{C}_{2g+2}$ is homotopic to the classifying space of the group of homeomorphisms of $D \setminus \underline{a}$ — the disc with $2g+2$ interior points removed — which fix the boundary of the disc pointwise but are allowed to permute the points of $\underline{a}$:

$$\mathcal{C}_{2g+2} \simeq B\mathrm{Homeo}^+(D \setminus \underline{a}) \simeq B\mathfrak{Br}_{2g+2};$$

the associated mapping class group is the braid group $\mathfrak{Br}_{2g+2}$ on $2g+2$ strings. Thought of as the fundamental group of $\mathcal{C}_{2g+2}$ based at $\underline{a}$, the group $\mathfrak{Br}_{2g+2}$ is generated by the braids σ_i, $i = 1, \dots, 2g+1$, where σ_i interchanges the points a_i and a_{i+1}.

We shall now determine the map that Φ induces on fundamental groups. Write $\pi : \Sigma_{\underline{a}} \to D$ for the double covering map branched at $a_1, \dots, a_{2g+2}$. Let D_i be a subdisc in the interior of D containing the two points a_i and a_{i+1}. Then $\pi^{-1}(D_i)$ is an annulus contained in $\Sigma_{\underline{a}}$. To see this, note that $\pi^{-1}(D_i)$ has two boundary components and that its Euler characteristic is given by

$$\chi(\pi^{-1}(D_i)) = 2\chi(D_i) - 2 = 0.$$

Proposition 2.1. *On fundamental groups, the map Φ takes σ_i to the Dehn twist corresponding to the annulus $\pi^{-1}(D_i)$ in $\Sigma_{\underline{a}}$.*

Proof. Homeomorphisms of the base of a cyclic branched covering of a surface that permute the branch points can be lifted, and this lift is unique up to covering transformations, cf. Lemma 5.1 of [BH]. The braid σ_i corresponds to a homeomorphism h_i of D supported in the subdisc

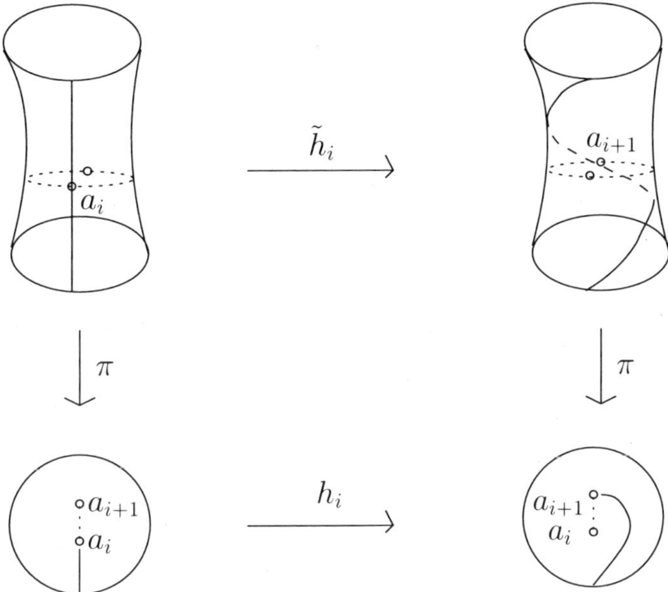

Fig. 1. Lifting h_i

D_i, and hence lifts to a homeomorphism $\tilde{h}_i$ of $\Sigma_{\underline{a}}$ supported in the annulus $\pi^{-1}(D_i)$. There is only one non-trivial covering transformation for π, given by the hyper-elliptic involution J, which interchanges the boundary components of the annulus $\pi^{-1}(D_i)$. Thus the lift $\tilde{h}_i$ is unique if we ask for the boundary circles to be fixed.

The mapping class group of an annulus is generated by the Dehn twist around its waist. By analysing the effect of $\tilde{h}_i$ on a path connecting the two boundary components of $\pi^{-1}(D_i)$, one checks the claim. See Figure 1. Q.E.D.

Thus, on fundamental groups, Φ induces the group homomorphism

$$\phi : \mathfrak{Br}_{2g+2} \to \Gamma_{g,2} \to \Gamma_g$$

where the generator σ_i is mapped to the Dehn twist around the curve α_i in Figure 2. This is the group homomorphism studied in [ST].

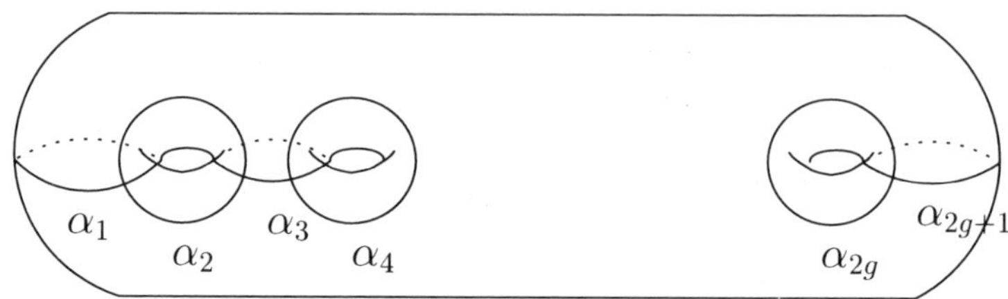

Fig. 2. The image of ϕ

§3. Action of the framed little 2-discs operad

Let $\mathcal{D}_k$ be the space of smooth embeddings of k disjoint, ordered copies of the disc D into its interior which restrict on each disc to the composition of a translation and a multiplication by an element of $\mathbb{C}^\times$. The spaces $\{\mathcal{D}_k\}$ form an operad $\mathcal{D}$, the framed little 2-discs operad, with structure maps

$$\gamma : \mathcal{D}_k \times (\mathcal{D}_{m_1} \times \cdots \times \mathcal{D}_{m_k}) \longrightarrow \mathcal{D}_{\Sigma m_i}$$

given by composition of embeddings. The operad $\mathcal{D}$ acts naturally on the configuration spaces $\mathcal{X} = \coprod_{m \geq 0} \mathcal{X}_m$ with $\mathcal{X}_m = \mathcal{C}_{2m}$. This action is defined by maps

$$\gamma_{\mathcal{X}} : \mathcal{D}_k \times (\mathcal{X}_{m_1} \times \cdots \times \mathcal{X}_{m_k}) \longrightarrow \mathcal{X}_{\Sigma m_i}$$

which take a point $(f; \underline{a}_1, \dots, \underline{a}_k)$ to the image of the union $\underline{a}_1 \cup \cdots \cup \underline{a}_k$ under the embedding f.

Now put $\mathcal{Y}_m = \mathcal{M}_{m-1,2}$ for $m \geq 1$, and $\mathcal{Y}_0 = \mathcal{M}_{0,1} \sqcup \mathcal{M}_{0,1}$. Each surface $\Sigma \in \mathcal{Y}_m$ has two ordered parametrised boundary circles, $\partial_1 \Sigma$ and $\partial_2 \Sigma$. For $f \in \mathcal{D}_k$, let $P_f = D \setminus f(D \cup \cdots \cup D)$. We think of P_f as a Riemann surface with $k+1$ parametrised boundary components. There is an operad action

$$\gamma_{\mathcal{Y}} : \mathcal{D}_k \times (\mathcal{Y}_{m_1} \times \cdots \times \mathcal{Y}_{m_k}) \longrightarrow \mathcal{Y}_{\Sigma m_i}$$

defined by

$$(f; \Sigma_1 \cup \cdots \cup \Sigma_k) \mapsto (P_f \cup P_f \cup \Sigma_1 \cup \cdots \cup \Sigma_k)/\equiv$$

where the union of the first boundary components $\cup_i \partial_1 \Sigma_i$ is identified with the interior boundary circles of the first P_f, and $\cup_i \partial_2 \Sigma_i$ is identified with the interior boundary circles of the second P_f using the parametrisations, cf. Figure 3. The resulting surface has a well-defined complex structure that restricts to the given complex structures on the subsurfaces, cf. [S].

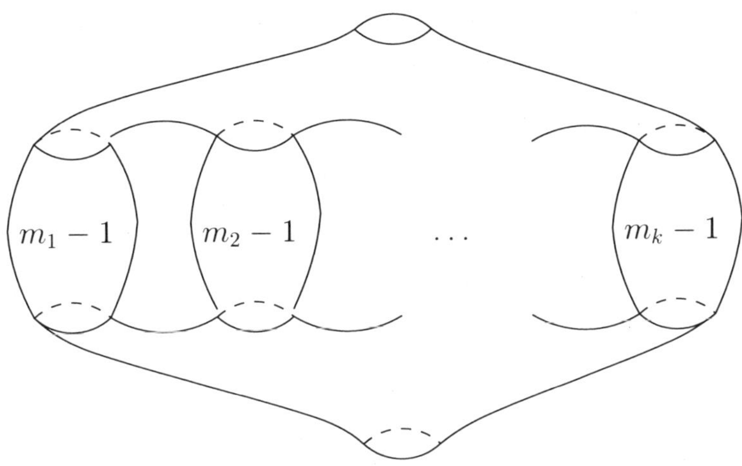

Fig. 3. $\mathcal{D}$ acting on moduli spaces

Proposition 3.1. *The map Φ defines a map $\mathcal{X} \to \mathcal{Y}$ of algebras over the framed little 2-discs operad.*

Proof. The index m was chosen so that Φ takes $\mathcal{X}_m$ to $\mathcal{Y}_m$. Both $\gamma_\mathcal{X}$ and $\gamma_\mathcal{Y}$ are defined by gluing of surfaces. The composed surface has a unique conformal structure that restricts to the given conformal structures on the subsurfaces. Furthermore, the covering map $\pi : \Sigma_{\underline{a}} \to D$ is conformal away from the set $\underline{a}$. It therefore follows that the assignment $\underline{a} \mapsto \Sigma_{\underline{a}}$ commutes with the gluing operations of $\gamma_\mathcal{X}$ and $\gamma_\mathcal{Y}$. Q.E.D.

Remark 3.2. We can also work on the level of mapping class groups and their classifying spaces. For this, replace $\mathcal{D}_k$ by the homotopy equivalent space $B\Gamma_{0,k+1}$, the classifying space of the mapping class group of D with k interior discs removed. Similarly, we also replace $\mathcal{X}_m$ by $B\mathfrak{Br}_{2m}$ and $\mathcal{Y}_m$ by $B\Gamma_{m-1,2}$, for $m > 0$. The actions of $\gamma_\mathcal{X}$ and $\gamma_\mathcal{Y}$, reinterpreted in these terms, are now induced by maps of the underlying mapping class groups. By definition they commute with Φ, or more precisely with $B\phi$. This corrects the statement and proof of 2.6 in [C].

§4. The induced map in homology

Proposition 3.1 has immediate consequences for the map that Φ induces on homology. Let p be a prime. The idea for the following is due to F. Cohen [C]. The case $p = 2$ is the (stronger) unstable version of Harer's conjecture, cf. [ST].

Corollary 4.1. *For $g > 2$ and $* > 0$,*

$$\Phi_* : H_*(\mathcal{C}_{2g+2}; \mathbb{Z}/p\mathbb{Z}) \to H_*(\mathcal{M}_{g,2}; \mathbb{Z}/p\mathbb{Z}) \text{ is zero }.$$

Proof. Recall that the first homology group of the mapping class group $\Gamma_{g,2}$, and hence that of $\mathcal{M}_{g,2}$ is zero for $g > 2$. On the other hand, $H_*(\mathcal{C}_{2g+2}, \mathbb{Z}/p\mathbb{Z})$ is generated by the first homology group under products and the operations induced from the action of $\mathcal{D}$, cf. [CLM]. The result now follows by Proposition 3.1. Q.E.D.

This implies that also the map in homology with rational coefficients (and hence with any field coefficients) is trivial as can be seen as follows. If the image of a class in rational homology is non-trivial then there must be a class in integer homology which is mapped by Φ_* to a class of infinite order. As all homology groups of the moduli spaces are finitely generated, this image class is divisible by at most finitely many primes and would therefore be non-zero in homology with coefficients in $\mathbb{Z}/p\mathbb{Z}$ for any prime p coprime to the maximal divisor. But this cannot happen by Corollary 4.1.

We now deduce the main result of [ST]. Maps of algebras over the little framed 2-discs operad group-complete to maps of double loop spaces. In particular, Φ induces a map of double loop spaces from the group-completion of $\mathcal{X}$ to that of $\mathcal{Y}$,

$$\Phi^+ : 2\mathbb{Z} \times B\mathfrak{Br}_\infty^+ \longrightarrow \mathbb{Z} \times B\Gamma_\infty^+;$$

here X^+ denotes Quillen's plus construction on X, while $\mathfrak{Br}_\infty := \lim_{k\to\infty} \mathfrak{Br}_k$ and $\Gamma_\infty := \lim_{g\to\infty} \Gamma_{g,2}$ are the infinite braid and stable mapping class groups. Note that here the zero component in $\mathbb{Z} \times B\Gamma_\infty^+$ corresponds to surfaces of Euler characteristic 2.

Corollary 4.2. *$\phi_* : H_*(\mathfrak{Br}_\infty; k) \to H_*(\Gamma_\infty; k)$ is zero for $* > 0$, and any constant coefficient system k.*

Proof. As the plus construction does not alter the homology, this follows because on connected components the map Φ^+ is homotopic to the constant map. Indeed, any double loop space map from $B\mathfrak{Br}_\infty^+ \simeq \Omega^2 S^3$ is determined by its restriction to $S^1 \subset \Omega^2\Sigma^2(S^1)$. For Φ^+ this restriction has to be homotopic to the constant map as $B\Gamma_\infty^+$ is simply connected. Q.E.D.

Recall that by the Harer-Ivanov homology stability theorem, the homology of the mapping class group is independent of the genus g and the number of boundary components in dimensions $* < g/2$. Thus Corollary 4.2 can be rephrased as follows.

Corollary 4.3. *For* $g > 2* > 0$,

$$\Phi_* : H_*(\mathcal{C}_{2g+2}; k) \to H_*(\mathcal{M}_{g,2}; k) \text{ is zero .}$$

§5. Appendix: Relation to hyper-elliptic mapping class groups

By definition Φ factors through the subspace of hyper-elliptic curves in $\mathcal{M}_{g,2}$. Note that the hyper-elliptic involution J interchanges the boundary components of $F_{g,2}$ and is only an element of the extension $\Gamma_{g,(2)}$ of $\Gamma_{g,2}$, the mapping class group associated to homeomorphisms that may exchange the boundary components as long as the parametrisations are preserved. The hyper-elliptic mapping class groups $\triangle_{g,(2)}$ and $\triangle_g$ are the commutants of J in $\Gamma_{g,(2)}$ and Γ_g. They are extensions of the braid groups of the disc D and the sphere S^2 respectively:

$$\begin{array}{ccccccccc}
0 & \longrightarrow & <J> & \longrightarrow & \triangle_{g,(2)} & \longrightarrow & \mathfrak{Br}_{2g+2} & \longrightarrow & 0 \\
 & & \downarrow{=} & & \downarrow & & \downarrow & & \\
0 & \longrightarrow & <J> & \longrightarrow & \triangle_g & \longrightarrow & \mathfrak{Br}_{2g+2}(S^2) & \longrightarrow & 0.
\end{array}$$

The kernel of the middle and right vertical maps is an extension of the free group on $2g+1$ generators by $\mathbb{Z}$ for $g > 0$. The homomorphism ϕ provides a splitting of the top row. As J maps the curves α_i, $i = 1, \ldots, 2g+1$, in Figure 2 to themselves, J commutes with the corresponding Dehn twists, cf. Lemma 4.6.7 of [B]. Hence

$$\triangle_{g,(2)} \simeq \mathfrak{Br}_{2g+2} \times <J> .$$

At the conference, R. Hain raised the question whether Corollary 4.2 can be extended to the hyper-elliptic mapping class group. This is not the case, as we shall explain now.

Consider the universal surface bundle E over $B\Gamma_g \simeq B\text{Homeo}^+(F_g)$ and the associated pull-back bundle $incl^*(E)$ over $B(< J >) = BC_2$. The vertical tangent bundle on E is classified by a map $\omega : E \to \mathbb{C}P^\infty$. Consider the composition

$$\theta : BC_2 \xrightarrow{incl} B\Gamma_g \xrightarrow{tr} Q(E_+) \xrightarrow{\omega} Q(\mathbb{C}P^\infty_+) = Q(\mathbb{C}P^\infty) \times Q(S^0);$$

here tr denotes the Becker-Gottlieb transfer map of the bundle E, while X_+ stands for the union of X with a disjoint base point, and $Q = \lim_{n\to\infty} \Omega^n S^n$ is the free infinite loop space functor. A straightforward computation of transfer maps, cf. Lemma 5.2 of [GMT], gives

$$\theta \simeq (2g+2)\psi + (-2g)\hat{t},$$

where

$$\psi : BC_2 \longrightarrow \mathbb{C}P^\infty \longrightarrow Q(\mathbb{C}P^\infty)$$

is induced by the inclusion $C_2 \to S^1$, and $\hat{t}$ is the transfer associated to the universal bundle $EC_2 \to BC_2$. The map ψ is non-trivial in homology with $\mathbb{Z}/2\mathbb{Z}$-coefficients, and hence so is *incl*. We have proved

Proposition 5.1. *The inclusion incl* $:< J >= C_2 \to \Gamma_g$ *is non-trivial on stable homology with* $\mathbb{Z}/2\mathbb{Z}$*-coefficients.*

References

[B] J. Birman, Braids, Links, and Mapping Class Groups, Ann. of Math. Stud., **82**, Princeton Univ. Press, 1975.

[BH] J. Birman and H. Hilden, On isotopies of homeomorphisms of Riemann surfaces, Ann. of Math. (2), **97** (1973), 424–439.

[C] F. R. Cohen, Homology of mapping class groups for surfaces of low genus, The Lefshetz Centennial conference, Part II, Contemp. Math., **58** II, Amer. Math. Soc., 1987, pp. 21–30.

[CLM] F. R. Cohen, T. J. Lada and J. P. May, The homology of iterated loop spaces, Lecture Notes in Math., **533**, Springer-Verlag, 1976.

[GMT] S. Galatius, I. Madsen and U. Tillmann, Divisibility of the stable Miller-Morita-Mumford classes, J. Amer. Math. Soc., **19** (2006), 759–779.

[S] G. Segal, The Definition of Conformal Field Theory, In: Topology, Geometry and Quantum Field Theory, (ed. U. Tillmann), London Math. Soc. Lecture Note Ser., **308**, Cambridge Univ. Press, 2004.

[ST] Y. Song and U. Tillmann, Braids, mapping class groups, and categorical delooping, Math. Ann., **339** (2007), 377–393.

Graeme Segal
All Souls College
Oxford University
Oxford OX1 4AL, UK
E-mail address: graeme.segal@all-souls.ox.ac.uk

Ulrike Tillmann
Mathematical Institute
Oxford University
Oxford OX1 3LB, UK
E-mail address: tillmann@maths.ox.ac.uk

Advanced Studies in Pure Mathematics 52, 2008
Groups of Diffeomorphisms
pp. 479–490

New examples of elements in the kernel of the Magnus representation of the Torelli group

Masaaki Suzuki

Abstract.

From our previous paper, it is known that the Magnus representation of the Torelli group is not faithful. In this paper, we show a certain kind of elements in the kernel of the representation.

§1. Introduction

Let Σ_g be an oriented closed surface of genus g and $\Sigma_{g,1}$ a compact surface removed an open disk from Σ_g. We denote by $\mathcal{M}_{g,1}$ the mapping class group of $\Sigma_{g,1}$ relative to the boundary, that is the group of isotopy classes of orientation preserving diffeomorphisms of $\Sigma_{g,1}$ which restrict to the identity on the boundary. Let $\mathcal{I}_{g,1}$ be the Torelli group of $\Sigma_{g,1}$, that is, the normal subgroup of $\mathcal{M}_{g,1}$ consisting of all the elements which act trivially on the first homology group of $\Sigma_{g,1}$.

A group is said to be linear, if it admits a finite dimensional faithful representation. In [6] and [1], it is proved that the mapping class group of a closed surface of genus two is linear. However, the linearity of the mapping class group of a surface of higher genera is still not known.

It is shown in [13] that the Magnus representation of the Torelli group

$$r_1 : \mathcal{I}_{g,1} \to GL(2g; \mathbb{Z}[H])$$

is not faithful for $g \geq 2$, where $H = H_1(\Sigma_{g,1}; \mathbb{Z})$. Then this representation cannot determine the linearity of the Torelli group $\mathcal{I}_{g,1}$. In [16], we characterized which commutators of two BSCC maps lie in the kernel of r_1, where the Dehn twist along a bounding simple closed curve is called BSCC map.

Received April 30, 2007.
Revised January 10, 2008.
The research is supported in part by the Grant-in-Aid for Scientific Research (No. 18840008), the Ministry of Education, Culture, Sports, Science and Technology, Japan.

Let $\mathcal{K}_{g,1}$ be the normal subgroup of $\mathcal{I}_{g,1}$ generated by all BSCC maps. Morita constructed a homomorphism

$$d_1 : \mathcal{K}_{g,1} \longrightarrow \mathbb{Z},$$

which is a secondary characteristic class of surface bundles. See [8], [10], [11] and [12] for details. The following open problem is mentioned in [12]: determine whether the Magnus representation of the Torelli group detects d_1 or not. In other words, does there exist some $\varphi \in \ker r_1$ so that $d_1(\varphi) \neq 0$.

In order to investigate the above problem, we will present new examples of elements in the kernel of the Magnus representation of the Torelli group in this paper. In particular, we will describe a relation between the kernel of $\mathcal{I}_{g,1} \to \mathcal{I}_g$, which is denoted by G, and the kernel of r_1, where $\mathcal{I}_g$ is the Torelli group of Σ_g. More precisely, the following is proved in this paper:

Main Theorem . (1) *Let ψ be a BSCC map and $\tilde{\gamma} \in G$. Then*

$$[\psi, \tilde{\gamma}\psi\tilde{\gamma}^{-1}] \in \ker r_1.$$

(2) *The 3-rd derived subgroup of G is contained in the kernel of r_1.*
(3) *Let φ_1, φ_2 be elements of the commutator subgroup of G. Then*

$$[\varphi_1, \varphi_2][\varphi_1, \varphi_2^{-1}] \in \ker r_1.$$

However, the value of d_1 for each element in the kernel of r_1 which is shown in this paper is zero. Then these elements are not useful to answer the above problem of Morita. We note that all the elements mentioned in [13], [16] map to zero by d_1.

In Section 2, we review the definition and the irreducible decomposition of the Magnus representation of the Torelli group. In Section 3, we study some properties and examples of the Magnus representation, which are needed to prove the main results. In Section 4, the main results of this paper are shown, that is, we present a certain kind of elements in the kernel of r_1, which do not appear in the previous papers [13], [16].

§2. Definition and irreducible decomposition of the Magnus representation of the Torelli group

In this section, we recall the definition of the Magnus representation of the Torelli group and its irreducible decomposition from [9], [13] and [14].

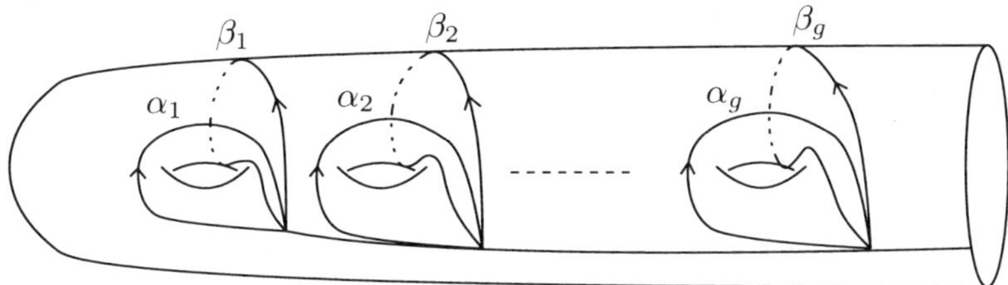

Fig. 1. Generators of Γ_0

We fix a system of generators $\alpha_1, \ldots, \alpha_g, \beta_1, \ldots, \beta_g$ of the free group $\Gamma_0 = \pi_1(\Sigma_{g,1})$ as shown in Figure 1. Let us simply write $\gamma_1, \ldots, \gamma_{2g}$ for them.

Definition 2.1. We call the mapping

$$\begin{array}{rccc} r: & \mathcal{M}_{g,1} & \longrightarrow & GL(2g;\mathbb{Z}[\Gamma_0]) \\ & \varphi & \longmapsto & \left(\overline{\frac{\partial \varphi(\gamma_j)}{\partial \gamma_i}}\right)_{i,j} \end{array}$$

the Magnus representation of the mapping class group, where $\frac{\partial}{\partial \gamma_i}$: $\mathbb{Z}[\Gamma_0] \to \mathbb{Z}[\Gamma_0]$ is the Fox derivation of the integral group ring $\mathbb{Z}[\Gamma_0]$ and $^- : \mathbb{Z}[\Gamma_0] \to \mathbb{Z}[\Gamma_0]$ is the antiautomorphism induced by the mapping $\gamma \mapsto \gamma^{-1}$.

The mapping r is not a homomorphism but a crossed homomorphism. In other words, we have the following.

Proposition 2.2 (Morita [9]). *For any elements $\varphi, \psi \in \mathcal{M}_{g,1}$, we have*

$$r(\varphi\psi) = r(\varphi) \cdot {}^{\varphi}r(\psi)$$

where ${}^{\varphi}r(\psi)$ denotes the matrix obtained from $r(\psi)$ by applying the automorphism $\varphi : \mathbb{Z}[\Gamma_0] \to \mathbb{Z}[\Gamma_0]$ on each entry.

Since the Torelli group acts trivially on H, we obtain a genuine representation of the Torelli group as follows. We restrict r to the Torelli group $\mathcal{I}_{g,1}$ and reduce the coefficients to $\mathbb{Z}[H]$:

$$r_1 : \mathcal{I}_{g,1} \longrightarrow GL(2g;\mathbb{Z}[H]).$$

That is to say, this homomorphism r_1 is the composite $r^{\mathfrak{a}}$ of the mapping r by the abelianization $\mathfrak{a} : \Gamma_0 \to H$. We call r_1 the Magnus representation of the Torelli group.

The Magnus representation of the Torelli group is reducible, but not completely reducible. More precisely, we have the following irreducible decomposition.

Theorem 2.3 ([14]). *For $g \geq 2$ there exists a non-singular matrix $P \in GL(2g; R)$ such that for any element $\varphi \in \mathcal{I}_{g,1}$*

$$P^{-1}\, r_1(\varphi)\, P = \left(\begin{array}{c|ccc|c} 1 & & \rho_{b_1}(\varphi) & & \rho_{b_3}(\varphi) \\ \hline 0 & & & & \\ \vdots & & \rho_B(\varphi) & & \rho_{b_2}(\varphi) \\ & & & & \\ \hline 0 & 0 & \cdots & 0 & 1 \end{array}\right).$$

Moreover, ρ_B is a $(2g-2)$-dimensional irreducible representation of $\mathcal{I}_{g,1}$. Here $R = \mathbb{Z}[{x_i}^{\pm 1}, {y_i}^{\pm 1}, \frac{1}{1-y_i}]$ $(\supset \mathbb{Z}[H])$ where x_i, y_i $(i = 1, \ldots, g)$ are obtained by abelianizing α_i, β_i respectively.

Though the actual value of P is slightly complicated, we express it explicitly according to [14]. That is to say,

$$P = \begin{pmatrix} P_{11} & 0 \\ P_{21} & I_g \end{pmatrix} \cdot \left(I_1 \oplus \left(\begin{pmatrix} 0 & -{}^tP_{11} \\ I_{g-1} & 0 \end{pmatrix}^{-1} {Q_2}^{-1} \right) \right)$$

where I_k is the identity matrix of degree k and

$$P_{11} = -\begin{pmatrix} 1-{y_1}^{-1} & & & 0 \\ 1-{y_2}^{-1} & 1-{y_2}^{-1} & & \\ \vdots & \vdots & \ddots & \\ 1-{y_g}^{-1} & 1-{y_g}^{-1} & \cdots & 1-{y_g}^{-1} \end{pmatrix}$$

$$P_{21} = \begin{pmatrix} 1-{x_1}^{-1} & & & 0 \\ 1-{x_2}^{-1} & 1-{x_2}^{-1} & & \\ \vdots & \vdots & \ddots & \\ 1-{x_g}^{-1} & 1-{x_g}^{-1} & \cdots & 1-{x_g}^{-1} \end{pmatrix}$$

$$Q_2 = \begin{pmatrix} 0 & 1 & & \\ & \ddots & \ddots & \\ & & \ddots & 1 \\ 1 & & & 0 \end{pmatrix}.$$

In this paper, we write $r_1'(\varphi)$ for $P^{-1}r_1(\varphi)P$ for simplicity.

§3. Some properties of $\rho_B, \rho_{b_1}, \rho_{b_2}, \rho_{b_3}$

In this section, we show the product formulas of $\rho_B, \rho_{b_1}, \rho_{b_2}, \rho_{b_3}$ and calculate them for elements of the kernel of the projection $\mathcal{I}_{g,1} \to \mathcal{I}_g$, where $\mathcal{I}_g$ is the Torelli group of Σ_g.

Lemma 3.1. *For any elements $\varphi_1, \varphi_2 \in \mathcal{I}_{g,1}$, we have*

(1) $\rho_B(\varphi_1\varphi_2) = \rho_B(\varphi_1)\rho_B(\varphi_2)$,
(2) $\rho_{b_1}(\varphi_1\varphi_2) = \rho_{b_1}(\varphi_2) + \rho_{b_1}(\varphi_1)\rho_B(\varphi_2)$,
(3) $\rho_{b_2}(\varphi_1\varphi_2) = \rho_B(\varphi_1)\rho_{b_2}(\varphi_2) + \rho_{b_2}(\varphi_1)$,
(4) $\rho_{b_3}(\varphi_1\varphi_2) = \rho_{b_3}(\varphi_2) + \rho_{b_1}(\varphi_1)\rho_{b_2}(\varphi_2) + \rho_{b_3}(\varphi_1)$.

Proof. By $r'_1(\varphi_1\varphi_2) = r'_1(\varphi_1)r'_1(\varphi_2)$, we get

$$\left(\begin{array}{c|c|c} 1 & \rho_{b_1}(\varphi_1\varphi_2) & \rho_{b_3}(\varphi_1\varphi_2) \\ \hline 0 & & \\ \vdots & \rho_B(\varphi_1\varphi_2) & \rho_{b_2}(\varphi_1\varphi_2) \\ 0 & & \\ \hline 0 & 0 \cdots 0 & 1 \end{array}\right) =$$

$$\left(\begin{array}{c|c|c} 1 & \rho_{b_1}(\varphi_1) & \rho_{b_3}(\varphi_1) \\ \hline 0 & & \\ \vdots & \rho_B(\varphi_1) & \rho_{b_2}(\varphi_1) \\ 0 & & \\ \hline 0 & 0 \cdots 0 & 1 \end{array}\right) \left(\begin{array}{c|c|c} 1 & \rho_{b_1}(\varphi_2) & \rho_{b_3}(\varphi_2) \\ \hline 0 & & \\ \vdots & \rho_B(\varphi_2) & \rho_{b_2}(\varphi_2) \\ 0 & & \\ \hline 0 & 0 \cdots 0 & 1 \end{array}\right).$$

Q.E.D.

Next, we consider the kernel of the projection $\mathcal{I}_{g,1} \to \mathcal{I}_g$. The kernel of $\mathcal{I}_{g,1} \to \mathcal{I}_g$, denoted by G, is isomorphic to the fundamental group $\pi_1(T_1\Sigma_g)$, where $T_1\Sigma_g$ is the unit tangent bundle of Σ_g. It is known that G is generated by $\tau_\zeta, \tilde{\alpha}_i, \tilde{\beta}_i$, $(i = 1, \ldots, g)$, see [4], [7] for details. These generators $\tau_\zeta, \tilde{\alpha}_i, \tilde{\beta}_i$ are given as follows. Let $\alpha_{i+}, \alpha_{i-}, \beta_{i+}, \beta_{i-}, \zeta$ be the curves in Figure 2, 3 and 4. Then $\tilde{\alpha}_i \in \mathcal{I}_{g,1}$ (respectively $\tilde{\beta}_i \in \mathcal{I}_{g,1}$), which is a lift of α_i (respectively β_i) to $\pi_1(T_1\Sigma_g)$, is a product of the Dehn twist along α_{i+} (respectively β_{i+}) and the inverse of the Dehn twist along α_{i-} (respectively β_{i-}). Besides, τ_ζ is the Dehn twist along a curve ζ which is parallel to the boundary of $\Sigma_{g,1}$.

Lemma 3.2. *For any element $\tilde{\gamma} \in G = \ker(\mathcal{I}_{g,1} \to \mathcal{I}_g)$, we have*

$$\rho_B(\tilde{\gamma}) = [\gamma]^{-1}I_{2g-2}, \quad \rho_{b_1}(\tilde{\gamma}^{-1}) = -[\gamma]\rho_{b_1}(\tilde{\gamma}), \quad \rho_{b_2}(\tilde{\gamma}^{-1}) = -[\gamma]\rho_{b_2}(\tilde{\gamma}).$$

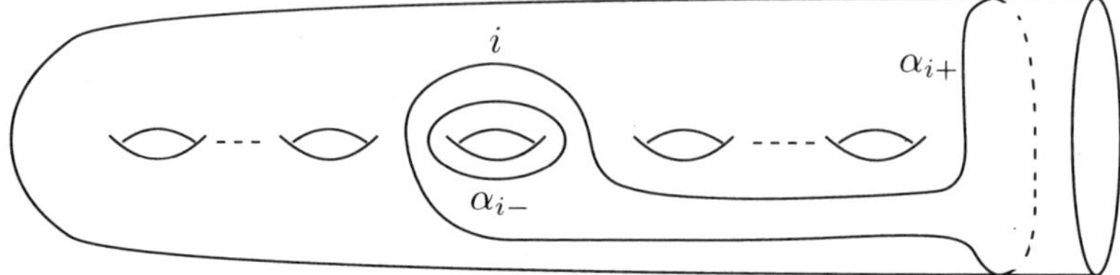

Fig. 2. Simple closed curves α_{i+} and α_{i-}

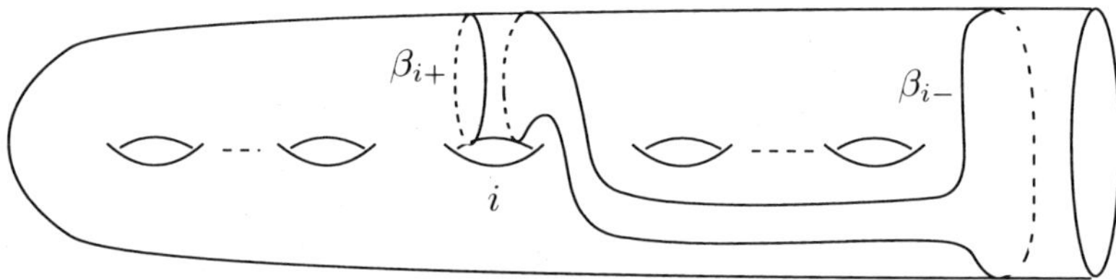

Fig. 3. Simple closed curves β_{i+} and β_{i-}

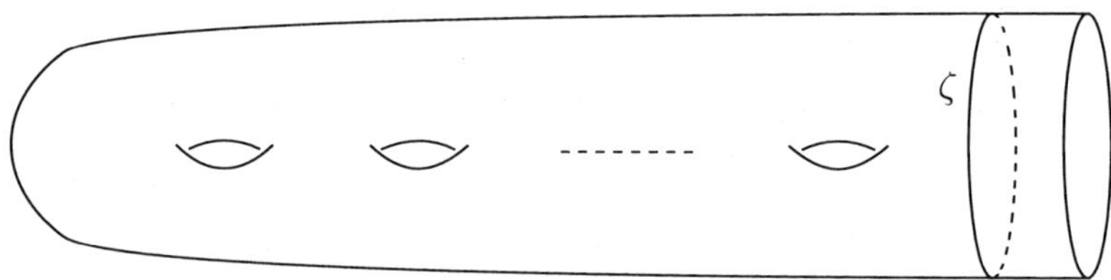

Fig. 4. Simple closed curve ζ

where $[\gamma]$ represents the homology class of γ whose lift to $\pi_1(T_1\Sigma_g)$ is $\tilde{\gamma}$. In particular, if φ belongs to the commutator subgroup of G, then

$$\rho_B(\varphi) = I_{2g-2}, \quad \rho_{b_1}(\varphi^{-1}) = -\rho_{b_1}(\varphi), \quad \rho_{b_2}(\varphi^{-1}) = -\rho_{b_2}(\varphi).$$

Proof. By direct calculations, we have

$$\tilde{\beta_1}(\alpha_j) = \begin{cases} [\alpha_1, \beta_1] \cdots [\alpha_g, \beta_g] \beta_1 \alpha_1 \bar{\beta}_1 & j = 1 \\ \beta_1 [\alpha_1, \beta_1] \cdots [\alpha_g, \beta_g] \alpha_j [\beta_g, \alpha_g] \cdots [\beta_1, \alpha_1] \beta_1 & 2 \leq j \leq g \end{cases},$$

$$\tilde{\beta_1}(\beta_j) = \begin{cases} \beta_1 & j = 1 \\ \beta_1 [\alpha_1, \beta_1] \cdots [\alpha_g, \beta_g] \beta_j [\beta_g, \alpha_g] \cdots [\beta_1, \alpha_1] \beta_1 & 2 \leq j \leq g \end{cases}.$$

Then the Magnus matrix $r_1(\tilde{\beta_1})$ is given by the following equations:

$$\mathfrak{a}\left(\overline{\frac{\partial \tilde{\beta_1}(\alpha_j)}{\partial \alpha_i}}\right) = \begin{cases} 1 - \bar{y}_i + \delta_{1,i} \bar{y}_1 & j = 1 \\ \delta_{j,i} \bar{y}_1 + \bar{y}_1 (1 - \bar{x}_j)(1 - \bar{y}_i) & 2 \leq j \leq g \end{cases}$$

$$\mathfrak{a}\left(\overline{\frac{\partial \tilde{\beta_1}(\beta_j)}{\partial \alpha_i}}\right) = \begin{cases} 0 & j = 1 \\ \bar{y}_1 (1 - \bar{y}_i)(1 - \bar{y}_j) & 2 \leq j \leq g \end{cases}$$

$$\mathfrak{a}\left(\overline{\frac{\partial \tilde{\beta_1}(\alpha_j)}{\partial \beta_i}}\right) = \begin{cases} \delta_{1,i}(1 - \bar{x}_1) - (1 - \bar{x}_i) & j = 1 \\ \delta_{1,i}(1 - \bar{x}_j) - \bar{y}_1 (1 - \bar{x}_i)(1 - \bar{x}_j) & 2 \leq j \leq g \end{cases}$$

$$\mathfrak{a}\left(\overline{\frac{\partial \tilde{\beta_1}(\beta_j)}{\partial \beta_i}}\right) =$$
$$\begin{cases} \delta_{1,i} & j = 1 \\ \delta_{j,i} \bar{y}_1 + \delta_{1,i}(1 - \bar{y}_j) - \bar{y}_1 (1 - \bar{x}_i)(1 - \bar{y}_j) & 2 \leq j \leq g \end{cases}.$$

Here $\bar{x}_i = x^{-1}, \bar{y}_i = y^{-1}$. Moreover, the conjugation this matrix by P is described as

$$\rho_B(\tilde{\beta_1}) = \bar{y}_1 I_{2g-2}, \quad \rho_{b_1}(\tilde{\beta_1}) = (-\bar{y}_1, \underbrace{0, \ldots, 0}_{2g-3}),$$

$$\rho_{b_2}(\tilde{\beta_1}) = {}^t(\underbrace{0, \ldots, 0}_{2g-2}), \quad \rho_{b_3}(\tilde{\beta_1}) = 0.$$

Similarly, we can compute the other matrices $\rho_B(\tilde{\alpha_i})$ and $\rho_B(\tilde{\beta_i})$:

$$\rho_B(\tilde{\alpha_i}) = \bar{x}_i I_{2g-2}, \quad \rho_B(\tilde{\beta_i}) = \bar{y}_i I_{2g-2}.$$

In addition, more simple computations show

$$\rho_B(\tau_\zeta) = I_{2g-2}.$$

These proved the first equation of the statement. Furthermore, the first equation and Lemma 3.1 follow that

$$\begin{aligned}0 &= \rho_{b_1}(\tilde{\gamma}^{-1}\tilde{\gamma}) = \rho_{b_1}(\tilde{\gamma}) + [\gamma]^{-1}\rho_{b_1}(\tilde{\gamma}^{-1})\\ 0 &= \rho_{b_2}(\tilde{\gamma}\tilde{\gamma}^{-1}) = [\gamma]^{-1}\rho_{b_2}(\tilde{\gamma}^{-1}) + \rho_{b_2}(\tilde{\gamma})\quad .\end{aligned}$$

This completes the proof. Q.E.D.

Lemma 3.3. *Let ψ be a BSCC map, that is, the Dehn twist along a 0-homologous simple closed curve. Then we obtain*

(1) $\rho_{b_1}(\psi) = \rho_{b_1}(\psi)\rho_B(\psi)$,
(2) $\rho_{b_2}(\psi) = \rho_B(\psi)\rho_{b_2}(\psi)$,
(3) $\rho_{b_1}(\psi)\rho_{b_2}(\psi) = 0$,
(4) $(\rho_B(\psi) - I_{2g-2})^2 = 0$.

Proof. It is shown in [15] that the Magnus matrix $r_1(\psi)$ of a BSCC map ψ can be written as

$$r_1(\psi) = I_{2g} + uv$$

where $u \in \mathbb{Z}[H]^{2g}, v \in {}^t\mathbb{Z}[H]^{2g}$ and $vu = 0$. Then we get

$$(1)\quad (r_1'(\psi) - I_{2g})^2 = (P^{-1}r_1(\psi)P - I_{2g})^2 = P^{-1}uvPP^{-1}uvP = O.$$

On the other hand, by Theorem 2.3 the following equation holds:

$$\begin{aligned}&(r_1'(\psi) - I_{2g})^2\\ &= \left(\begin{array}{c|c|c} 0 & \rho_{b_1}(\psi) & \rho_{b_3}(\psi) \\ \hline 0 & & \\ \vdots & \rho_B(\psi) - I & \rho_{b_2}(\psi) \\ \cline{2-3} 0 & 0\cdots 0 & 0 \end{array}\right)\left(\begin{array}{c|c|c} 0 & \rho_{b_1}(\psi) & \rho_{b_3}(\psi) \\ \hline 0 & & \\ \vdots & \rho_B(\psi) - I & \rho_{b_2}(\psi) \\ \cline{2-3} 0 & 0\cdots 0 & 0 \end{array}\right)\\ &= \left(\begin{array}{c|c|c} 0 & \rho_{b_1}(\psi)\rho_B(\psi) - \rho_{b_1}(\psi) & \rho_{b_1}(\psi)\rho_{b_2}(\psi) \\ \hline 0 & & \\ \vdots & (\rho_B(\psi) - I)^2 & \rho_B(\psi)\rho_{b_2}(\psi) - \rho_{b_2}(\psi) \\ \cline{2-3} 0 & 0\cdots 0 & 0 \end{array}\right).\end{aligned}$$

By comparing (1) with the right hand side of the above equation, we arrive at the statements of Lemma 3.3. Q.E.D.

§4. Main Theorem

In this section, we present several types of elements in the kernel of the Magnus representation of the Torelli group.

Theorem 4.1. *Let ψ be a BSCC map and $\tilde{\gamma} \in G = \ker(\mathcal{I}_{g,1} \to \mathcal{I}_g)$. Then*

$$[\psi, \tilde{\gamma}\psi\tilde{\gamma}^{-1}] \in \ker r_1.$$

Proof. If ψ' denotes $\tilde{\gamma}\psi\tilde{\gamma}^{-1}$, it is sufficient to prove $r_1'(\psi\psi') = r_1'(\psi'\psi)$. Thus we will show that

(1) $\rho_B(\psi\psi') - \rho_B(\psi'\psi) = 0$,
(2) $\rho_{b_1}(\psi\psi') - \rho_{b_1}(\psi'\psi) = 0$,
(3) $\rho_{b_2}(\psi\psi') - \rho_{b_2}(\psi'\psi) = 0$,
(4) $\rho_{b_3}(\psi\psi') - \rho_{b_3}(\psi'\psi) = 0$.

First, Lemma 3.2 says that $\rho_B(\tilde{\gamma})$ is central, so we have

$$\rho_B(\tilde{\gamma})\rho_B(\psi)\rho_B(\tilde{\gamma})^{-1} = \rho_B(\psi). \tag{2}$$

Then we obtain the following by the above equation.

$$\begin{aligned}
&\rho_B(\psi\psi') - \rho_B(\psi'\psi)\\
&\quad = \rho_B(\psi)\rho_B(\tilde{\gamma})\rho_B(\psi)\rho_B(\tilde{\gamma})^{-1} - \rho_B(\tilde{\gamma})\rho_B(\psi)\rho_B(\tilde{\gamma})^{-1}\rho_B(\psi)\\
&\quad = 0 \quad \text{by (2)}.
\end{aligned}$$

Next, we calculate ρ_{b_1}:

$$\begin{aligned}
&\rho_{b_1}(\psi\psi') - \rho_{b_1}(\psi'\psi)\\
&\quad = \rho_{b_1}(\psi') + \rho_{b_1}(\psi)\rho_B(\psi') - \rho_{b_1}(\psi) - \rho_{b_1}(\psi')\rho_B(\psi)\\
&\qquad\qquad \text{by Lemma 3.1 (2)}\\
&\quad = -[\gamma]\rho_{b_1}(\tilde{\gamma}) + [\gamma]\rho_{b_1}(\psi) + [\gamma]\rho_{b_1}(\tilde{\gamma})\rho_B(\psi) + \rho_{b_1}(\psi)\rho_B(\psi)\\
&\qquad -\rho_{b_1}(\psi) + \{[\gamma]\rho_{b_1}(\tilde{\gamma}) - [\gamma]\rho_{b_1}(\psi) - [\gamma]\rho_{b_1}(\tilde{\gamma})\rho_B(\psi)\}\rho_B(\psi)\\
&\qquad\qquad \text{by Lemma 3.2}\\
&\quad = -[\gamma]\rho_{b_1}(\tilde{\gamma})\{I - 2\rho_B(\psi) + \rho_B(\psi)^2\} \quad \text{by Lemma 3.3}\\
&\quad = -[\gamma]\rho_{b_1}(\tilde{\gamma})(I - \rho_B(\psi))^2\\
&\quad = 0 \quad \text{by Lemma 3.3}
\end{aligned}$$

Similarly, we get $\rho_{b_2}(\psi\psi') - \rho_{b_2}(\psi'\psi) = 0$. Finally, we obtain

$$
\begin{aligned}
&\rho_{b_3}(\psi\psi') - \rho_{b_3}(\psi'\psi) \\
&= \rho_{b_3}(\psi') + \rho_{b_1}(\psi)\rho_{b_2}(\psi') + \rho_{b_3}(\psi) \\
&\quad -\rho_{b_3}(\psi) - \rho_{b_1}(\psi')\rho_{b_2}(\psi) - \rho_{b_3}(\psi') \quad \text{by Lemma 3.1 (4)} \\
&= \rho_{b_1}(\psi)\{-\rho_B(\psi)\rho_{b_2}(\tilde{\gamma}) + [\gamma]^{-1}\rho_{b_2}(\psi) + \rho_{b_2}(\tilde{\gamma})\} \\
&\quad -\{-[\gamma]\rho_{b_1}(\tilde{\gamma}) + [\gamma]\rho_{b_1}(\psi) + [\gamma]\rho_{b_1}(\tilde{\gamma})\rho_B(\psi)\}\rho_{b_2}(\psi) \\
&\qquad\qquad \text{by Lemma 3.1 and 3.2} \\
&= 0 \quad \text{by Lemma 3.3.}
\end{aligned}
$$

This completes the proof. Q.E.D.

Suppose φ be an element of $\mathcal{I}_{g,1}$ such that $\rho_B(\varphi) = I_{2g-2}$, $\rho_{b_1}(\varphi) = 0$ and $\rho_{b_2}(\varphi) = 0$. Then the Magnus matrix $r_1'(\varphi)$ commutes with that of any element of $\mathcal{I}_{g,1}$. This means that $[\varphi, \mathcal{I}_{g,1}] \subset \ker r_1$. One of such elements is τ_ζ. We show other elements satisfying these properties.

We denote by $G^{(k)}$ the k-th term in the derived series of G so that $G^{(0)} = G$ and $G^{(k+1)} = [G^{(k)}, G^{(k)}]$.

Theorem 4.2. *If φ belongs to $G^{(2)}$, then $\rho_B(\varphi) = I_{2g-2}$, $\rho_{b_1}(\varphi) = 0$ and $\rho_{b_2}(\varphi) = 0$.*

Proof. Let g_1, g_2, g_3, g_4 be elements of G. It is sufficient to prove the statement for the case $\varphi = [[g_1, g_2], [g_3, g_4]]$. By Lemma 3.2, for $g, g' \in G$, $\rho_B([g, g'])$ is the identity matrix. Then we get $\rho_B(\varphi) = I_{2g-2}$. Moreover, $\rho_{b_1}(\varphi)$ is computed as

$$
\begin{aligned}
\rho_{b_1}(\varphi) &= \rho_{b_1}([g_1, g_2][g_3, g_4][g_2, g_1][g_4, g_3]) \\
&= \rho_{b_1}([g_1, g_2]) + \rho_{b_1}([g_3, g_4]) + \rho_{b_1}([g_2, g_1]) + \rho_{b_1}([g_4, g_3]) \\
&\qquad\qquad \text{by Lemma 3.1 and 3.2} \\
&= \rho_{b_1}([g_1, g_2]) + \rho_{b_1}([g_3, g_4]) - \rho_{b_1}([g_1, g_2]) - \rho_{b_1}([g_3, g_4]) \\
&\qquad\qquad \text{by Lemma 3.2} \\
&= 0.
\end{aligned}
$$

A similar calculation follows that $\rho_{b_2}(\varphi) = 0$. This completes the proof. Q.E.D.

Corollary 4.3. *The commutator $[G^{(2)}, \mathcal{I}_{g,1}]$ is contained in* $\ker r_1$. *In particular,*

$$G^{(3)} \subset \ker r_1.$$

Proof. This follows directly from Theorem 4.2 and the argument before Theorem 4.2. Q.E.D.

Corollary 4.4. *Let φ_1, φ_2 be elements of $G^{(1)}$. Then*

$$[\varphi_1, \varphi_2][\varphi_1, \varphi_2^{-1}] \in \ker r_1.$$

Proof. Since $[\varphi_1, \varphi_2]$ and $[\varphi_1, \varphi_2^{-1}]$ belong to $G^{(2)}$, it is sufficient to show

$$\rho_{b_3}([\varphi_1, \varphi_2][\varphi_1, \varphi_2^{-1}]) = 0.$$

We recall that $\rho_{b_1}([\varphi_1, \varphi_2]) = 0$, $\rho_{b_2}([\varphi_1, \varphi_2^{-1}]) = 0$, $\rho_B(\varphi_i) = I_{2g-2}$, $\rho_{b_1}(\varphi_i^{-1}) = -\rho_{b_1}(\varphi_i)$, $\rho_{b_2}(\varphi_i^{-1}) = -\rho_{b_2}(\varphi_i)$ by Lemma 3.2.

$$\begin{aligned}
&\rho_{b_3}([\varphi_1, \varphi_2][\varphi_1, \varphi_2^{-1}]) \\
&\quad= \rho_{b_3}([\varphi_1, \varphi_2]) + \rho_{b_3}([\varphi_1, \varphi_2^{-1}]) + \rho_{b_1}([\varphi_1, \varphi_2])\rho_{b_2}([\varphi_1, \varphi_2^{-1}]) \\
&\quad= \rho_{b_3}(\varphi_1\varphi_2) + \rho_{b_3}(\varphi_1^{-1}\varphi_2^{-1}) + \rho_{b_1}(\varphi_1\varphi_2)\rho_{b_2}(\varphi_1^{-1}\varphi_2^{-1}) \\
&\qquad+\rho_{b_3}(\varphi_1\varphi_2^{-1}) + \rho_{b_3}(\varphi_1^{-1}\varphi_2) + \rho_{b_1}(\varphi_1\varphi_2^{-1})\rho_{b_2}(\varphi_1^{-1}\varphi_2) \\
&\quad= \rho_{b_3}(\varphi_1) + \rho_{b_3}(\varphi_2) + \rho_{b_1}(\varphi_1)\rho_{b_2}(\varphi_2) \\
&\qquad+\rho_{b_3}(\varphi_1^{-1}) + \rho_{b_3}(\varphi_2^{-1}) + \rho_{b_1}(\varphi_1^{-1})\rho_{b_2}(\varphi_2^{-1}) \\
&\qquad+\{\rho_{b_1}(\varphi_1) + \rho_{b_1}(\varphi_2)\}\{-\rho_{b_2}(\varphi_1) - \rho_{b_2}(\varphi_2)\} \\
&\qquad+\rho_{b_3}(\varphi_1) + \rho_{b_3}(\varphi_2^{-1}) + \rho_{b_1}(\varphi_1)\rho_{b_2}(\varphi_2^{-1}) \\
&\qquad+\rho_{b_3}(\varphi_1^{-1}) + \rho_{b_3}(\varphi_2) + \rho_{b_1}(\varphi_1^{-1})\rho_{b_2}(\varphi_2) \\
&\qquad+\{\rho_{b_1}(\varphi_1) - \rho_{b_1}(\varphi_2)\}\{-\rho_{b_2}(\varphi_1) + \rho_{b_2}(\varphi_2)\} \\
&\quad= 2\{\rho_{b_3}(\varphi_1) + \rho_{b_3}(\varphi_1^{-1}) - \rho_{b_1}(\varphi_1)\rho_{b_2}(\varphi_1)\} \\
&\qquad+2\{\rho_{b_3}(\varphi_2) + \rho_{b_3}(\varphi_2^{-1}) - \rho_{b_1}(\varphi_2)\rho_{b_2}(\varphi_2)\} \\
&\quad= 2\{\rho_{b_3}(\varphi_1) + \rho_{b_3}(\varphi_1^{-1}) + \rho_{b_1}(\varphi_1)\rho_{b_2}(\varphi_1^{-1})\} \\
&\qquad+2\{\rho_{b_3}(\varphi_2) + \rho_{b_3}(\varphi_2^{-1}) + \rho_{b_1}(\varphi_2)\rho_{b_2}(\varphi_2^{-1})\} \\
&\quad= 0
\end{aligned}$$

This completes the proof. Q.E.D.

§5. Acknowledgements

The author would like to express his gratitude to Prof. Shigeyuki Morita for helpful suggestions and encouragements. He also would like to thank Prof. Teruaki Kitano and Prof. Takayuki Morifuji for valuable discussions and advices. Finally, he also would like to thank the referee for useful comments.

References

[1] S. Bigelow and R. Budney, The mapping class group of a genus two surface is linear, Algebr. Geom. Topol., **1** (2001), 699–708.

[2] J. Birman, Braids, Links and Mapping Class Groups, Ann. of Math. Stud., **82**, Princeton Univ. Press, 1975.

[3] R. H. Fox, Free Differential Calculus I, Ann. of Math. (2), **57** (1953), 547–560.

[4] D. Johnson, Conjugacy relations in subgroup of the mapping class group and group-theoretic description of the Rochlin invariant, Math. Ann., **249** (1980), 243–263.

[5] D. Johnson, A survey of the Torelli group, Contemp. Math., **20** (1983), 165–179.

[6] M. Korkmaz, On the linearity of certain mapping class groups, Turkish J. Math., **24** (2000), 367–371.

[7] S. Morita, Families of Jacobian manifolds and characteristic classes of surface bundles I, Ann. Inst. Fourier (Grenoble), **39** (1989), 777–810.

[8] S. Morita, On the structure of the Torelli group and the Casson invariant, Topology, **30** (1991), 603–621.

[9] S. Morita, Abelian quotients of subgroups of the mapping class group of surfaces, Duke Math. J., **70** (1993), 699–726.

[10] S. Morita, Casson invariant, signature defect of framed manifolds and the secondary characteristic classes of surface bundles, J. Differential Geom., **47** (1997), 560–599.

[11] S. Morita, Structure of the mapping class groups of surfaces: a survey and a prospect, In: Proceedings of the Kirbyfest, Geom. Topol. Monogr., **2**, Geom. Topol. Publ., 1999, pp. 349–406.

[12] S. Morita, Cohomological structure of the mapping class group and beyond, In: Problems on Mapping Class Groups and Related Topics, Proc. Sympos. Pure Math., **74**, Amer. Math. Soc., Providence, RI, 2006, pp. 329–354.

[13] M. Suzuki, The Magnus representation of the Torelli group $\mathcal{I}_{g,1}$ is not faithful for $g \geq 2$, Proc. Amer. Math. Soc., **130** (2002), 909–914.

[14] M. Suzuki, Irreducible decomposition of the Magnus representation of the Torelli group, Bull. Austral. Math. Soc., **67** (2003), 1–14.

[15] M. Suzuki, A class function on the Torelli group, Kodai Math. J., **26** (2003), 304–316.

[16] M. Suzuki, On the kernel of the Magnus representation of the Torelli group, Proc. Amer. Math. Soc., **133** (2005), 1865–1872.

[17] M. Suzuki, A geometric interpretation of the Magnus representation of the mapping class group, Kobe J. Math., **22** (2005), 39–47.

Department of Mathematics
Akita University
1-1 Tegata-Gakuenmachi, Akita, 010-8502
Japan

Advanced Studies in Pure Mathematics 52, 2008
Groups of Diffeomorphisms
pp. 491–504

On the simplicity of the group of contactomorphisms

Takashi Tsuboi

Abstract.

We consider the group $\mathrm{Cont}^r_c(M^{2n+1},\alpha)$ of C^r contactomorphisms with compact support of a contact manifold (M^{2n+1},α) of dimension $(2n+1)$ with the C^r topology. We show that the first homology group of the classifying space $B\overline{\mathrm{Cont}}^r_c(M^{2n+1},\alpha)$ for the C^r foliated M^{2n+1} products with compact support with transverse contact structure α is trivial for $1 \leq r < n+(3/2)$. This implies that the identity component $\mathrm{Cont}^r_c(M^{2n+1},\alpha)_0$ of the group $\mathrm{Cont}^r_c(M^{2n+1},\alpha)$ of contactomorphisms with compact support of a connected contact manifold (M^{2n+1},α) is a simple group for $1 \leq r < n+(3/2)$.

§1. Introduction

The groups of diffeomorphisms play an important role in the theory of foliations. This relationship is clear in the theory developed by Mather and Thurston, which asserts the relationship between the topology of the classifying space for C^r foliations of codimension n and that of the group $\mathrm{Diff}^r_c(\boldsymbol{R}^n)$ of C^r diffeomorphisms of $\boldsymbol{R}^n$ with compact support: $H_*(B\overline{\mathrm{Diff}}^r_c(\boldsymbol{R}^n);\boldsymbol{Z}) \cong H_*(\Omega^n B\overline{\Gamma}^r_n;\boldsymbol{Z})$. Here, $B\overline{\mathrm{Diff}}^r_c(\boldsymbol{R}^n)$ is the classifying space for the C^r foliated $\boldsymbol{R}^n$ products with compact support, $B\overline{\Gamma}^r_n$ is the classifying space for the Γ^r_n structures (C^r foliations of codimension n) with trivialized normal bundles, and Ω^n means the n-fold loop space.

Received March 22, 2007.
Revised February 9, 2008.
2000 *Mathematics Subject Classification*. Primary 57R50, 57R32; Secondary 57R17, 57R52.
Key words and phrases. contactomorphisms, classifying space, foliations.
The author is supported by Grant-in-Aid for Scientific Research 16204004 and 17104001, Grant-in-Aid for Exploratory Research 18654008, Japan Society for Promotion of Science, and by the 21st Century COE Program at Graduate School of Mathematical Sciences, the University of Tokyo.

The homology of $H_*(B\overline{\mathrm{Diff}}{}^r_c(\boldsymbol{R}^n);\boldsymbol{Z})$ is known for several cases and this gives the information on the connectivity of $B\overline{\Gamma}{}^r_n$: $B\overline{\Gamma}{}^r_n$ is n-connected; $B\overline{\Gamma}{}^\infty_n$ is $(n+1)$-connected (Mather [9], Thurston [12]); $B\overline{\Gamma}{}^r_n$ is $(n+1)$-connected ($r \neq n+1$) (Mather [10]); $B\overline{\Gamma}{}^0_n$ and $B\overline{\Gamma}{}^L_n$ (L stands for Lipschitz) are contractible (Mather [8]); $B\overline{\Gamma}{}^1_n$ is contractible (Tsuboi [17]); The connectivity of $B\overline{\Gamma}{}^r_n$ increases as $r \searrow 1$ (Tsuboi [15]); $\pi_{2n+1}(B\overline{\Gamma}{}^r_n) \geq \boldsymbol{R}$ ($r > 2-1/(n+1)$) (by using the characteristic classes for foliations). This connectivity information is closely related to the construction of foliations (Haefliger [3], [4], Thurston [13], [14]).

It is also known that if $H_1(B\overline{\mathrm{Diff}}{}^r_c(\boldsymbol{R}^n);\boldsymbol{Z}) = 0$, that is, if $r = 0$, $r = L$, $1 \leq r < n+1$, or $n+1 < r \leq \infty$, the identity component $\mathrm{Diff}^r_c(M)_0$ of the group $\mathrm{Diff}^r_c(M)$ of C^r diffeomorphisms with compact support of a connected n-dimensional manifold M is a simple group (Thurston [12], see also Banyaga [2]).

Now we are interested in the group of volume preserving diffeomorphisms, the group of symplectomorphisms of a symplectic manifold and the group of contactomorphisms of a contact manifold. The homology of the group of volume preserving diffeomorphisms is studied by McDuff ([7]). The homology of the group of symplectomorphisms is studied by Banyaga ([1]). The book [2] by Banyaga is a good reference for these groups.

In this paper, we apply the techniques of Mather [10] and Tsuboi [15] to the group of contactomorphisms to show that the first homology group of the classifying space $B\overline{\mathrm{Cont}}{}^r_c(\boldsymbol{R}^{2n+1},\alpha_{\mathrm{st}})$ for the C^r foliated $\boldsymbol{R}^{2n+1}$ products with compact support with transverse contact structure α_{st} is trivial for $1 \leq r < n+(3/2)$ (Theorem 5.1 for the standard contact manifold $(\boldsymbol{R}^{2n+1},\alpha_{\mathrm{st}})$). Then by the fragmentation technique (Thurston [12], Banyaga [2]), $H_1(B\overline{\mathrm{Cont}}{}^r_c(M^{2n+1},\alpha);\boldsymbol{Z})$ is trivial for any contact manifold (M^{2n+1},α) (Theorem 5.1), and the identity component $\mathrm{Cont}^r_c(M^{2n+1},\alpha)_0$ of the group $\mathrm{Cont}^r_c(M^{2n+1},\alpha)$ of C^r contactomorphisms with compact support of a connected contact manifold (M^{2n+1},α) is a simple group for $1 \leq r < n+(3/2)$ (Theorem 5.2).

The author is very grateful to the referee for his careful and detailed reading and valuable comments.

§2. Contactomorphisms of class C^r

First we need to define what is a C^r contactomorphism for $1 \leq r < \infty$.

Let (M^{2n+1},α) be a contact manifold of dimension $(2n+1)$, where α is a C^∞ 1-form on M^{2n+1} such that $\alpha \wedge (d\alpha)^n$ is a volume form. A (positive) C^r contactomorphism φ of (M^{2n+1},α) is a C^r diffeomorphism of M^{2n+1} such that $\varphi^*\alpha = w_\varphi \alpha$ where w_φ is a positive C^r function on M^{2n+1} depending on φ. In other words, a contactomorphism φ is a diffeomorphism whose tangent map $T\varphi$ preserves the C^∞ contact hyperplane field $\xi = \ker \alpha$ with transverse orientation.

Note that the action of the C^r diffeomorphism on the tangent bundle TM^{2n+1} or on the cotangent bundle T^*M^{2n+1} is usually of class C^{r-1}, however, it is well defined that the function w_φ is of class C^r and such diffeomorphisms form a group. For, the equations ${\varphi_j}^*\alpha = w_{\varphi_j}\alpha$ ($j = 1, 2$) imply

$$(\varphi_1\varphi_2)^*\alpha = {\varphi_2}^*{\varphi_1}^*\alpha = {\varphi_2}^*(w_{\varphi_1}\alpha) = ({\varphi_2}^* w_{\varphi_1}){\varphi_2}^*\alpha = ({\varphi_2}^* w_{\varphi_1})w_{\varphi_2}\alpha.$$

Hence $w_{\varphi_1\varphi_2} = ({\varphi_2}^* w_{\varphi_1})w_{\varphi_2}$, and if w_{φ_1} and w_{φ_2} are of class C^r, then so is $w_{\varphi_1\varphi_2}$. We see also that $w_{\varphi^{-1}} = \varphi^{-1*}w_\varphi$. Thus C^r contactomorphisms of (M^{2n+1}, α) form a group.

Note also that C^r contactomorphisms of (M^{2n+1}, α) are determined by the C^∞ contact hyperplane field ξ. For, if $\xi = \ker\alpha$ is defined by another C^∞ 1-form $\beta = k\alpha$ ($k > 0$), then $\varphi(k\alpha) = (\varphi^*k)w_\varphi\alpha = \dfrac{\varphi^*k}{k}w_\varphi(k\alpha)$, hence if w_φ is of class C^r, then so is $\dfrac{\varphi^*k}{k}w_\varphi$.

§3. Local contractibility and fragmentation property

Let $\mathrm{Cont}_c^r(M^{2n+1}, \alpha)$ or $\mathrm{Cont}_c^r(M^{2n+1}, \xi)$ denote the group of contactomorphisms of (M^{2n+1}, α) or of (M^{2n+1}, ξ) with compact support.

An element φ of this group, C^1 close to the identity, corresponds to a C^{r+1} function on $M = M^{2n+1}$. Consider the contact form $w\alpha_1 - \alpha_2$ on $M \times M \times \boldsymbol{R}_{>0}$, where w is the coordinate of $\boldsymbol{R}_{>0}$, $\alpha_i = p_i^*\alpha$ and $p_i : M \times M \times \boldsymbol{R}_{>0} \longrightarrow M$ is the projection to the i-th component ($i = 1, 2$). A C^r diffeomorphism φ of M belongs to $\mathrm{Cont}_c^r(M, \xi)$ (i.e., $\varphi^*\alpha = w_\varphi\alpha$ with w_φ being a positive C^r function) if and only if the graph of $(\varphi, w_\varphi) : M \longrightarrow M \times \boldsymbol{R}_{>0}$;

$$\{(u, \varphi(u), w_\varphi(u)) \mid u \in M\}$$

is a Legendrian submanifold of $(M \times M \times \boldsymbol{R}_{>0}, w\alpha_1 - \alpha_2)$ of class C^r (see Lychagin [6], Banyaga [2]). There is a C^∞ contactomorphism from a neighborhood of the graph of (id, 1) to the space $J^1(M, \boldsymbol{R})$ of 1-jets of functions on M. A C^r Legendrian submanifold of $J^1(M, \boldsymbol{R})$, C^1 close to 0 is the prolongation of a C^{r+1} function f on M, C^2 close to 0. In this way, a neighborhood of the identity in $\mathrm{Cont}_c^r(M^{2n+1}, \alpha)$ is diffeomorphic to a neighborhood of 0 in the space $C_c^{r+1}(M^{2n+1})$ of C^{r+1} functions on M^{2n+1} with compact support. This shows the local contractibility of $\mathrm{Cont}_c^r(M^{2n+1}, \alpha)$.

Note here that the identity of M^{2n+1} corresponds to the zero function, and we have a canonical C^∞ path to the identity for an element of $\mathrm{Cont}_c^r(M^{2n+1},\alpha)$, C^1 close to the identity. In fact, for an element φ C^1 close to the identity, we obtain the C^{r+1} function f_φ. Then tf_φ is the canonical C^∞ path in the space $C_c^{r+1}(M^{2n+1})$ of C^{r+1} functions and this corresponds to a C^∞ path in the space of graphs, that is, to a C^∞ path in $\mathrm{Cont}_c^r(M^{2n+1},\alpha)$ to the identity.

We also see the fragmantation property of $\mathrm{Cont}_c^r(M^{2n+1},\alpha)$. If we have a contactomorphism φ with compact support C^1 close to the identity, we choose finitely many Darboux coordinate neighborhoods $\{U_i\}$ which cover the support of φ. We choose a covering by smaller neighborhoods $W_i \subset \overline{W}_i \subset V_i \subset \overline{V}_i \subset U_i$, and using the C^r function f_φ which is associated to φ and a bump function with support in U_i, we obtain a C^r function with support in U_i which coincides with f_φ on V_i. Thus we have a contactomorphism with support in U_i which coincides with φ on W_i. In this way, φ can be fragmented. See Banyaga [2].

§4. Contactmorphisms for the standard structure on $\boldsymbol{R}^{2n+1}$

For the standard contact structure $(\boldsymbol{R}^{2n+1},\alpha_{\mathrm{st}})$, we give an explicit correspondence between a neighborhood of the identity in $\mathrm{Cont}_c^r(\boldsymbol{R}^{2n+1},\alpha_{\mathrm{st}})$ and a neighborhood of 0 in $C_c^{r+1}(\boldsymbol{R}^{2n+1})$.

Let $\alpha_{\mathrm{st}} = \mathrm{d}z - \sum_{i=1}^n y_i \mathrm{d}x_i$ be the standard contact form on $\boldsymbol{R}^{2n+1}$. Let $\mathrm{Cont}_c^r(\boldsymbol{R}^{2n+1},\alpha_{\mathrm{st}})$ denote the group of C^r contactomorphisms of $(\boldsymbol{R}^{2n+1},\alpha_{\mathrm{st}})$ with compact support.

When $\varphi \in \mathrm{Cont}_c^r(\boldsymbol{R}^{2n+1},\alpha_{\mathrm{st}})$ is C^1 close to the identity, we write the graph

$$\{(u,\varphi(u),w_\varphi(u)) \mid u \in \boldsymbol{R}^{2n+1}\} \subset \boldsymbol{R}^{2n+1} \times \boldsymbol{R}^{2n+1} \times \boldsymbol{R}_{>0}$$

in the form of 1-jets of a function of $2n+1$ variables as follows: For

$$u = (x^{(1)},y^{(1)},z^{(1)}) = (x_1^{(1)},\dots,x_n^{(1)},y_1^{(1)},\dots,y_n^{(1)},z^{(1)}),$$

we write $\varphi(u) = (x_1^{(2)}(u),\dots,x_n^{(2)}(u),y_1^{(2)}(u),\dots,y_n^{(2)}(u),z^{(2)}(u))$, where

$$\begin{aligned}
x_i^{(2)} &= x_i^{(2)}(x_1^{(1)},\dots,x_n^{(1)},y_1^{(1)},\dots,y_n^{(1)},z^{(1)}) \quad (i=1,\dots,n),\\
y_i^{(2)} &= y_i^{(2)}(x_1^{(1)},\dots,x_n^{(1)},y_1^{(1)},\dots,y_n^{(1)},z^{(1)}) \quad (i=1,\dots,n),\\
z^{(2)} &= z^{(2)}(x_1^{(1)},\dots,x_n^{(1)},y_1^{(1)},\dots,y_n^{(1)},z^{(1)}) .
\end{aligned}$$

Then the graph $\{(u, \varphi(u), w_\varphi(u)) \mid u \in \boldsymbol{R}^{2n+1}\}$ satisfies

$$\mathrm{d}z^{(2)} - \sum_{i=1}^{n} y_i^{(2)} \mathrm{d}x_i^{(2)} = w_\varphi(u)\{\mathrm{d}z^{(1)} - \sum_{i=1}^{n} y_i^{(1)} \mathrm{d}x_i^{(1)}\}.$$

That is,

$$\mathrm{d}\big(z^{(2)}(u) - z^{(1)} - \sum_{i=1}^{n} y_i^{(2)}(u)(x_i^{(2)}(u) - x_i^{(1)})\big) - (w_\varphi(u) - 1)\mathrm{d}z^{(1)}$$
$$+ \sum_{i=1}^{n} \left(w_\varphi(u) y_i^{(1)} - y_i^{(2)}(u)\right)\mathrm{d}x_i^{(1)} + \sum_{i=1}^{n} (x_i^{(2)}(u) - x_i^{(1)})\mathrm{d}y_i^{(2)} = 0.$$

Since the graph is close to $\{(u, u, 1) \mid u \in \boldsymbol{R}^{2n+1}\}$, the graph can be written as a graph with respect to the variables

$$(x^{(1)}, y^{(2)}, z^{(1)}) = (x_1^{(1)}, \dots, x_n^{(1)}, y_1^{(2)}, \dots, y_n^{(2)}, z^{(1)})$$

and the graph is written as

$$\{(x^{(1)}, y^{(1)}(x^{(1)}, y^{(2)}, z^{(1)}), z^{(1)},$$
$$y^{(2)}, x^{(2)}(x^{(1)}, y^{(2)}, z^{(1)}), z^{(2)}(x^{(1)}, y^{(2)}, z^{(1)}), w(x^{(1)}, y^{(2)}, z^{(1)}))\},$$

where $y^{(1)}(x^{(1)}, y^{(2)}, z^{(1)})$, $x^{(2)}(x^{(1)}, y^{(2)}, z^{(1)})$ and $z^{(2)}(x^{(1)}, y^{(2)}, z^{(1)})$ are C^r functions by the inverse function theorem, and

$$w(x^{(1)}, y^{(2)}, z^{(1)}) = w_\varphi(x^{(1)}, y^{(2)}(x^{(1)}, y^{(1)}, z^{(1)}), z^{(1)}).$$

If φ is a C^r contactomorphism with compact support, then the C^r function

$$f(x^{(1)}, y^{(2)}, z^{(1)}) = z^{(2)}(x^{(1)}, y^{(2)}, z^{(1)}) - z^{(1)}$$
$$- \sum_{i=1}^{n} y_i^{(2)} \left(x_i^{(2)}(x^{(1)}, y^{(2)}, z^{(1)}) - x_i^{(1)}\right)$$

has the derivatives

$$\frac{\partial f}{\partial x_i^{(1)}} = -w(x^{(1)}, y^{(2)}, z^{(1)})\ y_i^{(1)}(x^{(1)}, y^{(2)}, z^{(1)}) + y_i^{(2)} \quad (i = 1, \dots, n),$$
$$\frac{\partial f}{\partial y_i^{(2)}} = x_i^{(1)} - x_i^{(2)}(x^{(1)}, y^{(2)}, z^{(1)}) \quad (i = 1, \dots, n),$$
$$\frac{\partial f}{\partial z^{(1)}} = w(x^{(1)}, y^{(2)}, z^{(1)}) - 1 \quad ,$$

which are C^r functions, hence $f(x^{(1)}, y^{(2)}, z^{(1)})$ is a C^{r+1} function with compact support. Note again that the identity of $\boldsymbol{R}^{2n+1}$ corresponds to the zero function.

Conversely for a C^{r+1} function $f(x^{(1)}, y^{(2)}, z^{(1)})$ with compact support, C^2 close to the zero, by putting

$$
\begin{aligned}
w(x^{(1)}, y^{(2)}, z^{(1)}) &= \frac{\partial f}{\partial z^{(1)}} + 1 &&, \\
x_i^{(2)}(x^{(1)}, y^{(2)}, z^{(1)}) &= -\frac{\partial f}{\partial y_i^{(2)}} + x_i^{(1)} && (i = 1, \dots, n), \\
y_i^{(1)}(x^{(1)}, y^{(2)}, z^{(1)}) &= \frac{-\dfrac{\partial f}{\partial x_i^{(1)}} + y_i^{(2)}}{\dfrac{\partial f}{\partial z^{(1)}} + 1} && (i = 1, \dots, n), \\
z^{(2)}(x^{(1)}, y^{(2)}, z^{(1)}) &= f + z^{(1)} - \sum_{i=1}^{n} y_i^{(2)} \frac{\partial f}{\partial y_i^{(2)}} ,
\end{aligned}
$$

the graph is a Legendrian submanifold of $(\boldsymbol{R}^{2n+1} \times \boldsymbol{R}^{2n+1} \times \boldsymbol{R}_{>0}, w\alpha_1 - \alpha_2)$, C^1 close to the identity. Hence it is a graph of a C^r contactomorphism φ with compact support.

§5. Statement of result

Let $\mathrm{Cont}_c^r(M^{2n+1}, \alpha)$ be the group of C^r contactomorphisms with compact support of the contact manifold (M^{2n+1}, α) with the C^r topology, and $\mathrm{Cont}_c^r(M^{2n+1}, \alpha)^\delta$, the same group with the discrete topology. Let $B\overline{\mathrm{Cont}_c^r}(M^{2n+1}, \alpha)$ denote the homotopy fiber of the map between their classifying spaces: $B\mathrm{Cont}_c^r(M^{2n+1}, \alpha)^\delta \longrightarrow B\mathrm{Cont}_c^r(M^{2n+1}, \alpha)$. $B\overline{\mathrm{Cont}_c^r}(M^{2n+1}, \alpha)$ is the classifying space for the C^r foliated M^{2n+1} products with compact support with transverse contact structure defined by α. Here is our main result.

Theorem 5.1. *For* $1 \le r < n+(3/2)$, $H_1(B\overline{\mathrm{Cont}_c^r}(M^{2n+1}, \alpha); \boldsymbol{Z}) = 0$.

For $G = \mathrm{Cont}_c^r(M^{2n+1}, \alpha)$, $B\overline{G}$ is constructed as the realization of the semi-simplicial set $S_*^\infty(G)/G$. Here $S_i^\infty(G)$ is the set of smooth singular simplices of G. C^r diffeomorphism groups have a smooth structure such that the composition

$$(g_1, g_2) \longmapsto g_1 g_2 : G \times G \longrightarrow G$$

is smooth with respect to g_1 (but it is not smooth with respect to g_2 if r is finite). An element $\sigma : \Delta^i \longrightarrow G$ $(\in S_i^\infty(G))$ defines a foliated M^{2n+1}

product over Δ^i such that the leaf passing through $(t, x) \in \Delta^i \times M^{2n+1}$ is given by

$$\{(s, \sigma(s)\sigma(t)^{-1}(x)) \in \Delta^i \times M^{2n+1} \mid s \in \Delta^i\}.$$

A C^r foliated product has a natural C^r (semi-)norm ([15]).

We note here that for a topological group G, the classifying space $B\overline{G}$ is constructed as the realization of $S_*(G)/G$, where $S_*(G)$ is the set of singular simplices of G. When G has a smooth structure, the inclusion $|S_*^\infty(G)/G| \subset |S_*(G)/G|$ is usually a homotopy equivalence. The homotopy equivalence $|S_*^\infty(G)/G| \subset |S_*(G)/G|$ is shown by approximating a singular simplex by a smooth singular simplex and by constructing a canonical homotopy between them. The construction of the canonical homotopy uses the fact explained in Section 3 that there is a canonical path to the identity for an element of $\mathrm{Cont}_c^r(M^{2n+1}, \alpha)$, C^1 close the identity.

Let $\mathrm{Cont}_c^r(M^{2n+1}, \alpha)_0$ denote the connected component of the identity of $\mathrm{Cont}_c^r(M^{2n+1}, \alpha)$. Theorem 5.1 implies the simplicity of this group (Thurston [12], Banyaga [2]).

Theorem 5.2. *Let (M^{2n+1}, α) be a connected contact manifold. For $1 \leq r < n + (3/2)$, $\mathrm{Cont}_c^r(M^{2n+1}, \alpha)_0$ is a simple group.*

Proof. By the argument in Thurston [12] or Banyaga [2], $H_1(B\overline{\mathrm{Cont}_c^r}(M^{2n+1}, \alpha); \boldsymbol{Z}) = 0$ implies that the universal covering group $\widetilde{\mathrm{Cont}_c^r}(M^{2n+1}, \alpha)_0$ is a perfect group and then $\mathrm{Cont}_c^r(M^{2n+1}, \alpha)_0$ is a perfect group. As we explained in Section 3, $\mathrm{Cont}_c^r(M^{2n+1}, \alpha)_0$ has the fragmentation property. Again by the argument in Thurston [12] or Banyaga [2] using the fact that $\mathrm{Cont}_c^r(\boldsymbol{R}^{2n+1}, \alpha_{\mathrm{st}})_0$ is a perfect group, $\mathrm{Cont}_c^r(M^{2n+1}, \alpha)_0$ is a simple group. Q.E.D.

To prove Theorem 5.1, by the fragmentation technique (Thurston [12], Banyaga [2]), it suffices to prove Theorem 5.1 for the standard contact manifold $(\boldsymbol{R}^{2n+1}, \alpha_{\mathrm{st}})$. In other words, by taking a Darboux coordinate neighborhood $U \cong \boldsymbol{R}^{2n+1}$, by the fragmentation technique of Section 3, and the fact that for any Darboux coordinate neighborhood U_i in M, there is a contact isotopy sending U_i into U, the induced map $H_1(B\overline{\mathrm{Cont}_c^r}(\boldsymbol{R}^{2n+1}, \alpha_{\mathrm{st}}); \boldsymbol{Z}) \longrightarrow H_1(B\overline{\mathrm{Cont}_c^r}(M^{2n+1}, \alpha); \boldsymbol{Z})$ is surjective. Hence $H_1(B\overline{\mathrm{Cont}_c^r}(\boldsymbol{R}^{2n+1}, \alpha_{\mathrm{st}}); \boldsymbol{Z}) = 0$ implies $H_1(B\overline{\mathrm{Cont}_c^r}(M^{2n+1}, \alpha); \boldsymbol{Z}) = 0$.

In order to show Theorem 5.1 for $(\boldsymbol{R}^{2n+1}, \alpha_{\mathrm{st}})$, we need to construct several elements of $\mathrm{Cont}_c^\infty(\boldsymbol{R}^{2n+1}, \alpha_{\mathrm{st}})$ which coincide with contractions or translations on $[-1, 1]^{2n+1}$. We do this in the next section. In Section

7, we give the main construction and prove Theorem 5.1 for $(\boldsymbol{R}^{2n+1}, \alpha_{\mathrm{st}})$ in a way similar to that in Tsuboi [15].

§6. Lie algebra for the group of smooth contactomorphisms

For $r = \infty$, the Lie algebra of $\mathrm{Cont}_c^\infty(M^{2n+1}, \alpha)$ or $\mathrm{Cont}_c^\infty(M^{2n+1}, \xi)$ is described as follows: Let $\mathcal{X}_{\mathrm{cont}}^\infty(M^{2n+1}, \alpha)$ be the space of C^∞ vector fields X such that $L_X\alpha = k_X\alpha$ for a C^∞ function k_X depending on X.

If $L_{X_j}\alpha = k_{X_j}\alpha$ ($j = 1, 2$), then

$$\begin{aligned}
L_{[X_1,X_2]}\alpha &= (L_{X_1}L_{X_2} - L_{X_2}L_{X_1})\alpha = L_{X_1}(k_{X_2}\alpha) - L_{X_2}(k_{X_1}\alpha) \\
&= X_1(k_{X_2})\alpha + k_{X_2}k_{X_1}\alpha - X_2(k_{X_1})\alpha - k_{X_1}k_{X_2}\alpha \\
&= (X_1(k_{X_2}) - X_2(k_{X_1}))\alpha
\end{aligned}$$

Hence $k_{[X_1,X_2]} = X_1(k_{X_2}) - X_2(k_{X_1})$ and such vector fields form a Lie algebra. For the C^r case, however, the C^r vector fields X such that $L_X\alpha = k_X\alpha$ for a C^r function k would not form a Lie algebra, because $[X_1, X_2]$ and $k_{[X_1,X_2]}$ are usually of class C^{r-1}. This reflects the fact that the composition $(\varphi_1, \varphi_2) \longmapsto \varphi_1\varphi_2$ is smooth with respect to φ_1 but not smooth with respect to φ_2 in the group of C^r diffeomorphisms.

We need later several contactomorphisms of $(\boldsymbol{R}^{2n+1}, \alpha_{\mathrm{st}})$ for our construction. We write them as the time 1-maps of C^∞ contact vector fields.

For the standard contact form $\alpha_{\mathrm{st}} = \mathrm{d}z - \sum_{i=1}^n y_i \mathrm{d}x_i$, an element $X = \zeta\frac{\partial}{\partial z} + \sum_{i=1}^n \xi_i \frac{\partial}{\partial x_i} + \sum_{i=1}^n \eta_i \frac{\partial}{\partial y_i}$ of $\mathcal{X}_{\mathrm{cont}}^\infty(\boldsymbol{R}^{2n+1}, \alpha_{\mathrm{st}})$ is written by the function $a = \iota(X)\alpha$ on $\boldsymbol{R}^{2n+1}$ as follows:

$$X = (a - \sum_{j=1}^n y_j \frac{\partial a}{\partial y_j})\frac{\partial}{\partial z} - \sum_{i=1}^n \frac{\partial a}{\partial y_i}\frac{\partial}{\partial x_i} + \sum_{i=1}^n (\frac{\partial a}{\partial x_i} + y_i\frac{\partial a}{\partial z})\frac{\partial}{\partial y_i}.$$

For this contact vector field X, $k_X = \dfrac{\partial a}{\partial z}$.

The reason is as follows: Since

$$\begin{aligned}
L_X\alpha &= \iota(X)\mathrm{d}\alpha + \mathrm{d}\iota(X)\alpha = \sum_{i=1}^n (\xi_i \mathrm{d}y_i - \eta_i \mathrm{d}x_i) + \mathrm{d}(\zeta - \sum_{i=1}^n y_i\xi_i) \\
&= -\sum_{i=1}^n \eta_i \mathrm{d}x_i + \sum_{i=1}^n \frac{\partial \zeta}{\partial x_i}\mathrm{d}x_i + \sum_{i=1}^n \frac{\partial \zeta}{\partial y_i}\mathrm{d}y_i + \frac{\partial \zeta}{\partial z}\mathrm{d}z \\
&\quad - \sum_{i,j=1}^n y_i \frac{\partial \xi_i}{\partial x_j}\mathrm{d}x_j - \sum_{i,j=1}^n y_i \frac{\partial \xi_i}{\partial y_j}\mathrm{d}y_j - \sum_{i=1}^n y_i \frac{\partial \xi_i}{\partial z}\mathrm{d}z,
\end{aligned}$$

it satisfies

$$\begin{cases} \dfrac{\partial \zeta}{\partial x_i} - \eta_i - \displaystyle\sum_{j=1}^{n} y_j \dfrac{\partial \xi_j}{\partial x_i} = -y_i k_X \\ \dfrac{\partial \zeta}{\partial y_i} - \displaystyle\sum_{j=1}^{n} y_j \dfrac{\partial \xi_j}{\partial y_i} = 0 \\ \dfrac{\partial \zeta}{\partial z} - \displaystyle\sum_{j=1}^{n} y_j \dfrac{\partial \xi_j}{\partial z} = k_X \end{cases}$$

By differentiating $\dfrac{\partial \zeta}{\partial y_i} = \displaystyle\sum_{j=1}^{n} y_j \dfrac{\partial \xi_j}{\partial y_i}$ by y_k, we obtain

$$\frac{\partial^2 \zeta}{\partial y_i \partial y_k} = \frac{\partial \xi_k}{\partial y_i} + \sum_{j=1}^{n} y_j \frac{\partial^2 \xi_j}{\partial y_i \partial y_k}$$

which is symmetric in i and k. Hence we see that $\dfrac{\partial \xi_k}{\partial y_i} = \dfrac{\partial \xi_i}{\partial y_k}$. This implies that there is a function $a(x, y, z)$ such that $\xi_i = -\dfrac{\partial a}{\partial y_i}$. With this a,

$$\begin{aligned} \zeta(x, y, z) &= \int_{-\infty}^{y_i} \frac{\partial \zeta}{\partial y_i} \mathrm{d}y_i = \int_{-\infty}^{y_i} \sum_{j=1}^{n} y_j \frac{\partial \xi_j}{\partial y_i} \mathrm{d}y_i \\ &= \left[\sum_{j=1}^{n} y_j \xi_j \right]_{y_i=-\infty}^{y_i} - \int_{-\infty}^{y_i} \xi_i \mathrm{d}y_i \\ &= \sum_{j=1}^{n} y_j \xi_j + \int_{-\infty}^{y_i} \frac{\partial a}{\partial y_i} \mathrm{d}y_i \\ &= \sum_{j=1}^{n} y_j \xi_j + a = a - \sum_{j=1}^{n} y_j \frac{\partial a}{\partial y_j} \end{aligned}$$

Then

$$\begin{aligned} k_X(x, y, z) &= \frac{\partial \zeta}{\partial z} - \sum_{j=1}^{n} y_j \frac{\partial \xi_j}{\partial z} \\ &= \frac{\partial a}{\partial z} - \sum_{j=1}^{n} y_j \frac{\partial^2 a}{\partial y_j \partial z} + \sum_{j=1}^{n} y_j \frac{\partial^2 a}{\partial y_j \partial z} = \frac{\partial a}{\partial z} \end{aligned}$$

and

$$\begin{aligned}\eta_i(x,y,z) &= \frac{\partial \zeta}{\partial x_i} - \sum y_j \frac{\partial \xi_j}{\partial x_i} + y_i k_X \\ &= \frac{\partial a}{\partial x_i} - \sum_{j=1}^n y_j \frac{\partial^2 a}{\partial y_j \partial x_i} + \sum_{j=1}^n y_j \frac{\partial^2 a}{\partial y_j \partial x_i} + y_i \frac{\partial a}{\partial z} \\ &= \frac{\partial a}{\partial x_i} + y_i \frac{\partial a}{\partial z}\end{aligned}$$

Here are the contact vector fields we are interested in.

For the function $a(x,y,z) = 1$, the vector field $X = \dfrac{\partial}{\partial z}$, and its time t map is θ_t, where $\theta_t(x^{(1)}, y^{(1)}, z^{(1)}) = (x^{(1)}, y^{(1)}, z^{(1)} + t)$.

For the function $a(x,y,z) = x_i$, the vector field $X = x_i \dfrac{\partial}{\partial z} + \dfrac{\partial}{\partial y_i}$, and its time t map is ψ_i^t, where $\psi_i^t(x^{(1)}, y^{(1)}, z^{(1)}) = (x^{(1)}, y^{(1)} + t1_i, z^{(1)} + tx_i)$.

For the function $a(x,y,z) = y_i$, the vector field $X = -\dfrac{\partial}{\partial x_i}$, and its time t map is φ_i^{-t}, where $\varphi_i^t(x^{(1)}, y^{(1)}, z^{(1)}) = (x^{(1)} + t1_i, y^{(1)}, z^{(1)})$.

For the function $a(x,y,z) = 2z - \sum_{i=1}^n x_i y_i$, the vector field $X = 2z\dfrac{\partial}{\partial z} + \sum_{i=1}^n x_i \dfrac{\partial}{\partial x_i} + \sum_{i=1}^n y_i \dfrac{\partial}{\partial y_i}$, and its time t map is ε^t, where $\varepsilon^t(x^{(1)}, y^{(1)}, z^{(1)}) = (e^t x^{(1)}, e^t y^{(1)}, e^{2t} z^{(1)})$.

The supports of these vector fields are not compact. We take the product of the function a and a bump function and make the time t maps θ_t, ε_t, φ_i^t, ψ_i^t be with compact support and coincide with the original ones for $|t| \leq 1$ on a given compact subset of $\boldsymbol{R}^{2n+1}$.

§7. Main construction and the proof of Theorem 5.1

Using the above contractions and translations, we perform a construction similar to that in [15].

For a positive real number A, take the rectangle $R = [-1,1]^n \times [-A/n, A/n]^n \times [-1,1]$. Then

$$\begin{aligned}&\varepsilon^{-\log(2+A)}(R) \\ &= [-\frac{1}{2+A}, \frac{1}{2+A}]^n \times [-\frac{A/n}{2+A}, \frac{A/n}{2+A}]^n \times [-\frac{1}{(2+A)^2}, \frac{1}{(2+A)^2}].\end{aligned}$$

For $\gamma = (\gamma_1, \dots, \gamma_n)$, $(\gamma_i = \pm 1)$,

$$R_\gamma = \psi_1^{\gamma_1 \frac{(1+A)A/n}{2+A}} \dots \psi_n^{\gamma_n \frac{(1+A)A/n}{2+A}} (\varepsilon^{-\log(2+A)}(R))$$

is a parallelepiped contained in

$$\begin{aligned}&[-\frac{1}{2+A},\frac{1}{2+A}]^n\times J_{\gamma_1}\times\cdots\times J_{\gamma_n}\\&\times[-\frac{1}{(2+A)^2}-|\sum_{i=1}^n\gamma_i|\cdot\frac{(1+A)A/n}{2+A},\\&\qquad\frac{1}{(2+A)^2}+|\sum_{i=1}^n\gamma_i|\cdot\frac{(1+A)A/n}{2+A}],\end{aligned}$$

which intersects a line parallel to z axis in a segment of length $\frac{2}{(2+A)^2}$ or not at all. Here, $J_{-1}=[-A/n,-\frac{A^2/n}{2+A}]$ and $J_{+1}=[\frac{A^2/n}{2+A},A/n]$.

Now for $k=\pm1,\pm3$, the parallelepipeds $\theta_{\frac{k}{(2+A)^2}}(R_\gamma)$ are disjoint. Since

$$\begin{aligned}&\frac{1}{(2+A)^2}+|\sum_{i=1}^n\gamma_i|\cdot\frac{(1+A)A/n}{2+A}+\frac{3}{(2+A)^2}\\&\leq\frac{1}{(2+A)^2}+\frac{(1+A)A}{2+A}+\frac{3}{(2+A)^2}\\&=\frac{4+2A+3A^2+A^3}{(2+A)^2}=1-\frac{A(2-2A-A^2)}{(2+A)^2}\leqq 1\end{aligned}$$

for small A ($<\frac{1}{2}$), they are contained in R.

For $\delta=(\delta_1,\ldots,\delta_n)$, ($\delta_i=\pm1$), the images

$$\varphi_1^{\delta_1\frac{1+A}{2+A}}\ldots\varphi_n^{\delta_n\frac{1+A}{2+A}}(\theta_{\frac{k}{(2+A)^2}}(R_\gamma))$$

are contained in $I_{\delta_1}\times\cdots\times I_{\delta_n}\times J_{\gamma_1}\times\cdots\times J_{\gamma_n}\times[-1,1]$, where $I_{-1}=[-1,-\frac{A}{2+A}]$ and $I_{+1}=[\frac{A}{2+A},1]$.

Thus we obtain $2^n\times2^2\times2^n=2^{2n+2}$ contractions

$$\varphi_1^{\delta_1\frac{1+A}{2+A}}\ldots\varphi_n^{\delta_n\frac{1+A}{2+A}}\theta_{\frac{k}{(2+A)^2}}\psi_1^{\gamma_1\frac{(1+A)A/n}{2+A}}\ldots\psi_n^{\gamma_n\frac{(1+A)A/n}{2+A}}\varepsilon^{-\log(2+A)}$$

with images of R being disjointly contained in R. See Figure 1.

For $G=\mathrm{Cont}_c^r(\boldsymbol{R}^{2n+1},\alpha)$, let $\sigma\in S_1^\infty(G)$ represent a foliated product over the interval Δ^1 with support in R.

We divide the foliated product into 2^{2n+2} segments of length $\frac{1}{2^{2n+2}}$ in the direction of Δ^1. Then $\sigma=\sum_{j=1}^{2^{2n+2}}\sigma_j$ in $H_1(B\overline{G};\boldsymbol{Z})$. Because of the

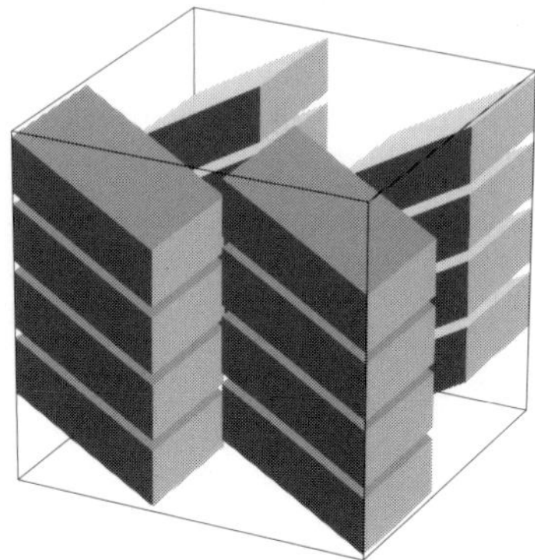

Fig. 1. The images of 2^4 contractions for $n = 1$

reparametrization in the direction of the interval Δ^1, the C^r norm of σ_j is $\dfrac{1}{2^{2n+2}}$ of that of σ.

Then we map each foliated product σ_j by the action of one of the above 2^{2n+2} contractions.

By the action of $\varepsilon^{-\log(2+A)}$, the C^r norm of the foliated product is multiplied by $\dfrac{1}{2+A}(2+A)^{2r}$. By the action of $\psi_1^{\gamma_1 \frac{(1+A)A/n}{2+A}} \cdots \psi_n^{\gamma_n \frac{(1+A)A/n}{2+A}}$, the C^r norm is multiplied by $(1 + \dfrac{(1+A)A}{2+A})^{r+1}$. By the action of the translations $\theta_{\frac{k}{(2+A)^2}}$ and $\varphi_1^{\delta_1 \frac{1+A}{2+A}} \cdots \varphi_n^{\delta_n \frac{1+A}{2+A}}$, the norm is unchanged.

Thus the foliated products obtained have the C^r norm

$$\frac{1}{2^{2n+2}}\frac{1}{2+A}(2+A)^{2r}(1+\frac{(1+A)A}{2+A})^{r+1}$$

times that of σ.

If $A = 0$, the factor is $2^{-2n-2}2^{2r-1} = 2^{2(r-n-\frac{3}{2})}$. Hence if $r < n + \dfrac{3}{2}$, by taking A small, the factor is smaller than 1.

Now we can show Theorem 5.1 for $(\boldsymbol{R}^{2n+1}, \alpha_{\mathrm{st}})$.

Proof of Theorem 5.1 *for* $(\boldsymbol{R}^{2n+1}, \alpha_{\mathrm{st}})$. Take $\sigma \in S_1^\infty(G) = S_1^\infty(\mathrm{Cont}_c^r(\boldsymbol{R}^{2n+1}, \alpha_{\mathrm{st}}))$ representing a foliated product over the interval Δ^1 with support in $[-\dfrac{A}{2+A}, \dfrac{A}{2+A}]^n \times [-A/n, A/n]^n \times [-1, 1]$. We show that it is written as a boundary in $B\overline{G} = B\overline{\mathrm{Cont}_c^r}(\boldsymbol{R}^{2n+1}, \alpha_{\mathrm{st}})$.

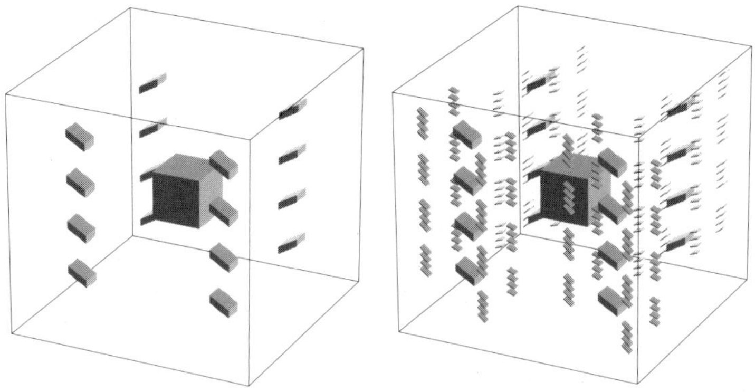

Fig. 2. The support of first and second iterates for σ with support in $[-\frac{A}{2+A}, \frac{A}{2+A}]^n \times [-A/n, A/n]^n \times [-1, 1]$

We perform the construction described above for the foliated product σ and repeat it infinitely many times. Since the supports of the resultant foliated products are disjoint, we take the union of all of them. See Figure 2. Let $I'\sigma$ be the (infinite) union without σ, and $I\sigma$, the (infinite) union with σ.

Then if we do the construction again, we see that $I\sigma = I'\sigma$ in $H_1(B\overline{G}; \boldsymbol{Z})$. Since $I\sigma = \sigma + I'\sigma$ in $H_1(B\overline{G})$, $\sigma = 0$ in $H_1(B\overline{G}; \boldsymbol{Z})$. See [15]. Q.E.D.

References

[1] A. Banyaga, Sur la structure du groupe des difféomorphismes qui préservent une forme symplectique, Comment. Math. Helv., **53** (1978), 174–227.

[2] A. Banyaga, The structure of classical diffeomorphism groups, Math. Appl., **400**, Kluwer Academic Publishers Group, Dordrecht, 1997.

[3] A. Haefliger, Feuilletage sur les variétés ouvertes, Topology, **9** (1970), 183–194.

[4] A. Haefliger, Homotopy and integrability, Lecture Notes in Math., **197**, Springer, 1971.

[5] M. Herman, Simplicité du groupe des difféomorphismes de classe C^∞, isotopes à l'identité, du tore de dimension n, C. R. Acad. Sci. Paris, **273** (1971), 232–234.

[6] V. V. Lychagin, On sufficient orbits of a group of contact diffeomorphisms, Math. USSR-Sb., **33** (1977), 223–242.

[7] D. McDuff, Local homology of groups of volume preserving diffeomorphisms, Ann. Sci. École Norm. Sup. (4), **15** (1982), 609–648.
[8] J. N. Mather, The vanishing of the homology of certain groups of homeomorphisms, Topology, **10** (1971), 297–298.
[9] J. N. Mather, Integrability in codimension 1, Comment. Math. Helv., **48** (1973), 195–233.
[10] J. N. Mather, Commutators of diffeomorphisms I, II and III, Comment. Math. Helv., **49** (1974), 512–528, **50** (1975), 33–40, and **60** (1985), 122–124.
[11] J. N. Mather, On the homology of Haefliger's classifying space, C.I.M.E., Differential Topology, 1976, pp. 71–116.
[12] W.Thurston, Foliations and group of diffeomorphisms, Bull. Amer. Math. Soc., **80** (1974), 304–307.
[13] W. Thurston, The theory of foliations of codimension greater than one, Comment. Math. Helv., **49** (1974), 214–231.
[14] W.Thurston, Existence of codimension-one foliations, Ann. of Math. (2), **104** (1976), 249–268.
[15] T. Tsuboi, On the homology of classifying spaces for foliated products, In: Foliations, Adv. Stud. Pure Math., **5**, North-Holland, 1985, pp. 37–120.
[16] T. Tsuboi, Foliations and homology of the group of diffeomorphisms, Sugaku, **36** (1984), 320–343; Sugaku expositions, **3** (1990), 145–181.
[17] T. Tsuboi, On the foliated products of class C^1, Ann. of Math. (2), **130** (1989), 227–271.
[18] T. Tsuboi, On the connectivity of the classifying spaces for foliations, Contemp. Math., **96** (1989), 319–331.

Graduate School of Mathematical Sciences
the University of Tokyo
Komaba Meguro, Tokyo 153-8914, Japan
E-mail address: tsuboi@ms.u-tokyo.ac.jp

Advanced Studies in Pure Mathematics 52, 2008
Groups of Diffeomorphisms
pp. 505–524

On the uniform perfectness of diffeomorphism groups

Takashi Tsuboi

Abstract.

We show that any element of the identity component of the group of C^r diffeomorphisms $\mathrm{Diff}^r_c(\boldsymbol{R}^n)_0$ of the n-dimensional Euclidean space $\boldsymbol{R}^n$ with compact support ($1 \leqq r \leqq \infty$, $r \neq n+1$) can be written as a product of two commutators. This statement holds for the interior M^n of a compact n-dimensional manifold which has a handle decomposition only with handles of indices not greater than $(n-1)/2$. For the group $\mathrm{Diff}^r(M)$ of C^r diffeomorphisms of a compact manifold M, we show the following for its identity component $\mathrm{Diff}^r(M)_0$. For an even-dimensional compact manifold M^{2m} with handle decomposition without handles of the middle index m, any element of $\mathrm{Diff}^r(M^{2m})_0$ ($1 \leqq r \leqq \infty$, $r \neq 2m+1$) can be written as a product of four commutators. For an odd-dimensional compact manifold M^{2m+1}, any element of $\mathrm{Diff}^r(M^{2m+1})_0$ ($1 \leqq r \leqq \infty$, $r \neq 2m+2$) can be written as a product of six commutators.

§1. Introduction

For a manifold M, let $\mathrm{Diff}^r_c(M)$ denote the group of C^r diffeomorphisms of M with compact support ($1 \leqq r \leqq \infty$). The *support* of a diffeomorphism f of M is defined to be the *closure* of $\{x \in M \mid f(x) \neq x\}$. Let $\mathrm{Diff}^r_c(M)_0$ denote the identity component of $\mathrm{Diff}^r_c(M)$. Here $\mathrm{Diff}^r_c(M)$ is equipped with the C^r topology. By the results of Mather and Thurston

---

Received January 8, 2008.
Revised April 19, 2008.
2000 *Mathematics Subject Classification.* Primary 57R52, 57R50; Secondary 37C05.
Key words and phrases. diffeomorphism group, uniformly perfect, commutator subgroup.
The author is partially supported by Grant-in-Aid for Scientific Research 16204004, 17104001, Grant-in-Aid for Exploratory Research 18654008, Japan Society for Promotion of Science, and by the 21st Century COE Program at Graduate School of Mathematical Sciences, the University of Tokyo.

([7], [8], [12]), for an n-dimensional manifold M^n, $\mathrm{Diff}_c^r(M^n)_0$ is a perfect group if $r = 0$ or $1 \leqq r \leqq \infty$ and $r \neq n+1$. A group is perfect if it coincides with its commutator subgroup.

We study in this paper the uniform perfectness of $\mathrm{Diff}_c^r(M^n)_0$. A group is uniformly perfect if any element can be written as a product of a bounded number of commutators. In [7], Mather showed that any element of $\mathrm{Homeo}_c(\boldsymbol{R}^n)$ can be written as a commutator. Hence any element of $\mathrm{Homeo}(S^n)_0$ can be written as a product of two commutators. In [14], $\mathrm{Diff}_c^r(\boldsymbol{R}^n)_0$ $(1 \leqq r < n+1)$ is shown to be uniformly perfect. Hence $\mathrm{Diff}^r(S^n)_0$ $(1 \leqq r < n+1)$ is also uniformly perfect. By the result of Herman [5], any element of $\mathrm{Diff}^\infty(S^1)_0$ can be written as a product of two commutators.

We show in this paper that any element of $\mathrm{Diff}_c^r(\boldsymbol{R}^n)_0$ $(1 \leqq r \leqq \infty$, $r \neq n+1)$ can be written as a product of two commutators (Theorem 2.1). The same technique applies to showing that for the interior M^n of a compact n-dimensional manifold which has a handle decomposition only with handles of indices not greater than $(n-1)/2$, any element of $\mathrm{Diff}_c^r(M^n)_0$ $(1 \leqq r \leqq \infty$, $r \neq n+1)$ can be written as a product of two commutators (Theorem 4.1). The handle decomposition of a compact manifold is summarized in Section 3.

For compact manifolds M^n, we show (Theorem 5.1) that if M^n has a handle decomposition without handles of middle indices, then any element of $\mathrm{Diff}^r(M^n)_0$ can be written as a composition of elements to which Theorem 4.1 is applicable. Then we show that for an even-dimensional compact manifold M^{2m} which has a handle decomposition without handles of the middle index m, any element of $\mathrm{Diff}^r(M^{2m})_0$ $(1 \leqq r \leqq \infty$, $r \neq 2m+1)$ can be written as a product of four commutators (Theorem 5.2). For an odd-dimensional compact manifold M^{2m+1}, Theorem 5.2 asserts that if there are no handles of indices m and $m+1$, any element of $\mathrm{Diff}^r(M^{2m+1})_0$ $(1 \leqq r \leqq \infty$, $r \neq 2m+2)$ can be written as a product of four commutators, but we have a stronger result for odd-dimensional compact manifolds. By using the idea of the paper [2] by Burago, Ivanov and Polterovich, we can prove that for any odd-dimensional compact manifold M^{2m+1}, any element of $\mathrm{Diff}^r(M^{2m+1})_0$ $(1 \leqq r \leqq \infty$, $r \neq 2m+2)$ can be written as a product of six commutators (Theorem 6.1).

The topology of the manifold may prevent the group $\mathrm{Diff}^r(M)_0$ from being uniformly perfect. We thought that if an element of $\mathrm{Diff}^r(M)_0$ could be connected to the identity only by a very long isotopy, then the number of commutators to write this element would be long. What we show here is the following. Unless the manifold is even-dimensional and having a handle decomposition with handles of the middle index, we can replace the isotopy by a nicer one to write a diffeomorphisms as a product of bounded number of commutators.

In the proof of Theorem 2.1, we use the result on the perfectness of the group $\mathrm{Diff}_c^r(\boldsymbol{R}^n)_0$ ($1 \leqq r \leqq \infty$, $r \neq n+1$) by Mather and Thurston ([8], [12]) and construct necessary diffeomorphisms. The author got the idea of this construction when he was studying the paper [6] of Dieter Kotschick remembering some discussion with him during his stay at the University of Tokyo in 2006. The author is very grateful to him. The author also thank Shigenori Matsumoto for several valuable comments.

While the author was preparing a preliminary version of this paper, Danny Calegari informed him the existence of the paper [2] by Burago, Ivanov and Polterovich, which the author overlooked. In [2] they proved the results corresponding to our Theorems 2.1, 4.1, and 5.2 in the case of spheres. Moreover they made an excellent observation of tracing the isotopy of a graph after intersecting another graph and showed the uniform perfectness of $\mathrm{Diff}^r(M^3)_0$ for closed 3-dimensional manifolds M^3. The proof of the uniform perfectness of $\mathrm{Diff}^r(M^{2m+1})_0$ for odd-dimensional manifolds M^{2m+1} is rather straight forward after the idea of their paper and our Theorem 5.2. Leonid Polterovich pointed out to the author that these groups treated in this paper are meager in their terminology (Remark 6.6). The author is very grateful to Danny Calegari and Leonid Polterovich for their valuable comments. The author is also grateful to the referee for the suggestions for improving the exposition of this paper.

§2. Diffeomorphisms of the Euclidean space

First we give the proof of the following theorem.

Theorem 2.1. *Let* $\mathrm{Diff}_c^r(\boldsymbol{R}^n)$ *be the group of diffeomorphisms of the* n*-dimensional Euclidean space* $\boldsymbol{R}^n$ *with compact support and let* $\mathrm{Diff}_c^r(\boldsymbol{R}^n)_0$ *be its identity component. If* $1 \leqq r \leqq \infty$ *and* $r \neq n+1$*, then any element of* $\mathrm{Diff}_c^r(\boldsymbol{R}^n)_0$ *can be written as a product of two commutators.*

Proof. Take an element $f \in \mathrm{Diff}_c^r(\boldsymbol{R}^n)_0$ ($r \neq n+1$). By the result of Mather and Thurston ([7], [8], [12]), f can be written as a product of commutators.

$$f = [a_1, b_1] \cdots [a_k, b_k], \quad a_1,\, b_1,\, \ldots,\, a_k,\, b_k \in \mathrm{Diff}_c^r(\boldsymbol{R}^n)_0,$$

where $[a_i, b_i] = a_i b_i {a_i}^{-1} {b_i}^{-1}$. Let U be an open ball in $\boldsymbol{R}^n$ such that the supports of a_i, b_i as well as the supports of the isotopies $\{a_{it}\}_{t\in[0,1]}$ ($a_{i0} = \mathrm{id}$ and $a_{i1} = a_i$), $\{b_{it}\}_{t\in[0,1]}$ ($b_{i0} = \mathrm{id}$ and $b_{i1} = b_i$) are contained in U. Let $g \in \mathrm{Diff}^r_c(\boldsymbol{R}^n)_0$ be an element such that $g^i(U)$ $(i \in \boldsymbol{Z})$ are disjoint. Put

$$F = \prod_{i=1}^{k} g^{k-i}([a_1, b_1] \cdots [a_i, b_i]) g^{i-k}.$$

Then F is an element of $\mathrm{Diff}^r_c(\boldsymbol{R}^n)_0$. Now the conjugate of F by g is as follows:

$$\begin{aligned} gFg^{-1} &= \prod_{i=1}^{k} g^{k-i+1}([a_1, b_1] \cdots [a_i, b_i]) g^{i-k-1} \\ &= \prod_{i=0}^{k-1} g^{k-i}([a_1, b_1] \cdots [a_{i+1}, b_{i+1}]) g^{i-k}. \end{aligned}$$

Hence

$$\begin{aligned} F^{-1}gFg^{-1} &= ([a_1, b_1] \cdots [a_k, b_k])^{-1} \prod_{i=0}^{k-1} g^{k-i}[a_{i+1}, b_{i+1}] g^{i-k} \\ &= f^{-1} \prod_{i=0}^{k-1} g^{k-i}[a_{i+1}, b_{i+1}] g^{i-k} \\ &= f^{-1} \Big[\prod_{i=0}^{k-1} g^{k-i} a_{i+1} g^{i-k}, \prod_{i=0}^{k-1} g^{k-i} b_{i+1} g^{i-k} \Big]. \end{aligned}$$

Put

$$A = \prod_{i=0}^{k-1} g^{k-i} a_{i+1} g^{i-k} \quad \text{and} \quad B = \prod_{i=0}^{k-1} g^{k-i} b_{i+1} g^{i-k},$$

then A and B are elements of $\mathrm{Diff}^r_c(\boldsymbol{R}^n)_0$. Thus f can be written as a product of two commutators: $f = [A, B][g, F^{-1}]$. Q.E.D.

The proof uses only the fact that there is an open set U which contains the support of given finitely many diffeomorphisms and a compact support diffeomorphism g such that $g^i(U)$ $(i \in \boldsymbol{Z})$ are disjoint. Hence we have the following corollary.

Corollary 2.2. *Let M^n be an n-dimensional manifold diffeomorphic to $N^p \times \boldsymbol{R}^q$ $(q \geqq 1,\ p + q = n)$ for a compact manifold N^p, then any element of $\mathrm{Diff}^r_c(M^n)_0$ $(1 \leqq r \leqq \infty,\ r \neq n + 1)$ can be written as a product of two commutators.*

§3. Review of the Morse theory and handle decompositions

In our theorems, the assumptions are given in terms of handle decompositions. We review in this section several facts on the Morse theory for manifolds and handle decompositions ([10], [11]).

A function $f : M^n \longrightarrow \boldsymbol{R}$ on a compact n-dimensional manifold M^n without boundary is called a Morse function if the critical points are nondegenerate, that is, the Hessian matrices of f at the critical points are nondegenerate. For such a function f, the set of critical points is a finite set. The index of the Hessian matrix of f at a critical point is called the index of the critical point.

Any compact connected n-dimensional manifold M^n without boundary admits a Morse function $f : M^n \longrightarrow \boldsymbol{R}$ such that $f(M^n) = [0, n]$, the set of critical points of index k is contained in $f^{-1}(k)$ ($k = 0, \ldots, n$) and $f^{-1}(0)$ and $f^{-1}(n)$ are one point sets ([11]).

Put $W_k = f^{-1}([0, k+1/2])$, and then this W_k is a compact manifold with boundary $\partial W_k = f^{-1}(k + 1/2)$. Let c_k be the number of critical points of index k. Then the manifold W_k is diffeomorphic to the manifold obtained from W_{k-1} by attaching c_k handles of index k ($k = 0, \ldots, n$). This means the following.

Let $D^k \times D^{n-k}$ be the product of the k-dimensional disk D^k and the $(n-k)$-dimensional disk D^{n-k}. Let $\varphi_i : (\partial D^k) \times D^{n-k} \longrightarrow \partial W_{k-1}$ ($i = 1, \ldots, c_k$) be diffeomorphisms with disjoint images. Let

$$W'_k = W_{k-1} \cup_{\bigsqcup_{i=1}^{c_k} \varphi_i} \bigsqcup_{i=1}^{c_k} (D^k \times D^{n-k})_i$$

be the space obtained from the disjoint union $W_{k-1} \sqcup \bigsqcup_{i=1}^{c_k} (D^k \times D^{n-k})_i$ by identifying $x \in (\partial D^k) \times D^{n-k} \subset (D^k \times D^{n-k})_i$ with $\varphi_i(x) \in \partial W_{k-1} \subset W_{k-1}$.

For given triangulations of W_{k-1} and of $(D^k \times D^{n-k})_i$, we have subdivisions of them such that φ_i after isotoped are piecewise linear isomorphisms to the images. Thus W'_k has a triangulation as a piecewise linear manifold.

On the other hand, by smoothing along the corner which is the image $\bigsqcup_{i=1}^{c_k} \varphi_i((\partial D^k) \times (\partial D^{n-k}))$, W'_k has a differentiable structure. This manifold W'_k is the manifold obtained from W_{k-1} by attaching c_k handles of index k ($k = 0, \ldots, n$) which we stated. The image of $D^k \times D^{n-k}$ is called a handle of index k. We will simply write the handle of index k as $(D^k \times D^{n-k})_i$.

Then the manifold W_k is diffeomorphic to the manifold W'_k with boundary. It is better to say that the manifold W_k is obtained from the manifold W'_k by adding the collar of the boundary $\partial W'_k$.

By using the sequence of submanifolds $D^n \cong W_0 \subset W_1 \subset \cdots \subset W_n = M^n$ and the diffeomorphisms $W_k \cong W'_k$, M^n is decomposed into the union of the handles $(D^k \times D^{n-k})_i$ ($i = 1, \ldots, c_k$; $k = 0, \ldots, n$). This decomposition into handles is called a handle decomposition of M. We write the handle decomposition as $D^n \cong W_0 \subset W_1 \subset \cdots \subset W_n = M^n$. This handle decomposition represents a piecewise linear structure as well as a differentiable structure. We call the image of $D^k \times \{0\}$ the core disk of the handle $(D^k \times D^{n-k})_i$ of index k, and the image of $\{0\} \times D^{n-k}$ its co-core disk.

For the above Morse function $f : M^n \longrightarrow \boldsymbol{R}$ and the constant function n, the function $n - f$ is a Morse function, and the critical points of the Morse function f of index k are nothing but the critical points of the Morse function $n - f$ of index $n - k$. Hence this gives rise to a handle decomposition of M^n called the dual handle decomposition. A handle decomposition and its dual handle decomposition can be considered identical as a decomposition of M^n into subsets. The handles of index k of the original handle decomposition corresponds to the handles of index $n - k$ of the dual handle decomposition. This duality switches the roles of core disks and co-core disks.

By choosing a Riemannian metric on the manifold M^n, the Morse function f defines the gradient vector field and the gradient flow Ψ_t. The singular points of the gradient vector field are precisely the critical points of f. The core disk and the co-core disk of a handle of a handle decomposition of M^n correspond to the local stable manifold and the local unstable manifold of the corresponding singular point p of the gradient flow Ψ_t, respectively ([10], [11]). Let e^k_i and e'^{n-k}_i denote the global stable manifold and the global unstable manifold, respectively, for the singular point p which is a critical point of index k of f. Then e^k_i and e'^{n-k}_i are diffeomorphic to $\boldsymbol{R}^k$ and $\boldsymbol{R}^{n-k}$, respectively. Then we know that the global stable manifolds and the global unstable manifolds of critical points give the cell decomposition $\bigcup_{k=0}^n \bigcup_{i=1}^{c_k} e^k_i$ and the dual cell decomposition $\bigcup_{k=0}^n \bigcup_{i=1}^{c_k} e'^{n-k}_i$ of M^n, respectively ([10]). The dual cell decomposition is the cell decomposition for the Morse function $n-f$. Consider the k-skeleton $X^{(k)}$ of the cell decomposition and the $(n-k-1)$-skeleton $X'^{(n-k-1)}$ of the dual cell decomposition:

$$X^{(k)} = \bigcup_{j \leqq k} \bigcup_{i=1}^{c_j} e^j_i \quad \text{and} \quad X'^{(n-k-1)} = \bigcup_{j \geqq k+1} \bigcup_{i=1}^{c_j} e'^{n-j}_i.$$

The boundary ∂W_k of W_k is transverse to the gradient flow Ψ_t, and hence $M \setminus (X^{(k)} \cup X'^{(n-k-1)})$ is diffeomorphic to $\partial W_k \times \boldsymbol{R}$ by the map

$$\partial W_k \times \boldsymbol{R} \ni (x, t) \longmapsto \Psi_t(x) \in M \setminus (X^{(k)} \cup X'^{(n-k-1)}).$$

Moreover $\Psi_t(\partial W_k)$ converges to $X^{(k)}$ as $t \longrightarrow -\infty$ and to $X'^{(n-k-1)}$ as $t \longrightarrow \infty$. Hence, $M \setminus X'^{(n-k-1)}$ is diffeomorphic to the interior $\mathrm{int}(W_k)$

of W_k and $X^{(k)}$ is a deformation retract of both W_k and $M \setminus X'^{(n-k-1)}$:

$$X^{(k)} \subset \operatorname{int}(W_k) \subset W_k \subset M \setminus X'^{(n-k-1)}.$$

Using the flow Ψ_t, for any neighborhood V of $X^{(k)}$ and for any compact subset A in $\operatorname{int}(W_k)$, we can construct an isotopy $\{G_t : \operatorname{int}(W_k) \longrightarrow \operatorname{int}(W_k)\}_{t\in[0,1]}$ with compact support such that $G_0 = \mathrm{id}_{\operatorname{int}(W_k)}$, $G_t|X^{(k)} = \mathrm{id}_{X^{(k)}}$ $(t \in [0,1])$ and $G_1(A) \subset V$. A similar statement is true for $X^{(k)} \subset M \setminus X'^{(n-k-1)}$.

By careful choices of the Morse function and the Riemannian metric on M, the cell complexes $X^{(k)}$ and $X'^{(n-k-1)}$ become differentiably embedded simplicial complexes. Since we use this fact, we give here a sketch of the proof.

Proposition 3.1. *Let L be an ℓ-dimensional simplicial complex differentiably embedded in ∂W_k $(\ell \geqq k)$. Then there is an $(\ell+1)$-dimensional simplicial complex $\widehat{L}$ differentiably embedded in W_k such that $\partial W_k \cap \widehat{L} = L$ and $\widehat{L}$ is a deformation retract of W_k.*

Sketch of the proof. The proof is roughly as follows: It is shown by the induction on k. For $k = 0$, we take the cone of L as $\widehat{L}$. We assume that the assertion is true for $k-1$ and we construct $\widehat{L}$ for W_k. First, for the handles $(D^k \times D^{n-k})_i$ of index k $(i = 1, \ldots, c_k)$, we can deform the co-core disks $(\{0\} \times D^{n-k})_i$ so that the belt spheres $S_i^{n-k-1} = (\{0\} \times (\partial D^{n-k}))_i$ are in general position to L. In a neighborhood of a belt sphere S_i^{n-k-1}, L is isomorphic to the product of a small k-dimensional disk B^k and $S_i^{n-k-1} \cap L$. We can subdivide $L'_i = S_i^{n-k-1} \cap L$ so that L'_i becomes an $(\ell-k)$-dimensional simplicial complex. Then we subdivide L and the triangulation of $(D^k \times D^{n-k})_i$ $(\subset W_k)$ so that $B^k \times (\{0\} * L'_i)$ $(\subset (B^k \times D^{n-k})_i)$ becomes an $(\ell+1)$-dimensional subcomplex of the subcomplex $(B^k \times D^{n-k})_i$ of $(D^k \times D^{n-k})_i \subset W_k$ $(i = 1, \ldots, c_k)$ after isotoping the triangulation of the handle $(D^k \times D^{n-k})_i$. Here, $\{0\}$ is the center of the co-core disk $(\{0\} \times D^{n-k})_i$ and we regard B^k as a small disk embedded in D^k. If remove $(\operatorname{Int}(B^k) \times D^{n-k})_i$ $(i = 1, \ldots, c_k)$ from W_k, we obtain a piecewise linear manifold W''_{k-1} isomorphic to W_{k-1} and an ℓ-dimensional simplicial complex

$$L_1 = (L \setminus \bigcup_{i=1}^{c_k} B^k \times L'_i) \cup \bigcup_{i=1}^{c_k} (\partial B^k) \times (\{0\} * L'_i)$$

on $\partial W''_{k-1}$. By the induction hypothesis, we have an $(\ell+1)$-dimensional simplicial complex $\widehat{L}_1$ in W''_{k-1} such that $\partial W''_{k-1} \cap \widehat{L}_1 = L_1$ and $\widehat{L}_1$ is a deformation retract of W''_{k-1}. Since $W''_{k-1} \cup \bigcup_{i=1}^{c_k} B^k \times (\{0\} * L'_i)$ is a deformation retract of W_k, $\widehat{L} = \widehat{L}_1 \cup \bigcup_{i=1}^{c_k} B^k \times (\{0\} * L'_i)$ is the desired $(\ell+1)$-dimensional simplicial complex. Q.E.D.

As for this proposition, the case where L is the empty set corresponds to the construction of a k-dimensional simplicial complex K^k in W_k which is a deformation retract of W_k. In this case, the complex K^k is constructed from the core disks $(D^k \times \{0\})_i$ $(\subset (D^k \times D^{n-k})_i)$.

Corollary 3.2. *There is a k-dimensional simplicial complex K^k differentiably embedded in W_k which is a deformation retract of W_k.*

We note here that by careful choices of the Morse function and the Riemannian metric on M^n, we can make K^k be differentiably embedded and be the union of the stable manifolds of the gradient flow. Hence we have the following proposition.

Proposition 3.3. *Let $D^n \cong W_0 \subset W_1 \subset \cdots \subset W_n = M^n$ be a handle decomposition.*

(1). *There is a k-dimensional simplicial complex K^k differentiably embedded in W_k such that, for any neighborhood V of K^k and for any compact subset A in $\mathrm{int}(W_k)$, there is an isotopy $\{G_t : \mathrm{int}(W_k) \longrightarrow \mathrm{int}(W_k)\}_{t\in[0,1]}$ with compact support such that $G_0 = \mathrm{id}_{\mathrm{int}(W_k)}$, $G_t|K^k = \mathrm{id}_{K^k}$ $(t \in [0,1])$ and $G_1(A) \subset V$.*

(2). *There is an $(n-k-1)$-dimensional simplicial complex K'^{n-k-1} differentiably embedded in $M \setminus W_k$ such that, for any neighborhood V of K^k and for any compact subset A in $M \setminus K'^{n-k-1}$, there is an isotopy $\{G_t : M \setminus K'^{n-k-1} \longrightarrow M \setminus K'^{n-k-1}\}_{t\in[0,1]}$ with compact support such that $G_0 = \mathrm{id}_{M\setminus K'^{n-k-1}}$, $G_t|K^k = \mathrm{id}_{K^k}$ $(t \in [0,1])$ and $G_1(A) \subset V$.*

Remark 3.4. For a compact connected n-dimensional manifold M^n with boundary ∂M^n, we have a handle decomposition of the form $D^n \cong W_0 \subset W_1 \subset \cdots \subset W_k = M^n$ for some $k < n$. Then Proposition 3.3 (1) holds.

§4. Diffeomorphisms of manifolds with small spines

We study the group of diffeomorphisms of open manifolds to which the idea of the proof of Theorem 2.1 applies.

Theorem 4.1. *Let M^n be the interior of a compact n-dimensional manifold with handle decomposition with handles of indices not greater than $(n-1)/2$, then any element of $\mathrm{Diff}^r_c(M^n)_0$ $(1 \leqq r \leqq \infty,\ r \neq n+1)$ can be written as a product of two commutators.*

Proof. This theorem is a corollary to the following Proposition 4.2. For, by Proposition 3.3 (Remark 3.4), we can construct a k-dimensional simplicial complex K^k ($k \leqq (n-1)/2$) differentiably embedded in M^n such that, for any compact set A in M^n and for any neighborhood V of K^k, there is an isotopy $\{G_t\}_{t\in[0,1]}$ required in Proposition 4.2. Q.E.D.

Proposition 4.2. *Let M^n be an n-dimensional manifold. Assume that $2k+1 \leqq n$ and there is a finite k-dimensional simplicial complex K^k differentiably embedded in M^n such that for any compact set A in M^n and any neighborhood V of K^k, there is an isotopy $\{G_t : M^n \longrightarrow M^n\}_{t\in[0,1]}$ such that $G_0 = \mathrm{id}_{M^n}$, $G_t|K^k = \mathrm{id}_{K^k}$ ($t \in [0,1]$) and $G_1(A) \subset U$. Then any element of $\mathrm{Diff}^r_c(M^n)_0$ ($1 \leqq r \leqq \infty$, $r \neq n+1$) can be written as a product of two commutators.*

For the proofs of this proposition and the theorems for diffeomorphisms of compact manifolds we need the following lemmas. These lemmas should be well-known but we include their proofs for the completeness.

Lemma 4.3. *Let M^n be a compact n-dimensional manifold. Let K^k and L^ℓ be k-dimensional and ℓ-dimensional finite simplicial complexes, respectively. Let $f : K^k \longrightarrow M^n$ and $g : L^\ell \longrightarrow M^n$ be differentiable maps and assume that f is an embedding. If $k+\ell+1 \leqq n$ then there is an isotopy $\{F_t : M^n \longrightarrow M^n\}_{t\in[0,1]}$ ($F_0 = \mathrm{id}$) such that $F_1(f(K^k)) \cap g(L^\ell) = \emptyset$.*

Proof. We construct the isotopy F_t, skeleton by skeleton. We consider that $K^k \subset M^n$ and let $K^{(m)}$ be the m skeleton of K^k ($m = 0, \ldots, k$). Assume that there is an isotopy $\{F^m_t\}_{t\in[0,1]}$ such that $F^m_1(K^{(m)}) \cap g(L^\ell) = \emptyset$. Assume also that the number of $(m+1)$-dimensional simplices of K^k is N_{m+1} and for $0 \leqq u < N_{m+1}$, we obtained an isotopy $\{F^{(m+1),u}_t\}_{t\in[0,1]}$ such that

$$F^{(m+1),u}_1(K^{(m)} \cup (\sigma^{m+1}_1 \cup \cdots \cup \sigma^{m+1}_u)) \cap g(L^\ell) = \emptyset.$$

For the $(m+1)$-dimensional simplex σ^{m+1}_{u+1} of K^k, $F^{(m+1),u}_1(\sigma^{m+1}_{u+1})$ is differentiably embedded in M^n. We take the normal bundle ν of $F^{(m+1),u}_1(\sigma^{m+1}_{u+1})$, and take the image U under the exponential map of a small disk bundle in ν. Let $\pi : U \longrightarrow F^{(m+1),u}_1(\sigma^{m+1}_{u+1})$ be the projection. We may assume that for neighborhoods V_0 and V_1 of $\partial\sigma^{m+1}_{u+1}$ in σ^{m+1}_{u+1} such that $\partial\sigma^{m+1}_{u+1} \subset V_0 \subset \overline{V_0} \subset V_1$, $\pi^{-1}(V_1) \cap L^\ell = \emptyset$ and $\pi^{-1}(\sigma^{m+1}_{u+1} \setminus V_0)$ does not intersect other $(m+1)$-dimensional simplices of $F^{(m+1),u}_1(K^k)$. Since this normal bundle is trivial, we have a projection $p : U \longrightarrow \boldsymbol{R}^{n-m-1}$ (of rank $n-m-1$) such that $p^{-1}(0) = F^{(m+1),u}_1(\sigma^{m+1}_{u+1})$. Note that $p(g(L^\ell) \cap U)$ is a finite union of the images under differentiable maps of simplices of dimension not greater than ℓ. Since $\ell \leqq n-k-1 \leqq n-(m+1)-1$, $p(g(L^\ell)\cap U)$ is a nowhere dense subset of $\boldsymbol{R}^{n-m-1}$. Take a point q close to 0 in $p(U) - p(g(L^\ell)) \subset p(U) \subset \boldsymbol{R}^{n-m-1}$. Let $\{F_t^{(m+1),u+1}\}_{t\in[0,1]}$ be an isotopy with support in U such that $\pi(F_t^{(m+1),u+1}(x)) = x$, $p(F_t^{(m+1),u+1}(x)) = t\mu(x)q$ for $x \in \sigma_{u+1}^{m+1}$, where $\mu : \sigma_{u+1}^{m+1} \longrightarrow [0,1]$ is a C^∞ function such that $\mu(x) = 1$ for $x \in \sigma_{u+1}^{m+1} \setminus V_1$ and $\mu(x) = 0$ for $x \in V_0$. Then $F_1^{(m+1),u+1}(\sigma_{u+1}^{m+1}) \cap g(L) = \emptyset$.

Thus we obtain an isotopy $\{F_t^{m+1}\}_{t\in[0,1]}$ such that $F_1^{m+1}(K^{(m+1)})\cap g(L^\ell) = \emptyset$ as the composition of $\{F_t^{(m+1),N_k-1}\}_{t\in[0,1]}, \ldots, \{F_t^{(m+1),0} = F_t^m\}_{t\in[0,1]}$.

Note that the support of the isotopy $\{F_t\}_{t\in[0,1]}$ can be made to be an arbitrarily small compact neighborhood of K^k. Q.E.D.

Remark 4.4. Under the notation of Lemma 4.3, if $k+\ell = n$, then we obtain F_t such that $F_1(f(K^{(k-1)})) \cap g(L^\ell) = \emptyset$, $F_1(f(K^k)) \cap g(L^{(\ell-1)}) = \emptyset$ and the intersection $F_1(f(\sigma^k)) \cap g(\tau^\ell)$ is transverse for each k-dimensional simplex σ^k of K^k and each ℓ-dimensional simplex τ^ℓ of L^ℓ, where $L^{(\ell-1)}$ denotes the $(\ell-1)$ skeleton of L^ℓ. For, we can proceed as in the proof of Lemma 4.3, and for the modification with respect to a k-dimensional simplex σ_{u+1}^k of K^k, we can use a regular value of the projection $p : U \longrightarrow \boldsymbol{R}^\ell$ to the fiber of the normal bundle of $F_1^{(k),u}(\sigma_{u+1}^k)$.

Lemma 4.5. *Let M^n be an n-dimensional manifold. Let K^k be a k-dimensional finite simplicial complex differentiably embedded in M^n. If $2k+1 \leqq n$, then there are an isotopy $\{F_t : M^n \longrightarrow M^n\}_{t\in[0,1]}$ with compact support ($F_0 = \mathrm{id}$) and an open neighborhood U of K^k such that $(F_1)^\ell(U)$ ($\ell \in \boldsymbol{Z}$) are disjoint.*

Proof. There is a neighborhood V of K^k such that for any neighborhood W of K^k ($K^k \subset W \subset \overline{W} \subset V$) and a compact subset $A \subset V$, there is an isotopy $\{G_t : M^n \longrightarrow M^n\}_{t\in[0,1]}$ with support in V such that $G_0 = \mathrm{id}$ and $G_1(A) \subset W$. The neighborhood U is defined by using the structure of normal bundles of each simplex of K^k used in the proof of Lemma 4.3. Then the isotopy is defined skeleton by skeleton by using the normal bundle projections.

By Lemma 4.3 applied to $g(L^\ell) = K^k$, there is an isotopy $\{h_t\}_{t\in[0,1]}$ such that $h_0 = \mathrm{id}$ and $h_1(K^k) \cap K^k = \emptyset$. We may assume that the support of the isotopy $\{h_t\}_{t\in[0,1]}$ is contained in V.

Take a neighborhood W_0 of K^k and W_1 of $h_1(K^k)$ such that $W_0 \cap W_1 = \emptyset$. Then $W = W_0 \cap (h_1)^{-1}(W_1)$ is a neighborhood of K^k such that

$W \cap h_1(W) = \emptyset$. Here we can take W_0 and W_1 such that their closures $\overline{W_0}$ and $\overline{W_1}$ are compact, and then $\overline{W}$ is compact.

For W and $h_1(\overline{W})$, we have an isotopy $\{G_t : M^n \longrightarrow M^n\}_{t\in[0,1]}$ with support in V such that $G_0 = \mathrm{id}$ and $G_1(h_1(\overline{W})) \subset W$.

Let F_t be the composition of G_t and h_t. Then $F_1(W) \subset W$. Since $F_1(W)\cap K^k = \emptyset$, we can take a neighborhood U of K^k such that $U \subset W$ and $U \cap F_1(W) = \emptyset$. Then $U \subset W \setminus F_1(W)$ and for $\ell > 0$, $(F_1)^\ell(U) \subset (F_1)^\ell(W) \setminus (F_1)^{\ell+1}(W)$. Hence $(F_1)^\ell(U)$ ($\ell \in \boldsymbol{Z}$, $\ell \geqq 0$) are disjoint. Then $(F_1)^\ell(U)$ ($\ell \in \boldsymbol{Z}$) are disjoint. Q.E.D.

Proof of Proposition 4.2. By Lemma 4.5, there are a neighborhood U of K^k and an element g of $\mathrm{Diff}_c^r(M^n)_0$ such that $g^i(U)$ ($i \in \boldsymbol{Z}$) are disjoint. For any element $f \in \mathrm{Diff}_c^r(M^n)_0$, by the assumption of the proposition, there is an isotopy $\{G_t : M^n \longrightarrow M^n\}_{t\in[0,1]}$ such that $G_0 = \mathrm{id}$, $G_t|K^k = \mathrm{id}_{K^k}$ and $G_1(\mathrm{supp}(\{f_t\}_{t\in[0,1]})) \subset U$. Then by the argument of Theorem 2.1, $G_1 \circ f \circ {G_1}^{-1}$ can be written as a product of two commutators in $\mathrm{Diff}_c^r(M^n)_0$. Hence f can be written as product of two commutators in $\mathrm{Diff}_c^r(M^n)_0$. Q.E.D.

§5. Diffeomorphisms of compact manifolds

If a compact manifold M has a decomposition into nice pieces, we can show that any element of $\mathrm{Diff}^r(M)_0$ can be written as a composition of diffeomorphisms to which we can apply Theorem 4.1, and that any element of $\mathrm{Diff}^r(M)_0$ can be written as a product of a bounded number of commutators.

Theorem 5.1. *Let M^n be a compact n-dimensional manifold. Let P^p and Q^q be p-dimensional and q-dimensional finite simplicial complexes differentiably embedded in M^n, respectively. Assume that $p+q+2 \leqq n$ and that $P^p\cap Q^q = \emptyset$. Then any element $f \in \mathrm{Diff}^r(M^n)_0$ ($1 \leqq r \leqq \infty$) can be written as a product $f = g\circ h$ such that $g \in \mathrm{Diff}_c^r(M^n\setminus k(Q^q))_0$ and $h \in \mathrm{Diff}_c^r(M^n \setminus P^p)_0$, where $k \in \mathrm{Diff}^r(M^n)_0$, $k(Q^q) \cap P^p = \emptyset$, and $\mathrm{Diff}_c^r(M^n \setminus k(Q^q))_0$ and $\mathrm{Diff}_c^r(M^n \setminus P^p)_0$ are considered as subgroups of $\mathrm{Diff}^r(M^n)_0$, respectively.*

By using Theorems 5.1 and 4.1, we obtain the following theorem.

Theorem 5.2. *Let M^n be a compact n-dimensional manifold. If M^n has a handle decomposition without handles of middle indices, that is, if there is a handle decomposition with p-handles, where $2p + 2 \leqq n$ or $2p - 2 \geqq n$, then any element of $\mathrm{Diff}^r(M^n)_0$ ($1 \leqq r \leqq \infty$, $r \neq n + 1$) can be written as a product of four commutators. In particular, if M^{2m} is a $(2m)$-dimensional compact manifold with a handle decomposition* *without handles of index m, any element of* $\mathrm{Diff}^r(M^{2m})_0$ $(1 \leqq r \leqq \infty$, $r \neq 2m+1)$ *can be written as a product of four commutators.*

Proof. We look at the handle decomposition and the dual handle decomposition of M^n. Then by using the core disks of the handles of indices not greater than $q = (n-2)/2$ of the handle decomposition and the dual handle decomposition, we obtain q-dimensional simplicial complexes P^q and Q^q such that $P^q \cap Q^q = \emptyset$. Since $q + q + 2 \leqq n$, any element $f \in \mathrm{Diff}^r(M^n)_0$ can be written as a product $f = g \circ h$ such that $g \in \mathrm{Diff}^r_c(M^n \setminus k(Q^q))_0$ and $h \in \mathrm{Diff}^r_c(M^n \setminus P^q)_0$ by Theorem 5.1, where $k \in \mathrm{Diff}^r(M^n)_0$. By Proposition 3.3, $M^n \setminus P^q$ and $M^n \setminus Q^q$ as well as $M^n \setminus k(Q^q)$ satisfy the assumption of Theorem 4.1. Hence g and h can be written as product of two commutators in $\mathrm{Diff}^r(M^n)_0$. Thus Theorem 5.2 is proved. Q.E.D.

Proof of Theorem 5.1. Let $\{f_t\}_{t\in[0,1]}$ be the isotopy such that $f_0 = \mathrm{id}$ and $f_1 = f$. Let $F : [0,1] \times M^n \longrightarrow M^n$ be the trace of the isotopy: $F(t,x) = f_t(x)$.

We look at the image $F([0,1] \times P^p) \subset M^n$. Since $p + 1 + q \leqq n - 1$, by Lemma 4.3, there is an isotopy $\{k_s\}_{s\in[0,1]}$ $(k_0 = \mathrm{id}$, $k_1 = k)$ such that $F([0,1] \times P^p) \cap k(Q^q) = \emptyset$.

Let U be a neighborhood of $F([0,1] \times P^p)$ and V be a neighborhood of $k(Q^q)$ such that $U \cap V = \emptyset$.

Let ξ be the vector field on $[0,1] \times M^n$ given by

$$\frac{\partial}{\partial t} + \Big(\frac{\mathrm{d}f_{t+s}(x)}{\mathrm{d}s}\Big)_{s=0}$$

at $(t, f_t(x))$. This ξ generates the isotopy f_t. Let η be a vector field on $[0,1] \times M^n$ with support in $[0,1] \times U$ such that $\eta = \xi$ on a neighborhood of $\{(t, f_t(x_0)) \mid x_0 \in P^p,\ t \in [0,1]\}$. Then $\eta = \partial/\partial t$ on $[0,1] \times V$ which is a neighborhood of $[0,1] \times k(Q^q)$. Then η generates an isotopy $\{g_t\}_{t\in[0,1]}$ such that g_t is the identity on the neighborhood V of $k(Q^q)$ and $g_t(x) = f_t(x)$ for x in a neighborhood of P^p. Put $h = {g_1}^{-1} f_1$, then h is the identity in a neighborhood of P^p, and it is isotopic to the identity as an element of $\mathrm{Diff}^r(M^n)$. Put $h_t = {g_t}^{-1} \circ f_t$. Then h_t is the identity on a neighborhood of P^p.

Thus $f = g \circ h$ and $g \in \mathrm{Diff}^r_c(M^n \setminus k(Q^q))_0$ and $h \in \mathrm{Diff}^r_c(M^n \setminus P^p)_0$. Q.E.D.

Remark 5.3. In the proof of Theorem 5.1, the decomposition of a diffeomorphism uses only the fact that $F([0,1] \times P^p) \cap k(Q^q) = \emptyset$.

Remark 5.4. For a compact manifold M we have a handle decomposition. For a compact odd-dimensional manifold M^{2m+1}, M^{2m+1} is covered by two open sets U_1 and U_2 which are neighborhoods of the union of handles of indices not greater than m and the union of dual handles of indices not greater than m. Then by the fragmentation lemma ([1]), there is a neighborhood $\mathcal{N}$ of the identity in $\mathrm{Diff}^r(M^{2m+1})_0$ such that every element f of $\mathcal{N}$ can be written as a product $f = g \circ h$, where $g \in \mathrm{Diff}^r_c(U_1)_0$ and $h \in \mathrm{Diff}^r_c(U_2)_0$. Hence by Theorem 4.1, every element f of $\mathcal{N}$ can be written as a product of four commutators of elements of $\mathrm{Diff}^r(M^{2m+1})_0$ ($1 \leqq r \leqq \infty$, $r \neq 2m+2$). For a compact even-dimensional manifold M^{2m}, M^{2m} is covered by three open sets U_1, U_2 and U_3. Here, U_1 and U_2 are neighborhoods of the union of handles of indices not greater than $m-1$ and the union of dual handles of indices not greater than $m-1$, and U_3 is a disjoint union of open balls which is a neighborhood of the union of m handles. Then by the fragmentation lemma, there is a neighborhood $\mathcal{N}$ of the identity in $\mathrm{Diff}^r(M^{2m})_0$ such that every element f of $\mathcal{N}$ can be written as a product $f = a \circ g \circ h$, where $g \in \mathrm{Diff}^r_c(U_1)_0$, $h \in \mathrm{Diff}^r_c(U_2)_0$ and $a \in \mathrm{Diff}^r_c(U_3)_0$. Hence by Theorem 4.1, every element f of $\mathcal{N}$ can be written as a product of six commutators of elements of $\mathrm{Diff}^r(M^{2m})_0$ ($1 \leqq r \leqq \infty$, $r \neq 2m+1$).

§6. Diffeomorphisms of odd-dimensional compact manifolds

In [2], Burago, Ivanov and Polterovich proved that for a closed 3-dimensional manifold M^3, any element of $\mathrm{Diff}^r(M^3)_0$ can be written as a product of ten commutators. Their method together with the general position argument in the previous sections gives the following theorem.

Theorem 6.1. *Let M^{2m+1} be a compact $(2m+1)$-dimensional manifold. Then any element of $\mathrm{Diff}^r(M^{2m+1})_0$ ($1 \leqq r \leqq \infty$, $r \neq 2m+2$) can be written as a product of six commutators.*

Proof. We look at the handle decomposition and the dual handle decomposition of M^{2m+1}. Then by using the core disks of the handles of indices not greater than m of the handle decomposition and the dual handle decomposition, we obtain m-dimensional simplicial complexes P^m and Q^m such that $P^m \cap Q^m = \emptyset$. By Proposition 3.3, $M^{2m+1} \setminus P^m$ and $M^{2m+1} \setminus Q^m$ satisfy the assumption of Theorem 4.1. Then the theorem follows from the following theorem and Theorems 2.1 and 4.1. Q.E.D.

Theorem 6.2. *Let M^{2m+1} be a compact $(2m+1)$-dimensional manifold. Let P^m and Q^m be m-dimensional finite simplicial complexes differentiably embedded in M^{2m+1}, respectively. Assume that $P^m \cap Q^m = \emptyset$. Then any element $f \in \mathrm{Diff}^r(M^{2m+1})_0$ ($1 \leqq r \leqq \infty$) can be written as a product $f = a \circ g \circ h$ such that $a \in \mathrm{Diff}^r_c(\bigsqcup_i U_i)_0$, $g \in \mathrm{Diff}^r_c(M^{2m+1} \setminus$*

$k(Q^m))_0$ and $h \in \mathrm{Diff}^r_c(M^{2m+1} \setminus k'(P^m))_0$, where $\bigsqcup_i U_i$ is a disjoint union of $(2m+1)$-dimensional open balls U_i embedded in M^{2m+1}, k, $k' \in \mathrm{Diff}^r(M^{2m+1})_0$, and $\mathrm{Diff}^r_c(\bigsqcup_i U_i)_0$, $\mathrm{Diff}^r_c(M^{2m+1} \setminus k(Q^m))_0$ and $\mathrm{Diff}^r_c(M^{2m+1} \setminus k'(P^m))_0$ are considered as subgroups of $\mathrm{Diff}^r(M^{2m+1})_0$, respectively.

For the proof of Theorem 6.2, we need several lemmas.

Lemma 6.3. *Let $P^{(m-1)}$ and $Q^{(m-1)}$ be the $m-1$ skeletons of P^m and Q^m, respectively. Then any element $f \in \mathrm{Diff}^r(M^{2m+1})_0$ can be written as a product $f = g \circ h$ such that $g \in \mathrm{Diff}^r_c(M^{2m+1} \setminus k(Q^m))_0$ and $h \in \mathrm{Diff}^r_c(M^{2m+1} \setminus P^{(m-1)})_0$, where $k \in \mathrm{Diff}^r(M^{2m+1})_0$ and $k(Q^m) \cap P^m = \emptyset$. Moreover there is an isotopy $\{h_t\}_{t\in[0,1]}$ such that $h_0 = \mathrm{id}$, $h_1 = h$, h_t is the identity in a neighborhood of $P^{(m-1)}$, and for $H(t,x) = h_t(x)$, $H([0,1] \times P^m) \cap k(Q^{(m-1)}) = \emptyset$ and, for m-dimensional simplices τ^m of P^m and σ^m of Q^m, the intersection $H([0,1] \times \tau^m) \cap k(\sigma^m)$ is transverse. Thus $H([0,1] \times P^m) \cap k(Q^m)$ is a finite set.*

Proof. Let $\{f_t\}_{t\in[0,1]}$ be the isotopy such that $f_0 = \mathrm{id}$ and $f_1 = f$. Let $F : [0,1] \times M^{2m+1} \longrightarrow M^{2m+1}$ be the trace of the isotopy: $F(t,x) = f_t(x)$. As in the proof of Theorem 5.1, we look at the image $F([0,1] \times P^m) \subset M^{2m+1}$.

Since the dimension of the manifold is $2m+1$, by Lemma 4.3 and Remark 4.4, there is an isotopy $\{k_s\}_{s\in[0,1]}$ ($k_0 = \mathrm{id}$, $k_1 = k$) such that $F([0,1] \times P^m) \cap k(Q^{(m-1)}) = \emptyset$, $F([0,1] \times P^{(m-1)}) \cap k(Q^m) = \emptyset$ and $k(\sigma^m)$ is transverse to $F([0,1] \times \tau^m)$ for each pair of m-dimensional simplices σ^m of Q^m and τ^m of P^m. Hence, $F([0,1] \times P^m) \cap k(Q^m)$ is a finite set:

$$F([0,1] \times P^m) \cap k(Q^m) = \{F(t_i, u_i) \mid i = 1, \ldots, r\} \subset M^{2m+1}.$$

We proceed as in the proof of Theorem 5.1. We can take an isotopy g_t fixing a neighborhood of $k(Q^m)$ and $g_t = f_t$ in a small neighborhood of $P^{(m-1)}$. Then for $H(t,x) = h_t(x) = g_t{}^{-1} \circ f_t(x)$, h_t is the identity on a neighborhood of $P^{(m-1)}$. Thus the intersection $H([0,1] \times P^m) \cap k(Q^m)$ is transverse. Since $H(t_i, u_i) = F(t_i, u_i)$,

$$\begin{aligned} H([0,1] \times P^m) \cap k(Q^m) &= \{F(t_i, u_i) \mid i = 1, \ldots, r\} \\ &= \{H(t_i, u_i) \mid i = 1, \ldots, r\} \subset M^{2m+1}. \end{aligned}$$

Q.E.D.

We would like to decompose an element $\overline{h}$ close to h as a composition of an element $a \in \mathrm{Diff}^r_c(\bigsqcup_i U_i)_0$, where $\bigsqcup_i U_i$ is a disjoint union of

$(2m+1)$-dimensional open balls U_i embedded in M^{2m+1}, an element $\overline{g} \in \mathrm{Diff}^r_c(M^{2m+1} \setminus k(Q^m))_0$ and an element $\overline{h}' \in \mathrm{Diff}^r_c(M^{2m+1} \setminus k'(P^m))_0$:

$$\overline{h} = a \circ \overline{g} \circ \overline{h}'.$$

By the classical result of Whitney [18], we have the following lemma.

Lemma 6.4. *Let $\{h_t\}_{t\in[0,1]}$ be an isotopy which is the identity in a neighborhood of $P^{(m-1)}$ and put $H(t,x) = h_t(x)$. Let $V \subset P$ be the complement of a neighborhood of $P^{(m-1)}$ where $h_t = \mathrm{id}$. Then there is an isotopy $\{\overline{h}_t\}_{t\in[0,1]}$ fixing a neighborhood of $P^{(m-1)}$ such that its trace $\overline{H} : [0,1]\times M^{2m+1} \longrightarrow M^{2m+1}$ is close to $H : [0,1]\times M^{2m+1} \longrightarrow M^{2m+1}$ and $\overline{H}|[0,1] \times V$ is an immersion outside of a finite subset. Moreover the image $\overline{H}([0,1] \times V) \subset M^{2m+1} \setminus (P^{(m-1)} \cup k(Q^{(m-1)}))$ has finitely many double point curves which is in general position with respect to the curves $\overline{H}([0,1] \times \{v\})$ $(v \in V)$. If $m \geqq 2$ these double point curves are disjoint, and if $m = 1$, there are at most finitely many triple points and cusps.*

Lemma 6.5. *For generic $\overline{h} = \overline{h}_1 \in \mathrm{Diff}^r_c(M^{2m+1} \setminus P^{(m-1)})_0$ given by Lemma 6.4, $\overline{h}$ can be decomposed as $\overline{h} = a \circ \overline{g} \circ \overline{h}'$, where $a \in \mathrm{Diff}^r_c(\bigsqcup_i U_i)_0$, $\bigsqcup_i U_i$ is a disjoint union of $(2m+1)$-dimensional open balls U_i embedded in M^{2m+1}, $\overline{g} \in \mathrm{Diff}^r_c(M^{2m+1} \setminus k(Q^m))_0$ and $\overline{h}' \in \mathrm{Diff}^r_c(M^{2m+1} \setminus P^m)_0$.*

For the proof of Lemma 6.5, we need to find the open balls U_i. These balls are neighborhoods of embedded arcs or embedded trees in $M^{2m+1} \setminus (P^m \cup k(Q^{(m-1)}))$. This is a construction essentially due to Burago, Ivanov and Polterovich ([2])

Let

$$\overline{H}([0,1] \times P^m) \cap k(Q^m) = \{\overline{H}(s_i, v_i) \mid i = 1, \ldots, r\} \subset M^{2m+1} \setminus (P^m \cup k(Q^{(m-1)})).$$

We look at $\overline{H}([s_i,1] \times \{v_i\})$. For generic $\overline{H}$, $\overline{H}([s_i,1] \times \{v_i\})$ does not intersect $P^m \cup k(Q^m)$ other than $\overline{H}(s_i, v_i) \in k(Q^m)$

If $m \geqq 2$, then for generic $\overline{H}$, $\overline{H}([s_i,1] \times \{v_i\})$ does not intersect the double point curves.

If $m = 1$, then $\overline{H}([s_i,1] \times \{v_i\})$ may intersect the double point curves. For generic $\overline{H}$, the intersection consists of finitely many points $\overline{H}(s_{i,i_1}, v_i) = \overline{H}(s'_{i,i_1}, v'_{i,i_1})$ $(i_1 = 1, \ldots, j_i)$, where we only take the double points such that $s'_{i,i_1} > s_{i,i_1}$. For the double points where $s'_{i,i_1} > s_{i,i_1}$, we look at the curve $\overline{H}([s'_{i,i_1}, 1] \times \{v'_{i,i_1}\})$. For generic $\overline{H}$, $\overline{H}([s'_{i,i_1}, 1] \times \{v'_{i,i_1}\})$ does not intersect $P^1 \cup k(Q^1)$ but may intersect the double point curves at finitely many points again. $\overline{H}(s'_{i,i_1,i_2}, v'_{i,i_1}) = \overline{H}(s''_{i,i_1,i_2}, v''_{i,i_1,i_2})$ $(i_2 = 1, \ldots, j_{i,i_1})$. Then for $s''_{i,i_1,i_2} > s'_{i,i_1,i_2}$, we look at the curve $\overline{H}([s''_{i,i_1,i_2}, 1] \times \{v''_{i,i_1,i_2}\})$.

We continue this process and obtain trees consisting of arcs of the form $\overline{H}([s, 1] \times \{v\})$ starting at the points of the intersection $\overline{H}([0, 1] \times P^1) \cap k(Q^1)$ bifurcating at the double points which are the intersections of the arcs and the forward image $\overline{h}_t(P^1)$ of P^1 under the isotopy. Note that the branches of the trees are finitely many. It is because outside of small neighborhoods of the tangencies of the double point curves and the curves $\overline{H}([0, 1] \times \{v\})$ $(v \in V)$ and outside of small neighborhoods of triple points and cusps, there exists a positive real number δ such that two intersecting points $\overline{H}(s_0, v)$, $\overline{H}(s_1, v)$ of the double point curves and $\overline{H}([0, 1] \times \{v\})$ satisfy $|s_0 - s_1| > \delta$. Thus we obtain final branches which look like $\overline{H}([s^{(k)}_{i,i_1,i_2,\ldots,i_k}, 1] \times \{v^{(k)}_{i,i_1,i_2,\ldots,i_k}\})$. Note also that the tree intersect $P^1 \cup k(Q^1)$ only at the starting point $\overline{H}(s_i, v_i) \in k(Q^1)$.

Proof of Lemma 6.5 *for* $m \geqq 2$. If $m \geqq 2$, using the curves $\overline{H}([s_i, 1] \times \{v_i\})$, we can define an isotopy $\{a_t\}_{t \in [0,1]}$ $(a_0 = \mathrm{id})$ with support in neighborhoods of $\overline{H}([s_i, 1] \times \{v_i\})$ such that $(a_1 \circ \overline{h})(P^m) \cap k(Q^m) = \emptyset$ and there is an isotopy $\{h'_t\}_{t \in [0,1]}$ such that $h'_0 = \mathrm{id}$, $h'_1 = a_1 \circ \overline{h}$ and $h'_t(P^m) \cap k(Q^m) = \emptyset$ $(t \in [0, 1])$.

We take a small neighborhood U_i of $\overline{H}([s_i, 1] \times \{v_i\})$ diffeomorphic to the $(2m+1)$-dimensional ball. We take these U_i to be disjoint and the intersection of U_i and $\overline{H}([0, 1] \times P^m)$ or $k(Q^m)$ is described as follows.

We put a coordinate $(x_1, x_2, \ldots, x_{m+1}, x_{m+2}, \ldots, x_{2m+1}) \in (-2, 2)^{2m+1}$ on U_i such that, for $\varepsilon_i > 0$,

$$\begin{aligned} &k(Q^m) \cap U_i = \{0\} \times \{0\}^m \times (-2, 2)^m, \\ &\overline{H}((s_i - 2\varepsilon_i(1 - s_i), 1] \times \{v_i\}) \cap U_i = (-2, 1] \times \{0\}^{2m}, \\ &\overline{h}_{s_i + t(1-s_i)}(P^m) \cap U_i = \{t\} \times (-2, 2)^m \times \{0\}^m \quad (t \in [-\varepsilon_i, 1]). \end{aligned}$$

Take an isotopy $\{a_t\}_{t \in [0,1]}$ with support in $\bigsqcup_{i=1}^r U_i$ such that on each U_i, $a_0 = \mathrm{id}$ and, for $(x_1, x_2, \ldots, x_{2m+1}) \in [-\varepsilon_i, 1] \times [-1, 1]^{2m} \subset (-2, 2)^{2m+1}$,

$$a_t(x_1, x_2, \ldots, x_{2m+1}) = (x_1 - (1 + \varepsilon_i)t, x_2, \ldots, x_{2m+1}).$$

Now $(a_1 \circ \overline{h}_1)(P^m) \cap k(Q^m) = \emptyset$. Moreover there is an isotopy $\{h'_t\}_{t \in [0,1]}$ from the identity to $a_1 \circ \overline{h}_1$ such that $h'_t(P^m) \cap k(Q^m) = \emptyset$ $(t \in [0, 1])$.

The reason is that we can modify $\overline{h}_t$ on U_i by replacing by

$$a_{(u_i + \varepsilon_i)/(1 + \varepsilon_i)} \circ \overline{h}_{s_i + u_i(1 - s_i)}$$

for $t = s_i + u_i(1 - s_i) \in [s_i - \varepsilon_i(1 - s_i), 1]$, i.e., $u_i \in [-\varepsilon_i, 1]$. Then

$$\begin{aligned}
&(a_{(u_i+\varepsilon_i)/(1+\varepsilon_i)} \circ \overline{h}_{s_i+u_i(1-s_i)})(\{-\varepsilon_i\} \times [-1,1]^m \times \{0\}^m) \\
&= a_{(u_i+\varepsilon_i)/(1+\varepsilon_i)}(\{u_i\} \times [-1,1]^m \times \{0\}^m) \\
&= \{u_i - (u_i + \varepsilon_i)\} \times [-1,1]^m \times \{0\}^m \\
&= \{-\varepsilon_i\} \times [-1,1]^m \times \{0\}^m
\end{aligned}$$

Thus there is an isotopy $\{h'_t\}_{t\in[0,1]}$ such that $h'_0 = \mathrm{id}$, $h'_1 = a_1 \circ \overline{h}_1$ and $h'_t(P^m) \cap k(Q^m) = \emptyset$ $(t \in [0,1])$.

Then by the proof of Theorem 5.1 (Remark 5.3), $a_1 \circ \overline{h}$ can be written as a composition $a_1 \circ \overline{h} = \overline{g} \circ \overline{h}'$, where $\overline{g} \in \mathrm{Diff}^r_c(M^{2m+1} \setminus k(Q^m))_0$ and $\overline{h}' \in \mathrm{Diff}^r_c(M^{2m+1} \setminus P^m)_0$. Thus $\overline{h} = (a_1)^{-1} \circ \overline{g} \circ \overline{h}'$. Since $(a_1)^{-1} \in \mathrm{Diff}^r_c(\bigsqcup_{i=1}^r U_i)_0$, Lemma 6.5 for $m \geqq 2$ is proved. Q.E.D.

Proof of Lemma 6.5 *for* $m = 1$. If $m = 1$, then we will take U_i considering the intersection with the double point curves.

First take a small neighborhood U_i of $\overline{H}([s_i, 1] \times \{v_i\})$ as in the case where $m \geqq 2$. U_i has the coordinate $(-2,2)^3$ as before. We will modify U_i by using several isotopies.

We also take small neighborhoods $U_{i,i_1}, U_{i,i_1,i_2}, \ldots$ of the branches $\overline{H}([s'_{i,i_1}, 1] \times \{v'_{i,i_1}\})$ $(s'_{i,i_1} > s_i)$, $\overline{H}([s''_{i,i_1,i_2}, 1] \times \{v''_{i,i_1,i_2}\})$ $(s''_{i,i_1,i_2} > s'_{i,i_1})$, ... We put a coordinate $(x_1, x_2, x_3) \in (-2,3) \times (-2,2)^2$ on U_{i,i_1} such that

$$\begin{aligned}
&\overline{H}([s'_{i,i_1} - 2\varepsilon_{i,i_1}(1 - s'_{i,i_1}), 1] \times \{v'_{i,i_1}\}) \cap U_{i,i_1} = (-2, 1] \times \{(0,0)\}, \\
&\overline{h}_{s'_{i,i_1}+t(1-s'_{i,i_1})}(P^1) \cap U_{i,i_1} = \{t\} \times (-2,2) \times \{0\} \quad (t \in [-\varepsilon_{i,i_1}, 1]),
\end{aligned}$$

and coordinates on $U_{i,i_1,i_2}, \ldots$ are taken in a similar way.

We take isotopies $\{a_t^{i,i_1}\}_{t\in[0,1]}$ with support in U_{i,i_1} such that $a_0^{i,i_1} = \mathrm{id}$ and, for $(x_1, x_2, x_3) \in [-\varepsilon_{i,i_1}, 1] \times [-1,1]^2 \subset (-2,3) \times (-2,2)^2$,

$$a_t^{i,i_1}(x_1, x_2, x_3) = (x_1 + t(1 + \varepsilon_{i,i_1}), x_2, x_3).$$

We also take isotopies $\{a_t^{i,i_1,i_2}\}_{t\in[0,1]}, \ldots$ with support in $U_{i,i_1,i_2}, \ldots$ in a similar way. Then we take U_i very thin so that

$$\Bigl(\bigl(\prod_{i_1,i_2,\ldots,i_k} a_1^{i,i_1,i_2,\ldots,i_k} \bigr) \circ \cdots \circ \bigl(\prod_{i_1,i_2} a_1^{i,i_1,i_2} \bigr) \circ \bigl(\prod_{i_1} a_1^{i,i_1} \bigr) \Bigr)(U_i)$$

does not intersect $\overline{H}([s_i, 1] \times P^1)$ outside of a neighborhood of $\overline{H}([s_i, 1] \times \{v_1\})$, where $\{a_t^{i,i_1,i_2,\ldots,i_k}\}_{t\in[0,1]}$ is the isotopy with support in a neighborhood $U_{i,i_1,i_2,\ldots,i_k}$ of the final branch $\overline{H}([s^{(k)}_{i,i_1,i_2,\ldots,i_k}, 1] \times \{v^{(k)}_{i,i_1,i_2,\ldots,i_k}\})$

defined in a similar way. Let

$$\overline{a} = \prod_{i=1}^{r} \Big(\prod_{i_1, i_2, \ldots, i_k} a_1^{i, i_1, i_2, \ldots, i_k} \Big) \circ \cdots \circ \Big(\prod_{i_1, i_2} a_1^{i, i_1, i_2} \Big) \circ \Big(\prod_{i_1} a_1^{i, i_1} \Big).$$

Then $\overline{a} \circ a_t \circ \overline{a}^{-1}$ is isotopic to the identity by the isotopy with support in the disjoint union of 3-dimensional open balls $\overline{a}(\bigsqcup_{i=1}^{r} U_i)$. By the construction, $((\overline{a} \circ a_1 \circ \overline{a}^{-1}) \circ \overline{h}_1)(P^1) \cap k(Q^1) = \emptyset$. We show that there is an isotopy $\{h'_t\}_{t \in [0,1]}$ from the identity to $(\overline{a} \circ a_1 \circ \overline{a}^{-1}) \circ \overline{h}_1$ such that $h'_t(P^1) \cap k(Q^1) = \emptyset$ $(t \in [0,1])$.

For the construction of h'_t, we define the local time $u_i \in [-\varepsilon_i, 1]$ on U_i $(1 \leqq i \leqq r)$ by $t = s_i + u_i(1 - s_i)$ as in the case where $m \geqq 2$. We can modify $\overline{h}_t$ on the union

$$U_i \cup \bigcup_{i_1} U_{i, i_1} \cup \bigcup_{i_1, i_2} U_{i, i_1, i_2} \cup \cdots \cup \bigcup_{i, i_1, i_2, \ldots, i_k} U_{i, i_1, i_2, \ldots, i_k}$$

for $t = s_i + u_i(1 - s_i) \in [s_i - \varepsilon_i(1 - s_i), 1]$ $(u_i \in [-\varepsilon_i, 1])$ and define h'_t there by

$$h'_t = (\overline{a} \circ a_{(u_i + \varepsilon_i)/(1 + \varepsilon_i)} \circ \overline{a}^{-1}) \circ \overline{h}_{s_i + u_i(1 - s_i)}.$$

Then this isotopy $\{h'_t\}_{t \in [0,1]}$ satisfies that $h'_0 = \mathrm{id}$, $h'_1 = (\overline{a} \circ a_1 \circ \overline{a}^{-1}) \circ \overline{h}_1$ and $h'_t(P^1) \cap k(Q^1) = \emptyset$ $(t \in [0,1])$.

Then by the proof of Theorem 5.1 (Remark 5.3), $(\overline{a} \circ a_1 \circ \overline{a}^{-1}) \circ \overline{h}_1$ can be written as a composition $(\overline{a} \circ a_1 \circ \overline{a}^{-1}) \circ \overline{h}_1 = \overline{g} \circ \overline{h}'$, where $\overline{g} \in \mathrm{Diff}_c^r(M^3 \setminus k(Q^1))_0$ and $\overline{h}' \in \mathrm{Diff}_c^r(M^3 \setminus P^1)_0$. Thus $\overline{h} = (\overline{a} \circ {a_1}^{-1} \circ \overline{a}^{-1}) \circ \overline{g} \circ \overline{h}'$. Since $\overline{a} \circ {a_1}^{-1} \circ \overline{a}^{-1} \in \mathrm{Diff}_c^r(\overline{a}(\bigsqcup_{i=1}^{r} U_i))_0$, Lemma 6.5 for $m = 1$ is proved. Q.E.D.

Proof of Lemma 6.5 *for* $m = 0$. This is an (easy) exceptional case. The only compact connected 1-dimensional manifold is the circle S^1. For $f \in \mathrm{Diff}^r(S^1)_0$ and $p \in S^1$, we take a point q distinct from p and $f(p)$, Let g be a C^r diffeomorphism of S^1 which coincides with f on a neighborhood of p and with the identity on a neighborhood of q. Then $h = g^{-1} \circ f$ is the identity on a neighborhood of p. Since g is isotopic to the identity as an element of $\mathrm{Diff}_c^r(S^1 \setminus \{q\})_0$, and h is isotopic to the identity as an element of $\mathrm{Diff}_c^r(S^1 \setminus \{p\})_0$, $f = g \circ h$ in a desired way. (Then any element of $\mathrm{Diff}^r(S^1)_0$ $(1 \leqq r \leqq \infty,\ r \neq 2)$ can be written as a product of four commutators as in Theorem 5.2.) Note that in this case the original isotopy for f is different from the composition of the isotopies for g and h. Q.E.D.

Proof of Theorem 6.2. For $h \in \mathrm{Diff}^r(M^{2m+1})$, let $\overline{h}$ be the diffeomorphism obtained by Lemma 6.4. By Lemma 6.5, $\overline{h}$ can be written as $\overline{h} = a \circ \overline{g} \circ \overline{h}'$, where $a \in \mathrm{Diff}^r_c(\bigsqcup_i U_i)_0$, $\overline{g} \in \mathrm{Diff}^r_c(M^{2m+1} \setminus k(Q^m))_0$ and $\overline{h}' \in \mathrm{Diff}^r_c(M^{2m+1} \setminus P^m)_0$. Then $h = a \circ \overline{g} \circ \overline{h}' \circ (\overline{h}^{-1}h)$. Since $\overline{h}^{-1}h$ is close to the identity, by Remark 5.4, $\overline{h}^{-1}h$ can be written as the product $\overline{h}^{-1}h = \widehat{h} \circ \widehat{g}$, where $\widehat{g} \in \mathrm{Diff}^r_c(M^{2m+1} \setminus k(Q^m))_0$ and $\widehat{h} \in \mathrm{Diff}^r_c(M^{2m+1} \setminus P^m)_0$. Then

$$h = a \circ \overline{g} \circ \overline{h}' \circ \widehat{h} \circ \widehat{g} = a \circ (\overline{g} \circ \widehat{g}) \circ \widehat{g}^{-1} \circ (\overline{h}' \circ \widehat{h}) \circ \widehat{g}.$$

Here $a \in \mathrm{Diff}^r_c(\bigsqcup_i U_i)_0$, $\overline{g} \circ \widehat{g} \in \mathrm{Diff}^r_c(M^{2m+1} \setminus k(Q^m))_0$ and $\widehat{g}^{-1} \circ (\overline{h}' \circ \widehat{h}) \circ \widehat{g} \in \mathrm{Diff}^r_c(M^{2m+1} \setminus \widehat{g}^{-1}(P^m))_0$. Thus Theorem 6.2 is shown. Q.E.D.

Remark 6.6. In Corollary 2.2 and Theorem 4.1, there is an open subset U of M^n and there is an element g of $\mathrm{Diff}^r_c(M^n)_0$ such that any element f of $\mathrm{Diff}^r_c(M^n)_0$ is conjugate to an element of $\mathrm{Diff}^r_c(U)_0$ and $g(U) \cap U = \emptyset$. Then any commutator $[a, b]$ in $\mathrm{Diff}^r_c(U)_0$ can be written as a product of 4 conjugates of g or g^{-1}. For, if a, $b \in \mathrm{Diff}^r_c(U)_0$, then by putting $c = g^{-1}ag$, $cb = bc$ and

$$\begin{aligned} aba^{-1}b^{-1} &= gcg^{-1}bgc^{-1}g^{-1}b^{-1} \\ &= gcg^{-1}c^{-1}cbgc^{-1}b^{-1}bg^{-1}b^{-1} \\ &= g(cg^{-1}c^{-1})(bcgc^{-1}b^{-1})(bg^{-1}b^{-1}). \end{aligned}$$

Thus for an n-dimensional manifold M^n satisfying the assumption of Corollary 2.2 or Theorem 4.1, any element f of $\mathrm{Diff}^r_c(M^n)_0$ can be written as a product of 8 conjugates of g or g^{-1} $((1 \leqq r \leqq \infty, r \neq n+1)$. By this observation, Theorem 5.2 implies that for an even-dimensional compact manifold M^{2m} which has a handle decomposition without handles of the middle index m, there is an element g such that any element f of $\mathrm{Diff}^r(M^{2m})_0$ can be written as a product of 16 conjugates of g or g^{-1} $(1 \leqq r \leqq \infty, r \neq 2m+1)$. Here g is taken so that g maps a neighborhood U of the union of the simplicial complexes P^k and Q^k in Theorem 5.2 to an open set $g(U)$ with $U \cap g(U) = \emptyset$. In a similar way, Theorem 6.1 implies that for an odd-dimensional compact manifold M^{2m+1}, there is an element g such that any element f of $\mathrm{Diff}^r(M^{2m+1})_0$ can be written as a product of 24 conjugates of g or g^{-1} $(1 \leqq r \leqq \infty, r \neq 2m+2)$. This implies that these groups are meager in the terminology of the paper [2] as Polterovich pointed out to the author. Note that, for a perfect group, if there is an element g with the above property, then it is uniformly perfect.

References

[1] A. Banyaga, The structure of classical diffeomorphism groups, Math. Appl., **400**, Kluwer Academic Publishers Group, Dordrecht, 1997, xii+197 pp.
[2] D. Burago, S. Ivanov and L. Polterovich, Conjugation-invariant norms on groups of geometric origin, in this volume, pp. 221–250.
[3] D. B. A. Epstein, The simplicity of certain groups of homeomorphisms, Compositio Math., **22** (1970), 165–173.
[4] S. Haller and J. Teichmann, Smooth perfectness through decomposition diffeomorphisms into fiber preserving ones, Ann. Global Anal. Geom., **23** (2003), 53–63.
[5] M. Herman, Sur la conjugaison différentiable des difféomorphismes du cercle à des rotations, Inst. Hautes Études Sci. Publ. Math., **49** (1979), 5–234.
[6] D. Kotschick, Stable length in stable groups, in this volume, pp. 401–413.
[7] J. Mather, The vanishing of the homology of certain groups of homeomorphisms, Topology, **10** (1971), 297–298.
[8] J. Mather, Commutators of diffeomorphisms I, II and III, Comm. Math. Helv., **49** (1974), 512–528, **50** (1975), 33–40 and **60** (1985), 122–124.
[9] S. Matsumoto and S. Morita, Bounded cohomology of certain groups of homeomorphisms, Trans. Amer. Math. Soc., **94** (1985), 539–544.
[10] J. Milnor, Morse theory, Ann. of Math. Stud., **51**, Princeton Univ. Press, Princeton, NJ, 1963, vi+153 pp.
[11] J. Milnor, Lectures on the h-cobordism theorem, Princeton Univ. Press, Princeton, NJ, 1965, v+116 pp.
[12] W. Thurston, Foliations and groups of diffeomorphisms, Bull. Amer. Math. Soc., **80** (1974), 304–307.
[13] T. Tsuboi, On 2-cycles of $B\mathrm{Diff}(S^1)$ which are represented by foliated S^1-bundles over T^2, Ann. Inst. Fourier (Grenoble), **31** (1981), 1–59.
[14] T. Tsuboi, On the homology of classifying spaces for foliated products, In: Foliations, Adv. Stud. Pure Math., **5**, North-Holland, 1985, pp. 37–120.
[15] T. Tsuboi, On the foliated products of class C^1, Ann. of Math. (2), **130** (1989), 227–271.
[16] T. Tsuboi, On the perfectness of groups of diffeomorphisms of the interval tangent to the identity at the endpoints, In: Proceedings of Foliations: Geometry and Dynamics, Warsaw, 2000, World Scientific, Singapore, 2002, pp. 421–440.
[17] T. Tsuboi, On the group of foliation preserving diffeomorphisms, Foliations 2005, Lodz, World Scientific, Singapore, 2006, pp. 411–430.
[18] H. Whitney, The singularities of a smooth n-manifold in $(2n-1)$-space, Ann. of Math. (2), **45** (1944), 247–293.

Graduate School of Mathematical Sciences
the University of Tokyo
Komaba Meguro, Tokyo 153-8914, Japan
E-mail address: tsuboi@ms.u-tokyo.ac.jp

Advanced Studies in Pure Mathematics

A SERIES OF UP-TO-DATE GUIDES OF LASTING INTEREST
TO ADVANCED MATHEMATICS

Volume 1 Algebraic Varieties and Analytic Varieties.
Edited by S. Iitaka. February, 1983

Volume 2 Galois Groups and their Representations.
Edited by Y. Ihara. December, 1983

Volume 3 Geometry of Geodesics and Related Topics.
Edited by K. Shiohama. June, 1984

Volume 4 Group Representations and Systems of Differential Equations.
Edited by K. Okamoto. March, 1985

Volume 5 Foliations.
Edited by I. Tamura. February, 1986

Volume 6 Algebraic Groups and Related Topics.
Edited by R. Hotta. March, 1985

Volume 7 Automorphic Forms and Number Theory.
Edited by I. Satake. February, 1986

Volume 8 Complex Analytic Singularities.
Edited by T. Suwa and P. Wagreich. February, 1987

Volume 9 Homotopy Theory and Related Topics.
Edited by H. Toda. February, 1987

Volume 10 Algebraic Geometry, Sendai, 1985.
Edited by T. Oda. · July, 1987

Volume 11 Commutative Algebra and Combinatorics.
Edited by M. Nagata and H. Matsumura.
October, 1987

Volume 12 Galois Representations and Arithmetic Algebraic Geometry.
Edited by Y. Ihara. November, 1987

Volume 13 Investigations in Number Theory.
Edited by T. Kubota. March, 1988

Volume 14 Representations of Lie Groups, Kyoto, Hiroshima, 1986.
Edited by K. Okamoto and T. Oshima. February, 1989

Volume 15 Automorphic Forms and Geometry of Arithmetic Varieties.
Edited by K. Hashimoto and Y. Namikawa. July, 1989

Volume 16 Conformal Field Theory and Solvable Lattice Models.
Edited by M. Jimbo, T. Miwa and A. Tsuchiya.
June, 1988

Volume 17 Algebraic Number Theory—in honor of K. Iwasawa.
Edited by J. Coates, R. Greenberg, B. Mazur and I. Satake. August, 1989

Volume 18 I—Recent Topics in Differential and Analytic Geometry.
Edited by T. Ochiai. November, 1990
II—Kähler Metric and Moduli Spaces.
Edited by Edited by T. Ochiai. November, 1990

Volume 19 Integrable Systems in Quantum Field Theory and Statistical Mechanics.
Edited by M. Jimbo, T. Miwa and A. Tsuchiya.
November, 1989

Volume 20 Aspects of Low Dimensional Manifolds.
Edited by Y. Matsumoto and S. Morita.
December, 1992

Volume 21 Zeta Functions in Geometry.
Edited by N. Kurokawa and T. Sunada.
December, 1992

Volume 22 Progress in Differential Geometry.
Edited by K. Shiohama. September, 1993

Volume 23 Spectral and Scattering Theory and Applications.
Edited by K. Yajima. February, 1994

Volume 24 Progress in Algebraic Combinatorics.
Edited by E. Bannai and A. Munemasa. May, 1996

Volume 25 CR-Geometry and Overdetermined Systems.
Edited by T. Akahori, G. Komatsu, K. Miyajima, M. Namba and K. Yamaguchi. July, 1997

Volume 26 Analysis on Homogeneous Spaces and Representation Theory of Lie Groups, Okayama-Kyoto.
Edited by M. Kashiwara, T. Kobayashi, T. Matsuki, K. Nishiyama and T. Oshima. April, 2000

Volume 27 Arrangements – Tokyo 1998.
Edited by M. Falk and H. Terao. August, 2000

Volume 28 Combinatorial Methods in Representation Theory.
Edited by M. Kashiwara, K. Koike, S. Okada, I. Terada, and H.-F. Yamada. December, 2000

Volume 29 Singularities – Sapporo 1998.
Edited by J.-P. Brasselet and T. Suwa.
December, 2000

Volume 30 Class Field Theory – Its Centenary and Prospect.
Edited by K. Miyake. April, 2001

Volume 31 Taniguchi Conference on Mathematics Nara '98.
Edited by M. Maruyama and T. Sunada. May, 2001

Volume 32 Groups and Combinatorics — in memory of Michio Suzuki.
Edited by E. Bannai, H. Suzuki, H. Yamaki and T. Yoshida. October, 2001

Volume 33 Computational Commutative Algebra and Combinatorics.
Edited by T. Hibi. February, 2002

Volume 34 Minimal Surfaces, Geometric Analysis and Symplectic Geometry.
Edited by K. Fukaya, S. Nishikawa and J. Spruck.
June, 2002

Volume 35 Higher Dimensional Birational Geometry.
Edited by S. Mori and Y. Miyaoka. July, 2002

Volume 36 Algebraic Geometry 2000, Azumino.
Edited by S. Usui, M. Green, L. Illusie, K. Kato, E. Looijenga, S. Mukai and S. Saito. November, 2002

Volume 37 Lie Groups, Geometric Structures and Differential Equations — One Hundred Years after Sophus Lie —.
Edited by T. Morimoto, H. Sato and K. Yamaguchi.
December, 2002

Volume 38 Operator Algebras and Applications.
Edited by H. Kosaki. January, 2004

Volume 39 Stochastic Analysis on Large Scale Interacting Systems.
Edited by T. Funaki and H. Osada. March, 2004

Volume 40 Representation Theory of Algebraic Groups and Quantum Groups.
Edited by T. Shoji, M. Kashiwara, N. Kawanaka, G. Lusztig and K. Shinoda. March, 2004

Volume 41 Stochastic Analysis and Related Topics in Kyoto
In honour of Kiyosi Itô.
Edited by H. Kunita, S. Watanabe and Y. Takahashi.
March, 2004

To be continued